Methods in Enzymology

Volume 89
CARBOHYDRATE METABOLISM
Part D

METHODS IN ENZYMOLOGY

EDITORS-IN-CHIEF

Sidney P. Colowick   Nathan O. Kaplan

Methods in Enzymology

Volume 89

Carbohydrate Metabolism

Part D

EDITED BY

Willis A. Wood

DEPARTMENT OF BIOCHEMISTRY
MICHIGAN STATE UNIVERSITY
EAST LANSING, MICHIGAN

1982

ACADEMIC PRESS

A Subsidiary of Harcourt Brace Jovanovich, Publishers

New York   London
Paris   San Diego   San Francisco   São Paulo   Sydney   Tokyo   Toronto

ACADEMIC PRESS, INC.
111 Fifth Avenue, New York, New York 10003

United Kingdom Edition published by
ACADEMIC PRESS, INC. (LONDON) LTD.
24/28 Oval Road, London NW1 7DX

Library of Congress Cataloging in Publication Data
Main entry under title:

Carbohydrate metabolism.

(Methods in enzymology ; v. 89)
Includes index.
1. Carbohydrates--Metabolism--Research--Methodology.
I. Wood, Willis A., Date. II. Series.
QP601.M49 vol. 89 [QP701] 574.19'25s 82-6769
ISBN 0-12-181989-2 [574.19'254] AACR2

PRINTED IN THE UNITED STATES OF AMERICA

82 83 84 85   9 8 7 6 5 4 3 2 1

Table of Contents

Section I. Analytical Methods

Section II. Enzyme Assay Procedures

Section III. Preparation of Substrates and Effectors

Section IV. Oxidation–Reduction Enzymes

Section V. Isomerases, Epimerases, and Mutases

Contributors to Volume 89

Article numbers are in parentheses following the names of contributors.
Affiliations listed are current.

KAREN E. ACKERMANN (100), *Department of Psychiatry, Washington University School of Medicine, St. Louis, Missouri 63110*

OSAO ADACHI (25, 26, 33, 34, 50, 76, 82), *Department of Applied Microbiology, Faculty of Agriculture, Yamaguchi University, Yamaguchi 753, Japan*

S. H. GEORGE ALLEN (64), *Department of Biochemistry, Albany Medical College, Albany, New York 12208*

MINORU AMEYAMA (4, 22, 24, 25, 26, 31, 32, 33, 34, 50, 76, 82), *Department of Applied Microbiology, Faculty of Agriculture, Yamaguchi University, Yamaguchi 753, Japan*

RICHARD L. ANDERSON (15, 42, 47, 95), *Department of Biochemistry, Michigan State University, East Lansing, Michigan 48824*

LARS ANDERSSON (74), *Pure and Applied Biochemistry, Chemical Center, University of Lund, S-220 07 Lund 7, Sweden*

FRANK B. ARMSTRONG (51), *Department of Biochemistry, North Carolina State University, Raleigh, North Carolina 27650*

ARTHUR R. AYERS (20), *Department of Cellular and Developmental Biology, Harvard University, Cambridge, Massachusetts 02138*

R. BARKER (11, 12, 13, 14), *Section of Biochemistry, Molecular and Cell Biology, Division of Biological Sciences, Cornell University, Ithaca, New York 14853*

GLENN C. BEWLEY (51), *Department of Genetics, North Carolina State University, Raleigh, North Carolina 27650*

DONALD L. BISSETT (15, 95), *Department of Biochemistry, Michigan State University, East Lansing, Michigan 48824*

ENRICO BOCCÙ (49), *Institute of Pharmaceutical Chemistry, University of Padova, Padova I-35100, Italy*

WINFRIED BOOS (10), *Department of Biology, University of Konstanz, D-7750 Konstanz, Federal Republic of Germany*

RICHARD B. BRANDT (6), *Department of Biochemistry, Medical College of Virginia, Virginia Commonwealth University, Richmond, Virginia 23298*

JUSTINO BURGOS (87, 88), *Department of Food Biochemistry and Food Technology, Faculty of Veterinary Science, University of Leon, Leon, Spain*

F. BUTZ (48), *Chemisches Laboratorium der Universität Freiburg, D-7800 Freiburg i Br., Federal Republic of Germany*

FLOYD M. BYERS (5), *Department of Animal Science, Texas A & M University, College Station, Texas 77840*

LARRY D. BYERS (57), *Department of Chemistry, Tulane University, New Orleans, Louisiana 70118*

E. CADMAN (14), *Section of Biochemistry, Molecular and Cell Biology, Division of Biological Sciences, Cornell University, Ithaca, New York 14853*

DINA F. CAROLINE (66), *Department of Radiology, Hospital of the University of Pennsylvania, Philadelphia, Pennsylvania 19104*

JULIA Y. CHAN (36), *Department of Biochemistry, University of Toronto, and Research Institute, Hospital for Sick Children, Toronto, Ontario M5G 1X8, Canada*

E. V. CHANDRASEKARAN (19, 38), *Department of Biochemistry, University of Georgia, Athens, Georgia 30602*

E. L. CLARK (13), *Section of Biochemistry, Molecular and Cell Biology, Division of Biological Sciences, Cornell University, Ithaca, New York 14853*

T. H. CLAUS (17), *Lederle Laboratories, Pearl River, New York 10965*

J. A. CROMLISH (40, 84), *Department of Biochemistry, Queen's University, Kingston, Ontario K7L 3N6, Canada*

D. A. CUMMING (17), *Department of Molecular Biology, Vanderbilt University School of Medicine, Nashville, Tennessee 37232*

SHAWKY M. DAGHER (54), *Faculty of Agricultural Sciences, American University of Beirut, Beirut, Lebanon*

A. STEPHEN DAHMS (37), *Department of Chemistry and Molecular Biology Institute, San Diego State University, San Diego, California 92182*

W. S. DAVIDSON (84), *Department of Biochemistry, Memorial University of Newfoundland, St. John's, Newfoundland A1C 5S7, Canada*

MARK DAVILA (38), *Department of Biochemistry, University of Georgia, Athens, Georgia 30602*

WILLIAM C. DEAL, JR. (54), *Department of Biochemistry, Michigan State University, East Lansing, Michigan 48824*

DAVID T. DENNIS (56), *Department of Biology, Queen's University, Kingston, Ontario K7L 3N6, Canada*

RONALD G. DUGGLEBY (56), *Department of Biochemistry, University of Queensland, St. Lucia, Brisbane, Queensland 4067, Australia*

JOHN H. ECKFELDT (79), *Department of Clinical Chemistry and Toxicology, Veterans Administration Medical Center, Minneapolis, Minnesota 55417, and Department of Laboratory Medicine and Pathology, University of Minnesota, Minneapolis, Minnesota 55455*

R. D. EICHNER (62), *Department of Immunology, John Curtin School of Medical Research, The Australian National University, Canberra City, A.C.T. 2601 Australia*

ALAN D. ELBEIN (92), *Department of Biochemistry, University of Texas Health Science Center, San Antonio, Texas 78284*

M. R. EL-MAGHRABI (17), *Department of Physiology, Vanderbilt University School of Medicine, Nashville, Tennessee 37232*

HARRY G. ENOCH (91), *Graduate Center for Toxicology, University of Kentucky, and Kentucky Department of Energy, Lexington, Kentucky 40578*

KARL-ERIK ERIKSSON (20), *Chemistry Department, Swedish Forest Products Research Laboratory, Stockholm, Sweden*

M. P. ESNOUF (98), *Nuffield Department of Clinical Biochemistry, Radcliffe Infirmary, Oxford OX2 6HE, England*

G. FERRI (55), *Istituto di Chimica Biologica, Università di Pavia, 27100 Pavia, Italy*

T. G. FLYNN (40, 84), *Department of Biochemistry, Queen's University, Kingston, Ontario K7L 3N6, Canada*

ANGELO FONTANA (49), *Institute of Organic Chemistry, University of Padova, Padova I-35100, Italy*

BARBARA FREIMÜLLER (52), *Zentrum Biochemie, Medizinische Hochschule, D-3000 Hanover 1, Federal Republic of Germany*

SHINOBU C. FUJITA (58), *Department of Pharmacology, Gunma University, Maebashi-shi, Gunma 371, Japan*

T. FUKASAWA (99), *Research Unit for Molecular Genetics, Keio University School of Medicine, Shinanomachi, Shinjuku-ku, Tokyo, Japan*

SHUICHI FURUTA (70), *Department of Biochemistry, Shinshu University School of Medicine, Matsumoto, Nagano 390, Japan*

PETER GIEROW (8), *Department of Biochemistry, Chemical Centre, University of Lund, S-220 07 Lund, Sweden*

ERWIN GOLDBERG (61), *Department of Biological Sciences, Northwestern University, Evanston, Illinois 60201*

ROBERT W. GRACY (93), *Department of Biochemistry, North Texas State University/Texas College of Osteopathic Medicine, Denton, Texas 76203*

L. GUERRILLOT (81), *Societé Rhone-Poulenc, Fabrications Biochimiques, 94400 Vitry sur Seine, France*

TAKAO HAMA (78), *Laboratory of Physiological Chemistry, Faculty of Pharmacy, Kobe-Gakuin University, Tarumi-ku, Kobe, Japan*

RAGY HANNA (19), *Department of Biochemistry, University of Georgia, Athens, Georgia 30602*

T. HARADA (71), *Kobe Women's College, Suma-ku, Kobe-shi 654, Japan*

ROY W. HARDING (66), *Radiation Biology Laboratory, Smithsonian Institution, Rockville, Maryland 20852*

R. P. HARRIS (98), *Oxford Enzyme Group, Preparation Laboratory, Department of Zoology, Oxford OX1 3PS, England*

TAKASHI HASHIMOTO (70), *Department of Biochemistry, Shinshu University School of Medicine, Matsumoto, Nagano 390, Japan*

MARK W. C. HATTON (28), *Department of Pathology, McMaster University, Hamilton, Ontario L8N 3Z5, Canada*

M. L. HAYES (11, 14), *Section of Biochemistry, Molecular and Cell Biology, Division of Biological Sciences, Cornell University, Ithaca, New York 14853*

FRITZ HEINZ (52, 83), *Arbeitsbereich Enzymologie, Zentrum Biochemie, Medizinische Hochschule, D-3000 Hannover 1, Federal Republic of Germany*

ALAN J. HILLIER (63), *Russell Grimwade School of Biochemistry, University of Melbourne, Parkville, Victoria 3052, Australia*

PAUL P. HIPPS (100), *Department of Psychiatry, Washington University School of Medicine, St. Louis, Missouri 63110*

T. HIRABAYASHI (71), *Suntory Co., Ltd., Minato-ku, Tokyo-to 107, Japan*

TH. HÖPNER (90), *Fachbereich Biologie der Universität, D-2900 Oldenburg, Federal Republic of Germany*

SABURO HOSOMI (39), *Faculty of Pharmaceutical Sciences, Osaka University, Yamada-oka, Suita 565, Japan*

MIYOSHI IKAWA (23), *Department of Biochemistry, University of New Hampshire, Durham, New Hampshire 03824*

KAZUTOMO IMAHORI (58), *Tokyo Metropolitan Institute of Gerontology, 35-2 Sakae-cho, Itabashi, Tokyo 173, Japan*

KEN IZUMORI (92), *Department of Food Science, Kagawa University, Miki-cho, Kagawa-ken, Japan*

G. RICHARD JAGO (63), *Russell Grimwade School of Biochemistry, University of Melbourne, Parkville, Victoria 3052, Australia*

BENGT JERGIL (8), *Department of Biochemistry, Chemical Centre, University of Lund, S-220 07 Lund, Sweden*

KANDIAH JEYASEELAN (68), *Department of Botany, University of Jaffna, Jaffna, Thirunelvely, Sri Lanka*

J. G. JOSHI (101), *Department of Biochemistry, University of Tennessee, Knoxville, Tennessee 37996*

HARRY KESTER (68), *Department of Genetics, Agricultural University, Wageningen 6703 BM, The Netherlands*

KINUKO KIMURA (44), *Laboratory of Biochemistry, Rikkyo (St. Paul's) University, Nishi-Ikebukuro, Toshima-ku, Tokyo 171, Japan*

JEREMY R. KNOWLES (18), *Department of Chemistry, Harvard University, Cambridge, Massachusetts 02139*

G. A. KOCHETOV (3, 7), *A. N. Belŏzersky Laboratory of Molecular Biology and Bioorganic Chemistry, Moscow State University, Moscow 117234, USSR*

LEONARD D. KOHN (59), *Section on Biochemistry of Cell Regulation, Laboratory of Biochemical Pharmacology, National Institute of Arthritis, Diabetes, and Digestive and Kidney Diseases, National Institutes of Health, Bethesda, Maryland 20205*

DANIEL J. KOSMAN (27), *Department of Biochemistry, State University of New York, Buffalo, New York 14214*

MARIA-REGINA KULA (72, 89), *Gesellschaft für Biotechnologische Forschung, D-3300 Braunschweig, Federal Republic of Germany*

G. KURZ (29, 45, 48), *Chemisches Laboratorium der Universität Freiburg, D-7800 Freiburg i. Br., Federal Republic of Germany*

PER-OLOF LARSSON (77), *Pure and Applied Biochemistry, Chemical Center, University of Lund, S-220 07 Lund 7, Sweden*

ALAN L. LEAVITT (1, 2), *Graduate Program in Civil Engineering, Department of Civil and Environmental Engineering, Cornell University, Ithaca, New York 14853*

CHI-YU LEE (43, 51, 61, 75, 94), *Andrology Laboratory, Obs/Gyn, Acute Care Hospital, The University of British Columbia, Vancouver, British Columbia V6T 2B5, Canada*

WOLFGANG LEICHT (83), *Sparte Pflanzenschutz, Bayer AG, 5090 Leverkusen-Bayerwerk, Federal Republic of Germany*

NANCY LEISSING (21), *Department of Safety Assessment, Travenol Laboratories, Morton Grove, Illinois 60053*

D. LESSMANN (45), *Chemisches Laboratorium der Universität Freiburg, D-7800 Freiburg i. Br., Federal Republic of Germany*

ROBERT L. LESTER (91), *Department of Biochemistry, University of Kentucky, Lexington, Kentucky 40536*

STEPHEN D. MCCURRY (9, 18), *Research Department, Kellogg Company, Battle Creek, Michigan 49016*

R. D. MACELROY (97), *Extraterrestrial Research Division, Ames Research Center, Moffett Field, California 94035*

E. T. MCGUINNESS (21), *Department of Chemistry, Seton Hall University, South Orange, New Jersey 07079*

J. D. MCVITTIE (98), *Department of Clinical Biochemistry, John Radcliffe Hospital, Headington, Oxford OX3 9DU, England*

E. MAIER (29), *Chemisches Laboratorium der Universität Freiburg, D-7800 Freiburg i. Br., Federal Republic of Germany*

MATS-OLLE MÅNSSON (77), *Pure and Applied Biochemistry, Chemical Center, University of Lund, S-220 07 Lund 7, Sweden*

JOHN P. MARKWELL (47), *Department of Biochemistry, Michigan State University, East Lansing, Michigan 48824*

KAZUNOBU MATSUSHITA (24, 31), *Department of Applied Microbiology, Faculty of Agriculture, Yamaguchi University, Yamaguchi 753, Japan*

P. MAURER (45), *Chemisches Laboratorium der Universität Freiburg, D-7800 Freiburg i. Br., Federal Republic of Germany*

H. PAUL MELOCHE (16), *Papanicolaou Cancer Research Institute, Miami, Florida 33136*

JOSEPH MENDICINO (19, 38), *Department of Biochemistry, University of Georgia, Athens, Georgia 30602*

C. R. MIDDAUGH (97), *Department of Biochemistry, University of Wyoming, Laramie, Wyoming 82071*

KLAUS MOSBACH (74, 77), *Pure and Applied Biochemistry, Chemical Center, University of Lund, S-220 07 Lund 7, Sweden*

U. MÜLLER (90), *Berstrasse 81, D-6900 Heidelburg, Federal Republic of Germany*

R. MICHAEL MULLIGAN (9), *Department of Biochemistry, Michigan State University, East Lansing, Michigan 48864*

DAVID W. NIESEL (51), *Department of Microbiology, The University of Texas at Austin, Austin, Texas 78712*

Y. NOGI (99), *Research Unit for Molecular Genetics, Keio University School of Medicine, Shinanomachi, Shinjuku-ku, Tokyo, Japan*

H. A. NUNEZ (11), *Michigan Department of Public Health, Lansing, Michigan 48909*

NGOZI A. NWOKORO (36), *Department of Biochemistry, University of Toronto, and Research Institute, Hospital for Sick Chil-*

dren, Toronto, Ontario M5G 1X8, Canada

KAZUKO ÔBA (60), Laboratory of Biochemistry, Faculty of Agriculture, Nagoya University, Chikusa-ku, Nagoya 464, Japan

E. L. O'CONNELL (16), Papanicolaou Cancer Research Institute, Miami, Florida 33136

TAIRO OSHIMA (58), Mitsubishi-Kasei Institute of Life Sciences, Machida-shi, Tokyo 194, Japan

FLORA H. PETTIT (65), Clayton Foundation Biochemical Institute, The University of Texas at Austin, Austin, Texas 78712

J. PIERCE (12, 14), Department of Biochemistry, University of Wisconsin, Madison, Wisconsin 53706

JOHN W. PIERCE (9), Department of Biochemistry, Michigan State University, East Lansing, Michigan 48864

REGINA PIETRUSZKO (73), Center for Alcohol Studies, Rutgers University, New Brunswick, New Jersey 08903

J. PILKIS (17), Department of Physiology, Vanderbilt University School of Medicine, Nashville, Tennessee 37232

S. J. PILKIS (17), Department of Physiology, Vanderbilt University School of Medicine, Nashville, Tennessee 37232

ALFRED POLLAK (18), 135 Marlee Avenue, Toronto, Ontario M6B 406, Canada

RAYMOND PORTALIER (35), Laboratoire de Microbiologie et de Génétique Moléculaire, Université Claude Bernard de Lyon, 69622 Villeurbanne, France

J. RODNEY QUAYLE (96), Microbiology Department, University of Sheffield, Sheffield S10 2TN, England

DOUGLAS D. RANDALL (69), Department of Biochemistry, University of Missouri–Columbia, Columbia, Missouri 65211

LESTER J. REED (65), Clayton Foundation Biochemical Institute and Department of Chemistry, The University of Texas at Austin, Austin, Texas 78712

ERWIN REGOECZI (28), Department of Pathology, McMaster University, Hamilton, Ontario L8N 3Z5, Canada

U. RUSCHIG (90), Fachbereich Chemie der Universität, D-2900 Oldenburg, Federal Republic of Germany

JOHN RUSSO (37), Department of Chemistry and Molecular Biology Institute, San Diego State University, San Diego, California 92182

HERMANN SAHM (46, 72, 89), Institut für Biotechnologie, KFA Jülich, D-5170 Jülich 1, Federal Republic of Germany

ROBERTO MARTÍN SARMIENTO (87, 88), Department of Food Biochemistry and Food Technology, Faculty of Veterinary Science, University of Leon, Leon, Spain

SHOJI SASAKI (80), Department of Botany, Faculty of Science, Hokkaido University, Sapporo 060, Japan

HARRY SCHACHTER (36), Department of Biochemistry, University of Toronto, and Research Institute, Hospital for Sick Children, Toronto, Ontario M5G 1X8, Canada

R. M. SCHEEK (53), Department of Physical Chemistry, University of Groningen, 9747 AG Groningen, The Netherlands

HORST SCHÜTTE (46, 72, 89), Gesellschaft für Biotechnologische Forschung, D-3300 Braunschweig, Federal Republic of Germany

T. SEGAWA (99), Research Unit for Molecular Genetics, Keio University School of Medicine, Shinanomachi, Shinjuku-ku, Tokyo, Japan

A. S. SERIANNI (11, 12, 13, 14), Department of Chemistry, University of Notre Dame, Notre Dame, Indiana 46556

WILLIAM R. SHERMAN (1, 2, 100), Department of Psychiatry, Washington University School of Medicine, St. Louis, Missouri 63110

EMIKO SHINAGAWA (22, 31, 32), Department of Applied Microbiology, Faculty of Agriculture, Yamaguchi University, Yamaguchi 753, Japan

RONALD A. SIMKINS (42), Department of Biochemistry, Michigan State University, East Lansing, Michigan 48824

E. C. SLATER (53), *Laboratory for Biochemistry, B.C.P. Jansen Institute, University of Amsterdam, 1018 TV Amsterdam, The Netherlands*

M. L. SPERANZA (55), *Istituto di Chimica Biologica, Università di Pavia, 27100 Pavia, Italy*

ROLF A. STEINBACH (46), *Gesellschaft für Biotechnologische Forschung, D-3300 Braunschweig, Federal Republic of Germany*

FRANÇOIS STOEBER (35), *Laboratoire de Microbiologie, Institut National des Sciences Appliquées de Lyon, 69621 Villeurbanne, France*

C. STOURNARAS (48), *Biochemical Department, Hellenic Pasteur Institute, Athens, Greece*

MARIJKE STRATING (67), *Department of Genetics, Agricultural University, Wageningen 6703 BM, The Netherlands*

JULIA M. SUE (18), *Department of Chemistry, Harvard University, Cambridge, Massachusetts 02139*

YASUTAKE SUGAWARA (80), *Department of Regulation Biology, Faculty of Science, Saitama University, Urawa 338, Japan*

NANAYA TAMAKI (78), *Laboratory of Nutritional Chemistry, Faculty of Nutrition, Kobe-Gakuin University, Tarumi-ku, Kobe, Japan*

O. TERADA (41), *Tokyo Research Laboratory, Kyowa Hakko Kogyo Company, Machida-shi, Tokyo 194, Japan*

N. E. TOLBERT (9), *Department of Biochemistry, Michigan State University, East Lansing, Michigan 48864*

RANDY TOPP (68), *Department of Toxicology, Agricultural University, Wageningen 6703 BM, The Netherlands*

PAUL S. TRESSEL (27), *Department of Biochemistry, State University of New York, Buffalo, New York 14214*

KIHACHIRO UEHARA (39), *College of General Education, Setsunan University, Ikedanakamachi, Neyagawa, Osaka 572, Japan*

SUSUMU UJITA (44), *Laboratory of Biochemistry, Rikkyo (St. Paul's) University, Nishi-Ikebukuro, Toshima-ku, Tokyo 171, Japan*

IKUZO URITANI (60), *Laboratory of Biochemistry, Faculty of Agriculture, Nagoya University, Chikusa-ku, Nagoya 464, Japan*

JOHN M. UTTING (59), *Department of Natural Science, Bernard Baruch College of CUNY, New York, New York, 10010*

T. UWAJIMA (41), *Tokyo Research Laboratory, Kyowa Hakko Kogyo Company, Machida-shi, Tokyo 194, Japan*

J. P. VANDECASTEELE (81), *Institut Français du Petrole, Direction Environnement et Biologie Pétroliere, 92500 Rueil-Malmaison, France*

DAVID L. VANDER JAGT (86), *Department of Biochemistry, The University of New Mexico School of Medicine, Albuquerque, New Mexico 87131*

FRANCESCO M. VERONESE (49), *Institute of Pharmaceutical Chemistry, University of Padova, Padova I-35100, Italy*

JAAP VISSER (67, 68), *Department of Genetics, Agricultural University, Wageningen 6703 BM, The Netherlands*

JEAN-PIERRE VON WARTBURG (30, 85), *Medizinisch-chemisches Institut der Universität, 3000 Bern 9, Switzerland*

ROBERT P. WAGNER (66), *313 Los Arboles Drive, Sante Fe, New Mexico 87501*

WILLIAM C. WENGER (15), *Department of Biochemistry, Michigan State University, East Lansing, Michigan 48824*

BENDICHT WERMUTH (30, 85), *Medizinisch-chemisches Institut der Universität, 3000 Bern 9, Switzerland*

ANDREA WESTERHAUSEN (83), *Arbeitsbereich Enzymologie, Zentrum Biochemie, Medizinische Hochschule, D-3000 Hannover 1, Federal Republic of Germany*

GEORGE M. WHITESIDES (18), *Department of Chemistry, Harvard University, Cambridge, Massachusetts 02139*

P. WILLNOW (90), *Boehringer-Mannheim GmbH, D-8132 Tutzing, Federal Republic of Germany*

CHI-HUEY WONG (18), *Department of Chemistry, Harvard University, Cambridge, Massachusetts 02139*

TAKASHI YONETANI (79), *Department of Biochemistry and Biophysics, School of Medicine, University of Pennsylvania, Philadelphia, Pennsylvania 19174*

JAMES H. YUAN (61), *Department of Chemical Sciences, Old Dominion University, Norfolk, Virginia 23508*

Preface

Volumes 89 and 90 of *Methods in Enzymology* contain new procedures that have been published since 1974 for the enzymes involved in the conversion of monosaccharides to pyruvate or to the breakdown products of pyruvate. During the seven years since the last volumes on this subject were published, the development of new chromatographic separations utilizing affinity and hydrophobic properties of enzymes have resulted in improved procedures differing radically from those available previously. Hence, many of the enzyme purifications now involve fewer steps, are easier to carry out, and often allow separation of isozymes. As with Volumes 41 and 42, preparation of the same enzyme from a number of sources has been included. This is in recognition of the increasing spectrum of interest of investigators in, for instance, the comparison properties from a variety of sources and phylogenetic relationships.

Because of the large amount of material to be presented in this well-defined region of metabolism, division of the chapters into two volumes was necessary. The placement of sections in each volume is arbitrary and is for the most part the pattern found in Volumes 41 and 42.

W. A. WOOD

METHODS IN ENZYMOLOGY

EDITED BY

Sidney P. Colowick and Nathan O. Kaplan

VANDERBILT UNIVERSITY
SCHOOL OF MEDICINE
NASHVILLE, TENNESSEE

DEPARTMENT OF CHEMISTRY
UNIVERSITY OF CALIFORNIA
AT SAN DIEGO
LA JOLLA, CALIFORNIA

METHODS IN ENZYMOLOGY

EDITORS-IN-CHIEF

Sidney P. Colowick   Nathan O. Kaplan

VOLUME XXXV. Lipids (Part B)
Edited by JOHN M. LOWENSTEIN

VOLUME XXXVI. Hormone Action (Part A: Steroid Hormones)
Edited by BERT W. O'MALLEY AND JOEL G. HARDMAN

VOLUME XXXVII. Hormone Action (Part B: Peptide Hormones)
Edited by BERT W. O'MALLEY AND JOEL G. HARDMAN

VOLUME XXXVIII. Hormone Action (Part C: Cyclic Nucleotides)
Edited by JOEL G. HARDMAN AND BERT W. O'MALLEY

VOLUME XXXIX. Hormone Action (Part D: Isolated Cells, Tissues, and Organ Systems)
Edited by JOEL G. HARDMAN AND BERT W. O'MALLEY

VOLUME XL. Hormone Action (Part E: Nuclear Structure and Function)
Edited by BERT W. O'MALLEY AND JOEL G. HARDMAN

VOLUME XLI. Carbohydrate Metabolism (Part B)
Edited by W. A. WOOD

VOLUME XLII. Carbohydrate Metabolism (Part C)
Edited by W. A. WOOD

VOLUME XLIII. Antibiotics
Edited by JOHN H. HASH

VOLUME XLIV. Immobilized Enzymes
Edited by KLAUS MOSBACH

VOLUME XLV. Proteolytic Enzymes (Part B)
Edited by LASZLO LORAND

VOLUME XLVI. Affinity Labeling
Edited by WILLIAM B. JAKOBY AND MEIR WILCHEK

VOLUME XLVII. Enzyme Structure (Part E)
Edited by C. H. W. HIRS AND SERGE N. TIMASHEFF

VOLUME XLVIII. Enzyme Structure (Part F)
Edited by C. H. W. HIRS AND SERGE N. TIMASHEFF

VOLUME XLIX. Enzyme Structure (Part G)
Edited by C. H. W. HIRS AND SERGE N. TIMASHEFF

VOLUME 80. Proteolytic Enzymes (Part C)
Edited by LASZLO LORAND

VOLUME 81. Biomembranes (Part H: Visual Pigments and Purple Membranes, I)
Edited by LESTER PACKER

VOLUME 82. Structural and Contractile Proteins (Part A: Extracellular Matrix)
Edited by LEON W. CUNNINGHAM AND DIXIE W. FREDERIKSEN

VOLUME 83. Complex Carbohydrates (Part D)
Edited by VICTOR GINSBURG

VOLUME 84. Immunochemical Techniques (Part D: Selected Immunoassays)
Edited by JOHN J. LANGONE AND HELEN VAN VUNAKIS

VOLUME 85. Structural and Contractile Proteins (Part B: The Contractile Apparatus and the Cytoskeleton)
Edited by DIXIE W. FREDERIKSEN AND LEON W. CUNNINGHAM

VOLUME 86. Prostaglandins and Arachidonate Metabolites
Edited by WILLIAM E. M. LANDS AND WILLIAM L. SMITH

VOLUME 87. Enzyme Kinetics and Mechanism (Part C: Intermediates, Stereochemistry, and Rate Studies)
Edited by DANIEL L. PURICH

VOLUME 88. Biomembranes (Part I: Visual Pigments and Purple Membranes, II)
Edited by LESTER PACKER

VOLUME 89. Carbohydrate Metabolism (Part D)
Edited by WILLIS A. WOOD

VOLUME 90. Carbohydrate Metabolism (Part E) (in preparation)
Edited by WILLIS A. WOOD

Section I

Analytical Methods

[1] Resolution of DL-*myo*-Inositol 1-Phosphate and Other Sugar Enantiomers by Gas Chromatography[1]

By ALAN L. LEAVITT and WILLIAM R. SHERMAN

Principle

Enantiomeric amino acids, 2-amino-1-phenylethanols, and lactic acid can be separated as simple derivatives by gas chromatography (GC) on a glass capillary column coated with a chiral liquid phase.[2] The phase, named Chirasil-Val,[2] consists of N-t-butyl-L-valinamide in α-amide linkage with a copolymer of dimethylsiloxane and a carboxyalkyl-methylsiloxane. Glass capillary columns coated with Chirasil-Val are commercially available,[3] thus giving a potentially straightforward general method for separating enantiomeric mixtures or performing quantitative analyses with chiral specificity. The original applications of Chirasil-Val were with nitrogen-containing substances and stressed the role of the nitrogen atoms in the achievement of chiral separations.[2] It appears from our work that this chiral phase also has the ability to separate non-nitrogen-containing optical isomers. Using this column we have resolved enantiomeric mixtures of several carbohydrates in the form of simple derivatives produced by generally applicable methods.

Derivatization Procedures

Trimethylsilyl–methylphosphate (*TMS-Me*) derivatives of inositol phosphates were prepared by complete trimethylsilylation of the molecules followed by methanolysis of the phosphate trimethylsilyl (TMS) ester moiety and then reaction with diazomethane.[4]

Heptafluorobutyric (*HFB*) *esters* of *chiro*-inositol[5] and of arabitol, of several aldoses and of two aldonolactones were prepared by heating a

[1] W. R. Sherman and A. L. Leavitt, "29th Conf. on Mass Spectrometry and Allied Topics," p. 49. Minneapolis, Minn., May 24–29, 1981. A. L. Leavitt and W. R. Sherman, *Carbohyd. Res.* in press, 1982. We thank Dr. B. E. Phillips and J. Zwicker for their contributions to the development of the inositol separations. This work was supported by NIH Grants NS-05159, NS-13781, AM-20579, and RR-00954.
[2] H. Frank, G. J. Nicholson, and E. Bayer, *J. Chromatogr. Sci.* **15**, 174 (1977); *J. Chromatogr.* **146**, 197 (1978); *Angew. Chem. Int. Ed. Engl.* **17**, 363 (1978).
[3] Applied Science Laboratories, State College, Pennsylvania.
[4] W. R. Sherman, A. L. Leavitt, M. P. Honchar, L. M. Hallcher, and B. E. Phillips, *J. Neurochem.* **36**, 1947 (1981); see also this volume [2].
[5] Sources of sugars: arabinose, D, Calbiochem; L, Sigma; fucose; D, Sigma; L, Pfanstiehl; mannose, D, Fisher; L, Sigma; *chiro*-inositol, D, Calbiochem; L, gift of Professor Laurens

TABLE I
ENANTIOMERIC SEPARATION OF DL-2,3,4,5,6-
PENTAKIS(O-TRIMETHYLSILYL)
myo-INOSITOL 1-DIMETHYLPHOSPHATE[a]

Retention time (min)			Column	
L-Isomer	D-Isomer	Resolution[b]	Temperature (°C)	Pressure (psi)
6.30	6.46	0.8	220	10
14.44	15.10	1.14	185	20
40.60	43.38	2.08	160	26

[a] Chromatographed on a 20-meter Chirasil-Val capillary column.
[b] $R = 2d/w_1 + w_2$, where d is the peak-to-peak separation and w_1, w_2 is the width at the baseline of the idealized (by triangulation) peak.

suspension of the sugar (0.1–0.5 mg) in 100 μl of heptafluorobutyrylimidazole[6] at 60° for 30–60 min with occasional mixing. The resulting solution may be chromatographed directly; however, condensing reaction by-products solidify and clog splitter exit lines. To remove excess reagent and by-products, the sample may be diluted with hexane and chilled to −20°, which precipitates the reagent. The supernatant is removed, and the precipitate is washed with cold hexane. The combined hexane extracts can be concentrated prior to GC.

Cyclic alkaneboronic esters of carbohydrates with an even number of hydroxyls (e.g., *chiro*-inositol[7] and arabinose[8]) were prepared by allowing 1 mg of the sugar to react with 10 mg of methaneboronic acid[9] or butaneboronic acid[9] in 1 ml of pyridine at room temperature overnight, or at 100° for 30 min. The completely dissolved samples are then chromatographed directly.

Carbohydrates with an odd number of hydroxyls (e.,g., aldohexoses) require a coderivative for the remaining hydroxyl in order to confer suitable chromatographic properties. Subsequent to the alkaneboronic acid reaction described above, the partially derivatized substrate can be

Anderson, University of Wisconsin; D- and L-xylose, Pierce; glucose, D, Fisher; L, Sigma; DL-glyceraldehyde, Sigma; D- and L-arabitol, Pierce; gulono-γ-lactone, D, Pfanstiehl; L, Nutritional Biochemicals Corp.; galactono-γ-lactone, D, Pierce; L, General Biochemicals, Inc.

[6] Pierce Chemical Co., Rockford, Illinois.
[7] J. Wiecko and W. R. Sherman, *J. Am. Chem. Soc.* **101,** 979 (1979).
[8] F. Eisenberg, Jr., this series, Vol. 28 [11].
[9] Alfa Products, Danvers, Massachusetts.

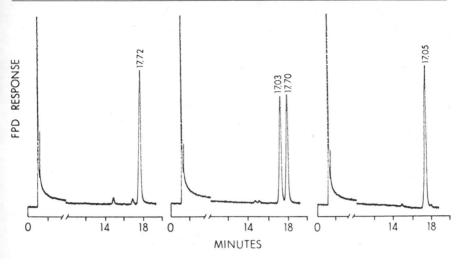

FIG. 1. Gas chromatogram of D-*myo*-inositol 1-P from phosphatidylinositol (left), synthetic DL-*myo*-inositol-1-P (center), and L-*myo*-inositol-1-P formed from bovine L-*myo*-inositol-1-P synthase (right). Each of the enantiomers is chromatographed as the TMS$_5$-Me$_2$ derivative[13] on a 25-meter Chirasil-Val glass capillary column. Each peak represents about 12 ng of the sugar acid. Conditions: flame photometric detection (FPD), phosphorus-selective mode, split injection, He carrier (20 psi), 190°.

trimethylsilylated[10] by adding 100 μl of *N*,*O*-bis(trimethylsilyl)trifluoro-acetamide containing 10% trimethylsilyl chloride.[11] Alternatively, the partial derivative can be acetylated[12] (50 μl of acetic anhydride), or the HFB derivative prepared using 50 μl of heptafluorobutyrylimidazole. Each of the latter reactions is essentially complete after an additional 30 min, and the sample is then ready for chromatography.

myo-Inositol Phosphates

DL-*myo*-Inositol 1-phosphate TMS$_5$ME$_2$[13] is well resolved into its enantiomers by a 20-meter Chirasil-Val column (Table I). An application of this separation is shown in Fig. 1, where D-*myo*-inositol-1-P TMS$_5$Me$_2$ (D-M1P TMS$_5$Me$_2$) from phosphatidylinositol[14] (left) is compared with synthetic[15] DL-M1P-TMS$_5$Me$_2$ (center) and M1P-TMS$_5$Me$_2$

[10] V. N. Reinhold, F. Wirtz-Peitz, and K. Biemann, *Carbohydr. Res.* **37**, 203 (1974).
[11] Regis Chemical Co., Morton Grove, Illinois.
[12] J. Wiecko and W. R. Sherman, *J. Am. Chem. Soc.* **98**, 7631 (1974).
[13] See footnote 16 of this volume [2] for abbreviations.
[14] F. L. Pizer and C. E. Ballou, *J. Am. Chem. Soc.* **81**, 915 (1959); C. E. Ballou and F. L. Pizer, *J. Am. Chem. Soc.* **81**, 4745 (1959).
[15] D. E. Kiely, G. L. Abruscato, and V. Baburao, *Carbohydr. Res.* **34**, 307 (1974).

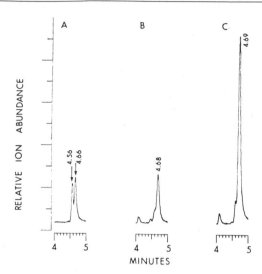

FIG. 2. Gas chromatography–mass spectrometry with the chiral capillary using selected ion monitoring. Each trace is that of m/z 649 (MH$^+$) of M1P TMS$_5$-Me$_2$ generated by ammonia chemical ionization. Trace A represents 20 ng of racemic M1P, the L-enantiomer eluting first; trace B was obtained by derivatization of a 20-mg (dry weight) sample of rat cerebral cortex and chromatography of an aliquot. Trace C is from cerebral cortex of a rat treated with LiCl (intraperitoneally, 3.6 meq/kg body weight daily for 9 days). The chromatograms show the D-enantiomer to be the major inositol phosphate in control and the enantiomer mainly affected by lithium in rat cerebral cortex. Conditions: column 210°, 20 meter, He carrier (10 psi).

from the reaction of bovine myo-inositol-1-P synthase[16] (right) which is thus shown to be the L-enantiomer.

The Chirasil-Val column is suitable for GC-MS use also. When installed in a Finnigan Model 3300 chemical ionization instrument, using ammonia as reagent gas, selected ion monitoring with enantiomeric separation can be performed on the M1P isomers. Figure 2 shows chromatograms of M1P-TMS$_5$Me$_2$ using m/z 649 (MH$^+$) as the detector ion. Figure 2A shows racemic M1P, and 2B shows M1P from cerebral cortex of a control rat. Figure 2C shows M1P from cerebral cortex of a rat treated with LiCl, where it can be seen that the D-enantiomer is elevated by the lithium.[4] Notice the short retention time, which, while degrading resolution somewhat, still allows unambiguous identification of the optical isomer. Chemical ionization GC-MS is about 10-fold less sensitive than phosphorus-selective flame photometric detection. These tissue results were obtained by direct silylation.[4]

[16] See this series, Volume 90 [50].

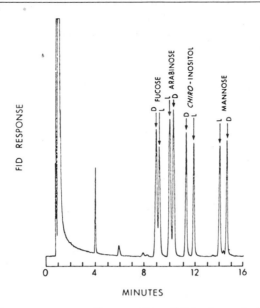

MINUTES

FIG. 3. A gas chromatogram, on a 25-meter Chirasil-Val column of mixed enantiomers of fucose, arabinose, *chiro*-inositol, and mannose, each as the per(heptafluorobutyric) esters. Each peak represents about 10 ng of each sugar, prior to derivatization. The peak at 4 min is an unknown related to derivatization. No aldose anomers are seen. Conditions: He carrier (15 psi); temperature program from 95° (isothermal to 8 min) then 3°/min; flame ionization detection (FID).

(±)-*chiro-Inositol.* This is the only unsubstituted inositol that occurs in enantiomeric form. The enantiomers are best resolved as the hexakis(heptafluorobutyryl) (HFB) derivative (Fig. 3; Table II). The *chiro*-inositol enantiomers also separate as the tris(methaneboronate) (Table III). The TMS derivative of (±)-*chiro*-inositol did not give enantiomeric resolution.

Aldoses. Figure 3 shows the separation of the enantiomers of fucose, arabinose, and mannose on a 25-meter Chirasil-Val column, each as the HFB derivative. Table II lists their retention times and the resolution achieved at a single temperature/pressure. Table II shows xylose and glucose to be only poorly separated as their HFB derivatives.

DL-Glucose is best separated on a 20-meter column as the methaneboronate coderivatized with a trimethylsilyl group or as the butaneboronate coderivatized with TMS or HFB moiety (Table III). The alkaneboronate-acetate derivatives are less well resolved. DL-Arabinose is only poorly resolved as the butaneboronate.

Sugars That Could Not Be Resolved. No enantiomeric separation was achieved on a 20-meter column with the following sugar derivatives:

TABLE II
ENANTIOMERIC SEPARATION AS THE HEPTAFLUOROBUTYRATE DERIVATIVES[a]

per(HFB) sugar	Retention time (min)		Resolution[b]	Column temperature (°C)
	L-Isomer	D-Isomer		
Fucose	7.01	6.76	0.50	100
Arabinose	6.12	6.41	0.71	105
chiro-Inositol	3.72	3.40	1.03	115
Mannose	3.13	3.38	1.19	125
Xylose	14.84	14.38	0.3[c]	95
Glucose	5.93	5.96	0.2[c]	120

[a] Gas chromatography on a 20-meter Chirasil-Val capillary column at 15 psi as the per(heptafluorobutyryl) (HFB) derivative.

[b] Resolution, except where noted, as defined in Table I, footnote b.

[c] Estimated resolution; i.e., the valley depth between two peaks divided by the average peak height.

DL-myo-inositol-4-P TMS$_5$Me$_2$; DL-myo-inositol cyclic-1,2-P TMS$_4$Me; any of the inositol phosphates as the per(TMS) derivatives; arabitol TMS or HFB derivative; DL-galactono- and gulono-1,4-lactones as the HFB derivatives, and DL-α-glycerophosphate as the butaneboronate TMS ester.

TABLE III
ENANTIOMERIC SEPARATION OF ALKANEBORONATE DERIVATIVES[a]

Boronate	Retention time (min)		Resolution[b]	Column[c] temperature (°C)
	L-Isomer	D-Isomer		
Glucose MeB-TMS	5.64	5.75	0.8	130
Glucose BuB-TMS	10.18	10.35	0.6	170
Glucose BuB-HFB[d]	6.09	6.22	0.8	180
Glucose MeB-Ac	14.65	15.01	0.2	110
Glucose BuB-Ac	15.04	15.28	0.2	170
Arabinose BuB	11.64	11.82	0.3	140
chiro-Inositol MeB	2.92	3.00	0.4	150

[a] Gas chromatography on a 25-meter Chirasil-Val capillary column as methane boronates (MeB) or butane boronates (BuB). Glucose is coderivatized with a trimethylsilyl (TMS), heptafluorobutyryl (HFB), or acetyl (Ac) group.

[b] Estimated resolution as defined in Table II, footnote c.

[c] Column head pressure, 20 psi.

[d] This derivative gives a complex chromatogram, although the enantiomers are the principal peaks.

Comments. The aldose enantiomers of Fig. 3 each chromatograph on the chiral column as a single peak, i.e., without evidence of resolved anomers. On a nonpolar capillary column >90% of each of these sugars also chromatographs as a single peak. Glucose, however, did evidence anomeric separation (β anomer R_t 1.7 that of α-anomer) on a nonpolar capillary column. The anomeric mixture from glucose chromatographed, however, as a single peak on the Chirasil-Val capillary.

While no general guidelines for separating optical isomers of other sugars can be suggested, our results indicate that this method has promise in a general sense. We have examined only a few derivative types and only one chiral phase (at least one other is available[17,18]); our findings suggest that an empirical approach using simple derivatives may prove to be worthwhile in other instances.

[17] Alltech Associates, Deerfield, Illinois.
[18] Additional studies on the separation of aldose enantiomers include: W. A. König, I. Benecke, and H. Bretting, *Angew. Chem.* (English Ed.) **20**, 693 (1981); W. A. König, I. Benecke, and S. Sievers, *J. Chromatogr.* in press (1982). Dr. H. Frank, of the Universität Tübingen, has performed similar separations (personal communication).

[2] Determination of Inositol Phosphates by Gas Chromatography[1]

By ALAN L. LEAVITT and WILLIAM R. SHERMAN

Background

The separation of inositol phosphates has been performed by paper[2] and thin-layer chromatography[3] as well as by electrophoresis[4] and ion exchange chromatography.[2] *myo*-Inositol 1- and 2-phosphates have been separated by high-pressure liquid chromatography.[5] When quantitative analyses have been performed by these methods, the sensitivity has generally been limited by that available with molybdate-based methods of phosphate analysis. Nevertheless, these techniques have great value in experiments involving ^{32}P.

[1] This work was supported by NIH Grants NS-05159, NS-13781, AM-20579, and RR-00954.
[2] C. Grado and C. E. Ballou, *J. Biol. Chem.* **236**, 54 (1961).
[3] J. K. Hong and I. Yamane, *Soil Sci. Plant Nutr.* (*Tokyo*) **26**, 391 (1980).
[4] U. B. Seiffert and B. W. Agranoff, *Biochim. Biophys. Acta* **98**, 574 (1965).
[5] L. M. Hallcher and W. R. Sherman, *J. Biol. Chem.* **255**, 10,896 (1980).

The gas chromatography (GC) of sugar phosphates followed quickly after the discovery of the utility of the trimethylsilyl (TMS) derivative for the GC of sugars.[6,7] The first report utilized the TMS derivatives of sugar phosphates, where the phosphate moiety was converted to the dimethyl ester with diazomethane.[8] Later it was found that both the alcoholic and acidic hydroxyl functions could be derivatized with TMS groups.[9-12] More recently another type of derivative for sugar phosphates has been described, the alkaneboronate dimethyl phosphate,[13] which is of less general utility.

Whereas quantitative analysis of simple sugars by GC is common and presents few methodological problems, sugar phosphates, by virtue of their extremely polar nature, are difficult to gas chromatograph quantitatively, especially in the small amounts often encountered in biological experiments. This report describes our experiences with the quantitative analysis of a small group of sugar phosphates, the isomers of *myo*-inositol phosphate. Although limited in scope, the principles should apply in the more extended case of sugar phosphates in general.

Detectors, Sensitivity

The three GC detectors of use for the measurement of small amounts of phosphorylated sugars are the flame ionization detector (FID), the flame photometric detector (FPD), and the mass spectrometer via the gas chromatographic inlet (GC–MS).

The FID is a general detector with a response proportional to the carbon content of the substrate. It is reliable and has a wide linear dynamic range and a usable sensitivity to about 50 pmol with typical sugar TMS derivatives.

The FPD in use in our instruments (Varian Model 3700 GC) is reported to have a phosphorus : carbon response ratio of 5×10^5 (on a gram-atom basis) when a 530-nm filter is used.[14] This detector responds principally to HPO emission generated by phosphorylated substances in a hydrogen

[6] R. Bentley, C. C. Sweeley, M. Makita, and W. W. Wells, *Biochem. Biophys. Res. Commun.* **11**, 14 (1963).

[7] C. C. Sweeley, R. Bentley, M. Makita, and W. W. Wells, *J. Am. Chem. Soc.* **85**, 2497 (1963).

[8] W. W. Wells, T. Katagi, R. Bentley, and C. C. Sweeley, *Biochim. Biophys. Acta* **82**, 408 (1964).

[9] T. Hashizume and Y. Sasaki, *Anal. Biochem.* **15**, 346 (1966).

[10] F. Eisenberg, Jr., and A. H. Bolden, *Anal. Biochem.* **29**, 284 (1969).

[11] W. R. Sherman, S. L. Goodwin, and M. Zinbo, *J. Chromatogr. Sci.* **9**, 363 (1971).

[12] D. J. Harvey and M. G. Horning, *J. Chromatogr.* **76**, 51 (1973).

[13] J. Wiecko and W. R. Sherman, *Org. Mass Spectrom.* **10**, 1007 (1975).

[14] P. L. Patterson, R. L. Howe, and A. Abu-Shumays, *Anal. Chem.* **50**, 339 (1978).

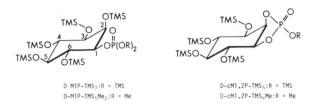

FIG. 1. Derivatives of inositol phosphates used for gas chromatography.

flame. The sensitivity of the FPD with our samples is 5–10 times that of the FID, with a comparable signal : noise ratio. Non-phosphorus-containing substances are virtually unseen. Most of our work with tissue inositol phosphates is done with the FPD.

GC–MS presents a greater problem with phosphorylated sugars than with many other substances owing to adsorption of these polar molecules by exposed metal surfaces of many instruments. The qualitative GC–MS of sugar phosphates has been reported,[12,15] but there are no reports of quantitative GC–MS analysis of these substances. We have found the measurement of chemical standards of TMS *myo*-inositol-1-P (M1P-TMS$_7$) and TMS *myo*-inositol cyclic-1,2-phosphate (cM1,2P-TMS$_5$) to be linear to about 1 pmol when using TMS-*chiro*-inositol-3-P as an internal standard (J. H. Allison and W. R. Sherman, unpublished observation). In this experiment, we used a Finnigan Model 3200 electron ionization instrument and ions *m/z* 389 for TMS *myo*-inositol-1-P and *chiro*-inositol-3-P; *m/z* 398 was used for *myo*-inositol-cyclic-1,2-P. The advantages of GC–MS over the FPD are principally in specificity: by using appropriate ions from a mass spectrum, near-absolute molecular specificity can be achieved, often with little compromise in sensitivity. Exposed metal surfaces and the complexity and cost of these instruments are the main drawbacks.

Derivatives

We have used two derivatives for quantitative work: the completely trimethylsilylated inositol phosphates (M1P-TMS$_7$, M2P-TMS$_7$, M4P-TMS$_7$, M5P-TMS$_7$, cM1,2P-TMS$_5$); and the TMS dimethylphosphorylinositols (M1P-TMS$_5$Me$_2$, etc., as well as cM1,2P-TMS$_4$Me), (Fig. 1).[16]

[15] M. Zinbo and W. R. Sherman, *J. Am. Chem. Soc.* **92**, 2105 (1970).
[16] Correct nomenclature for these ether esters is as follows: M1P-TMS$_7$, *myo*-inositol-2,3,4,5,6-pentakis(*O*-trimethylsilyl)-1-bis(trimethylsilyl)phosphate; M1P-TMS$_5$Me$_2$, *myo*-inositol-2,3,4,5,6-pentakis(*O*-trimethylsilyl)-1-dimethylphosphate; cM1,2P-TMS$_5$, *myo*-inositol-3,4,5,6-tetrakis(*O*-trimethylsilyl) cyclic-1,2-(trimethylsilyl)phosphate; etc.

RETENTION TIMES OF INOSITOL AND SORBITOL PHOSPHATES ON PACKED AND CAPILLARY COLUMNS

	Per(trimethylsilyl) derivative		Trimethylsilyl ether-dimethylphosphate derivative		
	3% OV-17[a]	CP^{tm}Sil 5[b]	CP^{tm}Sil 5[b]	DB-1[c]	SP-2255[d]
myo-Inositol-1-P	1.0 (8.5 min)	1.0 (7.8 min)	1.0 (5.9 min)	1.0 (7.1 min)	1.0 (6.3 min)
myo-Inositol cyclic-1,2-P	—	0.48	0.51	0.46	—
myo-Inositol-5-P	0.71	0.82	0.95	—	0.96
Sorbitol-6-P	—	0.88	0.75	—	—
myo-Inositol-4-P	0.87	1.03	0.90[e]	0.88[f]	0.86
myo-Inositol-2-P	1.37	1.17	—	1.10	—

[a] Packed glass column (6 foot × ¼ in. o.d.), 180°, 90 ml/min He carrier.
[b] Glass capillary (14 meters), 200°, 20 psi He carrier (Chrompack, Netherlands).
[c] Fused silica capillary (10 meters), 180°, 16 psi, He carrier (J & W Scientific, Rancho Cordova, California).
[d] Glass capillary (5 meters), 175°, 15 psi He carrier (Supelco, Bellefonte, Pennsylvania).
[e] A peak at 1.19 relative to M1P-TMS$_5$Me$_2$ has been identified by GC–MS as M1P-TMS$_6$Me; its formation is temperature dependent.
[f] A peak at 1.08 relative to M1P-TMS$_5$Me has been identified by GC–MS as M4P-TMS$_6$Me; its formation is temperature dependent.

Derivative selection in the case of the inositol phosphates is made on the basis of the separation desired. The table shows the relative retention times of these derivatives on four chromatography columns. As can be seen, the separation of M1P from M4P on nonpolar capillary columns must be performed with the TMS dimethylphosphorylinositols. No consistent differences between the two derivative types have been seen with respect to adsorptive losses, but the high reactivity of the TMS esters may result in losses due to covalent interactions with some liquid phases.

Per(trimethylsilyl) derivatives are prepared in our laboratory in two ways: either direct silylation of lyophilized tissues or water extraction of frozen tissue in a boiling water bath followed by centrifugation, removal of an aliquot, lyophilization, and silylation. The direct method has been used the most and, for the inositol phosphates of brain tissue, gives consistently reliable results.

Direct trimethylsilylation has been used in this laboratory for the analysis of both nonphosphorylated[17,18] and phosphorylated[19] inositols in small

[17] M. A. Stewart, V. Rhee, M. M. Kurien, and W. R. Sherman, Biochim. Biophys. Acta 192, 361 (1969).
[18] W. R. Sherman, P. M. Packman, M. H. Laird, and R. L. Boshans, Anal. Biochem. 78, 119 (1977).
[19] J. H. Allison, M. E. Blisner, W. H. Holland, P. P. Hipps, and W. R. Sherman, Biochem. Biophys. Res. Commun. 71, 664 (1976).

sections of brain tissue. Rats are decapitated into liquid nitrogen, and the heads are either stored at $-70°$ or rapidly dissected in a $-20°$ cold room. For M1P analysis in control tissue (0.3 mmol/kg dry tissue) 15–20 mg of tissue, cut into pieces of 2–3 mg, are placed in a cooled 1-ml conical centrifuge tube. The tubes are transferred, without warming above $-20°$ and within the space of 2 hr, to a lyophilizing flask in a freezer chest ($-37°$) and freeze-dried for 36 hr at a pressure of 5 μm or less. While still under vacuum, the jar is warmed to room temperature and further dried for 2–3 hr. The dry tissue is then freed of adventitious material and quickly weighed at room temperature on a Roller–Smith balance. Enough material for several analyses is obtained by transferring 3 mg of dry tissue to a 1-ml conical interior screw-capped reaction vial and adding 125 μl of trimethylsilylating reagent. The reagent consists of equal parts of dry pyridine[20] and N,O-bis(trimethylsilyl)trifluoroacetamide containing 10% trimethylsilyl chloride. The preparation is allowed to stand at room temperature for 48 hr; it is then analyzed or it may be stored at $-70°$ in a screw-capped jar containing Drierite. Jars stored at $-70°$ must be allowed to warm to room temperature before being opened because moisture condensing in the derivatized preparation can make the sample useless for analysis.

A sample concentration of 25 μg of dry tissue per microliter of derivatizing reagent provides adequate sensitivity for M1P analyses. Although more concentrated preparations would increase sensitivity, the greater viscosity makes accurate microsyringe injection difficult.

Trimethylsilyl ether methyl esters such as M1P-TMS$_5$Me$_2$ are best prepared from the per(trimethylsilyl) derivative. When tissues or chemical standards are allowed to react directly with diazomethane, migration of the phosphate ester occurs, giving a mixture of phosphate isomers. The isomers are sometimes minor by-products but always complicate the analysis.

In the procedure we have developed, a measured volume of trimethylsilylated product is transferred into a presilanized (with trimethylsilylating reagent) 2-ml reaction vial using a constriction pipette, leaving behind as much tissue as possible. The reagent is then removed under a stream of N_2 at room temperature and further dried under mechanical pump vacuum; the residue is taken up in about 1 ml of 10% absolute methanol in diethyl ether at $0°$. This procedure removes the TMS ester moieties on the phosphate, giving the free acid. Low speed centrifugation of samples at this stage to sediment tissue residues effects a substantial cleanup of the preparation. An aliquot is then removed, and diazomethane is either bubbled through the solution or added dropwise in methanolic ether until the yel-

<hr />

[20] Purchased or prepared by storing reagent grade pyridine over molecular sieve 5A that has been heated to 450° for 1 hr, cooled, and stored in a desiccator.

low color persists. After 5 min at 0°, the solvents and excess diazo-methane are removed in a stream of nitrogen and the residue is taken up in a measured volume of trimethylsilylating reagent, prepared as before. After about 30 min, the sample is ready for analysis. This procedure has been found to give quantitative recovery of the inositol phosphomono-esters when volumetric accuracy is preserved.

Tissue Extractions. We have limited experience with extraction of tis-sues to remove inositol phosphates. Kidney tissue, for example, gives a much more complex mixture on direct silylation than does brain. Extrac-tion with water for 5 min in a boiling water bath and lyophilization of an aliquot followed by trimethylsilylation has given preparations in which M1P is more easily measured. There appears to be no loss of M1P from standards or tissues by this method (L. Y. Munsell and W. R. Sherman, unpublished observations).

Gas Chromatography

Packed GC Columns. These columns have been used in our laboratory with variable success for several years. The packing that has been used the most[19] is 3% OV-17 on Gas Chrom Q (Applied Science Laboratories, State College, Pennsylvania). Another packing found to be satisfactory is 3% SP-2250 Clinical Packing (1-1767 Supelco, Inc., Bellefonte, Pennsyl-vania). Both of these packings use 50% phenylmethylsilicone as a liquid phase. This polar phase effectively resolves the inositol phosphate iso-mers. However, the inertness of the OV-17 packing varies from lot to lot, many batches being too adsorptive for M1P analyses. We have limited, but successful, experience with the SP-2250 clinical grade packing; it is pretested by the manufacturer for another application and may, therefore, be reliable for sugar phosphate analyses. The presilanized packed glass columns are generally 1.8 meters × 4 mm i.d. and are operated at 180° (isothermal) with helium carrier gas at a flow rate of 90 ml/min. The column extends into the injection port, but the packing does not. Injec-tions are made into a plug of silanized glass wool that is changed daily. The table gives retention times of several sugar phosphates. The selection of OV-17 is based on the fact that, on packed SE-30 columns, M1P-TMS$_7$ and β-glucose-6-P-TMS$_6$ are not separated.[11]

A useful column will give a linear dilution response to standards over the sample range of interest. Some columns, however, will adsorb sam-ples as large as 10 nmol. To start an analysis, it is usually necessary to inject several 10-μl samples of derivatized brain tissue (prepared for this purpose from lyophilized whole brain tissue) and to allow this material to chromatograph while temperature-programming from 180° to 280°. When a column is operating well, approximately 10 sample analyses can be run, followed by standards, before broad peaks from less volatile components begin to elute from the column. At this point, the column should be cleaned out by again temperature-programming to 280° for about 15 min. After the baseline has stabilized, additional analytical samples may be chromatographed as before. Eventually, the glass wool plug becomes contaminated by carbonaceous deposits. These charred areas tend reversibly to adsorb sample components, which may elute from the column during subsequent GC runs, creating "ghost" peaks. Deposits of divalent metals, such as zinc from deproteinizing procedures, will completely adsorb the sugar phosphates and thus must be avoided.

Capillary Columns. Because of the difficulties in using packed columns, we have turned to glass and fused silica capillary columns. Although dependable columns are not yet made with polar liquid phases such as OV-17 and SP-2250, the high efficiency of nonpolar capillary columns is such that α-Glc-6-P TMS$_6$ and β-Glc-6-P-TMS$_6$ can be separated from M1P-TMS$_7$. Although we have found column-to-column variability in quality, open tubular wall-coated columns that have been used successfully include the following: fused silica, 0.2 mm i.d., 10-meter, coated with 0.2 μm of OV-101 (Quadrex Corp., New Haven, Connecticut); glass, 0.25 mm i.d., 10-meter coated with 0.2 μm of CP^{t-m}Sil 5 (Chrompack, The Netherlands); and fused silica, 0.25 mm i.d., 10-meter, coated with 0.25 μm of a chemically bonded liquid phase, DB-1 (J & W Scientific, Rancho Cordova, California). These are all nonpolar polymethylsiloxane liquid phases similar to SE-30, but that give somewhat different relative retention times than packed SE-30 columns. Each of these columns chromatographs M1P-TMS$_7$ successfully with minimal priming before analyses. Some columns were obtained in the length described, and others were shortened by breaking them into two sections.

Our capillary GC uses a glass-lined inlet splitter (Varian Instruments) to control the amount of sample entering the column. A glass-lined adapter (Varian) also connects the effluent of the column to the detector. This equipment permits glass capillary column installation without the need for straightening column ends. With the appropriate injection port and detector adapters (Varian), fused silica columns may be installed so that the front end of the column serves as its own split tip and the detector end is positioned at the flame tip. Results are comparable to the all-glass system. GC–MS installation of the glass capillaries using glass-lined stainless steel tubing (Scientific Glass and Engineering, Austin, Texas) in the GC oven (to avoid column end-straightening) did lead to some deterioration of performance. With fused silica, the columns are led from the injection splitter tip directly into the mass spectrometer source (Finnigan 3300 chemical ionization instrument) with no loss in column resolution or obvious adsorptive losses.

Columns were operated at head pressures of about 8 psi giving flows of about 2–5 ml/min of helium carrier. In GC–MS experiments NH$_3$ was introduced, as ionizing reagent gas, at the effluent end of the column. All injections (1–2 μl) were made into a silanized injection port which had either a silanized glass wool plug or an integral glass frit below the injection point. About a 4 : 1 split was utilized. Operation without either a glass wool plug or a frit led to gross contamination of the capillary column with residues. In general, samples and standards are chromatographed on capillaries using temperature programming, although the data in the table were obtained by operating isothermally. Using a 5°/min temperature program of from 160° to 200° (hold 2 min) followed by a 2–3 min cool-down, samples can be run all day on a 4 m column without contaminant interference. Samples are interspersed with standards to monitor the adsorptivity of the injection port insert, and the insert is replaced if the response of the standard diminishes. During replacement, the column oven *must* be cooled to room temperature to avoid oxygen damage to the hot capillary column. Occasionally, removal of the first few cm of column is necessary to restore column inertness.

Separations on the capillary columns are not easily controlled, with this group of compounds, by selection of a suitable liquid phase because at this time only the nonpolar phases are available as highly deactivated columns. Instead, we change from the per(trimethylsilyl) derivative to the TMS$_5$Me$_2$ ether esters to achieve the M1P–M4P separation. Using an SP-2250 capillary column (analogous to an OV-17 packed column) separates M1P- and M4P-TMS$_7$—however, with the complete loss of cM1,2P-TMS$_5$, apparently through adsorption.

myo-Inositol 1,2-cyclic phosphate as the per(trimethylsilyl) derivative is the most avidly adsorbed substrate we have encountered. The underivatized substance may be prepared from *myo*-inositol-2-P (Sigma Chemical Co., St. Louis, Missouri) by reaction with dicyclohexylcarbodiimide.[21] We have found it to be useful in the assessment of the adsorptivity of columns. A series of dilutions to the limit of sensitivity will show a rapid loss of cM1,2P-TMS$_5$ from a defective column while M1P- or M2P-TMS$_7$ will give a constant response : weight ratio. This is illustrated by Fig. 2.

Figure 3 shows another effect that we have encountered with these derivatives—however, only on capillary columns. Phosphate migration, in this case from M2P-TMS$_7$ to M1P-TMS$_7$, can occur during chromatography or during injection. The migration is temperature-dependent; thus it can be controlled to a degree. In Fig. 3 the column temperature was reduced, while maintaining comparable retention times, by using hydro-

[21] F. L. Pizer and C. E. Ballou, *J. Am. Chem. Soc.* **81**, 915 (1959).

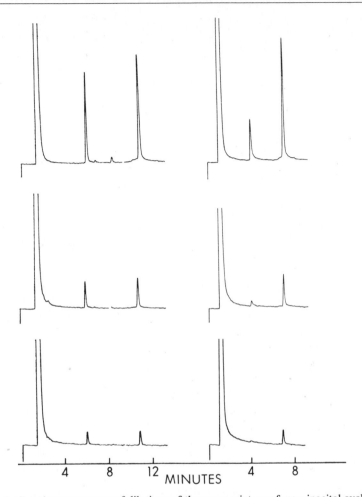

FIG. 2. Gas chromatograms of dilutions of the same mixture of *myo*-inositol cyclic-1,2-P TMS$_5$ (first peak) and *myo*-inositol-1-P TMS$_7$ (second peak) on two new commercial capillary columns. The column on the left is satisfactory; that on the right is not. Both columns were coated with a nonpolar polymethylsiloxane. The column on the left is 29 meters long (0.2 mm i.d., 0.25 μm film); that on the right is a 25-meter column (0.25 mm i.d., 0.2 μm film). In each case the top chromatograms represent 41 pmol of cM1,2P and 50 pmol of M1P on the column. The two center chromatograms are 12 and 14 pmol of cM1,2P and M1P; the bottom chromatogram is 6 and 7 pmol, respectively. The M1P chromatographs quantitatively on each column, but the cM1,2P is highly absorbed by the column on the right. Adsorption is a major difficulty in sugar phosphate gas chromatography; it can have a deleterious effect on the quantitation of M1P as well, although this is not seen here. The use of cM1,2P-TMS$_5$ as a test for column performance is recommended. Detection was by flame photometric detector in phosphorus-selective mode, split injection, He carrier.

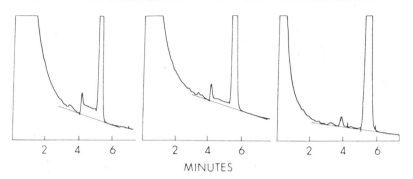

MINUTES

FIG. 3. Gas chromatograms of *myo*-inositol-2-P TMS$_7$ (at about 5.5 min) wherein *myo*-inositol-1-P TMS$_7$ (at about 4 min) is generated during passage through the column in a temperature-dependent manner. The chromatogram on the left was run at 195° (He carrier, 10 psi); that in the center at 180° (H$_2$ carrier, 9 psi); that on the right at 165° (H$_2$ carrier, 20 psi). The "bridging" between the two peaks is due to migration of the TMS phosphate from the inositol oxygen-2 to oxygen-1 during passage.

gen as carrier gas (elution times are shorter with hydrogen than with helium) and by increasing the carrier velocity.[22] When this occurs in the injection port the sample appears to be contaminated with the isomer. Reduction of the injection port temperature reduces or eliminates the formation of the isomeric phosphate. This is not observed in the reverse direction, i.e., from M1P to M2P, since the equatorial 1-phosphate is thermodynamically more stable.

In summary, a complete analysis of the inositol monophosphates is best done with two derivatives: cM1,2P is measured as the TMS$_5$ derivative and M1P in the absence of significant amounts of M4P also may be analyzed as the per(trimethylsilyl) derivatives. However, if analysis of M1P, M4P, and M5P is to be performed on a nonpolar capillary column, the TMS$_5$Me$_2$ derivative must be used.

Our experiences with the inositol phosphates offer general guidelines for the GC analysis of sugar phosphates. The extremely inert capillary columns installed in an all-glass system permit very polar compounds to be chromatographed with minimal adsorption. Although these columns often provide a high degree of resolution, substances as chemically similar as the inositol phosphate isomers may require derivative modification to attain a particular separation. By using an internal standard, quantitative accuracy can be preserved even if volumetric changes accompany such manipulations. The resulting analysis can be highly specific and sensitive.

[22] Caution to ensure leak-free capillary connections is especially important with hydrogen as carrier. With polyimide-strengthened fused silica columns, especial care in handling must be exercised because scratching the polyimide sheath can make the column susceptible to unobserved breakage. Commercial hydrogen alarms are available (e.g., from J & W Scientific) that warn the operator and shut off the instrument.

[3] Enzymic Method for Determination of Fructose 6-Phosphate, Xylulose 5-Phosphate, and Sedoheptulose 7-Phosphate

By G. A. KOCHETOV

Principle. Transketolase (D-sedoheptulose-7-phosphate : D-glyceraldehyde-3-phosphate glycolaldehydetransferase, EC 2.2.1.1) catalyzes the cleavage of keto sugars to form an "active glycolaldehyde," which is then transferred to the respective substrate.[1] In the presence of ferricyanide, the active glycolaldehyde is oxidized to form glycolic acid.[2] The scheme of the reaction (with, e.g., fructose 6-phosphate as the substrate) is as follows:

Fructose 6-phosphate + 2 $Fe(CN)_6^{3-}$ + H_2O →
$\qquad$ glycolic acid + erythrose 4-phosphate + 2 $Fe(CN)_6^{4-}$ + 2 H^+

The reaction is irreversible, and 2 mol of ferricyanide are reduced per mole of the consumed substrate.[3] The reduction of ferricyanide is accompanied by a decrease in the absorbance at 420 nm.

Knowing the molar extinction coefficient of ferricyanide, one can calculate its consumption (and consequently, the quantity of fructose 6-phosphate) in the reaction mixture. Other donor substrates of transketolase, e.g., xylulose 5-phosphate and sedoheptulose 7-phosphate, behave similarly.

Procedure.[4] Two spectrophotometric cuvettes, the experimental and reference, are filled with the following components (final concentrations are given): 0.05 M glycylglycine buffer, 2 mM $MgCl_2$, 0.1 mM thiamine pyrophosphate, and 1.4 mM $K_3Fe(CN)_6$. The experimental cuvette is, in addition, supplemented with 0.04 units of transketolase from bakers' yeast (for definition of unit, see this volume [7]). The total volume of each cuvette is 2.2 ml, and the pH is 7.6.

The samples are mixed, and the recorder is zeroed by adjusting the width of the slit. If the starting position is stable, 0.02 ml of fructose 6-phosphate (0.03–0.12 μmol) is added to each cuvette; the contents are mixed, and the decrease of absorbance at 420 nm is measured until the

---

[1] E. Racker, *in* "The Enzymes" (P. D. Boyer, H. Lardy, and K. Myrbäck, eds.), 2nd ed., Vol. 5, p. 397. Academic Press, New York, 1961.

[2] J. W. Bradbeer and E. Racker, *Fed. Proc., Fed. Am. Soc. Exp. Biol.* **20**, 88 (1961).

[3] P. Christen and A. Gasser, *Abstr. FEBS Meet., 10th* No. 604 (1975).

[4] R. A. Usmanov and G. A. Kochetov, *Biochem. Int.* **3**, 33 (1981).

reaction is complete at 20°. The time is usually within 5–15 min, depending on the quantity of fructose 6-phosphate in the sample. The measurements are carried out in a double-beam automatic spectrophotometer in 1-cm cuvettes; the 0–0.1 scale is used.

Calculation. The molar extinction coefficient for ferricyanide at 420 nm is $1000 \, M^{-1} \, cm^{-1}$.

The quantity of fructose 6-phosphate (μmol) is calculated from the formula

$$M = A_{420/2} \cdot V$$

where V is the volume of the reaction mixture.

Quantitative determination of other keto compounds (that can be donor substrates of transketolase) is similar to that of fructose 6-phosphate.

[4] Enzymic Microdetermination of D-Glucose, D-Fructose, D-Gluconate, 2-Keto-D-gluconate, Aldehyde, and Alcohol with Membrane-Bound Dehydrogenases

By MINORU AMEYAMA

D-Glucose + 2 ferricyanide $\xrightarrow{\text{D-glucose dehydrogenase}}$ D-glucono-δ-lactone + 2 ferrocyanide

D-Fructose + 2 ferricyanide $\xrightarrow{\text{D-fructose dehydrogenase}}$ 5-keto-D-fructose + 2 ferrocyanide

D-Gluconate + 2 ferricyanide $\xrightarrow{\text{D-gluconate dehydrogenase}}$ 2-keto-D-gluconate + 2 ferrocyanide

2-Keto-D-gluconate + 2 ferricyanide $\xrightarrow{\text{2-keto-D-gluconate dehydrogenase}}$

2,5-diketo-D-gluconate + 2 ferrocyanide

Aldehyde + 2 ferricyanide $\xrightarrow{\text{aldehyde dehydrogenase}}$ carboxylic acid + 2 ferrocyanide

Alcohol + 2 ferricyanide $\xrightarrow{\text{alcohol dehydrogenase}}$ aldehyde + 2 ferrocyanide

Enzymic microdetermination of sugars, in many cases, is based on measuring reduced pyridine nucleotides, such as alcohol, with NAD-dependent alcohol dehydrogenase and D-glucose with a coupling reaction composed of hexokinase, hexosephosphate isomerase, and D-glucose-6-phosphate dehydrogenase. Sometimes the reduced pyridine nucleotide is amplified by dyes, such as nitro blue tetrazolium; this also allows activity measurement in the visible region. Although these assay methods are practical for the purposes, the most important and critical

METHODS IN ENZYMOLOGY, VOL. 89

point is the reaction equilibrium. For example, in the determination of trace amounts of ethanol with NAD-dependent alcohol dehydrogenase from yeast or mammals, the reaction equilibrium lies toward alcohol formation; thus, it is difficult to assay trace amounts of ethanol accurately. Similar situations are generally seen in enzymic microdetermination of other substrates, such as D-glucose, D-fructose, or D-gluconate. Such assay methods also require several purified enzymes and, in addition, ATP and NAD or NADP, and they are rather expensive. Also, D-glucose cannot be assayed independently of D-fructose owing to the substrate specificity of hexokinase. From these points of view, sugar and related oxidases have some superiority to pyridine nucleotide-dependent enzymes.

For enzymic microdetermination of various substrates, use of individual membrane-bound microbial dehydrogenases is recommended. The acetic acid bacteria have many dehydrogenases on the outer surface of cytoplasmic membrane, and these catalyze the oxidation of various carbohydrates and accumulate a large amount of oxidation product in the culture medium, for instance, D-gluconate, 2-keto-D-gluconate, 5-keto-D-fructose, 2,5-diketo-D-gluconate, acetate, dihydroxyacetone, 5-keto-D-gluconate, and L-sorbose. The enzymes that have been developed for the assays to be described are listed in the table.[1-12] Since these enzymes are tightly bound to the cytoplasmic membrane, solubilization with the aid of detergents is necessary before further purification can be

[1] O. Adachi, K. Tayama, E. Shinagawa, K. Matsushita, and M. Ameyama, *Agric. Biol. Chem.* **42**, 2045 (1978).
[2] O. Adachi, E. Miyagawa, E. Shinagawa, K. Matsushita, and M. Ameyama, *Agric. Biol. Chem.* **42**, 2331 (1978).
[3] O. Adachi, K. Tayama, E. Shinagawa, K. Matsushita, and M. Ameyama, *Agric. Biol. Chem.* **44**, 503 (1980).
[4] M. Ameyama, K. Osada, E. Shinagawa, K. Matsushita, and O. Adachi, *Agric. Biol. Chem.* **45**, 1889 (1981).
[5] M. Ameyama, E. Shinagawa, K. Matsushita, and O. Adachi, *J. Bacteriol.* **145**, 814 (1981).
[6] K. Matsushita, E. Shinagawa, O. Adachi and M. Ameyama, *J. Biochem.* (*Tokyo*) **85**, 1173 (1979).
[7] K. Matsushita, E. Shinagawa, and M. Ameyama, this volume [31].
[8] E. Shingawa, K. Matsushita, O. Adachi, and M. Ameyama, *Agric. Biol. Chem.* **42**, 2355 (1978).
[9] E. Shinagawa, K. Matsushita, O. Adachi, and M. Ameyama, *Agric. Biol. Chem.* **45**, 1079 (1981).
[10] E. Shinagawa, K. Matsushita, O. Adachi, and M. Ameyama, *Agric. Biol. Chem.* **46**, 135 (1982).
[11] M. Ameyama, E. Shinagawa, K. Matsushita, and O. Adachi, *Agric. Biol. Chem.* **45**, 851 (1981).
[12] K. Matsushita, Y. Ohno, E. Shinagawa, O. Adachi, and M. Ameyama, *Agric. Biol. Chem.* **44**, 1505 (1980).

MEMBRANE-BOUND DEHYDROGENASES USED FOR ENZYMIC MICRODETERMINATION

Dehydrogenase	Strain	Solubilizing conditions	References[a]
Alcohol	*Gluconobacter*	0.1–1% Triton X-100	1
Alcohol	*Acetobacter*	0.1–1% Triton X-100	2
Acetaldehyde	*Gluconobacter*	1% Triton X-100	3
Acetaldehyde	*Acetobacter*	1% Triton X-100	4
D-Fructose	*Gluconobacter*	1% Triton X-100	5
Glycerol	*Gluconobacter*	1% Cholate	c
D-Gluconate	*Gluconobacter*	1% Tween 80	c
D-Gluconate	*Pseudomonas*	0.2% DOC[b] + 0.5% cholate	6
D-Gluconate	*Klebsiella*	2% Triton X-100	7
D-Gluconate	*Serratia*	2% Triton X-100	8
2-Keto-D-gluconate	*Gluconobacter*	2% Cholate + 0.2 M KCl	9
D-Sorbitol	*Gluconobacter*	1% Triton X-100 + 0.1 M KCl	10
D-Glucose	*Gluconobacter*	1% Triton X-100	11
D-Glucose	*Pseudomonas*	1% Triton X-100 + 1 M KCl	12

[a] Numbers refer to text footnotes.
[b] DOC, deoxycholate.
[c] M. Ameyama, unpublished.

undertaken. Oxidation of individual substrates is linked to the respiratory chain, and the reaction runs to completion. Since these enzymes are at the culture interface, where a large amount of sugar acids accumulates, they are fairly stable at acidic pH and the optimum pH is often in the acidic regions.

Assay Methods

Basic Ferricyanide Procedure

Enzymic microdetermination of substrate utilizes the method of Wood *et al.*[13] using ferricyanide as an electron acceptor with some modifications. The method is based on the initial reduction rate. When the method is calibrated using the rates observed with standards, the reduction rates with unknowns give the quantitative amounts of substrate present. One unit of enzyme activity is defined as the amount of enzyme catalyzing the oxidation of 1 μmol of substrate per minute under the assay conditions, and 4.0 absorbance units equal 1 μmol of substrate oxidized. When only a trace amount of substrate is present, increased sensitivity is obtained by reduction of total reaction mixture to 1.0 ml including the ferric-Dupanol reagent and omission of any further addition of water. Also an end-point

[13] W. A. Wood, R. A. Fetting, and B. C. Hertlein, this series, Vol. 5, p. 287.

version of the method can be used. Substrate oxidation can be assayed with other methods using phenazine methosulfate or 2,6-dichlorophenolindophenol as the electron acceptor. Although such assay methods show a good agreement with each other, the ferricyanide method is mainly employed in this chapter owing to its simplicity for routine use. Further, with 2,6-dichlorophenolindophenol, the molecular extinction coefficient is variable with pH, and this makes the assay less reliable for quantitative determination.

Reagents. These reagents are common to all methods described.

Potassium ferricyanide, 0.1 M, in distilled water
Triton X-100, 10% solution
Ferric sulfate–Dupanol reagent containing 5 g of $Fe_2(SO_4)_3 \cdot n\ H_2O$, 3 g of Dupanol (sodium lauryl sulfate), 95 ml of 85% phosphoric acid, and distilled water to 1 liter
Buffer solution (shown under individual enzymes)
Standard substrate solution for individual assays

Procedure Common to All Methods Described. The reaction mixture contains 0.1 ml of potassium ferricyanide, 0.5 ml of buffer, enzyme solution (0.03–0.05 unit of enzyme activity), standard substrate or samples to be assayed, and Triton X-100 to the final concentration of 0.5 to 1.0% in a total volume of 1.0 ml. After preincubation for 5 min at the indicated temperature for individual enzymes, the reaction is initiated by the addition of potassium ferricyanide. The reaction is terminated by adding 0.5 ml of ferric sulfate–Dupanol reagent; then, 3.5 ml of water are added to the reaction mixture. The resulting Prussian blue color is measured with a spectrophotometer at 660 nm after standing for 20 min at the same temperature as that of the reaction. Standard calibration assays are carried out at the same time.

Individual Determinations

D-Glucose

D-Glucose is determined with D-glucose dehydrogenase in the presence of dye. The enzyme is solubilized and purified from the membrane of *Pseudomonas* or *Gluconobacter* species (this volume [24]).

Other Reagents

Sodium acetate buffer, 0.1 M, pH 4.5, for the enzyme from *Pseudomonas fluorescens*
McIlvaine buffer, pH 3.0, for the enzyme from *Gluconobacter suboxydans*. This buffer solution is prepared by mixing 15.89 ml of 0.1 M

citric acid with 4.11 ml of 0.2 M Na$_2$HPO$_4$. During mixing, the pH of the solution is verified by a pH meter.

Standard D-glucose (0.5 μmol or less) or samples to be assayed.

D-Glucose dehydrogenase, dissolved in 0.01 M potassium phosphate, pH 6.0, containing 1% Triton X-100

Procedure. The enzyme reaction is performed at 25°. After preincubation for 5 min at 25°, the reaction is initiated by the addition of potassium ferricyanide. After termination of the reaction, the resulting Prussian blue color is measured after the preparation has been left standing for 20 min at 25°.

Sensitivity and Specificity of the Assay. The enzyme is highly specific to D-glucose, and the following sugars are also oxidized by the enzyme from *P. fluorescens,* but at considerably lower rates: D-mannose (8.6% of the rate of D-glucose oxidation), D-galactose (6.5%), L-rhamnose (7.5%), D-xylose (13%), L-arabinose (2.8%), and maltose (3.2%) when assayed at a substrate concentration of 0.1 M under the standard assay conditions. The enzyme from *G. suboxydans* oxidizes maltose to 5% of the rate of D-glucose. D-Glucose oxidation is not repressed by any other substrate analogs, even when such compounds are added to the reaction mixture at 10 times higher concentration than D-glucose. D-Fructose, D-glucose 6-phosphate, D-fructose 6-phosphate, D-fructose 1,6-diphosphate, sugar alcohols, D-ribose, sucrose, D-arabinose, L-sorbose, α-methylglucoside, hexonates, and ketohexonates have no effect on D-glucose oxidation. The minimum amount of D-glucose that can be assayed with 5% deviation is 10 nmol. If the total volume of the reaction mixture is reduced to 1.0 ml, including ferric-Dupanol reagent, 2 nmol can be assayed.

D-*Fructose*

D-Fructose dehydrogenase catalyzes the oxidation of D-fructose to 5-keto-D-fructose *in vivo,* and the enzyme activity is linked to the electron transport chain of *Gluconobacter* species. D-Fructose is oxidized *in vitro* in the presence of dye, and the rate of D-fructose oxidation can be followed by the reduction of the dye.

Other Reagents

McIlvaine buffer, pH 4.5. This buffer solution is prepared by mixing 11.0 ml of 0.1 M citric acid with 9 ml of 0.2 M Na$_2$HPO$_4$. The pH of the solution is adjusted, using a pH meter, by adding either solution.

Standard D-fructose solution (0.5 μmol or less) or samples to be assayed

D-Fructose dehydrogenase (this volume [25]), dissolved in 20-fold diluted McIlvaine buffer, pH 6.0, containing 0.1% Triton X-100 and 1 mM 2-mercaptoethanol. McIlvaine buffer, pH 6.0, is prepared by mixing 7.37 ml of 0.1 M citric acid with 12.63 ml of 0.2 M Na_2HPO_4.

Procedure. The reaction is performed at 25° using McIlvaine buffer, pH 4.5. After preincubation for 5 min at 25°, the reaction is initiated by the addition of potassium ferricyanide. The resulting Prussian blue color is measured after the preparation has stood for 20 min at 25°.

Sensitivity and Specificity of the Assay. D-Fructose dehydrogenase catalyzes the oxidation of only D-fructose in the presence of ferricyanide. When D-fructose (100 μmol) is oxidized in the presence of equimolar substrate analogs (100 μmol each), such as D-glucose, D-mannose, D-fructose 6-phosphate, D-glucose 1-phosphate, D-gluconate, 5-keto-D-fructose, D-glucose 6-phosphate, D-fructose 1,6-diphosphate, 2-keto-D-gluconate and 5-keto-D-gluconate, the rate of D-fructose oxidation is not affected appreciably. Trace amounts of D-fructose that are difficult to assay by measuring the initial reaction rate can be determined by end-point measurement and shows a good agreement with the theoretical values. The method is capable of estimating D-fructose accurately down to 10 nmol.

D-*Gluconate*

D-Gluconate dehydrogenase catalyzes the oxidation of D-gluconate to 2-keto-D-gluconate *in vivo,* and the enzyme activity is linked to the respiratory chain of the oxidative bacteria. D-Gluconate is oxidized *in vitro* in the presence of dye, and the rate of D-gluconate oxidation can be followed by the reduction of the dye.

Other Reagents

Sodium acetate, 0.1 M, pH 5.0, for the enzyme preparations from *Pseudomonas, Serratia,* and *Gluconobacter* species

McIlvaine buffer, pH 4.0, for the enzyme from *Klebsiella* species. This buffer solution is prepared by mixing 12.29 ml of 0.1 M citric acid with 7.71 ml of 0.2 M Na_2HPO_4. The pH of the buffer solution is ensured by adding either solution.

Standard D-gluconate solution, sodium salt (0.5 μmol or less) or samples to be assayed

D-Gluconate dehydrogenase (this volume [31]), dissolved in 0.01 M potassium phosphate, pH 6.0, containing 0.1% Trixon X-100

Procedure. The reaction is performed at pH 5.0 or 4.0 at 25°. After preincubation for 5 min at 25°, the reaction is initiated by the addition of potassium ferricyanide. The resulting Prussian blue color is measured after the preparation has stood for 20 min at 25°.

Sensitivity and Specificity of the Assay. The enzyme catalyzes oxidation only of D-gluconate in the presence of dyes such as ferricyanide, 2,6-dichlorophenolindophenol, or phenazine methosulfate. The following compounds are not oxidized by the enzyme: D-glucose, D-fructose, D-galactose, D-mannose, D-arabinose, D-xylose, maltose, L-rhamnose, 2-keto-D-gluconate, 5-keto-D-gluconate, 5-keto-D-fructose, D-galactonate, D-mannonate, 6-phospho-D-gluconate, L-idonate, D-arabonate, D-xylonate, L-sorbose, D-sorbitol, D-mannitol, alcohols, and aldehydes. When D-gluconate (100 μmol) is oxidized in the presence of equimolar substrate analogs (100 μmol each), the reaction rate of D-gluconate oxidation is not affected appreciably by any of them except for oxalate, oxamate, pyruvate, and 2-keto-D-gluconate (see this volume [31]). It is characteristic of the membrane-bound dehydrogenases to have a relatively low optimum pH and to catalyze a complete oxidation of the substrate. Trace amounts of D-gluconate that are difficult to assay by measuring the initial reaction rate can be determined by end-point measurement and show a good agreement with the theoretical values. The method is capable of estimation of D-gluconate accurately down to 10 nmol.

2-Keto-D-Gluconate

2-Keto-D-gluconate dehydrogenase catalyzes the oxidation of 2-keto-D-gluconate to 2,5-diketo-D-gluconate *in vivo*, and the enzyme activity is linked to the electron transport system of dark pigment-producing strains of *Gluconobacter* species. 2-Keto-D-gluconate is oxidized *in vitro* in the presence of dye, and the rate of oxidation can be followed by the reduction of the dye.

Other Reagents

    Sodium acetate buffer, 0.1 M, pH 4.0
    Standard 2-keto-D-gluconate, sodium salt (0.5 μmol or less) or samples to be assayed
    2-Keto-D-gluconate dehydrogenase (this volume [32]), dissolved in 0.01 M potassium phosphate, pH 6.0, containing 0.1% Triton X-100

Procedure. After preincubation for 5 min at 37°, the reaction is initiated by the addition of enzyme solution and incubation is carried out at 37°.

Sensitivity and Specificity of the Assay. The enzyme catalyzes oxidation only of 2-keto-D-gluconate in the presence of dye. The enzyme is completely inert toward 5-keto-D-gluconate, 2-keto-D-galactonate, and 2-keto-D-gluconate. The following compounds are not oxidized by the purified enzyme preparation: hexoses, pentoses, hexonates, sugar alcohols, hexose phosphates, alcohols, aldehydes, and other kinds of organic acids. Oxidation of 2-keto-D-gluconate is not affected by these substrate analogs even when such compounds are present at 10 times excess in the assay mixture except for succinate, citrate, and oxamate (see this volume [32]). The optimum pH of 2-keto-D-gluconate oxidation is 4.0, and the enzyme shows its catalytic properties as a typical membrane-bound dehydrogenase as mentioned in the preceding paragraphs. The method is capable of estimating 2-keto-D-gluconate accurately down to 10 nmol.

Aldehyde

Membrane-bound aldehyde dehydrogenase catalyzes the oxidation of aldehydes to corresponding acids *in vivo,* and the enzyme activity is linked to the electron transport system of the acetic acid bacteria. Aldehyde is oxidized *in vitro* in the presence of dye and the rate of oxidation can be followed by the reduction of the dye.

Other Reagents

McIlvaine buffer, pH 4.0. For preparation of this buffer, see D-gluconate determination by D-gluconate dehydrogenase mentioned above.

Standard acetaldehyde (0.5 μmol or less) or samples to be assayed. Redistillation of the commercial product of acetaldehyde is recommended in preparing the standard acetaldehyde solution; store in the cold until use.

Aldehyde dehydrogenase (this volume [82]) is dissolved in 0.01 M sodium acetate, pH 5.3, containing 0.5% Triton X-100, 25 mM benzaldehyde, and 10% sucrose for the enzyme from *Gluconobacter* species. The enzyme from *Acetobacter* species is dissolved in 0.01 M potassium phosphate, pH 6.0, containing 5% Triton X-100, 2% cetylpyridinium chloride, and 10 mM benzaldehyde.

Procedure. After preincubation for 5 min at 25°, the reaction is initiated by the addition of potassium ferricyanide and terminated by adding 0.5 ml of the ferric-Dupanol reagent. The resulting stabilized Prussian blue color is measured after standing for 20 min at 25°. The effect of benzaldehyde, which exists in the enzyme solution and is oxidized by the enzyme at 5% of the rate for acetaldehyde, is subtracted by running the assay without substrate.

Sensitivity and Specificity of the Assay. The enzyme from both genera of the acetic acid bacteria catalyzes oxidation of aliphatic aldehydes with a straight carbon chain length of 2 to 6. Formaldehyde is not oxidized. The following aldehydes are oxidized: glutaraldehyde (64% that of acetaldehyde), isobutyraldehyde (60%), cinnamaldehyde (17%), p-anisaldehyde (10%), and benzaldehyde (5%). No appreciable interference in aldehyde determination with the purified enzyme is observed when equimolar or excess amounts of other substances that may occur in sample solution are present in the assay mixture. Sugars, sugar alcohols, hexonates, ketohexonates, and alcohols show no effect in aldehyde determination. The optimum pH is found at 4.0, and the catalytic properties are typical of membrane-bound dehydrogenases. The method is capable of estimation of acetaldehyde accurately down to 10 nmol.

Although benzaldehyde, which is added as a stabilizing agent of the purified enzyme, reacts as a substrate, it can be ignored, if the initial reaction rate method is used, because of its very low reaction rate. However, in case of the end-point measurement, the enzyme should be freed from benzaldehyde by adsorption and desorption of the enzyme onto hydroxyapatite followed by extensive washing and desorption of the enzyme with concentrated salt solution in the presence of 1% Triton X-100 and 10% sucrose. In the absence of benzaldehyde, the enzyme can be preserved for over a year without appreciable loss in enzyme activity when stored at $-20°$.

Alcohol

Membrane-bound alcohol dehydrogenase of the acetic acid bacteria catalyzes oxidation of ethanol *in vivo,* and the enzyme activity is linked to the electron transport system of the organisms. The enzyme catalyzes the reaction *in vitro,* and the rate of alcohol oxidation is measured by following the reduction of dye.

Other Reagents

McIlvaine buffer, pH 4.0, for the enzyme from *Acetobacter* species. This buffer is prepared as shown in D-gluconate determination.

McIlvaine buffer, pH 6.0, for the enzyme from *Gluconobacter* species. This buffer is prepared as mentioned above (D-fructose determination).

Standard ethanol solution (0.5 μmol or less) or samples to be assayed

Alcohol dehydrogenase (this volume [76]), dissolved in 0.01 M potassium phosphate, pH 6.0, containing 0.1% Triton X-100

Procedure. The reaction is performed at 25°. After preincubation for 5 min at 25°, the reaction is initiated by the addition of potassium ferricyanide and terminated by adding 0.5 ml of ferric-Dupanol reagent. The stabilized Prussian blue color is measured after the preparation has stood for 20 min at 25°.

Sensitivity and Specificity of the Assay. The enzyme from both genera of the acetic acid bacteria catalyzes oxidation of aliphatic primary alcohols with a straight carbon chain length of 2 to 6. Since the other species of membrane-bound dehydrogenases are separated from alcohol dehydrogenase during the purification steps, the enzyme preparation with a specific activity over 150 units per milligram of protein is sufficient. No appreciable interference in alcohol determination is observed when equimolar or excess amounts of other substances, which may occur in sample solution, are present in the assay mixture. Unlike the NAD(P)-linked enzymes, the reaction of alcohol dehydrogenase of the acetic acid bacteria is irreversible and trace amounts of ethanol can be assayed with the enzyme by end-point measurement. This method is capable of estimation of alcohol accurately down to 10 nmol.

[5] Automated Enzymic Determination of L(+)- and D(−)-Lactic Acid

By FLOYD M. BYERS

Principle

These procedures are based on methods previously described[1] and entail automated micromethods for rapid assay of L(+)- and D(−)-lactic acid. Applications include cell-containing fluids and other biological extracts. Methods utilize an autosampler pump system and UV-Vis spectrophotometer, L(+)- and D(−)-LDH enzymes, and reduction of NAD^+ to $NADH_2$, which is measured at 340 nm. These enzymic methods are highly specific for either isomer, run at a rate of 60 samples per hour, use minimal quantities of reagents, and require only 40–80 μl of sample.

Other procedures previously used for measuring L(+)- and/or estimating D(−)-lactic acid in complex biological fluids are variable, unspecific, time-consuming, and cumbersome. The development of simple, rapid, specific assays for both L(+)- and D(−)-lactate isomers in biological fluids involving cellular extracts greatly facilitates lactate metabolism research. Although a chemical analysis for total lactic acid based on Barker and Summerson's[2] procedure has been automated by Hochella and Weinhouse,[3] it has the same disadvantage as the manual assay, in that it is

[1] S. R. Goodall and F. M. Byers, *Anal. Biochem.* **89**, 80 (1978).
[2] S. B. Barker and W. H. Summerson, *J. Biol. Chem.* **138**, 535 (1941).
[3] N. J. Hochella and S. Weinhouse, *Anal. Biochem.* **10**, 304 (1965).

METHODS IN ENZYMOLOGY, VOL. 89

variable and not completely specific for lactic acid.[2,4] While other modifications of the original colorimetric method are widely used[4-6] to eliminate pyruvic acid and other interfering substances, it is interesting that no color development occurs if a high-purity grade of H_2SO_4 is used,[4] suggesting a catalytic effect of trace impurities. Enzymic methods that are specific for L(+)-lactic acid, and thus offer more potential, have been automated[3,5,7] for analyses of serum and tissue extracts. Two of these methods[5,7] involve the measurement of the reduction of NAD^+ to $NADH_2$ with either UV or fluorometric procedures. The other method[3] uses a simple colorimetric assay by coupling the $NAD^+ \rightarrow NADH_2$ reaction via diaphorase to an INT dye reaction permitting measurement within the visible range on serum samples.

The methodology developed included adaptations of these procedures to allow automated analyses of complex biological fluids containing cellular extracts.

Materials and Methods

Reagents, Equipment, and Sample Preparation

Instrumentation includes a Technicon autoanalyzer pump and Sampler II and a Gilford Model 222 spectrophotometer with Beckman DU optics with a 15-mm continuous flowcell cuvette or an equivalent system. Output is traced on a strip chart recorder. The L(+) assay is based on the manual methods of Rosenburg and Rush,[8] and the D(−) method is patterned after the manual method of Boehringer Mannheim.[9] In both procedures lactate is oxidized to pyruvate by the appropriate LDH enzyme, and the corresponding reduction of NAD^+ to $NADH_2$ is measured directly at 340 nm. Hydrazine hydrate and semicarbazide hydrochloride are used in D(−) and L(+) assays, respectively, to trap pyruvate and prevent end-product inhibition. Rumen fluid samples are preserved and deproteinized with 25% metaphosphoric acid (5% final concentration) and centrifuged (12,000 g for 15 min). This procedure removes potential competitors for NAD^+, sample enzymes, and other proteins and stabilizes the lactic acid. The clarified rumen liquor is diluted with the appropriate buffer (Table I). For each assay, 40 or 80 μl of sample [for L(+) and D(−) procedures, respectively] are added to 800 μl of appropriate glycine buffers and placed in disposable

[4] R. J. Pennington and R. M. Sutherland, *Biochem. J.* **63**, 353 (1965).
[5] M. K. Schwartz, G. Kessler, and O. Bodansky, *J. Biol. Chem.* **236**, 1207 (1946).
[6] J. F. Speck, J. W. Moulder, and E. A. Evans, *J. Biol. Chem.* **164**, 119 (1946).
[7] C. S. Apstein, E. Puchner, and N. Brachfeld, *Anal. Biochem.* **38**, 20 (1970).
[8] J. C. Rosenburg and B. F. Rush, *Clin. Chem.* **12**, 299 (1969).
[9] Boehringer Mannheim, *Biochem. Bull.* **120**, 121 (1973).

TABLE I

LACTIC ACID ASSAY REAGENTS

Reagent No.	Procedure	
	L(+)-Lactic acid	D(−)-Lactic acid
1	0.2 M glycine + NaOH buffer (pH 9.6) with 1 g of semicarbizide hydrochloride and 1 ml of aqueous Brij[a] per liter	0.4 M glycine, 0.3 M hydrazine buffer (pH 9.0) with 1 ml of aqueous Brij per liter
2	0.85% NaCl with 1 ml of aqueous Brij per liter	0.85% NaCl with 1 ml of aqueous Brij per liter
3	0.25 M phosphate buffer (pH 7.4) with 0.15% albumin (bovine fraction V)	0.25 M phosphate buffer (pH 7.4) + 0.15% albumin (bovine fraction)
4	Rabbit muscle L(+)-LDH,[b] approximately 475 units of activity per milligram of protein, in 2.1 M ammonium sulfate suspension (diluted at 0.170 ml/20 ml of reagent 3, providing 800 units of activity)	*L. leichmannii* D(−)-LDH,[c] approximately 325 units of activity per milligram of protein, in 3.2 M ammonium sulfate suspension (diluted at 0.50 ml in 20 ml of reagent 3, providing 800 units of activity)
5	250 mg of NAD$^+$ dry powder[d] (ethanol free) added to reagent 4 immediately prior to assay	500 mg of NAD$^+$ dry powder[d] (ethanol free) added to reagent 4 immediately prior to assay

[a] Brij from Fisher Scientific Co.
[b] Miles Laboratories, EC 1.1.1.28, D(−)-lactic dehydrogenase.
[c] Sigma Chemical Co., EC 1.1.1.27, L(+)-lactic dehydrogenase.
[d] ICN Biochemical Co.

1-ml sample cups in autosampler trays. Standardized solutions of the respective lithium lactate salts are also preserved with 25% metaphosphoric acid and diluted similarly with their respective buffers. Standards used range in concentration from 0.50 to 10 mg of lactate per milliliter. Blanks are prepared in the same fashion by adding 40 or 80 μl of a 5.0% metaphosphoric acid solution to glycine buffers used in each assay. Actual rumen fluid blanks determined by running samples with all reagents except the LDH enzyme resulted in absorbance values similar to those of the blanks prepared as listed. All reagents are refrigerated until ready for use. The enzyme preparation is kept in ice water as aspirated into the assay to eliminate stability problems. The sampler is set with a 60–2 : 1 per hour cam and an 0.86 mm probe with 0.51 mm i.d. sample tubing. The wash solution is physiological saline with a wetting agent (Brij) added, which provides a more effective washout than distilled water. The samples are run at 60 per hour with a 2 : 1 sample-to-wash ratio. Optimum separation of sample peaks is obtained at this rate, and there is little difficulty with carryover between samples except when an extremely high sample is followed by a very low sample. In this situation, the low sample is rerun preceded by either a blank or another low sample. The maximum positive bias due to carryover effects even in this situation is less than 5% in any assay.

L(+)-Lactate Apparatus

Sample, air, and buffer (0.60, 0.42, and 2.5 ml/min) are combined by means of a "cactus" bubbler, and this combined stream is passed through a 28-turn mixing coil. The enzyme is then added (0.10 ml/min) via the capillary arm of a sidearm "h-cactus"; and the reaction mixture is then passed through another 28-turn mixing coil and a standard time-delay coil maintained at 37° in a heating bath. A portion of the reaction stream (1.2 ml/min) then enters the spectrophotometer via a 15-mm quartz continuous flowcell cuvette, and the fraction of full-scale absorbance is measured at 340 nm. Full scale is set at 0.500 absorbance, with the baseline set with blank samples.

D(−)-Lactate Apparatus

Buffer and enzyme (2.5 and 0.10 ml/min) are first combined by a sidearm "h-cactus" and passed through a 14-turn mixing coil. The sample (0.60 ml/min) is then combined via a "cactus" bubbler with the enzyme mixture and air (0.42 ml/min), and the reaction stream is passed through two 28-turn mixing coils, as complete mixing is crucial in this assay owing to the lower K_m for the D(−)-LDH enzyme. The reaction stream is then passed through three standard-length time-delay coils (needed because of the long incubation time) in a water bath set at 25°. The optimum temperature for the D(−) enzyme is lower, probably because it is of bacterial rather than mammalian origin. The greatest sensitivity (Δ absorbance/Δ lactate) is achieved at 25°. Sensitivity at 20, 30, and 35° was 3.3, 9.04, and 22.0% lower. A portion of the reaction mixture stream (1.2 ml/min) is then passed through the same continuous flowcell of the spectrophotometer, and the fraction of full-scale absorbance is measured at 340 nm. Full-scale absorbance is set at 0.750 absorbance with the baseline set to zero with the sample blank.

Development of Methodology and Protocol for Use

The reagent concentrations, flow rates, pH of the buffer solutions, and incubation temperatures were selected after preliminary testing to optimize conditions for maximal absorbance, precision, and optimal peak separation. Initially, a colorimeter was used with the INT Dye reduced by

NADH$_2$ through diaphorase. However, because of the large quantities of reducing agents in the rumen fluid extracts, this method gave unsatisfactory results. Rumen fluid samples containing essentially no lactic acid and samples introduced with no LDH in the system both resulted in very extensive dye reduction and color development. Thus this procedure was unacceptable, and in the further development of these assays a spectrophotometer was used at 340 nm with a blue filter to measure NADH$_2$ directly. With this method, rumen fluid samples acidified with metaphosphoric acid and run without LDH in the system caused no baseline shift above that of the metaphosphoric acid–water blank and no interfering substances were found to bias absorbance in any selected samples tested.

The reactions on which these assays are based are as follows:

$$\text{L(+)-Lactate} + \text{NAD}^+ \underset{}{\overset{\text{L(+)-LDH}}{\rightleftharpoons}} \text{pyruvate} + \text{NADH}_2$$
$$\text{D(−)-Lactate} + \text{NAD}^+ \underset{}{\overset{\text{D(−)-LDH}}{\rightleftharpoons}} \text{pyruvate} + \text{NADH}_2$$

Semicarbazide hydrochloride or hydrazine hydrate are added to the glycine buffer to trap the pyruvate initially in the sample, and that produced in the reaction, to drive the lactate to pyruvate reaction toward completion. Since the equilibrium of the reaction normally greatly favors the reverse reaction, the pyruvate from the sample and produced in the assay inhibits this reaction unless removed from the system. Very little interference exists from other substances, even at equimolar concentrations, owing to the specificity of the LDH enzymes.[10] Lactate dehydrogenase, other enzymes, and proteins endogenous to the biological fluids being assayed, are effectively eliminated by the metaphosphoric acid precipitation. Both the L(+) and D(−) enzymes are very specific for the respective isomers, and thus no cross-reactivity exists. When tested, the D(−)-LDH enzyme caused no reaction with L(+)-lactate and the L(+)-LDH enzyme did not react with D(−)-lactate; these findings confirm previous research.

In the L(+) assay, absorbance follows a slightly curvilinear function (Table II) of lactate level. Greater linearity, while not deemed necessary and of no practical significance since the standard curve is very reproducible, can be achieved by adding an additional time-delay coil to increase incubation time. With the D(−) procedure, absorbance changes linearly from 0 to 8.0 mg of lactic acid per milliliter. The minimum detectable level of lactate in either assay using the stated quantities of samples is 0.10 mg/ml, or 1% of the total range. Larger quantities of acidified sample need to be used if levels are below this range and will require more buffering, and thus the strength of the appropriate glycine buffer in which the sample

[10] L. Lundholm, E. Mohme-Lundhold, and N. Svedmyr, *Scand. J. Clin. Lab Invest.* **15**, 311 (1963).

TABLE II
STANDARD CURVES FOR LACTATE ASSAYS

Lactate concentration (mg/ml)	L(+) assay Fraction of full-scale absorbance (0.50) ± SD ($N = 5$)	CV^a (%)	D(−) assay Fraction of full-scale absorbance (0.75) ± SD ($N = 5$)	CV (%)
0.5	0.1425 ± 0.0055	3.86	0.0556 ± 0.0020	3.60
1.0	0.2330 ± 0.0084	3.61	0.1135 ± 0.0046	4.41
2.0	0.3570 ± 0.0045	1.26	0.2225 ± 0.0027	1.23
4.0	0.5760 ± 0.0114	1.98	0.4370 ± 0.0055	1.25
6.0	0.7400 ± 0.0110	1.49	0.6555 ± 0.0049	0.75
8.0	0.8850 ± 0.0087	0.99	0.8600 ± 0.0065	0.76

^{a} CV, coefficient of variation.

is diluted must be increased. As is evident, the sample buffer used in the D(−) assay is twice as concentrated as the buffer used in the L(+) assay. This is necessary because 80 μl of sample are used in the D(−) assay compared to 40 μl for the L(+) assay and the buffer must be adequate to bring the sample stream to pH 7.0 to allow the enzyme to function. Thus, the range of the assay can be easily modified by altering the degree of dilution of the sample in the buffer solution and making corresponding changes in buffer strength to accommodate changes in quantities of acid to be buffered. More concentrated samples are best handled by dilution.

Standard curves (Table II) for the L(+) and D(−) assays indicate that the precision of both assays as measured by standard deviations and coefficients of variation is very acceptable and variance is similar to that of manual enzymic assays for lactate. Precision in lactate analyses of rumen fluid samples parallels that of the lactate standards for either isomer. The recovery of lactic acid measured by adding incremental aliquots of standardized lithium lactate to deproteinized rumen fluid samples varying in level of lactic acid present initially, averaged 98.9% ± 3.6 (SD). Recovery of D(−) lactate was similar to that of L(+).

The application of these automated methods for assays of rumen fluid or similar clarified cellular extracts offers several advantages. They provide a specific and reliable measurement of total, L(+)- and D(−)-lactic acid directly, are easily adapted to commonly used autoanalyzer systems with 340 nm spectrophotometric capability, and can assay up to 60 samples per hour. Finally, the quantities of reagents used by this system are less than by previous methods, manual or automated, making the system more cost effective to run.

[6] Determination of D-Lactate in Plasma

By Richard B. Brandt

Early in this century, it was known that the predominant form of lactic acid in mammals was the L(+) isomer.[1] Eventually lactic acid was found to be mainly from glycolytic reactions under conditions where pyruvic acid was reduced.[2] The formation of the less predominant isomer, D-lactate, was reported in 1913 from homogenates of mammalian tissue when methylglyoxal (MeG) was added as a substrate.[3,4] The catalysis occurs through a dual-enzyme system, glyoxalase I [S-D-lactoyl-glutathione methylglyoxal-lyase (isomerizing); EC 4.4.1.5] and glyoxalase II (S-2-hydroxyacylglutathione hydrolase; EC 3.1.2.6)[5] with glutathione as a coenzyme.[6] The substrate for the system is the nonenzymically formed hemimercaptal between MeG and glutathione (GSH).[7] Interest in this system continues, since MeG is growth inhibitory, possibly by impairment of protein synthesis.[8,9] Therefore, data on the formation of D-lactate from MeG catabolism assume some importance. The synthesis of MeG in mammalian tissues is at least in part nonenzymic both during amino acid catabolism[10] and from dihydroxyacetone phosphate.[11] Data have been presented for the enzymic synthesis of MeG in liver.[12] D-Lactate formation occurs presumably from MeG in human blood, or other tissues where the glyoxalase system is active. *In vitro,* in whole blood when glycolysis was inhibited by fluoride or when substrates were added in quantities in excess of normal levels, D-lactate was synthesized at 0.1–0.5 mM/hr.[13] Bacterial formation of D-lactate through an MeG synthetase[14] appears to have been responsible for a lactic acidosis in a human case.[15]

[1] A. R. Mandel and G. Lusk, *Am. J. Physiol.* **16,** 129 (1906).

[2] J. S. Fruton, "Molecules and Life," p. 363. Wiley (Interscience), New York, 1972.

[3] H. D. Dakin and H. W. Dudley, *J. Biol. Chem.* **14,** 423 (1913).

[4] C. Neuberg, *Biochem. Z.* **51,** 484 (1913).

[5] E. Racker, *J. Biol. Chem.* **190,** 685 (1951).

[6] K. Lohmann, *Biochem. Z.* **254,** 332 (1932).

[7] M. Jowett, and J. H. Quastel, *Biochem. J.* **27,** 486 (1933).

[8] L. G. Egyud, and A. Szent-Györgyi, *Proc. Natl. Acad. Sci. U.S.A.* **56,** 203 (1966).

[9] M. Litt, *Biochemistry* **8,** 3249 (1969).

[10] W. H. Elliot, *Nature (London)* **185,** 467 (1960).

[11] S. A. Siegel and R. B. Brandt, *Va. J. Sci.* **29,** 99 (1978).

[12] J. Sato, Y. Wang, and J. Van Eys, *J. Biol. Chem.* **255,** 2046 (1980).

[13] R. B. Brandt and S. A. Siegel, *Submolec. Biol. Cancer, Ciba Found. Symp.* **67,** 211 (1979).

[14] R. A. Cooper and A. Anderson, *FEBS Lett.* **11,** 273 (1970).

[15] M. S. Oh, K. R. Phelps, M. Traube, J. L. Barbosa-Saldivar, C. Boxhill, and H. J. Carroll, *N. Engl. J. Med.* **301,** 249 (1979).

METHODS IN ENZYMOLOGY, VOL. 89

The metabolism of D-lactate in mammals is unclear although urinary excretion[16,17] and catabolism to CO_2 have been reported,[18] perhaps involving D-lactic acid dehydrogenase.[19] Rat tissue homogenates (brain, liver, heart, and kidney) have been shown to catalyze the oxidation of D-[^{14}C]lactate to [14]CO_2 at various rates different from those for L-lactate.[20]

In order to determine D-lactate in plasma, a reliable, stereospecific method has been developed that allows for measurement of D-lactate in concentrations encountered in humans.[21]

Assay Principle

The procedure is based on the spectrophotometric determination of NADH using a specific D-lactic dehydrogenase (D-LDH) to catalyze the formation of pyruvate, coupled to the reduction of NAD^+ to NADH at alkaline pH, with hydrazine as a trap for pyruvate to drive the reaction to completion[21] [Eq. (1)].

$$\text{D-Lactate} + \text{NAD}^+ \xrightarrow{\text{D-LDH}} \text{pyruvate} + \text{NADH} + \text{H}^+ \tag{1}$$

Equipment and Reagents

Equipment. A Beckman DU modified with a Gilford Model 252 system, automatic cuvette positioner, and a Lauda K-2/R constant-temperature bath was used for measuring NADH absorbance at 340 nm. Quartz semimicrocuvettes with 10-mm path length and 1.4-ml maximum volume were used for the measurements. Gilson Pipetman adjustable digital microliter pipettes were used for volume transfers. Amicon Centriflo membrane cones, type CF50A, were used for plasma deproteinization. The cones were rinsed with water and allowed to soak for 2 hr with several changes of water. Excess water was removed by centrifugation at 2000 rpm in an SS-34 rotor in a Sorvall RC2B for 10 min.

Reagents. All biochemicals were from Sigma Chemical Company except as noted. The buffer system used in the assay was glycine–hydrazine buffer, pH 9.5, prepared by dissolving 18.8 g of glycine and 8.8 ml of 95% hydrazine (Eastman Organic Chemical) in water to a total volume of 500

[16] M. Duran, J. P. Van Biervliet, J. P. Kamerling, and S. K. Wadman, *Clin. Chim. Acta* **74**, 297 (1977).
[17] M. A. Judge, and J. Van Eys, *J. Nutr.* **76**, 310 (1962).
[18] D. Giesecke and A. Fabritius, *Experientia* **30**, 1124 (1974).
[19] P. K. Tubbs and G. D. Greville, *Biochim. Biophys. Acta* **34**, 290 (1959).
[20] M. G. Waters, M. J. Rispler, E. S. Kline, and R. B. Brandt, *Va. J. Sci.* **32**, 138 (1981).
[21] R. B. Brandt, S. A. Siegel, M. G. Waters, and M. H. Bloch, *Anal. Biochem.* **102**, 39 (1980).

ml. The assay stock solution for D-lactate determination was prepared daily and contained 50 mg of NAD^+, 10 ml of glycine–hydrazine buffer, and 20 ml of distilled water. D-Lactate standard stock solution was 0.94 mg/ml Li lactate (9.75 mM D-lactate), stored at 4° and diluted for use in distilled water. D-Lactic dehydrogenase (D-LDH) from *Lactobacillus leichmannii* (D-lactate : NAD^+ oxidoreductase; EC 1.1.1.28) was from Sigma or Boehringer-Mannheim Biochemicals and was diluted in water to give about 600 units/ml. A correction for the interference of L-lactate on the apparent D-lactate concentration was found to be necessary for some D-LDH preparations. To determine whether corrections are required, the D-LDH used was tested for L-LDH activity by addition of L-lactate, equivalent to a final concentration of 35 mg/100 ml to the assay tube. If a significant increase in absorbance was found, then a correction formula using a linear regression standard curve within the range of physiological values for L-lactate was determined and the true D-lactate concentration was calculated.[21] These corrections for some D-LDH preparations were small but significant for the low concentration of D-lactate present in plasma as affected by the L-lactate present in each plasma sample. Sigma assay kit 826-UV was used for L-lactate analysis. Blood was collected in Becton-Dickinson heparinized Vacutainers.

General Method

Samples of either plasma or standard solutions of D-lactate were centrifuged in the Amicon cones, and the filtrate was added to the NAD^+, glycine–hydrazine assay solution. The reaction was initiated by addition of D-LDH; the sample was allowed to incubate to complete the reaction, and the absorbance of the formed NADH was determined. The concentration of D-lactate in plasma was determined from a linear regression formula from a standard solution. The previously published procedure[21] used a protein-free plasma assay for D-lactate where perchloric acid and alkali was used. That procedure required several sample centrifugations and dilution factors. The deproteinization introduced falsely high values for D-lactate when some glycolytic intermediates were present. It has been reported that various compounds, including many glycolytic intermediates, when added at 80–800 times their red blood cell or plasma concentration did not give significant falsely high values for D-lactate.[21]

The pH of the buffer system in the alkaline range was selected to aid in overcoming the equilibrium in favor of lactate formation. Hydrazine in the buffer, by reacting with pyruvate formed in the reaction, forces the reaction to completion. Varying pH from 7.5 to 10.0 showed a decreasing half-time ($t_{1/2}$) of reaction with increasing pH. At pH 9.5 and 25° the $t_{1/2}$

was 6.4 min, which was used to estimate a 99.9% completion of reaction of 64 min. Incubation at 25° gave a minimum $t_{1/2}$ in a range from 15° to 37°. The reaction time of 90 min allows for some loss of activity of D-LDH. Adding known amounts of D-lactate standard to plasma in the range of 0.015 to 0.4 mM showed recoveries of 92–107%.

Assay Protocol

Assay for Standard Curve for D-*Lactate*. About 2 ml of standard D-lactate, varying in concentration from 0.015 to 2.0 mM was added to the Amicon cones and centrifuged at 2000 rpm for 40 min. An aliquot of 0.8 ml of the filtrate was added to 2.4 ml of the NAD$^+$, glycine–hydrazine buffer and mixed by vortexing. Samples of 1.0 ml of each assay solution were added to 100 × 13 mm culture tubes followed by 0.05 ml of D-LDH to two tubes and 0.05 ml of water to the third tube. The tubes were vortexed and incubated 90 min at 25°. The absorbance was measured at 340 nm, and the reagent blank value was subtracted to determine the absorbance due to the reduction of NAD$^+$. A standard curve regression line for D-lactate was constructed to cover a range of 0.006 to 1.2 mM. The linear regression equation developed to determine plasma D-lactate is shown in Eq. (2).

Lactate conc. (mg/100 ml)
$$= \frac{\text{change in absorbance } - \text{ absorbance intercept}}{\text{slope}} \times \text{dilution factor} \quad (2)$$

The actual values determined to express the plasma concentration of D-lactate are given in Eq. (3).

$$\text{Lactate conc.} = \frac{\text{change in absorbance } + \text{ } 0.0045}{0.595} \times 4 \quad (3)$$

Plasma D-*Lactate Assay*. The procedure was the same as for the standard curve except that heparinized blood was centrifuged at 3200 rpm (1400 g) for 10 min in a clinical centrifuge and 2 ml of the plasma was added to the Amicon cone. A second reagent blank was also used with water in place of the filtered plasma. The final absorbance was corrected for the absorbance contributed by the D-LDH in the NAD$^+$, glycine–hydrazine solution.

For a typical analysis from a normal individual, the data from the table were used, and the calculation was as follows:

ΔA = 0.178 (mean of sample absorbance) − 0.146 (absorbance of
buffer, plasma filtrate, and separate D-LDH) = 0.032

DATA FOR PLASMA D-LACTATE

Tube	D-LDH[a]	Sample[b]	Absorbance[c]
1	−	−	0.113
2	−	−	0.113
3	+	−	0.117
4	+	−	0.115
5	−	+	0.143
6	+	+	0.179
7	+	+	0.177

[a] +, D-LDH added; −, water added in place of D-LDH.
[b] +, Plasma filtrate in buffer; −, water added in place of plasma filtrate.
[c] At 340 mm against a cuvette containing only water.

Using Eq. (3), the calculation is

$$\text{D-lactate} = \frac{0.032 + 0.0045}{0.595} \times 4$$

The calculated concentration (mg/100 ml) was 0.25. To express millimolar concentration, this value was multiplied by 0.111 and is 0.027 mM. The values reported for seven healthy males, ages 22–28, were 0.21 mg/100 ml ± 0.014 SEM (0.023 mM).[21] These values are in the range of 1–2% of the concentration of L-lactate.

Discussion

The assay described here was directed toward D-lactate concentrations that would be below those expected for the normal glycolytic product, L-lactate. The assay reported here is valid for D-lactate concentrations from 0.006 to 1.2 mM. The assay procedure of Gawehn and Bergmeyer[22] was modified, since their procedure was for food or microbiological samples of higher D-lactate concentration. An automated procedure for D- and L-lactate[23] was also directed to 200–500 times the concentration found in plasma. In order to measure D- and L-lactate in plasma, it is recommended that separate blood samples be collected for

[22] K. Gawehn and H. U. Bergmeyer in "Methods of Enzymatic Analysis" (H. U. Bergmeyer, ed.), 2nd Engl. ed. Vol. 3, p. 1492. Academic Press, New York, 1974.
[23] S. R. Goodall and F. M. Byers, Anal. Biochem. 89, 80 (1978).

each since an assay for L-lactate requires inhibition of glycolysis with an inhibitor such as fluoride, otherwise significant amounts of L-lactate will be formed. D-Lactate concentration in whole blood, however, does not increase significantly during storage over several hours, but addition of fluoride causes a rapid and significant increase as explained elsewhere.[13]

Acknowledgment

This work was supported by funds from the National Foundation for Cancer Research and was carried out with the aid of M. G. Waters and many medical students.

Section II

Enzyme Assay Procedures

[7] Determination of Transketolase Activity via Ferricyanide Reduction

By G. A. KOCHETOV

Principle. In addition to the transferase reaction, transketolase (EC 2.2.1.1) can catalyze, in the presence of ferricyanide, the oxidation of donor substrates.[1,2] For example, if fructose 6-phosphate is used as a substrate, the following reaction takes place:

Fructose 6-phosphate + 2 Fe(CN)$_6^{3-}$ + H$_2$O →
$$\text{glycolic acid + erythrose 4-phosphate + 2 Fe(CN)}_6^{4-} + 2\text{ H}^+$$

Two moles of ferricyanide are reduced per mole of the consumed substrate.[2] The reaction is followed spectrophotometrically by the decrease in absorbance at 420 nm, which is due to the reduction of ferricyanide. For calculations, the molar extinction coefficients for ferricyanide of 1000 M^{-1} cm^{-1} were used.

Procedure.[3] The reaction mixture contains (final concentrations): 50 mM glycylglycine buffer, 2 mM MgCl$_2$, 0.1 mM thiamine pyrophosphate, 1.4 mM K$_3$Fe(CN)$_6$, 3 mM fructose 6-phosphate, 0.01–0.02 units of transketolase; final volume 2 ml, pH 7.6. The reaction is initiated by addition of the enzyme. The decrease in the absorbance is measured during 20 sec at 420 nm against a reference sample containing all the components but transketolase. The measurements are carried out in a double-beam automatic spectrophotometer in 1-cm light pathway cuvettes; the 0–0.1 scale was used.

Definition of the Enzyme Unit. One unit of enzyme is the amount that catalyzes the conversion of 1 μmol of fructose 6-phosphate per minute under the above conditions.

Remarks. The rate of the transketolase reaction measured with ferricyanide is constant for 20–30 sec and then gradually decreases; the decrease is due to the inactivation of the enzyme,[4] the reason for which is obscure. There is a clear-cut linear dependence between the concentration of the enzyme (in the range indicated above) and the apparent reaction rate.

[1] J. N. Bradbeer and E. Racker, *Fed. Proc., Fed. Am. Soc. Exp. Biol.* **20,** 88 (1965).
[2] P. Christen and A. Gasser, *Abstr. FEBS Meet., 10th* No. 604 (1975).
[3] R. A. Usmanov and G. A. Kochetov, *Biochem. Int.,* **3,** 33 (1981).
[4] P. Christen, M. Cogoli-Greutor, M. J. Healy, and D. Lubini, *Eur. J. Biochem.* **63,** 223 (1976).

METHODS IN ENZYMOLOGY, VOL. 89

The above procedure for the transketolase activity assay requires a relatively large amount of the enzyme, as the initial rate of the transketolase-catalyzed oxidation is many times lower than the rate of the transketolase-catalyzed transferase reaction based on the standard substrates of transketolase (xylulose 5-phosphate and ribose 5-phosphate). The advantage of this method is that it requires no auxiliary enzymes. It differs favorably from other such methods[5,6] in that one can work with any of the donor substrates of transketolase. Finally, the ferricyanide method allows one to study the intermediate steps of the transferase reaction catalyzed by transketolase, as only one substrate, the donor, is used. No acceptor substrate is required.

[5] G. A. Kochetov and P. P. Philippov, *Anal. Biochem.* **48**, 286 (1972).
[6] G. A. Kochetov and R. A. Usmanov, *Anal. Biochem.* **88**, 296 (1978).

[8] Spectrophotometric Method for Glucose-6-Phosphate Phosphatase[1]

By PETER GIEROW and BENGT JERGIL

Glucose 6-phosphate + H_2O → glucose + P_i

Glucose-6-phosphatase (EC 3.1.3.9) is located mainly in the endoplasmic reticulum and is commonly used as a marker enzyme for this organelle. The assay of both glucose 6-phosphate hydrolysis and other reactions catalyzed by the enzyme[2] have been described in this series.[3,4] Assay methods of the former reaction are usually based on phosphate determination by the ammonium molybdate reaction.[5-7] Other methods utilize the release of radiolabeled glucose[8,9] or phosphate.[10] The procedure described here is based on a coupled enzymic reaction that can be followed spectrophotometrically.[11]

[1] This work was supported by a grant from the Swedish Natural Science Research Council.
[2] R. C. Nordlie, *Curr. Topics Cell. Regul.* **8**, 33 (1974).
[3] M. A. Swanson, this series, Vol. 2, p. 541.
[4] R. C. Nordlie and W. J. Arion, this series, Vol. 9, p. 619.
[5] P. A. Chen, T. Y. Toribara, and H. Warner, *Anal. Chem.* **28**, 1756 (1956).
[6] K. Itaya and M. Ui, *Clin. Chim. Acta* **14**, 361 (1966).
[7] C. L. Penney, *Anal. Biochem.* **75**, 201 (1976).
[8] H. Negishi, Y. Morishita, S. Kodama, and T. Matsuo, *Clin. Chim. Acta* **53**, 175 (1974).
[9] S. A. Kitcher, K. Siddle, and J. P. Luzio, *Anal. Biochem.* **88**, 29 (1978).
[10] J. W. Arion, B. K. Wallin, P. W. Carlson, and A. J. Lange, *J. Biol. Chem.* **247**, 2558 (1972).
[11] P. Gierow and B. Jergil, *Anal. Biochem.* **101**, 305 (1980).

Assay Method

Principle. The assay method is an extension of a method previously used for determination of glucose[12] and cholesterol.[13] Glucose released by the enzyme is oxidized by glucose oxidase (EC 1.1.3.4) and peroxidase (EC 1.11.1.7):

$$\text{Glucose 6-phosphate} + H_2O \xrightarrow{\text{glucose-6-P phosphatase}} \text{glucose} + P_i$$

$$\text{Glucose} + O_2 \xrightarrow{\text{glucose oxidase}} \text{gluconolactone} + H_2O_2$$

$$2\ H_2O_2 + \text{phenol} + \text{4-aminoantipyrine} \xrightarrow{\text{peroxidase}} \text{quinoneimine} + 4\ H_2O$$

Reagents

D-Glucose 6-phosphate, 0.1 M

Glucose oxidase (*Aspergillus niger;* Sigma Chemical Co., St. Louis, Missouri, type V), 1500 units/ml, purified as described below

Peroxidase (horseradish; Sigma type II), 800 units/ml, dissolved in water

Phenol, 0.2 M, + 4-aminoantipyrine, 8 mM; stable for a month when kept in the dark at 4°

Sodium cacodylate–HCl buffer, 0.1 M, pH 6.5

Procedure. The reaction mixture contains 125 μl of buffer, 50 μl of phenol–4-aminoantipyrine solution, 50 μl of glucose 6-phosphate, 10 μl of peroxidase, 10 μl of glucose oxidase, and the sample (usually 25–100 μg of microsomal protein)[14] in a final volume of 1 ml. The cuvette is thermostatted at 25°. The reaction is started by the addition of sample and is allowed to proceed for 2–3 min before analysis to avoid a slight lag phase. The formation of quinoneimine is then followed for 1–2 min at 510 nm in a recording spectrophotometer. A blank value has to be measured to account for a background activity caused by a phosphatase that contaminates glucose oxidase. The molar extinction coefficient for quinoneimine at 510 nm is 6.66 mM^{-1} cm^{-1}. It should be noted that two glucose 6-phosphate molecules are hydrolyzed for each molecule of quinoneimine formed.

[12] P. Trinder, *Ann. Clin. Biochem.* **6**, 24 (1969).

[13] C. C. Allain, L. S. Poon, C. S. G. Chan, W. Richmond, and P. C. Fu, *Clin. Chem.* **20**, 470 (1974).

[14] Microsomes were prepared as described by R. Ohlsson and B. Jergil, *Eur. J. Biochem.* **72**, 595 (1977). Fed male Sprague-Dawley rats were killed by decapitation, and the livers were quickly transferred to ice-cold 0.35 M sucrose in 10 mM Tris-HCl, pH 7.4. The livers were homogenized in a Potter–Elvehjem homogenizer. Microsomes were obtained as the 100,000 g pellet of the 9000 g supernatant. Smooth and rough microsomes were isolated, respectively, at the 0.9/1.3 M and 1.3/1.9 M sucrose interfaces after centrifugation of unfractionated microsomes in discontinuous sucrose gradients (95,000 g for 15 hr). The fractions were washed and suspended in the Tris buffer before use.

Purification of Glucose Oxidase

Commercially available glucose oxidase is contaminated by a phosphatase that hydrolyzes glucose 6-phosphate at low pH values. It is, therefore, desirable to purify glucose oxidase before use, particularly when low glucose-6-phosphatase activities are to be measured. The purification can be done by isoelectric focusing.[15] Since the isoelectric points of glucose oxidase and the phosphatase are rather close (4.0 and 4.4, respectively), a narrow pH range Ampholine (pH 3–5) should be used. The focusing of glucose oxidase in the pH gradient and its elution from the focusing column can be followed easily owing to the yellow color of this enzyme.

If no commercial apparatus is available for isoelectric focusing, the following procedure can be followed.[16] Two vertical glass tubes (1.4 × 30 cm) connected through a Y-shaped glass tube equipped with a three-way valve are used as electrode vessel and separation chamber. The latter is equipped with a water jacket and is cooled at 0° during the run. Approximately 2000 units of glucose oxidase (1.35 ml) are mixed with 9.6 ml of 10% sucrose and 50 μl of Ampholine (pH range 3–5). This mixture together with 10.8 ml of 40% sucrose, 200 μl of Ampholine (pH range 3–5), and 50 μl of Ampholine (pH range 3.5–10) is used to form a linear sucrose gradient in the separation chamber. The bottom of this chamber and the electrode vessel are filled with 1% H_2SO_4 in 50% sucrose (anode solution). The cathode solution (20 mM NaOH) is layered on top of the sucrose gradient. The focusing is allowed to proceed for 40 hr at 4° with an initial voltage of 600 V, which is increased to 1200 V after 1–2 hr (maximum current 3 mA).

After the run, fractions of 0.5–1 ml are collected and analyzed for glucose oxidase (using the conditions for glucose-6-P phosphatase but with glucose as substrate) and acid phosphatase.[17] Fractions containing glucose oxidase are combined. They are virtually free from phosphatase (less than 0.5% of the original activity). Glucose oxidase can be stored at 4° for several weeks.

Properties

Sensitivity and Reproducibility. The reaction rate is proportional to the amount of sample within wide limits (3–200 μg of microsomal protein), and the reaction is linear for at least 10 min with 125 μg of microsomal

[15] O. Vesterberg, this series, Vol. 22, p. 389.
[16] M. Sommarin and B. Jergil, *Eur. J. Biochem.* **88**, 49 (1978).
[17] H. U. Bergmeyer, "Methoden der Enzymatische Analyse," 2nd German ed., Vol. 1, p. 457. Verlag Chemie, Weinheim, 1970.

protein. Each single experiment is completed within a few minutes and can be followed directly in the spectrophotometer. It is recommended to use purified glucose oxidase at low glucose-6-P phosphatase concentrations. The activities recorded for glucose-6-P phosphatase are the same as those obtained with an ammonium molybdate method,[18] and the kinetic parameters do not differ significantly from those obtained with other methods.[9,10,19]

The coupled enzyme reaction described is only slightly inhibited by inorganic phosphate. Thus, the specific activity decreases less than 10% in 5 mM, and approximately 30% in 20 mM, sodium phosphate buffer, pH 6.5. Polymer systems of dextran and polyethylene glycol that may be used for the separation of subcellular particles[20] interfere with the ammonium molybdate methods but not with the present method. Crude tissue homogenates or isolated cytosol usually contain glucose that will interfere with the method. The glucose can be removed by gel filtration or dialysis.

[18] J. A. Lewis and J. R. Tata, *J. Cell. Sci.* **13**, 447 (1973).
[19] W. J. Arion, A. J. Lange, and L. M. Ballas, *J. Biol. Chem.* **251**, 6784 (1976).
[20] P.-Å. Albertsson, B. Andersson, C. Larsson, and H.-E. Åkerlund, *Methods Biochem. Anal.* **28** (in·press).

[9] Activation and Assay of Ribulose-1,5-bisphosphate Carboxylase/Oxygenase

By JOHN W. PIERCE, STEPHEN D. McCURRY,
R. MICHAEL MULLIGAN, and N. E. TOLBERT

Ribulose-1,5-bisphosphate (ribulose-P$_2$) carboxylase-oxygenase (EC 4.1.1.39) is a bifunctional enzyme that catalyzes the addition of CO$_2$ or O$_2$ to C-2 of ribulose-P$_2$. The products of the carboxylase reaction are two molecules of D-glycerate-3-P, and this enzyme activity is systematically classified as 3-phospho-D-glycerate carboxy-lyase (dimerizing). The products of the oxygenase reaction are one molecule of glycolate-2-P and D-glycerate-3-P. The oxygenase activity is of the internal monooxygenase type, since the reaction of ribulose-P$_2$ with O$_2$ results in the incorporation of one oxygen atom into each P-glycolate and H$_2$O. The carboxylase activity is generally assayed by the rate of formation of acid-stable ^{14}C product ([1-^{14}C]glycerate-3-P) formed from NaH^{14}CO$_3$.[1,2] The oxygenase

[1] M. Wishnick and M. D. Lane, this series, Vol. 23, p. 570.
[2] J. M. Paulsen and M. D. Lane, *Biochemistry* **5**, 2350 (1966).

METHODS IN ENZYMOLOGY, VOL. 89

activity is usually monitored by measuring the rate of O_2 consumption with an oxygen electrode.[3-5] Carboxylase activity has also been determined spectrophotometrically by measurement of NADH oxidation enzymically coupled to the rate of glycerate-3-P production.[6,7] The rate of disappearance of ribulose-P_2 has been examined.[8] Oxygenase activity has also been determined by measuring the amount of P-glycolate or glycolate produced after phosphatase treatment.[9-11] In addition, methods have been described for the simultaneous assay of both activities that depend on the use of specifically labeled radioactive ribulose-P_2 and $^{14}CO_2$.[12,13] In addition to enzymic activity, the amount of ribulose-P_2 carboxylase/oxygenase protein can be measured by antibody precipitation[14,15] and by gel electrophoresis.[16] The number of active sites can be quantitatively determined by binding carboxyarabinitol-P_2.[17,18] The simultaneous assay procedures have not been routinely used, and in this chapter only the standard ^{14}C and O_2 consumption assays will be detailed.

An appreciation of the protocols used in properly assaying ribulose-P_2 carboxylase/oxygenase activity requires an understanding of the manner in which various forms of the enzyme interconvert. The forms of ribulose-P_2 carboxylase/oxygenase to be discussed in this chapter are shown in Scheme 1.

The first two reactions of the scheme are referred to as activation and involve the slow, reversible addition of CO_2 ($^ACO_2$ = activator CO_2) to a lysyl residue on the enzyme to form a carbamate, followed by rapid, reversible interaction of the carbamate form of the enzyme with Mg^{2+} to form the active, ternary complex.[19] The adjective, "slow," used in describing the formation of the enzyme · $^ACO_2$ complex, is meant to indicate that the rate of formation and breakdown of this binary complex is much

[3] F. J. Ryan and N.E. Tolbert, *J. Biol. Chem.* **250**, 4229 (1975).
[4] T. J. Andrews, G. H. Lorimer, and N. E. Tolbert, *Biochemistry* **12**, 11 (1973).
[5] G. H. Lorimer, M. A. Badger, and T. J. Andrews, *Anal. Biochem.* **78**, 66 (1977).
[6] E. Racker, this series, Vol. 5, p. 266.
[7] R. M. Lilley and D. A. Walker, *Biochim. Biophys. Acta* **358**, 226 (1974).
[8] S. C. Rice and N. G. Pon, *Anal. Biochem.* **87**, 39 (1978).
[9] W. A. Laing, W. L. Ogren, and R. H. Hageman, *Plant Physiol.* **54**, 678 (1974).
[10] R. Chollet and L. L. Anderson, *Arch. Biochem. Biophys.* **176**, 344 (1976).
[11] J. T. Christeller and W. A. Laing, *Biochem. J.* **183**, 747 (1979).
[12] D. B. Jordan and W. L. Ogren, *Plant Physiol.* **67**, 237 (1981).
[13] S. S. Kent and J. D. Young, *Plant Physiol.* **65**, 465 (1980).
[14] V. A. Wittenbach, *Plant Physiol.* **64**, 884 (1979).
[15] B. Week, *in* "Manual of Quantitative Immunoelectrophoresis. Methods and Application," *Scand. J. Immunol* **2**, Suppl. 1, 37 (1973).
[16] H. Bauwe, *Biochem. Physiol. Pflanz.* **174**, 246 (1979).
[17] J. Pierce, N. E. Tolbert, and R. Baker, *Biochemistry* **19**, 934 (1980).
[18] S. D. McCurry, J. W. Pierce, and N. E. Tolbert, *J. Biol. Chem.* **256**, 6623 (1981).
[19] G. H. Lorimer, *Annu. Rev. Plant Physiol.* **32**, 349 (1981).

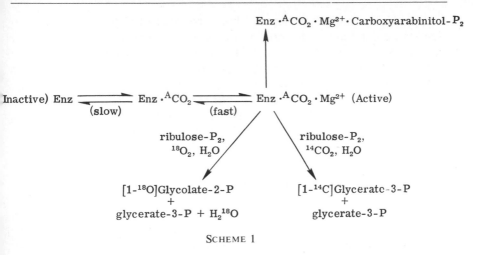

SCHEME 1

slower than the rate of catalysis. A large number of kinetic properties of the enzyme are dictated by this large difference in relative rates of activation and catalysis, and proper assay of the enzyme requires taking these differences into account.

The importance of carefully defining the activation state of the enzyme cannot be overemphasized. Much confusion in the literature regarding this enzyme can be related to improper activation or omission altogether of statements concerning the activation conditions that were used. Of course, it is critical in kinetic studies to start with the same amount of active enzyme. It has become customary to preincubate the enzyme with CO_2 and Mg^{2+} for a period of time (>10 min) prior to assaying initial catalytic activity over a short time period (<1 min) under desired conditions of substrate concentration, pH, etc. When this protocol is followed, fully activated enzyme is assayed over a period of time short enough so that inactivation during the time course of the assay is negligible. Thus, the large difference in the rates of catalysis and activation has been utilized to separate activation experimentally from catalytic phenomena.

The multiple effects of CO_2 on enzyme activity require careful consideration. Mention has already been made of the role of $^ACO_2$ in enzyme activation. In addition, a different molecule of CO_2 ($^SCO_2$ = substrate CO_2)[19,20] is used as the substrate for the carboxylase reaction. $^SCO_2$ and O_2 are competitive inhibitors with respect to the oxygenase and carboxylase reactions.[9,21] Proper assay of the oxygenase activity, therefore, requires activation of the enzyme with $^ACO_2$ and Mg^{2+} at high protein concentrations followed by delivery of the activated enzyme into "CO_2-

[20] G. H. Lorimer, *J. Biol. Chem.* **254**, 5599 (1979).
[21] M. Badger and T. J. Andrews, *Biochem. Biophys. Res. Commun.* **60**, 204 (1974).

free'' buffer with high dilution so that carry-over of CO_2 from the activation step is minimized. Similar precautions are required for the accurate determination of $K_m(CO_2)$ where fully activated enzyme is assayed under variable CO_2 concentrations with or without competition from O_2. In both experiments, allowance must be made for the unavoidable, though small, carry-over of CO_2 and O_2 from the activation step or with buffers.

During the carboxylase assay, the rate of conversion of HCO_3^- to CO_2 can quickly become a limiting factor especially at low HCO_3^- concentrations and/or high enzyme concentrations. The addition of carbonic anhydrase to the assay mixture has been used to enhance the rate of conversion of HCO_3^- to CO_2.[22,23] Alternatively, short assay times with low enzyme concentrations may be used, so that the initial rate of enzyme activity does not change during the assay period because of CO_2 depletion. Also, since O_2 is a competitive inhibitor of the enzyme with respect to CO_2, carboxylase assays should be run anaerobically to determine $K_m(CO_2)$, which is 10–15 μM. However, for reasons of experimental convenience, most assays are run aerobically, and the higher apparent $K_m(CO_2)$ of about 26 μM reflects the competition by O_2.[23] The apparent $K_m(CO_2)$ for the enzyme, however, is closer to the physiological K_m *in vivo*.

The oxygenase reaction is generally run in buffers in equilibrium with air levels of CO_2. The concentration of dissoved O_2 at 30° is about 250 μM under these conditions, and during the short time of the oxygenase assay this level of O_2 is not depleted enough to alter the rate. However, the amount of O_2 in buffers in equilibrium with air is far below the saturation level for O_2 of the oxygenase activity. $K_m(O_2)$ has been reported to be about 400 μM.[19] Therefore the oxygenase rate measured with air-equilibrated buffers is much less than maximal because of the low concentration of O_2 and the presence of competing CO_2. Thus, observed $K_m(O_2)$ values should be treated as apparent K_m values that relate only to the specified assay conditions unless extraordinarily great care is taken in removing contaminating CO_2.

The following kinetic parameters pertain to fully activated spinach enzyme at pH 8.2 and 30° (see tabulation below).

$K_{act}(CO_2)$	100 μM
$K_{act}(Mg^{2+})$	1000 μM
$K_m(CO_2)$	10–15 μM
$K_m(CO_2)$ apparent in air	26 μM
$K_m(O_2)$ apparent in air	400 μM
$K_m(ribulose-P_2)$	20–25 μM

[22] K. Okabe, A. Linklar, M. Tsuzuki, and S. Miyachi, *FEBS Lett.* **114**, 142 (1980).
[23] I. F. Bird, M. J. Cornelius, and A. J. Keys, *J. Exp. Bot.* **31**, 365 (1980).

The following sections of this article outline the protocols used in activation and assay of ribulose-P$_2$ carboxylase/oxygenase. In addition, a protocol is described for determining the amount of activated enzyme (Enz · ^{A}CO$_2$ · Mg^{2+}) in a given enzyme solution that will bind carboxyarabinitol-P$_2$.

Activation

For routine assays, the enzyme must be fully activated by CO$_2$ and Mg^{2+} before initiation of either the carboxylase or oxygenase reaction.[19] After isolation and/or storage as a frozen (NH$_4$)$_2$SO$_4$ slurry, a concentrated enzyme preparation (10–40 mg/ml) is prepared as described by McCurry et al.[24] If the enzyme has been stored in the cold, as is generally the case, incubation or dialysis at room temperature overnight is best to restore activity. The enzyme is activated at 2–16 mg/ml in a buffer with final concentrations of 100 mM Na$^+$-Bicine buffer (pH 8.2), 10 mM NaHCO$_3$, 20 mM MgCl$_2$, 0.2 mM EDTA, and 1 mM dithiothreitol. This can be achieved by starting with double-strength concentration of this activating buffer. Incubation should be for at least 10 min at 30°. For the carboxylase assay, the enzyme is activated at a protein concentration of $\geq$ 2 mg/ml and diluted 25- to 100-fold into the assay mixture in order to minimize the carry-over of activating NaHCO$_3$ into the assay solution. The preparations may be filter-sterilized to eliminate bacterial contamination. Protein concentrations are measured by A_{280}.[24]

Activation and assay of the crystalline tobacco enzyme is similar to that for the spinach enzyme except that it is also heat-activated at 50° for 20 min. Tobacco enzyme that has been crystallized is 30–50% more active after 50° heat activation, and the tobacco enzyme that has not been crystallized is inhibited by activation at the higher temperature.

^{14}C Ribulose-P$_2$ Carboxylase Assay

Reagents

Bicine, 200 mM, pH 8.2, containing 0.4 mM EDTA and 1 mM dithiothreitol
Ribulose-P$_2$, 12.5 mM, pH 6.5.[25] Store at pH 6.5 in a $-20°$ freezer.
NaH^{14}CO$_3$, 250 mM ($\geq$ 0.14 Ci/mol)
Activated enzyme, 2 mg/ml
MgCl$_2$, 2 M
HCl, 2 N

[24] S. D. McCurry, R. Gee, and N. E. Tolbert, this series, Vol. 90 [82].
[25] B. L. Horecker, J. Hurwitz, and A. Weissbach, *Biochem. Prep.* 6, 83 (1958).

It is convenient to run these assays in the same 8-ml glass scintillation vials to be used for counting the ^{14}C. Each vial contains 250 μl of Bicine buffer, 20 μl of ribulose-P_2 solution, 5 μl of $MgCl_2$ solution, 20 μl of NaH$^{14}CO_3$ solution, and 150 μl of H_2O.

For routine assays, all the reagents except ribulose-P_2 and NaH$^{14}CO_3$ may be premixed to reduce the number of pipetting steps. The vials are stoppered with rubber serum caps to prevent loss of $^{14}CO_2$ or changes in the specific radioactivity due to exchange with atmospheric CO_2. The stoppered vials may also be gassed with N_2. The reaction is initiated after temperature equilibration by the addition of activated enzyme (i.e., 20 μl) through the serum cap. The vial must be immediately swirled to mix. Vigorous mixing with a vortex should be avoided because protein denaturation may occur. After 15 sec to 1 min, the reaction is stopped by the addition of 200 μl of 2 N HCl. The vials, which contain the acidified samples, are slowly heated to dryness in an oven at 95° (in a hood) to remove excess $^{14}CO_2$ and acid. This drying step must be performed carefully to avoid caramelization of the samples, as the color will cause severe quenching during the scintillation counting. After cooling, 0.5 ml of H_2O is added followed by 4.5 ml of scintillation cocktail. The cocktail we use is composed of 2 liters of toluene, 1 liter of Triton X-100, 12 g of PPO, and 0.15 g of POPOP. The use of 8-ml vials saves on the cost of reagents, and the results are the same as with 20-ml vials.

Carboxylase sp. act (μmol/min/mg protein)

$$= \frac{^{14}C \ (dpm)}{(dpm \ ^{14}C/\mu mol \ CO_2) \cdot time \ (min) \cdot mg \ protein \ in \ assay}$$

When low CO_2 concentrations are used, the concentration of enzyme must be correspondingly decreased, or carbonic anhydrase added. Otherwise, the CO_2 concentration (about 14 μM at 1 mM HCO$_3^-$, pH 8) will change drastically owing to its utilization by the enzyme. Additionally, when various concentrations of CO_2 are used in the assay, contaminating CO_2 in the buffers can lead to serious and variable errors. It is convenient to determine the specific ^{14}C radioactivity of each solution by adding a small amount (about 5 nmol) of ribulose-P_2 to an aliquot and converting it quantitatively to [1-^{14}C]glycerate-3-P. The observed acid stable radioactivity divided by the amount of ribulose-P_2 added gives the specific radioactivity of the $^{14}CO_2$. A separate determination of the total radioactivity in the solution with excess ribulose-P_2 can then be used to determine the concentration of HCO$_3^-$(CO_2).

Notes. The enzyme solutions should be handled at room temperature, as the enzyme is labile in the cold. Ribulose-P_2 solutions are unstable, especially under alkaline conditions,[26] and should not be used longer than 2–4 weeks. Properties of these sugar bisphosphates are discussed elsewhere in this volume.[27] Ribulose-P$_2$ solutions are stored frozen at a pH of 5–6.5, at which it is most stable, and the assay buffer is relied upon to maintain the pH at 8.2 upon addition of this substrate.

Oxygenase Assay

Because of its relative convenience and availability, the most widely used assay of the oxygenase activity is based on the polarographic measurement of O$_2$ uptake with an oxygen electrode.

Reagents

Na$^+$-Bicine, 200 mM, pH 8.2, containing 0.4 mM EDTA. For careful work, this buffer should be prepared from freshly boiled, distilled H$_2$O. Addition of Bicine to the degassed water causes the pH to drop to about 3, and further CO$_2$ removal by drawing a vacuum over the solution is desirable. The pH is adjusted to 8.2 with 50% NaOH in which Na$_2$CO$_3$ is insoluble. Nitrogen gas is bubbled through the solution for 24 hr, and the solution is stored under nitrogen.

Ribulose-P$_2$, 12.5 mM, pH 6.5

MgCl$_2$, 2 M

Distilled H$_2$O, freshly boiled

Activated enzyme, 8–16 mg/ml

This high concentration of activated enzyme is used for maximum dilution into the assay mixture. If both oxygenase and carboxylase assays are to be run on the same preparation, the enzyme is activated at 8–16 mg/ml; a portion is used for the oxygenase assay, and a portion is diluted to 2 mg/ml with activation buffer for the carboxylase assay.

Assays are run in a final volume of 1 ml in a Rank Brothers (Cambridge, England) oxygen electrode at 30° and pH 8.2. The chamber of the oxygen electrode contains 500 μl of Bicine buffer, 40 μl of ribulose-P$_2$ solution, 10 μl of MgCl$_2$ solution, and 400–410 μl of water.

The solution is bubbled with humidified, CO$_2$-free air (or gas at the desired O$_2$ concentration) for 2–3 min while awaiting temperature equilibration. The chamber is closed with care taken to exclude bubbles. The solution of activated enzyme (10–20 μl) is added through the port, and O$_2$ uptake is measured over a period of 1 min.

[26] C. Paech, J. Pierce, S. D. McCurry, and N. E. Tolbert, *Biochem. Biophys. Res. Commun.* **83**, 1084 (1978).

[27] A. S. Serianni, J. Pierce, and R. Barker, this volume [12].

The rates obtained from this assay should be nearly linear for more than 1 min, but after this time the rate slows dramatically owing to enzyme deactivation in the CO_2-depleted medium. Tangential lines drawn from the first 15–30 sec of the reaction are used to determine the initial rate of the reaction.

$$\text{Oxygenase sp. act } (\mu\text{mol/min/mg protein}) = \frac{\mu\text{mol } O_2 \text{ uptake/min}}{\text{mg protein in assay}}$$

An inherent lag exists in this assay because of the time required for O_2 to diffuse across the membrane of the electrode. This unavoidable lag should be less than 5 sec, although longer lag times may result if the electrode is dirty or the membrane is too thick. Dithiothreitol must not be present in any significant concentration because of its reaction with O_2. The background rate of O_2 consumption, when either enzyme or ribulose-P_2 is omitted, should be negligible.

Measurement of the Amount of the Activated Enzyme (Enzyme $\cdot$ $^ACO_2$ $\cdot$ Mg^{2+})

The finding that carboxyarabinitol-P_2 binds most tightly ($K_D < 10$ pM) only to the ternary enzyme $\cdot$ $^ACO_2$ $\cdot$ Mg^{2+} complex[17] and that the resulting quaternary complex is essentially inert to exchange of any of its components[18,28] can be used as a basis to assess the amount of the activated enzyme $\cdot$ $^ACO_2$ $\cdot$ Mg^{2+} form in a given solution in an assay independent of measuring enzyme activity.[18] One determines the amount of activated enzyme by measuring either the amount of tightly bound [2'-^{14}C]carboxyarabinitol-P_2 or by measuring the amount of $^{14}CO_2$ that is rendered nonexchangeable by treatment with unlabeled carboxyarabinitol-P_2. Theoretically, there are eight active sites per mole of enzyme. Carboxyarabinitol-P_2 may be synthesized with ribulose-P_2 and cyanide.[17,28]

$^{14}CO_2$ Method

Ribulose-P_2 carboxylase/oxygenase (1–2 mg/ml) is activated with 20 mM $MgCl_2$ and 10 mM NaH$^{14}CO_3$ (0.5 Ci/mol) for at least 1 hr as described above under Activation. Longer activation is required to assure complete exchange with the added $^{14}CO_2$ with any CO_2 already in the preparation and bound to the enzyme. A small aliquot may be taken for measuring enzyme activity by fixation of substrate $^{14}CO_2$ with ribulose-P_2. A larger aliquot (0.4 ml) is delivered into 4.1 ml of a carboxyarabinitol-P_2 solution (0.1 mM carboxyarabinitol-P_2, 10 mM NaHCO$_3$, 20 mM $MgCl_2$

---

[28] H. M. Miziorko and R. L. Sealy, *Biochemistry* **19**, 1167 (1980).

in Bicine buffer) to form the enzyme $\cdot$ $^ACO_2$ $\cdot$ Mg^{2+} $\cdot$ carboxyarabinitol-P$_2$ complex. The presence of 10-fold excess of NaHCO$_3$ reduces the specific radioactivity of any unbound or loosely bound $^{14}CO_2$. After 30 min, 4.5 ml of a solution containing 40% polyethylene glycol, 10 mM NaHCO$_3$, and 20 mM MgCl$_2$ in 0.1^r M Bicine buffer is added to precipitate the enzyme complex.[29] The precipitate is collected by centrifugation (27,000 g for 10 min) and twice washed with 4 ml of 20% polyethylene glycol in the previous solution. The twice washed precipitate is dissolved in 0.6 ml of 50 mM Bicine buffer (pH 8.0) containing 5 mM NaHCO$_3$, and an aliquot is taken for determination of radioactivity. An aliquot may also be taken for determination of protein. The amount of radioactivity then is a measure of the number of enzyme sites that contain $^ACO_2$ and Mg^{2+}, that is, the number of activated sites.

[^{14}C]Carboxyarabinitol-P$_2$ Method

The method is similar to the $^{14}CO_2$ variant described above. Enzyme (1 mg) is incubated for 15 min in 1 ml of 100 mM Bicine buffer (pH 8.0) containing 10 mM NaHCO$_3$ and 20 mM MgCl$_2$ to form the activated enzyme $\cdot$ $^ACO_2$ $\cdot$ Mg^{2+}. A 20-μl aliquot of 1 mM [2'-^{14}C]carboxyarabinitol-P$_2$ (0.81 Ci/mol)[17] (a 1.5-fold excess of the [2'-^{14}C]carboxyarabinitol-P$_2$ over potential active sites) is added, and the solution is kept for 45 min. A 50-fold excess of nonradioactive carboxyarabinitol-P$_2$ is then added, and any exchange of the radioactive ligand with the nonradioactive ligand is allowed to proceed for 5 hr. A 2-ml aliquot of 20% polyethylene glycol containing 20 mM MgCl$_2$ is added to precipitate the protein complex. The precipitate is collected by centrifugation, washed twice with 3-ml portions of the 20% polyethylene glycol with 20 mM MgCl$_2$; the final precipitate is resuspended, and its radioactivity is determined.

[29] N. P. Hall and N. E. Tolbert, *FEBS Lett.* **96,** 167 (1978).

Section III

Preparation of Substrates and Effectors

[10] Synthesis of (2R)-Glycerol-o-β-D-Galactopyranoside by β-Galactosidase

By WINFRIED BOOS

(2R)-Glycerol-o-D-β-galactopyranoside (RGG) with its two glyceryl hydroxyl groups acylated by fatty acids occurs in large amounts in chloroplasts of plants and algae as one type of galactolipid (monogalactosyldiglyceride).[1-3] Of the two possible diastereoisomers (in respect to the optically active carbon 2 of glycerol) only the R form is found. After deacylation of the two fatty acids of the galactolipid, RGG is now an excellent carbon source for enteric bacteria such as Escherichia coli, endowed with the lac operon-encoded proteins. In fact, by comparing the utilization by E. coli of lactose and RGG, one is tempted to propose that RGG, not lactose, is the natural substrate of the lac operon-encoded metabolic functions of enteric bacteria.

1. With few exceptions lactose is found only in the milk of mammals whereas RGG can be released by lipases in large amounts in the gut of plant-eating animals.

2. RGG, but not lactose, is an inducer of the lac operon[4]; the latter has first to be altered into an inducer (allolactose) by the galactosyl transferase activity of β-galactosidase.[5]

3. Cells growing on lactose exhibit a lower β-galactosidase activity than those growing on RGG, probably because of the observed phenomenon that lactose after being taken up by the lactose transport system releases its split products glucose and galactose quantitatively into the medium.[6] Thus, growth on lactose is dependent on recapturing and metabolizing external glucose and galactose. Under these conditions glucose must exert a strong catabolite effect on the transcription of the lac operon, and growth will proceed effectively only when glucose is used up and the cells are growing on galactose. As yet it is not clear whether growth on RGG also proceeds via the release of glycerol and galactose

[1] P. S. Sastry and M. Kates, Biochemistry 3, 1271 (1964).
[2] A. A. Benson, J. F. Winterman, and R. Wiser, Plant. Physiol. 34, 315 (1959).
[3] H. E. Carter, K. Ohno, S. Noijma, C. L. Tipton, and N. Stanacev, J. Lipid Res. 2, 215 (1961).
[4] C. Burstein, M. Cohn, A. Kepes, and J. Monod, Biochim. Biophys. Acta 95, 634 (1965).
[5] A. Jobe and S. Bourgeois, J. Mol. Biol. 69, 397 (1972).
[6] R. E. Huber, R. Pisko-Dubienski, and K. L. Huber, Biochem. Biophys. Res. Commun. 96, 656. (1980).

METHODS IN ENZYMOLOGY, VOL. 89

into the medium. However, both glycerol and galactose exert only a mild catabolite effect, and the *lac* operon will remain fully induced under these conditions.

Thus, in *E. coli* RGG is an excellent substrate for β-galactosidase and the lactose transport system as well as the *lac* repressor.[7] Unlike lactose RGG also serves as substrate, albeit poorly for the third *lac* operon encoded enzyme, thiogalactoside transacetylase.[7a] RGG retains its inducing capacities for the *lac* operon even in strains that lack the lactose transport system.[8] The reason for this inducer ability is the presence of an additional transport system that can recognize RGG with a high affinity and can concentrate it inside the cells several thousandfold.[9] This system is a galactose transport system, is dependent on a periplasmic galactose binding protein, and is coded for by the *mgl* genes.[10] RGG has been used to isolate mutants in this system,[11] and, owing to the affinity of the galactose binding protein toward RGG, it has been used to identify the galactose binding protein as the chemoreceptor for galactose and RGG chemotaxis.[12] In the absence of a functional lactose transport system (*lac Y* mutants), growth of *Escherichia coli* on RGG is dependent on β-galactosidase and the *mgl*-coded galactose transport system.

As mentioned before, glycerol-*o*-β-D-galactopyranoside contains an asymmetric carbon at position 2 of the glycerol moiety. The 2*R* and 2*S* forms are diastereoisomers and can be distinguished by slight differences in their physical properties, such as melting point, optical rotation, and infrared spectra.[13] The naturally occurring 2*R* form, but not the 2*S* form, is recognized by the *mgl*-encoded galactose transport system and the galactose binding protein. However, both the 2*R* and the 2*S* forms are hydrolyzed by β-galactosidase and are transported by the lactose transport system.[7]

The isolation of RGG can be accomplished by extracting galactolipids from chloroplasts, deacylating by alkaline hydrolysis, and chromatographic separation of the glycerol mono- and digalactosides followed by crystallization of RGG.[1-3,13] Alternatively, one can make use of the transgalactosylating activity of β-galactosidase[14,15] with *o*-NPGal or another

[7] T. J. Silhavy and W. Boos, *J. Biol. Chem.* **248**, 6571 (1973).
[7a] R. E. Musso and I. Zabin, *Biochemistry* **12**, 553 (1973).
[8] W. Boos, P. Schaedel, and K. Wallenfels, *Eur. J. Biochem.* **1**, 382 (1967).
[9] W. Boos and K. Wallenfels, *Eur. J. Biochem.* **3**, 360 (1968).
[10] W. Boos, *Curr. Top. Membr. Transp.* **5**, 51 (1974).
[11] T. J. Silhavy and W. Boos, *J. Bacteriol.* **120**, 424 (1974).
[12] G. L. Hazelbauer and J. Adler, *Nature (London) New Biol.* **230**, 101 (1971).
[13] B. Wickberg, *Acta Chem. Scand.* **12**, 1187 (1958).
[14] K. Wallenfels and R. Weil, *in* "The Enzymes" (P. D. Boyer, ed.), 3rd ed., Vol. 7, p. 617. Academic Press, New York, 1972.
[15] R. E. Huber, G. Kurz, and K. Wallenfels, *Biochemistry* **15**, 1994 (1976).

suitable β-galactoside as galactosyl donor and glycerol as acceptor.[16,17] Despite the fact that β-galactosidase can hydrolyze both the 2R and 2S form of glycerol-o-β-D-galactopyranoside, only the 2R form is synthesized in the transgalactosylating reaction.[16] The disadvantage of the latter method is the necessary separation of the formed RGG from the split products of β-galactosidase hydrolysis, mainly galactose, and the small amounts of a second galactosyl transfer product that is formed by transfer of galactose to the hydroxyl group on the carbon 2 of glycerol. It is difficult to separate the latter compound from RGG by chromatography. Only repeated crystallization after seeding with RGG has been given successful purification.[17]

Described below is a simple and inexpensive method[7] for synthesizing RGG that utilizes the transgalactosylating properties of β-galactosidase and the chloroform solubility of a derivative of RGG that is formed by the transfer of galactose onto isopropylidene glycerol, in which the hydroxyl groups of carbons 2 and 3 of glycerol are blocked by acetone. Again, as in the case of glycerol, also with racemic isopropylidene glycerol as acceptor, only the corresponding 2R diastereoisomer is synthesized. From all products formed during the enzymic action of galactosidase, as well as from the starting material lactose, only the galactosyl transfer product together with isopropylidene glycerol is extracted with chloroform. The chloroform can be removed by vacuum distillation. From the remaining transfer product, nearly pure RGG can be obtained by mild hydrolysis with acidic acid. It can be crystallized from ethanolic solutions.

Preparation of (2R)-Glycerol-o-β-D-galactopyranoside

Racemic isopropylidene glycerol (150 g, Sigma Chemical Co., St. Louis, Missouri) and lactose (50 g) are added to 850 ml of buffer (10 mM ammonium bicarbonate, 1 mM MgCl$_2$, pH 7.2). *Escherichia coli* β-galactosidase (about 10,000 units of a crude preparation[18]) is added, and the solution is incubated at room temperature with occasional shaking. Lactose hydrolysis and galactosyl transfer are monitored by thin-layer chromatography. For this purpose silica gel-coated glass plates with a layer thickness of 0.25 mm can be used. The solvent n-butanol–pyridine–water (1:1:1, v/v/v) separates lactose and allolactose, as well as the monosaccharides glucose and galactose, from the transfer product. The spots are developed by spraying with 20% sulfuric acid and heating to

[16] W. Boos, J. Lehmann, and K. Wallenfels, *Carbohydr. Res.* **1**, 419 (1966).

[17] W. Boos, J. Lehmann, and K. Wallenfels, *Carbohydr. Res.* **7**, 381 (1968).

[18] One unit of enzyme activity is defined as the amount of enzyme that will hydrolyze 1 μmol of o-nitrophenyl galactoside in 1 min under the above conditions.

120° for 20 min. Isopropylidene glycerol evaporates from the plates and is not visible.

Under the given conditions for the enzymic synthesis, the enzyme is slowly inactivated. Therefore, the same amount of enzyme should be added after 3 and 6 hr. After 9 hr most of the lactose is used up. The solution is heated to 100° for 5 min. After cooling, the solution is concentrated by rotary evaporation until a slightly yellow syrup is obtained. The syrup is extracted three times with 200 ml of chloroform. The combined chloroform extracts are freed of chloroform by rotary evaporation. The remaining syrup contains only the transfer product and isopropylidene glycerol. The latter is removed by vacuum distillation (80°, 10–15 mm Hg). After distillation a thick somewhat yellowish syrup remains; 200 ml of 50% acetic acid are added, and the syrup is brought into solution by gentle stirring; the solution is kept at room temperature for 15 hr. After this time the transfer product is quantitatively hydrolyzed to RGG as checked by thin-layer chromatography. As solvent n-propanol–water (7 : 1, v/v) is used. In this solvent RGG and galactose can be separated from each other. This procedure may be important to test for any hydrolysis of RGG to galactose and glycerol. The acetic acid is removed by rotary evaporation. To remove the residual acetic acid, the syrup is dissolved twice in 200 ml of methanol and brought to dryness again by rotary evaporation. The residue is dissolved in 200 ml of hot ethanol. At room temperature the solution is seeded with authentic RGG, and crystallization is allowed to proceed. Crystallization from methanol is easier but yields less product since RGG is more soluble in methanol than in ethanol. In a typical experiment the yield is 8.35 g of RGG (24%, based on the initial amount of lactose), m.p. 124–128°, and $[\alpha]_D^{20} - 4.76°$ (water, $c = 10$). The product was twice recrystallized from methanol, m.p. 137–138°, $[\alpha]_D^{20} - 4.81°$ (water, $c = 10$). For comparison, the following values are reported in the literature: m.p. 140.5–141.5°, $[\alpha]_D^{20} - 7°$ [13]; m.p. 139–140°, $[\alpha]_D^{27} + 3.77°$.[19] A mixed melting point with authentic (2R)-glycerol-o-β-D-galactopyranoside shows no depression. The product comigrates with authentic RGG on thin-layer chromatography in both solvents. In addition, the pertrimethylsilylated derivative of the product cochromatographs in the gas chromatograph with authentic RGG.

The enzyme necessary for the preparation may conveniently be isolated from strains that contain high levels of this enzyme.[20] Purification of the enzyme is not necessary. In the above preparation the ammonium sulfate precipitate of crude extracts obtained from an overproducing strain was used. The slow inactivation of enzymic activity dur-

[19] H. E. Carter, R. H. McCluer, and E. D. Seifer, J. Am. Chem. Soc. 78, 3735 (1956).
[20] A. V. Fowler, J. Bacteriol. 112, 856 (1972).

ing the enzymic synthesis can be counteracted somewhat by using freshly distilled isopropylidene glycerol and recrystallized lactose (from water).

The initial crystallization of RGG will occur faster if the syrup containing RGG is passed over a BioGel P-2 (400 mesh) column (5 × 100 cm) after dissolving in 50 ml of water and elution in water. The fractions eluting from this column can be monitored by the anthrone reaction.[21] For crystallization the RGG containing fractions are concentrated by rotary evaporation followed by vacuum desiccation. The very viscous syrup is then dissolved in the smallest amount of boiling ethanol and left at room temperature.

Preparation of Radioactively Labeled RGG

For the preparation of [^{14}C]RGG, labeled in the glycerol moiety, the direct transfer of the galactosyl moiety from o-NPGal onto ^{14}C-labeled glycerol followed by chromatography can be used.[17] A typical preparation is the following: 0.45 ml of buffer (0.1 M potassium phosphate, 1 mM MgSO$_4$, pH 7.2) 0.2 mM [^{14}C]glycerol (12.1 mCi/mmol), and 50 mM o-NPGal are incubated at room temperature. β-Galactosidase (50 μl; 10 units, crude preparation) is added. After 1–2 min (50–70% hydrolysis of o-NPGal) the mixture is heated in boiling water for 3 min. Initially, o NPGal may not dissolve entirely at a concentration of 50 mM. This problem may be overcome by heating the solution prior to the addition of enzyme. After o-NPGal is dissolved, the solution is quickly cooled to room temperature and enzyme is added. Alternatively, the enzyme reaction can be carried out at 40° for a correspondingly shorter time. Under these conditions, o-NPGal remains in solution.

The entire mixture is transferred onto chromatography paper (Whatman 3 MM) as a 20 cm-long streak and chromatographed in phenol–water (4:1, v/v). This somewhat inconvenient solvent is used because it separates RGG well from contaminating galactose. The chromatogram is thoroughly dried in air. By autoradiography four bands containing radioactivity can be seen. The fastest migrating band contains glycerol. The next band consists 95% of [^{14}C]RGG. The yield of ^{14}C-labeled RGG in relation to the initial [^{14}C]glycerol in this experiment is 55%; 10–15% of the supplied glycerol remains ungalactosylated. The remaining two bands represent products that contain more than one galactosyl moiety. The RGG-containing band is cut out from the chromatogram and eluted with water. Using a descending chromatography setup, the radioactivity will migrate with the water front and is eluted quantitatively in less than 10 drops (~0.5 ml).

To remove the residual phenol the solution can be chromatographed

---

[21] F. J. Viles and L. Silverman, *Anal. Chem.* **21**, 950 (1949).

again in a different solvent, for instance *n*-butanol–pyridine–water (6 : 4 : 3, v/v/v). Autoradiography and elution from the dried paper yields a preparation that is free of galactose and glycerol but is about 5% 2-glyceryl-*o*-β-D-galactopyranoside.[17] This substance can be removed only by cocrystallization from ethanol of this preparation with unlabeled RGG. For this purpose a small amount of RGG is added to the radioactive solution and evaporated to dryness; 0.2–0.5 ml of ethanol is added to the solution of syrup and heated. After cooling, a very small RGG crystal is added at room temperature. With some skill, as little as 2 mg of RGG can be crystallized twice. In this case it is advantageous if the crystallizing RGG grows and remains attached to the glass walls, where it can be washed with cold ethanol quite easily. After two crystallizations, the specific radioactivity remains constant and the compound is chemically and radiochemically pure.

[11] Chemical Synthesis of Monosaccharides Enriched with Carbon Isotopes

By A. S. SERIANNI, H. A. NUNEZ, M. L. HAYES, and R. BARKER

This method[1,2] utilizes the facile condensation of cyanide anion with an aldehyde (or aldose) at controlled pH to form a pair of 2-epimeric aldononitriles. The aldononitriles are stabilized at pH <5.0 and hydrogenated over $Pd/BaSO_4$ to produce a pair of 2-epimeric aldditylimines that hydrolyze spontaneously to the corresponding aldoses. The epimers can be separated by chromatography on Dowex 50 in the Ba^{2+} or Ca^{2+} form (Scheme 1). In addition to the simple aldoses, deoxyaldoses and other water-soluble aldose derivatives may be used as acceptors.

Aldoses enriched at C-1 are prepared when cyanide enriched with ^{13}C or ^{14}C is used. Yields exceed 85%, making serial application of the synthesis feasible for the preparation of aldoses with multiple enrichment or with isotopes at positions other than C-1. For the preparation of glucose enriched at C-4 or C-5, however, it is better to use enzymic methods.[3]

The formation of aldononitriles from stoichiometric amounts of reactants is essentially quantitative at pH 8.0 when the starting aldose has four or fewer carbons. The presence of pyranose forms of the reactant aldose

[1] A. S. Serianni, H. A. Nunez, and R. Barker, *Carbohydr. Res.* **72**, 71 (1979).
[2] A. S. Serianni, E. L. Clark, and R. Barker, *Carbohydr. Res.* **72**, 79 (1979).
[3] A. S. Serianni, E. Cadman, J. Pierce, M. L. Hayes, and R. Barker, this volume [14].

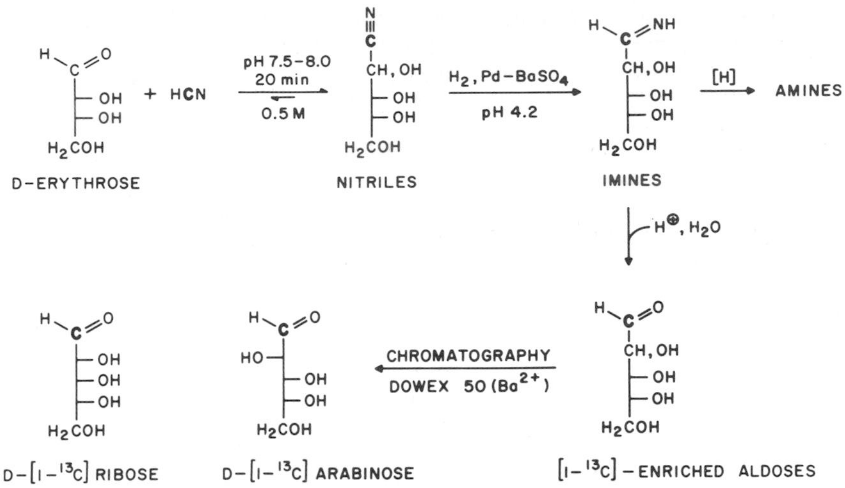

$$
\begin{array}{l}
\text{H}-\text{C}=\text{O} \\
\quad\text{—OH} \\
\quad\text{—OH} \\
\text{H}_2\text{COH}
\end{array}
\; + \; \text{HCN}
\xrightleftharpoons[\text{0.5 M}]{\substack{\text{pH 7.5–8.0}\\ \text{20 min}}}
\begin{array}{l}
\text{N} \\
\text{|||} \\
\text{C} \\
\text{|} \\
\text{CH, OH} \\
\quad\text{—OH} \\
\quad\text{—OH} \\
\text{H}_2\text{COH}
\end{array}
\xrightarrow[\text{pH 4.2}]{\text{H}_2, \text{Pd–BaSO}_4}
\begin{array}{l}
\text{H}-\text{C}=\text{NH} \\
\text{|} \\
\text{CH, OH} \\
\quad\text{—OH} \\
\quad\text{—OH} \\
\text{H}_2\text{COH}
\end{array}
\xrightarrow{[\text{H}]} \text{AMINES}
$$

D-ERYTHROSE   NITRILES   IMINES

$\xrightarrow{\text{H}^{\oplus}, \text{H}_2\text{O}}$

$$
\begin{array}{l}
\text{H}-\text{C}=\text{O} \\
\quad\text{—OH} \\
\quad\text{—OH} \\
\quad\text{—OH} \\
\text{H}_2\text{COH}
\end{array}
\qquad
\begin{array}{l}
\text{H}-\text{C}=\text{O} \\
\text{HO—} \\
\quad\text{—OH} \\
\quad\text{—OH} \\
\text{H}_2\text{COH}
\end{array}
\xleftarrow[\text{DOWEX 50 (Ba}^{2+})]{\text{CHROMATOGRAPHY}}
\begin{array}{l}
\text{H}-\text{C}=\text{O} \\
\text{|} \\
\text{CH, OH} \\
\quad\text{—OH} \\
\quad\text{—OH} \\
\text{H}_2\text{COH}
\end{array}
$$

D-[1-^{13}C] RIBOSE   D-[1-^{13}C] ARABINOSE   [1-^{13}C]–ENRICHED ALDOSES

SCHEME 1

reduces the equilibrium concentration of aldononitriles to ~80%. In these cases, a threefold excess of cyanide is used to increase the yield to >95%. Excess cyanide is recovered quantitatively prior to hydrogenation by aeration of the acidified reaction mixture and entrapment of hydrogen cyanide in methanolic potassium hydroxide. Removal of excess cyanide is essential to ensure efficient hydrogenation of the product aldononitriles.

General Materials and Methods

Glycolaldehyde and the pentoses can be obtained from commercial sources in high purity. D-Glyceraldehyde,[4] D-erythrose,[5] and D-threose[6] can be prepared from D-fructose, 4,6-O-ethylidene-D-glucose, and 4,6-O-ethylidene-D-galactose, respectively.

Potassium [^{14}C]cyanide is generally impure as provided by commercial sources and may contain as little as 30% of its radioactivity in the form of cyanide. Potassium [^{13}C]cyanide, produced by the Los Alamos Scientific Laboratory, is available from commercial suppliers with 90+ atom percent ^{13}C.

Palladium on barium sulfate (5%) can be purchased from Sigma. Catalytic reductions are carried out at atmospheric pressure using vigorous stirring (magnetic stirrer), and hydrogen uptake is measured by water

[4] A. S. Perlin, *Methods Carbohydr. Chem.* **1**, 61 (1962).
[5] A. S. Perlin, *Methods Carbohydr. Chem.* **1**, 64 (1962).
[6] D. H. Ball, *J. Org. Chem.* **31**, 220 (1966).

CONDITIONS FOR ALDONONITRILE FORMATION, HYDROGENATION, AND PRODUCT PURIFICATION

Aldononitrile chain length	Temperature[a] (°C)	Concentration of starting aldose	[CN]	pH reduction	Purification scheme[b]
C_2, C_3, C_4	20°	Equimolar, 0.1–0.2 M		1.7 ± 0.1	I
C_5, C_6	20°	0.5	1.5	4.2 ± 0.1	II

[a] For cyanide condensation reaction.

[b] Scheme I: Dowex 50 (H$^+$), BaCO$_3$, Dowex 1 (OAc$^-$), Dowex 50 (H$^+$), chromatography on Dowex 50 (Ba^{2+}) for C$_4$ aldoses; scheme II: Dowex 50 (H$^+$), Dowex 1 (OAc$^-$), chromatography on Dowex 50 (Ba^{2+}).

displacement in a calibrated cylinder.[7] Reductions at 20–30 psi H$_2$ are performed using a Parr pressure reaction apparatus.[8]

Gas chromatographic analyses utilize a 1.8 m × 2 mm (i.d.) column of OV-17 (3%) on high-performance Chromosorb W-AW (Applied Science), with a temperature program of 100–230° at 4–6°/min. Aqueous samples (6 μl) are added to 150 μl dry pyridine, followed by 150 μl of N,O-bis(trimethylsilyl)trifluoroacetamide (BSTFA) containing 1% trimethylchlorosilane (TMCS); the mixture is incubated at 60° for 30 min prior to analysis.

Radioactivity in aqueous samples (0.2 ml) is measured using 2.3 ml of a cocktail prepared from Triton X-100 (1 liter), PPO (8 g), POPOP (0.2 g), and toluene (2 liters).

Preparation of Aldononitriles

The concentrations of reactants used for aldononitrile formation are listed in the table. In a typical preparation of [1-^{13}C]hexoses, a two-neck flask is immersed in a water bath at 20° and an aqueous solution of potassium cyanide (2.0 g, 30 mmol, 13 ml of H$_2$O) is added. A pH electrode, mounted in a rubber stopper, is inserted securely, and the sidearm is equipped with a rubber stopper through which a small slot (large enough for the passage of small needles or polyethylene tubing) has been cut. Acetic acid (17.4 M) is slowly added with efficient stirring until the pH of the cyanide solution is lowered from ~11.9 to 7.8. The pentose solution (1.5 g, 10 mmol, 4 ml of H$_2$O) is added, and the pH of the reaction mixture

[7] A. I. Vogel, "A Textbook of Practical Organic Chemistry," p. 471. Wiley, New York, 1956.

[8] A. I. Vogel, "A Textbook of Practical Organic Chemistry," p. 866. Wiley, New York, 1956.

is maintained at 7.5–8.0 for 15–20 min with the addition of 1 M acetic acid or NaOH. Excess HCN provides some buffering capacity, and pH changes are minimal. If desired, samples can be withdrawn for gas-liquid chromatography (GLC) analysis.

At pH 7.5–8.0 at 20°, condensation is complete in ≤20 min and the aldononitriles are relatively stable. The C_5 and C_6 aldononitriles are more labile than their C_2, C_3, and C_4 homologs.[9] After 20 min of reaction, C_6 aldononitriles will produce <5% hydrolysis products, which can be identified by GLC, since they are retained longer than the aldononitriles.[9]

When the reaction is complete, acetic acid (17.4 M) is added to give pH 4.2 ± 0.1 unless GLC analysis is to be performed. In this case, after the reaction has proceded for 15–20 min at pH 7.5–8.0, acetic acid is added to pH ~5.5, and samples are taken for analysis. At this pH, hydrolysis is inhibited. If GLC analysis shows incomplete condensation, the pH can be raised with NaOH. If complete, more acetic acid is added to give pH 4.2 ± 0.1.

Removal or Recovery of HCN

Loss of HCN from the reaction vessel is negligible even when excess reagent is used and excess HCN should be removed or recovered from the reaction mixture prior to hydrogenation. Unenriched HCN can be removed in a well-vented hood by N_2 aeration of the reaction solution (pH 4.2) with a sintered-glass gas diffuser. Isotopically enriched HCN can be recovered. Nitrogen from the reaction vessel is passed through two traps fitted with efficient diffusors, each containing 250 ml of methanolic potassium hydroxide to a depth of 30 cm. In the preparation of 10 mmol of hexoses, 20 mmol of HCN will remain unreacted. To recover this material, methanolic solutions of KOH (20 mmol) are added to each trap, and N_2 is passed through at ~50 ml/min. More than 19 mmol of HCN is recovered in the first trap after 5–6 hr of aeration. The contents of this trap can be concentrated to dryness under vacuum at 25° to give a residue of KCN slightly contaminated with KOH. Recovered cyanide can be titrated with $AgNO_3$,[10,11] and it can then be used for further synthesis of aldononitriles.

The aldononitriles are stable at pH 4.2 for the 5–6 hr required to remove HCN. The reaction mixture should be readjusted if necessary to pH 4.2 ± 0.1 after aeration and prior to hydrogenation.

[9] A. S. Serianni, H. A. Nunez, and R. Barker, *J. Org. Chem.* **45**, 3329 (1980).
[10] J. Liebig, *Justus Liebigs Ann. Chem.* **77**, 102 (1851).
[11] G. Déningés, *C. R. Hebd. Seances Acad. Sci.* **117**, 1078 (1893).

Hydrogenation of Aldononitriles

C_5 and C_6 aldonononitriles can be hydrogenated at pH 4.2 ± 0.1 and pressures up to 30 psi H_2. C_2, C_3, and C_4 aldononitriles must be hydrogenated at pH 1.7 and atmospheric pressure to inhibit amine formation. In this case, 3 M H_2SO_4 is used to adjust from pH 4.2 to 1.7. A suitable reduction apparatus is described by Vogel.[7,8]

In a typical hydrogenation reaction, a brown suspension of Pd/BaSO$_4$ (5%, 60 mg per millimole of nitrile) in 5–10 ml of H_2O is reduced until a gray suspension is produced (~10 min). The aldononitrile solution, adjusted to the proper pH, is added to the catalyst, and the reaction vessel is flushed thoroughly with H_2. Hydrogenation is initiated by vigorous stirring or mechanical shaking. The rate and extent of reduction are followed manometrically with a pressure gauge (Parr apparatus)[8] or through the use of a calibrated cylinder with correction for the partial pressure of water in the vessel.[7] Reaction times vary: glycolonitrile, glyceronitrile, erythrono- and threononitriles 8–10 hr; pentono- and hexononitriles, 2–4 hr.

For large-scale (>5 mmol) preparations of C_2, C_3, and C_4 aldoses, the pH of the suspension (initially pH 1.7) should be measured and adjusted with H_2SO_4 periodically (once every 20 min) to suppress primary amine formation.

When hydrogenation is complete, the catalyst is removed by filtration through Celite or glass fiber filters.

Purification of Product Aldoses

After removal of the catalyst, the filtrate is treated batchwise with excess Dowex 50-X8 (H$^+$) to remove cations, including the by-product 1-amino-1-deoxyalditols, and the resin is removed by filtration. For the preparation of pentoses and hexoses, the filtrate is treated batchwise with Dowex 1-X8 (OAc$^-$), filtered, and concentrated to a gum under vacuum at 30° to remove acetic acid. The gum is dissolved in a minimum amount of water for chromatography.

Dowex 50 (H$^+$) filtrates containing C_2, C_3, and C_4 aldoses contain H_2SO_4 that is removed by the *slow* addition of BaCO$_3$ with efficient stirring until the solution is pH ~4.3. The suspension is centrifuged (Sorvall, 7000 g) to remove most of the sticky precipitate, and the supernatant is filtered through Celite. The filtrate is treated batchwise with Dowex 1-X8 (OAc$^-$) and then with Dowex 50-X8 (H$^+$). The final filtrate is concentrated to a small volume (2–5 ml) for use or storage (C_2 or C_3 aldoses) or for chromatography (C_4 aldoses). Careful washing of precipitates and resins is essential to ensure good yields.

Mixtures of 2-epimeric aldoses are separated readily by chromatography on Dowex 50-X8 (200–400 mesh, Ba^{2+})[12] at room temperature using deionized water as the eluent. A 3.4 × 100 cm column can be used to separate millimole quantities of 2-epimeric aldoses. The following 2-epimers in each pair elute first: threose, arabinose, xylose, glucose, galactose. Chromatography on Dowex 50-X8 (200–400 mesh, Ca^{2+})[13] improves the separation of lyxose from xylose.

Elution profiles are determined by spotting samples from fractions on Whatman No. 1 paper and testing for reducing sugars with silver nitrate,[14] by radioactivity assay[1,2] when ^{14}CN is used, by refractive index monitoring or by reducing sugar assay.[15] Fractions containing the purified aldoses are pooled and concentrated under vacuum (water aspirator) at 30°. The resulting solution (~30–40 ml) is treated with Dowex 50 (H$^+$) batchwise, the resin is removed by filtration, and the filtrate is concentrated under vacuum at 30° to a thick gum (pentoses and hexoses only). For most applications, syrups or concentrated aqueous solutions of the aldoses are satisfactory, and yields can be estimated by weighing thoroughly desiccated products or by quantitative analysis for reducing sugars.[15] If crystals are desired, they can be obtained by seeding the gums with authentic samples or by following published procedures for specific sugars.

General Comments

Reaction mixtures of cyanide and starting aldose must be maintained at pH 7.8 ± 0.2. At higher pH values, the aldononitriles decompose to aldonates and a variety of undesirable products.[9] At lower pH values, condensation is slow and incomplete.

The conditions for hydrogenation-solvolysis depend on the chain length of the aldononitrile. For C$_2$, C$_3$, and C$_4$ aldononitriles, the best conditions are pH 1.7 over Pd/BaSO$_4$ at atmospheric pressure. At higher pH values or higher pressures, the production of primary amines is favored. C$_5$ and C$_6$ aldononitriles can be hydrogenated at pH 4.2 and pressures up to 30 psi. 1-Amino-1-deoxyalditols (~10%) are produced, but they are removed in the first step of product purification by treatment with Dowex 50-X8 (H$^+$).

Gas-chromatographic analysis of pertrimethylsilyl derivatives of al-

[12] J. K. N. Jones and R. A. Wall, Can. J. Chem. 38, 2290 (1960).
[13] S. J. Angyal, G. S. Bethell, and R. Beveridge, Carbohydr. Res. 73, 9 (1979).
[14] R. M. C. Dawson, D. Elliott, W. Elliott, and K. M. Jones, eds., "Data For Biochemical Research," 2nd ed., p. 541. Oxford Univ. Press, London and New York, 1969.
[15] J. T. Park and M. J. Johnson, J. Biol. Chem. 181, 149 (1949).

dononitriles and aldoses is a convenient method to determine the progress of reactions and the purity of products.[9] The use of $K^{14}CN$ in the condensation reaction facilitates the assay of column effluents. We have found that $K^{14}CN$ provided by commercial suppliers frequently is impure and that 50–60% of the radioactivity is not incorporated in the final product. For this reason, yields cannot be calculated on the basis of $K^{14}CN$ used, nor can ^{14}C-enriched monosaccharides be prepared in good yields based on apparent ^{14}CN input.

The course of reactions using ^{13}C-enriched cyanide is easily followed by ^{13}C NMR. Spectra show only the resonances of enriched products and of reagents present in substantial amounts. A series of spectra obtained from intermediates and products formed in the synthesis of D-[1-^{13}C]ribose and arabinose are shown in Fig. 1.

The proportion of 2-epimeric aldoses obtained by this method is determined by the proportion of 2-epimeric aldononitriles formed from the condensation reaction. The proportion of epimers is particularly unfavorable in the condensation of cyanide with D-arabinose, where 70% of the product hexoses have the D-*manno* configuration. Conversion of D-mannose to the biologically more useful D-glucose, however, can be accomplished enzymically[3] or chemically using molybdate (Mo^{VI}) as a catalyst.[16] The latter process can be applied to tetroses and higher carbon aldoses. It is accompanied by inversion of the C1–C2 fragment. For example, D-[1-^{13}C, ^{2}H]mannose is converted to D-[2-^{13}C, ^{2}H]glucose. The equilibrium proportions of 2-epimeric aldoses are determined by their relative thermodynamic stabilities. Two catalysts are effective. The most convenient is 85% molybdic acid in aqueous solution. However, under forcing conditions or with long reaction times secondary products (3-epimers) are formed. The formation of secondary products can be avoided using dioxobis (2,4-pentanedionato-O,O') molybdenum as a catalyst in dimethylformamide solutions.[16]

Comments on Specific Aldoses

Glycolaldehyde.[2] This aldose is produced by the addition of cyanide to formaldehyde (1 : 1). Purification, after hydrogenation, does not require chromatography. Concentration to a syrup results in oligomerization, but in dilute solution oligomers hydrolyze to the *gem*-diol form.

---

[16] M. L. Hayes, N. Pennings, A. S. Serianni, and R. Barker, submitted for publication (1982).

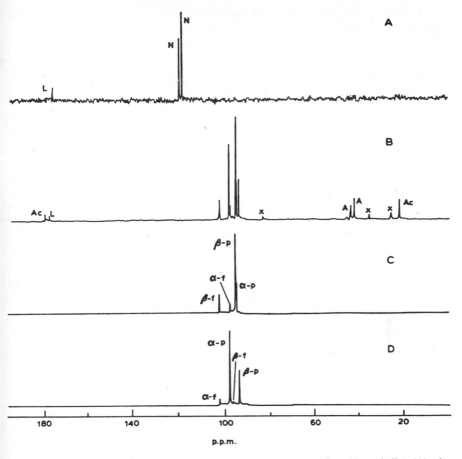

FIG. 1. Proton-decoupled ^{13}C NMR spectra of the intermediate aldononitriles (A), the reduction products (B), and the purified product aldoses (C and D) in the synthesis of D-[1^{13}C]ribose and arabinose from D-erythrose and 89.5%-enriched sodium [^{13}C]cyanide. All spectra show only the resonances of the enriched nuclei, unless otherwise indicated. (A) Cyanide condensation reaction mixture showing D-[1-^{13}C]ribono- and arabinononitriles (N) and D-[1-^{13}C]arabinono-γ-lactone (L). (B) Products from the hydrogenation of the mixture shown in (A): Ac = natural abundance resonances of acetic acid; A = D-[1-^{13}C]aminoalditols; L = D-[1-^{13}C]arabinono-γ-lactone; X = unidentified contaminants; the resonances between 95 and 103 ppm are due to product [1-^{13}C]aldoses. (C and D) Product aldoses after chromatography on Dowex 50-X8 (200–400 mesh) in the barium form: (C) = D-[1-^{13}C]ribose; (D) = D-[1-^{13}C]arabinose. Examination of the enriched aldoses by gas–liquid chromatography shows these products to be free of D-erythrose. The overall yield of the purified 1-^{13}C-enriched pentoses, using a 1 : 1 ratio of reactants, was 78%. From Serianni et al.[1]

DL-*Glyceraldehyde*.[2] This is produced in 80% yield as a racemic mixture from glycolaldehyde using a stoichiometric amount of cyanide. Oligomers form in concentrated solutions. Hydrolysis in dilute solution is slow, but eventually the *gem*-diol monomer is formed.

D-*Erythrose and* D-*Threose*.[2] These aldoses are formed in about equimolar amounts from D-glyceraldehyde and a stoichiometric amount of cyanide in 80% overall yield. Concentrated solutions contain oligomeric forms. Dilute solutions of D-erythrose at 30° contain β-furanose (63%), α-furanose (25%), *gem*-diol (12%), and aldehyde ($\sim$1%). Dilute solutions of D-threose contain α-furanose (51%), β-furanose (37%), *gem*-diol (12%), and aldehyde ($\sim$1%).

D-*Ribose and* D-*Arabinose*.[1] These aldoses are formed from D-erythrose and a stoichiometric amount of cyanide in 80% overall yield in a ratio of 1.4 : 1 *ribo* : *arabino*. Concentrated and dilute solutions contain only the expected monomeric pyranose and furanose forms.

D-*Glucose and* D-*Mannose*.[1] The nitriles are formed by the reaction of D-arabinose (0.5 M) with cyanide (1.5 M), pH 8.0, at 20° for 20 min. Excess cyanide is removed quantitatively by purging the acidic reaction mixture (pH 4.2) with N_2 for 5 hr. Hydrogenation gives the aldoses in 80% yield in a ratio 2.3 : 1 *manno* : *gluco*.

D-*Galactose and* D-*Talose*.[1] These are formed from D-lyxose using a threefold excess of cyanide. Overall yield is 85% with *galacto* : *talo* $\simeq$ 1 : 1.

Molybdate-Catalyzed Interconversion of Aldoses: Mannose and Glucose

Two methods are described:

A. An aqueous solution of D-[1-^{13}C]mannose (0.1 M), and 85% molybdic acid (2 mM) at pH 4.5, is incubated at 90 $\pm$ 2° for 5 hr at which time equilibrium is achieved with D-[1-^{13}C]mannose : D-[2-^{13}C]glucose = 1 : 2.5.[16] The reaction can be monitored by ^{13}C or ^{1}H NMR, or by GLC as described above. After treatment with Dowex 50 × 8 (H$^+$) and Dowex 1 × 8 (OAc$^-$) or (HCO$_3^-$) to remove NH$_4^+$ and MoO$_4^{2-}$ ions, the aldoses are purified by chromatography on Dowex 50 × 8 (Ba^{2+}) as described above.

The rate of epimerization is increased by higher temperatures, greater molybdate concentrations, and lower pH values. Extended reaction times or forcing conditions lead to the formation of secondary products. In the D-[1-^{13}C]mannose epimerization these are D-[1-^{13}C]altrose and D-[2-^{13}C]allose. Pentoses treated with 0.1 M molybdate at 80° for 12 hr yield an equilibrium mixture of all four epimers. Starting with D-[1-^{13}C]xylose, the equilibrium mixture consists of D-[1-^{13}C]xylose, 35%, D-[2-^{13}C]lyxose, 24%, D-[1-^{13}C]arabinose, 29%, and D-[2-^{13}C]ribose, 12%.

B. A solution of D-[1-^{13}C]mannose (0.1 M) and dioxobis(2,4-pentanedionato-O,O') molybdenum (5 mM) in dimethylformamide is incubated at 50° for 24 hr.[17] After cooling, 2 volumes of H_2O are added and the catalyst is extracted with CH_2Cl_2. The aqueous layer is deionized and the products separated as described above.

[17] Y. Abe, T. Takizawa, and T. Kunieda, *Chem. Pharm. Bull.* **28**, 1324 (1980).

[12] Chemical Synthesis of Aldose Phosphates Enriched with Carbon Isotopes

By A. S. SERIANNI, J. PIERCE, and R. BARKER

The synthesis of 1-^{13}C-enriched aldose phosphates can be accomplished by the condensation of ^{13}C-enriched cyanide with an appropriate aldose phosphate acceptor[1] as shown in Scheme 1. Condensation is rapid at pH 7.5–8.0, and aldononitrile phosphates are stable toward hydrolysis at this and lower pH values. 2-Epimeric aldononitrile phosphates are readily separated by ion-exchange chromatography on Dowex 1 (formate)[2] and hydrogenate smoothly at pH 1.7 ± 0.1 over Pd/BaSO$_4$. Overall yields exceed 80%. The method is preferred for the preparation of ^{13}C-enriched derivatives but can be applied successfully to the preparation of ^{14}C-enriched compounds provided that [^{14}C]cyanide of adequate purity is available. Commercial samples are often highly impure.

The availability of appropriate acceptors (glycolaldehyde phosphate, D-glyceraldehyde 3-phosphate, and D-erythrose 4-phosphate) is key to the application of this synthesis. The first two of these are readily prepared from commercially available and relatively inexpensive starting materials by oxidation with lead tetraacetate.[1,3] D-Erythrose 4-phosphate is most readily prepared by application of the synthesis described here using natural abundance cyanide and D-glyceraldehyde 3-phosphate as reactants.

[1] A. S. Serianni, J. Pierce, and R. Barker, *Biochemistry* **18**, 1192 (1979).
[2] The 2-epimeric aldose phosphates, on the other hand, do not separate well by ion-exchange chromatography, necessitating separation of the nitriles prior to hydrogenation.
[3] Sodium metaperiodate oxidation was also examined, but traces of iodate interfere with hydrogenation of the aldononitrile phosphates, and careful chromatographic purification was required.

METHODS IN ENZYMOLOGY, VOL. 89

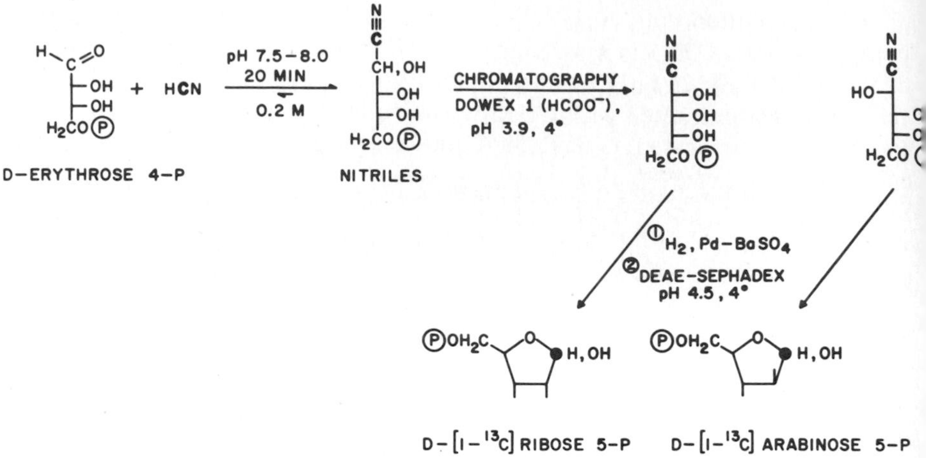

D-ERYTHROSE 4-P + HCN $\xrightarrow[\text{0.2 M}]{\substack{\text{pH 7.5-8.0} \\ \text{20 MIN}}}$ NITRILES $\xrightarrow[\text{pH 3.9, 4°}]{\substack{\text{CHROMATOGRAPHY} \\ \text{DOWEX 1 (HCOO⁻),}}}$

① H_2, Pd-BaSO$_4$
② DEAE-SEPHADEX pH 4.5, 4°

D-[1-^{13}C] RIBOSE 5-P   D-[1-^{13}C] ARABINOSE 5-P

SCHEME 1

Analytical Methods

Inorganic and organic phosphate are assayed by the method of Leloir and Cardini.[4] Radioactivity of aqueous samples (0.2 ml) is measured using 2.3 ml of a cocktail of Triton X-100 (1 liter), PPO (8.0 g), POPOP (0.2 g), and toluene (2 liters).

Aqueous solutions should be reduced in volume rapidly at approximately 30°. A good quality rotary evaporator with a water aspirator should remove 100 ml of water in 15 min.

Hydrogenation at atmospheric pressure is performed using the apparatus described by Vogel.[5]

Preparation of Aldose Phosphate Acceptors

Glycolaldehyde Phosphate. Disodium DL-glycerol 1-P hexahydrate (Sigma; 8.6 g, 27 mmol) is moistened with 5 ml of H_2O and dissolved in 400 ml of glacial acetic acid with efficient stirring. After dissolution of the salt and addition of 1.7 ml of 18 M sulfuric acid, lead tetraacetate (24 g, 54 mmol) is added during 15 min. After 2 hr, oxalic acid (4.5 g, 50 mmol) is added and stirring is continued for an additional 30 min. The suspension is filtered through Celite and the filtrate is concentrated at 30° under vacuum to approximately 30 ml. The filter cake is washed with 200 ml of H_2O, the concentrate and washings are combined, and barium acetate (13 g, 50

---

[4] L. F. Leloir and C. E. Cardini, this series, Vol. 3, p. 840.
[5] A. I. Vogel, "A Textbook of Practical Organic Chemistry," pp. 471–472. Wiley and New York, 1958.

mmol) is added with efficient stirring at 4° for 15 min. The white suspension is filtered through Celite, the precipitate is washed with H_2O, and the filtrate and washings are treated with excess Dowex 50-X8 (H^+). The suspension is filtered, the solution is concentrated as before to about 200 ml, and the concentrate is extracted overnight at 4° with diethyl ether in a continuous liquid–liquid extraction apparatus. The aqueous solution is recovered, concentrated as before to about 30 ml, and stored at $-20°$. The yield is 25 mmol (93%) by total phosphate with a trace of inorganic phosphate, and the purity is at least 95% by ^{13}C NMR.[1]

D-*Glyceraldehyde 3-P*. Disodium D-fructose-6-P dihydrate (Sigma; 3.4 g, 10 mmol) is treated with lead tetraacetate (18 g, 40.5 mmol) in the manner described for the preparation of glycolaldehyde-P, except that 1.1 ml of 18 M sulfuric acid (20 mmol) is added prior to the addition of the oxidant. After Dowex 50-X8 (H^+) treatment, the acidic solution of 2-*O*-glycoloyl-D-glyceraldehyde 3-P is concentrated to 10 ml and stored at 25° for 18 hr to yield D-glyceraldehyde 3-P and glycolic acid. Alternatively, hydrolysis can be carried out by incubating the acidic solution at 40° for 6 hr. The resulting solution is adjusted to pH 5.5 with 2 M NaOH and applied to a 1.2 × 50 cm column of DEAE-Sephadex A-25 (40–120 mesh, acetate) at 4°, washed with a small amount of H_2O, and eluted with a linear gradient of sodium acetate (1.5 l, 0.05–0.60 M, pH 5.5 ± 0.1). Glyceraldehyde 3-P elutes at 0.15 M sodium acetate and is preceded by glycolic acid. Fractions are assayed for D-glyceraldehyde 3-P by organic P analysis and for glycolic acid by the method of Lewis and Weinhouse.[6] Fractions containing D-glyceraldehyde 3-P are pooled, treated with excess Dowex 50-X8 (H^+), and concentrated twice under vacuum at 30° to approximately 5 ml to remove acetic acid. The yield is 8.1 mmol (81%) by organic phosphate analysis with a trace of inorganic phosphate, and the purity is at least 95% by ^{13}C NMR[1].

Preparation of Aldononitrile Phosphates

Cyanide condensations are carried out using apparatus described in this volume.[7] The K[13]CN solution (2.0 mmol, 14 ml H_2O) containing 10^7 cpm of K[14]CN[8] is added to the flask, which is fitted with pH electrode, and the solution is cooled to 5° with an ice bath prior to adjustment to pH 8.0 ± 0.1 with 4 M acetic acid. The solution of acceptor aldose phosphate (2.0 mmol, 4 ml of H_2O) at pH 7.5 is added with efficient stirring while maintaining the pH at 7.5–8.0 with the addition of 2 M acetic acid or 1 M

---

[6] K. F. Lewis and S. Weinhouse, this series, Vol. 3, p. 269.

[7] A. S. Serianni, H. A. Nunez, and R. Barker, this volume [11].

[8] K[14]CN is added to reaction mixtures mainly to assist in the detection of products during chromatography. For quantitation the phosphate assay is used, not ^{14}C incorporation, since commercial K[14]CN is frequently impure.

SEPARATION AND EPIMERIC DISTRIBUTION OF ALDONONITRILE
PHOSPHATES

Parent aldose phosphates	Chromatography on Dowex 1-X8 (formate)[a]		Ratio of epimers[c] and major forms
	Peak 1[b]	Peak 2[b]	
Glyceraldehyde 3-P	threo	erythro	1.3 : 1 erythro
Erythrose 4-P	arabino	ribo	1.4 : 1 ribo
Threose 4-P	xylo	lyxo	1.5 : 1 lyxo

[a] A 2.2 × 51 cm Dowex 1-X8 (200–400 mesh) column in the formate form was used. Solutions of aldononitrile phosphates were adjusted to pH 6.5–7.0 prior to application to the column. Conditions were as follows: gradients: for 4-carbon aldononitrile phosphates, 3 liters, 0.2–0.9 M sodium formate, pH 3.9; for 5-carbon aldononitrile phosphates, 3 liters, 0.05–0.8 M sodium formate, pH 3.9; temperature = 4°; flow rate = 0.5 ml/min; 7 ml per fraction.

[b] Aldononitrile-P configurations were determined by reduction to aldose phosphates, incubation with alkaline phosphatase, and characterization of the released aldose. The ^{13}C NMR spectra of the resulting aldoses were compared with those of standard pentoses.

[c] Ratios were determined by computerized integration of ^{13}C NMR spectra of epimeric mixtures and by organic phosphate analysis of purified preparations.

NaOH. Stoichiometric amounts of aldose phosphate and cyanide are used, and the final concentration of reactants is 0.05–0.1 M. After 15 min at 5° and pH 7.5–8.0, the ice bath is removed and the reaction mixture is allowed to warm to 25° during 30 min. The solution is adjusted to pH 4.0 ± 0.2 and 4 M acetic acid. Condensation is complete (>95%) when assayed by ^{13}C NMR using a short pulse width (10 μsec, 55°) and long delay time (10 sec) to facilitate aldononitrile detection.[1]

Preparation of DL-Glyceraldehyde 3-P

D-Glyceronitrile 3-P is adjusted to pH 1.7 ± 0.1 with Dowex 50-X8 (H$^+$) after cyanide condensation and hydrogenated directly to DL-glyceraldehyde 3-P as described below for C$_4$ and C$_5$ homologs.

Separation of Epimeric Tetrono- and Pentononitrile Phosphates

Epimeric mixtures of aldononitrile phosphates (4–6 mmol) are separated by ion-exchange chromatography on a 2.2 × 50 cm column of

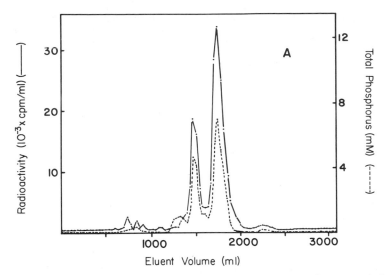

FIG. 1. Separation of DL-[1-^{13}C]xylononitrile 5-P and DL-[1-^{13}C]lyxononitrile 5-P. Chromatography of the 2-epimeric pentononitrile phosphates on a 2.2 × 50 cm Dowex 1-X8 (200–400 mesh) column in the formate form at 4° developed with a linear gradient of sodium formate (3 liters, 0.05 to 0.8 M, pH 3.9). The column effluent was assayed for radioactivity and total phosphate. The *xylo* epimer elutes before the *lyxo* epimer. From Serianni *et al.*[1]

Dowex 1-X8 (200–400 mesh, formate) at 4° using linear gradients of sodium formate/formic acid (see the table). The nitrile phosphate with *cis*-2,3-hydroxyl groups is the major product and is eluted last under these conditions (Fig. 1). Yields of the tetrono- and pentononitrile phosphates after cyanide condensation and chromatography are 85%. Fractions containing the aldononitrile phosphates are pooled and adjusted to pH 1.5 with Dowex 50-X8 (H$^+$). After filtration, the acidic solutions are concentrated to 100 ml under vacuum at 30° and extracted continuously with diethyl ether overnight at 4° to remove formic acid. The aqueous acidic solutions are recovered, concentrated under vacuum at 30° to approximately 10 ml, treated with 1 ml of 17.4 M acetic acid per millimole of nitrile-P, and adjusted to pH 1.7 ± 1 with Dowex 50-X8 (H$^+$) or 2 M NaOH prior to hydrogenation.

Hydrogenation of Aldononitrile Phosphates

Palladium-barium sulfate (5%, 60 mg per millimole of nitrile) is weighed into the reaction flask.[5] Water (5–10 ml) is added, and the suspension is reduced for 15–20 min at atmospheric pressure and 25° with efficient stirring. During this period, the catalyst changes from brown to light gray. The solution of aldononitrile phosphate, adjusted to pH 1.7 as described below, is added to the reaction vessel, which is filled and evacuated three times prior to a final charging with H_2. The concentration of aldononitrile phosphates can range from 40 to 100 mM in the hydrogenation reaction.

Typically, hydrogenation is complete in 6–8 hr at 25°. In a few instances, incomplete reduction was noted (by NMR analysis). In these cases, the spent catalyst is removed by filtration through glass fiber filters and a second reduction performed to complete the conversion to the aldose phosphate.

Hydrogenation products can be assayed by ^{13}C NMR to determine the extent of reduction to 1-amino-1-deoxyalditol phosphate and the amount of unreacted aldononitrile phosphate.[1] Yields of the three-, four-, and five-carbon aldose phosphates based on the analysis of ^{13}C NMR spectral peak areas in reactions with [1-^{13}C]aldononitrile phosphates are 85–95%.

After hydrogenation, the catalyst is removed on a glass fiber filter and the filtrate is treated batchwise with excess Dowex 50-X8 (H$^+$) for 10 min and then concentrated to 10 ml. Typically, the reaction mixture contains product aldose phosphate, 1-amino-1-deoxyalditol phosphate, and a small amount of aldononitrile phosphate. This solution is adjusted to pH 4.5 ± 0.1 with dilute NaOH and applied to a 1.2 × 50 cm DEAE-Sephadex A-25 (acetate) column at 4° which has been equilibrated with 0.05 M sodium acetate at pH 4.5 ± 0.1. The column is developed with a linear acetate-acetic acid gradient (1.5 liters, 0.05–0.8 M sodium acetate, pH 4.5 ± 0.1). Fractions (6 ml) are collected at a flow rate of 0.5 ml/min. The 1-amino-1-deoxyalditol phosphate elutes at the void volume, followed in order by aldose phosphate and aldononitrile phosphate. Fractions containing aldose phosphate are pooled, treated with excess Dowex 50-X8 (H$^+$), and concentrated under vacuum at 30° to approximately 10 ml. Recovery from DEAE-Sephadex chromatography based on phosphate assay is greater than 90%. Aldose phosphate solutions can be stored at pH 1.0–2.0 and −15° for many months.

General Comments on Ion-Exchange Chromatography and the Stability of Products

Epimeric tetrose and pentose phosphates are difficult to separate by ion-exchange chromatography. For this reason, the epimeric aldononitrile phosphates are separated on Dowex 1-X8 (formate) at 4° using linear formate/formic acid gradients at pH 3.9 ± 0.1. Epimeric mixtures of aldononitrile phosphates can also be separated on Dowex 1-X8 (chloride) using linear chloride gradients, but hydrogenation in the presence of chloride ion yields larger amounts of 1-amino-1-deoxyalditol phosphate ($\approx 45\%$).

It is important to maintain acidic conditions during the separation and handling of aldononitrile phosphates, since the reaction between the parent aldose and cyanide is reversible, and, at pH >8, purified aldononitrile phosphates revert to epimeric mixtures.[9]

Aldose phosphates, particularly the triose and tetrose phosphates, should be handled at low pH to avoid base-catalyzed isomerizations and β-elimination. In the pentose phosphate series, xylose 5-P and lyxose 5-P undergo isomerization to xylulose 5-P under mildly alkaline conditions (pH 7.5, 40°).

The acyclic triose and tetrose phosphates will isomerize to give mixtures that include keto compounds when chromatographed on Dowex 1-X8 (formate) at 4°. Purification of the alkali-sensitive aldose phosphates and the pentose phosphates can be achieved by anion-exchange chromatography on DEAE-Sephadex A-25 at 4° using linear gradients of acetic acid at pH 4.5 $\pm$ 0.1. The tetrose 4-phosphates frequently yield peaks with noticeable tailing, whereas triose and pentose phosphates yield symmetric peaks. Isomerization to keto compounds on DEAE-Sephadex was not observed under the conditions used.

It is usual to prepare acetal derivatives of the aldose phosphates having four or fewer carbons to protect the base-sensitive aldehydic function. The free aldose phosphates are stable, however, during long-term storage at pH 1.0–2.0. When stored at $-15°$ as 50 mM solutions, no changes are detectable after 2-months as determined by ^{13}C NMR analysis of 1-^{13}C-enriched compounds and by inorganic phosphate analysis. Storage at higher temperatures, however, results in degradation of these compounds even in acidic solution.

[9] A. S. Serianni, H. A. Nunez, and R. Barker, *J. Org. Chem.* **45**, 3329 (1980).

[13] Chemical Synthesis of Aldoses Enriched with Isotopes of Hydrogen and Oxygen

By A. S. SERIANNI, E. L. CLARK, and R. BARKER

The reaction sequence shown in Scheme 1 can be used to prepare a pair of 2-epimeric aldoses enriched with isotopes of hydrogen at H-1 and/or oxygen at O-2. The procedures for the incorporation of these

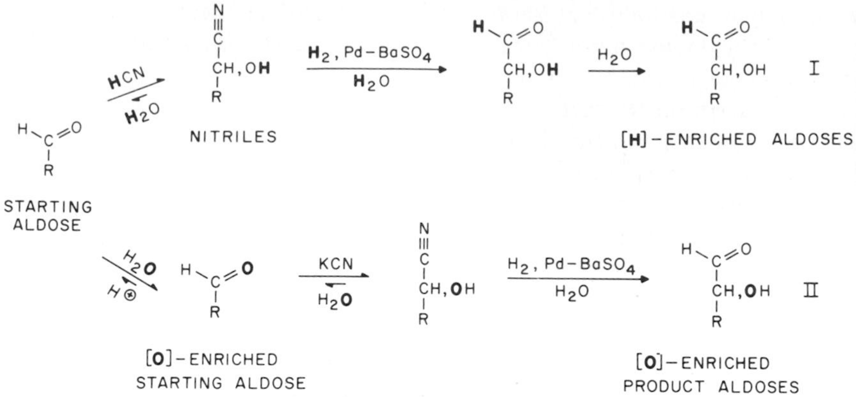

$$
H-\underset{R}{\overset{}{C}}=O \xrightarrow[H_2O]{HCN}
\begin{array}{c} N \\ \| \\ C \\ | \\ CH,OH \\ | \\ R \end{array}
\xrightarrow[H_2O]{H_2,\ Pd-BaSO_4}
H-\underset{R}{\overset{CH,OH}{C}}=O
\xrightarrow{H_2O}
H-\underset{R}{\overset{CH,OH}{C}}=O \quad I
$$

STARTING ALDOSE   NITRILES   [H]-ENRICHED ALDOSES

$$
H-\underset{R}{\overset{}{C}}=O \xrightarrow[H_2O]{KCN}
\begin{array}{c} N \\ \| \\ C \\ | \\ CH,OH \\ | \\ R \end{array}
\xrightarrow[H_2O]{H_2,\ Pd-BaSO_4}
H-\underset{R}{\overset{CH,OH}{C}}=O \quad II
$$

[O]-ENRICHED STARTING ALDOSE   [O]-ENRICHED PRODUCT ALDOSES

SCHEME 1

isotopes are adapted from those described in this volume [11] for the preparation of ^{13}C-enriched compounds. Aldose acceptors may include the simple aldoses,[1] aldose phosphates,[1] deoxyaldoses, and other water-soluble derivatives. The methods have been applied to the preparation of ^{2}H, ^{17}O, and ^{18}O derivatives, and should be applicable to the incorporation of ^{3}H with minor modification.

Hydrogen Isotopes

For the synthesis of ^{2}H-enriched compounds (Scheme 1, I), ^{2}H$_2$ gas is used to reduce aldononitriles over Pd/BaSO$_4$ in ^{2}H$_2$O at pH 1.7 ± 0.1 or 4.2 ± 0.1[2] and 25°. If the reduction is carried out in ^{1}H$_2$O, isotope exchange occurs and a significant amount of ^{1}H product is formed. In addition, water in the calibrated cylinder of the hydrogenation apparatus is replaced by mineral oil. Alternatively, a Parr pressure reaction apparatus may be employed.

Oxygen Isotopes

For the preparation of compounds enriched with oxygen isotopes, reactions are carried out on a semimicro scale (0.5–1.0 ml of O-enriched H$_2$O). The starting aldose is dried under vacuum to remove H$_2$O, dissolved in O-enriched H$_2$O and incubated under acidic conditions to pro-

---

[1] A. S. Serianni and R. Barker, *Can. J. Chem.* **57**, 3160 (1979).

[2] The appropriate pH of the reduction mixture is determined by the chain length of the aldononitrile, as described in this volume [11].

mote isotope exchange at the aldehydic oxygen[3] (Scheme 1, II). Finely ground potassium cyanide (dried under vacuum) is added at pH 7.5–8.0 to trap the oxygen isotope at O-2 of the aldononitriles. After cyanide condensation is complete (~20 min), the reaction mixture is adjusted to pH 3.5 and the oxygen-enriched H_2O is recovered by lyophilization. Reversal of cyanide addition is negligible under these conditions. The solid residue is dissolved in 1 M acetic acid, the solution is adjusted to pH 1.7 ± 0.1 or 4.2 ± 0.1[2] with H_2SO_4 or NaOH, and the mixture is reduced over $Pd/BaSO_4$. The reversibility of cyanide addition is slow at pH <4.5,[6] preventing loss of the oxygen label during reduction.

Materials

D-Glyceraldehyde[7] and D-erythrose[8] can be prepared from D-fructose and 4,6-O-ethylidene-D-glucose, respectively. Deuterium gas ($^2H_2$, 99.5 atom %), acetic acid-$^2H_4$ (99.5 atom %), and $H_2^{18}O$ (97 atom %) can be purchased from Merck, Sharpe and Dohme, Canada, Limited. Palladium–$BaSO_4$ (5%) and ^2H_2O (99.8 atom %) can be purchased from Sigma.

Reductions are carried out using apparatus described in this volume.[9]

Microelectrodes for microscale cyanide addition reactions (< 1 ml) can be purchased from Microelectrodes, Inc., Londonderry, New Hampshire. pH measurements in ^2H_2O solutions are corrected using the equation pH = pD − 0.4.[10]

General Method for the Preparation of Deuterated Carbohydrates: Synthesis of D-[1-^{13}C, ^{14}C, ^{2}H]Erythrose and Threose

A solution of $K^{13}CN$ (0.13 g, 2 mmol) in 13 ml of ^2H_2O at 20° containing $K^{14}CN$ (10^7 cpm) is added to a 25-ml flask[9] and adjusted to pH 8.0 ± 0.1 with 0.7 M acetic acid-$^2H_4$. D-Glyceraldehyde (2 mmol) is concentrated

---

[3] The rate of exchange of oxygen isotopes into O-1 depends on the structure of the starting aldose. For C_2–C_4 aldoses, hydrated aldehyde (gem-diol) forms are present in significant amounts (12–95%) in aqueous solution.[4] Exchange in the presence of acid is more facile than for the C_5 and C_6 aldoses, where gem-diolic forms are essentially absent. In the latter cases, heating may be required for adequate exchange.[5]

[4] A. S. Serianni, E. L. Clark, and R. Barker, Carbohydr. Res. 72, 79 (1979).

[5] D. Rittenberg and C. Graff, J. Am. Chem. Soc. 80, 3370 (1958).

[6] A. S. Serianni, H. A. Nunez, and R. Barker, J. Org. Chem. 45, 3329 (1980).

[7] A. S. Perlin, Methods Carbohydr. Chem. 1, 61 (1962).

[8] A. S. Perlin, Methods Carbohydr. Chem. 1, 64 (1962).

[9] A description of the reaction vessel can be found in this volume [11].

[10] R. Lumry, E. L. Smith, and R. R. Grant, J. Am. Chem. Soc. 73, 4330 (1951).

from 3 ml of ^2H_2O several times at 30° under vacuum. The residual gum is dissolved in 4–5 ml of ^2H_2O and added to the solution of $K^{13}CN$. The pH of the reaction mixture is maintained between 8.0 and 8.3 with additions of 0.7 M acetic acid-$^2H_4$ and/or 1.0 M NaO^2H. After 20–25 min, the pH is lowered to 4.0 ± 0.2 with 17 M acetic acid-$^2H_4$. A further adjustment of pH to 1.7 ± 0.2 is made with 6 M $^2H_2SO_4$.

Palladium–barium sulfate (5%, 60 mg per millimole of nitrile) is weighed into a 50 ml side-arm flask, 5 ml ^2H_2O is added, and the system is evacuated and charged three times with N_2. After the last evacuation, the system is charged with $^2H_2$ and the catalyst is reduced for 15–20 min at atmospheric pressure and 25° with efficient stirring. The ballast containing $^2H_2$ is filled with light mineral oil to prevent entry of H_2O into the reduction apparatus. The ^{13}C-enriched aldononitriles are then added, and the reduction is continued for 10 hr, or until the nitriles are completely reduced as determined by gas–liquid chromatography.[6] After removal of the catalyst by filtration, the filtrate is treated with $BaCO_3$ to remove SO_4^{2-} and, after removal of the precipitated salts, with Dowex 1-X8 (OAc$^-$) and then with Dowex 50-X8 (H$^+$) to deionize the solution. The filtrates and washings from these treatments are concentrated under vacuum at 30° to 2–5 ml and applied to a Dowex 50-X8 (200–400 mesh) column in the Ba^{2+} form[11] for separation as described in this volume [11]. Products are characterized by ^1H and ^{13}C NMR. The yield is 70% based on product weight as gums, cyanide assay,[6] and the recovery of radioactivity after separation. Incorporation of ^2H at H-1 is better than 95% based on analysis by ^1H NMR spectroscopy and mass spectrometry. Substitution at other positions was undetectable by ^1H and ^2H NMR spectroscopy.

General Method for the Preparation of Carbohydrates Containing Oxygen Isotopes: Synthesis of D-[2-^{18}O]Arabinose and Ribose

A solution of D-erythrose (0.2 mmol in 0.27 ml of H_2O) is concentrated to dryness under vacuum at 30° in a 5-ml pear-shaped flask, and the residue is desiccated overnight under vacuum over $MgClO_4$. The dried gum is dissolved in $H_2^{18}O$ (0.55 ml, 97 atom %), 17 μl of 18 M H_2SO_4 is added, and the flask is stoppered and incubated at 25° for 30 hr. A microelectrode is immersed in the solution, and, with efficient stirring, the pH is raised from ~1.0 to 7.0 with the careful addition of solid KCN (powdered and dried under vacuum over $MgClO_4$). Additional KCN (~15 mg) is added slowly while the pH is maintained below 7.5 with the addition of 18 M H_2SO_4 from a 5-μl syringe fitted with fine polyethylene tubing. The pH of the reaction mixture is maintained at 7.5–7.8 for 15 min and then

[11] J. K. N. Jones and R. A. Wall, *Can. J. Chem.* **38,** 2290 (1960).

lowered to pH 3.5 with 18 M H_2SO_4. The reaction mixture is frozen (CO_2–ethanol bath), and the $H_2^{18}O$ is recovered by lyophilization using an evacuated U-tube with the trapping flask cooled in a CO_2–ethanol bath. Recovery of $H_2^{18}O$ is >90%.

The dry, white residue is dissolved in 2.5 ml of 1 M acetic acid, the pH of the solution is adjusted from ~2.6 to 4.2 ± 0.1 with 1.5 M NaOH, and a sample is taken for analysis by gas–liquid chromatography.[6] The solution contains >95% aldononitriles and <5% hydrolysis products. Palladium–barium sulfate (5%, 42 mg) is added, and the mixture is reduced for 5–6 hr at atmospheric pressure as described in this volume [11].

After reduction is complete, the catalyst is removed by filtration and the solution is treated batchwise with Dowex 50-X8 (H^+). The resin is removed by filtration, and the filtrate is treated with $BaCO_3$ until neutral. The suspension is filtered through Celite, and the clear filtrate is treated with Dowex 1-X8 (OAc^-) batchwise and then with Dowex 50-X8 (H^+). The resin is removed by filtration, the deionized filtrate is concentrated at 30° under vacuum to a small volume (2–3 ml), and the mixture is applied to a 1.2 cm × 70 cm column of Dowex 50-X8 (200–400 mesh) in the Ba^{2+} form.[11] Fractions (3.5 ml) are collected at a flow rate of 0.3 ml/min. Two peaks elute as determined by $AgNO_3$ assay[12]: fractions 26–30, D-[2-^{18}O]arabinose; fractions 40–65, D-[2-^{18}O]ribose. The yield is ~70%. Incorporation of ^{18}O at O-2 is better than 80% as determined by mass spectral analysis of the pertrimethylsilylated alditols.

[12] R. M. C. Dawson, D. Elliott, W. Elliott, and K. M. Jones, eds., "Data for Biochemical Research," 2nd ed., p. 541. Oxford Univ. Press, London and New York, 1969.

[14] Enzymic Synthesis of ^{13}C-Enriched Aldoses, Ketoses, and Their Phosphate Esters

By A. S. SERIANNI, E. CADMAN, J. PIERCE, M. L. HAYES, and R. BARKER

A number of metabolically important carbohydrates with isotopic enrichment at specific sites can be prepared conveniently by enzymic modification of chemically synthesized precursors. For example, the general method for preparing isotopically enriched aldoses (this volume [11]) or aldose phosphates (this volume [12]), can be extended to prepare ketoses or ketose phosphates using commercially available enzymes and substrates such as ATP.

The conversions described below involve compounds enriched at specific sites with [13]C. In most cases, however, starting materials with other isotopic enrichments can be used, as can unenriched compounds. The only limitations are those due to enzymic mechanisms that exchange specific atoms with the solvent during the conversion.

Although several steps are involved in some syntheses, the substrates (apart from some enriched ones) and enzymes are commercially available. Dihydroxyacetone phosphate (DHAP) has been prepared enzymically from dihydroxyacetone and ATP by the action of glycerol kinase, since millimole quantities of DHAP are prohibitively expensive. Manipulations, such as chromatography, are straightforward, and yields are high.

Procedures for the following conversions are described.

DL-[1-^{13}C]Glyceraldehyde → D-[4-^{13}C]fructose → D-[4-^{13}C]glucose

D-[2-^{13}C]Ribose 5-phosphate → D-[2-^{13}C]ribulose 1,5-bisphosphate

D-[2-^{13}C]Glucose → D-[2-^{13}C] and [2,5-^{13}C]fructose 1,6-bisphosphate

D-[1-^{13}C]Mannose → D-[1-^{13}C]glucose

D-[1-^{13}C]Galactose → α-D-[1-^{13}C]galactose 1-phosphate

D-[4-^{13}C]Fructose and D-[4-^{13}C]Glucose

The ^{13}C-enriched starting material for this conversion is DL-[1-^{13}C]glyceraldehyde,[1] which is prepared by the condensation of K[13]CN with glycolaldehyde followed by hydrogenation and solvolysis of the nitrile product (see this volume [11]). The product is deionized by treatment with Dowex 50-X8 (H[+]) and Dowex 1-X8 (acetate) and is used as a substrate for muscle aldolase, as shown in Scheme 1. The second substrate for aldolase, dihydroxyacetone phosphate (DHAP), is prepared enzymically from dihydroxyacetone and ATP by the action of glycerolkinase.

The aldol-condensation products are D-[4-^{13}C]fructose 1-phosphate and L-[4-^{13}C]sorbose 1-phosphate. After treatment with acid phosphatase, the neutral sugars are separated by chromatography on a Dowex 50-X8 (200–400 mesh) column in the Ba[2+] form using distilled water as the eluent. The overall yield of [4-^{13}C]ketoses is 91%.

The conversion of D-[4-^{13}C]fructose to D-[4-^{13}C]glucose is accomplished in approximately 67% overall yield by the sequential action of hexokinase, phosphoglucose isomerase, and acid phosphatase. The lower yield is due to the isomerase equilibrium between fructose-6-P and glucose-6-P. Dephosphorylation of the mixture gives a 67/19 mixture of glucose and fructose. The latter is recovered in the final purification step and can be recycled. Overall recovery of ^{13}C isotope is 86%.

---

[1] D-[5-^{13}C]Glucose can be prepared by using DL-[2-^{13}C]glyceraldehyde in place of the 1-^{13}C derivative. This starting aldose is prepared from [1-^{13}C]glycolaldehyde and KCN as described in Chapter [11] of this volume.

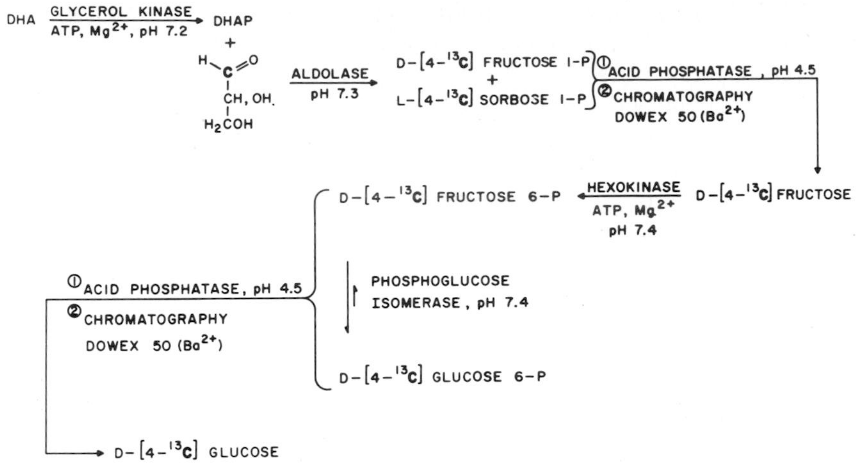

SCHEME 1

Preparation of Dihydroxyacetone Phosphate and DL-[*1-*^{13}*C*]*Glyceraldehyde.* Dihydroxyacetone phosphate (DHAP) is prepared in millimole quantities by the action of glycerol kinase on dihydroxyacetone and ATP. ATP (10 mmol), dihydroxyacetone (15 mmol), and $MgCl_2$ (10 mmol) are dissolved in H_2O (50 ml), and the pH of the solution is adjusted to 7.2 with 4 N NaOH. The solution is diluted to 100 ml with H_2O, glycerol kinase (EC 2.7.1.30) (~200 units) and myokinase (adenylate kinase, EC 2.7.4.3) (~100 units) are added and the pH of the reaction mixture is maintained at 7.0–7.2 with additions of 1 N NaOH over a period of 2 hr at room temperature. The reaction mixture is assayed for DHAP by alkaline lability of the phosphate group. Aliquots containing up to 1 μmol of P were treated with 100 μl of 1 N NaOH for 15 min, and the liberated P_i is assayed by the method of Leloir and Cardini.[2]

After 2 hr, the reaction mixture is treated with excess Dowex 50-X8 (H^+) batchwise, and the suspension is filtered. The filtrate is adjusted to pH 7.0 with 2 N NaOH, diluted to 500 ml with H_2O, and loaded on a DEAE-Sephadex column (2.5 cm × 74 cm) in the acetate form at 4°. The column is eluted with a 3-liter gradient of sodium acetate–acetic acid, 0.05 to 1.5 M at pH 4.5. Fractions (8 ml) are collected at 15-min intervals. DHAP elutes between fractions 130 and 160. The fractions are pooled and treated with Dowex 50-X8 (H^+) batchwise. The suspension is filtered, and the filtrate is concentrated at 30° under vacuum to 20 ml. The yield is 5.5 mmol, 37% based on P assay.[2]

[2] L. F. Leloir and C. E. Cardini, this series, Vol. 3, p. 840.

DL-[1-^{13}C]Glyceraldehyde is prepared from K^{13}CN and glycolaldehyde as described in this volume [11].

Preparation of D-[4-^{13}C]*Fructose and* L-[4-^{13}C]*Sorbose.* Dihydroxyacetone phosphate (5.5 mmol) and DL-[1-^{13}C]glyceraldehyde (6.5 mmol) are dissolved in 20 ml of H$_2$O, the pH is adjusted to 7.3 with 2 M NaOH, and the solution is diluted to 30 ml with H$_2$O. Aldolase (EC 4.1.2.13) (~100 units) is added, and the reaction is incubated for 5.5 hr at room temperature. Reaction is complete as determined by ^{13}C NMR. The mixture is diluted to 80 ml with H$_2$O and adjusted to pH 4.5 with 4 N acetic acid. Acid phosphatase (EC 3.1.3.2) (~100 units) is added, and the reaction is incubated at 37° overnight. Dephosphorylation is complete as determined by P assay.[2] An equal volume of absolute ethanol is added, and the mixture is incubated at 40° for 30 min. The suspended protein is removed by centrifugation at 7700 g for 30 min at 5° and the supernatant is collected and concentrated under vacuum at 30° to ~10 ml. The pH is adjusted to 6.0 with 1 N NaOH, and the solution is treated with Dowex 1-X8 (OAc$^-$) and Dowex 50-X8 (H$^+$), separately and batchwise. The deionized solution is concentrated to 2–3 ml and loaded on a Dowex 50-X8 (200–400 mesh) column (3.5 cm × 100 cm) in the Ba^{2+} form. The column is eluted with distilled H$_2$O, and fractions (5 ml) are collected at 0.4 ml/min. Fractions are assayed by AgNO$_3$.[3] Three peaks elute: fractions 86–106, L-[4-^{13}C]sorbose; fractions 110–127, DL-[1-^{13}C]glyceraldehyde; fractions 132–150, D-[4-^{13}C]fructose. Fractions containing the purified [^{13}C]hexoses are pooled and concentrated to dryness at 30° under vacuum. The yield is L-[4-^{13}C]sorbose, 0.5 g, 2.8 mmol; D-[4-^{13}C]fructose, 0.4 g, 2.2 mmol.

Preparation of D-[4-^{13}C]*Glucose.* D-[4-^{13}C]Fructose (0.5 mmol), D-[U-^{14}C]fructose (9.7 × 10^6 cpm), ATP (0.5 mmol), and MgCl$_2$ (0.5 mmol) are dissolved in 5 ml of H$_2$O. The pH is adjusted to 7.4 with 1 N NaOH and diluted to 8 ml with H$_2$O. Hexokinase (EC 2.7.1.1) (~50 units) and phosphoglucoisomerase (glucosephosphate isomerase, EC 5.3.1.9) (~25 units) are added and the pH is maintained at 7.3–7.5 for 30 min with additions of 1 N NaOH. The reaction is complete as determined by ^{13}C NMR. The reaction mixture is diluted to 25 ml and loaded on a DEAE-Sephadex column (1.5 cm × 86 cm) in the bicarbonate form. The column is eluted at room temperature with a 3-liter linear gradient of triethylammonium bicarbonate (0 to 0.4 M, pH 7.5). Fractions (10 ml) are collected at 0.5 ml/min and assayed for radioactivity. A single peak, containing D-[4-^{13}C]glucose 6-phosphate and D-[4-^{13}C]fructose 6-phosphate, elutes between fractions 134 and 160. The fractions are pooled and concentrated

[3] R. M. C. Dawson, D. Elliott, W. Elliott, and K. M. Jones, eds., "Data for Biochemical Research," 2nd ed., p. 541. Oxford Univ. Press, London and New York, 1969.

at 30° under vacuum to 40 ml. The solution is treated with Dowex 50 (H^+) batchwise, the resin is removed by filtration, and the filtrate is concentrated to 25 ml. The yield is 90%, based on radioactivity.

The solution of D-[4-^{13}C]glucose 6-phosphate and D-[4-^{13}C]fructose 6-phosphate is adjusted to pH 4.5 with 1 N NaOH, 50 units of acid phosphatase are added, and the reaction mixture is incubated at 37° overnight. The mixture is treated with an equal volume of absolute ethanol and centrifuged to remove protein as described above. The supernatants are pooled, concentrated to 25 ml, and deionized as described above with Dowex 1-X8 (OAc⁻) and Dowex 50-X8 (H^+). Volume is reduced at 30° under vacuum to 2–3 ml, and the sample is applied to a Dowex 50-X8 (200–400 mesh) column (3.5 cm × 100 cm) in the Ba^{2+} form. The column is eluted with distilled H_2O, and fractions (12 ml) are collected at 0.8 ml/min. Two peaks elute between fractions 30 and 50 (D-[4-^{13}C]glucose) and 50–70 (D-[4-^{13}C]fructose). Fractions containing the purified [^{13}C]hexoses are pooled, treated with Dowex 50 (H^+) batchwise, and concentrated to dryness. The yield is D-[4-^{13}C]glucose, 67%; D-[4-^{13}C]fructose, 19%; based on overall recovery of radioactivity and on weight as dried gums.

D-[2-^{13}C]Ribulose 1,5-Bisphosphate

The sequential action of phosphoriboisomerase and phosphoribulokinase on D-[2-^{13}C]ribose 5-phosphate gives D-[2-^{13}C]ribulose 1,5-bisphosphate in 85% overall yield. D-[2-^{13}C]Ribose 5-P can be prepared from D-[1-^{13}C]erythrose-4-P and KCN as described in this volume [12]. The preparation of the ^{13}C-enriched derivative is based on the preparation and purification described by Horecker et al.[4]

Preparation of D-[2-^{13}C]*Ribulose 1,5-Bisphosphate.* To 90 ml of a solution of $MgCl_2$ (10 mM), dithiothreitol (0.1 mM), and EDTA (0.1 mM) is added 1 mmol of ATP. After adjusting to pH 8 with 1 N NaOH, 100 units of phosphoriboisomerase (EC 5.3.1.6) and 50 units of phosphoribulokinase (EC 2.7.1.19) are added and the solution is readjusted to pH 8. The reaction is started by adding 10 ml of 0.1 M D-[2-^{13}C]ribose-5-phosphate (pH 8). Dilute NaOH is added periodically to maintain the pH between 7.6 and 8.0 as the reaction proceeds. The reaction is judged to be complete when no further addition of base is required to keep the pH approximately constant for 10 min. D-[2-^{13}C]Ribulose-1,5-bisphosphate may be isolated by Dowex 1 (Cl⁻) chromatography as described below for D-fructose-1,6-bisphosphate, or as follows.

The reaction solution is cooled to 4–10°, and protein is precipitated by

[4] B. L. Horecker, J. Hurwitz, and A. Weissbach, *J. Biol. Chem.* **218**, 785 (1956).

the addition of 0.1 volume of 50% trichloroacetic acid. Nucleotides are removed by successive addition of 4-g portions of acid- and alkali-washed activated Norit A charcoal until the absorbance at 257 nm (E_{257}[ADP] = 15,000 liters/mol-cm) is less than 1. After removal of the charcoal by filtration through Celite, the filter pad is washed, and the washings and filtrate are combined to give a solution that is adjusted to pH 7 with 1 N NaOH. A two-fold excess of barium acetate (1 g) is added, followed by 1.2 volumes of 95% ethanol. After storage overnight at 4°, the precipitated barium salts are collected by centrifugation. The precipitate is suspended in 50 ml of water, a few drops of glacial acetic acid are added, and the suspension is stirred with Dowex-50 (H^+) until a clear solution is obtained (~15 min). After filtering away the resin, the solution is adjusted to pH 6.5 and applied to a 2.5 × 40 cm Dowex 1-X8 column (200–400 mesh) in the chloride form at 4°. The column is developed with a 4-liter linear gradient (0–0.4 M LiCl in 10 mM HCl) at a flow rate of 1–2 ml/min. D-[2-^{13}C]Ribulose-1,5-bisphosphate is visualized by the orcinol–sulfuric acid colorimetric assay.[5] Fractions containing the sugars are pooled and adjusted to pH 7 with NaOH. For either isolation protocol, the compound is precipitated as its barium salt by adding a 1.2-fold excess (0.6 g) of barium acetate followed by 1.2 volumes of 95% ethanol. After storage at 4° for 16 hr the precipitate is collected and washed twice with ice-cold 75% ethanol, twice with ice-cold absolute ethanol, and dried under vacuum. The product is stored below −20°.

This protocol has been used to prepare up to 20 mmol of unenriched D-ribulose-1,5-P_2. For large-scale syntheses, nucleotide removal with charcoal avoids a tedious chromatographic separation. Other isolation procedures are available for smaller-scale syntheses.[6]

For use, a portion of the barium salt is suspended in water and stirred with Dowex 50 (H^+) until the solution is clear. Addition of a small volume of Na_2SO_4 solution will reveal incomplete removal of Ba^{2a} ion. After filtering, the solution is adjusted to pH 6 and assayed for total and inorganic phosphate[2]. The aqueous solution is stable for at least 1 month if stored at −20°.

D-Ribulose 1,5-bisphosphate is exceedingly unstable in alkaline solution and to long-term storage.[7] Either condition results in the formation of small amounts of D-xylulose-1,5-bisphosphate and monophosphates from phosphate elimination reactions. When high purity is required, D-xylulose-1,5-bisphosphate may be removed by the sequential action of

[5] B. L. Horecker, this series, Vol. 3, p. 105.
[6] S. S. Kent and J. D. Young, *Plant Physiol.* **65**, 456 (1980).
[7] C. Paech, J. Pierce, S. D. McCurry, and N. E. Tolbert, *Biochem. Biophys. Res. Commun.* **83**, 1084 (1978).

D-[2-^{13}C] GLUCOSE $\xrightarrow[\substack{\text{PFK, MYOKINASE,} \\ \text{ATP, Mg}^{2+}, \text{ pH 7.4}}]{\text{HEXOKINASE, PGI}}$ D-[2-^{13}C] FDP $\xrightarrow[\text{pH 7.4}]{\text{ALDOLASE, TPI}}$ D-[2,5-^{13}C] FDP

SCHEME 2

D-fructose bisphosphate aldolase and glycerolphosphate dehydrogenase. Ion-exchange chromatography may then be employed to separate D-ribulose-1,5-bisphosphate from the contaminating monophosphates.

D-[2-^{13}C]Fructose 1,6-Bisphosphate and D-[2,5-^{13}C]Fructose 1,6-Bisphosphate

The enzymes of the glycolytic pathway, hexokinase, phosphoglucose isomerase, and phosphofructokinase can be used to convert D-[2-^{13}C]glucose to D-[2-^{13}C]fructose 1,6-bisphosphate in 86% yield.[8] If myokinase is added, the conversion can be accomplished with less than 2 molar equivalents of ATP (Scheme 2). Product purification is achieved by ion-exchange chromatography.

The addition of aldolase and triosephosphate isomerase, or extended incubation with aldolase, which normally contains small amounts of the isomerase, gives D-fructose 1,6-bisphosphate in which the ^{13}C isotope is symmetrically distributed (Scheme 2).

Preparation of D-[2-^{13}C]*Fructose 1,6-Bisphosphate* (*FDP*). D-[2-^{13}C]Glucose (1.4 mmol) is dissolved in 17 ml of H_2O; ATP (2.1 mmol) and $MgCl_2$ (0.7 mmol) are added, and the pH is adjusted to 7.4 with 2 M NaOH. Myokinase (E.C. 2.7.4.3) (500 units), hexokinase (500 units), phosphoglucose isomerase (250 units), and phosphofructokinase (EC 2.7.1.11) (300 units) are added, and the mixture is incubated for 60 min at 34° while maintaining the pH of the reaction at pH 7.4 with additions of 2 M NaOH. Assay by ^{13}C NMR shows complete conversion to FDP.

The reaction mixture is diluted to 50 ml and applied to a 1.7 × 26 cm Dowex 1-X8 (200–400 mesh) column in the chloride form. The column is washed rapidly with 6 liters of 0.01 M HCl to remove AMP and ADP. The column is then eluted with 4 liters of 0.02 M HCl–0.02 M LiCl, and 11-ml fractions are collected at 0.5 ml/min. D-[2-^{13}C]FDP elutes between fractions 130 and 240 as determined by phosphate assay.[2] Fractions containing the bisphosphate are pooled and concentrated under vacuum at 30° to 200 ml. The solution is neutralized (pH 7.0) with 0.5 M NaOH, and Ba(OAc)$_2$ (5.6 mmol) is added with stirring, followed by 240 ml of ab-

---

[8] Midelfort *et al.*[9] have prepared D-[U-^{13}C]FDP by this method.

[9] C. F. Midelfort, R. K. Gupta, and I. A. Rose, *Biochemistry* **15**, 2178 (1976).

solute ethanol. The solution is stored at 4° overnight to precipitate of the barium salt of D-[2-^{13}C]fructose-1,6-bisphosphate. The precipitate is collected by centrifugation, suspended in H_2O, and treated with Dowex 50-X8 (H^+). The solution is filtered to remove the resin, concentrated to 3 ml at 30° under vacuum, adjusted to pH 4.0 with 0.1 M NaOH, and stored at 4°. The yield is 1.2 mmol (86%) based on phosphate assay.

D-[2-^{13}C]FDP can be converted to D-[2,5-^{13}C]FDP with partial enrichment at each site by adding aldolase and triosephosphate isomerase (~200 units each) to the reaction mixture. Purification of the 2,5-^{13}C derivative is accomplished as described above.

D-[1-^{13}C]Glucose from D-[1-^{13}C]Mannose

In the chemical synthesis of D-[1-^{13}C]glucose as described in this volume [11], the 2-epimer, D-[1-^{13}C]mannose, is formed as 70% of the final products. It is useful, therefore, to convert D-mannose to D-glucose and improve the yield of the more commonly encountered hexose.

This conversion is accomplished by the sequential action of hexokinase, phosphomannose isomerase, and phosphoglucose isomerase (Scheme 3). The equilibrium at 25° between the latter two enzymes yields the 6-phosphates of mannose, fructose, and glucose in the ratio 25 : 15 : 60. After separation from AMP, ADP, and other anions by chromatography on DEAE-Sephadex, the [4-^{13}C]hexose 6-phosphate mixture is treated with acid phosphatase and the hexoses are separated by chromatography on Dowex 50 (Ba^{2+}). Yields are high (>90%), and recovered mannose and fructose can be recycled. Approximately 60% of the starting mannose can be converted to glucose in one cycle.

Preparation of an Equilibrium Mixture of D-[1-^{13}C]*Mannose 6-Phosphate,* D-[1-^{13}C]*Fructose 6-Phosphate, and* D-[1-^{13}C]*Glucose 6-Phosphate.* D-[1-^{13}C]Mannose (0.2 mmol), D-[U-^{14}C]mannose (6.4 × 10^6 cpm), ATP (0.25 mmol), and MgCl$_2$ (0.25 mmol) are dissolved in 5 ml of H_2O, the pH is adjusted to 7.4 with 1 N NaOH, and the solution is diluted to 7 ml with H_2O. Hexokinase (50 units), phosphomannose isomerase (EC 5.3.1.8) (25 units), and phosphoglucose isomerase (25 units) are added,

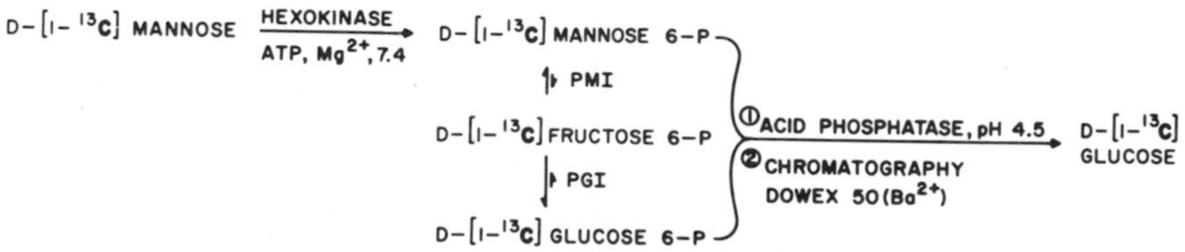

SCHEME 3

and the pH is maintained at 7.3–7.5 for 30 min with additions of 1 N NaOH. The reaction is judged to be complete by ¹³C NMR. The mixture is diluted to 25 ml with H_2O, adjusted to pH 6.5 with 1 N NaOH, and loaded on a DEAE-Sephadex column (1.5 cm × 90 cm) in the bicarbonate form. The column is eluted with a 4-liter linear gradient of triethylammonium bicarbonate (0 to 0.4 M, pH 7.5). Fractions (10 ml) are collected at 0.5 ml/min and assayed for radioactivity. A single peak, which contains D-[1-¹³C]mannose 6-phosphate, D-[1-¹³C]fructose 6-phosphate, and D-[1-¹³C]glucose 6-phosphate, elutes between fractions 200 and 230. These fractions are pooled, and the solution is concentrated at 35° under vacuum to 50 ml. The concentrate is treated with Dowex 50-X8 (H⁺) batchwise, the resin is removed by filtration, and the filtrate is concentrated to 30 ml. The yield is 90% as determined by radioactivity and P assay.[2]

Preparation of D-[1-¹³C]*Glucose.* One-half of the hexose 6-P mixture described above is diluted to 30 ml with H_2O and adjusted to pH 4.5 with 1 N NaOH. Acid phosphatase (~100 units) is added, and the reaction mixture is incubated at 36° overnight. An equal volume of absolute ethanol is added, and the mixture is incubated at 40° for 30 min. The suspension is centrifuged at 7700 g for 30 min at 5° to remove precipitated protein, and the supernatant is collected and concentrated at 35° under vacuum to ~10 ml. The solution is adjusted to pH 6.0 with 1 N NaOH followed by separate batchwise treatments with Dowex 1-X8 (OAc⁻) and Dowex 50-X8 (H⁺). The solution is concentrated to 10 ml and loaded on a Dowex 50-X8 (200–400 mesh) column (3.5 cm × 100 cm) in the Ba^{2+} form. The column is eluted with distilled H_2O, and fractions (10 ml) are collected at 0.5 ml/min and assayed for radioactivity. Three peaks elute: fractions 60–85, D-[1-¹³C]glucose; fractions 86–104, D-[1-¹³C]mannose; fractions 110–130, D-[1-¹³C]fructose. Fractions containing the purified hexoses are pooled and concentrated to 20–30 ml at 35° under vacuum. The yield is glucose, 65%; mannose, 22%; fructose, 11% as determined by radioactivity and weight as gums.

α-D-[1-¹³ C]Galactose 1-Phosphate

α-D-[1-¹³C]Galactose 1-phosphate is an important intermediate in the chemical synthesis of UDP-D-[1-¹³C]galactose.[10] D-[1-¹³C]Galactose can be converted chemically to the 1-phosphate by the orthophosphoric acid method of MacDonald.[10a] Alternatively, this conversion can be accomplished enzymically with the use of a crude galactokinase preparation

[10] H. A. Nunez and R. Barker, *Biochemistry* **19**, 489 (1980).
[10a] D. L. MacDonald, this series, Vol. 8, p. 121.

from yeast.[11] The enzyme preparation contains adenylyl kinase so that the conversion can be accomplished using slightly more than 50% of the stoichiometric amount of ATP. Product purification is achieved in high yield by ion-exchange chromatography on DEAE-Sephadex. Unphosphorylated galactose is recovered and can be recycled.

Preparation of α-D-[*1-*^{13}*C*]*Galactose 1-Phosphate.* Yeast galactokinase (EC 2.7.1.6) is partially purified through the DEAE-cellulose chromatography step of the procedure of Schell and Wilson.[12] Yeast used in this study has been galactose-adapted *Kluyveromyces fragilis* (Sigma). Crude galactokinase (40 ml) from 4 g of dried yeast is stored in 2-ml aliquots at $-20°$.

The reaction mixture contains 1.0 mmol of D-[1-^{13}C]galactose, 4.1 μCi of D-[1-^{3}H]galactose (specific activity, 2.8 Ci/mmol), 0.6 mmol of ATP, 0.2 mmol of $MgCl_2$, 33 μmol of dithiothreitol, 0.1 mmol of NaF, 1.7 mmol of triethanolamine–acetate (pH 8.0), and 2.0 ml (~7 units) of galactokinase in a total volume of 33 ml. Two drops of toluene are added, and the mixture is incubated at 30°. At suitable intervals, 50-μl samples are diluted to 0.25 ml and applied to 0.5-ml columns of Dowex 1-X8 (200–400 mesh) in the chloride form and washed with 1.0 ml of H_2O. The amount of radioactivity in the effluent is a direct measure of unreacted galactose. After 12 hr, 87% of the galactose is converted to α-D-[1-^{13}C]galactose 1-phosphate. The reaction mixture is diluted to 150 ml and applied at 2 ml/min to a Dowex 1-X2 (200–400 mesh) column (2.5 cm × 10 cm) in the bicarbonate form previously equilibrated with 0.02 M triethylammonium bicarbonate at pH 7.5. A 1-liter linear gradient of 0.02 to 0.4 M triethylammonium bicarbonate, pH 7.5, is started immediately, and 8-ml fractions are collected at a flow rate of 2 ml/min. α-D-[1-^{13}C]Galactose 1-phosphate elutes between fractions 81 and 93 and is obtained as its triethylammonium salt by repeated concentration under vacuum at 35° from 30-ml additions of H_2O. The yield is 80% based on radioactivity.

---

[11] D-Galactosamine is also phosphorylated by yeast galactokinase. The preparation of millimole quantities of α-D-galactosamine 1-P by this method has been described by D. M. Carlson and S. Roseman, this series, Vol. 28, p. 274.

[12] M. A. Schell and D. B. Wilson, *J. Biol. Chem.* **252,** 1162 (1977).

[15] D-Galactose 6-Phosphate and D-Tagatose 6-Phosphate

By RICHARD L. ANDERSON, WILLIAM C. WENGER, and DONALD L. BISSETT

D-Galactose 6-phosphate and D-tagatose 6-phosphate are intermediates in the catabolism of lactose and D-galactose in *Staphylococcus aureus*[1-5] and in group N streptococci.[6] D-Tagatose 6-phosphate is also an intermediate in galactitol catabolism in *Klebsiella pneumoniae*[7,8] and in *Escherichia coli*.[9]

D-Galactose 6-phosphate is used as a substrate in the assay of D-galactose-6-phosphate isomerase.[10] D-Tagatose 6-phosphate is used as a substrate in the assay of D-tagatose-6-phosphate kinase[11] and in the reverse reaction of D-galactose-6-phosphate isomerase[10] and also as a substrate-precursor in the assay of D-tagatose-1,6-bisphosphate aldolase.[12,13]

D-Galactose 6-Phosphate

The synthesis of D-galactose 6-phosphate generally follows the procedure used by Seegmiller and Horecker[14] for the synthesis of D-glucose 6-phosphate, but with modifications[15] that result in increased purity of the final product. For convenience, the entire procedure will be detailed here.

Principle. D-Galactose is multiply phosphorylated with polyphosphoric acid. All but the most resistant phosphate groups are then removed by acid hydrolysis, leaving primarily D-galactose 6-phosphate, but also

[1] D. L. Bissett and R. L. Anderson, *Biochem. Biophys. Res. Commun.* **52**, 641 (1973).
[2] D. L. Bissett and R. L. Anderson, *J. Bacteriol.* **119**, 698 (1974).
[3] D. L. Bissett, W. C. Wenger, and R. L. Anderson, *J. Biol. Chem.* **255**, 8740 (1980).
[4] D. L. Bissett and R. L. Anderson, *J. Biol. Chem.* **255**, 8745 (1980).
[5] D. L. Bissett and R. L. Anderson, *J. Biol. Chem.* **255**, 8750 (1980).
[6] D. L. Bissett and R. L. Anderson, *J. Bacteriol.* **117**, 318 (1974).
[7] J. Markwell, G. T. Shimamoto, D. L. Bissett, and R. L. Anderson, *Biochem. Biophys. Res. Commun.* **71**, 221 (1976).
[8] J. P. Markwell and R. L. Anderson, *Arch. Biochem. Biophys.* **209**, 592 (1981).
[9] J. Lengeler, *Mol. Gen. Genet.* **152**, 83 (1977).
[10] See this volume [95].
[11] R. L. Anderson and D. L. Bissett, this series, Vol. 90 [15].
[12] R. L. Anderson and D. L. Bissett, this series, Vol. 90 [34].
[13] R. L. Anderson and J. P. Markwell, this series, Vol. 90 [35].
[14] J. E. Seegmiller and B. L. Horecker, *J. Biol. Chem.* **192**, 175 (1951).
[15] W. C. Wenger and R. L. Anderson, *Carbohydr. Res.* **88**, 267 (1981).

other positional isomers of D-galactose monophosphate.[15] D-Galactose 6-phosphate is purified free of the positional isomers by ion-exchange chromatography.

Reagents

Polyphosphoric acid (Sigma Chemical Co.)
D-Galactose
Anhydrous sodium carbonate
Concentrated HBr
Barium carbonate
n-Octanol
Ethanol, 95%
Activated charcoal
Absolute ethanol
Diethyl ether
Dowex 50W-X8 resin (H⁺ form)
Ammonium hydroxide
Dowex 1-X4 (200–400 mesh) resin, washed in a column sequentially with 0.8 M potassium tetraborate until the chloride has been displaced, and 10 bed-volumes of water
Triethylammonium tetraborate
Methanol

Procedure. Water (2.3 ml) is carefully added to 23 g of polyphosphoric acid in an ice bath, and the mixture is mechanically stirred until the temperature is reduced to 5–10°. D-Galactose (10.0 g) is added immediately with vigorous stirring. The mixture is stirred gently for 16 hr at room temperature (24°), and then the reaction is terminated by the addition of 57 ml of water.

Anhydrous sodium carbonate (30 g) is added to the reaction mixture with stirring. To aid in the removal of CO_2, the mixture is warmed to about 60° and subjected to reduced pressure with a water aspirator for several minutes. The solution is diluted with 150 ml of water (the pH is now about 7.5) and cooled overnight in a cold room at 4°. (To reduce the rate of cooling, the solution is placed in a large water bath that was initially at 40°.) The resulting slurry of sodium polyphosphate crystals is cooled in an ice bath with stirring for 4 hr, then the crystals are removed by suction filtration.

To the filtrate, which contains multiply phosphorylated D-galactose, is added concentrated HBr (22 ml per 100 ml of filtrate). The mixture is refluxed for 16–24 hr, cooled to room temperature, treated with barium carbonate (20 g per 100 ml of filtrate), and stirred for 4 hr. Several drops of *n*-octanol are added to reduce foaming. The mixture, now at pH 6.4, is suction-filtered, and the residue is washed twice by suspending it in 10-ml portions of distilled water. The filtrate and washings are combined and treated with four volumes of 95% ethanol. The resulting flocculent precipitate is allowed to settle and is collected by centrifugation. The precipitate is extracted four times with 10-ml portions of water, the extracts are combined, and the amber solution is decolorized by treatment with 50 mg of activated charcoal for 2 hr with stirring. The mixture is suction-filtered and the filtrate is treated with four volumes of 95% ethanol. The resulting white precipitate is washed with absolute ethanol and diethyl ether and is allowed to dry. The product (1.9 g) is the barium salt of a mixture of D-galactose monophosphates.[15]

To obtain pure D-galactose 6-phosphate, a portion[16] of the barium D-galactose monophosphate mixture is dissolved in water and treated at room temperature with Dowex 50W-X8 resin (H$^+$ form) until the pH is 1.5. The resin is removed and the solution is titrated to pH 8.0 with ammonium hydroxide and applied to a column (1.0 × 30 cm) of Dowex 1-X4 resin (see Reagents). The column is washed with about 3 bed-volumes of water and eluted with a 400-ml linear gradient of 0.1 to 0.4 M triethylammonium tetraborate, followed by 100 ml of 0.4 M triethylammonium tetraborate.[15,17] Fractions (1.5 ml) are collected and assayed for carbohydrate.[18] The fractions comprising the largest peak (centering on fraction 220), which is D-galactose 6-phosphate, are combined, and the borate is removed by repeated evaporation with methanol.

Analysis of the Product.[15] The product, prior to its passage through the ion-exchange column, is only 77–80% enzymically active with D-galactose-6-phosphate isomerase in a coupled system. Analysis by combined gas-liquid chromatography–mass spectrometry and other techniques indicated the product to be a mixture of D-galactose 6-phosphate (80%), D-galactose 3-phosphate (7%), and D-galactose 5-phosphate (13%). After purification of the product by ion-exchange chromatography, pure D-galactose 6-phosphate is obtained that is 100% enzymically active.

D-Tagatose 6-Phosphate

The synthesis of D-tagatose 6-phosphate from 1,2:3,4-di-*O*-isopropylidene-D-tagatose has been described in this series by Totton and Lardy.[19] Here we expand on their procedure by detailing the synthesis of 1,2:3,4-di-

---

[16] A 100-mg sample was used in the procedure described here, but this could be scaled up.
[17] M. J. Lefebvre, N. J. Gonzales, and H. G. Pontis, *J. Chromatogr.* **15**, 495 (1964).
[18] M. Dubois, K. A. Gilles, J. K. Hamilton, P. A. Rebers, and F. Smith, *Anal. Chem.* **28**, 350 (1956).
[19] See this series, Vol. 3 [24].

O-isopropylidene-D-tagatose from D-galacturonic acid, based on two published syntheses,[20,21] and describe modifications introduced to increase the yield[22] and purity[3] of the final product, D-tagatose 6-phosphate.

Principle. The synthesis involves sequentially (*a*) the isomerization of D-galacturonic acid to D-tagaturonic acid; (*b*) the preparation of 1,2 : 3,4-di-O-isopropylidene-D-tagaturonic acid followed by its reduction to 1,2 : 3,4-di-O-isopropylidene-D-tagatose; (*c*) phosphorylation with diphenyl-chlorophosphonate to yield 1,2 : 3,4-di-O-isopropylidene-D-tagatose 6-diphenylphosphate followed by its reductive cleavage to 1,2 : 3,4-di-O-ispropylidene-D-tagatose 6-phosphate; (*d*) hydrolytic removal of the isopropylidene groups to yield D-tagatose 6-phosphate.

Preparation of D-*Tagaturonic Acid.*[20] D-Galacturonic acid (10 g, purified by recrystallization[23]) is dissolved in 2.5 liters of cold 0.15% (w/v) calcium hydroxide with vigorous stirring. The solution is allowed to stand at room temperature for 4–10 days, during which time crystals of calcium D-tagaturonate are formed. The crystals are collected by suction filtration, washed once with cold water, and dried over $CaSO_4$. The yield is 9.1 g.

*Preparation of 1,2 : 3,4-Di-O-isopropylidene-*D-*tagatose.*[21] Calcium D-tagaturonate (9.1 g) is stirred at room temperature in a solution of dry acetone (195 ml, distilled over Na_2SO_4) and concentrated H_2SO_4 (8 ml), in a reaction vessel equipped with a water-cooled condenser. After 4 hr, the solution is rapidly neutralized with an excess of calcium hydroxide suspended in water. The mixture is suction-filtered, and the pH of the filtrate is adjusted to 7.0 with H_2SO_4. The solution is lyophilized to give a white solid, which is the calcium salt of 1,2 : 3,4-di-O-isopropylidene-D-tagaturonic acid. The yield is 6.0 g.

To convert calcium 1,2 : 3,4-di-O-isopropylidene D-tagaturonate to the free acid, it is added with stirring to a mixture of 100 ml of water, 100 ml of diethyl ether, and 32.8 ml of 10% (w/v) H_2SO_4 at 4°. After 5 min, the ether layer is washed 4 times with 100-ml portions of cold water, dried over Na_2SO_4, and evaporated to dryness. The yield is 4.1 g of 1,2 : 3,4-di-O-isopropylidene-D-tagaturonic acid.

Ethereal diazomethane (prepared from Diazald,[24] Aldrich Chemical Co.) is added in excess (the solution will turn bright yellow at the end point) to 4.1 g of 1,2 : 3,4-di-O-isopropylidene-D-tagaturonic acid dissolved

[20] F. Ehrlich and R. Guttman, *Chem. Ber.* **67**, 573 (1934).

[21] P. A. J. Gorin, J. K. N. Jones, and W. W. Reid, *Can. J. Chem.* **33**, 1116 (1955).

[22] T. A. W. Koerner, R. J. Voll, A. L. E. Ashour, and E. S. Younathan, *J. Biol. Chem.* **251**, 2983 (1976).

[23] R. M. McCready, *Methods Carbohydr. Chem.* **2**, 27 (1963).

[24] N-Methyl-N-nitroso-p-toluenesulfonamide.

in a minimal amount of diethyl ether. (Appropriate safety precautions are to be followed in the generation and handling of diazomethane.[25]) The solution is allowed to stand at room temperature for 45 min and then is concentrated in a rotary evaporator to a syrup of the methyl ester of 1,2:3,4-di-*O*-isopropylidene-D-tagaturonic acid. The syrup is dissolved in 150 ml of anhydrous diethyl ether and stirred during the gradual addition of LiAlH$_4$ (3.7 g). The reaction mixture is stirred for 5 hr, then the excess LiAlH$_4$ is destroyed by the gradual addition of ethyl acetate; the end point is indicated by the cessation of bubbling. Water (150 ml) is added, and the organic solvents are removed by rotary evaporation. The solution is adjusted to near neutrality with 10% (v/v) acetic acid, filtered, neutralized, and extracted with an equal volume of chloroform. The chloroform layer is washed with a smaller volume of water, dried over Na$_2$SO$_4$, and evaporated to a syrup. Three recrystallizations from petroleum ether at 4° gives white needles of 1,2:3,4-di-*O*-isopropylidene-D-tagatose. The yield is 1.6 g.

Preparation of 1,2:3,4-Di-O-isopropylidine-D-tagatose 6-Phosphate. The synthesis of 1,2:3,4-di-*O*-isopropylidene-D-tagatose 6-diphenylphosphate and its reductive cleavage to 1,2:3,4-di-*O*-isopropylidene-D-tagatose 6-phosphate generally follow the procedures described by Totton and Lardy.[19] However, the yield of product in the reaction of diphenylchlorophosphonate with 1,2:3,4-di-*O*-isopropylidene-D-tagatose can be increased by incorporating the modifications of Koerner *et al.*[22] These modifications include the use of excess diphenylchlorophosphonate (1.5 molar equivalent) and a longer reaction time (60 hr at 5°).

Preparation of D-Tagatose 6-Phosphate.[3] 1,2:3,4-Di-*O*-isopropylidene-D-tagatose 6-phosphate (1.3 g) is dissolved in 15 ml of 0.05 N H$_2$SO$_4$ and heated on a steam bath for 30 min. Aqueous Ba(OH)$_2$ (CO$_2$ free) is then added until the pH is 10.2. The solution is filtered to remove suspended BaSO$_4$, and then is added to 4 volumes of 95% (v/v) ethanol to precipitate the barium salt of D-tagatose 6-phosphate. After standing at 4° for 24 hr, the precipitate is collected by centrifugation and washed with absolute ethanol, absolute ethanol–diethyl ether mixture (4:1, 1:1, and 1:4) and twice with anhydrous diethyl ether. The yield is about a gram of barium D-tagatose 6-phosphate.

Analysis of the Product.[3] Increased purity of the final product was achieved by carrying out the hydrolytic removal of the isopropylidene groups in 0.05 N H$_2$SO$_4$ at 100° for 30 min. If the hydrolysis was done at 100° in a neutral aqueous medium for 3 min,[19] we found the product to be

---

[25] T. J. DeBoer and H. J. Backer, *Org. Synth.* **36**, 16 (1956).

only 65% enzymically active with D-tagatose-6-phosphate kinase and other enzymes (compared to chemical analysis as ketohexose). Analysis by gas–liquid chromatography and natural-abundance ^{13}C NMR spectroscopy indicated the impurity to be the isopropylidene derivative. D-Tagatose 6-phosphate obtained by the use of $0.05 N$ H_2SO_4 in the final hydrolysis step is 100% enzymically active.

[16] Enzymic Synthesis of 2-Keto-3-Deoxygluconate 6-Phosphate Using 6-Phosphogluconate Dehydratase[1]

By E. L. O'CONNELL and H. PAUL MELOCHE

$$
\begin{array}{ccc}
\text{CO}_2\text{H} & & \text{CO}_2\text{H} \\
\text{H}-\text{C}-\text{OH} & & \text{C}=\text{O} \\
\text{HO}-\text{C}-\text{H} & \xrightarrow{\text{6-PGt dehydratase}} & \text{H}-\text{C}-\text{H} \\
\text{H}-\text{C}-\text{OH} & & \text{H}-\text{C}-\text{OH} \\
\text{H}-\text{C}-\text{OH} & & \text{H}-\text{C}-\text{OH} \\
\text{H}_2\text{C}-\text{OP} & & \text{H}_2\text{C}-\text{OP}
\end{array}
$$

6-Phosphogluconic dehydratase (EC 4.2.1.12) is an EDTA-sensitive enzyme, found in extracts of glucose-grown *Pseudomonas putida,* that converts 6-phosphogluconate to 2-keto-3-deoxygluconate 6-phosphate (KDPG).[2] Previously, KDPG had been routinely synthesized by the net condensation of pyruvate and D-glyceraldehyde 3-phosphate, and the product was isolated as the barium salt.[3] The major disadvantages of this method are the relative expense of the substrates for the condensation reaction, the separation of condensation product from the residual pyruvate and L-glyceraldehyde 3-phosphate as well as the difficulties of purifying the barium salt coupled to the inconvenience of working with the product in the barium form. These difficulties are circumvented in the procedure reported here in which the KDPG aldolase-pyruvate ketimine in extracts of the organism is reductively trapped, quantitatively inactivat-

[1] This work was supported in part by grants from the NSF, PCM 79-11565, and from the NIH GM, 24926 and RR 05690.
[2] H. P. Meloche and W. A. Wood, *J. Biol. Chem.* **239**, 3505 (1964).
[3] H. P. Meloche and W. A. Wood, this series, Vol. 9, p. 520.

ing the aldolase while leaving adequate dehydrase activity, which then is used to convert relatively inexpensive 6-phosphogluconate to KDPG. KDPG is then isolated as the lithium salt with a good degree of purity.

Reagents

Glucose-grown *Pseudomonas putida*
6-Phosphogluconate, or alternatively glucose 6-phosphate and bromine
Sodium cyanoborohydride (Aldrich Chemical Co.)
Protamine sulfate

Enzyme Preparation

Pseudomonas putida ATCC 12633 is grown as described for the submerged culture production of *P. saccharophila* with the exception that glucose is used in place of galactose. In a typical preparation, 25 g of packed cells are suspended in 50 ml of buffer composed of 50 mM potassium phosphate–0.5 mM manganese chloride, pH 7.5, and disrupted by sonic oscillation. Treatments, repeated as necessary for complete breakage of the cells, are carried out for 3-min intervals in an ice-bath to prevent heating of the extract. The preparation is then centrifuged at 5°. The supernatant contains 670 IU of the dehydratase and 1700 IU of KDPG aldolase. The aldolase is inactivated by treatment at 25° with 10 mM sodium pyruvate followed by 10 mM sodium cyanoborohydride for 30 min. All further steps are carried out at 5°.

With the phosphate–manganese chloride buffer described above, the extract is diluted to ca. 15 mg of protein per milliliter as estimated from the ratio of absorbancies at 280 and 260 nm.[4] The preparation is made 0.1 M by the addition of 13.35 mg of ammonium sulfate per milliliter, followed by 0.2 volume of 2% protamine sulfate pH 5.5. After centrifugation, the supernatant is made 1.5 M by the addition of 211 mg of ammonium sulfate per milliliter, followed by centrifugation. The supernatant from this step is then made 2 M by the further addition of 76.7 mg of ammonium sulfate per milliliter. After centrifugation, the pellet is dissolved in 10 ml of phosphate–manganese chloride buffer and dialyzed overnight against three changes of the same buffer. It should be noted that although manganese tends to precipitate from the buffer as used, its presence is critical to the success of the preceding and following steps from the standpoint of recovering the dehydratase, and this metal requirement cannot be replaced by, e.g., magnesium or other divalent cations. The dialyzed preparation contains 210 IU, or about one-third of the original dehydratase

[4] E. Layne, this series, Vol. 3, p. 447.

activity and no detectable KDPG aldolase activity. The dehydratase preparation is not stable, a characteristic of this enzyme,[2] and should be used for the synthesis of KDPG as soon as possible.

Substrate Preparation

Ten millimoles (3.34 g) of sodium glucose 6-phosphate (G6P) are dissolved in 7 ml of water and passed through a 2×15 cm column of Dowex 50 (hydrogen form). The column is eluted with water, and 5-ml fractions are collected. Acidic fractions, representing G6P, are pooled and adjusted to pH 5.9 using 2 M LiOH. Recovery is quantitative. Twelve millimoles of bromine (0.7 ml) are added with rapid stirring, and the solution is maintained between the pH limits 4.5 to 5.9 using LiOH. When the consumption of bromine is complete, as indicated by the pH stability, excess bromine is removed under reduced pressure and the mixture is concentrated to 50 ml using a rotary evaporator. The pH is adjusted to 7.6 using LiOH, followed by the addition of 10 volumes of absolute ethanol. The resulting precipitate is collected by centrifugation and washed with 100 ml of ethanol to remove lithium bromide and then washed twice with 50-ml portions of acetone. The solid is dried under vacuum and used in subsequent steps without further purification. The yield is 9.20 mmol of 6-phosphogluconate. The product is very hygroscopic and should be stored in a desiccator until used.

Assays Employed

Glucose 6-phosphate was assayed spectrophotometrically using NAD and glucose-6-phosphate dehydrogenase. 6-Phosphogluconate was assayed using NADP and 6-phosphogluconate dehydrogenase. 6-Phosphogluconate dehydratase activity was determined spectrophotometrically using a system coupled to KDPG adolase, lactic dehydrogenase, and NADH. The above assays have been described previously.[2]

2-Keto-3-deoxygluconate 6-Phosphate Synthesis

Lithium 6-phosphogluconate from above (9.20 mmol) is dissolved in 34 ml of water containing 250 μmol of reduced glutathione and 17 μmol of manganese chloride. The pH is adjusted to 7.5, 16 ml of the dehydratase preparation (200 IU) are added, and the sample is incubated at 25°. The progress of the reaction is estimated by assaying perchloric acid-treated samples of KDPG. After 120 min, no further product formation is evident. The reaction mixture is diluted to 200 ml with water, and then passed through a 2.2×16 cm column of Dowex 1 (Cl) and eluted with a linear gradient of 0 to $0.1 \, N$ HCl (total gradient 1 liter). Ten-milliliter fractions are collected. KDPG is found within the last third of the elution profile and is detected enzymically. (It should be noted that KDPG can also be detected using semicarbazide[5].) The pool of fractions containing the product is adjusted to pH 6.9 with LiOH and concentrated to 20 ml under vacuum. This solution is adjusted to pH 7.6 with LiOH, and 200 ml of absolute ethanol are added with rapid mixing. Any lumps are broken up, and the lithium salt is collected by centrifugation. The pellet is washed twice with 200-ml portions of absolute ethanol to remove lithium chloride, followed by two washes with 100-ml portions of acetone, and dried under vacuum. The yield is 2.53 g of material, which assay 90% pure calculated as $Li_3KDPG-2 \, H_2O$ of molecular weight 312. This corresponds to 7.3 mmol of KDPG, or a 79.3% yield. $Li_3KDPGal$ isolated in the same manner was found to be 100% pure calculated as the dihydrate, supporting the view that these lithium salts of sugar acid phosphate esters occur as dihydrates. Total organic phosphorus assay[6,7] corresponded to the KDPG present. In addition, no detectable 6-phosphogluconate, pyruvate, or glyceraldehyde 3-phosphate could be demonstrated in the preparation by appropriate enzymic assays.

[5] J. MacGee and M. Doudoroff, *J. Biol. Chem.* **210**, 617 (1954).
[6] L. F. Leloir and C. E. Cardini, this series, Vol. 3, p. 840.
[7] C. H. Fiske and Y. SubbaRow, *J. Biol. Chem.* **66**, 375 (1925).

[17] Chemical Synthesis of Fructose 2,6-Bisphosphate

By S. J. PILKIS, M. R. EL-MAGHRABI, D. A. CUMMING, J. PILKIS, and T. H. CLAUS

Principle. Fructose 2,6-bisphosphate is an allosteric activator of 6-phosphofructo-1-kinase and an inhibitor of fructose 1,6-bisphosphatase,[1-5] which was discovered in the course of studies on the regulation of hepato-

[1] E. Van Schaftingen and H. G. Hers, *Biochem. Biophys. Res. Commun.* **96**, 1524 (1980).
[2] T. H. Claus, J. Schlumpf, J. Pilkis, R. A. Johnson, and S. J. Pilkis, *Biochem. Biophys. Res. Commun.* **98**, 359 (1981).
[3] E. Van Schaftingen, L. Hue, and H. G. Hers, *Biochem. J.* **192**, 897 (1980).
[4] S. J. Pilkis, M. R. El-Maghrabi, J. Pilkis, T. H. Claus, and D. A. Cumming, *J. Biol. Chem.* **256**, 3171 (1981).
[5] S. J. Pilkis, M. R. El-Maghrabi, J. Pilkis, and T. H. Claus, *J. Biol. Chem.* **256**, 3619 (1981).

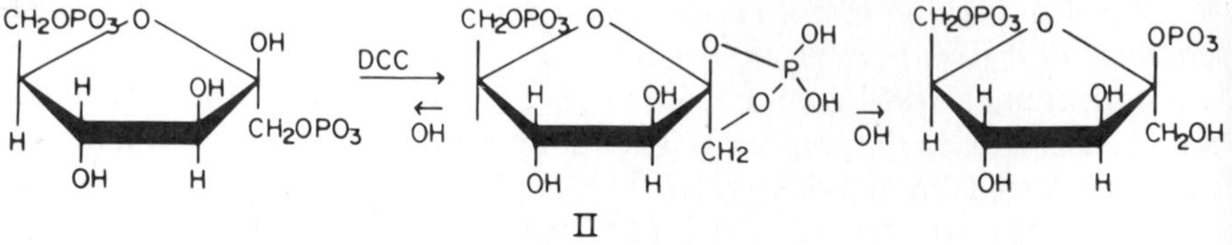

F 1,6-P_2 F 1,2 cyclic, 6-P_2 F 2,6-P_2

SCHEME 1

cyte 6-phosphofructo-1-kinase activity by glucagon.[6–10] The method of chemical synthesis of fructose 2,6-bisphosphate given below is based on the synthesis of fructose 2-phosphate first demonstrated by Pontis and Fischer.[11] They reported that treatment of fructose 1-phosphate with dicyclohexylcarbodiimide (DCC) in aqueous pyridine led to the formation of fructose cyclic 1,2-phosphate. Alkaline hydrolysis of that compound yielded fructose 2-phosphate as one product and fructose 1-phosphate as the other. We used this method to prepare fructose 2,6-bisphosphate except that we started with fructose 1,6-bisphosphate.[4] More recently, two other groups have reported the synthesis of fructose 2,6-bisphosphate by this procedure.[12,13] The reaction is given in Scheme 1.

Base-catalyzed ring opening of fructose 1,2-cyclic 6-bisphosphate (II) yields both fructose 1,6-bisphosphate and fructose 2,6-bisphosphate. Fructose 1,6-bisphosphate is destroyed by heating in alkali whereas the fructose 2,6-bisphosphate is stable to this treatment and can subsequently be purified by anion-exchange chromatography.[4]

Materials

D-fructose 1,6-bisphosphate (sodium salt) (Sigma)
Dicyclohexylcarbodiimide (Aldrich Chemical Company)
Pyridine (Aldrich), redistilled before use
Triethylamine (TEA) (Aldrich), redistilled before use

[6] S. J. Pilkis, J. Schlumpf, J. Pilkis, and T. H. Claus, *Biochem. Biophys. Res. Commun.* **88**, 3619 (1979).
[7] E. Furuya and K. Uyeda, *Proc. Natl. Acad. Sci. U.S.A.* **77**, 5871 (1980).
[8] T. H. Claus, J. Schlumpf, M. R. El-Maghrabi, J. Pilkis, and S. J. Pilkis, *Proc. Natl. Acad. Sci. U.S.A.* **77**, 6501 (1980).
[9] E. Van Schaftingen, L. Hue, and H. G. Hers, *Biochem. J.* **192**, 902 (1980).
[10] C. S. Richard and K. Uyeda, *Biochem. Biophys. Res. Commun.* **97**, 1535 (1980).
[11] H. G. Pontis and C. L. Fischer, *Biochem. J.* **89**, 452 (1963).
[12] K. Uyeda, E. Furuya, and A. D. Sherry, *J. Biol. Chem.* **256**, 8679 (1981).
[13] E. Van Schaftingen and H. G. Hers, *Eur. J. Biochem.* **117**, 319 (1981).

Assay of Phosphofructokinase and Its Activator. 6-Phosphofructo-1-kinase activity and activation of 6-phosphofructo-1-kinase by fructose 2,6-bisphosphate are determined with the aldolase-coupled assay.[2] Purified rat liver 6-phosphofructo-1-kinase is used in the activation assay.[4]

Carbohydrate Analysis. Carbohydrate is determined colorimetrically by the use of phenol in sulfuric acid.[4]

Reaction of D-Fructose 1,6-Bisphosphate with Dicyclohexylcarbodiimide

The sodium salt of D-fructose 1,6-bisphosphate is converted to the pyridinium salt by passage through a Dowex 50 (pyridinium form) column. The pyridinium salt (1.2 mmol) is then dissolved in 10 ml of pyridine containing 2 ml of H_2O and heated at 30° for 5 min. A similarly heated solution of DCC (12.0 mmol) and triethylamine (3.6 mmol) in 20 ml of pyridine is then added. The amount of reactants can be increased as desired. After 12 hr of incubation at 30° with constant shaking, the reaction mixture is diluted 10-fold with cold distilled H_2O, filtered to remove urea, and extracted with ethyl ether to remove excess DCC. The aqueous layer is evaporated to dryness under vacuum, dissolved in 10 mM triethylamine (TEA)–bicarbonate buffer, pH 8.4, and applied to a DEAE-Sephadex column (2.5 × 40 cm) equilibrated in the same buffer. The column is eluted with a linear gradient of 10 to 600 mM TEA-HCO$_3$ (total volume 1 liter). Two major peaks of reducing sugar are obtained: the first elutes with 300 mM TEA-HCO$_3$ and the second, which contains unreacted fructose 1,6-bisphosphate, elutes with 375 mM TEA-HCO$_3$ and is discarded. The material in the first peak does not activate phosphofructokinase unless aliquots of the material are first incubated with 1 N NaOH for 10 min at 37° in order to open cyclic phosphate ring structures. The material in this first peak is further purified prior to alkaline hydrolysis by rechromatography on DEAE-Sephadex (Fig. 1). Analysis of this material by ^{13}C NMR indicates that the compound is fructose 1,2-cyclic 6-bisphosphate (Fig. 2). The yield of fructose 1,2-cyclic 6-bisphosphate is 30–50%.

Preparation of Fructose 2,6-Bisphosphate by Alkaline Hydrolysis

A solution of 0.55 mmol of (II) is evaporated to dryness under reduced pressure and the residue is washed with H_2O three times to remove the TEA-HCO$_3$. The residue is dissolved in 5 ml of 1 N NaOH and kept at 37° for 15 min, diluted 4-fold with distilled water, and heated at 90° for 30 min in order to destroy fructose 1,6-bisphosphate that arises from the base-catalyzed ring opening. The ring opening is complete after 15 min, and the yield of fructose 2,6-bisphosphate from fructose 1,2-cyclic 6-bisphosphate

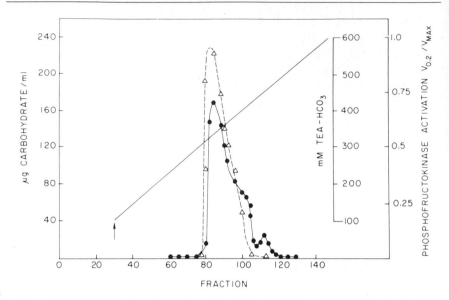

FIG. 1. DEAE-Sephadex chromatography of fructose 1,2-cyclic 6-bisphosphate. After separation of the products of the dicyclohexylcarbodiimide reaction with fructose 1,6-biphosphate on a DEAE-Sephadex column, the fractions eluting at 300 mM triethylamine (TEA)-HCO$_3$ and exhibiting ability to activate 6-phosphofructo-1-kinase after base-catalyzed ring opening were evaporated to dryness under reduced pressure. The residue was taken up in 10 ml of 100 mM TEA-HCO$_3$, pH 8.4, and applied to another DEAE-Sephadex column (0.9 × 50 cm) equilibrated with the same buffer. Elution was with a linear gradient of 100 to 600 mM TEA-HCO$_3$ (total volume 300 ml). The 2-ml fractions were collected and assayed for carbohydrate (●) and activating activity (△).

is 20%. The pH is then adjusted to 8.4 with acetic acid, and the sample is applied to a DEAE-Sephadex column. The product that activates 6-phosphofructo-1-kinase elutes at 375 mM TEA-HCO$_3$ and coincides with the major peak of carbohydrate (Fig. 3).[4] These fractions are pooled and evaporated to dryness; the residue is washed three times with H$_2$O. The compound can then be converted to the sodium salt by passing the solution through a Dowex 50 (Na$^+$ form) column.

Properties of the Product

This synthetic compound was identified as fructose 2,6-bisphosphate by ^{13}C NMR (Fig. 4).[4] By comparison of the chemical shifts obtained with those of Seymour et al.,[14] it was concluded that this compound is the β

[14] F. R. Seymour, R. D. Knapp, J. E. Zwerg, and S. H. Bishop, Carbohydr. Res. 72, 57 (1979).

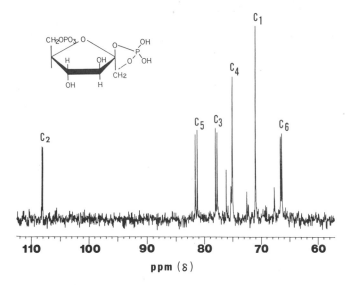

FIG. 2. Partial 22.5 mHz ^{13}C-NMR spectrum of fructose 1,2 cyclic 6-bisphosphate. The NMR conditions were as follows: 30 mg of the sample were dissolved in 0.45 ml of D_2O. The spectrum was obtained on a JEOL FX-90Q Fourier-transform NMR spectrometer. Data points (16,384) were accumulated over a 5000 Hz spectral width with a 5 μsec (45°) pulse width. The sample was completely proton decoupled, and a field-frequency lock was used. The acquisition time was 1.638 sec, and this spectrum is the result of 26,000 accumulations. Prior to Fourier transformation, a decreasing exponential apodization function was applied to the accrued free-induction decays. Zero-filling (16,000) was used during transformation. All resonances are measured downfield from an internal 2,2-dimethyl-2-silapentanesulfonic acid (Na^+ salt) set at -1.1 ppm. Resonance assignments were made according to Seymour et al.[14] The magnitude of the respective coupling constants (J) were as follows: C_2, $^2J_{31P-13C} = 3.1$ Hz; C_5, $^3J_{31P-13C} = 7.6$ Hz; C_3, $^3J_{31P-13C} = 6.7$ Hz; C_6, $^2J_{31P-13C} = 4.9$ Hz. Neither the C_4 nor the C_1 resonances showed any splitting; off-resonance ^{13}C-NMR experiments did confirm their relative assignments.

anomer.[4] The synthetic product was an inhibitor of fructose-1,6-bisphosphatase and activator of 6-phosphofructo-1-kinase with a K_i of 0.2 μM and a K_A of 0.05–0.1 μM for the respective rat liver enzyme.[4,5]

Comments

The overall yield of this method is limited primarily by the base-catalyzed ring opening reaction, but the key to purification is heating in alkali. This treatment results in complete hydrolysis and degradation of all fructose 1,6-bisphosphate as well as fructose 6-phosphate. The degradation products can then be separated from fructose 2,6-bisphosphate by

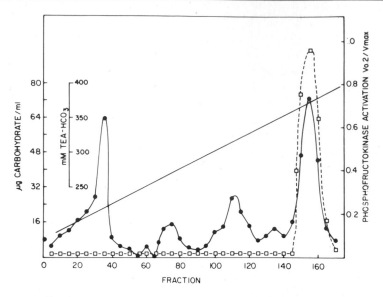

FIG. 3. Purification of fructose 2,6-bisphosphate by DEAE-Sephadex chromatography. After heat treatment of the products derived from alkaline hydrolysis of fructose 1,2 cyclic 6-bisphosphate, the pH was readjusted to 8.4 and the sample was applied to a DEAE-Sephadex column (0.7 × 16 cm) equilibrated with 10 mM triethylamine (TEA)-HCO$_3$, pH 8.4. Elution was with a linear gradient of 10 to 500 mM TEA-HCO$_3$ (total volume, 80 ml). The 0.05-ml fractions were collected and assayed for carbohydrate (●) and activating activity (□).

anion-exchange chromatography. Fructose 2,6-bisphosphate is stable to heating in alkali because it contains a phosphate group on the hemiketalic hydroxyl group at C-2.[4] The compound is acid labile. Incubation at pH 3 for 30 min at 22° produces equal amounts of fructose 6-phosphate and inorganic phosphate (see the table), and destroys its ability to activate 6-phosphofructo-1-kinase and to inhibit fructose 1,6-bisphosphatase. By measuring the amount of fructose 6-phosphate produced after mild acid

FIG. 4. The partial 22.5 mHz ^{13}C NMR spectrum of fructose 2,6-bisphosphate. The coupling constants, J, obtained for coupled resonance were: C$_2$, $^2J_{^{31}P-^{13}C}$ = 6.7 Hz; C$_3$, $^3J_{^{31}P-^{13}C}$ = 4.3 Hz; C$_5$, $^3J_{^{31}P-^{13}C}$ = 7.3 Hz; C$_6$, $J_{^{31}P-^{13}C}$ = 3.0 Hz. Resonance assignments were made according to Seymour *et al.*[14] The identification of this compound as fructose 2,6-bisphosphate was based on the particular resonances split, the magnitude of their coupling constants, and the absence of any splitting in the C$_4$ resonance.[4] Magnification of the C$_1$ resonance indicated the presence of slight splitting of the resonance with $^3J_{^{31}P-^{13}C}$ = 1.5 Hz. Such a small coupling constant is not inconsistent with the hexofuranosyl structure of this compound. The spectrum was obtained after 16,000 scans of 20 mg of fructose 2,6-bisphosphate. Spectroscopic conditions are those listed in the legend to Fig. 2.

ANALYSIS OF D-FRUCTOSE 2,6-BISPHOSPHATE[a]

	nmol/ml
Total phosphate	35.2
Acid-labile phosphate (pH 3.0, 30 min)	17.2
Acid-labile phosphate (1 N HCl, 100°, 30 min)	36.2
Carbohydrate content	17.3
Fructose 6-phosphate content after acid hydrolysis	18.0

[a] D-Fructose 2,6-bisphosphate, prepared by base-çata-lyzed ring opening of D-fructose 1,2-cyclic 6-bisphos-phate, was purified by anion-exchange chromatography and assayed for phosphate and carbohydrate content.

treatment, it is possible to determine the amount of fructose 2,6-bisphosphate originally present in the sample.[4]

Van Schaftingen and Hers[1] have described a method for preparing fructose 2,6-bisphosphate that involves mixing fructose-6-phosphate with phosphoric acid. This method appears to generate other phosphate esters besides fructose 1,6- and fructose 2,6-bisphosphate, and this makes it difficult to purify the fructose 2,6-bisphosphate.

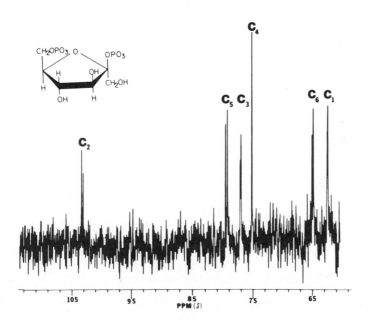

[18] Synthesis of Ribulose 1,5-Bisphosphate: Routes from Glucose 6-Phosphate (via 6-Phosphogluconate) and from Adenosine Monophosphate (via Ribose 5-Phosphate)[1]

By Chi-Huey Wong, Alfred Pollak, Stephen D. McCurry, Julia M. Sue, Jeremy R. Knowles, and George M. Whitesides

This procedure describes two routes to ribulose 1,5-bisphosphate (RuBP) (Fig. 1).[2] These routes provide procedures suitable for preparing several hundred grams of RuBP and of the major intermediates lying between starting materials and this metabolite (6-phosphogluconate, 6-PG; ribulose 5-phosphate, Ru-5-P; ribose 5-phosphate, R-5-P). The enzymes required are all used in immobilized form. These immobilized preparations have high stability, and the enzymes can be recovered conveniently and in good yields at the end of the reactions and re-used. Both routes require ATP cofactor recycling, and the procedure includes details of an improved synthesis of the acetyl phosphate used as the ultimate phosphate donor for this recycling. The route based on glucose 6-phosphate (G-6-P) also illustrates a convenient procedure for regeneration of NAD(P) from NAD(P)H under anaerobic conditions, based on conversion of α-ketoglutarate to glutamate.

The major technical problem in the preparation of RuBP is that of obtaining a product substantially free of xylulose 1,5-bisphosphate (XuBP). This analog is a powerful inhibitor[3] of ribulosebisphosphate carboxylase (RuBPC) ($K_{m,RuBP} = 20~\mu m$,[4] $K_{i,XuBP} = 3~\mu m$[5]). The RuBP isomerizes rapidly and spontaneously to XuBP ($\tau_{1/2} = 48$ hr at pH 8.0, 30°). Contamination of RuBP by 0.1% of XuBP results in significant inhibition of the enzyme. It is thus important to minimize exposure of the RuBP to isomerizing conditions during and after preparation. In particular,

[1] Supported by the National Institutes of Health, Grant GM 26543.

[2] C.-H. Wong, S. D. McCurry, and G. M. Whitesides, *J. Am. Chem. Soc.* **102**, 7938 (1980).

[3] S. D. McCurry, J. Pierce, N. E. Tolbert, and W. H. Orme-Johnson, *J. Biol. Chem.* **256**, 6623 (1981). Active RuBPC is a ternary complex (ECM) composed of enzyme (E), CO_2 (C), and Mg^{2+} (M). This complex is in equilibrium with an inactive form of the enzyme, and the interconversion of these forms is slow. XuBP combines very tightly with the inactive enzyme and acts as a noncompetitive inhibitor. It also binds less tightly to the ECM form of the enzyme as a competitive inhibitor.

[4] F. Dailey and R. S. Criddle, *Arch. Biochem. Biophys.* **201**, 594 (1980).

[5] J. M. Sue and J. R. Knowles, unpublished observation.

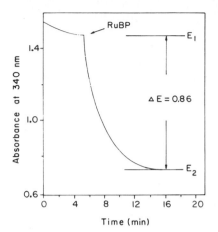

FIG. 1. A typical spectrophotometric trace obtained during analysis of ribulose 1,5-bisphosphate (RuBP) by the coupled-enzyme method. Methods are described in the procedure.

RuBP should not be exposed to pH > 8.[6] It also isomerizes as its solid barium salt if stored at room temperature; at $-20°$ it is stable for months. This procedure includes two procedures for removing small quantities of XuBP from RuBP.

The procedures for the preparation of RuBP summarized in Scheme 1 are the best available when large quantities (> 10 g) are required. Their ability to produce large quantities in a practical and economical procedure rests on the availability of good cofactor recycling schemes. The requirement for cofactor recycling necessarily makes these procedures more complex than smaller-scale procedures that do not involve cofactor recycling.[7-9] Of the two routes in Scheme 1, that starting from AMP is the more direct and more convenient. The route from G-6-P is, however, not much more complex. Which route is preferable in a particular laboratory depends upon the relative availability of the starting materials and relative desirability of the intermediates generated in each.

[6] C. Paech, J. Pierce, S. D. McCurry, and N. E. Tolbert, *Biochem. Biophys. Res. Commun.* **83,** 1084 (1978). The rate of degradation of RuBP at pH 8.3 and 30° was 1.25% per hour. We determined the stability of RuBP (0.5 mM, in 0.1 M Tris, pH 8.0, 30°) in solution by measuring the concentration of RuBP with the coupled enzymic method. The observed half-life for RuBP was 48 hr.

[7] G. D. Kuehn and T. C. Hsu, *Biochem. J.* **175,** 909 (1978).

[8] B. L. Horecker, J. Hurwitz, and A. Weissbach, *Biochem. Prep.* **6,** 83 (1958).

[9] J. Pierce, S. D. McCurry, R. M. Mulligan, and N. E. Tolbert, this volume [9].

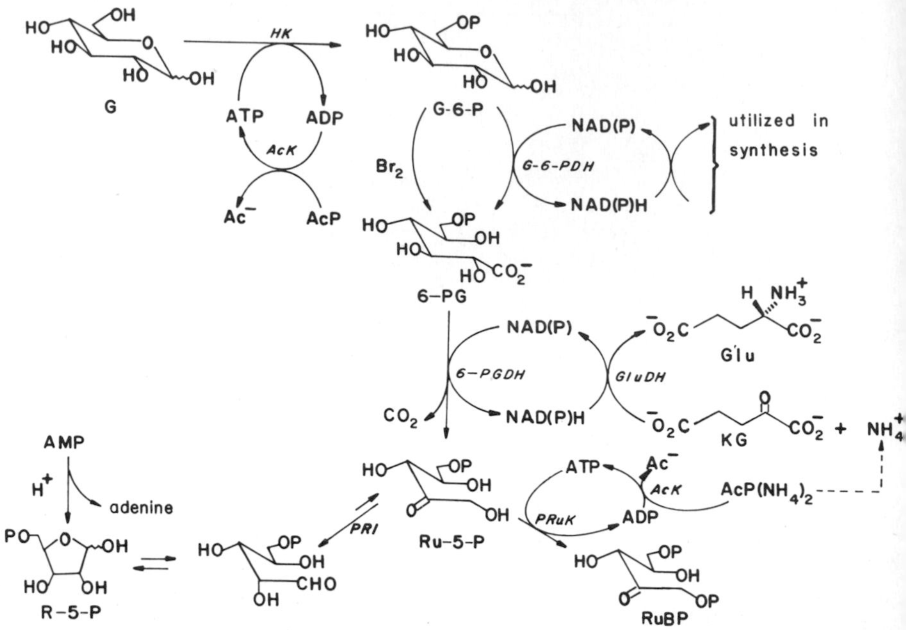

SCHEME 1. Enzymic syntheses of ribulose 1,5-bisphosphate (RuBP). Abbreviations: G-6-P, glucose 6-phosphate; 6-PG, 6-phosphogluconate; Ru-5-P, ribulose 5-phosphate; AcP, acetyl phosphate; Ac; acetate; 6-PGDH, 6-PG dehydrogenase; GluDH, glutamic dehydrogenase; PRuK, phosphoribulokinase; PRI, phosphoriboisomerase; AcK, acetate kinase.

Materials

The polymer used in the immobilizations, PAN-1000, is prepared using slight modifications of the procedure described previously.[10] A 1-liter round-bottom flask equipped with a Teflon-coated magnetic stirring bar is charged with acrylamide (55 g, 0.77 mol), N-acryloxysuccinimide (6 g, 35.6 mmol), azobis (isobutyronitrile) (0.35 g, 2.2 mmol), and 500 ml of tetrahydrofuran (THF) (AR grade, distilled from CaH_2). The solution is deoxygenated with argon for 30 min with vigorous stirring. The flask is capped with a Non-Air stopper and maintained in a constant-temperature water bath at 50° for 24 hr. Occasional relief of pressure inside the flask is necessary during the beginning stages of the reaction. The precipitated white polymer is separated by filtration on a large Büchner funnel (1-liter capacity). The polymer is washed on the funnel with four 200-ml portions of dry THF, transferred to a vacuum desiccator, and dried under vacuum

[10] A. Pollak, H. Blumenfeld, M. Wax, R. L. Baughn, and G. M. Whitesides, J. Am. Chem. Soc. 102, 6324 (1980).

(0.02 torr) at room temperature for 24 hr. The THF lost during the drying is trapped using a condenser kept at liquid nitrogen temperature to protect the pump. The white product (60 g, 99%) contains 912 μmol of active ester per gram.

Diammonium Acetyl Phosphate. This phosphate is prepared by reaction of acetic anhydride and phosphoric acid using a modification of a published procedure[11] that was designed to give better control over the concentration of ammonia in the methanol used in workup.

Ethyl acetate (AR grade 2000 mL) and 100% phosphoric acid (150 g, 1.5 mol, prepared by dissolving phosphorus pentoxide in 85% phosphoric acid, and stirring at ambient temperature for 2 hr) are transferred into a 3-liter, three-necked flask, equipped with a strong mechanical stirrer, a dropping funnel (500 ml), and a drying tube (Drierite) for ventilation. The flask is cooled in an ice bath (20 × 20 inches), and when the temperature of the reaction mixture had reached 0° (in about 30 min), the funnel is charged with cold (ca. 0°) acetic anhydride (275 g, 2.7 mol), which is then added dropwise (over 20 min) to the ethyl acetate–phosphoric acid mixture, with vigorous stirring. The resulting solution is stirred further at 0° for a total of 3.75 hr. *This interval is critical.* Major deviations (± 0.5 hr) result in significant decreases in yields.

A 5-liter, three-necked flask was placed in a cooling bath (25 × 25 inches, containing 2 gal of isopropanol and 1 gal of acetone). Pieces of Dry Ice are added to this bath at a rate that keeps the bath temperature at − 35 to − 40°). The reaction flask is fitted with a low-temperature thermometer, a drying (KOH) gas outlet tube, and a very strong overhead stirrer. Methanol (AR grade, 2000 ml) is added into the flask and stirred until the temperature reaches about − 30°; a rapid stream of gaseous ammonia (Matheson) is bubbled into the well-stirred methanol until its concentration reaches approximately 8 N as determined by titration. (To titrate, 1 ml of the solution is diluted with water to 25 ml and titrated with 1 N HCl to pH 6.0). The addition of ammonia is then stopped, and the gas inlet tube is replaced with a 3-liter funnel. The cold ethyl acetate–acetic anhydride–phosphoric acid mixture is quickly transferred to the funnei, from which it is dropped into the vigorously(!) stirred methanol–ammonia solution. This addition takes approximately 30–35 min, during, which time a fine, white suspension appears and the temperature rises to about − 10°. The white, precipitated material is quickly separated by filtration on a large Büchner funnel, washed with 1 liter of cold (0°) methanol, 1 liter of cold, anhydrous ether, and finally with 1 liter of cold, dry hexane (a mixture of hexanes). The product is immediately transferred into a large vacuum desiccator,

[11] J. M. Lewis, S. L. Haynie, and G. M. Whitesides, *J. Org. Chem.* **44,** 864 (1979).

which is attached to a good vacuum (ca. 2 torr). After about 6 hr of drying, 255 g of a white, fluffy product are obtained. An enzymic assay showed that this solid contains 89% diammonium acetyl phosphate (1.3 mol) by weight, corresponding to an 87% yield based on phosphoric acid.

Diammonium acetyl phosphate must be stored at $-15°$ in tightly stoppered containers to prevent its decomposition. If properly stored, it will decompose less than 5% per month.

Enzymes

The following enzymes are obtained from Sigma: yeast alcohol dehydrogenase (ADH, EC 1.1.1.1); aldehyde dehydrogenase (AldDH, EC 1.2.1.3); acetate kinase (AcK, EC 2.7.2.1); glucose-6-phosphate dehydrogenase (G-6-PDH, EC 1.1.1.49); 6-phosphogluconate dehydrogenase (6-PGDH, EC 1.1.1.44); α-glycerol-3-phosphate dehydrogenase (GDH, EC 1.1.1.8); glyceraldehyde-3-phosphate dehydrogenase (GAPDH, EC 1.2.1.12); 3-phosphoglycerate kinase (PGK, EC 2.7.2.3); glutamic dehydrogenase (GluDH, EC 1.4.1.3); hexokinase (HK, 2.7.1.1); fructose-1,6-bisphosphate aldolase (FDPA, EC 4.1.2.13); ribose-5-phosphate isomerase (PRI, EC 5.3.1.6); phosphoribulokinase (PRuK, EC 2.7.1.19); ribulose 1,5-bisphosphate carboxylase (RuBPC, EC 4.1.1.39); triosephosphate isomerase (TRI, EC 5.3.1.1); transaldolase (TA, EC 2.2.1.2). Ribulose-1,5-bisphosphate carboxylase (RuBPC) is isolated from spinach leaves[12,13] instead of using the commercially available source, which might contain PRuK.

For those enzymes suspended in ammonium sulfate, ammonium sulfate is removed by centrifuging the suspension at 15,000 g for 5 min (4°), and the precipitate is used directly in immobilizations. PRuK (Sigma, 250 U/313 mg) is dissolved in 10 ml of deoxygenated Tris buffer (50 mM; pH 7.5, containing 3 mM DTT) and dialyzed against 1 liter of the same buffer for 6 hr to remove impurities and the ammonium salts ($\sim$50% w/w), which would react with the active ester groups of PAN.

Glucose 6-Phosphate: Hexokinase-Catalyzed Phosphorylation of Glucose

Glucose 6-phosphate is prepared using HK-catalyzed phosphorylation and ATP regeneration.[14,15] A 3-liter solution containing glucose (1.4 mol), ATP 10 mmol), $MgCl_2$ (98 mmol), EDTA (4.8 mmol), and 1,3-

[12] F. J. Ryan and N. E. Tolbert, *J. Biol. Chem.* **250**, 4229 (1975).
[13] S. D. McCurry, R. Gee, and N. E. Tolbert, Vol. 90 [82].
[14] C.-H. Wong and G. M. Whitesides, *J. Am. Chem. Soc.* **103**, 4890 (1981).
[15] A. Pollak, R. L. Baughn, and G. M. Whitesides, *J. Am. Chem. Soc.* **99**, 2366 (1977).

dimercapto-2-propanol (18 mmol) is deoxygenated and maintained under argon. Immobilized hepokinase (1200 units, 6 ml of gel) and AcK (1200 units, 10 ml of gel) are added to this solution. Diammonium acetyl phosphate (1.4 mol, 90% purity) is added to the stirred reaction solution in 10 portions over 60 hr, and the solution is maintained at pH 7.5 by addition of 4 M KOH solution using a pH controller. The reaction is performed at 25°. After 3 days, 1.2 mol of G-6-P has been formed (0.364 M in 3.3 liters, yield 85%). The solution is separated from the polyacrylamide gels by decantation. Barium chloride (0.2 mol) is added to precipitate inorganic phosphate; this material is removed by filtration. Another portion of BaCl$_2$ (1.3 mol, 271 g) is added, followed by addition of ethanol (1.5 liters). The precipitated solid (677 g) contains 92% Ba G-6-P · 7 H$_2$O (1.2 mol). The turnover number (TN = moles of product per mole of enzyme or cofactor) for ATP during the synthesis is 140, and the recovered enzyme activities are: HK, 92%, AcK, 80%.

Enzyme Immobilization

Enzymes are immobilized in PAN gel using the general procedure described previously.[10] Each enzyme immobilization is carried out in the presence of substrates or products intended to occupy the active site and protect it against modification during immobilization. The concentration of substrates used to protect the enzymes, and the immobilization yields,[16] are PRI (R-5-P 200 mM and Ru-5-P 2 mM, 36%), PRuK (Ru-5-P 15 mM and ATP 15 mM, 30%), AcK (AcP 12.5 mM, and ADP 20 mM, 34%), 6-PGDH (6-PG 2.0 mM and NADP 0.5 mM, 38%), GluDH (2-ketoglutarate 15 mM, NADPH 0.5 mM, and NH$_4$Cl 20 mM, 40%).

A typical example of immobilization procedure of 10 mg of enzyme follows. PAN-1000 (1 g) was placed in a 30-ml beaker containing a stirring bar. HEPES buffer (4 ml, 0.3 M HEPES buffer) containing the concentrations given of active site-protective reagents was added. The mixture was brought into solution within 1 min by vigorous stirring. Aqueous solutions of dithiothreitol (DTT) (50 μl, 0.5 M) and triethylenetetramine (TET, 0.6 ml, 0.5 M) were added; 10 sec later, 1 ml of the enzyme solution in HEPES buffer was added. The mixture gelled within 3 min. The gel was allowed to stand under argon at room temperature for 1 hr to complete the coupling reaction. The gel was ground into fine particles with a pestle in a mortar for 2 min; 25 ml of deoxygenated HEPES buffer (50 mM, pH 7.5, containing 3 mM DTT and 50 mM (NH$_4$)$_2$SO$_4$) were added, and grinding was

---

[16] The yield obtained in the immobilization is defined by the equation

$$\text{Yield } (\%) = \frac{\text{units assayed in enzyme-containing gel}}{\text{units assayed in solution before immobilization}}$$

continued for an additional 2 min. The mixture was diluted with the same buffer solution (50 ml), stirred for 15 min to destroy the unreacted active esters, and separated by gentle centrifugation (3000 rpm). The washing procedure was repeated once with the same buffer containing no ammonium salt. The gel particles were then resuspended in the same volume of buffer for assay.

Enzyme Assays

Enzyme assays are carried out spectrophotometically following the standard procedures of Bergmeyer.[17] For the immobilized enzymes, aliquots of the solutions containing the suspended gel particles are taken by unexceptional procedures using volumetric pipettes: the gel particles have a density very similar to that of the solution, and the suspensions can be sampled in the same way as solutions provided that the particles are small and that the suspensions are shaken or stirred well immediately before sampling. During spectrophotometric assays, the cuvettes containing the gel suspensions are stoppered with Parafilm and shaken periodically for a few seconds to keep the suspensions well mixed. Almost linear assay reponses are obtained if the cuvettes are shaken 15–20 times over the course of assays lasting 1–5 min.

Assays for RuBP

Two methods are employed for RuBP determination. One is a coupled-enzyme method[18]; the other one is radiometric.[19,20] Scheme 2 shows the reactions involved in the coupled enzyme method. As shown in the reactions, one equivalent of RuBP reacts with four equivalents of NADH. Since the assay includes five enzymes, the possibility that impurities in one or several of them might influence the results is relatively large. In particular, we have found that RuBP carboxylase (RuBPC) must be prepared[19]; the commercially available sources of this enzyme seem to contain PRuK and PRI, both of which contribute to incorrect results.

The radiometric method involves only the first reaction in the coupled-enzyme analysis. The quantity of ^{14}C fixed in 3-phosphoglycerate is counted.

$$RuBP + {}^{14}CO_2 + H_2O \xrightarrow{\text{RuBPC}} D\text{-}[{}^{14}C]glycerate\ 3\text{-}P + D\text{-}glycerate\ 3\text{-}P$$

[17] H.-U. Bergmeyer, "Methods of Enzymatic Analysis," 2nd English ed., Vols. 1–4, Academic Press, New York, 1974.
[18] E. Racker, this series, Vol. 5, p. 266.
[19] J. Pierce, S. D. McCurry, R. M. Mulligan, and N. E. Tolbert, this volume [9].
[20] G. H. Lorimer, M. R. Badger, and T. J. Andrews, *Biochemistry* 15, 529 (1976).

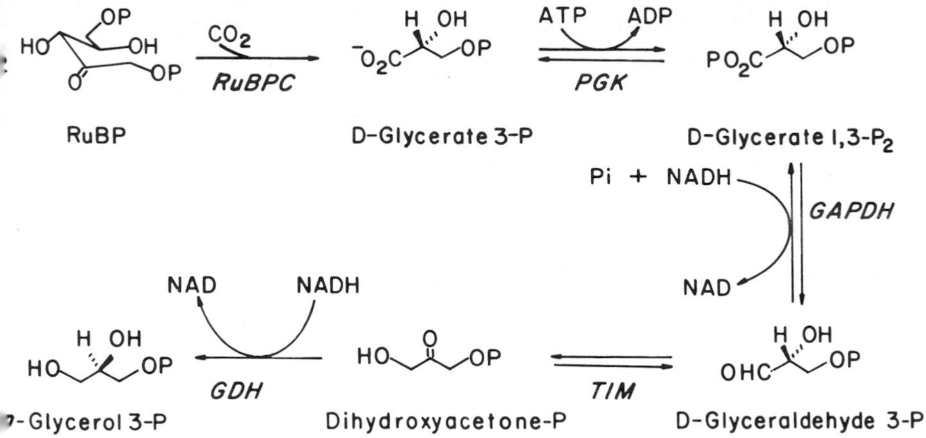

SCHEME 2

Solutions for the Coupled-Enzyme Assay

1. Tris-HCl buffer, 1.5 M, pH 8.0
2. Sodium bicarbonate, 0.5 M
3. Dithiothreitol (DTT), 0.5 M
4. Adenosine triphosphate, 0.1 M

Magnesium dichloride, 0.1 M

Triosephosphate isomerase–glycerolphosphate dehydrogenase (TIM–GDH), each 250 U/0.5 ml in 0.3 M $(NH_4)_2SO_4$

NADH-Na$_2$, 4 mM (7 mg/2 ml of H_2O)

Glyceraldehyde-3-phosphate dehydrogenase–phosphoglycerate kinase (GAPDH–PGK), each 250 units/ml in 0.3 M $(NH_4)_2SO_4$

RuBP, ~10 mM: dissolve 58.3 mg of Ba$_2$RuBP in 10 ml of 0.1 N HCl

RuBP carboxylase (70 U/ml) in 0.1 M Bicine buffer, pH 8, containing 10 mM NaHCO$_3$, 8 mM DTT, and 10 mM MgCl$_2$ (the buffer solution is deoxygenated with argon first), incubated at 30° for 30 min

The assay system is assembled from these solutions in a 3-ml cuvette: Tris (1.9 ml), NaHCO$_3$ (0.4 ml), DTT (30 μl), ATP (0.2 ml), MgCl$_2$ (0.2 ml), TIM–GDH (10 μl), GAPDH–PGK (20 μl), NADH (0.15 ml), and the RuBP sample (10 μl, obtained using a syringe).[21] Before adding RuBPC, the solution is mixed and read at 340 nm until no further change in absorbance occurs (Fig. 1, E_1, ca. 5 min). After adding RuBPC (30 units), the decrease of absorbance at 340 nm should be complete within 10 min (E_2). The content of RuBP in the sample (% w/w) is determined from the

---

[21] A. 10-μl Hamilton syringe ($\pm$0.1 μl) was used instead of a pipette to achieve an accurate volumetric transfer: a 1-μl error in sampling results in a 12% relative error in the result.

difference of absorbance ($\Delta E = E_1 - E_2$) and calculated according to the equation

$$\%Ba_2RuBP = \frac{\Delta E}{4\epsilon} \times \frac{V_1}{V_2} \times \frac{V_3}{1000W} \frac{MW}{} \times 100$$

where $\epsilon = 6.22$ mM^{-1} cm^{-1} (absorbance coefficient of NADH at 340 nm), V_1 = total volume of the assay solution (2.95 ml), V_2 = the volume of the RuBP sample solution taken (0.01 ml), V_3 = the volume of the original RuBP solution (10 ml), MW = the molecular weight of Ba$_2$RuBP (583.3), and W = the weight of Ba$_2$RuBp sample in V_3 (58.3 mg). Figure 1 is a typical curve obtained in this assay.

Radiometric Method. For the radiometric assay, aliquots from the RuBP stock solution are equilibrated in the assay buffer (0.1 M Bicine, pH 8.0) containing NaH^{14}CO$_3$ (20 mM, 0.16 Ci/mol), MgCl$_2$ (10 mM), and Na$_2$-EDTA (0.2 mM). The final concentration of RuBPC, determined by the Lowry method,[22] is 80 μg/ml (2 units/mg). The reaction is initiated by addition of RuBP (the total volume of solution is 7.5 ml) and allowed to run for 1 hr. It is then stopped by addition of acid, and the fixed ^{14}C is counted as described elsewhere.[20] The details of the procedure may be varied. The important feature is to keep the specific activity of the NaH^{14}CO$_3$ constant during the assay and to measure the amount of substrate added accurately. Each sample is assayed in triplicate at the estimated concentrations of 0.2, 0.4, and 0.9 mM. If the substrate is reasonably pure, the concentration as determined by either of the above assay methods should agree very well with determination of the concentration by purely chemical methods, e.g., by the orcinol test[23] and by organic phosphate determination.[24]

6-Phosphogluconate by Br_2 Oxidation of Glucose 6-Phosphate. To a stirred suspension of BaG-6-P · 7 H$_2$O [120 g (230 mmol) in H$_2$O, 912 ml], 24 ml of concentrated HCl are added. Sodium sulfate (34.5 g, 240 mmol) is added, and the resulting solution is stirred for about 5 min. The precipitate is removed by centrifugation, and the supernatant is adjusted to pH 5.4 with 4 N NaOH. Bromine (Br$_2$, 30 ml) is added. The solution is stirred for 30 min at 25°, and the pH is controlled at 5.0–6.0. The excess Br$_2$ is then removed by passing a stream of argon through the solution for 1 hr, and the resulting solution (containing ca. 200 mmol of 6-PG) is used without further purification.

6-Phosphogluconate can also be prepared by enzymic oxidation of glucose 6-phosphate.[14] This procedure is useful as a method for regenerat-

[22] A. Bensadoun and D. Weinstein, *Anal. Biochem.* **20,** 241 (1976).
[23] B. L. Horecker, this series, Vol. 3, p. 105.
[24] L. F. Leloir and C. E. Cardini, this series, Vol. 3, p. 840.

ing NAD(P)H from NAD(P), but is less convenient than bromine oxidation as a route to 6-PG.

Preparation of Ribose 5-Phosphate from 6-Phosphogluconate. To the 6-PG solution prepared above, (200 mmol of 6-PG in 1.3 liter of solution) is added sodium α-ketoglutarate (37 g, 0.22 mol), NH_4Cl (13.4 g, 0.25 mol), DTT (1.54 g, 10 mmol), and NADP (0.153 g, 0.2 mmol). The solution is deoxygenated by passing a stream of argon into the stirred solution for 30 min. To this solution is added 6-PGDH (800 units, 120 ml), GluDH (800 units, 100 ml), and PRI (800 units, 1 ml), each separately immobilized in PAN gel. The reaction mixture is stirred under argon, and the pH is controlled automatically at 7.5–7.8 by adding 3 N KOH through a peristaltic pump coupled to a pH controller. After 40 hr the stirring is stopped, the gel is allowed to settle, and the solution is decanted from the gel. The solution is treated with $BaCl_2$ (61 g, mol). Ethanol (the same volume as the aqueous phase) is added at 0°. The resulting precipitate is collected by filtration and washed with ethanol. After drying, the solid (58 g) contains R-5-P (52 g of BaR-5-P, 72% yield based on 6-PG, 90% purity).[17] The fractions of the starting enzymic activities recovered in the gel are: 6-PGDH, 92%; GluDH, 94%; PRI, 92%.

Preparation of Ribulose 1,5-Bisphosphate from 6-Phosphogluconate. A 3-liter solution containing 6-PG (0.2 mol), α-ketoglutarate (0.2 mol), NADP (0.2 mmol), ATP (1 mmol), DTT (10 mmol), and $MgCl_2$ (30 mmol) is degassed with argon. AcK (800 units, 5 ml), 6-PGDH (800 units, 120 ml), GluDH (800 units, 100 ml), and PRuK (800 units, 160 ml), each separately immobilized in PAN gel, are added, and the solution is blanketed with argon. Diammonium acetyl phosphate[7] (48 g of 90% pure materials, 0.25 mol) is added as a solid in 10 equal portions over 40 hr, and the pH is controlled at 7.6–7.9. The ammonium ion introduced with the acetyl phosphate suffices for glutamate synthesis. The solution is decanted from the gel; the enzyme activities recovered in the gel are: 6-PGDH, 92%; AcK, 92%; GluDH, 94%; PRuK, 92%. Acid-washed activated charcoal (30 g) is added to the decanted solution to remove colored impurities. The pH is adjusted to 4.5, and $BaCl_2$ (0.5 mol) is added. Most of the phosphate present in solution precipitates as barium phosphate, while the RuBP remains in solution. The pH is adjusted to 6.5, the ethanol (the same volume as the aqueous phase) is added at 0° to precipitate Ba_2RuBP. The product (98 g) is 78% pure when assayed enzymically (131 mmol, 66% yield based on 6-PG). The level of contamination by xylulose 1,5-bisphosphate (Ba_2XuBP) is 0.4% (determined using the assay based on GDH: see below). To increase the purity of RuBP, it is necessary to shorten the reaction time and minimize isomerization to XuBP by using larger quantities of enzymes (or smaller quantities of 6-PG). For example, conversion of 100 mmol of 6-PG to

RuBP in 2 liters of solution using the same quantities of enzymes described above is complete in 20 hr and gives Ba_2RuBP in 92% purity (Ba_2XuBP, 0.2%) and 80% yield (80 mmol) based on 6-PG.

Preparation of Ribose 5-Phosphate from AMP. The R-5-P is easily prepared by acid-catalyzed hydrolysis of AMP[25] following a slight modified literature procedure to facilitate large-scale preparation. To a suspension of 1 kg of Dowex 50W (H^+ form, 20–50 mesh), prewashed with $0.2 N$ HCl (two 2-liter portions) and H_2O (three 2-liter portions), in H_2O (with total volume of 2 liters) is added AMP-Na salt (100 g, 288 mmol). The pH of the suspension is 1.5–2.0. The suspension is then brought to a boil on a hot plate with stirring and heating is continued 30 min. The suspension is boiled for 5 min and then cooled immediately by pouring it into 300 g of ice. The mixture is filtered to separate the resin. Enzymic assay[17] indicates that the filtrate (1.2 liters) contains 250 mmol of R-5-P. The solution is stable at 4° for periods of months. When longer stability was required, R-5-P was isolated in the form of the barium salt. The above solution is mixed with $BaCl_2$ (56.2 g, 270 mmol) followed by addition of ethanol (1.2 liters) and cooling. The precipitated material (78 g) contains 94% by weight Ba · R-5-P (200 mmol).

Preparation of Ribulose 1,5-Bisphosphate from Ribose 5-Phosphate. A representative conversion of R-5-P to RuBP involved stirring R-5-P (0.2 mol, generated from its barium salt by adding 73.1 g (0.2 mol) of BaR-5-P to 1010 ml of $0.2 M$ H_2SO_4 with vigorous stirring for 30 min and centrifuging to remove the precipitate) under argon for 40 hr at pH 7.8 in a 3-liter reaction mixture containing PRI (100 units, 2 ml), PRuK (800 units, 160 ml), AcK (800 units, 5 ml), each immobilized separately in PAN gel, $MgCl_2$ (30 mmol), DTT (10 mmol), and ATP (1 mmol). Solid $AcP(NH_4)_2$ (0.25 mol) was added in 10 equal portions over this period. Isolation of Ba_2RuBP followed the procedure outlined above: 94 g of solid was obtained, containing 72% of Ba_2RuBP (116 mmol, 58% yield based on R-5-P), and 0.34% of Ba_2XuBP. The recovered enzyme activities were: PRI, 94%; PRuK, 92%; AcK, 92%. For further purification, 45 g of this material in 800 ml of H_2O was mixed with 300 g of Dowex 50W (H^+ form, 200–400 mesh) and stirred for 30 min at 25°, filtered, and washed with 100 ml of H_2O. The resulting solution was passed through Dowex 1 (800 g, 200–400 mesh, Cl^- form) supported in a 2-liter filter, and washed with 5 liters of 40 mM aqueous HCl to remove Ru-5-P and other impurities. RuBP was then desorbed by washing the resin with 2 liters of $0.15 M$ HCl–$0.1 M$ NaCl, and precipitated as described previously with 160 mmol of $BaCl_2$. The product (32 g) contained 94% Ba_2RuBP and 0.16% Ba_2XuBP by weight.

Determination of the Degree of Inhibition of Ribulosebisphosphate Car-

[25] J. R. Sokatch, *Biochem. Prep.* **12**, 2 (1968).

boxylase by Inhibitors Present in the Ribulose 1,5-Bisphosphate Using the Radiometric Assay Method. RuBP is a very unstable compound, particularly in basic solution, and the decomposition products, especially XuBP, are inhibitors of RuBPC.[26] The binding of XuBP to RuBPC is slow. Because the binding is slow, the presence of small amount of this potent inhibitor does not decrease the initial rate in an enzyme assay. In order to determine whether XuBP or other similar inhibitors were present, we incubated RuBPC with a solution of RuBP (600 μg of RuBPC and 2.25 μmol of RuBP in 7.5 ml of the solution described previously for the radiometric assay). Aliquots were taken periodically (approximately every 0.25–2 min), and the quantity of labeled phosphoglyceric acid was measured. When the reaction was complete (10 min), an additional 2.25 μmol of RuBP were added and the reaction again was followed. The difference between these rates measures the degree of inhibition of the second reaction by impurities present in the first: the rates will be the same in the two runs if there are no inhibitors present. A sample of RuBP that had been purified only to the extent that it had been precipitated as BaRuBP after the synthesis gave an inhibition in the second stage of 64%. If this inhibition were caused entirely by XuBP, it would represent a contamination in the range of 0.1%. Treatment of the RuBP sample by passage through a Dowex 1 Cl⁻ column (see below) to separate mono- and diphosphate resulted in an inhibition of 50%. The small amount of XuBP suggested by this assay can be removed by the aldolase–glycerolphosphate dehydrogenase purification described below. It is very difficult, if not impossible, to prevent the re-formation of XuBP. While the rate of epimerization of RuBP to XuBP during the purification under acidic conditions is not known, it is clear that under conditions used for the RuBPC assay (pH 8.0, 30°), RuBP undergoes β-elimination and rearrangement at the rate of 1.25% per hour.[6]

Purification of Ribulose 1,5-Bisphosphate. The level of contamination by XuBP present in these preparations would be unacceptable in many experiments. Two procedures are used for further purification.

In one, the sample (1 g, 72% purity with 0.4% XuBP) is treated with Dowex 50 (10 g, H⁺ form, 200–400 mesh), sonicated, filtered to remove resin, and purified on a Dowex 1 (200–400 mesh) column (1.5 × 20 cm) run with 30 mM HCl (100 ml), followed by HCl gradient formed by introducing 1 liter of 0.15 M HCl through a mixing chamber containing 200 ml of 30 mM HCl. Fractions of 20 ml are collected, and pentulose phosphates are detected using the orcinol test[24] or by enzymic assay for RuBP. The eluted RuBP (fractions 15–25) is isolated as the barium salt (0.6 g). It contains 96% Ba$_2$RuBP and 0.02% Ba$_2$XuBP.

[26] S. D. McCurry and N. E. Tolbert, *J. Biol. Chem.* **252**, 8344 (1977).

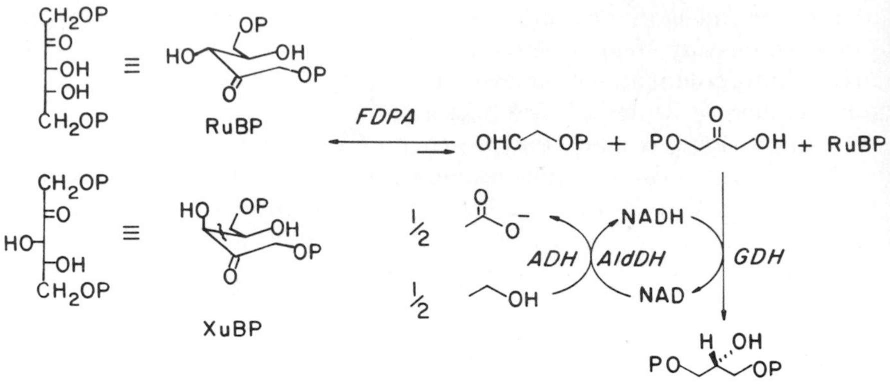

Scheme 3

A second purification is based on the selective enzymic destruction of XuBP: aldolase cleaves XuBP 20 ~ 30 times more rapidly than RuBP (see Scheme 3).[5] Incubation of a sample of RuBP contaminated with XuBP for a sufficiently long time (~10–20 min) effectively guarantees that all XuBP is digested by aldolase, provided that this reaction is coupled with glycerol-3-phosphate dehydrogenase (GDH) to convert the cleavage product dihydroxyacetone phosphate to glycerol phosphate. An NADH regeneration system based on coupled alcohol and aldehyde dehydrogenase (ADH–AldDH) is used for the GDH-catalyzed reaction. A sample of Ba_2RuBP (1 g, 72% purity, 0.4% XuBP) is suspended in 30 ml of H_2O at 4°, and treated with 10 g of Dowex 50 as described above. The solution is adjusted to pH 7.0 by slow addition of 1 N KOH at 0°. To the solution is added 50 units each of fructose-1,6-diphosphate aldolase (FDPA), glycerol-3-phosphate dehydrogenase (GDH), alcohol dehydrogenase (ADH), and aldehyde dehydrogenase (AldDH), followed by NAD (15 mg) and ethanol (0.1 ml). The mixture is stirred at 25° for 10 min, then acidified to pH 3.0. The RuBP is purified by anion-exchange chromatography as described previously and isolated as Ba_2RuBP. The resulting material (0.5 g, 94% purity) shows no (<0.01%) detectable contamination by XuBP.

Determination of Xylulose 1,5-Bisphosphate (XuBP). The same principle as described above is followed to determine the content of XuBP in a sample of RuBP. In the absence of ADH–AldDH for NADH regeneration, the decrease in the absorbance at 340 nm is measured in a period of ~10 min, and the change of absorbance is used to calculate the content of XuBP. Ba_2RuBP (20 mg) is dissolved in 0.1 N HCl (0.3 ml). Sodium sulfate (0.2 ml of 0.2 M Na_2SO_4) is added to precipitate the barium ion as barium sulfate. The supernatant, after centrifugation, is added to 2.5 ml of triethanolamine buffer (0.2 M, pH 7.6) containing 0.2 mM of NADH. Glycerol phosphate dehydrogenase (5 units) and fructose-1,6-diphosphate aldolase (5 units) are added to the mixture. The decrease of absorbance at 340 nm (ΔE) is recorded over a 10-min interval, and the content of XuBP is calculated according to the equation

$$\% Ba_2XuBP = \frac{\Delta E}{6.22} \times V \times \frac{MW}{W} \times 100$$

where V = total volume of assay solution in milliliters, MW = molecular weight of Ba_2XuBP (583.3), and W = weight of Ba_2RiBP sample in micrograms (ca. 20,000 μg). This method would detect 0.01% (0.005 μmol) of Ba_2XuBP in a 20-mg sample based on ΔE = 0.01.

[19] Isomers of α-D-Apiofuranosyl 1-Phosphate and α-D-Apiose 1,2-Cyclic Phosphate

By Joseph Mendicino, Ragy Hanna, and E. V. Chandrasekaran

The study of the biosynthetic reactions involving intermediates of D-apiose has been impeded by the lack of chemically pure phosphate esters of this branched-chain sugar. A more careful characterization of the small amounts of D-apiose derivatives formed in tissue extracts is possible if compounds of known structure and purity are available. Four monophosphate esters are formed when β-D-apiose tetraacetate is treated with anhydrous phosphoric acid. Two isomers of D-apiose, α-D-apio-L-furanosyl 1-phosphate and α-D-apio-D-furanosyl 1-phosphate, are obtained in the highest yield. Almost none of the corresponding β-D-apio-L-furanosyl 1-phosphate and β-D-apio-D-furanosyl 1-phosphate is found.[1] Several cyclic phosphate esters of D-apiose are also formed when β-D-apiose tetraacetate is treated with crystalline phosphoric acid. The α-D-1,2-cyclic D- and L-furanosylapiose phosphodiesters are formed from the corresponding α-D-apiose 1-phosphate esters when they are treated with alkali to remove the O-acetyl groups as shown in the following reactions.

$$\text{D-Apiose} \xrightarrow[\text{acetic anhydride}]{\text{pyridine}} \text{D-apiose tetraacetate}$$

$$\text{D-Apiose tetraacetate} \xrightarrow[\text{phosphoric acid}]{\text{crystalline}} \text{mixture of D-apiose 1-phosphate esters}$$

[1] J. Mendicino and R. Hanna, *J. Biol. Chem.* **245**, 6113 (1970).

Mixture of D-apiose 1-phosphate esters $\xrightarrow{\text{LiOH}}$ α-D-apio-L-furanosyl 1-phosphate,
α-D-apio-D-furanosyl 1-phosphate, α-D-apio-L-furanosyl 1,2-cyclic phosphate,
α-D-apio-D-furanosyl-1,2-cyclic phosphate

One of the cyclic phosphodiester derivatives has properties that are identical with those of a compound formed during the enzymic conversion of UDP-D-glucuronic acid to D-apiose derivatives in extracts of parsley and *L. minor*.[2,3] This compound arises from the breakdown of UDP-D-apiose, since hydrolysis of some nucleoside diphosphate sugars under alkaline conditions yields cyclic 1,2-phosphodiesters of the sugar moiety.[4] The cyclic 1,2-phosphodiester formed by the migration of the phosphate group of α-D-apio-D-furanosyl 1-phosphate is identical to the D-apiose cyclic phosphate intermediate formed from UDP-D-glucuronic acid in tissue extracts.[3]

Reagents

Anhydrous pyridine dried by distillation over calcium hydride
Acetic anhydride
Dry tetrahydrofuran
Crystalline phosphoric acid dried and stored over magnesium perchlorate
LiOH
LiCl
$K_2B_4O_7$

Synthesis of D-Apiose Tetraacetate

The diisopropylidene derivative[2] is prepared from D-apiose. A thoroughly dried syrup of D-apiose is then prepared from crystalline di-*O*-isopropylindeneapiose $[\alpha]_D^{25} + 58°$ (ethanol). The syrup (10 g) is dissolved in a mixture of 67 g of dry pyridine and 50 g of acetic anhydride, which has been previously cooled to 0°. After 13 days at 3° the reaction mixture is poured into 100 ml of ice water. The solution is extracted five times with 50-ml portions of dichloromethane, and the combined extracts are neutralized by shaking them with water and solid $NaHCO_3$ until effervescence ceases. The chloroform extract is washed with water and dried over anhydrous $MgSO_4$. The solvent is removed, and the dry syrup is stored under vacuum over magnesium perchlorate. The yield of product is 72% based on D-apiose tetraacetate, and the sample has $[\alpha]_D^{20} -25.8°$. A weighed sample is assayed by the Somogyi–Nelson procedure. It contains

[2] J. M. Picken and J. Mendicino, *J. Biol. Chem.* **242**, 1629 (1967).
[3] P. K. Kindel and R. R. Watson, **133**, 227 (1973).
[4] A. C. Paladini and L. F. Leloir, *Biochem. J.* **51**, 426 (1952).

46% by weight of D-apiose, which corresponds to the value expected for D-apiose tetraacetate.

Synthesis of the 1-Phosphate Esters of D-Apiose

Dry crystalline phosphoric acid is dissolved in tetrahydrofuran, and the D-apiose tetraacetate is dissolved in chloroform or tetrahydrofuran. The two solutions are mixed, and the solvent is removed under reduced pressure. This procedure ensures complete mixing of very small amounts of the two compounds in the resulting melt, and there is no reaction until the solvent is removed. In this manner, thoroughly dried D-apiose tetraacetate (10 mmol) is treated with 70 mmol of crystalline phosphoric acid in an evacuated reaction flask for 3 hr at 50°. The resulting dark syrup is dissolved in 50 ml of dry tetrahydrofuran, and it is then poured into 50 ml of ice-cold 1 N LiOH. The suspension is adjusted to pH 10.0 with LiOH, and it is kept at room temperature for 12 hr. Insoluble lithium phosphate is removed by filtration, and the precipitate is washed with 200 ml of 0.01 N LiOH. The filtrate and washes are combined and passed through a Dowex 50-NH$_4^+$ column (5 × 7 cm), and the column is washed with water until the eluate is free of acid-labile phosphate. The solution is then concentrated to 25 ml under reduced pressure. The yield of D-apiose-1-P, determined by analysis of acid-labile phosphate, is 30%.

The mixture of D-apiose 1-phosphate esters, approximately 2500 μmol, is applied to a Dowex 1 column (4.5 × 8 cm) in the bicarbonate form, and the column is washed with 10 liters of 10 mM K$_2$B$_4$O$_7$–2.5 mM NH$_4$OH. The phosphate esters are then eluted successively with solutions containing increasing concentrations of LiCl and decreasing concentrations of borate ion. Four peaks are eluted from the Dowex 1 column. Fraction I is eluted with 10 liters of 1 mM K$_2$B$_4$O$_7$–2.5 mM NH$_4$OH–30 mM LiCl. Fraction II is eluted with 5 liters of 0.1 mM K$_2$B$_4$O$_7$–2.5 mM NH$_4$OH–30 mM LiCl. Then fraction III is eluted with 10 liters of the same solution. Finally fraction IV is eluted with 10 liters of 0.5 M LiCl. The four peaks are collected and concentrated by evaporation under reduced pressure to 25 ml. Precipitated salt is removed by filtration. The samples of each fraction are then applied to Whatman 3 MM papers. Repeated development with a solvent containing 20% ethanol and 80% acetone causes LiCl and other salts to move off the paper, whereas the phosphate esters remain near the origin of the chromatogram. After chromatography, two thin strips one-third of the distance in from the sides of the paper are cut out and developed with the FeCl$_3$-sulfosaliciclic acid reagent[5] to reveal phosphate-containing areas. The developed strips are

[5] V. R. Runeckles and G. Krotkov, *Arch. Biochem. Biophys.* **70**, 442 (1957).

TABLE I
PROPERTIES OF ISOMERS OF D-APIOSE 1-PHOSPHATE ISOLATED
BY ION-EXCHANGE CHROMATOGRAPHY

| | Fraction | | | |
Property	I	II	III	IV
Yield (μmol)	190	80	1250	900
Assays (mole/mole of D-apiose)				
Reducing sugar after				
acid hydrolysis	1.0	1.0	1.0	1.0
Fructose-H_2SO_4 test as				
D-apiose	—	0.93	1.04	1.18
Acid-labile phosphate	1.06	1.03	0.97	0.99
Total phosphate	1.05	1.08	1.06	1.03
Hydrolysis rate constants (k min^{-1})				
at 26°				
H_2SO_4, 0.5 N	0.010	0.007	0.092	
H_2SO_4, 0.25 N	0.004	0.003	0.035	0.083

used as guides to cut out the appropriate areas of the untreated portions of the chromatogram. The phosphate esters are eluted from the paper with water, and the solutions are adjusted to pH 7.0 with LiOH and concentrated to 15 ml under reduced pressure. The yields and properties of each peak are shown in Table I.

The reaction mixture for the hydrolysis of the phosphate esters contains, in 5 ml, 0.5 N H_2SO_4 and 0.25% ammonium molybdate or 0.25 N H_2SO_4 and 0.125% ammonium molybdate, 0.1 ml of reducer, and 0.5 μmol of sugar phosphate. The reaction is carried out at 26°. The rate constants are calculated from slopes of the straight lines obtained by plotting the logarithm of ester concentration against time and applying the formula $K = 2.3 \log (ester_2/ester_1)/(t_2 - t_1)$.

Most of the D-apiose that is phosphorylated is recovered in fractions III and IV (2150 μmol compared to 2500 μmol applied to the column). Only about 10% of the total acid-labile sugar is present in fractions I and II.

The barium and cyclohexylammonium salts of the monophosphate esters in fractions III and IV are prepared from the lithium salt in greater than 95% yield.[1] The compounds in fractions I and II, which do not form insoluble barium salts, may be further purified by paper chromatography in solvent systems A and D shown in Table II.

Taken collectively, the analytical and chromatographic data show that the isolated products are isomeric monophosphate esters of D-apiose. The

TABLE II

PAPER CHROMATOGRAPHIC PROPERTIES OF THE FOUR ISOMERS OF D-APIOSE PHOSPHATE

	Solvent			
Compound	A Isopropanol– ammonia– water (7 : 1 : 2)	B Isopropanol– ammonia– water– borate[a]	C Methyl cellosolve– acetic acid– pyridine (8 : 4 : 1 : 1)	D Ethanol ammonium acetate, pH 7.5 (5 : 2)
D-Apio-L-furanosyl-1-P (IV)	0.42	0.38	0.64	0.41
D-Apio-D-furanosyl-1-P (III)	1.48	0.78	0.60	1.15
D-Apio-L-furanosyl-1,2-cyclic-P (II)	4.60	4.65	0.98	0.79
D-Apio-D-furanosyl-1,2-cyclic-P (I)	4.78	4.70	0.95	0.67
D-Glucose-1,2-cyclic-P	—	—	—	0.58
Fructofuranosyl-1-P	1.55	1.60	0.91	—
Glucopyranosyl-1-P	—	—	0.57	—

[a] Boric acid (0.618 g) is added to a solvent containing 700 ml of isopropanol, 100 ml of ammonium hydroxide (sp. gr. 0.896), and 200 ml of water. The results are expressed as the ratio of the distance traveled by the sugar phosphate to the distance traveled by P_i.

purified fractions from each peak contain no free reducing sugar or P_i, but during acid hydrolysis equivalent amounts of D-apiose and P_i are liberated. Support for the chemical identity of the synthesized phosphate esters is indicated by the stoichiometry between the total phosphate, acid-labile phosphate, and reducing sugar present in each compound. The isolated sugar phosphates have the same molar extinction coefficient and spectrum as D-apiose in the fructose–H_2SO_4 test.[1] On paper chromatography in three different solvent systems, the sugar released from each fraction after treatment with 0.1 N HCl at 100° for 15 min was recovered virtually quantitatively in a single well defined region, with an R_f in each case identical with that of D-apiose.[1] Although they did not form furfuraldehyde when treated with strong acid, D-apiose and the sugars formed by hydrolysis of the sugar phosphate esters reacted with the benzidine-trichloroacetic acid reagent on paper chromatograms to give a yellow spot with a white fluorescence under ultraviolet light. They gave a yellow-brown spot when the chromatograms were sprayed with p-anisidine hydrochloride. It is advised that correct protective procedures be used when spraying paper chromatograms with benzidine reagent because of its carcinogenicity.

Section IV

Oxidation– Reduction Enzymes

[20] Cellobiose Oxidase from *Sporotrichum pulverulentum*

By ARTHUR R. AYERS and KARL-ERIK ERIKSSON

$$\text{Cellobiose} + O_2 \rightarrow \text{cellobionic acid} + [0.5 \ O_2^-]^1$$

Assay Method

Principle. Oxidation of the assay substrate, lactose (4-O-β-galactosylglucose), is measured indirectly. The product of the reaction is lactobionic acid, which is quantitatively converted to an equimolar amount of glyoxylic acid. In the presence of lactate dehydrogenase, glyoxylic acid is reduced to glycolic acid, resulting in the oxidation of NADH and a decrease in the absorbance at 340 nm. Activity of cellobiose oxidase is determined by the decrease in absorbance at 340 nm.

Reagents

Sodium acetate buffer, 50 mM, pH 5.0
Lactose, 10 mg/ml
Sodium periodate, 25 mM, in 62 mM H$_2$SO$_4$
Ethylene glycol, 10 mg/ml
NADH, 2.8 mM, prepared fresh daily
Lactate dehydrogenase, 1.0 mg/ml

Procedure. Cellobiose oxidase activity can be measured with a modification[2] of the lactate dehydrogenase assay for aldonic acids.[3,4] Lactose is used as a substrate, since, unlike cellobiose and other cellodextrins, it is not hydrolyzed by the hydrolytic enzymes present in our crude preparations of cellobiose oxidase. To 1.0 ml of test sample diluted appropriately with 50 mM sodium acetate buffer, pH 5.0, add 0.10 ml of lactose solution (10 mg of lactose per milliliter in 50 mM sodium acetate buffer, pH 5.0) and incubate at 30° for 30 min. Terminate the reaction by the addition of 0.10 ml of 25 mM sodium periodate in 62 mM sulfuric acid, and incubate for 30 min at 30°. Add 0.05 ml of ethylene glycol solution (10 mg of ethylene glycol per milliliter) to consume excess periodate and incubate for 20 min at 30°. The glyoxylic acid produced by periodate oxidation of

---

[1] The final disposition of the second oxygen atom consumed in this reaction has not been determined. The superoxide anion has been detected. Assignment of the EC number and name of the enzyme await clarification of the reaction products.

[2] A. R. Ayers, S. B. Ayers, and K.-E. Eriksson, *Eur. J. Biochem.* **90,** 171 (1978).

[3] S. A. Barker, A. R. Law, P. J. Somers, and M. Stacey, *Carbohydr. Res.* **3,** 435 (1967).

[4] R. G. Jones, A. R. Law, and P. J. Somers, *Carbohydr. Res.* **42,** 15 (1975).

the lactobionic acid resulting from the action of cellobiose oxidase on lactose is measured by the addition of 0.100 ml of 2.8 mM NADH and 0.010 ml of purified lactate dehydrogenase (1.0 mg/ml, Sigma). The absorbance of the solution is read at 340 nm after 30 min of incubation at 30°. The amount of lactobionic acid produced in the sample incubation is proportional to the decrease in absorbance at 340 nm and is calibrated by comparison to a standard containing 0.10 μmol of hemicalcium lactobionate.[2]

Definition of Enzyme Unit and Specific Activity. One unit of cellobiose oxidase activity is the amount that will oxidize 1.0 μmol per minute of lactose to lactobionic acid. Specific activity is expressed as units per milligram of protein, as determined by the method of Lowry *et al.*[5] with bovine serum albumin as the protein standard.

Purification Procedure

Growth of the Organism. *Sporotrichum pulverulentum* (ATCC 32629) is cultured in 1-liter conical flasks containing 300 ml of a modified Norkrans medium.[2] After 10 days of culture the mycelia are removed by filtration through wire mesh (DIN 20) and any residual cellulose is eliminated by filtration through a sintered-glass funnel (L1) and a glass fiber filter (Sartorius, 13400). The culture filtrate is chilled to 5°, and purification is begun immediately. All steps are carried out at 0–5°. Each 15-liter batch of culture filtrate (about 60 flasks) contains approximately 40 units of cellobiose oxidase activity.

Step 1. SP-Sephadex Chromatography. The culture filtrate is diluted 10-fold with cold (5°) distilled water, and the pH is adjusted to 3.5 with glacial acetic acid. The diluted culture filtrate is passed rapidly (50 liters/ hr) through a shallow bed of SP-Sephadex C-50 (1.5 × 18 cm, Pharmacia Fine Chemicals) equilibrated with 20 mM sodium acetate buffer, pH 3.5, and retained on a sintered-glass funnel. The rapid flow rate is produced by applying decreased pressure to the funnel. A layer of nylon cloth (Monodur 50, AB Derma, Gothenburg, Sweden) held in place by stainless steel mesh protects the surface of the SP-Sephadex. After the passage of the culture filtrate, the SP-Sephadex is poured into a column (5 × 19 cm) sandwiched between two fresh layers (1 cm) of SP-Sephadex. The column is washed with 200 ml of 20 mM sodium acetate buffer, pH 3.5, and 200 ml of 20 mM sodium acetate buffer, pH 3.8. The cellobiose oxidase is eluted with a linear 4-liter gradient from 0.0 to 0.1 M NaCl in 20 mM sodium acetate buffer, pH 3.8. Fractions of 20 ml are collected at a flow rate of

[5] O. H. Lowry, N. J. Rosebrough, A. L. Farr, and R. J. Randall, *J. Biol. Chem.* **193,** 265 (1951).

4.3 ml hr^{-1} cm^{-2}. The active fractions in the cellobiose oxidase assay are pooled. Since the cellobiose oxidase is a hemoprotein, it may be convenient to follow the presence of the enzyme by its absorbance at 420 nm. The pooled fractions are concentrated to 10 ml by ultrafiltration.

Step 2. Agarose 0.5 M Chromatography. The concentrated enzyme from the SP-Sephadex column is applied to a column of agarose 0.5 M (2.6 × 91 cm, Bio-Rad Laboratories) equilibrated and eluted with 50 mM sodium acetate buffer, pH 5.0. Fractions of 9.5 ml are collected at a flow rate of 3.8 ml hr^{-1} cm^{-2}.

Step 3. Phenyl-Sepharose Chromatography. The active fractions from the agarose column are pooled, diluted to 1 liter with 50 mM sodium acetate buffer, pH 5.0, and made 1.5 M with respect to NaCl. The sample is then applied to a column of phenyl-Sepharose (1.7 × 25 cm, Pharmacia Fine Chemicals), washed with 100 ml of 1.5 M NaCl in 50 mM sodium acetate buffer, pH 5.0, and eluted with a linear 2-liter gradient simultaneously decreasing from 1.5 to 0.0 M NaCl and increasing from 0 to 50% ethylene glycol in 50 mM sodium acetate buffer, pH 5.0. Fractions of 20 ml are collected at a flow rate of 15 ml hr^{-1} cm^{-2}. Fractions corresponding to the major peak of absorbance at 420 nm are pooled and concentrated to 10 ml by ultrafiltration.

Step 4. Isoelectric Focusing. One-milliliter aliquots of the concentrated enzyme from the phenyl-Sepharose step are passed through PD-10 columns (disposable Sephadex G-25 columns, Pharmacia Fine Chemicals) equilibrated with 0.8% Ampholine ampholytes, pH 3.5–5.0 (LKB). The sample is collected in a total of 2.0 ml of ampholyte solution. Isoelectric focusing was performed in a pair of 0.5 × 15 cm Econocolumns (Bio-Rad Laboratories) connected with two 3-way nylon valves (see Fig. 1) as follows (pH 3.5–5.0 Ampholine ampholytes are used throughout):

1. Sucrose, 70% w/v, containing 1% v/v sulfuric acid is added to a height of 1 cm above the bottom frits of the columns. The passage between the columns is closed.

2. The balance column is filled to the top of the glass with 35% w/v sucrose containing 1% v/v sulfuric acid.

3. The reservoir is filled to within 1 cm of the top with 1% v/v sulfuric acid.

4. One milliliter of heavy protection solution (1.5 ml of 70% w/v sucrose, 0.1 ml of H$_2$O, 0.06 ml of ampholytes) is added to the sample column.

5. The sample is then added to the sample column in a gradient consisting of seven 0.5-ml steps prepared by mixing the following volumes of sample and heavy ampholyte solution (10 ml of 70% w/v sucrose, 0.4 ml of

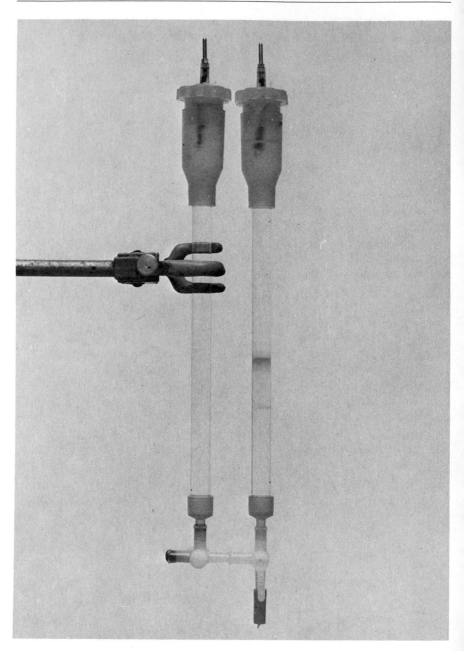

Fig. 1. Column isoelectric focusing of cellobiose oxidase. The sample column is at the right. Reproduced, with permission, from Ayers et al.[2]

PURIFICATION OF CELLOBIOSE OXIDASE FROM
Sporotrichum pulverulentum[a]

Step	Total enzyme (units)	Total protein (mg)	Specific activity (units/mg protein)	Yield (%)
Culture filtrate	41.8	13800	0.003	100
SP-Sephadex	27.2	578	0.047	65
Agarose, 0.5 M	5.7	49.3	0.115	13.6
Phenyl-Sepharose	4.4	18.0	0.244	10.5
Isoelectric focusing	1.7	4.2	0.417	4.2

[a] From Ayers *et al.*[2]

ampholytes): 0.10+0.40, 0.15+0.35, 0.20+0.30, 0.25+0.25, 0.30+0.20, 0.35+0.15, 0.40+0.10.

6. The sample column is filled to the top of the glass with light protection solution (1.3 ml of H_2O, 0.2 ml of 70% w/v sucrose, 0.02 ml of ampholytes).

7. The sample reservoir is filled to within 1 cm of the top with 0.4% w/v NaOH.

8. Platinum electrodes fitted in the upper closures are put in place [sample (−), balance (+), note: polarity given by Ayers *et al.*[2] is incorrect], and the passage between the columns is opened. The columns are submerged in an ice bath up to the upper closures, and power is applied: 400 V (ca 0.7 mA), 8 hr; 600 V (ca 0.8 mA), 16 hr; and 1000 V (ca 0.9 mA), 24 hr.

After focusing, a reddish band is clearly visible in the sample column (see Fig. 1, a minor reddish band of unknown identity is also sometimes observed at this stage). Fractions (3 drops = ca 0.15 ml) are collected into tubes containing 1 ml of ice by opening the valve on the sample side. One-milliliter of 50 mM sodium acetate buffer is then added to each fraction, and the absorbance at 420 nm is measured. The fractions of maximum absorbance are concentrated by ultrafiltration and placed in 50 mM sodium acetate buffer, pH 5.0, by passage through a PD-10 column. This preparation has a specific activity of over 0.4 units/mg, and only a single polypeptide is observed in analytical isoelectric focusing and sodium dodecyl sulfate gel electrophoresis.

Data from a typical preparation of cellobiose oxidase are summarized in the table.

Properties

General. The purified cellobiose oxidase is stable (less than 10% loss in activity per week) at 0 or $-20°$ in 50 mM sodium acetate buffer, pH 5.0. The enzyme has optimum activity at pH 5.0, an isolectric point of 4.2 at 20°, a maximum molecular weight estimated to be 102,000 by sodium dodecyl sulfate–polyacrylamide gel electrophoresis, and an $s_{20,w}$ of 5.6×10^{-13} S determined by sedimentation velocity centrifugation. The molecular weight as estimated by sedimentation equilibrium centrifugation is 93,000. The enzyme appears to be monomeric and binds to immobilized concanavalin A, which suggests that it is a glycoprotein.[2]

Spectral Properties. The oxidized form of the enzyme contains a Soret peak at 419 nm with shoulders at 355 and 465 nm and a small broad peak at 540 nm. The enzyme is reduced by substrate or dithionite, which shifts the Soret peak to 427 nm and yields peaks at 560 and 530 nm. The shoulder at 355 nm remains. The ratio of $A_{420 \text{ nm}}$ to $A_{280 \text{ nm}}$ of the oxidized form is 0.67. These spectral characteristics are those of a hemoprotein.

Flavin Content. Hydrolysis of the enzyme in 6 M HCl followed by spectrofluorimetric analysis suggests that there is one flavin per polypeptide. The flavin has not been identified.

Substrate Specificity. The enzyme has no detectable activity toward glucose or xylose, but galactose and mannose can be slowly oxidized. The disaccharides lactose and cellobiose are both readily oxidized, whereas 4-β-glucosylmannose is only slowly oxidized, and oxidation of both maltose and sophorose is not detected. Water-soluble cellodextrins (dp < 7) are oxidized with decreasing rapidity in the series from cellobiose to cellotetraose, but then with increasing size they are more rapidly oxidized. The activity of cellobiose oxidase is roughly equivalent on cellobiose and cellohexaose. Cellulose is not oxidized at a detectable rate.[2]

Oxygen Consumption. The enzyme consumes 1 mol of molecular oxygen per mole of substrate oxidized. One atom of oxygen is present in the onic acid reaction product, but the fate of the second atom is not fully known. Hydrogen peroxide has not been detected as a reaction product with cellobiose as substrate. However, it has been found that superoxide anion (O_2^-) is produced when cellobiose oxidation is catalyzed by cellobiose oxidase.[6]

Inhibitors. Cellobiose oxidase is inhibited by sodium azide, sodium cyanide, and 2,2-bipyridine.

Function in Cellulose Degradation. Several functions are proposed for cellobiose oxidase in cellulose degradation.

---

[6] K.-E. Eriksson, B. Pettersson, and S. Romanov, unpublished observation, 1981.

1. Since cellobiose is a competitive inhibitor for endo-1,4-β-glucanase and exo-1,4-β-glucanase activity,[7] the oxidation of cellobiose to its corresponding onic acid may facilitate cellulose degradation.

2. Oxidation of cellobiose by cellobiose oxidase generates the superoxide anion. This active radical species may attack the crystalline cellulose, thereby contributing to its conversion to amorphous cellulose, which is more susceptible to enzymic hydrolysis.

3. Cellobiose oxidase may oxidize the reducing glucosyl unit exposed by the action of endo-1,4-β-glucanases on cellulose. Oxidation would prevent subsequent re-formation of the glucosidic bond. The production of the reactive superoxide anion in close proximity to the cellulose may result in further disruption of the crystalline structure.

For a review of enzyme mechanisms involved in cellulose degradation by *S. pulverulentum*, see Eriksson.[8]

[7] T. M. Wood and S. I. McCrae, *Bioconversion Cellul. Subst. Energy, Chem. Microb. Protein Symp. Proc., 1st, 1977*, pp. 111–141 (1978).
[8] K.-.E. Eriksson, *Pure Appl. Chem.* **53**, 33 (1981).

[21] Sorbitol Dehydrogenase from Rat Liver[1]

By NANCY LEISSING and E. T. MCGUINNESS

$$\text{Sorbitol} + \text{NAD}^+ \rightleftharpoons \text{fructose} + \text{NADH} + \text{H}^+$$

Assay Method

Principle. The activity of sorbitol dehydrogenase (L-iditol : NAD$^+$ 5-oxidoreductase, EC 1.1.1.14) is determined from the forward direction of the reaction as shown. Initial velocity is determined by measuring the increase in absorbance at 340 nm due to the formation of NADH. Markedly elevated concentrations of sorbitol and/or NAD$^+$ ($>100 \times K_m$) should be avoided in the reaction cuvette because of the occurrence of substrate inhibition. Deviation from linearity occurs when more than 10% of either substrate has been converted to product.

Reagents

NAD$^+$, 92 mM in Tris buffer
Sorbitol, 400 mM in Tris buffer
Tris-HCl buffer, 50 mM, pH 9.6

[1] N. Leissing and E. T. McGuinness, *Biochim. Biophys. Acta* **524**, 254 (1978).

Procedure. The assay and blank cuvettes (1-cm light path) each contain 0.025 ml of NAD$^+$, 1.0 ml of Tris-HCl buffer, and 0.10 ml of enzyme solution, equilibrated to 25°. The enzyme is accurately diluted with 20 mM potassium phosphate buffer, pH 7.4, with 1 mM dithiothreitol, so that the assay cuvette contains an amount of enzyme in 0.10 ml that does not exceed a ΔA/min of 0.25 in the assay system. The reaction is initiated by adding 0.025 ml of sorbitol, and the increase in absorbance at 340 nm at 25° is monitored for at least 3 min using a chart recorder. Product inhibition is observed when about 10% of either substrate has been utilized. The blank reaction is initiated by adding 0.025 ml of Tris-HCl buffer and monitored over the same time period. The ΔA/min is calculated for each reaction, and the increase due to the enzymic activity in the assay cuvette is determined by the difference. One unit of enzymic activity is defined as the number of micromoles of NADH produced per minute per milligram of enzyme.

Purification Procedure

All procedures are carried out at 0–4° using 20 mM potassium phosphate buffer (pH 7.4) containing 1 mM dithiothreitol.

Step 1. Crude Extract. Rat livers (about 10 g each) that have been freshly excised from 200–250-g CR-CD Charles River male rats are homogenized for 2 min in the buffer at 4° using a Waring blender. The resulting homogenate is centrifuged at 10,000 g for 30 min.

Step 2. Protamine Sulfate Treatment. An 18-ml portion of 2% (w/v) protamine sulfate solution is gradually added with gentle mixing to the supernatant (100 ml) of the crude extract. After standing for 10 min, the solution is centrifuged at 10,000 g for 30 min, and the pellet is discarded.

Step 3. $(NH_4)_2SO_4$ Fractionation. The supernatant from the protamine sulfate step (~120 ml) is treated with $(NH_4)_2SO_4$ (21 g/100 ml, 35% saturation) by the slow addition of finely ground powder. The suspension is allowed to stand for 15 min, then centrifuged at 10,000 g for 10 min. This supernatant is treated with $(NH_4)_2SO_4$ by the further addition of 28 g/100 ml (75% saturation), allowed to stand for 30 min, and then centrifuged at 10,000 g for 20 min. The pellet is dissolved in buffer and either divided into 6-ml aliquots and stored at −20°, or immediately subjected to gel chromatography.

Step 4. Gel Chromatography. Alternative procedure, a or b, have been used at this point.

Procedure a. The enzyme solution is applied to an AcA-34 Ultrogel (LKB Instruments, Inc., Hicksville, New York 11801) column (2.6 × 30.5 cm) previously equilibrated with buffer. Elution is carried out in an upward direction using a flow rate of 9 ml/hr. Active fractions are pooled, divided into aliquots, and stored at $-20°$ until used. This procedure requires about 16 hr.

Procedure b. Alternatively, the enzyme fraction is applied to a Sephacryl G-200 (Pharmacia Fine Chemicals, Piscataway, New Jersey) column (2.6×25.0 cm) previously equilibrated with buffer. Elution is carried out in an upward direction using a flow rate of 60 ml/hr. Active fractions are pooled, divided into aliquots, and stored at $-20°$ until used. This procedure requires about 2.5 hr.

Step 5. AGNAD (Type 1) Affinity Chromatography. A 5.5-ml aliquot of enzyme solution is applied to an AGNAD (type 1, P-L Biochemicals Inc., Milwaukee, Wisconsin) column (2.2×2.6 cm). The column is then washed with 15–20 ml of buffer followed by 6 ml of 10 mM NADP solution. After an additional buffer wash (15–20 ml), 20 ml of 1.0 mM NAD$^+$ are passed through the column to elute the enzyme. A final buffer wash of 15–20 ml is then followed by 8 ml of 100 mM NAD$^+$ to remove the remaining dehydrogenase/proteins (flow rate throughout = 36 ml/hr). Twenty-nine percent of the activity and 98% of the protein applied to the column are recovered in the eluate. To determine the protein content, the pooled active fractions are first concentrated 19-fold using a B-15 Minicon concentrator (Amicon Corp., Lexington, Massachusetts) and then precipitated with trichloroacetic acid to remove interfering NAD$^+$.

The enzyme loses about 70% of its activity within 1 week when stored at $-20°$. The addition of 0.1 mg of bovine serum albumin per milliliter to the eluate stabilizes the preparation. When enzymic activity is too dilute for kinetic studies, the preparation is concentrated immediately before use at $4–8°$ with the B-15 Minicon concentrator.

When AGNAD type 3 replaces AGNAD type 1 under otherwise identical experimental conditions as in step 5, no enzymic activity is recovered even in the 100 mM NAD$^+$ eluent fractions.

A representative purification scheme, using procedure 4a is shown in Table I. This five-step purification procedure requires a total of 23 hr of elapsed time and about 6 hr of contact time.

Properties of the Enzyme

Substrate and Cofactor Specificity. The rates of polyol oxidation for the purified dehydrogenase (step 5) relative to sorbitol and NAD$^+$ were as follows: xylitol, 99; L-iditol, 97; D-arabinitol, 4; D-mannitol, 25; perseitol, 0; L-arabinitol, 0; galactitol, 0; ribitol, 50; sorbitol, 100. For the reverse reaction the rates of ketose reduction relative to D-fructose and NADH were as follows: D-fructose, 100; L-sorbose, 64. D-Mannose, D-ribose,

TABLE I
PURIFICATION OF RAT LIVER SORBITOL DEHYDROGENASE

Step	Volume (ml)	Total activity (units)	Total protein (mg)	Specific activity (units/mg protein)	Recovery (%)	Purification (fold)	Time (hr)
1. Crude extract	3.1	1.61	93.3	0.017	100	1.0	1
2. Protamine sulfate	3.1	1.73	47.1	0.037	108	2.1	1
3. Ammonium sulfate	1.1	1.16	28.8	0.040	72	2.3	2
4a. Ultrogel AcA-34[a]	5.5	0.92	10.4	0.088	57	5.1	16
5. AGNAD-type 1	18.0	0.30	0.04	7.47	18	439.0	3

[a] Steps 4a and 4b are alternative but equivalent methods.

D-xylose, and L-xylose showed no activity as substrates. Both oxidation and reduction were specific for NAD^+ and NADH, respectively. No activity was found with either NADP or NADPH as cofactors. These studies indicate that the enzyme is properly classified as L-iditol (sorbitol) dehydrogenase, based on the polyol configurations described by McCorkindale and Edson.[2] Comparison of the relative rates for polyol oxidation for steps 3, 4, and 5 showed similar ratios of activity.

Since 10 mM $NADP^+$ partially elutes (i.e., binds) sorbitol dehydrogenase from the AGNAD (type 1) affinity gel column, it was tested for its possible inhibitory effects. No inhibition was observed when $NADP^+$ was present in the reaction cuvette at 26 times the concentration of NAD^+ (8 mM $NADP^+$ and 0.3 mM NAD^+).

Profile of pH vs Activity. The influence of pH on the activity of the enzyme was studied in Tris-HCl, MES–NaOH, and glycine–NaOH buffer systems (50 mM) brought to a constant ionic strength of 0.1 with NaCl. The enzymic activity for the forward and reverse reactions over the pH range of 4.8–11.3 showed a pH optimum for D-fructose reduction at about 5.9, whereas sorbitol oxidation occurred over a broad range, with maximum activity at pH 8.1–8.5.

Thermal and Storage Stability. After the $(NH_4)_2SO_4$ fractionation step (Table I, step 3) the preparation remains stable with no loss of activity when stored at $-20°$ for at least 2 years. In this form no loss is seen after

[2] J. McCorkindale and N. Edson, *Biochem. J.* **57**, 518 (1954).

TABLE II
SUMMARY OF STABILITY DATA OF SORBITOL
DEHYDROGENASE (STEP 5)[a]

Addition	Concentration (mM)	Activity remaining after 1 week	
		4°	20°
None (control buffer)		0	78
Albumin	0.0154	103	103
Glycerol	6787	83	83
Ethylene glycol	5322	55	79
Sucrose	584	0	70
Maltose	27.8	21	93
Maltose	278	84	91
NaCl	100	0	0
Ammonium sulfate	3784	0	10
Phosphate buffer	55	0	78

[a] The assay system contained 2.0 mM NAD$^+$, 8.7 mM substrate, and 0.100 ml of enzyme solution in 50 mM Tris-HCl buffer, pH 9.6, in a total volume of 1.15 ml. The control buffer contained 10 mM potassium phosphate at pH 7.4.

5 hr at room temperature, and 35% activity remains after heating at 60° for 10 min.

The enzyme remains stable after further purification on either Ultrogel AcA-34 or Sephacryl G-200 with no apparent loss in activity after 5 hr at room temperature or at least 2 months at −20°. Stability problems are encountered, however, after AGNAD (type 1) affinity chromatography, about 70% loss occurring after storage at −20° for 1 week. Addition of bovine serum albumin, at a concentration of 0.1 mg/ml to the eluate fractions stabilized the activity both at −20° and at room temperature. This stability at room temperature allows very small amounts of enzyme to be detected by measuring its activity in an overnight reaction (Tris-HCl buffer, pH 9.6). The identical substrate specificity profile for steps 3, 4, and 5 demonstrates that the albumin does not alter the enzyme activity. Additional stability data are summarized in Table II.

Molecular Weight. This determination was carried out using gel permeation chromatography and disc gel electrophoresis. In the chromatographic method, standard marker proteins (chymotrypsinogen A, ovalbumin, bovine serum albumin, and aldolase), dissolved at a concentration of

10–15 mg/ml in the preparative buffer system, and the purified sorbitol dehydrogenase were applied to the Ultrogel AcA-34 column and eluted using the experimental conditions described above for the purification procedure. The eluate was collected at a constant flow rate of 9 ml/hr (4°). From the plot of the logarithms of the standard protein molecular weights against their K_{av} values,[3] sorbitol dehydrogenase, present as a single band, was judged to have a molecular weight of 95,000.

Polyacrylamide gel electrophoresis, in 3-cyclohexylamino-1,1-propanesulfonic acid (CAPS) buffer (pH 9.5), was performed according to the method of Hedrick and Smith.[4] Activity staining revealed a single band with an estimated molecular weight of 97,000.

Michaelis Constants. The Michaelis (K_m), constants determined in 50 mM N-2-hydroxylethylpiperazine-N'-2-ethanesulfonic acid (HEPES) buffer (pH 7.5), ionic strength 0.1 with NaCl, at 25° were as follows: 0.38 mM for sorbitol, 0.082 mM for NAD$^+$, 136 mM for fructose, and 0.067 mM for NADH. In this assay system, substrate inhibition was initially observed at 49 mM sorbitol, 7 mM NAD$^+$, 250 mM fructose, and 0.4 mM NADH.

Distribution. Preparation of sorbitol dehydrogenase from rat,[5] rabbit,[6] and horse[7] liver have been reported since the enzyme was last reviewed in this series.[8–10] Selected comparative properties from several mammalian sources have been examined.[11] The sheep[12] and rat[13] liver enzyme have been subjected to steady-state kinetic studies. Gabbay[14] has reviewed the role of this enzyme in the sorbitol pathway and the complications of diabetes mellitus.

[3] E. T. McGuinness, *J. Chem. Educ.* **50**, 826 (1973).
[4] J. L. Hedrick and A. J. Smith, *Arch. Biochem. Biophys.* **126**, 155 (1968).
[5] Y. Kida, *Nara Igaku Zasshi* **25**, 180 (1974).
[6] T. Moriyama, T. Nakano, T. Wada, H. Kikihana, K. Kida, Y. Kida, and K. Shimamoto, *Nara Igaku Zasshi* **24**, 356 (1973).
[7] J. P. Bailey, C. Renz, and E. T. McGuinness, *Comp. Biochem. Physiol.* **69B**, 909 (1981).
[8] J. B. Wolf, this series, Vol. 1 [46B].
[9] S. Horwitz, this series, Vol. 9 [31].
[10] T. E. King and T. Mann, this series, Vol. 9 [32].
[11] E. P. Walsall, S. A. Lyons, and R. P. Metzger, *Comp. Biochem. Physiol.* **59B**, 213 (1978).
[12] U. Christensen, E. Tüchsen, and B. Andersen, *Acta Chem. Scand. Ser. B* **B29**, 81 (1975).
[13] N. Leissing and E. T. McGuinness, Abstract of papers (Biol 49) 180th National Meeting, Amer. Chem. Soc., Las Vegas, August, 1980.
[14] K. H. Gabbay, *Annu. Rev. Med.* **26**, 251 (1975).

[22] D-Sorbitol Dehydrogenase from *Gluconobacter* *suboxydans,* Membrane-Bound

By EMIKO SHINAGAWA and MINORU AMEYAMA

D-Sorbitol + acceptor → L-sorbose + reduced acceptor

D-Sorbitol dehydrogenase occurs on the outer surface of the cytoplasmic membrane of *Gluconobacter* species. The enzyme is solubilized from the membrane and further purified to a homogeneous state.

Assay Method

Principle. The following assay systems are available: D-sorbitol is oxidized by D-sorbitol dehydrogenase, and the reaction rate is estimated (*a*) by spectrophotometry in the presence of 2,6-dichlorophenolindophenol and phenazine methosulfate; (*b*) by colorimetry in the presence of potassium ferricyanide; (*c*) by polarography with an oxygen electrode; or (*d*) by manometry in a conventional Warburg apparatus. In the following, the assay method with potassium ferricyanide is employed because of its simplicity for routine assay. This method is described by Wood *et al.*[1] in the assay of D-gluconate dehydrogenase of *Pseudomonas* species. Some modifications are made (see below).

Reagents

Potassium ferricyanide, 0.1 M, in distilled water
D-Sorbitol, 1 M, in distilled water
McIlvaine buffer, pH 4.5. This buffer solution is prepared by mixing 0.1 M citric acid and 0.2 M disodium phosphate
Triton X-100, 10%, in distilled water
Enzyme dissolved in 0.01 M sodium acetate, pH 5.8, containing 0.1% Triton X-100.
Ferric sulfate–Dupanol reagent containing 5 g of $Fe_2(SO_4)_3 \cdot n H_2O$, 3 g of Dupanol (sodium lauryl sulfate), 95 ml of 85% phosphoric acid, and distilled water to 1 liter

Procedure. The reaction mixture contains 0.1 ml of potassium ferricyanide, 0.5 ml of McIlvaine buffer, pH 4.5, 0.05 ml of Triton X-100, 0.1 ml of D-sorbitol, and enzyme solution in a final volume of 1.0 ml. After preincubation at 25° for 5 min, the reaction is initiated by the addition

[1] W. A. Wood, R. A. Fetting, and B. C. Hertlein, this series, Vol. 5, p. 287.

of potassium ferricyanide and stopped by adding 0.5 ml of ferric sulfate–Dupanol reagent. Then, 3.5 ml of water are added to the reaction mixture. The resulting Prussian blue color is measured at 660 nm after standing at 25° for 20 min. The rate of reduction of ferricyanide to fer-, rocyanide is correlated to the rate of D-sorbitol oxidation.

Definition of Unit and Specific Activity. One unit of enzyme activity is defined as the amount of enzyme catalyzing the oxidation of 1 μmol of D-sorbitol per minute under the conditions described above; 4.0 absorbance units equal 1 μmol of D-sorbitol oxidized. Specific activity (units per milligram of enzyme protein) is based on the protein estimation of modified Lowry method described by Dulley and Grieve.[2] In this method, sodium lauryl sulfate is added to the alkaline solution.

Source of Enzyme

Microorganism and Culture. Gluconobacter suboxydans var. α IFO 3254 can be obtained from the Institute for Fermentation, Osaka (17-85, Juso-honmachi 2-chome, Yodogawa-ku, Osaka 532, Japan). A stock culture is maintained on a potato–glycerol slant (see this volume [24]). The culture medium consists of D-sorbitol (10 g), yeast extract (3 g), and polypeptone (3 g) in 1 liter of tap water. The organism grown on the potato–glycerol slant is inoculated into 100 ml of the medium in a 500-ml shake flask, and cultivation is carried out at 30° for 24 hr with reciprocal shaking. In the case of large-scale culture, 25 liters of the medium are incubated in a 50-liter jar fermentor at 30° under vigorous agitation (500 rpm) and aeration (25–30 liters of air per minute). Bacterial cells are harvested at the late exponential phase.

Purification Procedure[3]

All purification steps are carried out at 0–5° unless otherwise stated. Centrifugations are carried out at 10,000 g for 10 min, and the buffer solution is sodium acetate. The following procedure is carried out with 50 g of wet cells of *G. suboxydans.*

Step 1. Preparation of Membrane Fraction. The cell paste is suspended in 0.01 M buffer, pH 5.0, and passed twice through a French press (American Instrument Co.) at 1000 kg/cm². After centrifugation to remove intact cells, the supernatant is centrifuged at 80,000 g for 90 min. The resultant precipitate is designated as the membrane fraction.

[2] J. R. Dulley and P. A. Grieve, *Anal. Biochem.* **64**, 136 (1975).
[3] E. Shinagawa, K. Matsushita, O. Adachi, and M. Ameyama, *Agric. Biol. Chem.* **46**, 135 (1982).

Step 2. Solubilization of Enzyme. The membrane fraction is suspended in 0.01 M buffer, pH 5.0, to a protein concentration of about 10 mg/ml. Triton X-100 (1%), KCl (0.1 M), and D-sorbitol (0.1 M) are added to the membrane suspension, and the suspension is stirred for 2 hr. The solubilized enzyme solution is obtained by centrifugation at 80,000 g for 90 min.

Step 3. Fractionation with Polyethylene Glycol. Polyethylene glycol 6000 is added to the supernatant to 20% (w/v) and left standing overnight. After centrifugation, the precipitate is dissolved with 0.01 M buffer, pH 5.0, containing 0.1% Triton X-100. A red solution is centrifuged to remove insoluble materials.

Step 4. DEAE-Cellulose Column Chromatography. The enzyme solution from step 3 is placed on a DEAE-cellulose column (3.5 × 20 cm) that has been equilibrated with 0.01 M buffer, pH 5.0, containing 0.1% Triton X-100 and the column is washed with the same buffer. The enzyme protein band having a red color migrates slowly down to the column and is eluted from the column far from the major impurities. Fractions having enzyme activity are combined, and the enzyme protein is precipitated by the addition of polyethylene glycol 6000, as above.

Step 5. CM-Cellulose Column Chromatography. The precipitate is collected by centrifugation and dialyzed against 0.01 M buffer, pH 5.8, containing 0.1% Triton X-100. The dialyzed enzyme is applied to a CM-cellulose column (3 × 10 cm) that has been equilibrated with the same buffer. Elution of the enzyme is made by a linear gradient of KCl up to 0.15 M in the same buffer. The enzyme activity is eluted at about 0.07 M KCl as a sharp protein peak accompanying enzyme activity and red color.

A typical enzyme purification is summarized in the table; over 200-fold purification is usually obtained using the method described above.

PURIFICATION OF D-SORBITOL DEHYDROGENASE FROM
Gluconobacter suboxydans var. α

Step	Total protein (mg)	Total activity (units)	Specific activity (units/mg)	Yield (%)
Membrane fraction	4940	9,120	2	100
Solubilized fraction	900	10,350	12	113
PEG[a] fraction	180	6,800	38	75
DEAE-cellulose	22	5,320	241	58
CM-cellulose	4	1,690	433	19

[a] PEG, polyethylene glycol 6000.

Properties[3]

Homogeneity. The enzyme shows a symmetrical peak with an apparent sedimentation constant of 5.8 S in the presence of 1% Triton X-100 (pH 5.0). When the purified enzyme having a specific activity of 430 units per milligram of protein is subjected to conventional polyacrylamide gel electrophoresis, one major protein band ($R_f = 0.60$) having enzyme activity and one minor protein band having no enzyme activity are found.

Absorption Spectra. The absorption spectrum of the purified enzyme reduced with sodium dithionite shows maxima at 551, 522, and 417 nm in the visible region, suggesting the presence of a cytochrome component in the purified enzyme preparation. The cytochrome is of the *c* type, based on the pyridine hemochromogen produced. The cytochrome is not reduced rapidly by the addition of D-sorbitol, until coenzyme Q_1 is added. When the buffer contains 10 mM D-sorbitol during the purification, the cytochrome is reduced about 70% relative to the reduction obtained with sodium dithionite.

Subunit Components. Upon gel electrophoresis in sodium dodecyl sulfate, the purified enzyme dissociates into three major protein bands with molecular weights (M_r) of 63,000, 51,000, and 17,000. Of the three components of D-sorbitol dehydrogenase, the largest (M_r 63,000) is a flavoprotein, since exposure of unstained gel to ultraviolet ray shows intense fluorescence. The largest component is probably the dehydrogenase protein. The second component (M_r 51,000), which shows red protein band on unstained gel and is stained by heme staining, is the cytochrome. The function of the smallest subunit remains unknown. The sum of these molecular weights is 131,000, the total molecular weight of D-sorbitol dehydrogenase.

Catalytic Properties. D-Sorbitol dehydrogenase can be assayed *in vitro* in the presence of any one of the following dyes as an electron acceptor: potassium ferricyanide, phenazine methosulfate, 2,6-dichlorophenolindophenol, nitro blue tetrazolium, or Wurster's blue. NAD, NADP, or oxygen are completely inactive as electron acceptors. This finding is the same as for other membrane-bound dehydrogenases described in this volume [24, 25, 31, 32, 82]. Of tested substrates, D-sorbitol is specifically oxidized and D-mannitol is also oxidized at 5% the D-sorbitol rate. Other sugar alcohols, such as D-arabitol, L-iditol, .*meso*-erythritol, dulcitol, ribitol, or xylitol, are completely inert to the enzyme. The optimum temperature for D-sorbitol oxidation is 25°. Oxidation of D-sorbitol proceeds most rapidly at pH 4.5. The apparent Michaelis constant for D-sorbitol is determined to be 30 mM at pH 4.5.

Flavin Determination. Extraction of the flavin prosthetic group of D-sorbitol dehydrogenase by acid precipitation by trichloroacetic acid did not release flavin, since the flavin is covalently bound to the dehydrogenase protein. When the enzyme is digested with Pronase and trypsin, a significant amount of flavin is released. Although a complete proteolytic digestion is not achieved, about 0.4 mol of flavin per mole of the enzyme was found.

[23] Hexose Oxidase from *Chondrus crispus*

By Miyoshi Ikawa

$$\text{D-Glucose} + O_2 \rightarrow \delta\text{-D-gluconolactone} + H_2O_2$$
$$\text{D-Galactose} + O_2 \rightarrow \gamma\text{-D-galactonolactone} + H_2O_2$$

The partial purification of a carbohydrate oxidase capable of oxidizing both glucose and galactose from the red alga *Iridophycus flaccidum* was first reported by Bean and Hassid.[1] The enzyme also occurs in the red alga *Chondrus crispus,* and the method for the isolation of the enzyme from this source reported by Sullivan and Ikawa[2] is described here.

Assay Method A

Principle. The hydrogen peroxide formed in the oxidation of the sugar reacts with the chromogen, *o*-dianisidine,[3] in the presence of peroxidase to form a dye, with absorbance at 402 nm.

Reagents

Sodium phosphate buffer, 0.1 M, pH 6.3
Glucose, 0.1 M in buffer
o-Dianisidine · 2 HCl, 3.0 mg/ml in water
Peroxidase, 0.1 mg/ml of peroxidase (Sigma Chemical Co.) in buffer

Procedure. To 1.5 ml of glucose are added 1.2 ml of buffer, 0.1 ml of *o*-dianisidine, 0.1 ml of peroxidase, and 0.1 ml of enzyme solution. The mixture is incubated at 25° for 15 min, after which 1 drop of concentrated HCl is added and the absorbance is read at 402 nm.

Definition of Unit and Specific Activity. A standard curve is constructed using hydrogen peroxide (0–3.0 μg/ml) in place of enzyme. One unit is

[1] R. C. Bean and W. Z. Hassid, *J. Biol. Chem.* **218**, 425 (1956).
[2] J. D. Sullivan, Jr., and M. Ikawa, *Biochim. Biophys. Acta* **309**, 11 (1973).
[3] A. St. G. Huggett and D. A. Nixon, *Biochem. J.* **66**, 12 P (1957).

defined as the amount of enzyme catalyzing the production of 10^{-3} μmol of H_2O_2 per minute at 25°, pH 6.3, and at a glucose concentration of 0.05 M. Specific activity is expressed as units per milligram of protein. Protein is determined by the Lowry et al.[4] procedure, using bovine serum albumin as the standard.

Assay Method B

Principle. The green alga *Chlorella pyrenoidosa* is very sensitive to hydrogen peroxide. A convenient semiquantitative method for detecting the enzyme consists of placing the extract on a paper disk and placing the disk on the surface of a *Chlorella*-seeded agar plate made up of glucose-containing growth medium. Activity is measured by the size of the rings of growth inhibition observed after incubation.[5]

Reagents

Chlorella pyrenoidosa Chick. Strain No. 395 from the Culture Collection of Algae, Department of Botany, University of Texas, Austin, Texas.

Chlorella growth medium. The growth medium contains glucose (10 g), K_2HPO_4 (2.5 g), KH_2PO_4 (2.5 g), KNO_3 (1 g), $MgSO_4 \cdot 7 H_2O$ (0.25 g), and trace element solution (0.4 ml) per liter. The trace element solution consists of $CaCl_2 \cdot 2 H_2O$ (1.25 g), $FeSO_4 \cdot 7 H_2O$ (0.25 g), $ZnSO_4 \cdot 7 H_2O$ (0.25 g), $MnCl_2 \cdot 4 H_2O$ (25 mg), H_3BO_3 (25 mg), NH_4VO_3 (2.5 mg), $(NH_4)_6Mo_7O_{24} \cdot 4 H_2O$ (2.5 mg), $CoCl_2 \cdot 6 H_2O$ (2.5 mg), $CuSO_4 \cdot 5 H_2O$ (2.5 mg), and sodium citrate dihydrate (5 g) per 250 ml.

Agar

Paper disks: Bacto-Disks-Sterile Blanks $\frac{1}{4}$ inch (6.5 mm) (Difco Laboratories, Detroit, Michigan)

Procedure. A thick *Chlorella* suspension is prepared by inoculating about 250 ml of sterile growth medium from a slant and allowing the culture to grow under artificial light until it becomes dark green. This culture may be kept in the refrigerator for several weeks.

To make up 20 plates, 50 ml of *Chlorella* culture is added to 50 ml of sterile medium. (If the cultures are especially heavy, the volume of culture can be decreased and the amount of sterile medium correspondingly increased.) This is added with gentle swirling to 100 ml of medium (containing 2 g of agar) that has been autoclaved and allowed to cool to 50°.

[4] O. H. Lowry, N. J. Rosebrough, A. L. Farr, and R. J. Randall, *J. Biol. Chem.* **193**, 265 (1951).

[5] M. Ikawa, D. S. Ma, G. B. Meeker, and R. P. Davis, *J. Agric. Food Chem.* **17**, 425 (1969).

The plates are poured before the agar has had a chance to solidify. The plates should be a light yellow-green and may be stored in a refrigerator for several days if not used immediately.

The paper disks are held in tweezers, dipped into the test solution (excess solution being removed by touching the disk to the side of the container), and placed on the surface of the *Chlorella*-seeded agar, usually 3–4 disks per plate. By this procedure approximately 20 μl of solution are soaked up by each disk. The plates are kept under fluorescent light, and the zone of growth inhibition around each disk is read after 24–48 hr.

Purification Procedure

All steps during the purification are carried out at 0–5° unless otherwise stated.

Source of Enzyme. *Chondrus crispus* (commonly known as Irish moss) occurs in abundance along the northern New England coast. Samples are collected year round in the intertidal zone at Rye Beach, New Hampshire. Freshly collected fronds are washed with cold tap water, blotted, and dried at room temperature, after which they are ground to a powder.

Extraction with Buffer. One liter of 0.1 M, pH 6.8, sodium phosphate buffer is added to 100 g of powdered fronds; the mixture is kept at 5° for 1–2 days with occasional shaking. The mixture is filtered by suction through cheesecloth, and the filtrate is centrifuged at 20,000 g for 30 min.

n-Butanol Treatment. The bright red-orange supernatant solution is shaken with an equal volume of *n*-butanol and, after standing for several minutes, is centrifuged at 10,000 g for 30 min. The upper butanol fraction and the phycocyanin pigments[6] at the interface are removed and discarded.

Ammonium Sulfate Precipitation. Ammonium sulfate is added slowly and with shaking to the butanol-treated extract at the rate of 45 g/100 ml. After several hours the mixture is centrifuged at 12,000 g for 20 min. The precipitate is dissolved in 50–100 ml of 0.01 M, pH 6.8, sodium phosphate buffer, and the solution is dialyzed against periodically changed distilled water for 2–3 days. The retentate is centrifuged at 10,000 g for 10 min. The supernatant is saved and sufficient pH 6.8 sodium phosphate is added to make it 0.1 M in phosphate.

DEAE-Cellulose Chromatography. A column (1.5 × 12 cm) is prepared by packing with 10 g of Whatman DE-52 ion-exchange cellulose and equilibrating with 0.1 M, pH 6.8, phosphate buffer. The buffered supernatant from the preceding step is placed on the column, and the column is

[6] S. P. Leibo and R. F. Jones, *Arch. Biochem. Biophys.* **106**, 78 (1964).

PURIFICATION OF HEXOSE OXIDASE FROM *Chondrus crispus*

Fraction	Volume (ml)	Total units	Total protein (mg)	Specific activity (units per mg protein)	Yield (%)
Aqueous extract	690	81,420	2277	35	100
(NH₄)₂SO₄ precipitate	425	69,700	468	149	85
DEAE-cellulose eluate	84	49,340	76	650	66
Sephadex G-200 eluate	59	8,190	2	4095	11

washed with 500 ml of the same buffer. Fractions (3–4 ml) are collected at a flow rate of 15–20 ml/hr. The enzyme is eluted from the column with $0.1 M$, pH 6.8, phosphate buffer, which is also $0.3 M$ in NaCl; the activity elutes in approximately the first 60 ml. The activity can be conveniently located using the *Chlorella* plates. Active fractions are pooled and dialyzed overnight against several changes of distilled water.

Pepsin-Trypsin Treatment. The retentate is adjusted to pH 3.5 with dilute HCl (final volume = approximately 80 ml), 20 mg of pepsin (3 × crystallized) are added and the mixture is incubated on a shaker at 37° for 5 hr. The mixture is then adjusted to pH 6.8 with dilute NaOH, and pH 6.8 sodium phosphate is added to make the mixture $0.01 M$ in phosphate. Trypsin (2 × crystallized) (20 mg) is added, the mixture is incubated with shaking at 37° for 5 hr, the digest is then lyophilized.

Gel Filtration. The lyophilized digest is suspended in distilled water and applied to a Sephadex G-200 column (2.5 × 96 cm), and the column is developed with $0.1 M$, pH 6.8, phosphate buffer. Fractions (3–4 ml) are collected at a flow rate of 10–12 ml/hr. Again activity can be located using the *Chlorella* plates. Activity is eluted at approximately 300–360 ml, where a slight 280 nm peak is observed. Active fractions are pooled, dialyzed against distilled water, and lyophilized.

A summary of the purification procedure is given in the table.

Properties

Hexose oxidase from *Chondrus crispus* is a glycoprotein of molecular weight about 130,000. It contains about 70% carbohydrate consisting mainly of galactose and xylose; it also contains about 12 gram atoms of copper per mole.

Specificity. The relative rates of oxidation of substrates are as follows: D-glucose, 100; D-galactose, 82; maltose, 40; cellobiose, 32; lactose, 22; D-glucose 6-phosphate, 10; D-mannose, 8; 2-deoxy-D-glucose, 8;

2-deoxy-D-galactose, 6; D-fucose, 2; D-glucuronic acid, 2; D-xylose, 1. L-Glucose, D-fructose, δ-D-gluconolactone, D-gluconic acid, γ-D-galactonolactone, dulcitol, D-arabinose, xylitol, and sucrose are not oxidized.

Inhibitors. The enzyme is strongly inhibited by sodium diethyldithiocarbamate and to a lesser extent by cyanide, hydroxylamine, azide, acetate, and pyruvate.

Other Properties. The enzyme has a pH optimum at 6.3 and a temperature optimum at 25°. Its K_m for glucose is 4 mM and for galactose 8 mM.

[24] D-Glucose Dehydrogenase from *Pseudomonas fluorescens*, Membrane-Bound

By KAZUNOBU MATSUSHITA and MINORU AMEYAMA

D-Glucose + acceptor → D-glucono-δ-lactone + reduced acceptor

D-Glucose dehydrogenase (EC 1.1.99.a) occurs on the outer surface of the cytoplasmic membrane of oxidative bacteria such as *Pseudomonas*[1] and *Gluconobacter*[2] species, and initiates a direct oxidation of D-glucose through an electron transport chain. The enzyme is solubilized from the membrane and further purified to a homogeneous state.

Assay Method

Principle. A spectrophotometric assay at 25° measures the decrease of absorbance at 600 nm of 2,6-dichlorophenolindophenol (DCIP) mediated with phenazine methosulfate (PMS). The activity can also be measured with PMS, DCIP, Wurster's blue (WB), ferricyanide, or coenzyme Q (CoQ) as an electron acceptor.[1]

Reagents

D-Glucose, 1 M, in distilled water
Tris-HCl, buffer, 50 mM, pH 8.75
DCIP, 6.7 mM, in distilled water
PMS, 20 mM, in distilled water

[1] K. Matsushita, Y. Ohno, E. Shinagawa, O. Adachi, and M. Ameyama, *Agric. Biol. Chem.* **44**, 1505 (1980).
[2] M. Ameyama, E. Shinagawa, K. Matsushita, and O. Adachi, *Agric. Biol. Chem.* **45**, 851 (1981).

Enzyme dissolved in 0.01 M potassium phosphate, pH 6.0, containing 5 mM MgCl$_2$ and 1% Triton X-100

Procedure. The cuvette with 1-cm light path contains 1 ml of buffer, 0.1 ml of DCIP, 0.1 ml of PMS, 0.1 ml of glucose, enzyme solution containing less than 0.1 unit, and water in a final volume of 3 ml. A reference cuvette contains all components except D-glucose. The reaction is initiated by the addition of D-glucose. Enzyme activity is measured as the initial reduction rate of DCIP. Potassium cyanide (1 mM) or sodium azide (4 mM) must be added in the reaction mixture when enzyme activity of the membrane is measured.

Definition of Unit and Specific Activity. One unit of enzyme oxidizes 1 μmol of D-glucose, or reduces 1 μmol of DCIP, per minute at 25° under the above conditions. The extinction coefficient of DCIP at pH 8.75 is taken as 15.1 mM^{-1}, so that 1 unit of enzyme activity correspond to a ΔA of 5.0 at 600 nm. Specific activity is expressed as units per milligram of protein. The protein content is determined by the modified Lowry method[3] because the sample contains Triton X-100.

Source of Enzyme

Microorganisms. Many strains of *Pseudomonas aeruginosa* and *P. fluorescens* are suitable sources of D-glucose dehydrogenase. Purification of the dehydrogenase of *P. fluorescens* FM-1 isolated in our laboratory is described below.

Cultures. The stock culture is maintained on a potato–glycerol slant prepared as follows: Freshly sliced potato (200 g) is boiled in 1 liter of tap water and autoclaved for 10 min at 2 kg/cm². The autoclaved gruel is centrifuged at 12,000 g for 20 min, and a light-yellow supernatant is obtained. To the supernatant are added dried yeast (10 g), glycerol (20 g), polypeptone (10 g), D-glucose (5 g), and agar powder (15 g) in 1 liter of tap water.

The organism is grown in a synthetic medium containing 5 g of sodium D-gluconate as the sole carbon source, 1 g of ammonium sulfate, 0.75 g of K$_2$HPO$_4$, 0.25 g of KH$_2$PO$_4$, 0.3 g of MgSO$_4$ · 7 H$_2$O, 0.03 g of FeSO$_4$ · 7 H$_2$O, and 0.5 g of yeast extract in 1 liter. Cultures are grown aerobically in a 50-liter jar fermentor containing 30 liters of the medium at 30° and harvested at the late exponential phase (about 7 hr).

Purification Procedure[1]

Potassium phosphate buffer (pH 6.0) containing 5 mM MgCl$_2$ is used throughout the purification process. All procedures are carried out at 0–5°.

[3] J. R. Dulley and P. A. Grieve, *Anal. Biochem.* **64,** 136 (1975).

Step 1. Preparation of Membrane Fraction. Cells are harvested by continuous-flow centrifugation at 10,000 g at <4°, washed with distilled water, and stored at −20° until use. About 100 g of wet cells are suspended in a sixfold volume of 0.01 M buffer, and DNase is added to 10 μg/ml. The cell suspension is twice passed through a French press (American Instrument Co.) at 1000 kg/cm² and centrifuged at 1800 g for 10 min to remove intact cells. The membrane fraction is collected by centrifugation at 68,000 g for 60 min and stored at −20° until use.

Step 2. Solubilization of Enzyme. The membrane fraction prepared from 100 g of wet cells is suspended in 0.01 M buffer at a final protein concentration of 10 mg/ml. Ten percent cholic acid (pH 7.4) is added to 1% concentration. After stirring for 2 hr, the suspension is centrifuged at 68,000 g for 60 min, and the precipitate is suspended in 0.01 M buffer to the original protein concentration. Triton X-100 and KCl are added to 1% and 1 M, respectively, and the suspension is stirred overnight. The suspension is centrifuged at 68,000 g for 60 min, and the supernatant containing D-glucose dehydrogenase is obtained.

Step 3. Polyethylene Glycol Fractionation. Polyethylene glycol 6000 is added to the supernatant, to 18% (v/v), and the mixture is stirred for 30 min. The supernatant is obtained by centrifugation at 10,000 g for 20 min.

Step 4. Ethanol Fractionation. Cold ethanol (−20°) is added to the supernatant up to 20% (v/v), and the precipitate formed by stirring for 30 min is removed by centrifugation at 10,000 g for 20 min. To the supernatant, ethanol is further added to 40% (v/v), and the precipitate is collected. The precipitate is suspended in 0.01 M buffer, and Triton X-100 is added to 1% in a final volume of 30 ml. The suspension is stirred overnight, then centrifuged at 10,000 g for 20 min to obtain a clear supernatant.

Step 5. Hydroxyapatite Column Chromatography. The supernatant is applied to a hydroxyapatite column (3.5 × 5 cm) preequilibrated with 0.001 M buffer containing 1% Triton X-100. The column is washed with 300 ml of the same buffer, and the enzyme is eluted with 0.05 M buffer containing 1% Triton X-100. The peak fractions are pooled as the purified enzyme, and the fractions in the tailing portion is discarded because of impurities. In case some impurities remain, especially when the membrane starting material has a specific activity of less than 2.0 units per milligram of protein, further purification can be achieved in step 6.

Step 6. Phenyl-Sepharose Column Chromatography. KCl is added to 2 M in the pooled fraction, and the solution is applied to a phenyl-Sepharose CL-4B (Pharmacia Fine Chemicals) column (1 × 4 cm) washed with 0.01 M buffer containing 2 M KCl. The column is washed with 100 ml of 0.01 M buffer containing 2 M KCl. The enzyme is eluted by a linear gradient consisting of 150 ml of 0.01 M buffer–0.05% Triton X-100, and 150 ml of 0.01 M buffer–1% Triton X-100.

TABLE I
PURIFICATION OF D-GLUCOSE DEHYDROGENASE FROM
Pseudomonas fluorescens

Fraction	Total protein (mg)	Total activity (units)	Specific activity (units/mg protein)	Yield (%)
Membrane fraction	1958	6814	3.5	100
Solubilized fraction	419	3749	8.9	55
PEG[a] supernatant	85	3388	40	50
Triton supernatant	11	2685	244	39
Hydroxyapatite	2.8	1080	386	16

[a] PEG, polyethylene glycol 6000.

A typical purification is summarized in Table I. The purified D-glucose dehydrogenase is nearly homogeneous in urea-sodium dodecyl sulfate (SDS) gel electrophoresis.

Properties

Molecular Properties.[1,4] The enzyme shows a single protein band with 87,000 ± 2400 of molecular weight by urea–SDS gel electrophoresis. The molecular weight of the enzyme is also estimated to be 93,000 by sucrose density gradient centrifugation in the presence of Triton X-100. These findings indicate that the enzyme is a single polypeptide and is present as a monomer in the presence of Triton X-100. By lowering the concentration of Triton X-100 in the enzyme solution, however, the enzyme aggregates and the activity decreases. The lost activity returns upon addition of detergent. This activation is also caused by adding phosphatidylglycerol, cardiolipin, or phospholipid extracted from *Pseudomonas*, but not phosphatidylethanolamine. Thus, D-glucose dehydrogenase may be highly hydrophobic. This fact is confirmed by amino acid analysis of the purified enzyme, which shows a polarity of 39.7%.

Prosthetic Group.[5] The absorption spectrum of the purified enzyme, with the Triton X-100 replaced by Brij 58, shows a peak at 285 nm and a shoulder at 275 nm. By the addition of D-glucose or sodium borohydride, an absorption peak at 340 nm appears owing to reduction of the prosthetic

[4] K. Matsushita, Y. Ohno, E. Shinagawa, O. Adachi, and M. Ameyama, *Agric. Biol. Chem.* **46,** 1007 (1982).

[5] M. Ameyama, K. Matsushita, Y. Ohno, E. Shinagawa, and O. Adachi, *FEBS Lett.* **130,** 179 (1981).

TABLE II
KINETIC PARAMETERS OF D-GLUCOSE DEHYDROGENASE FOR
VARIOUS ELECTRON ACCEPTORS

Electron acceptors[a]	K_m (mM)	V_{max}[b]	Optimum pH
DCIP	1.60	1220	(6.0)[c]
Ferricyanide	0.69	57	4.5
WB	0.56	290	8.75
PMS	0.13	258	8.75
CoQ_1	0.060	24	6.5
CoQ_2[d]	0.061	28	—
CoQ_4[d]	0.031	5.7	—
CoQ_6[d]	0.0097	2.0	—

[a] DCIP, dichlorophenolindophenol; WB, Wurster's blue; PMS, phenazine methosulfate; CoQ, coenzyme Q.
[b] Micromoles of D-glucose oxidized per minute per milligram of enzyme protein.
[c] It is impossible to assay below pH 5.5. The assay was performed at pH 6.0.
[d] The reaction mixture contains 5% ethanol.

group. The prosthetic group extracted from the purified enzyme shows an absorption spectrum having a peak at 246 nm and a shoulder at 326 nm. The peak at 246 nm is decreased and that at 326 nm is increased by the addition of sodium borohydride. The prosthetic group has a fluorescence maximum at 480 nm with excitation at 370 nm and excitation maxima at 370 nm, 330 nm, and 260 nm with emission at 480 nm. These characteristics indicate that the prosthetic group of the enzyme is pyrroloquinoline quinone, which is isolated from cells of methylotrophic bacteria[6] and has been shown to be a prosthetic group of both methanol dehydrogenase of *Hyphomicrobium* X and the D-glucose dehydrogenase of *Acinetobacter calcoaceticus*.[7] The same prosthetic group is found also in D-glucose dehydrogenase of *Gluconobacter*,[2] alcohol dehydrogenase of acetic acid bacteria (see this volume [76]), and aldehyde dehydrogenase of acetic acid bacteria (see this volume [82]).

Kinetic Properties.[1,4] D-Glucose dehydrogenase has a dual pH optimum depending on the electron acceptor used. The enzyme has an acidic optimum pH with DCIP or ferricyanide, and the activity with PMS or WB shows a maximum at pH 8.75. The Michaelis constant for D-glucose is

[6] S. A. Salisbury, H. S. Forrest, W. B. T. Crure, and O. Kennard, *Nature* (*London*) **280**, 843 (1979).
[7] J. A. Duine, J. Frank, Jr., and J. K. Van Zeeland, *FEBS Lett.* **108**, 443 (1979).

also different depending on pH: 0.47 mM at pH 6.0 with DCIP and 6.3 mM at pH 8.75 with PMS–DCIP. Kinetic parameters of the purified enzyme for various electron acceptors are shown in Table II. CoQ has the highest affinity for the enzyme. CoQ and its long-chain homologs are directly reduced by the enzyme. *Pseudomonas* species have CoQ$_9$ as a natural ubiquinone, and at a slower but significant rate the enzyme is able to reduce CoQ$_9$ solubilized with 5% ethanol or phospholipid vesicles. The enzyme activity is completely inhibited with p-benzoquinone (1.7 mM) or EDTA (3.3 mM).

Substrate Specificity and Reaction Product.[1] D-Glucose dehydrogenase possesses a fairly broad substrate specificity as follows: D-glucose, 100%; D-xylose, 13%; D-mannose, 8.6%; L-rhamnose, 7.5%; D-galactose, 6.5%; maltose, 3.2%; and L-arabinose, 2.8%. The enzyme does not oxidize D-fructose, L-sorbose, D-arabinose, α-methylglucoside, D-ribose, sucrose, D-gluconate, and D-galactonate. A reaction product of the enzyme is D-glucono-δ-lactone.

Immunological Properties. Antibody elicited with the purified D-glucose dehydrogenase shows a single or diffused precipitin line against the purified enzyme in a Ouchterlony double-diffusion analysis. Immunoprecipitates between the antibody and the purified enzyme or membrane-solubilized supernatant show only one polypeptide band in addition to the γ-globulin subunits in urea–SDS gel electrophoresis. The antibody inhibits the activity of the purified enzyme, but not that of the membrane.

[25] D-Fructose Dehydrogenase from *Gluconobacter industrius,* Membrane-Bound

By Minoru Ameyama and Osao Adachi

D-Fructose + acceptor → 5-keto-D-fructose + reduced acceptor

D-Fructose dehydrogenase (EC 1.1.99.11) occurs on the outer surface of the cytoplasmic membrane of *Gluconobacter* species. The enzyme is solubilized from the membrane and further purified to a homogeneous state.

Assay Method

Principle. The reaction rate can be estimated (*a*) by spectrophotometry in the presence of 2,6-dichlorophenolindophenol and phenazine methosulfate; (*b*) by colorimetry in the presence of potassium ferricyanide; (*c*) by polarography with an oxygen electrode; or (*d*) by manometry in a conventional Warburg apparatus. The ferricyanide method is employed in this purification because of its simplicity for routine use. This method is described by Wood *et al.*[1] in the assay of D-gluconate dehydrogenase. Some modifications are made as described below.

Reagents

Potassium ferricyanide, 0.1 *M*, in distilled water
D-Fructose, 1 *M*, in distilled water
McIlvaine buffer, pH 4.5, prepared by mixing 0.1 *M* citric acid and 0.2 *M* disodium phosphate
Ferric sulfate–Dupanol reagent containing 5 g of $Fe_2(SO_4)_3 \cdot n\ H_2O$, 3 g of Dupanol (sodium lauryl sulfate), 95 ml of 85% phosphoric acid, and distilled water to 1 liter
10% Triton X-100 in distilled water

Procedure. The reaction mixture contains 0.1 ml of potassium ferricyanide, 0.5 ml of McIlvaine buffer, 0.1 ml of Triton X-100, enzyme solution, and 0.1 ml of D-fructose in a total volume of 1.0 ml. After preincubation at 25° for 5 min, the reaction is initiated by the addition of potassium ferricyanide and stopped by adding 0.5 ml of ferric sulfate–Dupanol reagent. Then, 3.5 ml of water are added to the reaction mixture. The resulting Prussian blue color is measured at 660 nm after the mixture has stood at 25° for 20 min. The rate of reduction of ferricyanide to ferrocyanide is correlated to the rate of D-fructose oxidation.

Definition of Unit and Specific Activity. One unit of enzyme activity is defined as the amount of enzyme catalyzing the oxidation of 1 μmol of D-fructose per minute under the conditions described above; 4.0 absorbance units equal 1 μmol of D-fructose oxidized. Specific activity (units per milligram of protein) is based on the protein estimation by Lowry *et al.*[2] with bovine serum albumin as the standard. A modified method described by Dulley and Grieve[3] is employed for the samples that contain Triton X-100. In this method, sodium lauryl sulfate is added to the alkaline solution.

Source of Enzyme

Microorganism and Culture. Gluconobacter industrius IFO 3260 can be obtained from the Institute for Fermentation, Osaka (17-85, Juso-honmachi

[1] W. A. Wood, R. A. Fetting, and B. C. Hertlein, this series, Vol. 5, p. 287.
[2] O. H. Lowry, N. J. Rosebrough, A. L. Farr, and R. J. Randall, *J. Biol. Chem.* **193**, 265 (1951).
[3] J. R. Dulley and P. A. Grieve, *Anal. Biochem.* **64**, 136 (1975).

2-chome, Yodogawa-ku, Osaka 532, Japan). The stock culture is maintained on a potato–glycerol slant (see this volume [24]).

The culture medium consists of glycerol (4 g), sodium glutamate (6 g), 500 mg each of KH_2PO_4 and K_2HPO_4, 200 mg of $MgSO_4 \cdot H_2O$, and 10 mg each of $FeSO_4 \cdot 7 H_2O$, NaCl, and $MnSO_4 \cdot 7 H_2O$ in 1 liter of tap water. Also, 400 μg each of thiamine hydrochloride, nicotinic acid, and calcium pantothenate and 100 μg of p-aminobenzoic acid are added per liter as supplements. The organism grown on the potato–glycerol slant is inoculated into 100 ml of the medium in a 500-ml shake flask, and cultivation is carried out at 30° for 24 hr with reciprocal shaking. For large-scale culture, 25 liters of the medium are inoculated into 50-liter jar fermentor at 30° under vigorous agitation (500 rpm) and aeration (25–30 liters of air per minute). Bacterial cells are harvested at the late exponential phase.

Purification Procedure[4]

All operations are carried out at 0–5°, unless otherwise stated. McIlvaine buffer at various pH levels are used. The following purification is described for 50 g of wet cells of *G. industrius*.

Step 1. Preparation of Membrane Fraction. Cells are harvested from 25 liters of a 20-hr culture, and the cell paste is suspended in distilled water and passed twice through a French press (American Instrument Co.) at 1000 kg/cm². Intact cells are removed by centrifugation at 5000 g for 10 min. The resulting supernatant is designated the cell homogenate. The cell homogenate is then centrifuged at 68,000 g for 90 min; the sedimented pellet, designated the membrane fraction, can be stored at −20° for over 6 months without appreciable loss of enzyme activity.

Step 2. Solubilization of Enzyme. The membrane fraction is suspended in 20-fold diluted McIlvaine buffer, pH 6.0, and the protein concentration is adjusted to 30 mg/ml. To the suspension, 10% Triton X-100 and 2-mercaptoethanol are added to final concentrations of 1% and 1 mM, respectively. The suspension is gently stirred for 3 hr, solubilizing the enzyme from the membrane fraction. Cell debris is removed by centrifugation at 68,000 g for 60 min, and a clear rose-red supernatant is retained.

Step 3. DEAE-Cellulose Column Chromatography. The supernatant solution (340 ml) is applied to a column of DEAE-cellulose (2.5 × 20 cm) that has been equilibrated with 20-fold diluted buffer, pH 6.0, containing 0.1% Triton X-100 and 1 mM 2-mercaptoethanol. After the column is further washed with 300 ml of the same buffer to remove nonadsorbed protein, the enzyme is eluted with a descending pH gradient composed of

---

[4] M. Ameyama, E. Shinagawa, K. Matsushita, and O. Adachi, *J. Bacteriol.* **145**, 814 (1981).

PURIFICATION OF D-FRUCTOSE DEHYDROGENASE OF
Gluconobacter industrius

Fraction	Total protein (mg)	Total activity (units)	Specific activity (units/mg protein)	Yield (%)
Cell homogenate	8345	6955	0.8	100
Membrane fraction	5050	5115	1.0	73
Solubilized fraction	966	5940	6.2	85
DEAE-cellulose	121	4190	34.6	60
Hydroxyapatite	26	2500	96.4	40
Polyethylene glycol	13	1135	87.3	16
Final preparation	9	1550	172.0	22

500 ml of 20-fold diluted buffer, pH 6.0, and 500 ml of 20-fold diluted buffer, pH 4.5. Both buffer solutions are also supplemented with 0.1% Triton X-100 and 1 mM 2-mercaptoethanol. Ten-milliliter fractions are collected, and the D-fructose dehydrogenase elutes at about pH 5.2 as a rose-red material. Fractions containing enzyme activity are combined (300 ml) and concentrated by membrane filtration (Toyo ultrafilter UP-50) or dehydration with polyethylene glycol 6000. The concentrated enzyme solution is dialyzed against 20-fold diluted buffer, pH 6.0, containing 0.1% Triton X-100 and 1 mM 2-mercaptoethanol.

Step 4. Hydroxyapatite Column Chromatography. The dialyzed enzyme solution is then applied to a hydroxyapatite column (2 × 5 cm) that has been equilibrated with the same buffer as that used for dialysis. After the column is washed with 100 ml of the same buffer, elution of the enzyme is accomplished by a gradient of McIlvaine buffer, pH 6.0. One container is filled with 200 ml of 20-fold diluted buffer, and another is filled with 200 ml of undiluted buffer. Triton X-100 (0.1%) and 2-mercapto-ethanol (1 mM) are also present in both buffers. The enzyme activity is eluted at 200–250 ml. The active enzyme fractions of about 75 ml are placed in a dialysis tubing and concentrated to about 5 ml by dehydration with polyethylene glycol 6000. After removal of excess polyethylene glycol 6000 in the enzyme solution by extensive dialysis against 20-fold diluted buffer, pH 6.0, containing 0.1% Triton X-100 and 1 mM 2-mer-captoethanol, insoluble materials are removed by centrifugation at 12,000g for 20 min. The results of a typical purification are given in the table. The purified D-fructose dehydrogenase, prepared as above, usually has a specific activity of about 180 units per milligram of protein per minute under the standard assay conditions.

Properties[4]

Homogeneity. The enzyme shows a symmetrical sedimentation peak with an apparent sedimentation constant of 5.8 S in the presence of 1% Triton X-100 in 20-fold diluted McIlvaine buffer, pH 6.0. Gel filtration also shows a symmetrical elution peak of protein coincident with enzyme activity when analyzed by a column of Sephadex G-200 (1 × 115 cm) that has been equilibrated with 20-fold diluted buffer, pH 6.0, containing 0.1% Triton X-100 and 1 mM 2-mercaptoethanol. The apparent molecular weight was 140,000 by gel filtration. For the estimation of purity and homogeneity of the enzyme, the use of a conventional polyacrylamide gel electrophoresis is not recommended. The purified enzyme dissociates into subunit components during gel electrophoresis and gives three protein bands.

Absorption Spectra. The purified D-fructose dehydrogenase has a cytochrome c-like absorption spectrum. A partially reduced hemoprotein is present in the enzyme preparation, suggesting that the hemoprotein is more autoxidizable when compared with heme c in alcohol dehydrogenase of acetic acid bacteria (see this volume [76]). Absorption maxima at 553–550 nm, 523 nm, and 417 nm are observed for the reduced enzyme, and a single peak at 409 nm is observed for the oxidized enzyme.

Subunit Composition. The purified enzyme has three components, which dissociate in sodium dodecyl sulfate gel electrophoresis. The components have apparent molecular weights of 67,000, 50,800, and 19,700, and the sum of these components is 137,500 for the enzyme complex. When the purified enzyme is subjected to a conventional polyacrylamide gel electrophoresis prepared at pH 8.3 and run at pH 9.4, three protein bands are also stained. Enzyme activity is detected in the slowest-moving band (R_f = 0.25) the molecular weight of which is 67,000. When an unstained gel is irradiated by fluorescent light, intense fluorescence is observed only with this activity band. This indicates the presence of a flavin dehydrogenase in the D-fructose dehydrogenase complex. When an unstained gel is scanned at 550 nm, the middle protein band (R_f = 0.41), with an apparent molecular weight of 50,800, is found to be cytochrome c. The function of the smallest component of the enzyme (R_f = 0.75) remains unclear.

Stability. There is no significant loss activity when the dehydrogenase is stored at 0–4° in McIlvaine buffer, pH 4.0 to 5.0, containing 0.1% Triton X-100 and 1 mM 2-mercaptoethanol, for at least 2 weeks. Addition of sucrose, glycerol, or fructose to 5–10% has a stabilizing effect. The enzyme activity is completely preserved when the enzyme is frozen at −20° or below. The presence of Triton X-100 in the enzyme solution is essential for the preservation of activity of the purified enzyme, and it becomes labile and less active when the detergent is removed from the enzyme solution.

Catalytic Properties. The D-fructose dehydrogenase can be assayed *in vitro* in the presence of any one of the following dyes as an electron acceptor: potassium ferricyanide, phenazine methosulfate, nitro blue tetrazolium, or 2,6-dichlorophenolindophenol. NAD, NADP, or oxygen are completely inactive as electron acceptors. This finding applies to other membrane-bound dehydrogenases (see this volume [24], [31], [76]). Of tested substrates, only D-fructose is oxidized. When D-fructose (100 μmol) is oxidized in the presence of substrate analogs (100 μmol each), such as D-glucose, D-mannose, D-fructose 6-phosphate, D-glucose 1-phosphate, D-gluconate, 2-keto-D-gluconate, 5-keto-D-gluconate, the reaction rate is not affected. An apparent Michaelis constant for D-fructose at pH 4.5 is 10 mM. The pH optimum is pH 4.0, and optimum temperature is about 25°. The reaction product of D-fructose oxidation is identified as 5-keto-D-fructose by paper chromatography. The reaction product is specifically reduced to D-fructose in the presence of NADPH and 5-keto-D-fructose reductase, which can be crystallized from the cytosol of *G. industrius.*[5]

[5] M. Ameyama, K. Matsushita, E. Shinagawa, and O. Adachi, *Agric. Biol. Chem.* **45**, 863 (1981).

[26] D-Glucose Dehydrogenase from *Gluconobacter suboxydans*

By OSAO ADACHI and MINORU AMEYAMA

D-Glucose + NADP $\rightleftarrows$ D-glucono-δ-lactone + NADPH

Assay Method

Principle. D-Glucose dehydrogenase (EC 1.1.1.47) is measured spectrophotometrically by following the rate of NADPH formation at 340 nm.

Reagents

NADP, 1.5 mM, in distilled water; stored at $-20°$ until use
D-Glucose, 1 M, in distilled water
Tris-HCl buffer, 1 M, pH 8.0
Enzyme, dissolved in 0.01 M potassium phosphate, pH 7.5, containing 1 mM 2-mercaptoethanol

Procedure. The enzyme activity is measured by reading the increase of absorbance at 340 nm in a recording spectrophotometer thermostatted at 25° in a thermostatted room (25°). Prior to the assay, the buffer solution and distilled water are warmed to 25°. The complete reaction mixture contains 0.1 ml of NADP, 0.1 ml of D-glucose, 0.3 ml of Tris-HCl and enzyme solution in a total volume of 3.0 ml. The reaction is initiated by the addition of enzyme.

Definition of Unit and Specific Activity. One enzyme unit causes the formation of 1 μmol of NADPH per minute under the assay conditions described. Specific activity (units per milligram of protein) is based on the spectrophotometric protein estimation by measuring at 280 nm; $E_{cm}^{1\%}$ value of 15.0 is used throughout.

Source of Enzyme

Microorganism and Culture. Gluconobacter suboxydans IFO 12528 can be obtained from the Institute for Fermentation, Osaka (17-85, Juso-honmachi 2-chome, Yodogawa-ku, Osaka 532, Japan). This is the same strain as that used for preparation of crystalline 6-phospho-D-gluconate dehydrogenase (see this volume [50]). Cultivation of the organism is also performed under the same conditions as mentioned for 6-phospho-D-gluconate dehydrogenase [50].

Purification Procedure[1]

All operations are carried out at 0–5°, unless otherwise stated. Potassium phosphate buffer containing 1 mM 2-mercaptoethanol is used throughout. The procedure reported is for 150 g of wet cells of *G. suboxydans.* Centrifugations are performed at 12,000 g for 20 min.

Step 1. Cell-Free Extract. Cell paste of *G. suboxydans* harvested from 30 liters of a 24-hr culture in a 50-liter jar fermentor is washed twice with cold water and then suspended in 0.01 *M* buffer, pH 7.5. After cell disruption by a French press (American Instrument Co.) at 1000 kg/cm², the cell homogenate is centrifuged at 68,000 g for 90 min; D-glucose dehydrogenase is present in the supernatant (1600 ml).

Step 2. DEAE-Sephadex Column Chromatography (I). The enzyme solution is applied to a DEAE-Sephadex A-50 column (5 × 35 cm) that has been equilibrated with 0.01 *M* buffer, pH 6.0. About 50% of the total

---

[1] O. Adachi, K. Matsushita, E. Shinagawa, and M. Ameyama, *Agric. Biol. Chem.* **44,** 301 (1980).

enzyme activity passes through the column, and a large amount of impure protein is adsorbed.

Step 3. DEAE-Sephadex Column Chromatography (II). The enzyme solution from step 2 is adjusted to pH 7.5 with ammonia and applied to a second DEAE-Sephadex A-50 column (2 × 20 cm), which has been equilibrated with 0.01 M buffer, pH 7.5. Nonadsorbed proteins are washed through with the same buffer. Elution of the enzyme is affected by a linear gradient of NaCl formed from 500 ml of 0.01 M buffer, pH 7.5, and 500 ml of the same buffer containing 0.4 M NaCl. The dehydrogenase is eluted at about 0.15 M concentration of the salt. To the combined enzyme solution, ammonium sulfate is added to 3.9 M (51.5 g/100 ml). The precipitated enzyme is collected by centrifugation and dissolved in 0.01 M buffer, pH 7.5, and dialyzed thoroughly against the same buffer.

Step 4. Affinity Chromatography on Blue-Dextran Sepharose. After removal of insoluble materials by centrifugation, the dialyzed enzyme is applied to a Blue-Dextran Sepharose 4B column (2 × 20 cm) equilibrated with 0.01 M buffer, pH 7.5. The column is treated with the same buffer until the absorbance of the eluate at 280 nm decreases to 0.05. Elution of the enzyme is made with 0.01 M buffer containing 0.35 M KCl, and a sharp protein peak corresponding to the enzyme activity is eluted. The pooled fractions are dialyzed overnight against 0.01 M buffer, pH 7.5, containing 0.1 M NaCl.

Step 5. DEAE-Sephadex Column Chromatography (III). The dialyzed enzyme is applied to a DEAE-Sephadex A-50 column (1.5 × 10 cm) equilibrated with 0.01 M buffer, pH 7.5, containing 0.1 M NaCl. The column is washed with the same buffer and eluted by a linear gradient of NaCl formed by 200 ml of 0.01 M buffer, pH 7.5, containing 0.1 M NaCl and 200 ml of 0.01 M buffer, pH 7.5, containing 0.3 M NaCl. A major protein peak elutes around at 0.15 M NaCl is coincident with enzyme activity. Peak fractions with specific activities over 110 units per milligram of protein are combined and dialyzed against saturated ammonium sulfate containing 0.1 M D-glucose, pH 7.0, until the enzyme precipitates in the dialysis bag.

Step 6. Crystallization. The precipitate is collected by centrifugation and dissolved in a minimum volume of 0.1 M buffer, pH 7.5. After standing for a few hours in the cold, insoluble materials are removed by centrifugation. A few grains of solid ammonium sulfate are added to the enzyme solution, and the solution is kept in a refrigerator. Crystals of the enzyme appear as fine rods or needles.

The enzyme is purified about 1800-fold with a yield of 30% as shown in the table.

PURIFICATION OF D-GLUCOSE DEHYDROGENASE FROM
Gluconobacter suboxydans

Step	Total protein (mg)	Total activity (units)	Specific activity (units/mg protein)	Yield (%)
Cell-free extract	14,640	1160	0.08	100
DEAE-Sephadex A-50 (I)	5,640	570	0.10	50
DEAE-Sephadex A-50 (II)	1,290	435	0.34	38
Blue-Dextran Sepharose	15	403	26.62	35
DEAE-Sephadex A-50 (III)	4	390	97.50	34
Crystallization	2.5	350	140.00	30

Properties

Homogeneity. The sedimentation pattern of the enzyme shows a single symmetrical peak with an apparent sedimentation constant of 5.8 S. Upon gel electrophoresis, the enzyme shows a single protein band coincident with enzyme activity.

Molecular Properties. The absorption maximum is at 280 nm, and the $E_{cm}^{1\%}$ value is estimated to be 4.0 on the basis of absorbance and dry weight determinations. The molecular weight of the enzyme is determined to be 153,000 by gel filtration on Sephadex G-200 column. The enzyme dissociates into four identical subunits having a molecular weight of about 40,000 each.

Catalytic Properties. General catalytic properties of D-glucose dehydrogenase have been extensively studied with partially purified enzyme preparations.[2,3] A crystalline preparation as a suspension in ammonium sulfate is quite stable at 5° for several months. Even in the absence of ammonium sulfate, appreciable loss of activity is not observed after storage at 5° for 2 weeks. The optimum pH for D-glucose oxidation is 8.5–9.0, and a higher reaction rate is usually observed in Tris-HCl than in potassium phosphate. The optimum temperature is 50°. D-Glucose dehydrogenase is specific to NADP and completely inactive with NAD. The activity is potently inhibited by sulfhydryl reagents and divalent heavy metals such as *p*-chloromercuribenzoate, Hg^{2+}, Cu^{2+}, and Ni^{2+}. The enzyme possesses broad substrate specificity in the order D-glucose, 100%, D-mannose, 88%, mannose, 6%, D-glucose 6-phosphate, 1%, and

[2] K. Okamoto, *J. Biochem.* **53**, 348 (1963).
[3] G. Avigad, Y. Alroy, and S. Englard, *J. Biol. Chem.* **243**, 1936 (1968).

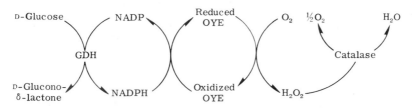

SCHEME 1. GDH, D-glucose dehydrogenase; OYE, old yellow enzyme.

D-galactose, 1%. The enzyme is inert to D-fructose, D-arabinose, D-xylose, and L-sorbose. The reaction product of the enzyme is D-glucono-δ-lactone. The apparent Michaelis constants for D-glucose and NADP are $5 \times 10^{-3}\ M$ and $1 \times 10^{-5}\ M$, respectively. In the reverse direction, only 4% of the reaction occurs at neutral pH.

Participation of Old Yellow Enzyme in Regeneration of NADP. It is indicated that NADPH formed in D-glucose oxidation is spontaneously oxidized to NADP by an enzyme, NADPH dehydrogenase, that exists predominantly in the cytoplasma of *Gluconobacter* species. The enzyme is similar to the old yellow enzyme in yeast.[4] An experiment similar to that described for 6-phospho-D-gluconate dehydrogenase (see this volume [50]) is conducted in a conventional Warburg flask: D-glucose dehydrogenase, old yellow enzyme, excess amounts of D-glucose, and limited amounts of NADP or NADPH are incubated in the presence of catalase. A linear oxygen uptake is observed showing that a cyclic coupling system is functioning to regenerate NADPH, as shown in Scheme 1.

[4] O. Adachi, K. Matsushita, E. Shinagawa, and M. Ameyama, *J. Biochem.* **86,** 699 (1979).

[27] Galactose Oxidase from *Dactylium dendroides*

By PAUL S. TRESSEL and DANIEL J. KOSMAN

Galactose oxidase (EC 1.1.3.9) is a Cu(II)-containing extracellular fungal enzyme that catalyzes the reaction (1).[1-6]

$$RCH_2OH + O_2 \rightarrow RCHO + H_2O_2 \tag{1}$$

Although the enzyme exhibits some activity toward a variety of primary alcohols[7-12] (and *only* primary alcohols[11]), among hexoses it is nearly specific for the C_6-OH of galactose.[3,13,14] Because of this high degree of

[1] J. A. D. Cooper, W. Smith, M. Bacila, and H. Medina, *J. Biol. Chem.* **234,** 445 (1959).

METHODS IN ENZYMOLOGY, VOL. 89

hexose specificity, galactose oxidase has been used successfully to oxidize such residues in sphingoglycolipids,[15] galactolipids,[16] and glycoproteins.[17] The resulting terminal C_6-aldehyde is subsequently reduced with [³H]NaBH₄, thus radiolabeling the biomolecule under extremely mild conditions.[17] The enzyme has also been immobilized in bioelectrodes that respond to galactose as nonreducing termini (lactose) and other primary alcohols (glycerol) in biological fluids. The oxidation of these metabolites is followed either by O_2-[18] or H_2O_2-sensitive[19] electrodes. This chapter presents the rather simple methodology for the growth of one[20] of the fungi that secrete the enzyme[21] and the purification of the enzyme from the fungal growth medium.[3,6,11,22] These procedures are, in part, based on extensive work by Horecker and his collaborators, who contributed an earlier chapter on this enzyme.[22]

Fungal Growth

Strain. Dactylium dendroides (NRRL 2903) is available from the USDA, Northern Regional Research Laboratories, Peoria, Illinois.[23]

[2] G. Avigad, C. Asensio, D. Amaral, and B. L. Horecker, *Biochem. Biophys. Res. Commun.* **4**, 474 (1961).

[3] G. Avigad, D. Amaral, C. Asensio, and B. L. Horecker, *J. Biol. Chem.* **237**, 2736 (1962).

[4] D. Amaral, L. Bernstein, D. Morse, and B. L. Horecker, *J. Biol. Chem.* **238**, 2281 (1963).

[5] F. Kelly-Falcoz, H. Greenberg, and B. L. Horecker, *J. Biol. Chem.* **240**, 2966 (1965).

[6] D. J. Kosman, M. J. Ettinger, R. Weiner, and E. J. Massaro, *Arch. Biochem. Biophys.* **165**, 456 (1974).

[7] G. Zancan and D. Amaral, *Biochim. Biophys. Acta* **198**, 146 (1970).

[8] D. J. Kosman, L. D. Kwiatkowski, M. J. Ettinger, and J. D. Brodie, *Fed. Proc., Fed. Am. Soc. Exp. Biol.* **32**, 550, Abs. (1973).

[9] L. D. Kwiatkowski and D. J. Kosman, *Biochem. Biophys. Res. Commun.* **53**, 715 (1973).

[10] G. Hamilton, J. DeJersey, and P. Adolf, *in* "Oxidases and Related Redox Systems" (T. E. King, H. S. Mason, and M. Morrison, eds.), p. 103. University Park Press, Baltimore, Maryland, 1973.

[11] P. Tressel and D. J. Kosman, *Anal. Biochem.* **105**, 150 (1980).

[12] P. Tressel, Ph.D. dissertation, State University of New York, Buffalo, 1980.

[13] R. Schlegel, C. Gerbeck, and R. Montgomery, *Carbohydr. Res.* **7**, 193 (1968).

[14] A. Maradufu and A. Perlin, *Carbohydr. Res.* **32**, 93 (1974).

[15] R. M. Bradley and J. N. Kanfer, *Biochim. Biophys. Acta* **84**, 210 (1964).

[16] C. A. Lingwood, *Can. J. Biochem.* **57**, 1138 (1979).

[17] A. G. Morell and G. Ashwell, this series, Vol. 28, p. 205.

[18] S. Dahodwala, M. Weibel, and A. Humphrey, *Biotechnol. Bioeng.* **18**, 1679 (1976).

[19] J. Johnson, YSI, Yellow Springs, Ohio, personal communication.

[20] M. Nobles and C. Madhosingh, *Biochem. Biophys. Res. Commun.* **12**, 146 (1963).

[21] J. Gancedo, C. Gancedo, and C. Asensio, *Arch. Biochem. Biophys.* **119**, 508 (1967).

[22] D. Amaral, F. Kelly-Falcoz, and B. L. Horecker, this series, Vol. 9, p. 87.

[23] F. Lombard, Chief Mycologist at the Forest Products Laboratory, Madison, Wisconsin,

Stationary Cultures. Fungal mycelia are maintained on 1.5% agar slants prepared using the liquid medium described under Starter Flasks (see below). A small tuft of mycelia is transferred to a slant and allowed to grow for 3–5 days at 25° in the dark. The tube is then covered with aluminum foil and stored at 4°. Such slants remain viable for ∼3 months. Dextrose Sabouraud agar slants (Difco Co., Detroit, Michigan) may also be used.[3,6,11]

Liquid (Shake Flask) Cultures. Although galactose oxidase has been successfully isolated from the medium of fermentor-grown *D. dendroides,*[24] growth in shake flask cultures is simpler and more efficient for the occasional or one-time preparation. Growth in liquid medium involves an initial 3-day "starter flask" or inoculum growth followed by a 6-day "culture flask" preparation. The media for both are prepared and autoclaved in three separate fractions, the only difference being the carbon source: dextrose [D(+)-glucose] is used in starter flasks, and L(−)-sorbose is used in culture flasks (Table I). This combination appears to maximize galactose oxidase production by the fungus.[11,24–26] The quantities in Table I are for 20 × 2 liter (Erlenmeyer) flasks (culture flasks), each containing 1250 ml of total medium; and 2 × 500 ml (Erlenmeyer) flasks (starter flasks) each containing 268 ml of total medium. Depending upon the level of enzyme production, which does vary, 30–60 mg of galactose oxidase can be isolated from a culture of this size.

Starter Flasks. Using sterile technique, to each of the 2 × 500 ml "salts" flasks (A), add one each of the 34-ml "metals" solution (B) and glucose solution (C), and 20 μl of thiamine stock solution. Five small tufts of mycelia are transferred from a stationary slant to each of the starter flasks. (A sealed 9-inch Pasteur pipette with a small hook bent in the tip is convenient for this transfer). The flasks are shaken at 300 rpm (1-inch radius) for 3 days at room temperature.

Stationary Slants. These slants are prepared by first dissolving 1.5 g of agar in 75 ml of solution A (salts) and autoclaving the mixture. While still hot, 12.5 ml each of sterile solutions B (metals) and C (glucose) and 10 μl of thiamine stock are added. This mixture is distributed using sterile technique into 10 capped, sterile culture tubes (16 × 150 mm) and allowed to cool. Either cotton plugs or "capits" may be used. The tubes are wrapped in foil and stored at 4°.

---

reports that this is the original Bacila strain[1]; personal communication to Daniel J. Kosman.

[24] Z. Marcus, G. Miller, and G. Avigad, *Appl. Microbiol.* **13**, 686 (1965).
[25] A. R. Shatzman and D. J. Kosman, *J. Bacteriol.* **130**, 455 (1977).
[26] A. R. Shatzman and D. J. Kosman, *Biochim. Biophys. Acta* **544**, 163 (1978).

TABLE I
FUNGAL GROWTH MEDIUM FOR *Dactylium dendroides*

Flask A: Salts, nitrogen source	Flask B: Trace metals	Flask C: Carbon source
Solution A	Solution B	Solution C
219 g Na$_2$HPO$_4$ (62 mM)^a	10.7 g MgSO$_4$ · 7 H$_2$O (1.6 mM)	Starter flask:
212 g KH$_2$PO$_4$ (62 mM)	50 mg MnSO$_4$ · H$_2$O (12 μM)	5.4 g of glucose
26 g (NH$_4$)$_2$NO$_3$ (13 mM)	77 mg ZnSO$_4$ · 7 H$_2$O (11 μM)	(55 mM) in
51 g (NH$_4$)$_2$SO$_4$ (15 mM)	40 mg CaCl$_2$ · H$_2$O (11 μM)	68 ml H$_2$O
19 g NaOH (15 mM)	40 mg FeSO$_4$ (11 μM)	(2 × 34 ml in
19 g KOH (14 mM)	67 mg CuSO$_4$ · 5 H$_2$O (11 μM)^{b,c}	50-ml flasks)
20.4 liters H$_2$O	2.57 liters H$_2$O	
		or
Autoclaved as 1.0 liter in	Autoclaved as 250 ml in	Culture flask:
20 × 2 liter flasks and	10 × 500 ml flasks and 34 ml	250 g of sorbose
200 ml in 2 × 500 ml	in 2 × 50 ml flasks	(55 mM), 2.5
flasks		liters H$_2$O
Add solution from flasks		Autoclaved as 250 ml
B and C		in 10 × 500 ml
		flasks
Add thiamine stock solution:		
337 mg/10 ml (5 μM), not		
autoclaved; prepared in		
sterile H$_2$O and kept		
frozen between uses		

^{a} Concentrations in final growth medium when prepared as described in the text.

^{b} Adding the equivalent ^{63}Cu as ^{63}CuNO$_3$ prepared by dissolving ^{63}Cu metal (powder or shot, Alfa Inorganics, Danvers, Massachusetts) in 1 N HNO$_3$ and diluting yields isotopically pure ^{63}Cu enzyme for magnetic resonance studies.

^{c} Use of Analar Salts (BDH, obtained from Gallard-Schlesinger, Carle Place, New York) or equivalent and omission of CuSO$_4$ yields a Cu-free medium. Fungal growth is not affected, and an equivalent amount of galactose oxidase is secreted as an apoprotein.[24,25]

Culture Flasks. For each pair of 20 × 2-liter flasks (salts, A) are mixed one each of the metals (B) and sorbose (C) flasks. This mixture is then divided equally between 2 × 2-liter flasks (A). To each of the 20 × 2-liter flasks are then added 100 μl of thiamine stock followed by no more than 15 ml of inoculum from the starter flasks. A light inoculum (~0.1% v/v or ~1g cells) appears to lead to greater enzyme production[24,25]; enzyme production decreases at high culture density.[25] The culture is grown at 20–25° for 6 days on a rotary shaker, 1–2-inch radius, 100–150 rpm (a larger radius requires a slower rate). Temperatures higher than 25° adversely affect enzyme production; below 18° little growth occurs. Enzyme production can be monitored during growth by removing ~0.5 ml medium from several of the flasks, pooling the samples, and assaying the pooled sample for enzyme activity (see below).

Enzyme Assay

Galactose oxidase has most commonly been assayed by monitoring the production of H_2O_2 (Reaction 1).[3] This is achieved by coupling this product to the horseradish peroxidase (HRP)-catalyzed oxidation of one of a number of chromogens, i.e., o-toluidine,[13] o-dianisidine,[3,22] o-cresol,[27] etc. The oxidized forms of these latter reagents absorb strongly in the visible region and thus provide a sensitive measure of galactose oxidase activity. There are at least two serious drawbacks to this assay; HRP activates galactose oxidase[28] whereas some of the chromogens inhibit.[27] They are also known or potential carcinogens. Thus, the coupled assay must be used with caution and only for qualitative purposes (as for monitoring enzyme production during culture growth).[29]

Coupled Assay

Assay Solution. To 100 ml of 0.1 M sodium phosphate buffer, pH 7.0, add:

  $D(+)$-Galactose, 500 mg (Sigma, "substantially glucose free," or equivalent)

  Horseradish peroxidase, 5 mg (Sigma, type III, mixture of basic isozymes)

  o-Dianisidine (3,3'-dimethoxybenzidine), 5 mg, dissolved in 0.5 ml of methanol

This solution should be kept in the cold and protected from the light when not in use. It will last ~2–4 weeks if these precautions are taken. Discard when the absorbance at 460 nm becomes greater than 0.1 versus buffer alone.

Method. A recording spectrophotometer operating at 460 nm and ~25° is used. The linear absorbance increase is followed for ~2 min, and the ΔA/min is calculated. The amount of enzyme added should be adjusted so that this value is between 0.2 and 0.6. Commonly, 5–50 μl of sample can be added to 1.0 ml of assay mixture. In a 1-cm cell, 1 absorbance unit at 460 nm is equivalent to 33 nmol of H_2O_2,[6] although an enzyme unit (EU) is that amount of enzyme giving ΔA/min $= 1.0$.[29]

---

[27] W. Fischer and J. Zapf, *Hoppe-Seyler's Z. Physiol. Chem.* **337**, 186 (1964).

[28] P. Tressel and D. J. Kosman, *Biochem. Biophys. Res. Commun.* **92**, 781 (1980).

[29] A coupled assay using an NADH-peroxidase may avoid at least some of these limitations [G. Avigad, *Anal. Biochem.* **86**, 470 (1978)]. Avigad notes that if a unit is defined as 1.0 A/min for all coupled assays, for a given quantity of enzyme, the relative "unit activity" is 1:6.4:28.8 (NADH:o-dianisidine:o-toluidine), i.e., o-toluidine is the most sensitive.

Direct Assay

Aside from manometric and polarographic techniques, which monitor O_2 consumption, other direct assays employ alcohol substrates that are themselves chromophoric. Of the several that have been used,[8,11,12] 3-methoxybenzyl alcohol appears to be the most reliable. The oxidation to the aldehyde is accompanied by a $\Delta\epsilon_{314} = 2691\ M^{-1}\ cm^{-1}$, and this can be readily monitored.[11]

Assay Solution. Use 0.06 M 3-methoxybenzyl alcohol (Aldrich Chemical Co.) in 0.1 M sodium phosphate buffer, pH 7.0.

Method. A recording spectrophotometer operating at 314 nm and ~25° is used. The linear absorbance increase is followed for 2–3 min as above. This assay is about one-tenth as sensitive as the coupled assay at the concentrations of alcohol substrate and the $\Delta\epsilon$ associated with each. Consequently, either more enzyme must be employed or a scale expansion used, if available. In an assay volume of 1.0 ml, a $\Delta A_{314\ nm} = 1.0$ is equivalent to 0.37 μmol of product aldehyde formed.

Enzyme Purification

Enzyme activity in the culture medium plateaus between day 5 and day 6 at 8–20 EU/ml (coupled assay, o-dianisidine). The origin of this variability has not been elucidated, although temperature, bacterial contamination, stationary slant viability, size of inoculum, and density of culture growth can all be contributing factors. In the 25 liters of a 20-flask preparation, this represents about 60–150 mg of galactose oxidase. Purification proceeds as follows. All operations are performed at 4°.

1. The contents of the culture flasks are filtered through 210-μm nylon mesh bags (CMN-210, Small Parts, Inc., Miami, Florida) (double layers of cheesecloth can substitute); the mycelia are squeezed dry and discarded. The filtrate is most readily collected in three 4-gallon plastic buckets.

2. The three buckets are stirred magnetically (3 × ½ inch bar, high-torque magnetic stirrer), and 50 g of microcrystalline cellulose (Sigma Cell-50, Sigma Chemical Co., St. Louis, Missouri) are added to each. The mixture is stirred for 30 min.

3. Ammonium sulfate (ACS reagent grade, strained to eliminate lumps) is added to 90% saturation (at 4°) over 45 min. This amounts to 5.3 kg/25 liters. This mixture is allowed to stir for an additional 45 min.

4. The precipitate (and cellulose) is collected by suction filtration on a Büchner funnel 33 cm in diameter using a double layer of grade 617 paper (VWR, Rochester, New York). The filter cake is washed with 60% satu-

rated $(NH_4)_2SO_4$ adjusted to pH 7.0 with NH_4OH, then suspended in ~500 ml of this same solution (total volume ~700 ml) and stirred magnetically to a smooth slurry.

5. This slurry is poured into a 5 × 50 cm chromatography column and allowed to settle; then the $(NH_4)_2SO_4$ is drained (1 drop/sec). Care should be taken to avoid trapping air bubbles. The protein is eluted from the packed column with ~500 ml of 0.1 M sodium phosphate buffer, pH 7.0. A reddish-brown front develops that contains the galactose oxidase. Approximately 200–300 ml of this colored eluate are collected and dialyzed against 6.5 liters of 10 mM sodium phosphate buffer, pH 7.0, containing 0.5 mM EDTA and 0.4 mM $CuSO_4$. The dialysis may proceed overnight, but all above operations must be completed in 1 day.

6. After an additional two changes of the dialysis buffer (omitting each time the EDTA and $CuSO_4$), the dialyzate is treated with 50 g of wet DEAE-cellulose (equilibrated with the 10 mM buffer). After stirring for 15 min, the DEAE is removed by gentle suction filtration (two layers of grade 617 paper, 7–9 cm are used), and, if not clear, the light amber mother liquor is centrifuged for 30 min at 20,000 g. This solution is then concentrated (Amicon PM-10 membrane, 76 mm at 40 psi) to ~20 ml. The concentrate is dialyzed against two changes of 1 liter of 5 mM sodium phosphate buffer, pH 7.8, and again clarified by centrifugation, if necessary. The above operations should be completed in 1 day.

7. The dialyzate is adsorbed onto a 2 × 3 cm phosphocellulose column (Bio-Rad Cellex-P equilibrated with the 5 mM buffer). This must be done carefully, as the avidity with which galactose oxidase binds to the phosphocellulose can cause the protein to "channel" into the cellulose. The column is washed with 25 ml of the 5 mM buffer. A salmon-colored fraction is eluted that contains ~5% of the total activity loaded on the column. The enzyme, which is readily apparent as a blue-green band at the top of the column, is eluted with 15 ml of 25 mM, followed by 15 ml of 30 mM, sodium phosphate buffers, pH 7.8. Since this column can be run quickly (~0.5 ml/min), a fraction collector is not essential. The enzyme can be collected in 4–5 fractions of 5 ml each. The peak tube(s) will be distinctly greenish blue. The presence of a yellow contaminant will also be apparent because of the yellowish caste of the fractions.

8. This yellow contaminant can be removed by either a second pass over the phosphocellulose column as outlined above[11] or by Sepharose 6B chromatography.[30] The latter involves pressure dialysis of the pooled fractions (~20 ml) against 0.1 M ammonium acetate buffer, pH 7.2, and concentration to 10–15 ml. The concentrate is loaded onto a 1.2 × 25 cm column of Sepharose 6B equilibrated with the same buffer which is also

[30] M. Hutton and E. Regoeczi, *Biochim. Biophys. Acta* **438**, 339 (1976).

TABLE II
RELATIVE SPECIFIC ACTIVITIES OF GALACTOSE OXIDASE DURING
PURIFICATION[11]

| | Relative activity | |
Step	Coupled assay[a]	Direct assay[b]
Media filtrate	1.0	1.0
(NH₄)₂SO₄ eluate	1.83	1.35
Post-DEAE batch absorption	18.55	13.96
First phosphocellulose	21.53	33.09
Second phosphocellulose or Sepharose 6B	24.45	72.17

[a] Galactose oxidation was coupled to the horseradish peroxidase–o-dianisidine reaction.

[b] The assay follows production of 3-methoxybenzaldehyde at 314 nm.[11] O_2-electrode results are similar.

used to elute the galactose oxidase (0.2 ml/min). The enzyme chromatographs as an essentially homogeneous peak following the elution of the yellow impurity.

These consecutive chromatographic steps are necessary to remove completely contaminating protein, which apparently binds to galactose oxidase, a highly basic protein (pI ~12).[6,31] Among these contaminants is one that acts as an inhibitor of turnover *in the absence of HRP*.[11,12] This is reflected by the data presented in Table II, which compares the relative specific activity of galactose oxidase during purification determined by the HRP-coupled and direct chromophoric substrate assays (as well as O_2-uptake measurements, which parallel the latter method). Thus, the degree of "purification" as measured by activity depends on the method of assay; it is ~24-fold measured by the coupled assay and ~72-fold measured by a direct assay.

Properties

The galactose oxidase so purified has a specific activity of 3600 EU/mg (coupled assay, o-dianisidine) or 166 μmol of 3-methoxybenzaldehyde per minute per milligram in the direct assay.[11] The turnover number at 100% O_2 ($K_m \geq 3$ mM[32]), 0.1 M galactose ($K_m = 175$ mM[32]), and pH 7.0 is 650 sec^{-1} using a Clark oxygen electrode. These represent the three most

[31] S. Bauer, G. Bauer, and G. Avigad, *Isr. J. Chem.* **5**, 126 p (1967).

[32] L. D. Kwiatkowski, M. Adelman, R. Pennally, and D. J. Kosman, *J. Bioinorg. Chem.*

readily available assay parameters. The protein is isozymic, consisting of a family of 5 or 6 isozymes as visualized in both starch[6] and pulsed power acrylamide gel electrophoresis.[11] (Continuous current is not sufficient for the electrophoresis of galactose oxidase in 5–7.5% acrylamide gels.[6,11]) Based on comparisons of protein (Coomassie Brilliant Blue G) and activity[33] staining, the isozymes do not differ markedly in specific activity.[11,12,34] The protein consists of a single polypeptide chain with apparently two intrachain disulfide bonds; $M_r = 68,000$[6]; ϵ at 280 nm $= 104,900$ M^{-1} cm^{-1}; $E_{1\,cm}^{1\%} = 15.4$ at the same wavelength.[6] The Cu(II) can be resolved employing diethyldithiocarbamate,[5,22] and the holo enzyme can be reconstituted by addition of Cu(II). No other metals have been reported to bind to or activate the apoprotein. The Cu(II)/Cu(I) $E^{\circ\prime} = 0.300 \pm 0.02$ mV at pH 7.0.[35]

Care must be taken in the storage and/or shipment of galactose oxidase. Lyophilization generally results in loss of (some) activity and irreversible alteration of the Cu(II)-associated spectral properties. (Such samples are essentially colorless, i.e., have very little of the visible absorbance of active enzyme.) Similar problems occur with samples that are repeatedly frozen and thawed. (Even samples in frost-free refrigerator freezer compartments are so affected.) Thus, the following procedure is recommended. The enzyme solution, ~5 mg/ml in 0.1 M sodium phosphate buffer, pH 7.0, is filtered through a sterile Millipore 0.1 μm filter (A010A025A) into a vessel rinsed well with 6 N HNO$_3$, distilled water, and then methanol, capped with aluminum foil, and oven dried. Refrigerated solutions prepared in this way retain all activity and Cu(II) spectral properties for at least 1 month. (Bacterial contamination is the prime concern in such storage; common bacteriostatic agents, e.g., N$_3^-$, are galactose oxidase inhibitors.) For long-term storage or shipment, such solutions are quick-frozen in liquid N$_2$ (or on crushed Dry Ice). Storage is then at $-70°$ or in liquid N$_2$. Samples are unchanged even after 1 year of storage at $-70°$. Shipment is on Dry Ice or in N$_2$-thermos vessels. Shipment or storage of galactose oxidase as an ammonium sulfate suspension has not been reported. However, it should be noted that ammonium sulfate can remove Cu(II) from plasma benzylamine oxidase,[36] another nonblue or type 2 Cu(II) protein.[37]

[33] J. Robinson and G. Lee, *Arch. Biochem. Biophys.* **120,** 428 (1967).

[34] The isozymes may exhibit somewhat different Cu(II) absorbance. A measure of the purity and "nativeness" of galactose oxidase is the ratio of absorbance, at 445/355 and 280/445 (in nm); the former is ≥1.6, and the latter is ≤27; i.e., the enzyme should exhibit a well defined 445 nm Cu(II)-associated transition.

[35] D. Melnyk and M. J. Ettinger, *J. Am. Chem. Soc.* (1982), in press.

[36] L. Schallinger and M. J. Ettinger, personal communication.

[37] T. Vänngård, *in* "Biological Applications of Electron Spin Resonance" (H. M. Swartz, J. R. Bolton, and D. C. Borg, eds.), p. 411. Wiley (Interscience), New York, 1972.

[28] Galactose Oxidase from Commercial Samples

By MARK W. C. HATTON and ERWIN REGOECZI

Since the report by Morell et al.[1] of a procedure to label the galactose-terminating glycans of desialylated glycoproteins, galactose oxidase (EC 1.1.3.9) has become a popular tool for protein biochemists and cell biologists. Briefly, the desialylated glycoprotein or galactose-terminating molecule is exposed to galactose oxidase, which specifically oxidizes C-6 of the terminating galactosyl residues to an aldehyde. After the enzyme has been removed, the oxidized substrate is reduced by NaB^3H_4 to yield a product containing ^3H-labeled galactosyl residues. One of the drawbacks of the method is the presence of proteolytic enzymes and other impurities in many commercial samples of galactose oxidase.[2] This chapter describes a simple affinity procedure that removes these contaminants from galactose oxidase. Furthermore, details concerning stability and storage of the enzyme, measurements of enzymic activity, and its purity are also given.

Principle

Agarose is a polysaccharide consisting of an alternating sequence of D-galactose and 3,6-anhydro-L-galactose with $\beta(1 \to 4)$ and $\alpha(1 \to 3)$ links. During column chromatography at room temperature, galactose oxidase partially recognizes agarose as a galactose substrate and binds weakly to the polysaccharide. Consequently, galactose oxidase will be retarded by the column and will emerge well separated from the contaminating proteins, which have no affinity for the matrix.

Procedure

Samples of galactose oxidase from several suppliers (Sigma Chemical Co., St. Louis, Missouri; Worthington Biochemical Corp., Freehold, New Jersey; P-L Biochemicals Inc., Milwaukee, Wisconsin) have been purified using the procedure to be described, and similar results were obtained with all samples.[2] A column (1 cm × 10 cm) of agarose (Sepharose 6B; Pharmacia Ltd., Uppsala, Sweden) is equilibrated at room temperature with a buffer at a neutral pH. Preferred eluents included 0.1 M sodium or

---

[1] A. G. Morrell, C. J. A. Van den Hamer, I. H. Scheinberg, and G. Ashwell, J. Biol. Chem. **241**, 3745 (1966).

[2] M. W. C. Hatton and E. Regoeczi, Biochim. Biophys. Acta **438**, 339 (1976).

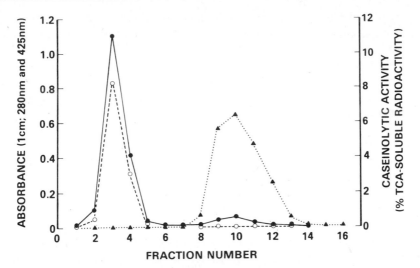

FIG. 1. Fractionation of a commercial sample of galactose oxidase (EC 1.1.3.9) on Sepharose 6B (1 cm × 10 cm) at room temperature. After dialysis against 0.1 M ammonium acetate, adjusted to pH 7.2 with NH₄OH, the crude enzyme (300 units according to the supplier) chromatographed through the column during elution with 0.1 M ammonium acetate, pH 7.2. The fraction volume was 5 ml. Absorbance measurements were made at 280 nm (●——●). Galactose oxidase activity (▲ · · · · ▲) was measured by the horseradish peroxidase–o-dianisidine procedure (as described by Hatton and Regoeczi[2]) using 10 μl of eluted fraction and an incubation time of 5 min at 37°. The samples were measured directly at 425 nm. Proteolytic activity (O――O) was determined, using ¹²⁵I-labeled casein as the substrate, by the following procedure: A volume, 100 μl, of each fraction was added to 0.4 ml of casein (5 mg/ml; trace-labeled with [¹²⁵I]casein) in 0.05 M sodium phosphate; pH 6.1. After 5 hr of incubation at 37°, 1 ml of trichloroacetic acid (20% w/v) was added and the precipitate was centrifuged to yield a clear supernatant. The extent of proteolysis, measured as the radioactivity present in the supernatant, was calculated as a percentage of protein-bound radioactivity originally present in the substrate.

ammonium acetate, pH 7.2, 0.1 M sodium phosphate, pH 7.2, or 0.02 M Tris-HCl (pH 7.5) containing 0.15 M NaCl. The commercial sample of enzyme (100–800 units, as given by the supplier) is dissolved in 1–2 ml of the equilibrating buffer and dialyzed against 1 liter of the same buffer at 4° for 20 hr. After loading the column, the protein is eluted at room temperature. Figure 1 shows a typical elution profile after measurement of the effluent fractions for protein content (absorbance at 280 nm), galactose oxidase, and proteolytic enzyme activity. The fractions containing galactose oxidase are pooled and pressure-dialyzed (N₂ back pressure, 10 psi; 8/32-inch Visking tubing) against 0.05 M sodium phosphate buffer, pH 6.5, at 4°. Concentrated preparations (0.2–0.5 mg/ml) of the enzyme are stored at −70°. At this temperature, the purified enzyme suffers no significant loss of activity over a period of at least 6 months. The enzyme also withstands freezing and thawing (37°) more than ten times without noticeable loss of activity.

The affinity of galactose oxidase for Sepharose is influenced by the nature of the equilibrating buffer. For the buffers described in Fig. 1, the elution volume of the enzyme is approximately four times that of the protein peak containing the proteolytic contaminants. By using 0.02 M Tris-HCl containing 0.05 M NaCl (pH 7.2), this ratio is increased to approximately six. However, a more acidic buffer, e.g., 0.1 M sodium phosphate at pH 6.5, or 0.1 M sodium acetate at pH 5.5, decreases the elution ratio to approximately three.

Measurement of Enzymic Activity

We have used two techniques to characterize galactose oxidase activity of the purified enzyme using galactose as the substrate. Galactose oxidase concentration is determined using an $E_{280\,nm}^{1\%,\,1\,cm}$ value of 15.4.[3] The technique that links galactose oxidase with a horseradish peroxidase–o-dianisidine system is based on that of Amaral et al.[4] and has been described before.[2] For quantities of galactose oxidase ranging from 60 to 200 ng in the assay, the horseradish peroxidase–o-dianisidine system gives values for the specific activity ranging from 300 units/mg to 390 units/mg [1 unit of enzymic activity produces an absorbance change (425 nm) of 1.00 per minute at 37°].

More recently, Avigad[5] has reported an alternative method that utilizes NADH with NADH peroxidase (EC 1.11.1.1) as the coupled enzyme. Using the NADH–NADH peroxidase assay at 37° a value of 72.7 (±5.3 SD) units/mg has been calculated for purified samples of galactose oxidase, where 1 unit is equivalent to the oxidation of 1 μmol of NADH/min at pH 7.0.

The absence of proteolytic and glycosidic enzymes is an important property of a galactose oxidase sample destined for use with desialylated glycoprotein substrates. Samples of the purified enzyme (containing up to 10 μg of protein) have been shown to be free from proteolytic activity using a variety of protein substrates including fibrinogen, casein, transferrin, and albumin.[2] Furthermore, using a procedure derived from that of Hugget and Nixon,[6] glucose oxidase activity has not been detected. As-

[3] D. J. Kosman, M. J. Ettinger, R. E. Weiner, and E. J. Massaro, Arch. Biochem. Biophys. **165**, 456 (1974).

[4] D. Amaral, F. Kelly-Falcoz, and B. L. Horecker, this series, Vol. 9, p. 87.

[5] G. Avigad, Anal. Biochem. **86**, 470 (1978).

[6] A. S. G. Hugget and D. A. Nixon, Biochem. J. **66**, 12P (1957).

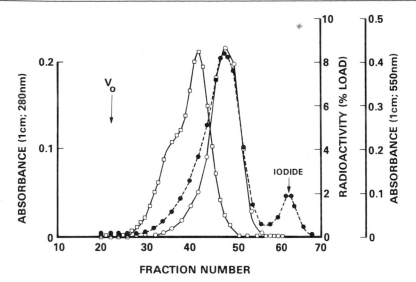

FIG. 2. Chromatography of 50 μg of purified galactose oxidase (traced with ^{125}I-labeled galactose oxidase) and human serum albumin (Behringwerke, A. G. Marburg, G. F. R.; 10 mg) on a column of Sephadex G-200 (50 cm × 2.2 cm) equilibrated with 0.01 M Tris-HCl, pH 7.2, containing 0.25 M NaCl. The fraction volume was 3.4 ml. □——□, Absorbance at 280 nm; ●---●, radioactivity; ○——○, galactose oxidase activity using the horseradish peroxidase–o-dianisidine method with the addition of H_2SO_4 to stop the reaction (550 nm).

says for various glycosidase enzymes have been undertaken using a technique[7] similar to that described for β-galactosidase by Wallenfels,[8] but with the appropriate nitrophenylated glycoside. Thus, using 0.1 M sodium acetate, pH 5.0, as the incubation buffer, β-galactosidase, α-fucosidase, and β-N-acetylglucosaminidase activities were not measurable using 10-μg quantities of galactose oxidase.

Other Properties of Galactose Oxidase

The gel filtration properties of the enzyme, purified by affinity chromatography on Sepharose, have been studied using Sephadex G-200.[2] In Fig. 2, the elution profile of galactose oxidase is contrasted with that of human serum albumin (M_r 69,000) and Na[125]I. Using the eluent 0.01 M Tris-HCl, pH 7.2, containing 0.25 M NaCl, the behavior of ^{125}I-labeled galactose oxidase closely matched the enzymic activity and behaved as a macromolecule of M_r approximately 39,000 as determined by the $V_e : V_o$

[7] M. W. C. Hatton, E. Regoeczi, and K.-L. Wong, *Can. J. Biochem.* **52**, 845 (1974).
[8] K. Wallenfels, this series, Vol. 5, p. 212.

ratio technique of Determann.[9] The value of 39,000 compares well with the measurement of 42,400 reported for galactose oxidase by Kelly-Falcoz *et al.*[10] using an equilibrium centrifugation technique. However, if the Sephadex G-200 column is equilibrated with 0.1 M ammonium acetate, pH 7.2, containing 0.1 M D-galactose, ^{125}I-labeled galactose oxidase and enzymic activity chromatographed with an M_r of approximately 52,000, which is comparable with the result (55,000) obtained by Bauer *et al.*[11] from sedimentation velocity and equilibration measurements.

Comments

The agarose affinity procedure is a simple method for purifying an impure sample of galactose oxidase. The isolated enzyme is very stable if stored frozen in 0.05 M phosphate buffer, pH 6.5, at low temperature. No measurable quantities of contaminating proteolytic or carbohydrate-specific enzymes have been detected. In addition, Sepharose chromatography offers an effective means to separate an oxidized galactose-terminating product from galactose oxidase after reaction.[12]

[9] H. Determann, "Gel Chromatography," p. 105. Springer-Verlag, Heidelberg, 1968.
[10] F. Kelly-Falcoz, H. Greenberg, and B. L. Horecker, *J. Biol. Chem.* **240**, 2966 (1965).
[11] S. Bauer, G. Blauer, and G. Avigad, *Isr. J. Chem.* **5**, 126p (1967).
[12] E. Regoeczi, P. A. Chindemi, M. W. C. Hatton, and L. R. Berry, *Arch. Biochem. Biophys.* **205**, 76 (1980).

[29] D-Galactose Dehydrogenase from *Pseudomonas fluorescens*

By E. MAIER and G. KURZ

β-D-Galactose + NAD$^+$ $\rightleftarrows$ D-galactono-1,5-lactone + NADH + H$^+$

D-Galactono-1,5-lactone $\rightleftarrows$ D-galactono-1,4-lactone

$$\searrow \qquad\qquad \swarrow$$
$$\text{D-galactonate} + \text{H}^+$$

D-Galactose dehydrogenase (β-D-galactose : NAD$^+$ 1-oxidoreductase; EC 1.1.1.48) catalyzes the pyridine nucleotide-dependent dehydrogenation of β-D-galactopyranose at C-1[1–3] with the formation of NADH and

[1] K. Wallenfels and G. Kurz, *Biochem. Z.* **335**, 559 (1962).
[2] K.-H. Ueberschär, E.-O. Blachnitzky, and G. Kurz, *Eur. J. Biochem.* **48**, 389 (1974).
[3] A. L. Cline and A. S. L. Hu, *J. Biol. Chem.* **240**, 4493 (1965).

D-galactono-1,5-lactone. Depending on the pH value, the immediately formed D-galactono-1,5-lactone rearranges intramolecularly to the corresponding 1,4-lactone or hydrolyzes to D-galactonate or both.[2]

Assay Method

Principle. The assay is based on the spectrophotometric determination of NADH formed with D-galactose as substrate at high pH.

Reagents

Bis(hydroxyethyl)amine-HCl buffer, 66 mM, pH 9.1
NAD$^+$, 30 mM, in bis(hydroxyethyl)amine-HCl buffer, pH 9.1
D-Galactose, 188 mM, in bis(hydroxyethyl)amine-HCl buffer, pH 9.1
Enzyme. The enzyme solution is diluted with the appropriate buffers to give a solution with an activity between 0.2 and 0.5 unit/ml.

Procedure. A mixture of 0.1 ml of NAD$^+$, 0.1 ml of D-galactose, and 2.7 ml of bis(hydroxyethyl)amine-HCl buffers is incubated at 30° in a spectrophotometer cell with a 1.0-cm light path. The reaction is started by the addition of 0.1 ml of the enzyme solution. The extinctions at 340 nm or 366 nm are measured at 15-sec intervals.

Definition of Unit and Specific Activity. One unit of enzyme is defined as the amount that catalyzes the reduction of 1 μmol of NAD$^+$ per minute under the above conditions. Specific activity is expressed as units per milligram of protein.

Organism and Cultivation

The strain of *Pseudomonas* used was originally obtained from Boehringer Mannheim GmbH (Mannheim, F.R.G.) and has been classified according to Stanier *et al.*[4] as *P. fluorescens* biotype E or F.[5] The bacteria were maintained in lyophilized stock cultures.

The bacteria are grown in a minimal medium containing 33 mM potassium sodium phosphate buffer, pH 6.8, and, per liter, 1 g of NH$_4$Cl, 0.5 g of MgSO$_4$ · 7 H$_2$O, 10 mg of CaCl$_2$ · 2 H$_2$O, 2.5 mg of FeSO$_4$ · 7 H$_2$O, and 2 g of D-galactose as carbon source at 30° in 1000-ml flasks containing 200 ml on a reciprocal shaker.[5,6] The cultivation of *P. fluorescens* on a pilot-plant scale is performed in a three-step cultivation scheme.[6]

---

[4] R. Y. Stanier, N. J. Palleroni, and M. Doudoroff, *J. Gen. Microbiol.* **43**, 159 (1966).
[5] D. Lessmann, K.-L. Schimz, and G. Kurz, *Eur. J. Biochem.* **59**, 545 (1975).
[6] F. Wengenmayer, K.-H. Ueberschär, G. Kurz, and H. Sund, *Eur. J. Biochem.* **40**, 49 (1973).

First Culture. Lyophilized cells are inoculated into 1000-ml flasks holding 200 ml of medium, and the cultures are incubated on a reciprocal shaker for 24 hr at 30°.

Second Culture. The content of one flask of the first culture is added to 10 liters of the same medium in a 14.5-liter fermentor (NBS Scientific, New Brunswick, New Jersey). The culture is allowed to grow under mild agitation (120 rpm) for 24 hr at 26° at an air supply of 1 volume per volume of culture per minute.

Pilot-Plant Culture. Minimal medium, 200 liters, containing 3 g of NH_4Cl instead of 1 g per liter are inoculated with 10 liters of the second culture in a fermentor (Wehrle-Werk AG, Emmendingen, F.R.G.). The culture is agitated (120 rpm) at 26° and an air flow of 0.6 volume per volume of culture per minute was applied. The pH must be controlled and automatically adjusted to pH 6.8 with a solution of concentrated NH_3. After 6 hr the culture is fed with 6 liters of a solution of D-galactose (200 g/liter) at a flow rate of 800 ml/hr. In the late log phase the culture is cooled and the bacteria are harvested with the aid of a continuous-flow centrifuge (type AS 16 RR, 15 500 rpm, Sharples, France). The cell paste must be slightly *reddish;* if the cells are white or an ugly gray-brown, the run should be discarded.

The cell paste is lyophilized. The lyophilized bacteria may be stored at −15° for several months without loss of activity.

Purification Procedure

The purification procedure for D-galactose dehydrogenase is a slight modification of the procedure described.[7] The last purification step, a relatively complex preparative electrophoresis, is replaced by a simpler ion-exchange chromatography step. All operations are carried out at 0–4° unless stated otherwise.

Step 1. Preparation of Cell-Free Extract. Lyophilized bacteria (500 g) are pulverized with the aid of a ball mill and suspended in 20 mM sodium acetate buffer, pH 5.0 (4800 ml); 9.6 ml of buffer are used per gram of cells. The suspension is homogenized with an Ultra Turrax type 45 (Janke und Kunkel KG, Staufen, F.R.G.) for 10 min. The temperature should not exceed 30° during this operation. Subsequently the pH is adjusted to 5.1. Ribonuclease A (100 mg) and deoxyribonuclease I (100 mg) are added, and the suspension is stirred at room temperature for 3 hr. The subsequent centrifugation at 20,000 g and 4° yields a yellowish, turbid solution (about 3800 ml). In the supernatant the acid-soluble absorbance at 260 nm was about 65% of the total absorbance at 260 nm.

[7] E.-O. Blachnitzky, F. Wengenmayer, and G. Kurz, *Eur. J. Biochem.* **47**, 235 (1974).

Step 2. Ammonium-Sulfate Fractionation. Solid $(NH_4)_2SO_4$ (182 g per liter of solution) is slowly added to the stirred cell-free extract. After addition, stirring is continued for 4 hr. The precipitate formed is removed by centrifugation at 20,000 g for 45 min, and additional $(NH_4)_2SO_4$ (105 g per liter of solution) is added to the supernatant. The suspension is kept overnight and subsequently centrifuged at 20,000 g for 45 min. The precipitate is washed with about 300 ml of a solution of 270 g $(NH_4)_2SO_4$ in 1 liter of 20 mM sodium acetate buffer, pH 5.1, and then collected by centrifugation. The washed precipitate is suspended in about 150 ml of 16 mM sodium acetate buffer, pH 5.1, and this suspension is dialyzed against 20 liters of the same buffer for at least 12 hr. Any precipitate is removed by centrifugation at 48,000 g for 30 min.

Step 3. Bulk Separation on CM-Sephadex. The solution of the preceding step (about 170 ml) is added under mild stirring to 1200 ml of CM-Sephadex C-50 equilibrated with 20 mM sodium acetate buffer, pH 5.0; 1 ml of CM-Sephadex C-50 gel is used per 7.5 mg of protein, corresponding to 75 units of D-galactose dehydrogenase. Stirring is continued for 10 min.[8] Under mild suction the CM-Sephadex C-50 gel is collected on a Büchner funnel and washed directly on the funnel three times with 1000 ml of 20 mM sodium acetate buffer, pH 5.0. Subsequently the gel is suspended in 1000 ml of 0.1 M sodium acetate buffer, pH 5.6, and this is stirred for 20 min.[8] The eluted protein is removed by means of the Büchner funnel. The elution process is repeated twice in exactly the same manner.[8] The eluates are combined and concentrated to 200 ml with the aid of an Amicon hollow-fiber concentrator CH 3 [Amicon GmbH, Witten (Ruhr), F.R.G.] equipped with a hollow-fiber cartridge type H1P 10.

Step 4. DEAE-Sephadex Chromatography. The concentrated protein solution is applied to a column (35 × 5 cm) of DEAE-Sephadex A-50 equilibrated with 0.1 M sodium acetate buffer, pH 5.0. After loading the protein on the DEAE-Sephadex A-50, the column is washed with three column-volumes of 0.1 M sodium acetate buffer, pH 5.0, containing 20 mM NaCl. Subsequently a linear gradient is applied: the starting buffer is 0.1 M sodium acetate buffer, pH 5.0, containing 20 mM NaCl; the limiting buffer is 0.1 M sodium acetate buffer, pH 5.0, containing 0.4 M NaCl; the total gradient volume is 3000 ml. Fractions of 15 ml are collected at a flow rate of 60 ml/hr. Fractions with an activity greater than 4 units/ml are combined (about 300 ml) and concentrated to a volume of about 100 ml with the aid of an Amicon ultrafiltration cell Model 402 [Amicon GmbH, Witten (Ruhr), F.R.G.], using a UM-10 membrane.

---

[8] In order to obtain the specified yields, it is absolutely necessary to follow the indicated times and pH values.

PURIFICATION OF D-GALACTOSE DEHYDROGENASE FROM *Pseudomonas fluorescens*[a]

Step	Volume (ml)	Activity (units/ml)	Protein (mg/ml)	Specific activity (units/mg protein)	Yield (%)	Purifica (fol(
1. Cell-free extract	3800	21.5	15.5	1.39	100	—
2. Ammonium sulfate fractionation	170	438	44.6	9.82	91	7.
3. Bulk separation on CM-Sephadex	200	309	3.6	85.83	76	61.
4. DEAE-Sephadex chromatography	100	398	1.18	337.28	49	242.
5. Hydroxyapatite chromatography	10	3659	4.31	849.0	45	611
6. DEAE-cellulose chromatography	10	2428	2.86	849.0	30	611

[a] Typical data obtained with 500 g of lyophilized bacteria.

Step 5. Hydroxyapatite Chromatography. The solution from the DEAE-chromatography is dialyzed for 12 hr against 10 liters of 10 mM potassium sodium phosphate buffer, pH 7.2, containing 0.1 M NaCl and applied to a hydroxyapatite column (20 × 3 cm) equilibrated with the same buffer. Instead of being dialyzed, the enzyme solution may be diluted with 2 mM sodium acetate buffer, pH 5.0, to a volume of 300 ml and then directly applied to the column. Subsequently the column is washed with four column-volumes of 17 mM potassium sodium phosphate buffer, pH 7.2, containing 0.1 M NaCl. The enzyme is eluted with 20 mM potassium sodium phosphate buffer, pH 7.2, containing 0.1 M NaCl. Fractions of 15 ml are collected at a flow rate of 60 ml/hr. Fractions that appear to be homogeneous in disc polyacrylamide gel electrophoresis are combined and concentrated to a final volume of about 10 ml with the aid of Amicon ultrafiltration cells Model 402 and Model 52, using UM-10 membranes.

Step 6. DEAE-Cellulose Chromatography.[9] The concentrated eluates from the hydroxyapatite chromatography are dialyzed for at least 12 hr against 2 liters of 40 mM sodium acetate buffer, pH 5.0, containing 58 mM NaCl with two changes of dialyzate. The dialyzed solution is placed onto a DEAE-column (80 × 2 cm) equilibrated with the same buffer. The elution of the enzyme is also performed with exactly the same buffer. Fractions of 4 ml are collected at a flow rate of 8.0 ml/hr. Fractions with an activity greater than 1 unit/ml are combined and concentrated to a volume of about 10 ml.

A typical purification is summarized in the table.

[9] This last purification step is necessary only to separate last traces of proteolytic enzymes.[7] It might be omitted if the enzyme is not to be kept for prolonged times.

Properties

Homogeneity. The purified enzyme with a specific activity of 849 units/mg is homogeneous by ultracentrifugation, by disc gel electrophoresis, by sodium dodecyl sulfate gel electrophoresis, and by gel isoelectric focusing.

Stability and pH Stability. The purified enzyme is stored at $-20°$ in 0.1 M phosphate buffer, pH 6.5–7.6, or in sodium acetate buffers, pH 5.0–5.6, containing enough glycerol (30–50%) to prevent freezing.[7] No decrease in enzyme activity could be observed within 18 months.

The enzyme is stable in the range from pH 5.0–9.0.[7]

Structure. D-Galactose dehydrogenase from *P. fluorescens* has a molecular weight of 64,000. The enzyme is composed of two identical subunits and has two binding sites.[7,10]

For a further description of the properties of D-galactose dehydrogenase from *P. fluorescens,* see Blachnitzky *et al.*[7], for its application for the determination of D-galactose, see Kurz and Wallenfels.[11,12]

[10] E. Maier and G. Kurz, unpublished observation.
[11] G. Kurz and K. Wallenfels, *in* "Methoden der enzymatischen Analyse" (H. U. Bergmeyer, ed.), 3rd German ed., p. 1324. Verlag Chemie, Weinheim/Bergstr., F.R.G., 1974.
[12] G. Kurz and K. Wallenfels, *in* "Methoden der enzymatischen Analyse" (H. U. Bergmeyer, ed.), 3rd German ed., p. 1225. Verlag Chemie, Weinheim/Bergstr., F.R.G., 1974.

[30] Aldose Reductase from Human Tissues

By BENDICHT WERMUTH and JEAN-PIERRE VON WARTBURG

$$R - CHO + NADPH + H^+ \rightleftarrows R - CH_2OH + NADP^+$$

Aldose reductase (alditol : NADP$^+$ 1-oxidoreductase, EC 1.1.1.21) and polyol dehydrogenase (L-iditol : NAD$^+$ 5-oxidoreductase, EC 1.1.1.14) constitute the sorbitol (polyol) pathway converting glucose to fructose in extrahepatic tissues. In addition to glucose, aldose reductase catalyzes the reduction of other sugar aldehydes and of several aliphatic and aromatic aldehydes. Another enzyme, aldehyde reductase[1] (alcohol : NADP oxidoreductase, EC 1.1.1.2), shows a similar substrate specificity. The two enzymes are distinguishable by relative substrate specificities and inhibitor sensitivities, by column chromatography, and immunologically.[2]

[1] J. P. von Wartburg and B. Wermuth, this volume [85].
[2] P. L. Hoffman, B. Wermuth, and J. P. von Wartburg, *J. Neurochem.* **35**, 354 (1980).

Assay Method

Principle. Aldose reductase is assayed spectrophotometrically by recording the decrease in NADPH absorbance at 340 nm. With crude enzyme solutions, blank reactions may occur with NADPH in the absence of any aldehyde substrate; consequently the rate of this reaction must be recorded before addition of exogenous aldehyde. Depending on the substrate and on the tissue analyzed, other dehydrogenases, notably aldehyde reductase, may interfere. Interference with aldehyde reductase is diminished by the addition of diphenylhydantoin or phenobarbitone to the assay medium. Moreover, the ratio of activities measured with D-xylose, DL-glyceraldehyde, and D-glucuronate may be used to distinguish between the two enzymes.[2]

Reagents

   Sodium phosphate buffer, 0.1 M, pH 7.0
   NADPH, 1.6 mM; made up daily in deionized water and kept on ice
   D-Xylose, 1 M
   DL-Glyceraldehyde, 0.1 M, dedimerized at 85° for 10 min
   D-Glucuronate, 0.1 M
   Diphenylhydantoin, 0.01 M, dissolved in 0.01 M NaOH

Procedure. To a 1-ml cuvette add 730 µl of assay buffer, 50 µl of NADPH, 20 µl of diphenylhydantoin (which inhibits aldehyde reductase by approximately 90%), 5–100 µl of enzyme solution, and deionized water to a total volume of 0.9 ml. Place cuvette in photometer and record blank reaction if necessary. Start enzyme reaction by the addition of 100 µl of substrate solution. If the absence of aldehyde reductase has been ascertained, diphenylhydantoin can be omitted.

Definition of Enzyme Unit and Specific Activity. One unit of enzyme is defined as the amount of protein catalyzing the oxidation of 1 µmol of NADPH per minute. The specific activity is expressed as milliunits of enzyme per milligram of protein. The concentration of protein is estimated from the absorbance at 260 and 280 nm.[3]

Preparation of Cibacron Blue-Sepharose. Sepharose containing a spacer arm with a free carboxyl group is formed by covalent linkage of 6-aminohexanoic acid to Sepharose 4B[4] using the cyanogen bromide method.[5] One liter of the derived Sepharose is suspended in an equal volume of water, and 4 g each of Cibacron Blue F3 G-A and N-cy-

---

[3] E. Layne, this series, Vol. 3 [73].
[4] Sepharose 4B containing covalently linked 6-aminohexanoic acid is commercially available from Pharmacia, Uppsala, Sweden.
[5] P. Cuatrecasas and C. B. Anfinsen, this series, Vol. 22 [31].

clohexyl-N'-(2-morpholinoethyl)carbodiimide metho-p-toluene sulfonate are added. The pH is adjusted to and maintained at 5 by the addition of NaOH or HCl, and the slurry is agitated for 20 hr. The Sepharose is then thoroughly washed on a sintered-glass funnel with 0.1 N NaOH, 0.1 N HCl, 2 M NaCl, and H_2O until the dye is undetectable in the effluent. Upon coupling, the dye changes its color from blue to purple.

Purification Procedure

A general procedure has been worked out in our laboratory to isolate and purify carbonyl reducing enzymes from human tissues, and the preparation of brain aldose reductase will be described in detail. Brains, either whole or without cortex and cerebellum, were obtained from legal medical autopsies. The brains were frozen 6–20 hr postmortem and stored at $-20°$. The whole purification is carried out at $4°$, and all buffers contain 50 μM EDTA and 0.5 mM 2-mercaptoethanol to stabilize the enzyme and 0.02% sodium azide to prevent bacterial growth.

Step 1. Extraction. Approximately 300 g of brain are homogenized with an equal volume (v/w) of 0.1 M sodium phosphate buffer, pH 7.4, in a Waring blender for 2–3 min. The homogenate is centrifuged for 2 hr at 45,000 g. The precipitate is reextracted with the same volume of buffer and recentrifuged. The precipitate is discarded. The pooled supernatants are dialyzed for 24 hr against three changes of 10 liters of 5 mM sodium phosphate buffer, pH 7.4, and centrifuged for 1 hr at 23,000 g.

Step 2. DEAE-Cellulose Chromatography. The supernatant fluid is applied to a column (2 × 40 cm) of DEAE-cellulose equilibrated with dialysis buffer. The column is washed with 500 ml of the same buffer before a linear gradient (2 × 300 ml) of 5 to 100 mM sodium phosphate buffer, pH 7.4, is applied. Aldose reductase activity emerges from the column in two fractions immediately before and after aldehyde reductase (Fig. 1). Both fractions are combined for further purification.

Step 3. Gel Filtration. The pooled enzyme (maximum 150 ml/run) is applied to a column (4 × 180 cm) of Sephadex G-100 equilibrated with 10 mM sodium phosphate buffer, pH 6.2. The column is washed at a flow rate of 90 ml/hr, and tubes containing aldose reductase activity are pooled.

Step 4. Affinity Chromatography. The pool of aldose reductase activity is applied to a column (1 × 25 cm) of Cibacron Blue-Sepharose equilibrated with 10 mM sodium phosphate buffer, pH 6.2. The column is washed with 50 ml of 100 mM sodium phosphate buffer, pH 6.2; elution of aldose reductase activity is accomplished by a linear gradient (2 × 12 ml) of 0 to 1 mM NADPH in the original column buffer. Fractions containing aldose

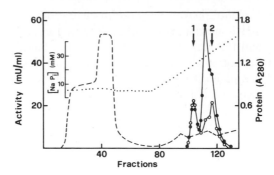

FIG. 1. Elution pattern of human brain aldose and aldehyde reductase activities from DEAE-cellulose. Brain extract was applied to a column equilibrated with 5 mM sodium phosphate, pH 7.4. Enzyme activities were eluted with a linear gradient of 5 to 100 mM phosphate buffer (····). Protein (---) was recorded with a Uvicord II spectrophotometer. Activity was assayed using either 0.1 M xylose (O——O) or 0.01 M DL-glyceraldehyde (●——●) as substrate. No diphenylhydantoin was included in the assay mixture. The elution positions of aldose reductase activity are indicated by arrows.

reductase activity are pooled, dialyzed against 10 mM sodium phosphate buffer, pH 7.0, and stored at 4°.

Table I summarizes typical purification data.

Alternative Procedure. An alternative purification procedure[6] including treatment with calcium phosphate gel, ammonium sulfate precipitation, chromatography over DEAE-cellulose and Sephadex G-100, and isoelectric focusing has been described for the enzyme from placenta. The purified enzyme had a specific activity of 1.68 units per milligram of protein with 400 mM D-xylose and 0.2 mM NADPH as substrates at pH 6.2.

Properties

Purity and Stability. The outlined purification procedure yields a homogeneous enzyme that is stable for several days in the pH range 6–9 at 4°. The activity rapidly decreases below pH 5.0.

Molecular Forms. DEAE-cellulose chromatography of brain extract yields two forms of aldose reductase. Polyacrylamide gel electrophoresis of the purified enzyme forms in the presence and absence of 2-mercaptoethanol indicates that they exist in a thiol-dependent equilibrium. The mechanism of the conversion, however, is not clear.

Physicochemical Properties. Aldose reductase consists of a single polypeptide chain with a molecular weight of 35,000–40,000. The isoelectric point has been reported to be 5.8–6.0.[2,6] The amino acid composition

[6] R. S. Clements, Jr. and A. I. Winegrad, *Biochem. Biophys. Res. Commun.* **47**, 1473 (1972).

TABLE I
PURIFICATION OF HUMAN BRAIN ALDOSE REDUCTASE

Fraction	Activity[a] (μmol min^{-1})	Protein (mg)	Specific activity (nmol min^{-1} mg^{-1})	Recovery (%)
Extract	7.25	5878	1.2	100
DEAE-cellulose	3.00	210	14	41
Sephadex G-100	2.70	25	108	37
Cibacron Blue-Sepharose	1.52	3.2[b]	475	21

[a] Activity was measured using D-xylose (0.1 M) as substrate.
[b] The enzyme contains tightly bound material absorbing at 260 nm that may interfere with the protein determination.

is shown in comparison with aldehyde reductase in Table III of this volume [85].

pH Optimum. The optimum pH for aldehyde reduction is between 6 and 7.

Substrate Specificity. Aldose reductase catalyzes the reduction of a number of physiological and xenobiotic aldehydes (Table II). Glycol- and polyolaldehydes are preferred substrates. The Michaelis constants for aldoses increase with increasing chain length, whereas the maximal rate of reduction is less dependent on the size of the substrate. Isocorticosteroids, intermediates in an alternative pathway of corticosteroid metabolism, in which the 17-ketol is replaced by a 17-aldol side chain, are the most efficiently reduced substrates known to date. The 20 α- and β-epimers are reduced at comparable rates. The reduction of biogenic aldehydes, derived from the biogenic amines by oxidative deamination, is little investigated but may be of some importance in view of the finding that in rat brain aldose reductase is responsible for the reduction of these aldehydes.[7]

The reverse reaction, oxidation of polyols to polyolaldehydes by NADP$^+$, proceeds very slowly. It can be used to detect reductase activity in polyacrylamide gels.[8]

Aldose reductase is specific for NADPH as coenzyme. With NADH (0.12 mM) about 10% of the NADPH-dependent activity is obtained. During catalysis the pro-4R hydrogen atom is transferred from the nicotinamide ring of the coenzyme to the substrate.[2,9]

[7] S. R. Whittle and A. J. Turner, *Biochim. Biophys. Acta* **657**, 94 (1981).
[8] M. M. O'Brien and P. J. Schofield, *Biochem. J.* **187**, 21 (1980).
[9] H. B. Feldman, P. A. Szczepanik, P. Havre, R. J. M. Corrall, L. C. Yu, H. M. Rodman, B. A. Rosner, P. D. Klein, and B. R. Landau, *Biochim. Biophys. Acta* **480**, 14 (1977).

TABLE II
SUBSTRATE SPECIFICITY OF HUMAN BRAIN ALDOSE REDUCTASE

Substrate	Concentration (mM)	Michaelis constant (mM)	Relative velocity (%)
DL-Glyceraldehyde	∞	0.06	100
D-Xylose	∞	16	78
D-Glucose	∞	90	45
D-Galactose	∞	110	45
D-Glucuronate	∞	4	75
11-Deoxyisocorticosterone[a]	∞	<0.001	150
Isocortisol[a]	∞	<0.001	150
Indole-3-acetaldehyde	1.2	—	77
4-Hydroxyphenylacetaldehyde[b]	0.4	—	113
4-Hydroxyphenylglycolaldehyde[b]	0.02	—	48
4-Nitrobenzaldehyde	∞	0.015	110
Caprinaldehyde	0.5	—	20
Butyraldehyde	0.5	—	12
NADPH	—	0.004	

[a] Isocorticosteroids are synthesized either enzymically [V. Lippman and C. Monder, *J. Steroid. Biochem.* **7**, 719 (1976)] or chemically [S. Oh and C. Monder, *J. Org. Chem.* **41**, 2477 (1976)]. The presented data are part of unpublished results of B. Wermuth and C. Monder (1981).

[b] Biogenic aldehydes are prepared by incubation of the appropriate amine with monoamine oxidase [B. Tabakoff, R. Anderson, and S. G. A. Alivisatos, *Mol. Pharmacol.* **9**, 428 (1973)].

Activators and Inhibitors. Ammonium sulfate activates aldose reductase, whereas chloride ions are inhibitory. Flavonoids, e.g., quercetin, quercitrin, and rutin, 3,3-tetramethyleneglutaric acid, Alrestatin (1,3-dioxo-1H-benz[de]isoquinoline-2-(3H)acetic acid), Sorbinil [S-6-fluorospiro(chroman-4,4'-imidazolidine)-2'',5''-dione], and the chromone, 7-hydroxy-4-oxo-4H-chromen-2-carboxylic acid, known inhibitors of animal lens aldose reductase, are also potent inhibitors of the human enzyme.[2,6,10] Differences exist in the susceptibility of the enzyme from various tissues. Aldose reductase from lens, for example, is about 10 times as susceptible to inhibition by quercetin ($IC_{50} = 0.6 \mu M$) as the enzymes from placenta ($IC_{50} = 7 \mu M$) or brain (IC_{50} = approximately 10 μM).

Tissue Distribution. Aldose reductase activity has been demonstrated in brain, kidney, placenta, testis, lens, lung, heart, and pancreas, but appears to be absent from liver. The enzyme is localized in the cytoplasm of the cell.

[10] P. F. Kador, J. H. Kinoshita, W. H. Tung, and L. T. Chylack, Jr., *Invest. Ophthalmol.* **19**, 980 (1980).

[31] D-Gluconate Dehydrogenase from Bacteria, 2-Keto-D-gluconate-Yielding, Membrane-Bound

By Kazunobu Matsushita, Emiko Shinagawa, and Minoru Ameyama

D-Gluconate + acceptor → 2-keto-D-gluconate + reduced acceptor

D-Gluconate dehydrogenase (EC 1.1.99.3) occurs on the outer surface of cytoplasmic membrane of oxidative bacteria, such as *Pseudomonas, Klebsiella, Serratia,* and acetic acid bacteria.[1,2] The enzyme activity is linked to the electron transport chain in the cytoplasmic membrane constituting a D-gluconate oxidase system.[3]

Assay Method

Principle. The assay is performed spectrophotometrically at 25° by measuring the decrease of absorbance at 600 nm of 2,6-dichlorophenolindophenol (DCIP) mediated with phenazine methosulfate (PMS). The activity can be also measured with PMS, DCIP, ferricyanide, or coenzyme Q (CoQ) as an electron acceptor.[4]

Reagents

D-Gluconate, sodium salt, 1 M, in distilled water
Potassium phosphate buffer, 0.1 M, pH 6.0
DCIP, 10 mM, in distilled water
PMS, 3 mM, in distilled water
Enzyme, dissolved in 0.01 M potassium phosphate, pH 6.0, containing 5 mM MgCl$_2$ and 0.1% Triton X-100

Procedure. A cuvette with 1-cm light path contains 1 ml of potassium phosphate buffer, 0.1 ml of DCIP, 0.1 ml of PMS, 0.1 ml of gluconate, enzyme solution (less than 0.1 unit), and water in a final volume of 3.0 ml. A reference cuvette contains all components except D-gluconate. The reaction is initiated by the addition of D-gluconate. Enzyme activity is measured as the initial reduction rate of DCIP.

[1] E. Shinagawa, T. Chiyonobu, O. Adachi, and M. Ameyama, *Agric. Biol. Chem.* **40**, 475 (1976).
[2] E. Shinagawa, T. Chiyonobu, K. Matsushita, O. Adachi, and M. Ameyama, *Agric. Biol. Chem.* **42**, 1055 (1978).
[3] K. Matsushita, M. Yamada, E. Shinagawa, O. Adachi, and M. Ameyama, *J. Bacteriol.* **141**, 389 (1980).
[4] K. Matsushita, E. Shinagawa, O. Adachi, and M. Ameyama, *J. Biochem. (Tokyo)* **86**, 249 (1979).

METHODS IN ENZYMOLOGY, VOL. 89

Definition of Unit and Specific Activity. One unit of enzyme is defined as the amount of enzyme that catalyzes the oxidation of 1 μmol of D-gluconate, or reduction of 1 μmol of DCIP, per minute at 25° under the above conditions. The extinction coefficient of DCIP at pH 6.0 is taken as 10 mM^{-1}, so that 1 unit of enzyme activity corresponds to 3.3 absorbance changes at 600 nm. Specific activity is expressed as the number of units per milligram of enzyme protein. Protein content is determined by the modified Lowry method[5] as the sample contains Triton X-100.

Source of Enzyme

Microorganisms. D-Gluconate dehydrogenase is obtained from the following strains: *Pseudomonas aeruginosa* IFO 3445,[6] *Pseudomonas fluorescens* FM-1 (see this volume [24]), *Klebsiella pneumoniae* IFO 3317, *Serratia marcescens* IFO 3054,[7] and *Gluconobacter sphaericus* IFO 12467. Bacterial strains having IFO numbers can be obtained from the Institute for Fermentation, Osaka (17-85, Juso-honmachi 2-chome, Yodogawa-ku, Osaka 532, Japan).

Cultures. The stock culture is maintained on a potato–glycerol slant (see this volume [24]). *Pseudomonas aeruginosa* and *P. fluorescens* are grown on a D-gluconate medium [24]. *Klebsiella pneumoniae* is grown on a medium consisting of 0.5% sodium D-gluconate, 0.5% D-glucose, 0.1% polypeptone, and 0.1% yeast extract. *Serratia marcescens* is grown on a medium consisting of 0.5% sodium D-gluconate, 0.5% D-glucose, 0.1% NH_4Cl, 0.3% KH_2PO_4, 0.04% Na_2SO_4, and 0.04% $MgSO_4 \cdot 7\ H_2O$. Cells are grown aerobically at 30° in a 50-liter jar fermentor containing 30 liters of medium. Cells are harvested at early stationary phase.

Purification Procedure

D-Gluconate dehydrogenase is purified from four strains, *P. aeruginosa,*[6] *P. fluorescens, K. pneumoniae,* and *S. marcescens.*[7] The enzyme of *P. aeruginosa* and *P. fluorescens* is solubilized from the membrane with 0.2% deoxycholate–0.5% cholate and 1% cholate, respectively, and then further purified through ammonium sulfate fractionation and column chromatography in the presence of Triton X-100. The enzyme from *K. pneumoniae* and *S. marcescens* is solubilized with 2% Triton X-100 and

---

[5] J. R. Dulley and P. A. Grieve, *Anal. Biochem.* **64**, 136 (1975).

[6] K. Matsushita, E. Shinagawa, O. Adachi, and M. Ameyama, *J. Biochem.* (*Tokyo*) **85**, 1173 (1979).

[7] E. Shinagawa, K. Matsushita, O. Adachi, and M. Ameyama, *Agric. Biol. Chem.* **42**, 2355 (1978).

purified by column chromatography in the presence of Triton X-100. In this chapter, purification procedures of the enzyme from *P. fluorescens* and *K. pneumoniae* are described.

Preparation from P. fluorescens

Potassium phosphate buffer, pH 6.0, containing 5 mM MgCl$_2$ is used throughout the purification. All procedures are carried out at 0–5°, unless otherwise stated.

Step 1. Preparation of Membrane Fraction. This step is performed essentially as described for preparation of the membrane fraction of D-glucose dehydrogenase (see this volume [24]).

Step 2. Solubilization of Enzyme. The membrane fraction prepared from 100 g of wet cells is suspended in 0.01 M buffer to a final protein concentration of 10 mg/ml. Ten percent cholic acid (pH 7.4) is added to the membrane suspension up to 1%. After stirring for 2 hr, the suspension is centrifuged at 68,000 g for 60 min, and the supernatant containing D-gluconate dehydrogenase is obtained.

Step 3. Ammonium Sulfate Fractionation. Ammonium sulfate is added to the supernatant to 13.4 g/100 ml, and the precipitate formed is removed by centrifugation at 10,000 g for 10 min. After further addition of ammonium sulfate to the supernatant to 11.5 g/100 ml, the solution is centrifuged at 10,000 g for 10 min. The precipitate is suspended in 0.01 M buffer containing 0.1% Triton X-100 and 0.01 M sodium D-gluconate. The suspension of about 50–60 ml is dialyzed overnight against two changes of 1 liter of the suspension medium. The dialyzed enzyme is centrifuged at 10,000 g for 10 min to obtain a clear supernatant.

Step 4. DEAE-Sephadex Column Chromatography. The supernatant is applied to a DEAE-Sephadex A-50 column (4 × 18 cm) that has been equilibrated with 0.01 M buffer containing 0.1% Triton X-100 and 0.01 M sodium D-gluconate. The column is washed with 400 ml of the same buffer. Elution of the enzyme is made with the same buffer containing 0.1 M KCl. The pooled fractions are concentrated to about 25 ml by ultrafiltration (Toyo UP-20 filter).

Step 5. Hydroxyapatite Column Chromatography. The concentrated enzyme solution is diluted 2-fold with 0.01 M buffer to reduce concentrations of salt and detergent, and placed on a hydroxyapatite column (3.5 × 3.5 cm) preequilibrated with 0.01 M buffer containing 0.01 M sodium D-gluconate. The column is washed stepwise with 100 ml each of 0.01 M buffer containing 0.01 M sodium D-gluconate, 0.1 M buffer containing 0.01 M sodium D-gluconate, and 0.01 M buffer containing 0.01 M sodium D-gluconate and 0.1% Triton X-100. The enzyme is eluted with 0.15 M

TABLE I
PURIFICATION OF D-GLUCONATE DEHYDROGENASE FROM *Pseudomonas fluorescens*

Fraction	Total protein (mg)	Total activity (units)	Specific activity (units/mg protein)	Yield (%)
Membrane fraction	2482	6876	2.8	100
Solubilized fraction	755	7079	9.4	103
Ammonium sulfate	236	8907	38	130
DEAE-Sephadex A-50	19	4505	237	66
Hydroxyapatite	9	4230	470	62

buffer containing 0.01 M sodium D-gluconate and 0.1% Triton X-100. The peak fractions showing both enzyme activity and a characteristic pink color are combined.

A typical purification is summarized in Table I.

Preparation from Klebsiella pneumoniae

Potassium phosphate buffer, pH 6.0, is used throughout the purification.

Step 1. Preparation of Membrane Fraction. About 400 g of frozen cell paste of *K. pneumoniae* is thawed and suspended in 0.01 M buffer. The cell suspension is passed twice through a French press (American Instrument Co.) at 1000 kg/cm^2 and centrifuged at 1800 g for 10 min to remove intact cells. The resulting supernatant is then centrifuged at 68,000 g for 60 min, and the pellet is collected as membrane fraction.

Step 2. Solubilization of Enzyme. The membrane fraction is suspended in 0.01 M buffer containing 2% Triton X-100 and 0.1 M sodium D-gluconate at a final protein concentration of about 40 mg/ml. The suspension is stirred gently for 4 hr and then MnCl$_2$ is added to 50 mM and stirred for an additional hour. The suspension is centrifuged at 68,000 g for 60 min, and the supernatant is dialyzed against two changes of 10 volumes of 0.01 M buffer containing 0.1% Triton X-100.

Step 3. DEAE-Cellulose Column Chromatography. The dialyzed enzyme is centrifuged at 12,000 g for 20 min to remove insoluble materials. The supernatant is applied to a DEAE-cellulose column (4.5 × 40 cm), equilibrated with 0.01 M buffer containing 0.1% Triton X-100. The column is washed with the same buffer containing 0.04 M NaCl. Elution is effected with 0.01 M buffer containing 0.1% Triton X-100 and 0.08 M NaCl. The pooled solution is concentrated with a ultrafilter (Toyo UP-20) to about 20 ml.

TABLE II

PURIFICATION OF D-GLUCONATE DEHYDROGENASE FROM *Klebsiella pneumoniae*

Fraction	Total protein (mg)	Total activity (units)	Specific activity (units/mg protein)	Yield (%)
Membrane fraction	22,500	23,250	1.0	100
Solubilized fraction	4,790	22,680	4.7	98
DEAE-Cellulose	187	19,660	105	85
DEAE-Sephadex A-50	59	14,110	239	61
Hydroxyapatite	11	5,643	513	24

Step 4. DEAE-Sephadex Column Chromatography. The concentrated enzyme is diluted with 0.01 M buffer containing 0.1% Triton X-100 and applied to a DEAE-Sephadex A-50 column (2 × 20 cm) preequilibrated with 0.01 M buffer containing 0.1% Triton X-100 and 0.04 M NaCl. After the column is washed with the same buffer, the enzyme is eluted with a linear gradient by increasing NaCl concentration to 0.08 M. The fractions having enzyme activity are combined and concentrated by ultrafiltration.

Step 5. Hydroxyapatite Column Chromatography. The concentrated enzyme solution is diluted twice with 0.01 M buffer and applied to a hydroxyapatite column (2 × 5 cm) equilibrated with 0.01 M buffer. The column is washed with the same buffer and then with 0.1 M potassium phosphate, pH 6.0. Protein impurities are removed by these treatments. The enzyme is eluted with 0.1 M buffer containing 0.1% Triton X-100. Fractions exhibiting an orange color are collected.

A typical purification is summarized in Table II.

Properties

Molecular Properties. D-Gluconate dehydrogenase of *P. aeruginosa* shows a single protein band on polyacrylamide gel electrophoresis,[6] but the enzyme preparations from *K. pneumoniae* and *S. marcescens* are separated into two protein bands.[7] The low-mobility band is yellow, is fluorescent under ultraviolet light, and shows enzyme activity. Another protein band with high mobility is red. Dissociation of enzyme from *P. fluorescens* that usually appears as a single protein band also sometimes occurs under other experimental conditions. In sodium dodecyl sulfate (SDS) gel electrophoresis, enzyme preparations from all four strains dissociate into three separate polypeptide bands (Table III). The sum of the molecular weight of the three components is 136,000–140,000, the total molecular

TABLE III
PROPERTIES OF D-GLUCONATE DEHYDROGENASE FROM OXIDATIVE BACTERIA

Properties	Pseudomonas aeruginosa[4]	P. fluorescens	Klebsiella pneumoniae	Serratia marcescens[6]
Molecular weight	66,000	66,500	66,000	68,000
of subunits	50,000	47,500	52,000	52,500
	22,000	21,700	22,000	25,500
Absorption peaks	554	554	554	554
of reduced	522	522	523	523
enzyme (nm)	417	418	416	418
Heme content (nmol/mg of protein)	14.7	14.6	16.8	—
Optimum pH with ferricyanide assay	5.0	5.0	4.0	5.0
K_m for D-gluconate	0.8 (pH 6)	0.3 (pH 6)	0.8 (pH 6)	0.8 (pH 6)
(mM)	3.2 (pH 5)	—	2.9 (pH 4)	2.3 (pH 5)
K_i (mM)				
Oxamate	0.6	—	0.8	1.0
Oxalate	2.9	—	2.8	2.0
Pyruvate	17.9	—	18.3	17.5
2KGA[a]	123.0	—	—	—

[a] 2-Keto-D-gluconate.

weight. The molecular weight of the dehydrogenase from P. aeruginosa is estimated to be 132,000–138,000 by polyacrylamide gel electrophoresis performed under different gel concentrations.[8] Sucrose density gradient centrifugation gives a molecular weight of 124,000–131,000 for the enzyme of P. aeruginosa in the presence of Triton X-100.[6] The molecular weight obtained by two analytical centrifugation methods were similar. The enzyme is monomeric in the presence of more than 0.1% Triton X-100, whereas in the absence of the detergent it becomes dimeric.[6] The removal of Triton X-100 also causes a decrease of enzyme activity. Activation of the enzyme is observed with phospholipids, especially cardiolipin, in the presence of Triton X-100.[6] Thus, D-gluconate dehydrogenase has a hydrophobic and phospholipid-interacting domain, and hence is an intrinsic membrane protein.

    Subunit Components. Purified D-gluconate dehydrogenase shows a visible absorption spectrum of the cytochrome c type having an asymmetrical α-peak (Table III). The cytochrome is a diheme cytochrome c of a low potential (+ 100 mV) as with the P. aeruginosa enzyme. The cytochrome is

[8] J. L. Hederick and A. J. Smith, *Arch. Biochem. Biophys.* **126,** 155 (1968).

reduced with D-gluconate via a dehydrogenase component. The prosthetic group of the enzyme is not released with trichloroacetic acid. A partial release of flavin is achieved when the enzyme is subjected to proteolytic digestion. The digested and hydrolyzed peptide shows the same pH-fluorescence curve as the flavin peptide. Thus, the prosthetic group is covalently bound to the enzyme. For the enzyme of *Pseudomonas fluorescens,* the prosthetic group is shown to be 8α-[N(1)-histidyl]-FAD. One of the three components dissociated upon SDS gel electrophoresis, having the largest molecular weight, is a flavoprotein with enzyme activity as a primary dehydrogenase. The second component, the middle protein band in SDS gel electrophoresis, is a cytochrome and migrates as a red band. The function of the third component, having the lowest molecular weight, has not been identified.

Catalytic Properties. D-Gluconate dehydrogenase is highly specific for D-gluconate and shows the optimum pH at 4.0–5.0 (Table III). The following compounds are not oxidized by the enzyme: D-glucose, D-fructose, D-galactose, D-mannose, D-xylose, 2-keto-D-gluconate, 5-keto-D-gluconate, 5-keto-D-fructose, D-galactonate, D-mannonate, 6-phospho-D-gluconate, L-idonate, D-arabonate, D-xylonate, and L-sorbose. Alcohols and aldehydes are also inert. The Michaelis constant for D-gluconate is 0.3–0.8 mM at pH 6.0 with PMS-DCIP, and 2.3–3.2 mM at pH 5.0 with ferricyanide (Table III). These kinetic properties are virtually the same for both the purified enzyme and the original membrane fraction. The purified enzyme can reduce PMS, DCIP, ferricyanide, and CoQ$_1$, but not menadione[4] to the same degree. NAD, NADP, or oxygen are completely inactive as an electron acceptor. The activity is inhibited competitively by pyruvate and 2-keto-D-gluconate and noncompetitively by oxamate. A mixed type of inhibition is observed with oxalate (Table III).

Reconstitution of D-Gluconate Oxidase.[4] In *P. aeruginosa,* a membrane preparation depleted D-gluconate dehydrogenase can be obtained through solubilization of the enzyme by detergent and subsequent extensive dialysis against alkaline buffer, pH 8.0. The membrane preparation thus obtained is depleted completely of D-gluconate dehydrogenase, but cytochrome oxidase activity is fully retained. When the purified enzyme is mixed with the enzyme-depleted membrane, however, little D-gluconate oxidase activity is restored. Restoration of the oxidase activity is achieved by adding CoQ to a mixture of the enzyme and the enzyme-depleted membrane. Of the CoQ homologs examined, CoQ$_2$ is the most effective, and CoQ$_1$ and CoQ$_4$ are also effective but less so. On addition of increasing amounts of the enzyme to a fixed amount of the enzyme-depleted membrane, the oxidase activity reaches a constant level indicating reconstitution of the D-gluconate oxidase system of the membrane.

[32] 2-Keto-D-gluconate Dehydrogenase from *Gluconobacter melanogenus*, Membrane-Bound

By EMIKO SHINAGAWA and MINORU AMEYAMA

2-Keto-D-gluconate + acceptor → 2,5-diketo-D-gluconate + reduced acceptor

2-Keto-D-gluconate dehydrogenase (EC 1.1.99.4), which catalyzes the oxidation of 2-keto-D-gluconate to 2,5-diketo-D-gluconate, is localized on the outer surface of the cytoplasmic membrane of *Gluconobacter* species; the enzyme is the primary dehydrogenase of the 2-keto-D-gluconate oxidizing system *in vivo*. Many bacteria have been examined for their capacity to oxidize 2-keto-D-gluconate, and the oxidizing activity has been found only in pigment-producing strains of *Gluconobacter,* such as *G. melanogenus, G. liquefaciens,* and *G. sphaericus.*[1]

Assay Method

Principle. 2-Keto-D-gluconate dehydrogenase catalyzes the reaction *in vitro* in the presence of dyes such as 2,6-dichlorophenolindophenol, phenazine methosulfate, ferricyanide, or ubiquinone derivatives. In the following, ferricyanide is used because of its simplicity for routine assay. This method is described by Wood *et al.*[2] in the assay of D-gluconate dehydrogenase. Some modifications are described here.

Reagents

Potassium ferricyanide, 0.1 *M*, in distilled water
Sodium 2-keto-D-gluconate, 1 *M*, in distilled water
Sodium acetate buffer, 0.1 *M*, pH 4.0
Triton X-100, 5% solution
Ferric sulfate–Dupanol reagent containing 5 g of $Fe_2(SO_4)_3 \cdot n\ H_2O$, 3 g of Dupanol (sodium lauryl sulfate), 95 ml of 85% phosphoric acid, and distilled water to 1 liter
Enzyme dissolved in 0.01 *M* potassium phosphate, pH 6.0, containing 0.1% Triton X-100

Procedure. The reaction mixture contains 0.1 ml of potassium ferricyanide, 0.5 ml of sodium acetate buffer, 0.1 ml of 2-keto-D-gluconate,

[1] E. Shinagawa, T. Chiyonobu, O. Adachi, and M. Ameyama, *Agric. Biol. Chem.* **40,** 475 (1976).
[2] W. A. Wood, R. A. Fetting, and B. C. Hertlein, this series, Vol. 5, p. 287.

0.1 ml of Triton X-100, and enzyme solution in a total volume of 1.0 ml. After preincubation at 37° for 5 min, the reaction is initiated by the addition of potassium ferricyanide and is carried out at 37° for 5 min. The reaction is stopped by adding 0.5 ml of ferric sulfate–Dupanol reagent. Then, 3.5 ml of water is added to the reaction mixture, and the resulting Prussian blue color is measured at 660 nm after standing at 37° for 20 min. The rate of reduction of ferricyanide is correlated to the rate of 2-keto-D-gluconate oxidation.

Definition of Unit and Specific Activity. One unit of enzyme activity is defined as the amount of enzyme catalyzing oxidation of 1 μmol of 2-keto-D-gluconate per minute under the above assay conditions, and 4.0 absorbance units equal 1 μmol of 2-keto-D-gluconate oxidized. Specific activity is expressed as units per milligram of protein. Protein concentration is determined according to modified Lowry method described by Dulley and Grieve.[3]

Source of Enzyme

Microorganism and Culture. Gluconobacter melanogenus IFO 3293 can be obtained from the Institute for Fermentation, Osaka (17-85, Juso-honmachi 2-chome, Yodogawa-ku, Osaka 532, Japan). The stock culture is maintained on a potato–glycerol slant. The preparation of this medium is shown in this volume [24].

The culture medium contains D-glucose (5 g), sodium D-gluconate (5 g), yeast extract (3 g), and polypeptone (3 g) in 1 liter of tap water. The organism grown on the potato–glycerol slant is inoculated into 100 ml of the medium in a 500-ml shake flask, and cultivation is carried out at 30° for 24 hr with reciprocal shaking. To 30 liters of the medium in a 50-liter jar fermentor, 2 liters of inoculum (20 shaking flasks) are inoculated, and incubation is carried out at 30° under vigorous agitation (500 rpm) and aeration (25–30 liters of air per minute). Bacterial cells are harvested at the late exponential phase. About 100 g of wet cells are obtained from two batches of cultures.

Purification Procedure[4]

All purification steps are performed at 0–5°, and centrifugations are at 10,000 g for 10 min unless otherwise stated.

Step 1. Preparation of Membrane Fraction. The cell paste (100 g) is suspended in 0.01 M sodium acetate buffer, pH 5.0, and passed twice

[3] J. R. Dulley and P. A. Grieve, *Anal. Biochem.* **64**, 136 (1975).
[4] E. Shinagawa, K. Matsushita, O. Adachi, and M. Ameyama, *Agric. Biol. Chem.* **45**, 1079 (1981).

through a French press at 1000 kg/cm². After centrifugation to remove intact cells, the supernatant is centrifuged at 80,000 g for 60 min. The resultant precipitate, the membrane fraction, can be stored at −20° without appreciable loss of enzyme activity for over several months.

Step 2. Solubilization of Enzyme. About 40 g of frozen membrane fraction are thawed and suspended in 400 ml of 0.01 M sodium acetate buffer, pH 5.0, containing 2% Brij 35 and 0.2 M KCl. The suspension is stirred for 3 hr and centrifuged at 80,000 g for 60 min. Substantial D-gluconate dehydrogenase (see this volume [31]) is solubilized, but 2-keto-D-gluconate dehydrogenase is retained in the precipitated membrane. The precipitate is suspended in 0.01 M Tris-HCl buffer, pH 8.0, containing 2% sodium cholate and 0.2 M KCl and stirred for 60 min. After standing overnight, the solubilized enzyme is completely recovered in the supernatant from centrifugation at 80,000 g for 60 min.

Step 3. Ammonium Sulfate Fractionation. To the solubilized enzyme, solid ammonium sulfate is added to 0.8 M (10.6 g/100 ml). After stirring for 60 min, the precipitate is removed by centrifugation and discarded. The ammonium sulfate concentration of the supernatant is increased to 1.7 M by adding 12 g of the salt per 100 ml. The precipitate is collected by centrifugation and dissolved in 0.01 M sodium acetate buffer, pH 5.5, containing 1% Triton X-100. The solution is dialyzed against the same buffer containing 0.1% Triton X-100.

Step 4. CM-Cellulose Column Chromatography. To the dialyzed solution, polyethylene glycol 6000 is added to 25% (w/v), to remove contaminating cholate from the preceding steps. The precipitate is collected by centrifugation and dissolved in 0.01 M sodium acetate buffer, pH 5.5, containing 1% Triton X-100. After centrifugation, the supernatant is placed on a CM-cellulose column (3.5 × 20 cm) that has been equilibrated with the same buffer. Elution of the enzyme is made by a linear gradient of KCl to 0.2 M in the same buffer. The fractions containing enzyme activity are combined, and polyethylene glycol 6000 is added to 10% (w/v). To the supernatant obtained by centrifugation, polyethylene glycol 6000 is added to 25% (w/v).

Step 5. Hydroxyapatite Column Chromatography. The precipitate obtained by centrifugation is dissolved in 0.01 M potassium phosphate buffer, pH 6.0, containing 0.1% Triton X-100. After centrifugation to remove insoluble materials, the resulting supernatant is applied to a hydroxyapatite column (3 × 6 cm) that has been equilibrated with the same buffer. The column is treated first with 0.05 M potassium phosphate buffer, pH 6.0, containing 0.1% Triton X-100 and then with 0.1 M potassium phosphate buffer, pH 6.0, containing 1% Triton X-100. The major part of the enzyme activity is eluted at 0.1 M potassium phosphate as a sharp protein

PURIFICATION OF 2-KETO-D-GLUCONATE DEHYDROGENASE FROM
Gluconobacter melanogenus

Fraction	Total protein (mg)	Total activity (units)	Specific activity (units/mg protein)	Yield (%)
Membrane fraction	12,470	37,400	3	100
Washed membrane	6,700	33,600	5	90
Solubilized enzyme	1,900	43,680	23	117
Ammonium sulfate fraction	380	37,130	98	99
CM-cellulose	40	30,240	750	81
PEG[a] fraction	34	27,720	828	74
Hydroxyapatite	12	15,260	1247	41

[a] PEG, polyethylene glycol 6000.

peak accompanied by enzyme activity, and fractions of specific activity over 1000 units per milligram of protein are combined.

As shown in the table, over 600-fold purification of the enzyme is usually obtained.

Properties[4]

Homogeneity. The purified enzyme is homogeneous as judged by analytical ultracentrifugation and polyacrylamide gel electrophoresis. The enzyme shows a symmetrical sedimentation peak with an apparent sedimentation constant of 5.9 S in 0.1 M potassium phosphate, pH 7.0, containing 1% sodium cholate. The purified enzyme also gives a single protein band with enzyme activity on disc gel electrophoresis.

Absorption Spectra. The purified enzyme has a characteristic deep rose-red color due to the cytochrome component. The typical reduced cytochrome c type of absorption spectrum has maxima at 554 nm, 523 nm, and 417 nm. The cytochrome component is reduced by the addition of either 2-keto-D-gluconate or sodium dithionite. The oxidized form of the enzyme shows a single peak at 411 nm. The cytochrome content is measured as pyridine hemochromogen, and 6.87 nmol of cytochrome per milligram (0.91 mol per mole of enzyme) are obtained with the purified enzyme.

Molecular Properties. The molecular weight of the enzyme is determined to be 133,000. The enzyme dissociates into three subunits with molecular weights of 61,000, 47,000, and 25,000 based on polyacrylamide gel electrophoresis in the presence of sodium dodecyl sulfate. The largest component is the dehydrogenase, which shows an intense fluorescence under ultraviolet light. The flavin moiety is bound to the enzyme covalently, and a partial release of the flavin is obtained by proteolytic digestion. The second subunit (M_r 47,000) is the cytochrome component. The function of the smallest component remains unclear.

Stability. The purified enzyme is stable at pH 5.0 through 8.0, and the stability of the enzyme is affected significantly by the concentration of Triton X-100 added to the enzyme solution. When the concentration of Triton X-100 is below its critical micelle concentration (CMC), the enzyme activity decreases rapidly. Thus, addition of Triton X-100 is essential to maintain the enzyme activity.

Catalytic Properties. 2-Keto-D-gluconate dehydrogenase can be assayed *in vitro* in the presence of any one of the following dyes as an electron acceptor: potassium ferricyanide, phenazine methosulfate, nitro blue tetrazolium, or 2,6-dichlorophenolindophenol; NAD, NADP, and oxygen are completely inert. Of tested substrates, only 2-keto-D-gluconate is oxidized. When 2-keto-D-gluconate is oxidized in the presence of excess amounts of substrate analogs such as sugars, sugar alcohols, hexonates, ketohexonates, and aldehydes, the reaction rate of 2-keto-D-gluconate oxidation is not affected. An apparent Michaelis constant for 2-keto-D-gluconate determined at pH 4.0 is found to be 50 mM with ferricyanide as the acceptor. The optimum pH of 2-keto-D-gluconate oxidation is pH 4.0, and the optimum temperature is 39°. The enzyme is inhibited noncompetitively by some kinds of organic acids, such as oxamate, succinate, citrate, and oxalate. The apparent K_i values for citrate and succinate are 22 mM and 175 mM, respectively. Sulfhydryl reagents and chelating agents were ineffective inhibitors.

[33] 5-Keto-D-gluconate Reductase from *Gluconobacter suboxydans*

By Minoru Ameyama and Osao Adachi

5-Keto-D-gluconate + NADPH $\rightleftharpoons$ D-gluconate + NADP

5-Keto-D-gluconate reductase (EC 1.1.1.69) catalyzes the reduction of 5-keto-D-gluconate to D-gluconate and occurs only in acetic acid bacteria.

Assay Method

Principle. 5-Keto-D-gluconate is reduced in the presence of NADPH to give D-gluconate and NADP. The reaction rate is measured by the decrease of absorbance at 340 nm.

Reagents

NADPH, 1.5 mM, in distilled water; store at $-20°$ until used
NADP, 1.5 mM, in distilled water; store at $-20°$ until used
5-Keto-D-gluconate, sodium salt, 0.2 M in distilled water
5-Keto-D-gluconate prepared according to DeMoss (this series, Vol. 3, p. 236)
D-Gluconate, sodium salt, 1 M, in distilled water
Potassium phosphate buffer, 0.1 M, pH 6.0
Glycine–NaOH buffer, 0.1 M, pH 10.0
Enzyme solution in 10 mM potassium phosphate, pH 6.0, containing 1 mM 2-mercaptoethanol

Procedure. NADPH oxidation is followed in a recording spectrophotometer at 25°. Prior to assay, the buffer solutions and distilled water are brought to 25°. The complete reaction mixture contains 0.1 ml of NADPH (or NADP), 0.05 ml of 5-keto-D-gluconate (or 0.1 ml of D-gluconate), 1.0 ml of potassium phosphate buffer (or 1.0 ml of glycine-NaOH buffer), and enzyme solution in a total volume of 3.0 ml. The reaction is initiated by the addition of 10 μl of enzyme. The decrease (or increase) of absorbance is followed by a recorder attached to the spectrophotometer.

Definition of Unit and Specific Activity. One unit of enzyme activity oxidizes of 1 μmol of NADPH per minute under the assay conditions described. Specific activity (units per milligram of protein) is based on the spectrophotometric protein estimation at 280 nm, an $E_{cm}^{1\%}$ value of 15.0 is used throughout.

Source of Enzyme

Microorganism and Culture. Gluconobacter suboxydans IFO 12528 can be obtained from the Institute of Fermentation, Osaka (17-85, Juso-honmachi 2-chome, Yodogawa-ku, Osaka 532, Japan). Stock cultures are maintained on a potato–glycerol slant (see this volume [24]). The culture medium consists of sodium D-gluconate (20 g), D-glucose (5 g), glycerol (3 g), yeast extract (3 g), polypeptone (2 g), and 200 ml of potato extract in 1 liter of tap water. The pH of the medium is 6.5. *Gluconobacter suboxydans* is transferred from the potato–glycerol slant to 100 ml of the medium in a 500-ml shake flask, and cultivation is carried out at 30° for 24 hr with reciprocal shaking. For large-scale cultures, 1.5 liters of inoculum (15 shake flasks) are transferred to 30 liters of the medium in a 50-liter jar fermentor. In this case, the potato extract is usually omitted from the medium, and cultivation is performed at 30° for 24 hr under vigorous aeration at 30 liters per minute with 500 rpm agitation.

Purification Procedure

All operations are carried out at 0–5° throughout the purification steps unless otherwise stated. Centrifugations are at 12,000 g for 20 min. The buffer solution used is potassium phosphate, pH 6.0, which is supplemented with 2-mercaptoethanol (1 mM), sodium D-gluconate (0.1 M), and glycerol (10%).

Step 1. Preparation of Cell-Free Extract. Cell paste of *G. suboxydans* (300 g wet weight) harvested from 60 liters (two large-scale runs) is washed twice with cold distilled water and suspended in 0.01 M buffer. Cell suspension is passed twice through a French press (American Instrument Co.) at 1000 kg/cm². The resulting cell homogenate is run at 68,000 g for 60 min in a preparative ultracentrifuge. About 360 ml of supernatant are obtained with protein concentration of about 30 mg/ml.

Step 2. DEAE-Sephadex Column Chromatography. The cell-free extract from the preceding step is applied to a DEAE-Sephadex A-50 column (5 × 35 cm) that has been equilibrated with 0.002 M buffer. After washing with 600 ml of the same buffer, elution of the enzyme is made stepwise with buffer containing 0.2 M NaCl. After elution of major protein peak, enzyme activity is eluted from the column. The enzyme fractions are combined (620 ml) and precipitated with ammonium sulfate at 3.6 M (47.5 g/100 ml). The precipitate collected by centrifugation is dialyzed against 0.002 M buffer overnight.

Step 3. Affinity Chromatography on Blue-Dextran Sepharose. After the dialyzed enzyme is centrifuged to remove insoluble materials, the resulting supernatant solution (32 ml) is applied to a Blue-Dextran Sepharose 4B column (2 × 20 cm) that has been equilibrated with 0.002 M buffer. The column is washed with the same buffer until the absorbance of the eluate at 280 nm decreases to 0.1. The column is further treated with the same buffer containing 0.1 M KCl to remove some impurities. Elution of the enzyme is made with 0.002 M buffer containing 0.3 M KCl. The enzyme fractions are combined (60 ml) and dialyzed against 0.002 M buffer overnight.

Step 4. pH Gradient Chromatography on DEAE-Sephadex. The dialyzed enzyme is applied to a DEAE-Sephadex A-50 column (1 × 20 cm) that has been equilibrated with the same buffer. After washing with 100 ml of the same buffer, elution of the enzyme is made by a descending

PURIFICATION OF 5-KETO-D-GLUCONATE REDUCTASE FROM *Gluconobacter suboxydans*

Step	Total protein (mg)	Total activity (units)	Specific activity (units/mg protein)	Yield (%)
Cell-free extract	10,360	584	0.05	100
DEAE-Sephadex A-50 chromatography	433	494	1.14	84
Blue-Dextran Sepharose chromatography	25	320	12.75	56
pH gradient chromatography (I)	13	245	18.60	42
pH gradient chromatography (II)	3.4	234	69.23	40
Crystallization	3.3	230	69.69	40

pH gradient of the buffer formed between two buffer systems. One system is 200 ml of 0.002 M buffer, pH 6.0, containing 2-mercaptoethanol (1 mM), sodium D-gluconate (0.1 M), and glycerol (10%). The second buffer system is of the same composition except that the pH is adjusted to 3.78 with 1 M D-gluconic acid. Two hundred milliliters of each of the buffers are used, and 6-ml fractions are collected. Two major protein peaks are eluted at pH 5.0 and 4.5. The enzyme activity is found in the second protein peak, which elutes at pH 4.5. The pooled enzyme solution (50 ml) is adjusted to pH 6.0 with dipotassium phosphate. Irreversible precipitation of protein impurities often occurs during pH readjustment, and it is removed by centrifugation.

When pH gradient chromatography is repeated in the same manner, the purified enzyme contains no critical impurity as judged by conventional disc gel electrophoresis, and the specific activity reaches at about 65.0 units per milligram of protein or more.

Step 5. Crystallization. Ammonium sulfate is added to 3.9 M (51.5 g/100 ml). The precipitate is collected by centrifugation and dissolved in a minimal volume of 0.01 M buffer. Insoluble materials are removed by centrifugation. Crystallization of the enzyme is performed by adding finely powdered ammonium sulfate. At approximately 2.5 M concentration, crystals of the enzyme appear as fine needles or rods, and growth of the crystals is complete within a month at 4°.

The enzyme is purified about 1200-fold with an overall recovery of 40%. A typical purification is summarized in the table.

Properties[1]

Substrate Specificity. 5-Keto-D-gluconate reductase is highly specific to 5-keto-D-gluconate, and other substrates are essentially inactive when

---

[1] O. Adachi, E. Shinagawa, K. Matsushita, and M. Ameyama, *Agric. Biol. Chem.* **43**, 75 (1979).

assayed at pH 6.0 in the presence of NADPH. In enzymic reduction of 5-keto-D-gluconate, the reaction product is identified to be exclusively D-gluconate. In the oxidation reaction, D-gluconate is most rapidly oxidized at pH 10.0 in the presence of NADP. The rate of D-gluconate oxidation is about 60% of the rate of 5-keto-D-gluconate reduction.

Stability. The enzyme activity is fairly unstable, over 75% being lost within a few days unless either D-gluconate or 5-keto-D-gluconate is added during storage or purification steps; 2-keto-D-gluconate has no effect.

Thermostability of the enzyme is also much enhanced by addition of either D-gluconate or 5-keto-D-gluconate, or both, but not by 2-keto-D-gluconate. The enzyme is heat labile and loses its activity completely when heated for a few minutes to more than 40°. When the enzyme solution is supplemented with D-gluconate or 5-keto-D-gluconate, no loss of enzyme activity is observed after heating at 55° for 5 min even at 0.1 μg/ml.

Catalytic Properties. The optimum pH for reduction is pH 5.5, and 80% of maximal activity is observed at pH 6.0. In the presence of NADP, D-gluconate oxidation is maximal at pH 10.0. The apparent equilibrium constant is approximately 10^{-12} M. Both reduction of 5-keto-D-gluconate and oxidation of D-gluconate proceed most rapidly at 50°. Apparent Michaelis constants for 5-keto-D-gluconate and NADPH at pH 6.0 are 0.9 mM and 6 μM, respectively. On the other hand, the Michaelis constants for D-gluconate and NADP assayed at pH 10.0 are 20 mM and 20 μM, respectively. The enzyme activity is inhibited by sulfhydryl reagents such as p-chloromercuribenzoate or mercuric ions. Divalent heavy metals also inhibit the enzyme activity.

Molecular Properties. The molecular weight of the enzyme is 100,000 by gel filtration on Sephadex G-200. The enzyme is composed of four identical subunits (M_r 25,000 each). The apparent sedimentation constant is 5.3 S in 0.1 M potassium phosphate, pH 6.0, containing 0.1 M D-gluconate and 5% glycerol at 25°.

Enzyme Formation and Physiological Role. Maximum enzyme formation is observed at the end of the exponential phase of growth. In the decreasing growth phase, about 65% of the maximum enzyme activity remains. The suggested physiological role is to reduce 5-keto-D-gluconate to D-gluconate to supply a carbon source and regenerate NADP in a manner similar to that for 2-keto-D-gluconate reductase (see this volume [34]).

[34] 2-Keto-D-gluconate Reductase from Acetic Acid Bacteria

By MINORU AMEYAMA and OSAO ADACHI

$$2\text{-Keto-D-gluconate} + NADPH \rightleftharpoons \text{D-gluconate} + NADP$$

2-Keto-D-gluconate reductase occurs in acetic acid bacteria and catalyzes the reduction of 2-keto-D-gluconate to D-gluconate.

Assay Method

Principle. 2-Keto-D-gluconate is reduced by the enzyme in the presence of NADPH to give D-gluconate and NADP. The reaction rate is measured by the decrease of optical density at 340 nm.

Reagents

NADPH, 1.5 mM, in distilled water
NADP, 1.5 mM, in distilled water
2-Keto-D-gluconate, sodium salt, 0.1 M, in distilled water
D-Gluconate, sodium salt, 1 M, in distilled water
Potassium phosphate buffer, 1 M, pH 6.0, for the enzyme from *Gluconobacter liquefaciens*
Potassium phosphate buffer, 1 M, pH 7.0, for the enzyme from *Acetobacter rancens*
Glycine–NaOH buffer, 1 M, pH 10.0, for the enzyme from *G. liquefaciens*
Glycine–NaOH buffer, 1 M, pH 12.0, for the enzyme from *A. rancens*
Enzyme in 1 mM potassium phosphate, pH 6.0, containing 1 mM 2-mercaptoethanol

Procedure. The enzyme activity is measured by reading the decrease (or increase) of absorbance of NADPH at 340 nm in a recording spectrophotometer with thermostatted water at 25° or a spectrophotometer in a thermostatted room (25°). Prior to the assay, the buffer solutions and distilled water are warmed to 25°. The complete reaction mixture contains 0.1 ml of NADPH (or NADP), 0.1 ml of 2-keto-D-gluconate (or 0.1 ml of D-gluconate), 0.1 ml of potassium phosphate buffer (or 0.1 ml of glycine-NaOH buffer), and enzyme solution in a total volume of 3.0 ml. The reaction is initiated by the addition of 10 μl of enzyme.

Definition of Unit and Specific Activity. One unit of enzyme activity causes the oxidation of 1 μmol of NADPH per minute under the assay conditions described above. Specific activity (units per milligram of protein) is based on the spectrophotometric protein estimation by measuring the absorbance at 280 nm. An $E_{cm}^{1\%}$ value of 15.0 is used throughout.

Source of Enzyme

Microorganisms and Cultures. Gluconobacter liquefaciens IFO 12388, *G. suboxydans* IFO 12528, *Acetobacter rancens* IFO 3298, and *A. ascendens* IFO 3299 are supplied from the Institute for Fermentation, Osaka (17-85, Juso-honmachi 2-chome, Yodogawa-ku, Osaka 532, Japan). Stock cultures are maintained on potato–glycerol slants (see this volume [24]). The culture medium for acetic acid bacteria consists of glycerol (3 g), D-glucose (5 g), sodium D-gluconate (20 g), polypeptone (5 g), and yeast extract (2 g) in 1 liter of tap water. The pH of the medium is 6.5. Aerobic cultures of bacteria are grown in a jar fermentor or in shake flasks for about 24 hr. About 150–200 g of wet cells are usually harvested from 30 liters of medium (fermentor).

Purification Procedure

Preparation from G. liquefaciens [1]

All operations are carried out at 0–5° unless otherwise stated. Centrifugations are at 12,000 g for 20 min. The buffer solution is potassium phosphate, pH 6.0, supplemented with 2-mercaptoethanol (1 mM).

Step 1. Preparation of Cell-Free Extract. Cell paste of *G. liquefaciens* (300 g wet weight) harvested from 60 liters of two 24-hour cultures in a 50-liter jar fermentor is washed twice with distilled water and suspended in 0.01 M buffer. The cell suspension is passed twice through a French press (American Instrument Co.) at 1000 kg/cm². The resulting cell homogenate is centrifuged at 68,000 g for 90 min in a preparative ultracentrifuge.

Step 2. First Ammonium Sulfate Fractionation. Solid ammonium sulfate is added to the supernatant to 1.2 M (16.4 g/100 ml, 0°). After standing in a cold, the precipitate is removed by centrifugation and discarded. The salt concentration of the supernatant is then increased to 3.3 M by adding 27.2 g/100 ml. The precipitate obtained by centrifugation is dissolved in a small volume of 0.01 M buffer.

Step 3. Heat Treatment. 2-Mercaptoethanol and D-gluconate are added to the enzyme solution to 0.05 M and 0.1 M, respectively. The solution is

[1] T. Chiyonobu, E. Shinagawa, O. Adachi, and M. Ameyama, *Agric. Biol. Chem.* **39**, 2263 (1975).

rapidly warmed to 55° in a 70–80° water bath with vigorous stirring and then held at 55° in a 55° bath for 5 min. The heat-treated enzyme solution is rapidly cooled in an ice bath, and the precipitate of denatured protein is removed by centrifugation. The precipitate of denatured protein is washed with 0.01 M buffer, and the wash is combined with the original supernatant. The enzyme solution is dialyzed against two changes of 10 liters of 0.01 M buffer.

Step 4. Second Ammonium Sulfate Fractionation. The dialyzed enzyme is brought to 1.2 M ammonium sulfate by the addition of 16.4 g of the salt per 100 ml of enzyme solution. After stirring for 30 min, the resulting precipitate is removed by centrifugation and discarded. The supernatant solution is brought to 2.2 M ammonium sulfate by the addition of 12.7 g of the salt per 100 ml of the original dialyzed enzyme solution and stirred for 30 min. The precipitate is collected by centrifugation and dissolved in 50 ml of 0.01 M buffer and dialyzed overnight against 3 liters of the same buffer.

Step 5. DEAE-Sephadex Column Chromatography. The dialyzed solution is applied to a DEAE-Sephadex A-50 column (3 × 35 cm) that has been equilibrated with 0.01 M buffer. The column is washed with 0.01 M buffer containing 0.1 M NaCl until the absorbance of the effluent at 280 nm decreases to 0.1 or less. The activity is eluted with 0.01 M buffer containing 0.2 M NaCl. Ammonium sulfate is added to the enzyme solution to 3.0 M, and the precipitate is collected by centrifugation. The precipitate is dissolved in 25 ml of 0.01 M buffer and dialyzed against 3 liters of the same buffer overnight. Insoluble material is removed by centrifugation.

Step 6. Hydroxyapatite Column Chromatography. The dialyzed enzyme solution is applied to a hydroxyapatite column (3 × 5 cm) that has been equilibrated with the same buffer as that used for dialysis in step 5. The column is treated with 0.01 M buffer until the absorbance of the effluent at 280 nm decreases to 0.1 or less. The activity is eluted from the column with 0.01 M buffer containing 0.2 M NaCl. The fractions with a specific activity over 55.0 units per milligram of protein are combined, and ammonium sulfate is brought to 3.0 M by the addition of 39.8 g/100 ml (0°).

Step 7. Crystallization. The precipitate is collected by centrifugation and dissolved in a minimum volume of 0.1 M buffer; insoluble materials are removed by centrifugation. To the supernatant, a few grains of solid ammonium sulfate are added with gentle stirring until a faint turbidity appears. At approximately 1.9 M ammonium sulfate, crystals of the enzyme come out as fine needles or fine rods and grow to a hexagonal thin plate within 2 weeks in a refrigerator (4°). The enzyme is purified 760-fold with a yield of 40% as summarized in Table I.

TABLE I
PURIFICATION OF 2-KETO-D-GLUCONATE REDUCTASE FROM *Gluconobacter liquefaciens*

Fraction	Total protein (mg)	Total activity (units)	Specific activity (units/mg protein)	Yield (%)
Cell-free extract	56,200	4500	0.08	100
Ammonium sulfate fraction, 1st	24,200	4200	0.17	94
Heat treatment	14,600	3700	0.25	82
Ammonium sulfate fraction, 2nd	2,000	3230	1.61	73
DEAE-Sephadex A-50	365	2740	7.50	61
Hydroxyapatite	36	2100	58.30	47
Crystalline enzyme	29	1770	61.03	40

Preparation from A. rancens[2]

The following purification starts with 400 g of wet cells of *A. rancens*. All operations are carried out at 0–5°, unless otherwise stated. Centrifugations are at 12,000 g for 20 min. The buffer is potassium phosphate, pH 6.0, which is supplemented with 2-mercaptoethanol to 1 mM.

Preparation of cell-free extract (step 1) and ammonium sulfate fractionation (step 2) are performed essentially as described above for *G. liquefaciens*.

Step 3. First DEAE-Sephadex Column Chromatography. The dialyzed enzyme solution (126 ml) is applied to a column of DEAE-Sephadex A-50 (5 × 30 cm) that has been equilibrated with 0.01 M buffer. The enzyme is eluted by a 2-liter linear gradient of 0 to 0.5 M NaCl in 0.01 M buffer. The enzyme activity is eluted at about 0.2 M NaCl, and fractions are combined to give 720 ml. Ammonium sulfate is added to 3.3 M, and the precipitate obtained by centrifugation is dissolved in a small volume of 0.01 M buffer and dialyzed overnight against the same buffer.

Step 4. First Hydroxyapatite Column Chromatography. The dialyzed enzyme (43 ml) is applied to a column of hydroxyapatite (3 × 7.5 cm) equilibrated with 0.01 M buffer. The column is washed with 300 ml of 0.01 M buffer, then treated with 0.075 M buffer; active fractions are combined to give 250 ml. After precipitation with ammonium sulfate to 3.3 M, the enzyme is dissolved in a small volume of 0.01 M buffer and dialyzed overnight against 0.01 M buffer.

Step 5. Second Hydroxyapatite Column Chromatography. The dialyzed enzyme (10 ml) is rechromatographed on a hydroxyapatite column (2.2 ×

[2] T. Chiyonobu, E. Shinagawa, O. Adachi, and M. Ameyama, *Agric. Biol. Chem.* **40**, 175 (1976).

TABLE II
PURIFICATION OF 2-KETO-D-GLUCONATE REDUCTASE FROM *Acetobacter rancens*

Fraction	Total protein (mg)	Total activity (units)	Specific activity (units/mg protein)	Yield (%)
Cell-free extract	41,531	3622	0.09	100
Ammonium sulfate fraction	5,065	2244	0.44	62
DEAE-Sephadex A-50 (I)	596	2107	3.54	58
Hydroxyapatite (I)	43	1161	27	32
Hydroxyapatite (II)	11	1045	95	29
DEAE-Sephadex A-50 (II)	2	671	336	19
Crystalline enzyme	2	671	336	19

5 cm) under the essentially the same conditions. The column is treated with 0.05 M and 0.1 M buffer, stepwise. The enzyme activity elutes at 0.1 M, and the fractions are combined to give 30 ml. The solution is dialyzed against 9 volumes of water containing 1 mM 2-mercaptoethanol to reduce the phosphate concentration to 0.01 M.

Step 6. Second DEAE-Sephadex Column Chromatography. The dialyzed enzyme solution is applied to a DEAE-Sephadex A-50 column (1.0 × 15 cm) as described in the step 3. The column is treated with 0.01 M buffer containing 0.1 M NaCl. Elution of the enzyme is made between 0.01 M buffer containing 0.1 M NaCl (250 ml) and 0.01 M buffer containing 0.3 M NaCl (250 ml). Protein peak and enzyme activity are eluted at about 0.15 M NaCl, and the fractions are combined to give 40 ml. The enzyme is precipitated with 3.3 M ammonium sulfate and collected by centrifugation. The precipitate is dissolved in a minimal volume of 0.1 M buffer, and insoluble materials are removed by centrifugation.

Step 7. Crystallization. Crystallization is performed by adding finely powdered ammonium sulfate until slight turbidity. At approximately 2.2 M, crystals appear as fine needles or thin plates.

A typical purification is summarized in Table II.

Variations in Procedures.[3] Purification and crystallization of the enzyme from *A. ascendens* IFO 3299 is performed using affinity chromatography on a Blue Dextran–Sepharose 4B column. After chromatography on DEAE-Sephadex A-50 and hydroxyapatite, as described above, the enzyme solution in 0.01 M buffer is applied to a column of Blue Dextran–Sepharose 4B that has been equilibrated with the same buffer. The column is washed with the same buffer until the absorbance at 280 nm decreases

[3] O. Adachi, T. Chiyonobu, E. Shinagawa, K. Matsushita, and M. Ameyama, *Agric. Biol. Chem.* **42**, 2057 (1978).

TABLE III
PURIFICATION OF 2-KETO-D-GLUCONATE REDUCTASE FROM *Acetobacter ascendens*

Fraction	Total protein (mg)	Total activity (units)	Specific activity (units/mg) protein)	Yield (%)
Cell-free extract	177,960	2566	0.01	100
Ammonium sulfate fraction	131,600	2180	0.02	85
DEAE-Sephadex A-50	3,580	1040	0.29	40
Hydroxyapatite	650	966	1.50	37
Ammonium sulfate fraction	265	688	2.60	26
Blue Dextran–Sepharose 4B	3.4	483	142.05	18
Crystalline enzyme	2	374	187.00	15

to 0.01. Elution is effected by adding KCl to 0.3 M to the buffer solution. Conventional disc gel electrophoresis shows homogeneity of the enzyme preparation at this stage. Crystallization of the enzyme is immediately performed as summarized in Table III.

Properties[1–3]

Stability. The crystalline enzyme as well as crude enzyme preparations of 2-keto-D-gluconate reductase are stable in 0.01 M potassium phosphate, pH 6.0, containing 1 mM 2-mercaptoethanol in the cold (5°). The enzyme is stable to heating at 55° for 5 min, and this is enhanced when 0.1 M D-gluconate or 2-keto-D-gluconate is added even at 0.1 μg of protein per milliliter; 5-keto-D-gluconate has no effect. Thus, 2-keto-D-gluconate reductase is a different protein from 5-keto-D-gluconate reductase (see this volume [33]).

Effect of pH. Enzyme activity for reduction of 2-keto-D-gluconate to D-gluconate is found at pH 6.0 with all enzyme preparations of acetic acid bacteria. Oxidation of D-gluconate to 2-keto-D-gluconate is found at pH 10.5 with the enzyme from *G. liquefaciens,* 12.0 with the enzyme from *A. rancens,* and 11.0 with the enzyme from *A. ascendens.*

Substrate Specificity. Table IV lists the substrate specificity of 2-keto-D-gluconate reductase of acetic acid bacteria. In addition to 2-keto-D-gluconate, 2-keto-D-galactonate and 2-keto-L-gulonate are rapidly reduced; some interesting differences are seen in the profile of substrate specificity of the enzymes between two genera of acetic acid bacteria. Activity of D-gluconate oxidation is weak or absent with the enzyme from *Acetobacter* and intense from *Gluconobacter.* 2-Keto-D-gluconate reductase of acetic acid bacteria requires NADPH as the coenzyme and shows no activity with NADH. An apparent Michaelis constant for NADPH is of the order of $10^{-6} M$, and for NADP, D-gluconate, and 2-keto-D-gluconate, of the order of $10^{-3} M$, when assayed at the optimum pH of reactions in both directions.

Molecular Properties. 2-Keto-D-gluconate reductase of acetic acid bacteria has a molecular weight of about 120,000 when estimated by gel filtration on a Sephadex G-200 column. Upon sodium dodecyl sulfate gel electrophoresis, the enzyme from *A. rancens* dissociates into a subunit having a molecular weight of 15,000, whereas the enzyme from *A. ascendens* gives a subunit of M_r 40,000. It seems likely that the enzyme from *G. liquefaciens* is composed of heterogeneous subunits—two larger subunits (M_r 43,000 each) and one smaller subunit (M_r 34,000).

Activators and Inhibitors. No activators in 2-keto-D-gluconate reduction and D-gluconate oxidation have been found. The enzyme is strongly inhibited by sulfhydryl reagents such as p-chloromercuribenzoate or mercuric ions.

Enzyme Formation and Physiological Role. The maximum enzyme formation is observed at the end of the exponential phase of bacterial growth for both genera. In the cells at decreasing growth phase, about 65% of the maximum enzyme activity remains. The physiological role appears to be

TABLE IV

SUBSTRATE SPECIFICITY OF 2-KETO-D-GLUCONATE REDUCTASE OF ACETIC ACID BACTERIA[a]

Substrate	Acetobacter ascendens[b]		A. rancens[c]		Gluconobacter liquefaciens[d]		G. suboxydans[e]	
	NADP	NADPH	NADP	NADPH	NADP	NADPH	NADP	NADPH
conate	7	0	9	0	67	0	48	0
spho-D-gluconate	0	0	0	0	0	0	0	0
nnonate	0	0	0	0	0	0	0	0
ate	3	0	11	0	94	0	83	0
actonate	1	0	6	0	47	0	39	0
bonate	0	0	0	0	0	0	0	0
onate	0	0	1	0	23	0	26	0
o-D-gluconate	0	100	0	100	0	100	0	100
o-D-gluconate	0	5	0	11	0	0	0	0
o-D-galactonate	0	95	0	79	0	79	0	80
o-L-gulonate	0	36	0	177	0	25	0	45

The data are expressed as relative reaction rates against 2-keto-D-gluconate as 100%.
Crystalline preparation with a specific activity of 187 units per milligram of protein is used.
Crystalline preparation with a specific activity of 336 units per milligram of protein is used.
Crystalline preparation with a specific activity of 61 units per milligram of protein is used.
Highly purified preparation with a specific activity of 111 units per milligram of protein is used.

to reduce 2-keto-D-gluconate to D-gluconate to supply a substrate for gluconokinase and a carbon source for growth and to regenerate NADP at the same time. When D-glucose and/or D-gluconate are depleted in the culture medium and converted to 2-keto-D-gluconate by membrane-bound dehydrogenase systems, gradual cell growth is still observed even at the late stationary phase due to 2-keto-D-gluconate utilization via the reductase.

[35] D-Mannonate and D-Altronate-NAD Dehydrogenases from *Escherichia coli*

By Raymond Portalier and François Stoeber

$$\text{D-Fructuronate} + \text{NADH} + \text{H}^+ \rightleftharpoons \text{D-mannonate} + \text{NAD}^+ \tag{1}$$
$$\text{D-Tagaturonate} + \text{NADH} + \text{H}^+ \rightleftharpoons \text{D-altronate} + \text{NAD}^+ \tag{2}$$

Mannonate dehydrogenase (D-mannonate : NAD$^+$ oxidoreductase, EC 1.1.1.57) and altronate dehydrogenase (D-altronate : NAD$^+$ oxidoreductase, EC 1.1.1.58) catalyzed the reduction of D-fructuronate (D-lyxo-5-hexulosonate) or D-tagaturonate (D-arabino-5-hexulosonate) into D-mannonate, Eq. (1), or D-altronate, Eq. (2), respectively. These inducible degradative enzymes are involved in glucuronate and galacturonate metabolism in bacteria. They were first identified in extracts from *E. coli* ATCC 9637 by Ashwell and his group.[1]

Assay Method

Principle. Mannonate and altronate dehydrogenase activities can be determined, in either the forward or reverse reaction, by measuring the initial rates of oxidation or reduction of NADH and NAD, respectively. Maximum initial rates in the forward reactions are 5-fold (mannonate dehydrogenase) to 10-fold (altronate dehydrogenase) higher than in the reverse reactions. Activities are measured spectrophotometrically by the decrease or increase in absorbance at 340 nm, in the presence of D-fructuronate or D-tagaturonate and D-mannonate or D-altronate, respectively.

[1] G. Ashwell, this series, Vol. 5, p. 190.

Reagents

For reduction of D-fructuronate or D-tagaturonate
  D-Fructuronate, sodium salt, 20 mM
    or D-tagaturonate, sodium salt, 10 mM
  NADH, 0.2 mM
  Na_2HPO_4–NaH_2PO_4 buffer, 50 mM, pH 6.3 at 22°
For oxidation of D-mannonate
  D-Mannonate, potassium salt, 20 mM
  NAD, 0.8 mM
  Glycylglycine–NaOH buffer, 50 mM, pH 8.4 at 22°
For oxidation of D-altronate
  D-Altronate, potassium salt, 20 mM
  NAD, 0.8 mM
  Glycylglycine-NaOH buffer, 50 mM, pH 8.9, at 22°

The solutions of reactants are made in phosphate or glycylglycine buffers. Stock solutions of substrate (0.1 M) or coenzyme (10 mM) are made in water. D-Fructuronate, D-tagaturonate, D-mannonate, and D-altronate are not commercially available and were prepared according to previously published procedures.[2-5]

Procedure. Assays are carried out in silica cells of 10-mm light path maintained at 37° and containing 0.7 ml of coenzyme, 0.1 ml of substrate, and 0.1 ml of buffer. Reactions are initiated by the addition of 0.1 ml of the various enzyme fractions diluted just before assay in 10 mM sodium phosphate buffer, pH 7.0, to obtain a concentration of about 5 milliunits of enzyme per milliliter in the assay mixture. With crude extracts, a blank without substrate is run to correct for NADH oxidase activity, when reduction of fructuronate or tagaturonate is estimated. Changes in absorbance at 340 nm are measured continuously with a multiple sample absorbance recording spectrophotometer.

Units. One unit of enzyme activity oxidizes NADH (reduction of fructuronate or tagaturonate) or reduces NAD (oxidation of mannonate or altronate) at the rate of 1 μmol per minute at 37°. Specific activity is the number of enzyme units per milligram of protein.

Protein is conveniently measured by direct ultraviolet spectrophotometry,[6] but with crude extracts and fractions high in nucleic acid content the method of Lowry *et al.*[7] is used.

[2] G. Ashwell, A. J. Wahba, and J. Hickman, *J. Biol. Chem.* **235**, 1559 (1960).
[3] F. Ehrlich and R. Guttmann, *Ber. Dtsch. Chem. Ges.* **67**, 1345 (1934).
[4] J. W. Pratt and N. K. Richtmyer, *J. Am. Chem. Soc.* **77**, 1906 (1955).
[5] C. S. Hudson, and H. S. Isbell, *J. Am. Chem. Soc.* **51**, 2225 (1929).
[6] See this series, Vol. 3 [73].
[7] O. H. Lowry, N. J. Rosebrough, A. L. Farr, and R. J. Randall, *J. Biol. Chem.* **193**, 265 (1951).

Purification Procedure

General Considerations. Galacturonate-induced cells of wild-type *E. coli* K-12 specifically contain altronate dehydrogenase activity whereas glucuronate-induced bacteria contain both altronate and mannonate dehydrogenase activities. The principal objective of these studies was to prove the individuality of each enzyme and to determine their specific properties. In the procedure given below, the enzymes are purified and separated from glucuronate-grown cells. Altronate dehydrogenase has also been purified from galacturonate-induced bacteria; results showed that the same enzyme is induced by both uronic acids.

Cultivation of Bacteria. *Escherichia coli* K-12 strain S 3000 (Hfr H) is grown aerobically on minimal salt medium "63"[8] supplemented with 0.5 μg of thiamine per milliliter in the presence of 16 mM (4 g/liter) D-glucuronate as the sole carbon source. Erlenmeyer flasks of 2-liter capacity, containing 400 ml of growth medium, are inoculated with $\frac{1}{50}$ the volume of an overnight bacterial suspension. Cells are harvested in the late-exponential phase of growth. The purification procedures given here are essentially those described by Portalier and Stoeber.[9,10] All steps are carried out at 0–4°.

Buffers. Buffers used for the purifications are the following: buffer I (sodium phosphate, 10 mM, pH 7.0); buffer II (sodium phosphate, 10 mM, pH 7.0, containing 10 mM 2-mercaptoethanol); buffer III (sodium phosphate, 10 mM, pH 7.5, containing 10 mM 2-mercaptoethanol).

Step 1. Preparation of Crude Extract. After 16 hr of incubation, cells are collected by centrifugation (20 min, 15,000 g) and washed twice with cold buffer I. Homogenized cell suspensions are disrupted by sonic action for 10 min in a Raytheon 250-W, 10 kHz, sonic oscillator, and the suspension is centrifuged for 1 hr at 38,000 g. Recovery of enzyme activities in the supernatant (100 ml) is 98%.

Step 2. Protamine Sulfate Treatment. Twenty-one milliliters of a 2% aqueous solution of protamine sulfate are added dropwise, with slow stirring, to 100 ml of the above supernatant (final concentration of 3.3 g/liter). After standing for 1 hr, the mixture is centrifuged for 30 min at 25,000 g, and the heavy white precipitate of protamine nucleic acid complex is discarded. Precipitation of nucleic acids with manganese chloride or streptomycin sulfate inactives the enzymes.

Step 3. Ammonium Sulfate Fractionation. To the clear supernatant solution from step 2 is added gradually 27 g of solid ammonium sulfate (1.6

---

[8] J. H. Miller, *in* "Experiments in Molecular Genetics," p. 431. Cold Spring Harbor Laboratory, Cold Spring Harbor, New York, 1972.

[9] R. C. Portalier and F. R. Stoeber, *Eur. J. Biochem.* **26,** 50 (1972).

[10] R. C. Portalier and F. R. Stoeber, *Eur. J. Biochem.* **26,** 290 (1972).

M). Equilibration is continued for an additional hour after the last addition of salt; the suspension is centrifuged, and the precipitate is discarded. The supernatant is then brought to 2.5 M salt by the further slow addition of 22 g of solid ammonium sulfate. The precipitate is collected and dissolved in 12 ml of buffer I. The residual salt is removed by two dialysis treatments for about 15 hr against 1 liter of buffer I, with stirring. A turbid precipitate, which occasionally appears on dialysis, is removed by centrifugation (1 hr, 38,000 g) before proceeding with the next step.

Step 4. DEAE-Sephadex Column Chromatography. The ammonium sulfate-free solution obtained in step 3 (14.5 ml) is applied to the top of a DEAE-Sephadex A-50 column (2 × 40 cm). The gel is equilibrated against buffer III. Fractions (1.8 ml) are eluted with a linear gradient extending between zero and 0.4 M NaCl in 1 liter of buffer III at a flow rate of 15 ml/cm² per hour. The enzymes are eluted at approximately 0.2 M NaCl: fractions 220–250, which contain 90% of mannonate dehydrogenase activity, and fractions 250–280, which contain altronate dehydrogenase activity, are pooled separately and precipitated by addition of ammonium sulfate (2.35 M). Precipitates are centrifuged, solubilized in 8 ml of buffer I, and dialyzed twice during 15 hr against 600 ml of the same buffer.

Step 5. Gel Filtration. The above dialyzates are chromatographed on columns (2.5 × 40 cm) of Sephadex G-200, for mannonate dehydrogenase, or Sephadex G-150, for the altronate dehydrogenase. Sephadex gels are equilibrated with buffer II containing 0.1 M NaCl. The enzymes (2-ml fractions) are eluted with the same buffer at a flow rate of 0.6 ml/cm² per hour. Mannonate dehydrogenase activity is recovered in fractions 48–62 of the G-200 column, and altronate dehydrogenase activity in fractions 48–61 of the G-150 column. A purification of approximately 50- to 60-fold is generally achieved for both enzymes. Recovery and specific activity data of a typical preparation are given in Table I.

Properties

Stability. The enzymes are quite stable during all stages of purification. Crude and purified preparations retain full activity for at least 3 weeks (mannonate dehydrogenase) and 2 months (altronate dehydrogenase) when stored at 4° in buffer I.

Effect of pH and Ionic Strength. Maximum rates of NADH oxidation for both enzymes occur at pH 6.3 in sodium phosphate buffer. Variation of buffer concentrations between 5 and 100 mM has no effect on enzyme activities. Maximum rates of NAD reduction in glycylglycine buffer occur at pH 8.4 for mannonate dehydrogenase and pH 8.9 for altronate dehydrogenase.

TABLE I

PURIFICATION OF MANNONATE AND ALTRONATE DEHYDROGENASES FROM *E. coli*

Step	Fraction	Mannonate dehydrogenase				Altronate dehydrogenase			
		Total units	Total protein (mg)	Specific activity (units/mg protein)	Yield (%)	Total units	Total protein (mg)	Specific activity (units/mg protein)	Yield (%)
1	Crude extract	6720	1130	6	100	7000	1130	6.2	100
2	Protamine sulfate supernatant	7760	525	15	115	8160	525	15.6	116
3	Ammonium sulfate	6950	360	19.5	103	6270	360	17.4	90
4	DEAE-Sephadex eluate	5320	86	62	79	4430	74	60	63
5	Sephadex G-200 eluate	4800	15	320	71	—	—	—	—
6	Sephadex G-150 eluate	—	—	—	—	4100	12.5	328	59

Specificity. Mannonate dehydrogenase oxidizes both D-fructuronate and D-tagaturonate in the forward reaction and reduces both D-mannonate and D-altronate in the reverse reaction; D-tagaturonate and D-altronate are 20% as active as D-fructuronate and D-mannonate, respectively. Altronate dehydrogenase is specific for D-tagaturonate and D-altronate. None of the following carbonyl-containing substrates, examined at a concentration of 5 mM, is reduced by the enzymes: D-glucuronate, D-galacturonate, D-mannuronate, D-fructose, D-tagatose, D-mannose, D-glucose, D-galactose, D-sorbose, D-arabinose, L-arabinose, D-ribose, L-lyxose, D-xylose, L-xylose, D-ribulose, D-*manno*-heptulose, and D-*altro*-heptulose. NADPH is about 5% as active as NADH for mannonate dehydrogenase and 10% for altronate dehydrogenase. No activity is detected with NADP in the reverse reactions.

Kinetic Properties. Kinetic studies show that the mannonate dehydrogenase-catalyzed reaction is consistent with "a rapid equilibrium random bi-bi + dead-end EBQ complex."[11] The K_m values for NADH, fructuronate, NAD, and mannonate are 0.033 mM, 0.5 mM, 0.42 mM, and 0.78 mM, respectively.[12] The forward reaction catalyzed by altronate dehydrogenase is that of "an ordered bi-bi mechanism,"[11] in which NADH is the first substrate to bind to the enzyme and tagaturonate is the last product to be released. The K_m values for NADH, tagaturonate, NAD, and altronate are 0.085 mM, 0.67 mM, 0.053 mM, and 0.075 mM, respectively.[13]

Inhibitors. Results from product, alternate product, and dead-end product inhibition studies of mannonate and altronate dehydrogenases are summarized in Table II. The enzymes are inhibited by sulfhydryl-inhibiting agents such as p-chloromercuribenzoate (PCMB). Mannonate dehydrogenase is completely inhibited by 5 μM PCMB whereas altronate dehydrogenase is not inhibited at this concentration. The apparent K_i values for PCMB are 0.1 μM for mannonate dehydrogenase and 1 mM for altronate dehydrogenase. Altronate dehydrogenase is inactivated by PCMB: at 1 mM PCMB, the half-life of the enzyme is 5.8 min; this inhibition is reversed by 2-mercaptoethanol. No metal requirement for the activities is observed.

Heat Stability. The half-life of mannonate dehydrogenase at 53° in buffer I is 10 min; enzyme substrates NADH, NADPH, fructuronate, tagaturonate, NAD, mannonate, and altronate protect the enzyme against heat inactivation. The half-life of altronate dehydrogenase at 58.5° in buffer I is 10.5 min; NADH, NADPH, altronate, and mannonate protect

[11] W. W. Cleland, *Biochim. Biophys. Acta* **67**, 104 (1963).
[12] R. C. Portalier, *Eur. J. Biochem.* **30**, 220 (1972).
[13] R. C. Portalier, *Eur. J. Biochem.* **30**, 211 (1972).

TABLE II

INHIBITION OF MANNONATE AND ALTRONATE DEHYDROGENASES BY THEIR PRODUCTS AND ANALOGS OF THESE PRODUCTS

Enzyme	Reaction	Inhibitor	Variable substrate	Type of inhibition[a]	Inhibition constants	
					K_i (slope) (mM)	K_i (intercept) (mM)
Mannonate dehydrogenase	Forward	Mannonate	NADH	C	2	—
			Fructuronate	C	2.5	—
		NAD	NADH	C	0.14	—
			Fructuronate	NC	1.85	2.72
		Altronate	NADH	NC	42	17
			Fructuronate	C	4	—
		ATP	NADH	NC	5.9	10
			Fructuronate	NC	2.9	7
	Reverse	Fructuronate	NAD	C	0.6	—
			Mannonate	C	0.44	—
		NADH	NAD	C	0.034	—
			Mannonate	NC	0.054	0.175
		Tagaturonate	NAD	C	0.63	—
			Mannonate	C	1.3	—

Enzyme	Direction	Inhibitor	Varied substrate	Type[a]	K_i	K_i
Altronate dehydrogenase	Forward	Altronate	NADH	NC	3.8	1.6
			Tagaturonate	NC	0.72	3.5
		NAD	NADH	C	1	—
			Tagaturonate	NC	21	4
		Fructuronate	NADH	C	4.5	21
			Tagaturonate	NC	4	—
		Mannonate	NADH		2	—
			Tagaturonate	C (hyperbolic)	—	6
		ATP	NADH	C	0.9	—
			Tagaturonate	NC	2.3	9.1
	Reverse	NADH	Altronate	NC	0.04	0.28
			NAD	C	0.02	0.2
		Tagaturonate	Altronate	NC	0.03	—
			NAD	C	0.61	1.3
		Mannonate	Altronate	None	2.9	—
			NAD (altronate at saturated concentration)		—	—

[a] C, competitive; NC, noncompetitive.

enzyme activity from heat denaturation, whereas NAD, tagaturonate, and fructuronate do not.

Genetics. Mannonate and altronate dehydrogenase structural genes are characterized: they map at minutes 97 and 52, respectively, on the *E. coli* genetic map.[14,15]

Other Properties. Mannonate and altronate dehydrogenases are used as coupling enzymes for a convenient direct assay of uronate isomerase[15] (EC 5.3.1.12). This enzyme catalyzes the isomerization of D-galacturonate and D-glucuronate into their corresponding keto analogs, D-tagaturonate and D-fructuronate, respectively.[1] Uronate isomerase activity is assayed spectrophotometrically at 37° by determining the decrease in NADH concentration at 340 nm. When coupling involves altronate dehydrogenase, the reaction mixture contains 45 mM sodium phosphate buffer, pH 7.6; 0.28 mM NADH; 10 mM sodium D-galacturonate; altronate dehydrogenase in excess; and the isomerase properly diluted. Altronate dehydrogenase extract (step 3 fraction) is prepared from galacturonate-induced cells of strain MH1, deficient in uronate isomerase activity.[15] Colorimetric assays of the enzymes, using phenazine methosulfate (PMS) as an intermediate acceptor for electron transfer from NADH to *p*-nitro blue tetrazolium (PNBT) were developed.[16] Enzyme activities are estimated at 37° by the increase in absorbance at 550 nm, which corresponds to the formation of diformazan. These new methods are mainly used as qualitative tests for rapid *in situ* characterization of mutant clones in the *E. coli* hexuronate system. The assay mixture for mannonate dehydrogenase activity contains 200 mM glycylglycine–NaOH buffer, pH 9.0; 1 mM NAD; 1 mM potassium D-mannonate; 0.5 mM PNBT; 0.033 mM PMS; 20 μg of gelatin per milliliter; and enzyme in a total volume of 1 ml. The assay system (1 ml) for altronate dehydrogense activity contains 200 mM glycylglycine–NaOH buffer, pH 9.0; 0.5 mM NAD; 0.5 mM potassium D-altronate; 0.25 mM PNBT; 0.033 mM PMS; 10 μg of gelatin per milliliter; and the enzyme extract.

[14] J. M. Robert-Baudouy and R. C. Portalier, *Mol. Gen. Genet.* **131**, 31 (1974).
[15] R. C. Portalier, J. M. Robert-Baudouy, and G. M. Nemoz, *Mol. Gen. Genet.* **128**, 301 (1974).
[16] R. C. Portalier and F. R. Stoeber, *Biochim. Biophys. Acta* **289**, 19 (1972).

[36] 2-Keto-3-deoxy-L-fuconate Dehydrogenase from Pork Liver

By HARRY SCHACHTER, JULIA Y. CHAN, and NGOZI A. NWOKORO

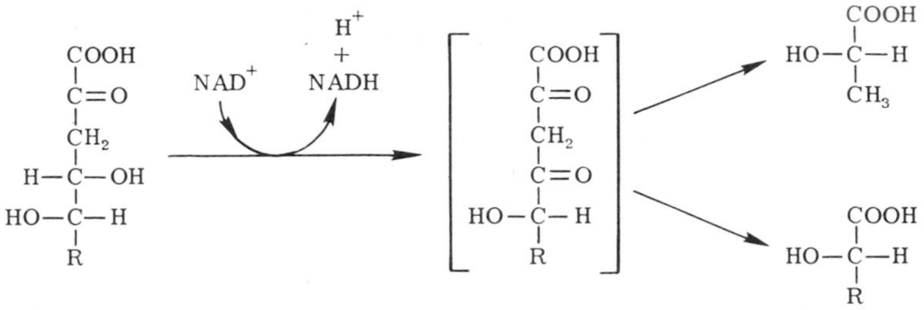

$$
\begin{array}{ccc}
\text{COOH} & & \text{COOH} \\
| & \text{NAD}^+ \quad \text{NADH} & | \\
\text{C}=\text{O} & & \text{C}=\text{O} \\
| & \xrightarrow{\hspace{2cm}} & | \\
\text{CH}_2 & & \text{CH}_2 \\
| & & | \\
\text{H}-\text{C}-\text{OH} & & \text{C}=\text{O} \\
| & & | \\
\text{HO}-\text{C}-\text{H} & & \text{HO}-\text{C}-\text{H} \\
| & & | \\
\text{R} & & \text{R}
\end{array}
$$

2-Keto-3-deoxy-L-fuconate →

$\qquad$ 2,4-diketo-3-deoxy-L-fuconate → 2 L-lactate $\qquad$ (1)

2-Keto-3-deoxy-L-galactonate →

$\qquad$ 2,4-diketo-3-deoxy-L-galactonate → L-lactate + glycerate $\qquad$ (2)

2-Keto-3-deoxy-D-arabonate →

$\qquad$ 2,4-diketo-3-deoxy-D-arabonate → L-lactate + glycolate $\qquad$ (3)

The enzyme 2-keto-3-deoxy-L-fuconate dehydrogenase [2-keto-3,6-dideoxy-L-galactonate:NAD^+ 4(or 5)-oxidoreductase, EC 1.1.1.-] has been purified from pork liver[1] and is a component of a pathway in pork liver capable of oxidizing 1 mol of L-fucose to 2 mol of L-lactate.[2] The pathway also acts on L-galactose and D-arabinose. The terminal reactions of this pathway for L-fucose (R = —CH_3), L-galactose (R = —CH_2OH) and D-arabinose (R = —H) are indicated above. The source of the two hydrogens removed by 2-keto-3-deoxy-L-fuconate dehydrogenase has not as yet been established to be at C-4, and therefore the product thought to be formed by the enzyme is indicated in brackets in the scheme above. The purified dehydrogenase does not cleave the 2,4-diketo-3-deoxyal-donates to the products indicated in the above scheme; this cleavage requires the action of one or more enzymes different from 2-keto-3-deoxy-L-fuconate dehydrogenase.

Assay Methods

Three assay methods have been used in various aspects of the work on 2-keto-3-deoxy-L-fuconate dehydrogenase.

[1] N. A. Nwokoro and H. Schachter, *J. Biol. Chem.* **250**, 6185 (1975).

[2] J. Y. Chan, N. A. Nwokoro, and H. Schachter, *J. Biol. Chem.* **254**, 7060 (1979).

Assay A [1]

Principle. The disappearance of the substrate 2-keto-3-deoxy-L-fuconate is followed spectrophotometrically by the periodate–thiobarbiturate reaction for 2-keto-3-deoxyaldonic acids,[3] which gives a pink color with a maximum at 549 nm. This assay can be used with nonradioactive substrates and with crude preparations of the enzyme and is therefore the most generally applicable assay method.

Reagents

Glycine–NaOH buffer, 0.5 *M*, pH 9.0
or
Tris-HCl buffer, 0.5 *M*, pH 9.0
NAD⁺, 10 m*M*
2-Keto-3-deoxy-L-fuconate, 10 m*M*. This material is not commercially available; we have synthesized it in our laboratory[1] by first oxidizing L-fucose to L-fuconate with bromine water followed by dehydration of L-fuconate to 2-keto-3-deoxy-L-fuconate with partially purified pork liver L-fuconate hydrolyase.[4] The 2-keto-3-deoxy-L-fuconate was purified from the enzyme mixture by chromatography on Dowex AG-1-X2.[1] The compound has been synthesized in the 1-¹⁴C, U-¹⁴C, and 1-¹⁴C, 6-³H forms by starting with the respective radioactive L-fucose compounds.[1,2]
Sodium periodate, 0.2 *M* in 9 *M* orthophosphoric acid
Sodium arsenite, 10% in 0.5 *M* sodium sulfate–0.1 *N* sulfuric acid .
2-Thiobarbituric acid, 0.6% in water
Sodium sulfate, saturated solution in water
Cyclohexanone
Trichloroacetic acid, 10%, ice cold

Procedure. The assay mixture contains 0.4 ml of buffer, 0.25 ml of NAD⁺, 0.1 ml of 2-keto-3-deoxy-L-fuconate, enzyme solution (at least 2 milliunits) and water to a final volume of 1.0 ml. Two control incubations are usually carried out, one lacking NAD⁺ and the other containing heat-inactivated enzyme (100° for 15 min). Incubations are carried out at 37° for 2 hr. The reaction can be terminated either by freezing or by the addition of 1.0 ml ice-cold 10% trichloroacetic acid. In the latter case, protein precipitate is removed after 10 min at 4° by centrifugation at 23,000 *g* for 10 min. To a 0.2-ml aliquot of the reaction mixture or of the supernatant after centrifugation is added 0.1 ml of sodium periodate. The mixture is shaken and left at room temperature for 20 min. Excess periodate is then

---

[3] L. Warren, *J. Biol. Chem.* **234**, 1971 (1959).
[4] R. Yuen and H. Schachter, *Can. J. Biochem.* **50**, 798 (1972).

destroyed by addition of 1.0 ml sodium arsenite. The tube is shaken until the yellow color disappears, and 3.0 ml of thiobarbituric acid and 0.5 ml of saturated sodium sulfate are added. The tubes are agitated, capped, and heated at 100° for 15 min. After cooling, the pink thiobarbiturate complex is extracted with 4.6 ml of cyclohexanone. The organic phase is separated by centrifugation in a clinical centrifuge for a few minutes, and the absorbance at 549 nm is determined.

Definition of Enzyme Unit. An enzyme unit is defined as the amount of enzyme catalyzing the disappearance of 1.0 μmol of substrate in 1.0 min at 37° under the above conditions. The amount of substrate destroyed can be calculated from the difference in absorbance between the standard assay mixture and the controls, assuming a molar extinction coefficient at 549 nm of 42,000 $cm^{-1} M^{-1}$ for the pink complex extracted into cyclohexanone after periodate–thiobarbiturate treatment of 2-keto-3-deoxy-L-fuconate.[1,2]

Assay B[1]

Principle. The enzyme is assayed by spectrophotometric determination of the appearance of NADH, which has an absorption maximum at 340 nm. This assay is rapid, but it is unreliable with crude enzyme preparations. It is used primarily for assaying enzyme preparations that have undergone the first four steps of purification (see below).

Procedure. The same incubations are used as are described above for assay A. The incubations are placed directly in 1-ml cuvettes, and the change in absorption at 340 nm is followed at 37° for 1 hr.

Assay C[2]

Principle. Radioactive 2-keto-3-deoxy-L-fuconate is used in this sensitive radiochemical assay. The incubation mixture is subjected to high voltage electrophoresis on paper to separate radioactive substrate from certain radioactive products. Radioactive components are quantitated by measurement of radioactivity. This assay is used primarily to detect contamination of the dehydrogenase with enzymes that destroy the product of the dehydrogenase, since the various components shown in the Eqs. (1), (2), and (3) are separable on high-voltage electrophoresis.[2]

Procedure. Incubations are as described for assay A except that 2-keto-3-deoxy-L-fuconate is radioactive (a suitable specific activity is about 2×10^4 dpm/μmol). Incubations are carried out at 37° for 1 hr, or less if the specific radioactivity of the substrate is made high enough for detection of product formation. The reaction is terminated by freezing, and the sample is lyophilized and reconstituted with water to a volume of

PURIFICATION OF PORK LIVER NAD$^+$-DEPENDENT 2-KETO-3-DEOXY-L-FUCONATE DEHYDROGEN.

Fraction	Protein concentration (mg/ml)	Enzyme activity (milliunits/ ml)	Specific enzyme activity (milliunits/ mg protein)	Purification factor	Recove (%)
23,000 g supernatant	26.7	31	1.17	1.0	100
Ammonium sulfate fraction	60.6	159	2.62	2.2	79
Concentrated Sephadex G-150 fraction	10.1	167	16.5	14.1	68
Concentrated isoelectric-focused fraction	2.9	118	40.8	35	58
Concentrated NAD$^+$-agarose affinity column fraction	~0.15[a]	593	~3950[a]	~3390[a]	27

[a] Protein concentration was estimated from absorption at 210 nm because of low protein recove in the final preparation.

0.1 ml. An aliquot of 0.05 ml is subjected to high-voltage paper electrophoresis on Whatman 3 MM paper saturated with formic acid–acetic acid–water (1.7 : 1.7 : 100, v/v/v) adjusted to pH 3.1 with pyridine, at 2 kV for about 2 hr. Both 2-keto-3-deoxy-L-fuconate and 2,4-diketo-3-deoxy-L-fuconate, the suggested product of the dehydrogenase, move rapidly to the anode in this electrophoretic system and are not separated; this assay cannot therefore be used to assay the purified dehydrogenase. However, L-lactate moves less rapidly toward the anode and is readily determined, thereby indicating the presence of an enzyme system that cleaves the dehydrogenase product to smaller molecules.

Purification Procedure

The dehydrogenase has been purified to apparent homogeneity by a 5-step procedure summarized in the table. Buffer A in the purification procedure is 0.05 M Tris-HCl, pH 8.0, containing 5 mM MgCl$_2$ and 1 mM dithiothreitol.

Step 1. Homogenization and Extraction. Pig livers were obtained within 20 min after slaughter and packed in ice until homogenized (within 2 hr after slaughter). All subsequent steps were carried out at 4°. Pieces were pooled from three separate livers, and 310 g were ground in a commercial meat grinder, suspended in 4 volumes of 0.25 M sucrose in buffer A, and homogenized at low speed in a large Waring blender with four bursts of 30-sec duration. The homogenate was filtered through four layers of surgical gauze and centrifuged at 23,000 g for 1 hr. The supernatant was dialyzed against buffer A prior to enzyme assay with assay A.

Step 2. Ammonium Sulfate Fractionation. Solid ammonium sulfate (23 g per 100 ml supernatant from step 1) was added slowly with stirring to the supernatant from step 1. After complete dissolution of the salt, the solution was centrifuged at 23,000 *g* for 1 hr and the pellet was discarded. Solid ammonium sulfate (12.2 g per 100 ml of original supernatant from step 1) was added slowly with stirring, and the resulting precipitate was collected by centrifugation as above and dissolved in buffer A to a final volume of 190 ml.

Step 3. Sephadex G-150 Chromatography. A Pharmacia type K100/100 column was packed with Sephadex G-150 (10 × 90 cm) and equilibrated with buffer A. A portion (85 ml) of the ammonium sulfate fraction (step 2) was passed through this column with buffer A as eluting medium. The flow rate was 180 ml/hr. The column was monitored by absorbance at 280 nm and by enzyme assay A. The fraction size was 15.2 ml. The major enzyme peak (fractions 180–198) was pooled (300 ml) and quickly concentrated to a volume of about 70 ml with a Diaflo ultrafiltration apparatus (Amicon) using a UM-10 membrane.

Step 4. Preparative Isoelectric Focusing. A 10-ml aliquot of the preceding fraction (step 3) was concentrated to 5 ml with a small Diaflo apparatus and subjected to isoelectric focusing in a 440 ml isoelectric focusing column (Column 8102, LKB Producter). The column was set up with a stabilizing gradient of 0 to 46% sucrose containing carrier ampholytes at a concentration of 1% in the pH range 5–8. The column was run at 400 V for the first 18 hr and 900 V for the subsequent 60 hr; the column was kept at 4° with a circulating bath. At the end of the run, 4.0-ml fractions were collected with a peristaltic pump at 96 ml/hr. Dehydrogenase activity could not be detected by enzyme assay A owing to interference from the ampholytes; the enzyme was detected by assay method B. Fractions 51–55 were pooled and subjected to repeated (at least eight times) dilution with buffer A and concentration with a Diaflo ultrafiltration apparatus equipped with a PM-10 membrane; this procedure reduced ampholyte and sucrose concentrations to levels that did not interfere with enzyme assay A or with polyacrylamide gel electrophoretic analysis. The final enzyme volume was 12 ml.

Step 5. Affinity Column Chromatography. In a typical preparation, a column (2.5 × 24 cm) of NAD$^+$-agarose (AGNAD, P-L Biochemicals) was equilibrated with 0.05 M Tricine [N-Tris(hydroxymethyl)methylglycine] buffer at pH 9.0, and 25 ml of step 4 fraction (pooled from two to three isoelectric focusing runs) were loaded on the column at a flow rate of 15 ml/hr. The column was then washed with 200 ml of 0.05 M Tricine buffer, pH 9.0, to remove unadsorbed material. The adsorbed protein was eluted with 150 ml of 5.0 mM NAD$^+$ in 0.05 M Tricine buffer, pH 8.0, at a flow rate of 100 ml/hr. Fraction size was 25 ml. The fractions were monitored for enzyme by assay B. Fractions 11–13 were pooled and dialyzed for 24 hr against three changes of 5 mM Tris-HCl, pH 8.0. The enzyme was concentrated to 10 ml with an Amicon Diaflo apparatus and then to 2.3 ml with a Schleicher and Schuell ultrafiltration membrane. Microliter volumes of concentrated enzyme for gel electrophoresis were obtained with a Minicon A25 concentrator (Amicon). The final purification is about 3000-fold with a yield of 27% (see the table).

Comments on Purification. The purification procedure has been carried out several times and is highly reproducible. The specific activity of the highly purified enzyme has varied between 4 and 8 units per milligram of protein. Sephadex G-150 column chromatography (step 3) may result in 2 or 3 peaks of enzyme activity; however, only the major peak has been studied, since the minor peaks are unstable. The most effective purification step is step 5 involving affinity chromatography on NAD$^+$-agarose; the dehydrogenase does not stick to agarose without the NAD$^+$ ligand, suggesting that a true affinity effect was responsible for the purification step.

Properties[1,5]

Stability. Enzyme preparations after steps 1–4 are stable at $-70°$ for at least 2 years if the protein concentration is at least 4 mg/ml and if 1 mM dithiothreitol is present. Step 5 enzyme was never obtained in large enough amounts to concentrate to a high protein concentration; this preparation lost 80% of its activity after 2 months at $-20°$.

Purity. Step 5 enzyme migrates as a single protein band on analytical polyacrylamide disc gel electrophoresis at pH 9.1; this band corresponds with dehydrogenase activity towards both 2-keto-3-deoxy-L-fuconate and 2-keto-3-deoxy-D-arabonate. A single major band was also obtained after electrophoresis at pH 7.1 or in sodium dodecyl sulfate–mercaptoethanol. The molecular weight of the enzyme was shown to be 89,500 by gel filtration on Sephadex G-150, and 99,600 by analytical ultracentrifugation. The sedimentation coefficient $S_{20,obs}$ for the step 5 enzyme is 5.2.

pH, Ionic Strength, and Cations. The dehydrogenase did not show a pH optimum. There was no activity at pH 6 or lower, and activity rose sharply as incubation pH was increased above pH 6; the activity was still increasing at pH 10.5, the highest pH studied. Increasing buffer concentration from 0.08 M to 0.2 M had no effect on activity. The enzyme showed no cation requirement, but Cd^{2+}, Co^{2+}, Pb^{2+}, Zn^{2+}, and Cu^{2+} at 10 mM concentrations inhibited the step 4 enzyme; this inhibition was prevented

[5] N. A. Nwokoro and H. Schachter, *J. Biol. Chem.* **250**, 6191 (1975).

by prior exposure of the enzyme to NAD^+. Enzyme activity was not affected by 10 mM concentrations of K^+, Na^+, Ca^{2+}, Fe^{3+}, and Mg^{2+}.

Sulfhydryl Requirements. Step 4 enzyme was inhibited by several sulfhydryl-reactive reagents, but this inhibition could be partially prevented by exposure of the enzyme to NAD^+.

Substrate Specificity. The dehydrogenase shows absolute specificity for NAD^+; no other coenzyme was found to be effective. The only effective hydrogen donors for the purified enzyme are 2-keto-3-deoxy-L-fuconate, 2-keto-3-deoxy-D-arabonate, and 2-keto-3-deoxy-L-galactonate.[2] Several other 2-keto-3-deoxyaldonates, as well as various other sugars and sugar acids, were ineffective substrates for the enzyme.[5] However, the 2-keto-3-deoxyaldonate derivatives of D-fucose, 6-deoxy-D-glucose, 6-deoxy-L-glucose, L-glucose, and L-arabinose were not available for testing.

Kinetics. The K_m for 2-keto-3-deoxy-L-fuconate is 0.15 mM, for 2-keto-3-deoxy-D-arabonate is 0.25 mM, and for NAD^+ is 0.22 mM.

Product Identification. Stoichiometry studies show that the purified enzyme transfers two atoms of hydrogen to NAD^+ for every molecule of 2-keto-3-deoxy-L-fuconate or 2-keto-3-deoxy-D-arabonate destroyed. The resulting product is a sugar acid with a mobility on high-voltage electrophoresis (assay C) identical to that of the substrate. The product is resistant to the periodate–thiobarbiturate reagent and is probably 2,4-diketo-3-deoxyaldonate, although removal of hydrogens from C-5 of the substrate cannot be ruled out.

Distribution. The enzyme has been detected in the 23,000 g supernatants from the livers of pig, calf, beef, lamb, mouse, and rat[5]; 12% or less of the total enzyme activity was detected in the 23,000 g pellets in these studies. These activities were all NAD^+ dependent. Tissue distribution studies[2] carried out in the pig showed the enzyme to be mainly in liver and kidney with relatively low activities in heart, stomach, lung, and brain, and undetectable activities in intestine and submaxillary gland. The distribution of two other enzymes in the L-fucose catabolic pathway (L-fucose dehydrogenase[6,7] and L-fuconate hydro-lyase[4]) was also studied, and it was evident that only liver and kidney had appreciable levels of all three enzymes required for the oxidation of L-fucose.

Acknowledgments

These studies were supported by the Medical Research Council of Canada.

[6] H. Schachter, J. Sarney, E. J. McGuire, and S. Roseman, *J. Biol. Chem.* **244,** 4785 (1969).
[7] P. W. Mobley, R. P. Metzger, and A. N. Wick, this series, Vol. 41, p. 173.

[37] D-Xylose Dehydrogenase[1]

By A. STEPHEN DAHMS and JOHN RUSSO

$$\text{D-Xylose} + \text{NAD}^+ \rightarrow \text{D-xylonate} + \text{NADH} + \text{H}^+$$

D-Xylose dehydrogenase (EC 1.1.1.175) functions in the metabolism of D-xylose in a pseudomonad.[2,3]

Assay Method

Principle. The continuous spectrophotometric assay measures the xylose-dependent rate of NADH production at 340 nm.

Reagents

N,N-Bis(2-hydroxyethyl)glycine (Bicine buffer), 0.1 M, pH 8.0
D-Xylose, 0.30 M
NAD$^+$, 50 mM

Procedure. The following are added to microcuvettes with a 1 cm light path: 30 μl of buffer, 10 μl of D-xylose, 10 μl of NAD$^+$, and 20 μl of D-xylose dehydrogenase. The reaction was initiated by the addition of enzyme. The rates are conveniently measured with a Gilford multisample absorbance recording spectrophotometer. The cuvette compartment was thermostatted at 25°.

Definition of Unit and Specific Activity. One unit is defined as the amount of enzyme that catalyzes the production of 1 μmol of NADH per minute. Specific activity is in terms of units per milligram of protein. Protein is conveniently measured by the method of Lowry *et al.*[4] and Warburg and Christian.[5]

Purification Procedure

The enzyme is purified from pseudomonad MSU-1 (ATCC 27855) grown as described elsewhere[6] except that the carbon source was D-xylose. The preparation of cells is suspended in 0.10 M Bicine buffer,

[1] This research was supported by United States Public Health Research Grant GM-22197, National Science Foundation Grant GB-38671, and the California Metabolic Research Foundation.
[2] A. S. Dahms, *Biochem, Biophys. Res. Commun.* **60**, 1933 (1974).
[3] A. S. Dahms, *Fed. Proc., Fed. Am. Soc. Exp. Biol.* **33**, 1900 (1974).
[4] O. H. Lowry, N. J. Rosebrough, A. L. Farr, and R. J. Randall, *J. Biol. Chem.* **193**, 265 (1951).
[5] O. Warburg and W. Christian, *Biochem. Z.* **310**, 384 (1941).
[6] A. S. Dahms, D. Sibley, W. Huismann, and A. Donald, this series, Vol. 90 [47].

PURIFICATION OF D-XYLOSE DEHYDROGENASE

Fraction	Volume (ml)	Total protein (mg)	Total activity (units)	Specific activity (units/mg protein)
Cell extract	175	4100	287	0.070
Protamine sulfate supernatant	310	3805	247	0.065
(NH$_4$)$_2$SO$_4$ precipitate, 35–55%	15	833	175	0.21
Sephadex G-200	35	108	87	0.81
Affinity step	12	24.1	60.3	2.5

pH 8.0, containing 0.14 mM 2-mercaptoethanol and 1 mM NAD$^+$ (buffer A). The following procedures are performed at 0–4°. The enzyme is substantially thermolabile but can be stabilized by the inclusion of xylose or NAD$^+$ in the processing buffers.

Protamine Sulfate Treatment. To a cell extract containing 0.2 M ammonium sulfate, an amount of 2% (w/v) protamine sulfate solution (pH 7.0) is added to give a concentration of 0.33%. After 30 min the precipitate is removed by centrifugation and discarded.

Ammonium Sulfate Fractionation. The protein in the supernatant from the protamine sulfate step is fractionated by the addition of crystalline ammonium sulfate. The protein precipitating between 1.42 M and 2.39 M is collected by centrifugation and dissolved in buffer A.

Chromatography on Sephadex G-200. The above fraction is chromatographed on a column (6 × 60 cm) of Sephadex G-200 equilibrated with buffer A. Fractions (7 ml) are collected during elution with the same buffer, and those that contain the highest specific activity are combined.

Chromatography on NAD$^+$–agarose. The pooled Sephadex G-200 fractions are dialyzed against 10 mM Bicine–0.14 mM mercaptoethanol, pH 8, for 6 hr. The sample is applied to a 5-ml column of agarose hexane–NAD$^+$ (P-L. Biochemicals, Milwaukee, Wisconsin) and eluted sequentially with 20 ml each of the same buffer containing 0.1 M NaCl, containing 1 mM AMP, and containing 5 mM NAD$^+$. D-Xylose dehydrogenase elutes in the final step. Bicine buffer is added to a final concentration of 0.1 M before storage at $-20°$. The protein is purified 50-fold with 21% overall recovery.

A typical purification is summarized in the table.

Properties

Substrate Specificity. Of 15 carbohydrates analyzed, only D-xylose served as a substrate (K_m = 0.5 mM). Compounds that did not serve as substrates (<2% of the rate) at 50 mM concentrations were: L-arabinose,

L-xylose, D-arabinose, D-ribose, D-lyxose, D-fucose, L-fucose, D-galactose, L-galactose, D-mannose, L-rhamnose, D-glucose, L-glucose, and D-lyxose. The enzyme is specific for NAD^+ ($K_m = 0.2$ mM); $NADP^+$ was ineffective. Attempts to isolate a stable γ-lactone using purified fractions (free of galactono-γ-lactone hydrolase activity) were unsuccessful, suggesting that the product is the unstable δ-lactone that spontaneously hydrolyzes. The enzyme is not affected by 1 mM EDTA or o-phenanthroline and was not stimulated by a number of divalent cations.

pH Optimum. Dehydrogenase activity as a function of pH is maximal at pH 8.0.

Effects of Thiols and Thiol Group Reagents. The purified enzyme was insensitive to thiols and thiol group reagents during assay, as none of the following compounds affected the reaction velocity: 3 mM 2-mercaptoethanol, reduced glutathione, dithiothreitol, iodoacetate, and p-mercuribenzoate. However, higher yields of the enzyme were consistently achieved when mercaptoethanol was included in the processing and storage buffers.

Stability. The protein is thermolabile. It possesses a half-life of 2 min at 32° and 15 min at 25°. It can be stabilized by its substrates individually or by high ionic strength. The enzyme is stable for several weeks in the frozen state.

[38] D-Apiitol Dehydrogenase from Bacteria

By E. V. Chandrasekaran, Mark Davila, and Joseph Mendicino

D-Apiose (3-C-hydroxymethyl-D-tetrose), one of the best known branched-chain sugars, is widely distributed in the cell wall polysaccharides of seaweeds, *Lemna minor,* and certain other plants. In parsley, it is present as a component of the flavone apiin. A gram-negative coccus, isolated from the surface of germinating parsley seeds by enrichment culture, oxidized D-apiose to CO_2 and utilized this sugar as a sole source of carbon for growth.

A specific inducible D-apiitol dehydrogenase[1] that is dependent on NAD^+ for activity is isolated from crude extracts of this bacterium.

$$
\begin{array}{ccc}
\text{CHO} & & \text{CH}_2\text{OH} \\
| & & | \\
\text{H}-\text{COH} & & \text{H}-\text{COH} \\
| & & | \\
\text{HOH}_2\text{C}-\text{COH} & + \text{NADH} + \text{H}^+ \rightleftarrows \text{HOH}_2\text{C}-\text{COH} + \text{NAD}^+ \\
| & & | \\
\text{CH}_2\text{OH} & & \text{CH}_2\text{OH} \\
\text{D-Apiose} & & \text{D-Apiitol}
\end{array}
$$

[1] R. Hanna, J. M. Picken, and J. Mendicino, *Biochim. Biophys. Acta* **315**, 259 (1973).

METHODS IN ENZYMOLOGY, VOL. 89

Reagents

1 M Glycylglycine–NaOH, pH 9.5
1 M Glycylglycine–NaOH, pH 7.5, and 0.2 M D-Apiitol
D-Apiose, 0.2 M
NAD$^+$, 0.01 M
NADH, 0.001 M
The four latter reagents are in distilled water.

Enzyme Assay

D-Apiitol dehydrogenase is assayed spectrophotometrically by measuring the absorbance change at 340 nm resulting from the reduction of NAD$^+$ in the presence of D-apiitol or the oxidation of NADH in the presence of D-apiose. The standard reaction mixture for measuring the rate of oxidation of D-apiitol contains, in 0.5 ml/ 80 mM glycylglycine–NaOH (pH 9.5), 20 mM D-apiitol, 1 mM NAD$^+$, and the enzyme. The reduction of D-apiose is measured in a 0.5-ml reaction mixture that contains 80 mM glycylglycine–NaOH (pH 7.5), 20 mM D-apiose, 0.1 mM NADH, and enzyme. The change in absorbance at 340 nm with time is linear with respect to the amount of enzyme added when less than 10% of the substrate is utilized. One unit of activity in these assays is defined as the amount of enzyme required to oxidize 1 μmol of D-apiitol per minute or the amount necessary to reduce 1 μmol of D-apiose per minute.

Purification Procedures

Isolation of Microorganism

Parsley seeds are germinated by placing them between two sheets of wet filter paper for 12 days. Bacteria grow on the top of the germinated seeds are suspended in minimal salts medium, and an organism capable of utilizing D-apiose is isolated by streaking this solution on 0.5% agar plates containing minimum salts medium composed of 7.0 g of K$_2$HPO$_4$, 3.0 g of KH$_2$PO$_4$, 1.0 g of MgSO$_4$, 1.0 g of (NH$_4$)$_2$SO$_4$, and 5.0 g of D-apiose per liter. A culture of the gram-negative micrococcus is maintained on nutrient agar slants containing 0.5% D-apiose.

Cultivation of Bacteria and Induction of D-Apiitol Dehydrogenase

The bacteria are grown on glucose, since they can later be incubated with D-apiose to induce the formation of the same amount of D-apiitol dehydrogenase as is present in cells grown on D-apiose. The bacteria are grown in a medium containing the following ingredients (grams per liter): K_2HPO_4, 7.0; KH_2PO_4, 3.0; $(NH_4)_2SO_4$, 1.0; $MgSO_4$, 1.0; glucose, 5.0. The flasks are shaken at 25° for 24 hr. The initial culture, which contains 250 ml of medium in a 500-ml Erlenmeyer flask, is used to inoculate six 2000-ml Erlenmeyer flasks, each of which contains 1200 ml of sterile medium. After 24 hr the cells are harvested by centrifugation at 30,000 g for 20 min, and they are washed twice with 10 volumes of the sterile medium lacking glucose. The final yield is 11.6 g, dry weight, of cells. Growth is measured by determining the absorbance at 660 nm on samples diluted to give readings of less than 0.5, and cell yield is estimated by drying a washed sample to constant weight of 104°. To induce the formation of D-apiitol dehydrogenase, the washed cells are suspended in an equal volume of minimal salts medium, and they are shaken for 24 hr at 30°. Then 0.4 volume of the starved cell suspension is mixed with 0.5 volume of minimal salts medium containing 3% casein hydrolyzate (Casamino acids, Difco) and 40 μmol of D-apiose per milliliter. This mixture is shaken for 24 hr at 30°. More D-apiose (40 μmol/ml) is added after 12 hr. The cells are harvested by centrifugation and washed three times with 10 volumes of distilled water. D-Apiose can be isolated in high yield from parsley seeds or *Lemna minor* by mild acid hydrolysis and paper chromatography, as described by Picken and Mendicino.[2]

Purification Steps

Step 1. Crude Extract. All of the procedures are carried out at 3°. The cells are suspended in 250 ml of 0.05 M Tris-HCl, pH 7.5, and they are broken in a French press. The resulting suspension is centrifuged at 34,000 g for 20 min, and the supernatant is centrifuged again at 100,000 g for 1 hr. The D-apiitol dehydrogenase activity is present in the supernatant solution.

Step 2. Streptomycin Sulfate Treatment. Streptomycin sulfate (20%; 11.3 ml) is added to 183 ml of the crude extract. The mixture is stirred for 15 min, and an inactive precipitate is removed by centrifugation and discarded. Excess streptomycin sulfate is removed by passing the supernatant solution through a small column (2.2 cm × 3 cm) of Rexyon 102 (H^+) (Rohm and Haas Co.).

Step 3. Calcium Phosphate Gel Absorption. Calcium phosphate gel[3] (0.5 mg per milligram of protein) that has been aged for 2 weeks is slowly added to the solution from step 2, and the suspension is stirred for 30 min and centrifuged.

Step 4. Ethanol Precipitation. The supernatant solution from step 3 is

[2] J. M. Picken and J. Mendicino, *J. Biol. Chem.* **242**, 1629 (1967).
[3] D. Keilin and E. F. Hartree, *Proc. R. Soc. London, Ser. B.* **124**, 397 (1938).

PURIFICATION OF D-APIITOL DEHYDROGENASE

Step	Volume (ml)	Protein (mg/ml)	Total activity (units)	Specific activity (units/mg) protein)
1. Crude extract	183	458	36.6	0.00043
2. Treatment with streptomycin sulfate	175	64	27.9	0.00248
3. Treatment with calcium phosphate gel	160	48	24.0	0.0032
4. Precipitation with ethanol	40	58	24.0	0.0103
5. Chromatography on DEAE-cellulose	35	23	19.5	0.024
6. Chromatography on Sephadex G-200	32	4	13.0	0.102

diluted with 2 volumes of $0.05\ M$ Tris-HCl, pH 7.5. This solution, 480 ml, is cooled to 0°, and 480 ml of 95% ethanol at $-20°$ are slowly added with stirring. Afterward the precipitate is collected by centrifugation, and it is dissolved in 35 ml of $0.05\ M$ Tris-HCl, pH 7.5. Some insoluble material is removed by centrifugation.

Step 5. DEAE-Cellulose Chromatography. The enzyme is adsorbed to a DEAE-cellulose column (2.2 cm × 7 cm) and eluted from the column with a solution formed as a linear gradient from 250 ml of 0.03 Tris-HCl, pH 7.5, in the mixing chamber and 250 ml of the buffer with $0.5\ M$ NaCl in the reservoir. The D-apiitol dehydrogenase emerges in a protein peak after about 100 ml of eluent has been collected. The fractions containing activity are pooled and concentrated by the use of a Diaflo membrane. The solution is dialyzed against 1 liter of $0.05\ M$ Tris-HCl, pH 7.5.

Step 6. Gel Filtration on Sephadex G-200. The enzyme is applied to a Sephadex G-200 column (5 cm × 100 cm) previously equilibrated against $0.05\ M$ Tris-HCl, pH 7.5. It is eluted with this buffer, and the activity is found in a protein peak after 1150 ml of buffer had passed through the column. The active fractions are combined, concentrated, and dialyzed as described above. The results obtained in a typical preparation are summarized in the table. The enzyme has been purified 200-fold from the crude extract with a yield of about 30%.

Properties

Stability, pH Optima, and Molecular Weight. The enzyme preparation remains active when frozen at $-20°$ for at least several weeks. The enzyme has a broad pH profile with an optimum at 7.5 in $0.08\ M$ Tris-HCl or $0.08\ M$ glycylglycine buffer when assayed in the direction of D-apiose reduction. A pH optimum of 10 is observed when the enzyme activity is measured in the direction of D-apiitol oxidation using 0.08 M glycine–NaOH buffer. A molecular weight of 115,000 has been determined by sucrose density centrifugation. The same value is obtained by gel chromatography on Sephadex G-200 according to the method of Andrews.[4]

Specificity and Kinetic Constants. The enzyme is highly specific for D-apiose. No activity is detected in the standard assay system when the same concentration of D-glucose, D-fructose, D-mannose, D-glucosamine, D-galactose, D-ribose, D-xylose, L-rhamnose, L-fucose, D-fucose, D-arabinose, L-arabinose, D-erythrose, or D-glyceraldehyde is used. The enzyme is also specific for D-apiitol with few possible exceptions. D-Ribitol, D-sorbitol, D-arabitol, L-arabitol, D-mannitol, D-xylitol, and D-galactitol are inactive when tested in the standard assay. The activity obtained with *myo*-inositol and *meso*-erythritol is 38 and 13%, respectively, of that obtained with D-apiitol. The dehydrogenase is specific for NAD$^+$ as the electron acceptor and NADH as the electron donor. NADP$^+$, NADPH, ascorbate, FAD, FADH$_2$, cytochrome c, and ferricyanide are all inactive. The apparent K_m for D-apiose is $7.14 \times 10^{-2} M$; D-apiitol, $1.16 \times 10^{-2} M$; NAD$^+$, 3.5×10^{-4}; NADH, $1.5 \times 10^{-5} M$. High concentrations ($1 \times 10^{-4} M$) of NADH inhibit the reaction.

[4] P. Andrews, *Biochem. J.* **91**, 222 (1964).

[39] D-Erythrulose Reductase[1] from Beef Liver

By KIHACHIRO UEHARA and SABURO HOSOMI

$$\text{D-Erythrulose} + \text{NADPH(NADH)} + \text{H}^+ \rightleftarrows \text{D-threitol}^{[2]} + \text{NADP}^+(\text{NAD}^+)$$

Assay Method

Principle. D-Erythrulose reductase activity is measured by following the decrease in absorption of NAD(P)H at 340 nm accompanying the reduction of D-erythrulose.

[1] D-Threitol : NADP$^+$ oxidoreductase, EC 1.1.1.162.

[2] In a previous paper,[3] the enzymic reaction product of D-erythrulose was incorrectly identified as erythritol. Gas–liquid and thin-layer chromatographic data have confirmed the product to be D-threitol,[4] not erythritol. For this reason the systematic name D-threitol : NADP$^+$ oxidoreductase[1] is used instead of erythritol : NADP$^+$ oxidoreductase, which has been designated by the IUB Enzyme Commission.

[3] K. Uehara, T. Tanimoto, and H. Sato, *J. Biochem.* (*Tokyo*) **75**, 333 (1974).

[4] K. Uehara, S. Mannen, S. Hosomi, and T. Miyashita, *J. Biochem.* (*Tokyo*) **87**, 47 (1980).

Reagents

Sodium phosphate buffer, 0.1 M, pH 6.25

D-Erythrulose, 50 mM, prepared by the isomerization of D-erythrose in pyridine[3]; stored at $-20°$

NAD(P)H, 2 mM, prepared daily

Procedure. To a cuvette with a 1-cm light path are added 2.7 ml of sodium phosphate buffer, 0.1 ml of NAD(P)H, 0.1 ml of D-erythrulose (replaced by water in blank), and 0.1 ml of enzyme.

The reaction is initiated at 28° by the addition of enzyme. A blank without D-erythrulose must be run to correct for any nonspecific oxidation of NAD(P)H.

Definition of Unit and Specific Activity. One unit of enzyme is defined as that amount which causes the oxidation of 1 μmol of NAD(P)H per minute under the assay conditions described above. Specific activity is expressed as units per milligram of protein. The concentration of protein is determined from absorbances at 280 nm and 260 nm as originally described by Warburg and Christian.[5] The protein concentration of purified preparations is measured by absorbance at 290 nm using the extinction coefficient $E_{1 \text{cm}}^{1\%} = 5.1$) of the crystalline enzyme. The reason for using the absorbance at 290 nm instead of 280 nm is given below.

Purification Procedure

The following is a modification of a previously published procedure.[3,6] All operations are performed at 0–6° unless otherwise stated.

Step 1. Homogenization. Three kilograms of fresh beef liver obtained immediately after slaughter (chilled in ice and brought to the laboratory) is cut into 2–3 cm cubes and washed thoroughly with cold water to remove blood and blood clots. The pieces of liver are divided into eight equal portions, each of which is then homogenized with 600 ml of cold 50 mM sodium phosphate buffer, pH 8.0, in a Waring blender for 4 min at high speed.

Step 2. Acetone Fractionation. To the homogenate obtained above, cold acetone (-5 to 0°) is slowly added with vigorous stirring, to a final concentration of 30% (v/v). After stirring for 2 min at 0°, the resulting precipitate is removed by centrifugation at 14,000 g for 5 min. To each liter of the supernatant, 400 ml of cold acetone are then added slowly with stirring. The suspension is centrifuged at 14,000 g for 2 min at 0°, and the resulting precipitate is discarded. More cold acetone is added to the

[5] O. Warburg and W. Christian, *Biochem. Z.* **310**, 384 (1941).
[6] K. Uehara and T. Tanimoto, *J. Biochem. (Tokyo)* **78**, 519 (1975).

supernatant to give a final concentration of 67% (v/v), and the suspension is stirred for 2 min. The precipitated protein is collected by centrifugation at 14,000 g for 2 min and dissolved in cold distilled water. The undissolved portion is discarded after centrifugation at 14,000 g for 20 min.

Step 3. First DEAE-Cellulose Chromatography. The enzyme fraction from step 2 is dialyzed overnight against two changes of 800 ml of 2.5 mM sodium phosphate buffer, pH 7.0, containing 1 mM dithiothreitol (DTT) and 10 μM NADP$^+$. The dialyzed enzyme is applied to a DEAE-cellulose column (3 × 40 cm) equilibrated with the dilute sodium phosphate buffer described above and washed with two column volumes of the same buffer. The enzyme is subsequently eluted with sodium phosphate buffer, pH 7.0, containing 1 mM DTT and 10 μM NADP$^+$, the concentration of the sodium phosphate being increased in a linear gradient from 2.5 mM to 100 mM in a volume of 1600 ml. The flow rate is 60 ml/hr, and 10-ml fractions are collected. The enzyme activity is usually eluted at the buffer concentration of 50 mM. Fractions containing more than 2 units per milliliter are pooled, adjusted to 4 mM DTT and 40 μM NADP$^+$, and concentrated to about 20 ml by ultrafiltration using an Amicon ultrafiltration cell equipped with a PM-10 membrane.

Step 4. Second DEAE-Cellulose Chromatography. The concentrated enzyme from step 3 is dialyzed overnight against 800 ml of 2.5 mM sodium phosphate buffer, pH 7.0, containing 1 mM DTT and 10 μM NADP$^+$, then applied to a DEAE-cellulose column (2 × 40 cm) equilibrated with the same buffer. The enzyme is eluted with sodium phosphate buffer, pH 7.0, containing 1 mM DTT and 10 μM NADP$^+$, the concentration of the sodium phosphate being increased in a linear gradient from 2.5 mM to 100 mM in a volume of 800 ml. The flow rate is 25 ml/hr, and 5-ml fractions are collected. The enzyme is eluted at about 50 mM buffer. Fractions containing enzyme activity are pooled, adjusted to 4 mM DTT and 40 μM NADP$^+$, and concentrated to about 5 ml by ultrafiltration using an Amicon ultrafiltration cell equipped with a PM-10 membrane.

Step 5. First Hydroxyapatite Chromatography. Hydroxyapatite, prepared by the method of Siegelman et al.,[7] is mixed with Hyflo Super-Cell (0.5 g per gram dry weight of gel) to permit a fairly rapid flow and washed twice with 2.5 mM sodium phosphate buffer, pH 7.0, containing 2 mM DTT and 20 μM NADP$^+$. The mixture is suspended in the same buffer and poured into a column (2 × 14 cm) to a height of 10 cm. The enzyme fraction from step 4 is dialyzed overnight against 500 ml of the same buffer and then applied to the column. The column is eluted successively with

[7] H. W. Siegelman, G. A. Wieczorek, and B. C. Turner, Anal. Biochem. 13, 402 (1965).

100-ml portions of 2.5 mM, 5.0 mM, 7.5 mM, and 10 mM sodium phosphate buffer, pH 7.0, containing 2 mM DTT and 20 μM NADP$^+$. The flow rate is 15 ml/hr, and 3-ml fractions are collected. Fractions containing enzyme activity are pooled, adjusted to 4 mM DTT and 40 μM NADP$^+$, and concentrated to about 2 ml by ultrafiltration, using a collodion bag.

Step 6. Second Hydroxyapatite Chromatography. The concentrated enzyme from step 5 is dialyzed overnight against 500 ml of 2.5 mM sodium phosphate buffer, pH 7.0, containing 2 mM DTT and 20 μM NADP$^+$. The dialyzed enzyme is applied to a hydroxyapatite column (1.5 × 7 cm) prepared as described above. The column is eluted with sodium phosphate buffer, pH 7.0, containing 2 mM DTT and 20 μM NADP$^+$, the concentration of the sodium phosphate being increased in a linear gradient from 2.5 mM to 15 mM in a volume of 160 ml. The flow rate is 15 ml/hr, and 3-ml fractions are collected. Fractions with high specific activity are pooled, adjusted to 4 mM DTT and 40 μM NADP$^+$, and concentrated to a final protein concentration of about 8 mg/ml by ultrafiltration using a collodion bag.

Step 7. Crystallization. The concentrated enzyme from step 6 is dialyzed in cellulose tubing against 30 ml of 1.03 M ammonium sulfate, pH 8.2, containing 15 mM DTT and 0.2 mM NADP$^+$ for 4 hr. Any precipitate appearing at this point is removed by centrifugation at 6000 g for 5 min. The clear supernatant is dialyzed against 30 ml of 1.15 M ammonium sulfate, pH 8.2, containing 15 mM DTT and 0.2 mM NADP$^+$ for 8 hr, and then against 30 ml of 1.31 M ammonium sulfate, pH 8.2, containig 15 mM DTT and 0.2 mM NADP$^+$. At this point, amorphous precipitate sometimes appears, which can be dissolved by dialysis against 1.03 M ammonium sulfate. Generally, crystals appear as needles after repeated cycles of dialysis against 1.03 M and 1.31 M ammonium sulfate, both containing 15 mM DTT and 0.2 mM NADP$^+$. To complete the crystallization, the ammonium sulfate concentration is raised to 1.43 M. For recrystallization, the crystals are dissolved in a minimal volume of 2.5 mM sodium phosphate buffer, pH 8.0, containing 20 mM DTT and 0.2 mM NADP$^+$, and dialyzed again in the same manner. The specific activity of the crystalline enzyme is 71.3 units per milligram of protein under the assay conditions described above, and does not increase upon recrystallization. The crystalline enzyme is homogeneous as judged by polyacrylamide gel electrophoresis[8] and ultracentrifugation.[3]

The purification procedure is summarized in the table.

[8] D. E. Williams and R. A. Reisfeld, *Ann. N. Y. Acad. Sci.* **121,** 373 (1964). It is essential that the electrophoresis be carried out with the electrode buffer containing 10 μM NADP$^+$ as a stabilizer of the enzyme.

PURIFICATION OF D-ERYTHRULOSE REDUCTASE FROM BEEF LIVER

Fraction	Total protein (mg)	Total activity[a] (units)	Specific activity (units/mg protein)	Purification (fold)	Yield (%)
1. Homogenate[b]	—	—	—	—	—
2. Acetone precipitate 50–67%	4060	3660	0.9	1	100
3. First DEAE-cellulose	254	3540	13.9	15	97.0
4. Second DEAE-cellulose	125	2200	17.6	20	60.1
5. First hydroxyapatite	36.7	1890	51.5	57	51.6
6. Second hydroxyapatite	29.0	2180	75.2	84	57.4
7. Crystalline enzyme	22.0	1570	71.3	79	42.9

[a] The enzyme activity is conveniently assayed with NADH as a coenzyme.

[b] An accurate estimate of the activity of homogenate is difficult to obtain owing to the NADH oxidase activity present.

Properties

Stability. The enzyme can be stored at 4° as a crystalline suspension in 1.43 M ammonium sulfate, pH 8.2, containing 15 mM DTT and 0.2 mM NADP$^+$ for several months without any loss of enzyme activity. D-Erythrulose reductase is inactivated by exposure to low temperature.[3,6] The cold inactivation is accelerated by increasing the salt concentration and decreasing the enzyme concentration.[6] Protection against cold inactivation is afforded by NADP$^+$ at low concentrations.[3] NAD$^+$, ATP, and other compounds with structures related to that of NADP$^+$ do not exert protection.[3] NADP$^+$ also protects the enzyme against inactivation by heat,[6] or rose bengal-sensitized photoinactivation.[9]

Physical and Chemical Properties. As normally isolated, D-erythrulose reductase contains 2–3 mol of bound NADP$^+$ per mole of enzyme,[9,10] so that the extinction coefficient of the enzyme at 280 nm is not constant. Thus, the concentration of pure enzyme protein is calculated from absorbance measurement at 290 nm, at which wavelength the contribution of NADP$^+$ to the absorbance is negligible.

The molecular weight of the enzyme is estimated to be 90,000 by sedimentation equilibrium analysis, Sephadex G-200 gel filtration and sucrose density gradient centrifugation.[3] The subunit molecular weight, determined by sodium dodecyl sulfate–polyacrylamide gel electrophoresis, is 22,000.[3] Thus, this enzyme is composed of four subunits of identical molecular weight. The isoelectric point of the enzyme determined by

[9] K. Uehara, S. Mannen, and S. Hosomi, *J. Biochem.* (*Tokyo*) **85**, 1003 (1979).

[10] K. Uehara, S. Mannen, and S. Hosomi, *Seikagaku* **50**, 1117 (1978) (in Japanese).

isoelectrofocusing is pH 6.75.[3] The amino acid composition of the beef liver enzyme is similar to that[4] of the chicken liver enzyme.

Substrate Specificity.[3] This enzyme is highly specific for D-erythrulose. The following derivatives of aldose and ketose do not serve as substrate: trioses (DL-glyceraldehyde and dihydroxyacetone), pentoses (D-arabinose, D-lyxose, D-ribose, D-xylose, D-ribulose, and D-xylulose), and hexoses (D-glucose, D-galactose, D-mannose, and D-fructose). This enzyme can use NADH as well as NADPH as a coenzyme, but the latter is more active.

Although the equilibrium of the reaction catalyzed by this enzyme strongly favors the reduction of D-erythrulose, the reverse reaction (dehydrogenation of D-threitol) is slightly detectable when carried out at pH 7.5 or 9.0 using NADP⁺ as a coenzyme.[9] Erythritol does not serve as a substrate under the same conditions.

Kinetic Properties. The pH optima for the reduction of D-erythrulose measured with NADH and NADPH are 6.25 and 5.85, respectively.[3] The K_m value for NADH is 0.22 mM, whereas that for NADPH is 6.8 μM.[3] The K_m value for D-erythrulose is 0.36 mM for the NADH-linked reaction.

NADP⁺, 2′,5′-ADP, and 2′-AMP are potent competitive inhibitors of the enzyme with respect to NADH and have the K_i values of 6.1 μM, 6.3 μM, and 13 μM, respectively.[9]

Distribution. D-Erythrulose reductase is present in the pancreas, heart, and kidneys, but is found especially in the liver of the beef. It is present in human liver[11] and has been crystallized from chicken liver.[4]

[11] S. Hosomi, Y. Yamamura, K. Higashino, H. Seike, K. Uehara, S. Mannen, and M. Inoue, *Seikagaku* **50**, 710 (1978) (in Japanese).

[40] Glycerol Dehydrogenase from Rabbit Muscle

By T. G. Flynn and J. A. Cromlish

$$\text{D-Glyceraldehyde} + \text{NADPH} + \text{H}^+ \rightleftharpoons \text{glycerol} + \text{NADP}^+$$
$$\text{D-Glucose} + \text{NADPH} + \text{H}^+ \rightleftharpoons \text{sorbitol} + \text{NADP}^+$$
$$\text{D-Xylose} + \text{NADPH} + \text{H}^+ \rightleftharpoons \text{xylitol} + \text{NADP}^+$$
$$p\text{-Nitrobenzaldehyde} + \text{NADPH} + \text{H}^+ \rightleftharpoons p\text{-nitrobenzylalcohol} + \text{NADP}^+$$

Glycerol dehydrogenase[1,2] (glycerol:NADP⁺ oxidoreductase, EC 1.1.1.72) activity has been found in several mammalian tissues.[1-5] The

[1] C. J. Toews, *Biochem. J.* **105**, 1067 (1967).
[2] A. W. Kormann, R. O. Hurst, and T. G. Flynn, *Biochim. Biophys. Acta* **258**, 40 (1972).

enzyme catalyzes the $NADP^+$-linked reduction of a wide variety of aldoses and aliphatic and aromatic aldehydes.[2] In this respect it is similar to both NADP L-hexonate dehydrogenase (EC 1.1.1.19) and aldose reductase (alditol: $NADP^+$ 1-oxidoreductase, EC 1.1.1.21). Glycerol dehydrogenase activity from rabbit muscle is due to two enzymes, both of which are different from aldehyde reductase. One of the rabbit muscle enzymes is immunologically identical with aldose reductase from rabbit lens.

Assay

Principle. Glycerol dehydrogenase activity is assayed spectrophotometrically by determining the decrease in absorbance of NADPH at 340 nm. In the following procedure the reaction is carried out at 25° in a Beckman 25 double-beam recording spectrophotometer. The scale is expanded to give a full scale expansion of 0.25.

Reagents

Sodium phosphate buffer, 100 mM, pH 6.5. All the reagents below are dissolved in this buffer.
NADPH, 2.5 mM (10.4 mg/2.5 ml)
DL-Glyceraldehyde, 62.5 mM (28.16 mg/5 ml)

Procedure. To a quartz cuvette of 1 cm path length are added 2.0 ml of buffer (final concentration 100 mM), 0.2 ml of NADPH (final concentration 200 μM), and 0.2 ml of DL-glyceraldehyde (final concentration 5 mM). The reaction is initiated by the addition of 0.1 ml of enzyme solution. The reaction is measured against a matched cuvette containing buffer. A control cuvette containing all reactants except substrate is set up in the same way and measured against a buffer blank. The difference in the rates of change of absorbance between sample and control cuvettes is then measured. The assay is performed in this way because of the considerable oxidation of NADPH that occurs in crude enzyme solutions in the absence of DL-glyceraldehyde. The reaction in the sample cuvette is not measured directly against a blank cuvette containing NADPH but no substrate, because, prior to addition of enzyme to start the reaction, the absorbance is effectively zero and the oxidation of NADPH would result in negative absorbance changes. When fractions obtained during purification of the enzyme are assayed, 0.4 M $(NH_4)_2SO_4$ (final concentration) is included in the reaction mixture. This is to counteract the inhibitory effect of

[3] H. P. Wolf and F. Leuthardt, *Helv. Chim. Acta* **36**, 1463 (1953).
[4] F. Leuthardt and H. P. Wolf, *Helv. Chim. Acta* **37**, 1732 (1954).
[5] B. W. Moore, *J. Am. Chem. Soc.* **81**, 5837 (1959).

2-mercaptoethanol, which is included routinely in all solutions during purification.

Enzyme Units and Specific Activity. One unit of enzyme activity is defined as the amount of enzyme that causes the oxidation of 1 μmol of NADPH per minute at 25°. Specific activity is expressed in units of activity per milligram of protein. Protein is determined by the Bio-Rad protein assay[6,7] using bovine γ-globulin as standard.

Purification Procedure

Buffers

Buffer A: 10 mM sodium phosphate buffer, pH 7.0, containing 5 mM 2-mercaptoethanol

Buffer B: 10 mM sodium phosphate buffer, pH 7.0, containing 5 mM 2-mercaptoethanol and 1 M NaCl

Buffer C: 5 mM sodium phosphate buffer, pH 7.4, containing 5 mM 2-mercaptoethanol

Buffer D: 200 mM sodium phosphate buffer, pH 7.4, containing 5 mM 2-mercaptoethanol

Step 1. Two male New Zealand white rabbits are killed by an overdose of phenobarbital or by cervical dislocation, and the muscle of the back legs is removed. The muscle is either processed immediately or stored frozen at $-20°$ until required. All stages of purification are performed at 0–4°. The excised muscle (approximately 500 g) is cut into small pieces and homogenized in a Waring blender for 3–4 min in buffer A at a 4:1, v/w, ratio. The homogenate is centrifuged at 13,200 g for 1 hr. The supernatant is filtered through glass wool and collected.

Step 2. Solid $(NH_4)_2SO_4$ (29.1 g/100 ml) is added over a 45-min period to give a saturation of 0.5. After stirring for an additional 15 min, the precipitate is removed by centrifugation and discarded. The supernatant is filtered through glass wool. To the supernatant is added additional solid $(NH_4)_2SO_4$ (19.4 g/100 ml) over a 45-min period with stirring to give a final $(NH_4)_2SO_4$ saturation of 0.8. After stirring for a further 15 min, the supernatant is centrifuged and the supernatant is discarded. The precipitate is dissolved in a minimal volume (about 130 ml) of buffer A.

Step 3. The dissolved precipitate from step 2 is applied to a column of Sephadex G-100 (10 × 90 cm) previously equilibrated with buffer A. The column is eluted with the same buffer and fractions (approximately 15 ml) are collected. Fractions with enzyme activity are pooled (approximately 1200 ml).

[6] Bio-Rad Technical Bulletin 1051, April 1977.
[7] M. M. Bradford, *Anal. Chem.* **72**, 248 (1976).

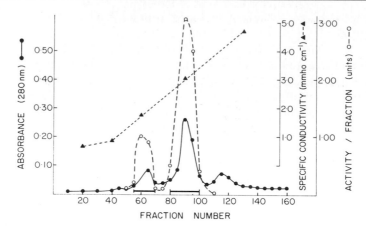

Fig. 1. DEAE-Sephacel chromatography of glycerol dehydrogenases from rabbits. Fractionation was achieved using a linear salt gradient (see text). The first activity peak is GDH₁, and the second is GDH₂.

Step 4. The collected fractions from step 3 are added to a suspension of Blue Sepharose C-16B[8] (18 g in sufficient buffer A to make a thick slurry), which has been previously equilibrated with buffer A, and the mixture is stirred for 12 hr. After cessation of stirring, the mixture is allowed to settle and the supernatant is poured off and discarded. A little of the supernatant is left so that the remaining Blue Sepharose can form a slurry. This slurry is poured into a column (2.5 × 12.5 cm) and washed with 1 liter of buffer A. The column is then developed with a linear salt gradient (750 ml of buffer A + 750 ml of buffer B). Fractions (approximately 5 ml) containing enzyme activity are pooled and concentrated to about 100 ml by ultrafiltration.[9] The concentrated solution is dialyzed overnight against 6 liters of buffer C. The specific conductivity of the enzyme solution should be less than 1 mho/cm after dialysis.

Step 5. The concentrated enzyme solution is applied to a DEAE-Sephacel column (2.5 × 30 cm) that has been previously equilibrated with buffer C. The column is washed with 500 ml of buffer C and then developed with a linear salt gradient (1000 ml of buffer C + 1000 ml of buffer D); 5 ml fractions are collected. Glycerol dehydrogenase activity elutes as two peaks (Fig. 1). the first peak is designated glycerol dehydrogenase₁ (GDH₁) and the second, GDH₂. The fractions comprising each peak of activity are collected, pooled, and reduced in volume to approximately 10 ml by ultrafiltration.

[8] Pharmacia (Canada) Ltd.
[9] Amicon Corp., Lexington, Massachusetts. Amicon stirred ultrafiltration cell with PM-10 membrane.

PURIFICATION OF GLYCEROL DEHYDROGENASE FROM RABBIT MUSCLE

Fraction	Volume (ml)	Total protein (mg)	Total activity (units)	Specific activity (units/mg protein)	Recovery (%)
›ernatant after st centrifugation	1820	13,200	175.6	0.013	100
›ernatant after 0% (NH$_4$)$_2$SO$_4$	2020	12,700	158.3	0.013	90
‹et 50–80% NH$_4$)$_2$SO$_4$	134	3,440	91.4	0.027	52
›hadex G-100	1200	1,650	101.3	0.061	58
‹e Sepharose ›ooled fractions	470	164.9	45.3	0.275	26
AE-Sephacel ¡DH$_1$	68.0	2.34	7.74	3.31	4.40
¡DH$_2$	90	6.15	32.4	5.26	18.5
›hadex G-75 ¡DH$_1$	10	2.00	7.54	3.83	4.30
¡DH$_2$	12	5.5	31.6	5.75	18.00

Step 6. The pooled concentrates of GDH$_1$ and GDH$_2$ are applied separately to a Sephadex G-75 column (2.5 × 80 cm) previously equilibrated in buffer A and eluted with buffer A. Fractions are again pooled and either dialyzed exhaustively against distilled water and lyophilized or dialyzed exhaustively against distilled water containing polyethylene glycol (30 mg/ml) and then frozen at −20°.

The purification procedure is summarized in the table. There are significant differences in the properties of the two glycerol dehydrogenases as described below.

Properties

Stability. Both GDH$_1$ and GDH$_2$ are stable when lyophilized or when stored frozen in 30% polyethylene glycol.

pH Optima. Using DL-glyceraldehyde as substrate the pH optimum of GDH$_1$ is 6.5 and that of GDH$_2$ is 7.0.

Specificity. Both GDH$_1$ and GDH$_2$ have a broad substrate specificity reducing D-glyceraldehyde, L-glyceraldehyde, glycolaldehyde, methylglyoxal, propionaldehyde, butyraldehyde, valeraldehyde, D-erythrose, D-xylose, and D-glucose, and p-nitrobenzaldehyde. Both enzymes have high K_m values for D-glucuronate (GDH$_1$, 42.2 mM; GDH$_2$, 186 mM) that distinguishes them from NADP L-hexonate dehydrogenase. GDH$_1$ is distinguishable from GDH$_2$ in its greater affinity for

D-glucose (K_m = 156 mM for GDH$_1$, 527 mM for GDH$_2$). Other distinguishing features are described below.

Apparent Thermodynamic Equilibrium Constant K'. K' has been calculated for the reaction

$$\text{Glycerol} + \text{NADP}^+ \rightleftharpoons \text{D-glyceraldehyde} + \text{NADPH} + \text{H}^+$$

Equilibrium was reached in approximately 2 hr. *K'* was calculated as 1.67 ± 0.023 × 10^{-6}, indicating that formation of NADP$^+$ and glycerol is favored.[2]

Electrophoresis Behavior. On polyacrylamide gel electrophoresis at pH 8.9, GDH$_2$ migrates faster than GDH$_1$ indicating that GDH$_2$ is more negatively charged.

Molecular Weights. On polyacrylamide gel electrophoresis in sodium dodecyl sulfate, GDH$_1$ and GDH$_2$ migrate as single bands; from appropriate molecular weight standards a molecular weight of 40,200 is calculated for GDH$_1$ and one of 41,500 for GDH$_2$. Molecular weights of 34,000 ± 4000 are obtained by gel filtration on Sephadex G-75.

Coenzyme Specificity. NADPH is the preferred coenzyme for GDH$_1$ and GDH$_2$, but both enzymes can utilize NADH to give about 10% of the activity exhibited with NADPH.

Activators and Inhibitors. Glycerol dehydrogenase activity is inhibited by 2-mercaptoethanol, but the effects of the mercaptan can be largely reversed by the presence of 0.4 M (NH$_4$)$_2$SO$_4$. The latter is an inhibitor of purified glycerol dehydrogenase, and the apparent activating effect of this salt in the presence of 2-mercaptoethanol is unexplained. Glycerol dehydrogenase activity is inhibited by p-chloromercuribenzoate[2] and by NaF.[1] GDH$_1$ and GDH$_2$ are also inhibited by phenobarbitol, diphenylhydantoin, and flavonoids, e.g., quercitin.

Immunological Data. Antibodies prepared in sheep to GDH$_2$ cross-react with absolute identity with rabbit muscle GDH$_1$. There is no cross-reaction between antisera to GDH$_2$ and NADP L-hexonate dehydrogenase from any source. Interestingly, antisera to GDH$_2$ give an immunological reaction of absolute identity with rabbit lens aldose reductase.

Comment. The pH optimum, molecular weight, and greater affinity for D-glucose suggest that GDH$_1$ is closely related to aldose reductase. The immunological identity of lens aldose reductase and GDH$_2$, and therefore also of GDH$_1$, adds support to this contention. The glycerol dehydrogenase purified from rabbit muscle by Kormann *et al.*[2] was probably a mixture of unresolved GDH$_1$ and GDH$_2$.

[41] Glycerol Oxidase from *Aspergillus japonicus*

By T. Uwajima and O. Terada

$$\text{Glycerol} + O_2 \rightarrow \text{glyceraldehyde} + H_2O_2$$

Glycerol oxidase is a novel enzyme that catalyzes the oxidation of glycerol in the presence of oxygen to glyceraldehyde and hydrogen peroxide without the requirement of any exogenous cofactors. The enzyme is formed in some strains of *Aspergillus, Neurospora,* and *Penicillium* grown on glycerol as a sole carbon source.[1] Protoheme IX and copper ions are contained in the enzyme as prosthetic groups. The enzyme, in combination with lipoprotein lipase, has been employed in a spectrophotometric assay for triglycerides in sera.

Assay Method

Principle. The enzyme assay is based on the measurement of hydrogen peroxide generated in the oxidation of glycerol. The hydrogen peroxide is oxidized with 4-aminoantipyrine and phenol in the presence of peroxidase to form a quinoneimine dye, according to Allain *et al.*[2]

Reagents

N-Tris(hydroxymethyl)methyl-2-aminoethanesulfonic acid (TES)–ammonia buffer, 0.1 M, pH 7.0
4-Aminoantipyrine, 2.4 mM
Phenol, 42.0 mM
Glycerol, 0.1 M
Peroxidase (specific activity 110), 0.2 mg/ml

Procedure. The assay system in a final volume of 3.0 ml contains 0.5 ml of glycerol, 1.0 ml of TES buffer, 0.5 ml of 4-aminoantipyrine, 0.5 ml of phenol, 0.1 ml of peroxidase, and 0.1 ml of glycerol oxidase. The mixture is incubated at 37° for 10 min with shaking. The reaction is initiated by the addition of glycerol oxidase. The control run contains all the components except substrate. The formation of the quinoneimine (EmM = 5.33 at 500 nm)[3] is estimated using a spectrophotometer.

[1] T. Uwajima, H. Akita, K. Ito, A. Mihara, K. Aisaka and O. Terada, *Agric. Biol. Chem.* **44,** 399 (1980).
[2] C. C. Allain, L. S. Poon, C. S. G. Chan, W. Richmond, and P. C. Fu, *Clin. Chem.* **20,** 470 (1974).
[3] W. Richmond, *Clin. Chem.* **19,** 1350 (1973).

METHODS IN ENZYMOLOGY, VOL. 89

Definition of Unit and Specific Activity. One unit of enzyme catalyzes the formation of 1 μmol of hydrogen peroxide per minute, under the assay conditions. Specific activity is expressed as units per milligram of protein, determined according to Lowry *et al.*[4]

Purification Procedure[1]

Culture of the Organism. Aspergillus japonicus AT 008 is grown in 15-liter volumes in a 30-liter fermentor at 30°, with an aeration rate of 15 liters per minute and an agitation speed of 350 rpm. The medium consists of 3% glycerol, 0.3% yeast extract, 0.1% meat extract, 0.1% KH_2PO_4, and 0.1% $MgSO_4 \cdot 7 H_2O$, adjusted to pH 7.2 with NaOH. The inoculum is 0.9 liter of an overnight culture in the same medium. After 20 hr of growth, the mycelia are harvested from the 15-liter broth by filtration and washed with distilled water. The yield is approximately 26.5 g of wet mycelial mat per liter of medium.

All subsequent purification steps are performed at 0–5°.

Step 1. Extraction of the Enzyme. The mycelial mats are suspended in 5.0 liters of 0.05 M borate buffer (H_3BO_3–KCl–Na_2CO_3), pH 10.0, and disrupted continuously by a Dyno Mill KDL (W. A. Bachofen, Basel, Switzerland) containing 0.25 mm in diameter glass beads at 3000 rpm. The broken mycelial suspension is centrifuged at 10,000 g for 20 min, and the pellet is discarded.

Step 2. Ammonium Sulfate Fractionation. The cell-free extract is brought to 0.40 (24.3 g/100 ml) saturation with ammonium sulfate, and the resulting precipitate is removed by centrifugation at 20,000 g for 20 min. The ammonium sulfate concentration of the supernatant is then increased to 0.70 (47.2 g/100 ml) saturation. The active precipitate is collected by centrifugation at 20,000 g for 20 min and dissolved in 80.0 ml of 0.05 M borate buffer, pH 10.0. The crude enzyme solution is dialyzed for 48 hr against four changes of 20 liters of 0.05 M borate buffer, pH 10.0.

Step 3. DEAE-Sephadex A-25 Column Chromatography. The dialyzed enzyme is subjected to column chromatography on DEAE-Sephadex A-25, and the enzyme is passed through the column (5.5 × 60 cm) previously equilibrated with 0.05 M borate buffer, pH 10.0. The column is then washed with 0.05 M borate buffer, pH 10.0, containing 0.05 M ammonium sulfate, which removes most of the inactive protein. The enzyme is eluted from the column with 0.05 M borate buffer, pH 10.0, containing 0.45 M ammonium sulfate. Fractions with a specific activity greater than 10.0 are combined and concentrated by the addition of ammonium sulfate to 0.70 saturation.

[4] O. H. Lowry, N. J. Rosebrough, A. L. Farr, and R. J. Randall, *J. Biol. Chem.* **193**, 265 (1951).

TABLE I
PURIFICATION OF GLYCEROL OXIDASE FROM *Aspergillus japonicus* AT 008

Fraction	Volume (ml)	Total protein (mg)	Total activity (units)	Specific activity (units/mg protein)	Yield (%)
1. Cell-free extract	4800	9600	16,510	1.72	100
2. Ammonium sulfate	102	2105	10,950	5.20	66.3
3. DEAE-Sephadex A-25	625	485	9,360	19.2	56.7
4. Hydroxyapatite	520	53	6,780	129.0	41.0
5. Sephadex G-200	210	16.6	5,345	322.0	32.3

Step 4. Hydroxyapatite Column Chromatography. The active precipitate is collected by centrifugation at 20,000 g for 20 min and dissolved in 15.0 ml of 0.05 M borate buffer, pH 10.0. The enzyme solution is dialyzed for 48 hr against four changes of 10 liters of 0.05 M borate buffer, pH 10.0. The dialyzed enzyme is loaded onto a column (4.5 × 25 cm) of hydroxyapatite equilibrated with 0.05 M borate buffer, pH 10.0. The column is washed with 0.05 M borate buffer, pH 10.0, containing 0.30 M ammonium sulfate, which removes the yellow pigment. The enzyme is eluted with 0.05 M borate buffer, pH 10.0, containing 0.75 M ammonium sulfate. Fractions with a specific activity greater than 60.0 are combined and precipitated by the addition of ammonium sulfate to 0.70 saturation.

Step 5. Sephadex G-200 Gel Filtration. The precipitate is collected by centrifugation at 20,000 g for 20 min and dissolved in 5.0 ml of 0.05 M borate buffer, pH 10.0. The enzyme solution is applid to gel filtration on a column (3.0 × 100 cm) of Sephadex G-200 equilibrated with 0.05 M borate buffer, pH 10.0. Two protein peaks emerge from the column, and the enzyme activity is entirely associated with the first peak. The active fractions are pooled and concentrated by the addition of ammonium sulfate to 0.70 saturation. The precipitate is collected by centrifugation at 20,000 g for 20 min and dissolved in 0.02 M TES buffer, pH 7.0, for investigation. A typical purification is summarized in Table I.

Properties[5]

Purity. The purified enzyme sediments as a single, symmetric schlieren peak on ultracentrifugation and gives a single band on the acrylamide gel electrophoresis at pH 8.3.[6]

[5] T. Uwajima and O. Terada, *Agric. Biol. Chem.* **44**, 2039 (1980).
[6] B. J. Davis, *Ann. N. Y. Acad. Sci.* **121**, 404 (1964).

Molecular Weight. The molecular weight is calculated as 400,000, by a sedimentation equilibrium method.[7]

Isoelectric Point. The isoelectric point is found to be 4.9, by isoelectric focusing.

Spectral Properties and Prosthetic Groups. The enzyme shows the typical absorption spectra of a hemoprotein; the absorption maxima are located at 557 nm and 430 nm in reduced form, and at 557 nm, 530 nm, 420 nm, 280 nm, and 238 nm in oxidized form. In the latter form, the absorption ratios $A_{557\,nm} : A_{280\,nm}$, $A_{530\,nm} : A_{280\,nm}$, and $A_{420\,nm} : A_{280\,nm}$ are 0.078, 0.083, and 0.41, respectively, and the extinction coefficient ($E_{1\,cm}^{1\%}$) of the enzyme at 280 nm is 14.7. The anaerobic addition of glycerol to the enzyme causes both a shift of the Soret band from 420 nm to 410 nm and decreases of the peaks at 557 nm and 530 nm. The pyridine ferrohemochrome of the enzyme shows an absorption spectrum almost identical to that of hematin, indicating that the iron porphyrin in the enzyme is protoheme IX. The enzyme contains 0.94 mol of heme per mole of enzyme protein as determined with the extinction coefficient of the pyridine ferrohemochrome (EmM = 32 at 557 nm).[8] On the other hand, analysis of the enzyme by atomic absorption spectrophotometry displays the presence of copper ions at a concentration of 2.04 gram atoms per mole of enzyme protein. Electron spin resonance signals are observed at g = 1.99, g = 2.00, and g = 2.02. The g = 1.99, and g = 2.02 signals are diminished by the anerobic addition of glycerol, and the three signals completely disappear on the addition of either a reducing agent, dithionite, or a copper chelating agent, diethyldithiocarbamate. These results indicate that the copper ions associated with the enzyme are present in the cupric form and directly involved in the electron transfer from the substrate.

Stability. The enzyme is stable at 5° between pH 7.0 and 10.5, and it can be stored at −80° for at least 6 months. In the alkaline pH region the enzyme solution is especially stable and can be preserved for a week without loss of activity. On heating, the enzyme is stable up to 45°, but loses approximately 40% of activity after treatment at 50° for 15 min.

pH Optimum. The pH optimum for the catalytic activity is 7.0 with glycerol as substrate.

Substrate Specificity. Table II illustrates the relative activities of the enzyme with a variety of hydroxyl compounds. Glycerol is oxidized most rapidly, and the K_m value and V_{max} are 10.4 mM and 935.6 μmol min^{-1} mg^{-1}, respectively. Dihydroxyacetone is also oxidized at a rate 58.9% of that for glycerol; however, glycerol 3-phosphate, methanol, ethanol, and polyvinyl alcohol are not oxidized at all. Among the saccharides, galactose and fructose are slightly oxidized.

[7] D. A. Yphantis, *Ann. N. Y. Acad. Sci.* **88**, 586 (1960).
[8] K. G. Paul, H. Theorell, and Å, Åkeson, *Acta. Chem. Scand.* **7**, 1248 (1953).

TABLE II
SUBSTRATE SPECIFICITY OF GLYCEROL OXIDASE

Substrate	Relative activity (%)
Glycerol	100
Dihydroxyacetone	58.9
1,3-Propanediol	6.73
1,3-Butanediol	6.19
1,4-Butanediol	1.19
Glycerol 3-phosphate	0
Dihydroxyacetone phosphate	0
D-Galactose	1.46
D-Fructose	0.47
D-Glucose	0
Saccharose	0
Methyl alcohol	0
Ethyl alcohol	0
n-Propyl alcohol	0
iso-Propyl alcohol	0
Ethylene glycol	0
Polyethylene glycol	0
Polyvinyl alcohol 2000	0
Phenol	0

Activators. Divalent metal ions, such as Zn^{2+}, Mg^{2+}, Ni^{2+}, Co^{2+}, and Mn^{2+}, activate the enzyme. Dialysis of the enzyme against $0.05\ M$ borate buffer, pH 10.0, for 18–24 hr results in stimulation of the enzyme activity.

Inhibitors. Sodium azide and hydroxylamine are potent competitive inhibitors, with K_i values of 2.0 μM and 1.2 μM, respectively. Diethyl-

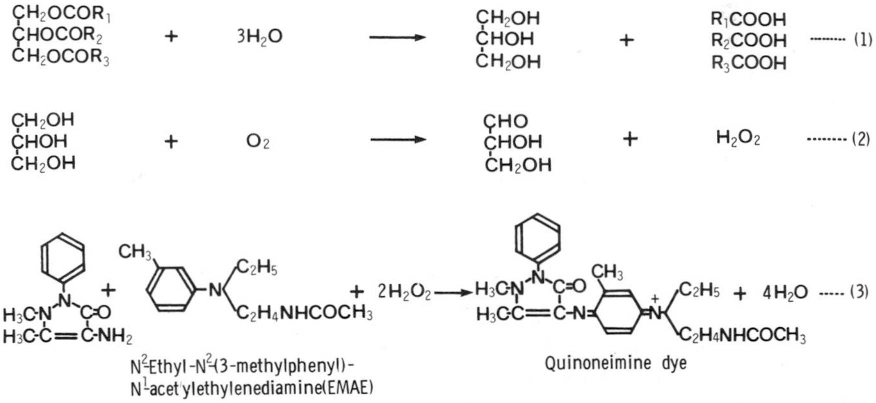

SCHEME 1

dithiocarbamate and 2 : 3-dimercapto-1-propanol also inhibit the enzyme competitively, with K_i values of 17.0 μM and 12.0 μM, respectively.

Application of the Enzyme to the Assay of Triglycerides. The enzyme, in combination with lipoprotein lipase, can be employed for the determination of serum triglycerides. The enzymic method is based on the formation of a quinoneimine dye in the sequence of reactions shown in Scheme 1.

A typical reaction medium in a final volume of 3.0 ml contains 45 units of lipoprotein lipase, 225 units of glycerol oxidase, 30 μmol of TES buffer (pH 6.8), 1.4 μmol of 4-aminoantipyrine, 4.1 μmol of EMAE, 30 units of peroxidase, 3 mg of Triton X-100, and 20 μl of serum. In addition, enzyme stabilizers, reaction stimulators, and ascorbic acid oxidase may be involved in the system. The reaction mixture is incubated at 37° for 10 min, and the resulting dye is read at 555 nm on a spectrophotometer. The enzymic method is more simple, sensitive, and far less susceptible to interference from bilirubin, heparin, ethylenediaminetetraacetic acid, and oxalic acid, compared with other enzymic and chemical methods.

[42] L-Sorbose-1-phosphate Reductase

By RICHARD L. ANDERSON and RONALD A. SIMKINS

L-Sorbose 1-phosphate + NAD(P)H + H$^+$ → D-glucitol 6-phosphate + NAD(P)$^+$
D-Fructose 1-phosphate + NAD(P)H + H$^+$ → D-mannitol 6-phosphate + NAD(P)$^+$

L-Sorbose-1-phosphate reductase functions in L-sorbose catabolism in wild-type *Klebsiella pneumoniae*, wherein it is induced by L-sorbose but not by D-fructose.[1] It also functions as an essential enzyme in D-fructose catabolism in a genetically reconstructed strain of this organism, wherein it is constitutive and catalyzes the second reaction of the pathway: D-fructose → D-fructose 1-phosphate → D-mannitol 6-phosphate → D-fructose 6-phosphate.[2]

Assay Method[3]

Principle. The rate of D-fructose 1-phosphate-dependent NADH oxidation is determined by measuring the rate of absorbance decrease at 340 nm.

[1] N. E. Kelker, R. A. Simkins, and R. L. Anderson, *J. Biol. Chem.* **247**, 1479 (1972).
[2] R. A. Simkins and R. L. Anderson, in preparation.
[3] R. A. Simkins and R. L. Anderson, in preparation.

Reagents

(2-N-Morpholino)ethane sulfonate (MES) buffer, 0.25 M, pH 6.2
NADH, 10 mM
MnCl$_2$, 30 mM
D-Fructose 1-phosphate, 0.10 M

Procedure. The following are added to a microcuvette with a 1.0-cm light path: 20 μl of MES buffer, 5 μl of NADH, 5 μl of MnCl$_2$, 5 μl of D-fructose 1-phosphate, enzyme, and water to a volume of 0.15 ml. The reaction is initiated by the addition of L-sorbose-1-phosphate reductase. With crude extracts, a control cuvette minus D-fructose 1-phosphate is included to correct for NADH oxidase. The reaction rates are conveniently measured with a Gilford multiple-sample absorbance-recording spectrophotometer. The cuvette compartment should be thermostatted at 30°.

Definition of Unit and Specific Activity. One unit of L-sorbose-1-phosphate reductase is defined as the amount that catalyzes the oxidation of 1 μmol of NADH per minute in the standard assay. Specific activity (units per milligram of protein) is based on protein determinations by the Lowry procedure or by the absorbance at 220 nm.[4]

Purification Procedure[3]

Organism and Growth Conditions. *Klebsiella pneumoniae* PRL-R3 (formerly designated *Aerobacter aerogenes* PRL-R3) was grown at 30° in a medium consisting of 0.71% Na$_2$HPO$_4$, 0.15% KH$_2$PO$_4$, 0.3% (NH$_4$)$_2$SO$_4$, 0.01% MgSO$_4$, 0.0005% FeSO$_4$ · 7 H$_2$O, and 0.5% L-sorbose (autoclaved separately). The cultures were grown overnight in 500-ml volumes in Fernbach flasks on a rotary shaker.

Preparation of Cell Extracts. The cells were harvested by centrifugation, suspended in buffer A [50 mM MES buffer (pH 6.1), 0.2% (v/v) 2-thioethanol, and 2 mM EDTA], and broken by sonic disruption with a Raytheon 250-W 10-kHz sonic oscillator, Model DF-101. The disruption chamber was cooled with circulating ice water. The cellular debris was removed by centrifugation, and the resulting supernatant fluid was designated the cell extract.

DEAE-Cellulose Chromatography I. The cell extract was applied to a column (1.6 × 8 cm) of DEAE-cellulose equilibrated with buffer B [25 mM MES buffer (pH 6.1), 0.2% (v/v) 2-thioethanol, and 2 mM EDTA]. The column was washed with 16 ml of buffer B, and then the protein was eluted with a 160-ml gradient of 0 to 0.3 M NaCl in the same buffer.

[4] M. P. Tombs, F. Souter, and N. F. MacLagan, *Biochem. J.* **73,** 167 (1959).

PURIFICATION OF L-SORBOSE-1-PHOSPHATE REDUCTASE

Fraction	Volume (ml)	Total protein (mg)	Total activity (units)	Specific activity (units/mg protein)	Recovery (%)
Cell extract	53	1083	76.5	0.071	(100)
DEAE-cellulose I	28	114	41.3	0.36	54
DEAE-cellulose II	19	44	29.3	0.67	38
NADP-affinity I[a]	26	9.9	23.4	2.4	31
NADP-affinity II[a]	26	3.8	11.2	2.9	15

[a] The values have been adjusted to correct for the proportion of the preceding step not applied to the column.

Fractions (1.7 ml) were collected, and those that contained most of the activity (fractions 31 through 46) were combined.

DEAE-Cellulose Chromatography II. The combined fractions from the above step were dialyzed for 18 hr at 4° against buffer B containing 20% (v/v) glycerol. This dialyzed preparation was then applied to a column (1.2 × 8 cm) of DEAE-cellulose equilibrated with buffer B. The column was washed with 9 ml of buffer B containing 0.05 M NaCl, and then the protein was eluted with a 90-ml linear gradient of 0.05 to 0.25 M NaCl in the same buffer. Fractions (1.4 ml) were collected, and those that contained most of the activity (fractions 25 through 38) were combined.

Affinity Chromatography I. The combined fractions from the second DEAE-cellulose chromatography step were dialyzed overnight at 4° against buffer B containing 20% (v/v) glycerol. A portion was applied to a column (1.0 × 6.3 cm) of NADP-Sepharose.[5,6] The column was then washed with 20 ml of 25 mM MES buffer (pH 6.1) containing 0.2% (v/v) 2-thioethanol, followed by 18 ml of this same solution containing in addition 1 mM MnCl$_2$, 2 mM D-fructose 1-phosphate, and 20 μM NADH. L-Sorbose-1-phosphate reductase was then eluted with 25 mM MES buffer (pH 6.1) containing 0.2% (v/v) 2-thioethanol, 2 mM EDTA, and 20 μM NADH. Fractions (1.1 ml) were collected, and those that contained L-sorbose-1-phosphate reductase activity were combined. This preparation, after dialysis against buffer B, was used for most of the studies reported. Examination of the enzyme by polyacrylamide gel electrophoresis revealed two minor contaminants that comprised about 5% of the total protein. When a more purified preparation was required, a second affinity chromatography step was performed as described below.

[5] S. C. March, I. Parikh, and P. Cuatrecasas, *Anal. Biochem.* **60**, 149 (1974).
[6] R. Lamed, Y. Levin, and M. Wilchek, *Biochim. Biophys. Acta* **304**, 231 (1973).

Affinity Chromatography II. A portion of the above enzyme preparation was applied to a column (1.0 × 6.3 cm) of NADP-Sepharose. The column was washed with 11 ml of 25 mM MES buffer (pH 6.1) containing 0.2% (v/v) 2-thioethanol and 2 mM EDTA, followed by 11 ml of this same solution containing in addition 1 mM NAD$^+$. L-Sorbose-1-phosphate reductase was then eluted with 25 mM MES buffer (pH 6.1) containing 0.2% (v/v) 2-thioethanol, 2.5 mM NAD$^+$, 2.5 mM D-fructose 1-phosphate, and 1.0 mM MnCl$_2$. Fractions (1.0 ml) were collected, and those that contained L-sorbose-1-phosphate reductase activity were combined. Examination of this preparation by polyacrylamide gel electrophoresis revealed one minor contaminant that comprised less than 1% of the total protein.

A typical purification is summarized in the table.

Properties[3]

Substrate Specificity and Kinetic Constants. L-Sorbose 1-phosphate (K_m = 0.6 mM), D-fructose 1-phosphate (K_m = 0.4 mM), and their reduction products (D-glucitol 6-phosphate and D-mannitol 6-phosphate, respectively) are the only known substrates. The equilibrium lies toward the hexitol phosphates. The V_{max} obtained with D-fructose 1-phosphate is about one-third that obtained with L-sorbose 1-phosphate. The reduction of L-sorbose, D-fructose, D-fructose 6-phosphate, or D-fructose 1,6-bisphosphate (all at 10 mM) could not be detected.

NADPH and NADH are about equally effective as coenzymes (K_m = 20 μM for both; same V_{max}).

Effect of Divalent Metal Ions. The enzyme (which is stored in an EDTA-containing buffer) exhibits no activity unless an activating divalent metal ion is included in the assay mixture. To examine the activating effect of low concentrations of metal salts, the EDTA was removed by passage of the enzyme through a column of BioGel P-6 that had been equilibrated with 25 mM MES buffer (pH 6.1), containing 0.2% (v/v) 2-thioethanol. The following restorations (%) of activity were then achieved with various metal salts (10 μM): MnCl$_2$, 100; ZnSO$_4$, 100; FeSO$_4$, 85; NiCl$_2$, 73; MgCl$_2$, 10; CaCl$_2$, 0; and no salt, 0.

Effect of pH. Activity as a function of pH in 67 mM MES buffer is maximal at pH 6.2, and 70% maximal at pH 5.6 and 7.0.

Molecular Weight. L-Sorbose-1-phosphate reductase has a molecular weight of 90,000 and is composed of two apparently identical 45,000 molecular weight subunits.

[43] Glucose-6-phosphate Dehydrogenase from Mouse

By CHI-YU LEE

Glucose 6-phosphate + NADP$^+$ $\rightleftharpoons$ 6-phosphogluconate + NADPH + H$^+$

Glucose-6-phosphate dehydrogenase (EC 1.1.1.49; D-glucose 6-phosphate : NADP$^+$ 1-oxidoreductase) catalyzes the oxidation of glucose 6-phosphate in the presence of NADP$^+$ and is the first enzyme in the pentose shunt pathway. This enzyme from many sources has been successfully purified mainly by general ligand affinity chromatography using derivatives of NADP$^+$ as the immobilized ligands.[1-4]

Assay Method

The activity of glucose-6-phosphate dehydrogenase is followed spectrophotometrically at 340 nm. In a total volume of 1 ml, the reaction mixture contains 0.1 M Tris-HCl, pH 8.0, 1 mM glucose-6-phosphate, 1 mM NADP$^+$, and a suitable amount of enzyme to cause an absorbance change of 0.05 to 0.1 A/min. One unit of enzyme activity is defined as the amount of enzyme that catalyzes the oxidation of 1 μmol of glucose 6-phosphate per minute.

Affinity Columns

The following derivatives of NADP$^+$ and its moieties immobilized on Sepharose are good ligands for the purification of glucose-6-phosphate dehydrogenase and several other NADP$^+$-dependent dehydrogenases[5]: (I) 8-(6-aminohexyl)amino-NADP$^+$; (II) 8-(6-aminohexyl)amino-2'-phosphoadenosine diphosphoribose; (III) 8-(6-aminohexyl)amino-2',5'-ADP; and (IV) N^6-(6-aminohexyl)amino-2',5'-ADP. In addition, mouse glucose-6-phosphate dehydrogenase can also be adsorbed and purified by an 8-(6-aminohexyl)amino-AMP-Sepharose column. Preparations of these affinity ligands have been reported[2,6-8] Ligands I and IV are commercially

[1] N. O. Kaplan, J. Everse, J. E. Dixon, F. E. Stolzenbach, C.-Y. Lee, C.-L. T. Lee, S. S. Taylor, and K. Mosbach, *Proc. Natl. Acad. Sci. U.S.A.* **71**, 3450 (1974).

[2] C.-Y. Lee and N. O. Kaplan, *Arch. Biochem. Biophys.* **168**, 665 (1975).

[3] C.-Y. Lee, C. H. Langley, and J. Burkhart, *Anal. Biochem.* **86**, 697 (1978).

[4] C.-Y. Lee, J. H. Yuan, D. Moser, and J. Kramer, *Mol. Cell. Biochem.* **24**, 67 (1979).

[5] C.-Y. Lee and A. F. Chen, *in* "The Pyridine Nucleotide Coenzymes" (J. Everse, K. S. You, and B. M. Anderson, eds.), Chapter 6, pp. 189–224. Academic Press, New York, 1982.

[6] J. Lopez-Barea, and C.-Y. Lee, *Eur. J. Biochem.* **98**, 487 (1979).

[7] C.-Y. Lee, and C.-J. Johanson, *Anal. Biochem.* **77**, 90 (1977).

[8] P. Brodelius, P.-O. Larsson, and K. Mosbach, *Eur. J. Biochem.* **47**, 81 (1974).

Copyright © 1982 by Academic Press, Inc.
All rights of reproduction in any form reserved.
ISBN 0-12-181989-2

available from P-L Biochemicals, Milwaukee, Wisconsin. Ligand IV is also available from Pharmacia, Piscataway, New Jersey, and from Sigma, St. Louis, Missouri. The details for the preparation of ligands II and III are presented here.

Preparation of 8-(6-Aminohexyl)-amino-2',5'-ADP

8-Bromo-5'-AMP is prepared by bromination of 5'-AMP according to the procedure described by Lee et al.[9] One gram of 8-bromo-5'-AMP is dissolved in 10 ml of distilled water and passed through a Dowex 50-X8 cation exchange column (50 ml) equilibrated in H^+ form. Fractions containing UV-absorbing material are pooled and neutralized with tri-n-butylamine and lyophilized to complete dryness.

One gram of 8-bromo-5'-AMP in tri-n-butylammonium salt is dissolved in 20 ml of triethyl phosphate (liquid) containing 20% $POCl_3$ (v/v). The solution is shaken vigorously for 24 hr at 4°. It is then slowly added to 100 ml of water in which the pH is maintained at 7.0 by proper additions of 4 M LiOH at 4°. After three successive extractions with an equal volume of ether, 20 g of 1,6-diaminohexane are added to the aqueous phase. After the removal of precipitated materials by centrifugation, the solution is heated to 60° for 3 hr. The conversion of 8-bromoadenine nucleotides to 8-(6-aminohexyl)amino derivatives is monitored by changes in λ_{max} from 263 nm to 278 nm. After the reaction is complete, the products are purified by two Dowex 1-X2 anion exchange columns connected in series (2 × 50 cm each) and equilibrated in formate ion. After adsorption, the columns are developed with a 0 to 0.3 M linear formic acid gradient (1 liter by 1 liter). Three main fractions containing UV-absorbing materials are recovered at a ratio of 3 : 3 : 4 (based on the absorbance at 278 nm). Each fraction is concentrated separately by rotary evaporation and precipitated with three volumes of cold ethanol at −20°. By thin-layer chromatography and NMR spectroscopy, these three fractions are identified as 8-(6-aminohexyl)amino derivatives of 5'-AMP, 2',5'-ADP, and 3',5'-ADP, respectively. About 150 mg of 8-(6-aminohexyl)amino-2',5'-ADP are prepared by this protocol.

Preparation of 8-(6-Aminohexyl)amino-2'-phosphoadenosine Diphosphoribose

$NADP^+$ (1 g, 1.2 mmol) is dissolved in 10 ml of 1 M sodium acetate buffer, pH 4.5. Liquid bromine (0.3 ml, 5.4 mmol) is then added dropwise to the rapidly stirring solution, and the pH is maintained at 4.5 with 1 M NaOH. After 30 min of reaction at room temperature, the unreacted

[9] C.-Y. Lee, J. H. Yuan, and E. Goldberg, this volume [61].

bromine is extracted three times with 10 ml each of carbon tetrachloride. Cold acetone (40 ml) is added to the aqueous phase, which contains 8-Br-NADP$^+$, and the preparation is stored at $-70°$ for at least 1 hr. The yellow precipitate is washed with 40 ml of cold acetone and redissolved in 5 ml of water. To the solution of 8-Br-NADP$^+$ is added 1,6-diaminohexane (3 g, 25 mmol). The reaction mixture is heated at 60° in a water bath. The progress of the replacement reaction is monitored spectrophotometrically, and after 3.5 hr the absorbance maximum shifts from 263 to 278 nm and the $A_{278} : A_{263}$ changes from 0.74 to 1.1. The reaction mixture contains mainly 8-(6-aminohexyl)amino-2'-phosphoadenosine diphosphoribose and presumably 2-hydroxynicotinaldehyde. The reaction mixture is diluted 100-fold to 1000 ml and loaded on a Dowex-1-X8 anion exchange column (260 ml bed), which has been washed with 1 M ammonium carbonate and equilibrated with water. The column is then washed with 5 column volumes of water, and the elution is made with a 0 to 0.1 M ammonium carbonate gradient (1 liter each). The fractions with λ_{max} at 278 nm are pooled and lyophilized.

Thin-layer chromatography is performed to follow the synthesis and to examine the purity of the final product. Eastman Kodak silica gel plates with a fluorescent indicator are used for all analyses in a solvent system containing 0.1 M sodium phosphate (pH 6.8)–ammonium sulfate–propan-1-ol (100 : 60 : 2). The chromatograms are visualized by UV light. Under these conditions, the R_f values are NADP$^+$, 0.56; 8-(6-amino-hexyl)amino-2'-phosphoadenosinediphosphoribose, 0.46; and 8-Br-NADP$^+$, 0.14. The molar extinction coefficient of 8-Br-NADP$^+$ and 8-(6-aminohexyl)amino-2'-phosphoadenosinediphosphoribose were determined to be 20,700 and 18,000 at 263 and 278 nm, respectively. The overall yield of the synthesis is 65%.

Coupling to Sepharose-4B by Cyanogen Bromide Activation

The affinity ligands are coupled to Sepharose by cyanogen bromide activation[10] as described in this volume[9] for the preparation of AMP- or ATP-Sepharose. They can also be coupled directly to commercially available CNBr-activated Sepharose (Pharmacia Chemical Co. and Sigma). Briefly, 30 ml of water-washed Sepharose-4B (Pharmacia) are suspended in 50 ml of 0.5 M sodium carbonate at pH 11.0 and 4° to which 10 g of cyanogen bromide dissolved in 20 ml of acetylnitrile are added dropwise over a period of 5 min. Constant stirring under the hood and occasional addition of ice chips to maintain low temperture are required. During the activation, the pH of the reaction mixture is maintained at 11.0 ± 0.1 by

[10] R. Axén, J. Porath, and S. Ernbäck, Nature (London) 214, 1302 (1967).

dropwise addition of 4 M NaOH. About 10 min after the addition of cyanogen bromide, the activated Sepharose is filtered on a Büchner funnel and washed with distilled water for 2 min until the pH is neutral. Within 5 min after activation, the packed and washed gel (30 ml) is added to 20 ml of 8-(6-aminohexyl)amino- 2',5'-ADP or 8-(6-aminohexyl)amino-2'- phosphoadenosinediphosphoribose (concentration 10 mg/ml in H_2O, pH 9.5 to 10.0 adjusted with NaOH). After constant shaking or stirring for 12–24 hr at 4°, the liganded Sepharose is filtered through a funnel. The first filtrate is saved and freeze-dried for the next coupling to freshly activated Sepharose. The affinity gel is then washed extensively with 300 ml of 2 M NaCl and then distilled water. The affinity gels can be stored in 50% glycerol at −20°.

According to the method described by Niesel *et al.*[11] the ligand density of the prepared affinity gel is estimated to be 0.7 to 1.0 μmol of Sepharose per milliliter. This is considered optimal for enzyme purification.

Purification Procedure

Unless otherwise specified, the enzyme purification is performed at 4°. The phosphate buffers used are in the potassium form. The mice are from the Jackson Laboratory, Bar Harbor, Maine.

Two different purification schemes are presented to obtain homogeneous glucose-6-phosphate dehydrogenase from mouse tissues.

Glucose-6-phosphate Dehydrogenase from Mouse Kidney

Step 1. 8-(6-Aminohexyl)amino-2',5'-ADP-Sepharose Affinity Chromatography. Frozen kidneys (80 g) from 400 DBA/2J mice are homogenized in 250 ml of 10 mM phosphate, pH 6.5, containing 10 mM 2-mercaptoethanol and 1 mM EDTA. After centrifugation at 27,000 g for 20 min, the supernatant is applied directly to the ADP-affinity column (2.5 × 10 cm) equilibrated with the same buffer; glucose-6-phosphate dehydrogenase is quantitatively adsorbed. After washing with two liters of equilibration buffer, the enzyme is eluted biospecifically with 0.2 mM $NADP^+$ included in the same buffer (flow rate ~30 to 50 ml/hr by gravity difference). Fractions containing glucose-6-phosphate dehydrogenase activity are pooled and concentrated to 2 ml by an Amicon unit fitted with a PM-10 membrane.

Step 2. DEAE-Sephadex Ion-Exchange Chromatography. After adjusting the pH to 8.0 with 1 N NH_4OH, the concentrate is loaded on a DEAE-Sephadex column equilibrated with 10 mM phosphate, pH 8.0 (column size 50 ml). After loading and washing with 100 ml of the same

[11] D. W. Niesel, G. C. Bewley, C.-Y. Lee and F. B. Armstrong, this volume [51].

TABLE I
PURIFICATION OF GLUCOSE-6-PHOSPHATE DEHYDROGENASE FROM
MOUSE KIDNEY

Fraction	Total activity (units)	Specific activity[a] (units/mg protein)	Purification (fold)
Crude homogenate	200	0.028	—
Affinity column 8-(6-aminohexyl) amino-2′,5′-ADP-Sepharose	150	10	356
DEAE-Sephadex	90	160	5715

[a] Protein was determined by fluorescamine assays of P. Böhlen, S. Stein, W. Dairman, and S. Udenfriend [*Arch. Biochem. Biophys.* **155**, 213 (1973)] using bovine serum albumin as the protein standard.

buffer, glucose-6-phosphate dehydrogenase is eluted with a linear NaCl gradient (0 to 0.5 M, 250 by 250 ml) in 10 mM phosphate, pH 8.0 (elution by gravity difference of 100 cm). The enzyme is expected to appear at a salt concentration of 0.2–0.3 M NaCl.

A summary of the purification of this enzyme from mouse kidneys is presented in Table I.

Glucose-6-phosphate Dehydrogenase from Mouse Testis or Kidney

An identical procedure is employed to purify glucose-6-phosphate dehydrogenase from testis and kidney of DBA/2J mice. The purification of this enzyme from testis homogenate is described here in detail.

Step 1. Frozen testes from DBA/2J mice (200 g) are homogenized in 200 ml of 10 mM phosphate, pH 6.5 at 4° followed by centrifugation at 27,000 g for 20 min. The supernatant is passed through an 8-(6-aminohexyl)amino-AMP-Sepharose column (5 × 12 cm) which has been equilibrated with the same buffer. After wash with 4 liters of 10 mM phosphate, pH 6.5, glucose-6-phosphate dehydrogenase is then eluted with 0.5 mM NADP$^+$ in the same buffer. The enzyme is eluted in fractions of about 200 ml.

Step 2. Fractions of active enzyme are concentrated to 3 ml and dialyzed extensively against 10 mM phosphate, pH 6.5, to remove the endogenous NADP$^+$. The dialyzate is loaded on a small 8-(6-aminohexyl)amino-2′,5′-ADP-Sepharose column (1 × 15 cm) in the same buffer. After quantitative adsorption, the affinity column is washed

TABLE II

PURIFICATION OF GLUCOSE-6-PHOSPHATE DEHYDROGENASE FROM MOUSE TESTIS

Fraction	Total protein[a] (mg)	Total activity (units)	Specific activity (units/mg protein)	Yield (%)	Purification (fold)
Crude homogenate	21,600	650	0.03	100	1
8-(6-Aminohexyl) amino-AMP-Sepharose	29.5	290	10	44.6	333
8-(6-Aminohexyl) amino-2',5'-ADP-Sepharose	0.95	150	158	23.1	5333

[a] Protein was determined by fluorescamine assays of P. Böhlen, S. Stein, W. Dairman, and S. Udenfriend [*Arch. Biochem. Biophys.* **155**, 213 (1973)] using bovine serum albumin as the protein standard.

with 50 ml of 50 mM phosphate, pH 6.5. The enzyme is eluted with a 0 to 1 mM NADP$^+$ linear gradient (50 ml by 50 ml) in the washing buffer. The flow rate is controlled at 15–20 ml/hr. Fractions of 2 ml are collected, and the enzyme activity appears at the end of the gradient. The purified enzyme has a specific activity of 158 units/mg.

The results of one such purification are summarized in Table II.

Properties

Glucose-6-phosphate dehydrogenase from mouse tissues can be purified to homogeneity by either procedure as judged by sodium dodecyl sulfate–acrylamide gel electrophoresis. These procedures should be applicable to the purification of this enzyme from other strains of mice.

Glucose-6-phosphate dehydrogenase from mouse kidney or testis is specific to NADP$^+$ and glucose 6-phosphate. Less than 1% activity was observed with NAD$^+$ as the coenzyme. Galactose 6-phosphate and 2-deoxyglucose 6-phosphate exhibit only 5% and 3%, respectively, of the activity with glucose 6-phosphate. K_ms for glucose 6-phosphate and NADP$^+$ are 50 and 10 μM, respectively. AMP and ATP show little or no inhibition at a nucleotide concentration of 5 mM. It remains to be established how the NADP$^+$-dependent glucose-6-phosphate dehydrogenase can be adsorbed to an AMP-affinity column. The mouse enzyme is a tetramer with a native molecular weight of 240,000. Antiserum raised in rabbits against purified mouse glucose-6-phosphate dehydrogenase partially cross-reacts with the human or guinea pig enzymes.

[44] Glucose-6-phosphate Dehydrogenase, Vegetative and Spore *Bacillus subtilis*

By Susumu Ujita and Kinuko Kimura

Assay Method

Principle. The spectrophotometoric assay can be used to measure glucose-6-phosphate dehydrogenase activity (EC 1.1.1.49) based on reduction of NADP+.

Reagents

Glycine (OH⁻) buffer, 200 mM, pH 9.4
MgCl$_2$, 200 mM
Glucose 6-phosphate, 8 mM
NADP+, 0.8 mM

Assay Procedure. Measurement is carried out in 1.0-cm light path cuvettes at 340 nm using a recording spectrophotometer. The complete assay mixture contains the following components at the final concentration: 20 mM glycine (OH⁻) buffer, pH 9.4, 20 mM MgCl$_2$, 0.8 mM glucose 6-phosphate, and 0.08 mM NADP+. All assay mixtures are incubated for 5 min at 25° before the reaction is initiated by adding a suitable amount of glucose-6-phosphate dehydrogenase.

Definition of Unit and Specific Activity. One unit of the enzyme activity is defined as that amount of enzyme causing the reduction of 1 μmol of NADP+ per minute. Specific activity is expressed as units per milligram of protein. Protein concentrations are estimated by the spectrophotometric method of Warburg and Christian or by the procedure of Lowry *et al.*

Cultivation and Harvest of Bacteria

Bacillus subtilis PCI 219 (ATCC 6633) is grown in a nutrient medium containing, per liter, glucose, 2 g; agar powder, 12 g; bouillon, 10 g at 37°. The media are sterilized by autoclaving 1 day before use. Vegetative cells at mid-exponential phase are harvested by using distilled water and by centrifugation and are washed with 0.1 M Tris (Cl⁻) buffer, pH 7.4, containing 1 mM EDTA (buffer I). Spores are harvested after about 2 weeks' cultivation. Collected spores are subjected to sonic treatment for 9 min and centrifuged for 30 min at 44,000 g. Vegetative cells and mother cells of sporangia are removed by this treatment, if present. The precipitated spores are stored at −20° until use.

Purification Procedure

Unless otherwise specified, all operations are carried out at 0–5° and centrifugation is performed at 20,000 g for 30 min.

Purification of Vegetative Cell Enzyme

Step 1. Preparation of the Crude Extract. The cell paste is suspended in 2.5 times its volume of buffer I and disrupted by sonication (20 kHz) for periods of 3 min with 5-min intervals for a total treatment time of 12 min, with ice-cooling. Unbroken cells and cell debris are removed by centrifugation.

Step 2. Protamine Sulfate Treatment. Under mechanical stirring, a 1% solution of protamine sulfate (an amount equivalent to 9% of the total protein present in the crude extract) is added dropwise to the crude extract. After an additional 15 min, the precipitate is removed by centrifugation.

Step 3. Ammonium Sulfate Fractionation. To the protamine sulfate supernatant, 28 g of solid ammonium sulfate per 100 ml (40%) is added. The suspension is centrifuged after stirring for 30 min at 0°. Another 14 g of ammonium sulfate (60%) is added, and the centrifugation is repeated after stirring for 30 min. The collected precipitate is dissolved in buffer I and dialyzed for 5 hr with four changes against 2 liters of 0.01 M Tris (Cl⁻) buffer, pH 7.4, containing 2 mM EDTA.

Step 4. Chromatography on DEAE-Cellulose. The dialyzed protein solution is applied to a DEAE-cellulose column (3.5 × 25 cm) equilibrated with buffer I. The column is washed with 500 ml of buffer I and then with 500 ml of buffer I containing 0.1 M NaCl. Enzyme elution is performed with a linear gradient from 0.1 to 0.4 M NaCl in a total volume of 1500 ml of buffer I.

Step 5. Chromatography on DEAE-Sephadex A-50. The combined active fraction from DEAE-cellulose is concentrated to 1–2 ml, and salt is also removed by means of a collodion membrane. This procedure is repeated by addition of buffer I, and the volume is finally adjusted to 30–40 ml. The concentrated enzyme solution is applied to a column (3 × 40 cm) of DEAE-Sephadex A-50 equilibrated with buffer I. The column is first developed with 700 ml of buffer I containing 0.1 M NaCl and then with 1000 ml of buffer I containing 0.175 M NaCl. Inactive proteins are almost completely removed by the second elution. The enzyme is subsequently eluted with 1500 ml of buffer I containing 0.2 M NaCl.

Step 6. Chromatography on Hydroxyapatite. The pooled fractions showing enzyme activity are concentrated, and the buffer system is changed to 0.001 M phosphate buffer, pH 6.8, using a collodion membrane. The concentrated solution is applied to a hydroxyapatite column (2 × 12 cm) equilibrated with 0.001 M phosphate buffer, pH 6.8, and is eluted stepwise with 150 ml each of 0.001 M, 0.0075 M, 0.015 M, 0.03 M, and 0.05 M phosphate buffer, pH 7.4. The bulk of the enzyme is eluted in 0.03 M buffer. The fractions containing high specific activity are concentrated, and the buffer system is changed to buffer I.

Step 7. Gel Filtration. The concentrated enzyme solution (2 ml) is applied to a Sephadex G-200 superfine column (3.5 × 48 cm) and eluted with buffer I at a rate of 5 ml/hr. Fractions of 2 ml are collected, and those containing high enzyme activity are pooled. This purified enzyme is homogeneous by SDS gel electrophoresis.

Partial Purification of Spore Enzyme

Step 1. Preparation of Crude Extract. Glass beads (mesh, 0.1 mm) are added to a spore suspension in a 200-ml beaker of a vibrogen cell mill (Edmund Bühler, Tübingen) (4 volumes of beads to 1 volume of suspended spores), and the spores are disrupted at the maximum speed for 10 min with water-cooling. The glass beads are removed by decantation and by centrifugation at 10,000 g for 15 min. Unbroken spores and spore debris are removed by centrifugation.

Procedure after Step 1. Further purification of spore enzyme is basically carried out as described for the purification of vegetative cell enzyme. The enzyme, which has a specific activity greater than 30, is prepared according to this method. But a high degree of purification similar to that obtained for the vegetative enzyme is difficult to attain, because spores contain less glucose-6-phosphate dehydrogenase than do vegetative cells.

A typical purification is summarized in the table.

PURIFICATION OF GLUCOSE-6-PHOSPHATE DEHYDROGENASE FROM VEGETATIVE CELLS

Fraction	Protein (mg)	Total activity (units)	Specific activity (units/mg protein)	Yield (%)	Purification (fold)
Crude extract[a]	30,000	850	0.028	100	1.0
Protamine sulfate	26,400	830	0.031	98	1.1
$(NH_4)_2SO_4$, 40–60%	11,200	800	0.071	94	2.5
DEAE-cellulose	969	786	0.81	93	28.9
DEAE-Sephadex A-50	43.4	620	14.3	73	511
Hydroxyapatite	7.5	370	49.3	44	1761
Sephadex G-200	3.8	255	67.1	30	2396

[a] From 500 g of wet cells.

Properties

K_m Values and Optimum pH. Both vegetative cell and spore enzymes have the same K_m values with respect to substrate and coenzyme, which is calculated to be $6.7 \times 10^{-6} M$ for NADP$^+$ and $7.5 \times 10^{-5} M$ for glucose 6-phosphate. The optimum pH is observed to be around pH 9.2 for both vegetative cell and spore enzymes as reported by Marquet and Dedonder.[1] This value is similar to values for the enzymes from *B. licheniformis.*[2]

Activation by Divalent Cations. Both vegetative cell and spore enzymes are activated by Mg^{2+}, Mn^{2+}, and Ca^{2+}. Maximum activation by these cations is attained at the concentration of 30–40 mM.

Coenzyme Specificity. This enzyme utilizes NADP$^+$ and NAD$^+$ as an electron acceptor. The NADP$^+$: NAD$^+$ activity ratio of this enzyme is about 1.5 in the absence of cations, and this ratio increases when cations are present.

Molecular Weight. The molecular weight of the minimum active molecular form is about 120,000 (measured by sucrose density gradient centrifugation), and both vegetative cell and spore enzymes are comparable. This enzyme may aggregate in the presence of a neutral salt [e.g., (NH$_4$)$_2$SO$_4$ or KCl] and after long periods of storage. The molecular weight measured by gel filtration appears larger than that obtained by sucrose density gradient centrifugation. Therefore, the molecular weight of 350,000, as previously reported[3] may be plausible. The molecular weights of subunits are 58,000 by SDS gel electrophoresis. This enzyme is assumed to be a dimer.

[1] M. Marquet and R. Dedonder, *C.R. Hebd. Seances Acad. Sci.* **241**, 1090 (1955).
[2] D. Opheim and R. W. Bernlohr, *J. Bacteriol.* **116**, 1150 (1973).
[3] S. Ujita and K. Kimura, *J. Biochem.* (*Tokyo*) **77**, 197 (1975).

[45] D-Glucose-6-phosphate Dehydrogenases from *Pseudomonas fluorescens*

By P. MAURER, D. LESSMANN, and G. KURZ

β-D-Glucose 6-phosphate + NAD(P)$^+$ $\rightleftarrows$

6-phospho-D-glucono-1,5-lactone + NAD(P)H + H$^+$

D-Glucose 6-phosphate is metabolized in *Pseudomonas fluorescens* by two pathways, the Entner–Doudoroff pathway and the hexose monophosphate pathway, both starting with the dehydrogenation of

METHODS IN ENZYMOLOGY, VOL. 89

D-glucose 6-phosphate. These different pathways are assumed to be regulated by two distinct D-glucose-6-phosphate dehydrogenases (D-glucose 6-phosphate : NAD(P)$^+$ 1-oxidoreductase, EC 1.1.1.49). The D-glucose-6-phosphate dehydrogenase assigned to the Entner–Doudoroff pathway (Entner–Doudoroff enzyme) is subject to complex metabolic regulation by both inductive and allosteric control,[1] whereas no regulatory properties have been demonstrated so far for the dehydrogenase that is part of the hexose monophosphate pathway (Zwischenferment).[2]

Evaluation of Enzyme Activities

Both D-glucose-6-phosphate dehydrogenases from *P. fluorescens* show specificity for NAD$^+$ as well as for NADP$^+$, but to a different extent. The exact portion which each of the D-glucose-6-phosphate dehydrogenases contributes to the activities measured with NAD$^+$ and NADP$^+$ may be calculated from the ratios of reaction rates with NAD$^+$ to NADP$^+$ determined for the isolated enzymes. In the presence of the two enzymes the measured activities are given by:

$$U_{NAD^+}^{meas} = U_{NAD^+}^{EDE} + U_{NAD^+}^{ZF} \tag{1}$$

$$U_{NADP^+}^{meas} = U_{NADP^+}^{EDE} + U_{NADP^+}^{ZF} \tag{2}$$

For the Entner–Doudoroff enzyme (EDE) the ratio of reaction rates with NAD$^+$ to NADP$^+$ is

$$U_{NAD^+}^{EDE}/U_{NADP^+}^{EDE} = 1.29 \tag{3}$$

and for the Zwischenferment (ZF):

$$U_{NAD^+}^{ZF}/U_{NADP^+}^{ZF} = 0.55 \tag{4}$$

These equations allow the derivation of expressions for the contributions of the two enzymes to the activities measured; for the Entner–Doudoroff enzyme:

$$U_{NAD^+}^{EDE} = 1.743\ U_{NAD^+}^{meas} - 0.959\ U_{NADP^+}^{meas} \tag{5}$$

$$U_{NADP^+}^{EDE} = 1.351\ U_{NAD^+}^{meas} - 0.743\ U_{NADP^+}^{meas} \tag{6}$$

and for the Zwischenferment:

$$U_{NAD^+}^{ZF} = 0.959\ U_{NADP^+}^{meas} - 0.743\ U_{NAD^+}^{meas} \tag{7}$$

$$U_{NADP^+}^{ZF} = 1.743\ U_{NADP^+}^{meas} - 1.351\ U_{NAD^+}^{meas} \tag{8}$$

[1] D. Lessmann, K.-L. Schimz, and G. Kurz, *Eur. J. Biochem.* **59**, 545 (1975).
[2] D. Lessmann and G. Kurz, unpublished observation.

Assay Method

Principle. The assays are based on the spectrophotometric determination of NAD(P)H formed with D-glucose 6-phosphate as substrate at high pH conditions.

Reagents

Tris-HCl buffer, 50 mM, pH 8.9
NAD$^+$, 30 mM, in Tris-HCl buffer, pH 8.9
NADP$^+$, 30 mM, in Tris-HCl buffer, pH 8.9
D-Glucose 6-phosphate, 200 mM, in Tris-HCl buffer, pH 8.9

Enzyme. The enzyme solution is diluted with the appropriate buffers to give a solution with an activity between 0.2 and 0.5 units/ml.

Procedure. A mixture of 0.1 ml of NAD$^+$ or NADP$^+$ solution, 0.1 ml of D-glucose 6-phosphate solution, and 0.7 ml of Tris-HCl buffer is incubated at 30° in a spectrophotometer cell with a 1.0-cm light path. The reaction is started by the addition of 0.1 ml of the enzyme solution. The extinctions at 340 nm or 366 nm are measured at 15-sec intervals.

Definition of Units and Specific Activity. One unit of enzyme is defined as the amount that catalyzes the reduction of 1 μmol of NAD$^+$ or NADP$^+$ per minute under the above conditions. Specific activity is expressed as units per milligram of protein.

Organism and Cultivation

The strain of *Pseudomonas* used was obtained from Boehringer Mannheim GmbH (Mannheim, F.R.G.) and has been classified as *P. fluorescens* biotype E or F.[1] The bacteria were maintained in lyophilized stock cultures.

The cultivation of *P. fluorescens* on a pilot-plant scale is performed on a three-step cultivation scheme.[1]

First Culture. One-cell cultures grown on yeast extract–peptone plates or lyophilized cells are inoculated into 1000-ml flasks holding 200 ml of minimal medium containing 33 mM potassium sodium phosphate buffer, pH 6.8, and, per liter, 1 g of NH$_4$Cl, 0.5 g of MgSO$_4$ · 7 H$_2$O, 10 mg of CaCl$_2$ · 2 H$_2$O, 2.5 mg of FeSO$_4$ · 7 H$_2$O, and 2 g of D-glucose as carbon source. The cultures are incubated on a reciprocal shaker for 15 hr at 30°.

Second Culture. The second culture is grown in a 14.5-liter fermentor (NBS Scientific, New Brunswick, New Jersey) holding 10 liters of a minimum medium containing, per liter, 6 g of Na$_2$HPO$_4$ · 12 H$_2$O, 2.3 g of KH$_2$PO$_4$, 1 g of NH$_4$Cl, 0.5 g of MgSO$_4$ · 7 H$_2$O, 0.1 g of ferric ammonium citrate, 10 mg of $CaCl_2 \cdot 2 H_2O$, and 3 g of glucose. The inoculum is 6%, and the culture is agitated (200 rpm) for 8 hr at 30° and supplied with an air flow of 1 volume per volume of culture per minute.

Pilot-Plant Culture. Two hundred liters of minimum medium containing, per liter, 6 g of $Na_2HPO_4 \cdot 12 H_2O$, 2.3 g of KH_2PO_4, 4 g of KCl, 3 g of NH_4Cl, 0.5 g of $MgSO_4 \cdot 7 H_2O$, 0.2 g of ferric ammonium citrate, 30 mg of $CaCl_2 \cdot 2 H_2O$, and 7.5 g of glucose are inoculated with 10 liters of the second culture in a fermentor (Wehrle-Werk AG, Emmendingen, F.R.G.). Under mild agitation (120 rpm) and an air supply of 0.6 volume per volume of culture per minute, the culture is grown at 30° for 15 hr. The pH is controlled and automatically adjusted to pH 6.8 with a solution of concentrated NH_3. The bacteria are harvested with the aid of a continuous-flow centrifuge (type AS 16 RR, 15,500 rpm, Sharples France).

The cell paste is lyophilized. The lyophilized bacteria may be stored at −15° for several months without loss of activity.

Entner–Doudoroff Enzyme

Isolation

The isolation procedure for the Entner–Doudoroff enzyme has been designed to allow the purification of the Zwischenferment from the same cell-free extract. In the first two steps both enzymes are enriched together and they are separated in the third step, the polyethylene glycol fractionation.

If only the isolation of the Entner–Doudoroff enzyme is planned, it is not necessary to perform the assays with NAD^+ as well as with $NADP^+$, because only about 3% of the activity measured with NAD^+ in the cell-free extract is due to the activity of the Zwischenferment[1] and will be separated after the third step. Subsequent to the digestion of nucleic acids (step 1) all solutions are made 20 mM in 2-mercaptoethanol, and all operations are performed at 0–4°.

Step 1. Preparation of Cell-Free Extract. Lyophilized bacteria (540 g) are ground in a ball mill and suspended in 50 mM potassium sodium phosphate buffer, pH 7.0 (5000 ml); 9.25 ml of buffer are used per gram of cells. The suspension is homogenized with an Ultra Turrax Type 45 (Janke und Kunkel KG, Staufen, F.R.G.) intermittently for a total of 5 min. The temperature should not exceed 30° during this operation. Ribonuclease A (100 mg) and deoxyribonuclease I (100 mg) are added, and the suspension is stirred at room temperature. The pH is controlled and readjusted to pH 7.0 with a solution of 4 M K_2HPO_4. After 4 hr the suspension is made 20 mM in 2-mercaptoethanol and subsequently centrifugated at 20,000 g and 4° for 90 min. A yellowish, turbid cell-free extract (about 4000 ml) is obtained.

Step 2. Ammonium-Sulfate Precipitation. Solid $(NH_4)_2SO_4$ (290 g/l solution) is added slowly to the cell-free extract, and the suspension is mildly stirred overnight. The precipitate is collected by centrifugation at 20,000 g for 60 min and subsequently dissolved in 2000 ml of 50 mM potassium sodium phosphate buffer, pH 7.0.

Step 3. Polyethylene Glycol Fractionation. The protein solution of the preceding step is adjusted to pH 7.0 by the addition of 4 M K_2HPO_4 and diluted with 50 mM potassium sodium phosphate buffer, pH 7.0, to a final D-glucose-6-phosphate dehydrogenase activity of 14 units/ml solution. Solid polyethylene glycol 6000 (80 g/l solution) is added to the stirred solution within 30 min, and stirring is continued for 1 hr. After removal of the precipitate by centrifugation at 20,000 g for 6 min, additional polyethylene glycol 6000 (80 g/l solution) is added within 30 min to the stirred supernatant. The suspension is stirred for 1 hr, then the precipitate is collected by centrifugation at 20,000 g for 60 min and subsequently dissolved in 300 ml of 10 mM potassium sodium phosphate buffer, pH 6.2, and dialyzed overnight against 20 liters of the same buffer.

Step 4. Bulk Separation on CM-Sephadex. The dialyzed enzyme solution is added, with mild stirring, to about 400 ml of CM-Sephadex C-50, equilibrated with 10 mM potassium sodium phosphate buffer, pH 6.2. One milliliter of CM-Sephadex C-50 gel is used per 100 units of D-glucose-6-phosphate dehydrogenase activity. By the addition of a solution of 10 mM KH_2PO_4, the pH is cautiously lowered, usually to pH between 5.9 and 6.0, until the adsorption of D-glucose-6-phosphate dehydrogenase is just complete. The CM-Sephadex C-50 gel is then collected under mild suction on a Büchner funnel and washed directly on the funnel with a total of 1500 ml of 10 mM potassium sodium phosphate buffer, pH 6.2, until the gel is nearly colorless. Subsequently, the washing procedure is repeated with 1000 ml of 20 mM potassium sodium phosphate buffer, pH 6.2. For elution of D-glucose-6-phosphate dehydrogenase, the gel is suspended in 200 ml of 20 mM potassium sodium phosphate buffer, pH 6.6, and this is stirred for 15 min. The eluted protein is removed by means of the Büchner funnel. This elution process is repeated until no more D-glucose-6-phosphate dehydrogenase is eluted, usually three times. The eluates are combined, and the enzyme is precipitated with solid $(NH_4)_2SO_4$ (300 g per liter of 1 solution). The precipitate is collected by centrifugation at 20,000 g for 30 min and dissolved in 25 ml of 20 mM imidazole–HCl buffer, pH 6.6, containing 0.2 M NaCl and 20 mM 2-mercaptoethanol. In order to remove the residual $(NH_4)_2SO_4$, the en-

TABLE I

PURIFICATION OF D-GLUCOSE-6-PHOSPHATE DEHYDROGENASE (ENTNER–DOUDOROFF ENZYME) FR
Pseudomonas fluorescens [a]

Step	Volume (ml)	Activity[b] (units/ml)	Protein (mg/ml)	Specific activity[b] (units/mg protein)	Yield (%)	Purific (fo]
1. Preparation of cell-free extract	4000	13.66	42.6	0.32	100	—
2. Ammonium-sulfate precipitation	2100	24.72	51.5	0.48	95	1
3. Polyethylene glycol fractionation	400	112.3	61.9	1.81	82.2	5
4. Bulk separation on CM-Sephadex	30	1024.8	11.2	91.5	56.3	286
5. DEAE-Sephacel chromatography	25	890.4	2.8	318	40.7	994

[a] Typical data obtained with 540 g of lyophilized bacteria.
[b] Related to the Entner–Doudoroff enzyme.

zyme solution is dialyzed for at least 12 hr against 5 liters of the same buffer with two changes of dialyzate.

Step 5. DEAE-Sephacel Chromatography. The dialyzed enzyme solution from the preceding step (about 30 ml) is concentrated with the aid of an Amicon ultrafiltration cell Model 52 [Amicon GmbH, Witten (Ruhr), F.R.G.] to a volume of about 20 ml, using a PM-30 membrane, and subsequently applied to a column (100 × 3 cm) of DEAE-Sephacel equilibrated with 20 mM imidazole–HCl buffer, pH 6.6, containing 0.2 M NaCl and 20 mM 2-mercaptoethanol. The elution of the enzyme is performed with exactly the same buffer at a flow rate of 20 ml/hr. Fractions of 20 ml are collected, and those revealing no impurities by disc polyacrylamide gel electrophoresis are combined and concentrated to a final volume of about 20 ml with the aid of Amicon ultrafiltration cells Model 402 and Model 52, using PM-30 membranes.

A summary of a typical purification procedure is given in Table I.

Properties of the Entner–Doudoroff Enzyme

Homogeneity. The purified enzyme with a specific activity of 318 units/mg is homogeneous by disc gel electrophoresis and by sodium dodecyl sulfate gel electrophoresis.

Stability and pH Stability. The purified enzyme is stored at 4° as a precipitate in 3.2 M ammonium sulfate solution, containing 20 mM 2-mercaptoethanol, in the neutral pH range. No decrease in enzyme activity could be observed within 12 months. The enzyme is stable in the range of pH 6.0 to 10.0.[1]

Structure. The Entner–Doudoroff enzyme from *P. fluorescens* has, in the neutral pH range in dilute solutions, a molecular weight of 220,000. This molecular-weight form is involved in an association–dissociation equilibrium with aggregates of higher molecular weight. The enzyme is composed of subunits (M_r 55,000 each) that are identical and not covalently linked.[3] By the active-enzyme centrifugation method,[4] the molecular weight of the smallest enzymically active unit has been determined to be only 110,000. This molecular form contains two subunits in the oligomer and has two binding sites for the substrate and two for the allosteric effector.[5]

Kinetic Properties. The Entner–Doudoroff enzyme uses NAD^+ and $NADP^+$ as cosubstrate. Under the conditions of the standard assay, the rate of the reaction with $NADP^+$ is only 77.5% of the rate with NAD^+, giving a ratio of reaction rates with NAD^+ to $NADP^+$ of 1.29. The dependence of the reaction velocity on the concentration of NAD^+ or $NADP^+$ in presence of D-glucose 6-phosphate shows hyperbolic curves. The respective values of the apparent Michaelis constants, in 50 mM Tris-HCl buffer, pH 8.9, with a constant concentration of 20 mM D-glucose 6-phosphate, at 30° for NAD^+ and $NADP^+$ are $K_{app} = 0.15$ mM and $K_{app} = 0.36$ mM.

The enzyme shows homotropic effects for D-glucose 6-phosphate and is strongly inhibited by ATP.[1]

For a further description of the properties of the Entner–Doudoroff enzyme from *P. fluorescens,* see Lessmann *et al.*[1]

Zwischenferment

Isolation

The Zwischenferment of *P. fluorescens* contains reactive sulfhydryl groups, and in order to avoid their oxidation all solutions are made 20 mM in 2-mercaptoethanol. All operations are performed at 0–4° unless stated otherwise.

The first two isolation steps are the same as for the Entner–Doudoroff enzyme.

Step 3. Polyethylene Glycol Fractionation. After removal of the precipitate containing the Entner–Doudoroff enzyme by centrifugation, polyethylene glycol 6000 is added within 60 min a third time to the stirred supernatant (530 g/l solution). After this addition, the syrupy solution is

---

[3] P. Maurer, D. Lessmann, and G. Kurz, unpublished observation.
[4] R. Cohen and M. Mire, *Eur. J. Biochem.* **23,** 267 (1971).
[5] P. Maurer and G. Kurz, unpublished observation.

stirred for at least 12 hr. The precipitate formed is collected by centrifugation at 20,000 g for 90 min and subsequently suspended in 300 ml of 10 mM potassium sodium phosphate buffer, pH 5.9, and dialyzed for 24 hr against 30 liters of the same buffer, with two changes of dialyzate. Any precipitate is removed by centrifugation at 48,000 g for 30 min.

Step 4. Bulk Separation on CM-Sephadex C-50. The clear solution from step 3 (about 540 ml) is added, with mild stirring, to about 500 ml of CM-Sephadex C-50, equilibrated with 10 mM potassium sodium phosphate buffer, pH 5.9; 1 ml of gel is used per 1.5 units of D-glucose-6-phosphate dehydrogenase activity. Subsequently the pH value is cautiously lowered to pH 5.4 by the careful addition of a 10 mM acetic acid solution. In order to obtain the complete adsorption of the enzyme, stirring is continued for 5–10 min. Then the CM-Sephadex C-50 gel is collected under mild suction on a Büchner funnel and washed directly on the funnel with 300 ml of 10 mM potassium sodium phosphate buffer, pH 5.9. For elution of D-glucose-6-phosphate dehydrogenase activity, the gel is suspended in 200 ml of 20 mM potassium sodium phosphate buffer, pH 7.0, and the suspension is stirred for 15 min. The eluted protein is removed by means of the Büchner funnel, and the elution process is repeated until no more D-glucose 6-phosphate activity is eluted, usually three times. The eluates are combined and concentrated to about 100 ml with the aid of an Amicon ultrafiltration cell Model 402 [Amicon GmbH, Witten (Ruhr), F.R.G.] using a PM-30 membrane.

Step 5. Hydroxyapatite Chromatography. The enzyme solution obtained from step 4 is equilibrated against 1.5 mM potassium sodium phosphate buffer, pH 6.8, containing 0.1 M NaCl, by chromatography on a column (40 × 5 cm) of Sephadex G-25 (medium grade) and subsequently applied to a hydroxyapatite column (40 × 3 cm) preequilibrated with the same buffer. The column is washed with four column volumes of 30 mM potassium sodium phosphate buffer, pH 6.8, containing 0.1 M NaCl. The elution of the enzyme is performed by a phosphate buffer gradient: starting buffer 30 mM potassium sodium phosphate buffer, pH 6.8, containing 0.1 M NaCl; limiting buffer 80 mM potassium sodium phosphate buffer, pH 6.8, containing 0.1 M NaCl; total gradient volume 4000 ml. Fractions of 25 ml are collected at a flow rate of 50 ml/hr. The fractions containing D-glucose-6-phosphate dehydrogenase activity are combined (about 1000 ml), and the solution is concentrated to a volume of about 100 ml, again with the aid of the Amicon ultrafiltration cell Model 402, using a PM-30 membrane.

Step 6. DEAE-Sephadex Chromatography. The enzyme solution is equilibrated against 10 mM piperazine-HCl buffer, pH 6.0, containing 0.15 M NaCl, by passage through a column (40 × 5 cm) of Sephadex G-25 (medium grade) and then applied to a column (40 × 5 cm) of DEAE-Sephadex A-50 preequilibrated with the same buffer. After four column volumes of buffer are passed through the column, the enzyme is eluted with a linear NaCl gradient, 0.15 to 0.40 M NaCl, in the same buffer; total gradient volume 5000 ml. Fractions of 25 ml are collected at a flow rate of 50 ml/hr. The Zwischenferment is completely separated by this procedure from the Entner–Doudoroff enzyme still present after the hydroxyapatite chromatography and emerges as the first peak of activity in the range of 0.22 to 0.24 M NaCl. The fractions containing the Zwischenferment are combined (about 500 ml) and concentrated to a volume of about 5 ml by ultrafiltration using PM-30 membranes.

Step 7. Preparative Electrophoresis. Preparative disc gel electrophoresis is performed in vertical slab gels with end elution using the gel system according to Davis.[6] With the equipment Ultraphor (Colora Messtechnik GmbH, Lorch/Württ., F.R.G.), a separation chamber for gels 137 mm wide and 7.5 mm thick is suitable. A separation gel with a height of 70 mm and a spacer gel of about 30 mm is used. For elution, 125 mM Tris-HCl buffer, pH 8.1, is employed, and 375 mM Tris-HCl buffer, pH 8.1, is used as a counter cell buffer. In order to exchange the buffer, the concentrated enzyme solution is passed through a column of Sephadex G-25 (medium grade) equilibrated with 58.5 mM Tris-H$_3$PO$_4$ buffer, pH 6.9, and the protein solution obtained (7–9 ml) is made 15% in glycerol. With the aid of a peristaltic pump, about 5 ml of 50 mM Tris-H$_3$PO$_4$ buffer, pH 6.9, made 5% in glycerol, is layered on the spacer gel to protect the enzyme solution against the electrode buffer. Subsequently, the protein solution, 15% in glycerol, is layered between the spacer gel and the protecting zone. During concentration the current is kept constant at 40 mA; during separation, at 50 mA. End elution is performed at a flow rate of 40 ml/hr, and fractions of 2 ml are collected. Enzyme fractions containing no impurities are combined and concentrated to a volume of about 5 ml by ultrafiltration using PM-30 membranes.

A summary of a typical purification procedure is given in Table II.

Properties of Zwischenferment

Homogeneity. The purified enzyme with a specific activity of 96 units/mg with NADP$^+$ as cosubstrate is homogeneous by disc gel electrophoresis and by sodium dodecyl sulfate electrophoresis.

Stability and pH Stability. The purified enzyme is stable for months as precipitate in 3.2 M ammonium sulfate solution, containing 20 mM 2-mercaptoethanol, in the neutral pH range at 4°. The enzyme is stable in the range from pH 6.5 to 10.0.

[6] B. J. Davis, *Ann. N. Y. Acad. Sci.* **121**, 404 (1964).

TABLE II

PURIFICATION OF D-GLUCOSE-6-PHOSPHATE DEHYDROGENASE (ZWISCHENFERMENT)
FROM *Pseudomonas fluorescens*[a]

Step	Volume (ml)	Activity (units/ml) measured with		Protein (mg/ml)	Specific activity[b] (units/mg protein) calculated for		Total activity[b] calculated for NADP+ (units)	Yield (%)
		NAD+	NADP+		NAD+	NADP+		
1. Preparation of cell-free extract	4000	14.0	11.21	42.6	0.008	0.0145	2473	100
2. Ammonium-sulfate precipitation	2100	25.08	20.08	51.5	0.012	0.021	2336	94.5
3. Polyethylene glycol fractionation	540	10.65	9.62	28.3	0.048	0.087	1329	53.7
4. Bulk separation on CM-Sephadex	100	35.30	32.91	24.2	0.22	0.40	967	39.1
5. Hydroxyapatite chromatography	100	28.46	26.57	3.1	1.40	2.54	796	32.2
6. DEAE-Sephadex chromatography	5	71.67	130.30	3.8	18.86	34.29	651	26.3
7. Preparative electrophoresis	5	49.6	90.25	0.94	52.80	96.02	455	18.4

[a] Typical data obtained with 540 g of lyophilized bacteria.
[b] Related to the Zwischenferment.

Structure. The Zwischenferment of *P. fluorescens* has a molecular weight of 265,000 and is composed of four subunits, each of a molecular weight of 65,000. No tendency to form aggregates of higher molecular weight has been observed.

Kinetic Properties. The Zwischenferment of *P. fluorescens* uses NAD+ as well as NADP+ as cosubstrate, preferring the latter. Under the conditions of the standard assay, the rate of reaction with NAD+ is 55% of the rate with NADP+, giving a ratio of NAD+ : NADP+ of 0.55. The reaction rate obeys with all substrates the Michaelis–Menten rate law. The limiting Michaelis constant for NAD+ is $K_a = 0.72$ mM, and for NADP+ $K_a = 0.63$ mM; the limiting Michaelis constant for D-glucose 6-phosphate in presence of NAD+ as the fixed cosubstrate is $K_b = 1.33$ mM; and in presence of NADP+, $K_b = 0.57$ mM.

For a comparison of the two D-glucose-6-phosphate dehydrogenases from *P. fluorescens* with enzymes from other sources, see the excellent review of Levy.[7]

[7] H. R. Levy, *Adv. Enzymol.* **48**, 97 (1979).

[46] Glucose-6-phosphate Dehydrogenase from *Methylomonas* M15

By ROLF A. STEINBACH, HORST SCHÜTTE, and HERMANN SAHM

D-Glucose 6-phosphate + NAD$^+$(NADP$^+$) $\rightleftharpoons$ D-gluconate 6-phosphate + NADH
(NADPH) + H$^+$

Studies have shown that in many methanol-utilizing bacteria the methanol is successively oxidized to carbon dioxide via formaldehyde and formate. However, in the obligate methanol-utilizing bacterium *Methylomonas* M15, formaldehyde dehydrogenase and formate dehydrogenase activities were not detected. In this bacterium formaldehyde is mainly oxidized to CO_2 via the dissimilatory ribulose monophosphate cycle[1]; therefore, *Methylomonas* M15 contains glucose-6-phosphate dehydrogenase of high specific activity. This enzyme and 6-phosphogluconate dehydrogenase together with the key enzymes of the ribulose monophosphate cycle—3-hexulosephosphate synthase and 3-hexulosephosphate isomerase—constitute a cyclic mechanism for oxidation of formaldehyde to CO_2.[2,3]

Assay Method

Principle. Glucose-6-phosphate dehydrogenase is measured spectrophotometrically by following the rate of NADH or NADPH formation at 340 nm in the presence of saturating amounts of glucose 6-phosphate and NAD or NADP.

Reagents

Tris-HCl buffer, 60 mM, adjusted to pH 9.0
Glucose 6-phosphate, 45 mM
NAD, 45 mM
NADP, 3 mM

Procedure. The assay is carried out at 30° in a recording spectrophotometer. The reaction mixture consists of 1.0 ml of buffer, 0.1 ml of NAD or NADP, limiting amounts of enzyme and distilled water to give a total volume of 2.9 ml. The reaction is carried out in a 3.0-ml quartz cell with a 1.0-cm light path. After sufficient warming of the solution, the

[1] R. A. Steinbach, H. Sahm, and H. Schütte, *Eur. J. Biochem.* **87**, 409 (1978).
[2] J. Colby and L. J. Zatman, *Biochem. J.* **148**, 513 (1975).
[3] T. Strøm, T. Ferenci, and J. R. Quayle, *Biochem. J.* **144**, 465 (1974).

reaction is initiated by the addition of 0.1 ml of glucose 6-phosphate and thorough mixing. The rate of absorbance change at 340 nm is followed for at least 2 min, and activities are calculated by using $\epsilon = 6.22$ cm^2/μmol for NADH or NADPH at 340 nm.

Definition of Enzyme Unit and Specific Activity. One unit of enzyme activity is defined as the amount of enzyme necessary to reduce 1 μmol of NAD or NADP per minute under the conditions of the assay. Specific activity is expressed as units per milligram of protein.

Production of Glucose-6-phosphate Dehydrogenase

Cultures of *Methylomonas* M15 (DSM 580) may be obtained from the German Collection for Microorganisms, Göttingen. This bacterium is an obligate methylotrophic strain that can grow only on methanol as a carbon and energy source. Cells are grown in the following basal medium: 3.75 g of KH_2PO_4, 2.5 g of Na_2HPO_4, 4.0 g of $(NH_4)_2SO_4$, 0.5 g of $MgSO_4 \cdot 7$ H_2O, 25 mg of $Ca(NO_3)_2 \cdot 4$ H_2O, 5 mg of $FeSO_4 \cdot 7$ H_2O, 5 mg of $ZnSO_4 \cdot 7$ H_2O, 10 ml of methanol in 1 liter of distilled water, pH 7.0.[4] For large-scale cultivation the organism is grown in a 200- or 500-liter fermentor at a stirrer speed of 1000 rpm, a temperature of 33°, and an aeration rate of 0.5 liter of air per liter of working volume per minute. The pH is maintained at 7.0 by automatic addition of 4 M NH_4OH. The cells are harvested at the end of the exponential growth phase by a continuous-flow centrifuge and stored at $-20°$ until use.

Purification Procedure

All operations are carried out in the cold room (4°).

Step 1. Preparation of Crude Extract. The frozen cells (250 g wet weight) are softened overnight and suspended with 1 liter of 0.01 M potassium phosphate buffer (pH 7.5) using a Waring blender. The cells are passed twice through a Manton–Gaulin homogenizer at 600 kp/cm^2. After 8 mg deoxyribonuclease are mixed into the solution for 30 min to reduce the viscosity, cell debris are removed by centrifugation at 40,000 g for 2 hr using a Sorvall centrifuge RC-2B and the SS 34 rotor.

Step 2. N-Cetyl-N,N,N-trimethylammonium Bromide Precipitation. To the crude extract obtained by step 1, a 10% solution of *N*-cetyl-*N*,*N*,*N*-trimethylammonium bromide in water is added with stirring to give a final concentration of 1.5% (v/v). The pH of the suspension is maintained at 7.5 by adding ammonia if necessary. The resulting precipi-

---

[4] H. Sahm and F. Wagner, *Eur. J. Appl. Microbiol.* **2**, 147 (1975).

tate is removed by centrifugation at 20,000 g for 1 hr. The clear supernatant is dialyzed against 10 mM Tris-HCl buffer, pH 7.5.

Step 3. DEAE-Cellulose Column Chromatography. A column (10 × 100 cm) is packed with DEAE-cellulose and equilibrated with 10 mM Tris-HCl buffer (7.5). The enzyme obtained by step 2 is applied to the column, and the column is washed with 15 liters of equilibration buffer. Elution is carried out by a linear gradient between 8 liters of equilibration buffer and 8 liters of the same buffer containing 0.3 M sodium chloride. The flow rate is 120 ml/hr, and 20-ml fractions are collected. The peak fractions are combined, concentrated by ultrafiltration through an Amicon hollow-fiber cartridge type H1P10 and dialyzed against 10 mM Tris-HCl buffer (pH 9.0).

Step 4. DEAE-Sephadex Column Chromatography. The concentrated enzyme solution from step 3 is applied to a column packed with DEAE-Sephadex A-50 (5 × 50 cm) and equilibrated with 10 mM Tris-HCl buffer (pH 9.0). The column is washed with 1.5 liters of equilibration buffer, and elution is carried out by a 4-liter linear gradient between 10 mM Tris-HCl buffer (pH 9.0) and 10 mM Tris-HCl buffer (pH 9.0) containing 0.3 M sodium chloride at a flow rate of 80 ml/hr. The enzyme is eluted with about 0.2 M sodium chloride concentration. The active fractions are concentrated by ultrafiltration using an Amicon PM-10 membrane.

Step 5. Affinity Chromatography on Blue Dextran–Sepharose Cl-6B. The enzyme of step 4 is dialyzed against 10 mM Tris-HCl buffer (pH 7.5) and applied to a Blue Dextran–Sepharose Cl-6B column (2.4 × 16 cm) equilibrated with 10 mM Tris-HCl buffer (pH 7.5). The column is washed with starting buffer until the protein concentration in the effluent becomes negligible, than a linear gradient is produced with 400 ml of starting buffer and 400 ml of 100 mM Tris-HCl buffer pH 7.5 containing 0.1 mM NADP. Fractions of 6.5 ml are collected and tested for glucose-6-phosphate dehydrogenase activity. The enzyme is eluted at approximately 0.02 mM NADP. The enzyme-containing fractions are combined, concentrated by ultrafiltration through an Amicon PM-10 membrane, and stored in portions at −20°.

The purification procedure is summarized in the table.

Remarks. DNase treatment in the first step can be omitted if a larger volume is used, since N-cetyl-N,N,N-trimethylammonium bromide is useful also for the removal of nucleic acids from the crude extract. Furthermore, it is possible to use streptomycin sulfate, protamine sulfate, manganous salts, or polyethyleneimine for this purpose, but the conditions for precipitation have to be adjusted. A sufficient purification of the enzyme is important prior to the affinity chromatography. The Blue Sepharose Cl-6B used here is a cross-linked agarose gel Sepharose Cl-6B

PURIFICATION OF GLUCOSE-6-PHOSPHATE DEHYDROGENASE FROM *Methylomonas* M15

Fraction	Volume (ml)	Total protein (mg)	Total activity (U)	Specific activity (units/mg of protein)	Yield (%)	Purifica factc
Crude extract	950	9560	6120	0.64	100	1
N-Cetyltrimethylammonium bromide fraction	230	4260	6080	1.43	99	2.
DEAE-cellulose	200	820	5110	6.2	83	9.
DEAE-Sephadex A-50	110	165	4365	26.5	71	41.
Blue Dextran–Sepharose CL-6B	18	32.5	3980	122.8	65	192

containing dextran covalently bound to Cibacron Blue F3G-A. This material binds preferentially proteins that have an affinity for NAD^+ and $NADP^+$. Several other, similar, media have become commercially available in the meantime and could be used. It may be necessary to adjust the elution buffer for optimal performance.

Properties

Molecular Weight and Subunit Structure. The molecular weight of the glucose-6-phosphate dehydrogenase was determined by the sedimentation equilibrium method. Assuming a partial specific volume of $0.750 \text{ cm}^3/\text{g}$, a molecular weight of $108,000 \pm 5000$ was calculated from the data. Polyacrylamide gel electrophoresis in the presence of sodium dodecyl sulfate indicated that the enzyme is a dimer, probably composed of two identical subunits with a molecular weight of $55,000 \pm 3000$.

Kinetic Properties. The enzyme exhibits activity with either NAD or NADP as electron acceptor. Under standard assay conditions the reaction rate is about 30% higher with NAD than with NADP. The binding of coenzymes is independent of glucose 6-phosphate and also the binding of glucose 6-phosphate is independent of the concentration of NAD or NADP. The K_m values are 0.2 mM for NAD, 0.014 mM for NADP, and with respect to glucose 6-phosphate, 0.29 mM and 0.11 mM, respectively.

Effect of pH. The enzyme is stable at a pH range of 7.5–9.0 and has its maximum activity at pH 9.0 and 45°.

Inhibitors. Since in *Methylomonas* M15 glucose-6-phosphate dehydrogenase is a key enzyme of the dissimulatory and the assimilatory pathways of the methanol metabolism, a possible regulatory effect of a number of metabolites on the enzyme activity was investigated.[1] Although the enzyme activity remains unaltered after the addition of acetyl-CoA, citrate, malate, phosphoenolpyruvate, or pyruvate, NADPH and NADH inhibit the enzyme completely at a final concentration of 1 mM when either NAD or NADP is used as electron acceptor. Furthermore, 1 mM ATP causes an enzyme inhibition of 35% with NAD and 16% with NADP as electron acceptor.

[47] D-Galactitol-6-phosphate[1] Dehydrogenase

By RICHARD L. ANDERSON and JOHN P. MARKWELL

D-Galactitol 6-phosphate[1] + NAD$^+$ → D-tagatose 6-phosphate + NADH + H$^+$

This inducible enzyme functions in the catabolism of galactitol in *Klebsiella pneumoniae*.[2,3]

Assay Method[3]

Principle. The rate of D-galactitol 6-phosphate-dependent NADH formation is determined by measuring the rate of absorbance increase at 340 nm.

Reagents

Tris-HCl buffer, 0.2 M, pH 8.5
NAD$^+$, 50 mM
D-Galactitol 6-phosphate,[4] 22.5 mM
MnCl$_2$, 2 mM

Procedure. The following components are added to a microcuvette with a 1.0-cm light path: 50 μl of buffer, 20 μl of NAD$^+$, 20 μl of D-galactitol 6-phosphate, 5 μl of MnCl$_2$, enzyme, and water to a volume of 0.15 ml. The reaction is initiated by the addition of D-galactitol-6-phosphate dehydrogenase. The reaction rates are conveniently measured with a Gilford multiple-sample absorbance-recording spectrophotometer. The cuvette compartment should be thermostatted at 30°.

Definition of Unit and Specific Activity. One unit of D-galactitol-6-phosphate dehydrogenase is defined as the amount that catalyzes the

---

[1] D-Galactitol 6-phosphate is sometimes referred to as L-galactitol 1-phosphate, which is the same compound.

[2] J. Markwell, G. T. Shimamoto, D. L. Bissett, and R. L. Anderson, *Biochem. Biophys. Res. Commun.* **71**, 221 (1976).

[3] J. P. Markwell and R. L. Anderson, *Arch. Biochem. Biophys.* **209**, 592 (1981).

[4] J. B. Wolff and N. O. Kaplan, *J. Biol. Chem.* **218**, 849 (1956).

PURIFICATION OF D-GALACTITOL-6-PHOSPHATE DEHYDROGENASE

Fraction	Volume (ml)	Protein (mg)	Activity (units)	Specific activity (units/mg protein)	Recovery (%)
Cell extract	21.8	296	15.9	0.053	100
DEAE-cellulose	28.0	35.0	11.8	0.34	74
Ammonium sulfate	2.2	16.3	9.85	0.61	62
Sephadex G-150	7.7	2.4	3.26	1.4	20

reduction of 1 μmol of NAD$^+$ per minute in the standard assay. Specific activity (units per milligram of protein) is based on protein determinations by the Lowry procedure.

Alternative Assay Procedure. The reverse reaction may also be measured spectrophotometrically.[3]

Purification Procedure[3]

Organism and Growth Conditions. The bacterial strain used is *Klebsiella pneumoniae* PRL-R3 (formerly designated *Aerobacter aerogenes* PRL-R3). It is grown aerobically at 30° in 1-liter volumes of a medium containing 0.15% KH_2PO_4, 0.71% Na_2HPO_4, 0.3% $(NH_4)_2SO_4$, 0.01% $MgSO_4$, 0.0005% $FeSO_4 \cdot 7 H_2O$, and 0.5% galactitol (autoclaved separately). The cells are harvested by centrifugation and washed with 0.85% (w/v) NaCl.

Preparation of Cell Extracts. Pelleted cells are suspended in buffer A [0.025 M Tris-HCl (pH 8.5), 10% (v/v) glycerol, and 0.5 mM dithiothreitol] and broken by sonic treatment (10,000 Hz) in a Raytheon Model DF-101 (250-W) sonic oscillator cooled with circulating ice water. The broken-cell suspension is centrifuged at 12,000 g for 10 min, and the resulting supernatant fluid is designated the cell extract.

General. The following procedures are performed at 0–4°. A summary of the purification is shown in the table.

DEAE-Cellulose Chromatography. The cell extract is applied to a column (1.2 × 6 cm) of DEAE-cellulose that has been equilibrated with buffer A. Protein is then eluted with a 200-ml linear gradient of 0 to 0.35 M KCl in buffer A. Fractions (1.8 ml) are collected, and those that contain the peak of the activity (fractions 28 through 43) are pooled.

Ammonium Sulfate Fractionation. To the above pooled fractions, ammonium sulfate (192 g/liter) is added slowly with stirring. After 15 min the resulting suspension is clarified by centrifugation and a second portion of ammonium sulfate (220 g/liter) is added. After 15 min, the precipitate is collected by centrifugation and dissolved in 2 ml of buffer A.

Sephadex G-150 Chromatography. The above fraction is chromatographed on a column (1.4 × 82 cm) of Sephadex G-150 equilibrated with buffer A. The protein is eluted with the same buffer. Fractions (1.3 ml) are collected, and those that contain the peak of the activity (fractions 55 through 60) are pooled. The enzyme is 26-fold purified with a 20% over-all recovery.

Properties[3]

Substrate Specificity. D-Galactitol 6-phosphate (K_m = 0.2 mM) and D-tagatose 6-phosphate (K_m = 0.1 mM) are the only known substrates.

Compounds (10 mM) that did not serve as substrates (<0.1% of the rate with D-galactitol 6-phosphate) in the presence of NAD^+ were D-mannitol 6-phosphate, D-glucitol 6-phosphate, D-glucose 6-phosphate, D-galactose 6-phosphate, and D-fructose 1-phosphate.

Compounds (10 mM) that did not serve as substrates (<0.1% of the rate with D-tagatose 6-phosphate) in the presence of NADH were D-fructose 6-phosphate, D-fructose 1-phosphate, D-fructose 1,6-bisphosphate, D-glucose 6-phosphate, D-galactose 6-phosphate, and L-sorbose 1-phosphate.

No activity was detected when $NADP^+$ and NADPH were substituted for NAD^+ and NADH, respectively, with any of the sugar phosphates tested.

Effect of Divalent Metal Ions. A divalent metal is required both for activity and stability. Treatment of the enzyme with 10 mM EDTA followed by removal of this chelating agent by passage of the enzyme through a column of Sephadex G-25 absolished all activity unless a divalent metal ion was included in the assay mixture. If the metal salt (67 μM) was added within 15 min of the EDTA treatment, the following restorations (%) of activity were achieved: $CdSO_4$, 91; $MnCl_2$, 71; $CoCl_2$, 61; $ZnCl_2$, 61. If the metal-ion addition was delayed for 150 min, only 4% of the activity could be restored.

pH Optimum. Activity as a function of pH is maximal at about pH 8.7 in Tris-HCl buffer.

Molecular Weight. An estimation of the molecular weight by Sephadex G-150 chromatography yielded a value of 86,000.

Stability. D-Galactitol-6-phosphate dehydrogenase in the Sephadex G-150 fractions retained most of its activity for at least 2 weeks when stored at 0° or −20°.

[48] 6-Phospho-D-gluconate Dehydrogenase from *Pseudomonas fluorescens*

By C. Stournaras, F. Butz, and G. Kurz

6-Phospho-D-gluconate + $NAD(P)^+ \rightarrow$ D-ribulose 5-phosphate + CO_2 + NAD(P)H + H^+

6-Phospho-D-gluconate is formed in *Pseudomonas* by the dehydrogenation of D-glucose 6-phosphate and subsequent hydrolysis of 6-phospho-D-glucono-1,5-lactone and by the hydrogenation of 2-oxo-6-phospho-D-gluconate.[1–5] It is further metabolized either by the 6-phospho-D-gluconate dehydratase in the Entner–Doudoroff pathway or by the 6-phospho-D-gluconate dehydrogenase (decarboxylating) (6-phospho-D-gluconate : $NAD(P)^+$ oxidoreductase (decarboxylating), EC 1.1.1.44) in the hexose monophosphate pathway.

Assay Method

Principle. The assay is based on the spectrophotometric determination of NADH formed with 6-phospho-D-gluconate as substrate at high pH conditions.

Reagents

Tris-HCl buffer, 100 mM, pH 8.20
NAD^+, 20 mM, in Tris-HCl buffer, pH 8.20
6-Phospho-D-gluconate, 50 mM, in Tris-HCl buffer, pH 8.20

Enzyme. The enzyme solution is diluted with the appropriate buffers to give a solution with an activity between 0.2 and 0.5 unit/ml.

Procedure. A mixture of 0.1 ml of NAD^+, 0.1 ml of 6-phospho-D-gluconate, and 0.7 ml of Tris-HCl buffer is incubated at 30° in a spectrophotometer cell with a 1.0-cm light path. The reaction is started by the addition of 0.1 ml of the enzyme. The extinctions at 340 nm or 366 nm are measured at 15-sec intervals.

Definition of Unit and Specific Activity. One unit of enzyme is defined as the amount that catalyzes the reduction of 1 μmol of NAD^+ per minute

[1] W. A. Wood, *Bacteriol. Rev.* **19**, 222 (1955).
[2] L. N. Ornston, *Bacteriol. Rev.* **35**, 87 (1971).
[3] R. C. Eisenberg, S. J. Butters, S. C. Quay, and S. B. Friedman, *J. Bacteriol.* **120**, 147 (1974).
[4] B. K. Roberts, M. Midgeley, and E. A. Dawes, *J. Gen. Microbiol.* **78**, 319 (1973).
[5] P. H. Whiting, M. Midgeley, and E. A. Dawes, *Biochem. J.* **154**, 659 (1976).

under the above conditions. Specific activity is expressed as units per milligram of protein.

Organism and Cultivation

The strain of *Pseudomonas* used was originally obtained from Boehringer Mannheim GmbH (Mannheim, F.R.G.) and has been classified as *P. fluorescens* biotype E or F.[6,7] The bacteria were maintained in lyophilized stock cultures.

The cultivation of the bacteria is performed exactly as described by Maurer *et al.*[8]

Purification Procedure

Lyophilized bacteria are used as enzyme source. All solutions are 20 mM in 2-mercaptoethanol, and all operations are performed at 0–4° unless otherwise stated.

Step 1. Preparation of Cell-Free Extract. The lyophilized and pulverized bacteria (300 g) are suspended in 20 mM sodium acetate buffer, pH 5.1 (4500 ml); 15 ml of buffer are used per gram of cells. The suspension is homogenized with an Ultra-Turrax type 45 homogenizer (Janke und Kunkel KG, Staufen, F.R.G.) intermittently for a total of 10 min, taking care that the working temperature does not exceed 25°. The suspension is stirred at room temperature for 4 hr and subsequently centrifuged at 16,000 g at 4° for 120 min. A yellow, turbid cell-free extract (about 4250 ml) is obtained.

Step 2. Bulk Separation on CM-Sephadex. The cell-free extract, having a pH value of about 6.0, is adjusted to pH 5.10 by adding cautiously a 1 M acetic acid solution. Subsequently the solution is added, with mild stirring, to 6000 ml of CM-Sephadex C-50 preequilibrated with 20 mM sodium acetate buffer, pH 5.10; 1 ml of CM-Sephadex C-50 is used per 10 mg of protein. In order to obtain nearly complete adsorption of the enzyme, the suspension is stirred for 30 min.[9] The CM-Sephadex C-50 gel is then collected under mild suction on a large Büchner funnel and washed directly on the funnel with 2000 ml of the same buffer. This washing-out procedure is repeated until the filtrate is clear. For elution of 6-phospho-D-gluconate dehydrogenase, the gel is suspended in 1500 ml of 20

---

[6] R. Y. Stanier, N. J. Palleroni, and M. Doudoroff, *J. Gen. Microbiol.* **43**, 159 (1966).

[7] D. Lessmann, K.-L. Schimz, and G. Kurz, *Eur. J. Biochem.* **59**, 545 (1975).

[8] P. Maurer, D. Lessmann, and G. Kurz, this volume [45].

[9] If necessary, the pH value is cautiously lowered until the adsorption of the enzyme is just complete.

mM sodium acetate buffer, pH 5.60, containing 150 mM NaCl, and this is stirred for 10 min. The eluted protein is removed by filtration on a Büchner funnel. This process is repeated until no more enzyme activity is eluted. The eluates are combined and concentrated to 1500 ml with the aid of an Amicon hollow-fiber concentrator CH 3 [Amicon GmbH, Witten (Ruhr), F.R.G.] equipped with a hollow-fiber cartridge type H1P-10.

Step 3. Ammonium Sulfate Precipitation. Solid $(NH_4)_2SO_4$ (409 g/l solution) is added slowly to the protein solution of step 2, and the suspension is gently stirred overnight. The precipitate is collected by centrifugation at 20,000 g for 60 min and is subsequently dissolved in about 65 ml of 50 mM imidazole-HCl buffer, pH 6.5, containing 110 mM NaCl. In order to remove the residual $(NH_4)_2SO_4$ the protein solution is dialyzed for at least 12 hr against 7500 ml of the same buffer with two changes of dialyzate.

Step 4. DEAE-Cellulose Chromatography. The dialyzed protein solution of step 3, containing about 25 mg of protein per milliliter, is applied to a column (100 × 5 cm) of DEAE-cellulose preequilibrated with 50 mM imidazole-HCl buffer, pH 6.5, containing 110 mM NaCl. After the protein is loaded on the column, the elution is performed with exactly the same buffer. Fractions of 13.3 ml are collected at a flow rate of 40.0 ml/hr. Fractions with an activity greater than 0.5 units/ml are combined and concentrated to a final volume of about 100 ml with the aid of an Amicon ultrafiltration cell Model 402 using a PM-30 membrane.

Step 5. Hydroxyapatite Chromatography. The enzyme solution obtained from step 4 is directly applied to a hydroxyapatite column (40 × 3 cm) preequilibrated with 5 mM potassium sodium phosphate buffer, pH 6.5, containing 110 mM NaCl. Subsequently the column is washed with four column-volumes of 48 mM potassium sodium phosphate buffer, pH 6.5, containing 110 mM NaCl. The enzyme is eluted with 51 mM potassium sodium phosphate buffer, pH 6.5, containing 110 mM NaCl. Fractions of 8.3 ml are collected at a flow rate of 25 ml/hr. Fractions with an activity greater than 0.2 units/ml are combined and concentrated to a final volume of about 4 ml by ultrafiltration, using UM-10 membranes.

Step 6. Preparative Electrophoresis. Preparative disc gel electrophoresis is performed in vertical gels with end elution using the equipment Ultraphor (Colora Messtechnik GmbH, Lorch/Württ., F.R.G.) with a separation chamber for gels 137 mm wide and 7.5 mm thick. The gel system according to Williams and Reisfeld[10] is adapted to the preparative requirements. A separation gel with a height of 70 mm and a spacer gel of about 30 mm is used. The total acrylamide concentration of the separation

[10] D. E. Williams and R. A. Reisfeld, *Ann. N. Y. Acad. Sci.* **121,** 373 (1964).

PURIFICATION OF 6-PHOSPHO-D-GLUCONATE DEHYDROGENASE (DECARBOXYLATING)
FROM *Pseudomonas fluorescens*[a]

Step	Volume (ml)	Activity (units/ml)	Protein (mg/ml)	Specific activity (units/mg protein)	Yield (%)	Purification (fold)
reparation of cell-free extract	4250	0.82	15.4	0.053	100	—
ulk separation on CM-Sephadex	1500	2.03	2.31	0.88	87	16.5
mmonium sulfate precipitation	75	36	24.5	1.47	77	27.5
EAE-cellulose chromatography	100	18.9	1.32	14.32	54	268
ydroxyapatite chromatography	4	352.5	7.28	48.42	40	906
reparative electrophoresis	3	302	2.5	121	26	2264

Typical data obtained with 300 g of lyophilized bacteria.

gel is 7.5% at a ratio of acrylamide : bisacrylamide of 95 : 5. For elution, a 56 mM Tris-HCl buffer, pH 7.2, is used as counter cell buffer.

The concentrated enzyme solution of step 5 is dialyzed for at least 12 hr against 1000 ml of 50 mM Tris-HCl buffer, pH 5.5, with two changes of dialyzate. The protein solution obtained (about 5 ml) is made 15% in glycerol. With the aid of a peristaltic pump, 5 ml of 50 mM Tris-H$_3$PO$_4$ buffer, pH 5.5, made 5% in glycerol, is layered on the spacer gel to protect the enzyme solution against the electrode buffer. Subsequently, the protein solution, 15% in glycerol, is layered between the spacer gel and the protecting zone. During concentration the current is kept constant at 40 mA; during separation, at 50 mA. Elution is performed at a flow rate of 42 ml/hr, and fractions of 1.4 ml are collected. Enzyme fractions containing no impurities are pooled and concentrated to a volume of about 3 ml by ultrafiltration using UM-10 membranes.

A typical purification is summarized in the table.

Properties

Homogeneity. The purified enzyme with a specific activity of 121 units/mg is homogeneous by disc gel electrophoresis, by sodium dodecyl sulfate electrophoresis, and by gel isoelectric focusing.

Stability and pH Stability. The purified enzyme shows no loss of activity over a period of months when stored under a solution of 3.2 M (NH$_4$)$_2$SO$_4$ in the neutral pH range. However, a slow formation of multiple enzyme forms has been observable, most likely by spontaneous deamidation.[11]

[11] C. Stournaras and G. Kurz, *Hoppe-Seyler's Z. Physiol. Chem.* **359**, 1115 (1978).

The enzyme is stable in the range from pH 4.8 to 10.8.

Structure. The dehydrogenase has a molecular weight of 128,000 and is composed of four subunits, each of molecular weight 32,000.[12,13]

Kinetic Properties. The dehydrogenase uses NAD^+ as well as $NADP^+$ as cosubstrate, exhibiting, under the conditions of the standard assay, a specific activity of 121 μmol for NADH and 23 μmol for NADPH formed per minute per milligram of protein, giving a ratio of reaction rates with NAD^+ to $NADP^+$ of 5.26.[13]

The limiting Michaelis constants, under the conditions of the assay are, for NAD^+, $K = 0.2$ mM and for 6-phospho-D-gluconate, $K = 0.13$ mM.

[12] C. Stournaras and G. Kurz, *Hoppe-Seyler's Z. Physiol. Chem.* **359**, 329 (1978).
[13] C. Stournaras, P. Maurer, and G. Kurz, unpublished observation.

[49] 6-Phosphogluconate Dehydrogenase from *Bacillus stearothermophilus*

By Francesco M. Veronese, Enrico Boccù, and Angelo Fontana

The enzyme 6-phosphogluconate dehydrogenase [6-PGDH; EC 1.1.1.44, 6-phospho-D-gluconate : $NADP^+$ 2-oxidoreductase (decarboxylating)] plays a key role in the pentose phosphate cycle, since it catalyzes the reversible oxidative decarboxylation of 6-phospho-D-gluconate to yield D-ribulose 5-phosphate and CO_2, with NADP reduced to NADPH. The enzyme has been extensively studied and has been isolated from a number of sources, including microorganisms.[1–11] More recently, the bacterial dehydrogenase has been isolated and characterized from *Streptococcus faecalis*,[12] *Escherichia coli*,[13] and *Bacillus stearothermophilus*.[14–17] This chapter describes a purification procedure for 6-PGDH from the thermophilic microorganism *B. stearothermophilus*, which shows an optimum growth temperature near 60°. The thermophilic 6-PGDH is of particular interest in view of its noteworthy stability to heat and protein denaturants.[13–16] The enzyme has been the subject of a detailed enzymo-

[1] S. Pontremoli, A. De Flora, E. Grazi, G. Mangiarotti, A. Bonsignore, and B. L. Horecker, *J. Biol. Chem.* **236**, 2975 (1961).
[2] M. Rippa, E. Grazi, and S. Pontremoli, *J. Biol. Chem.* **241**, 1632 (1966).
[3] M. Rippa, M. Signorini, and S. Pontremoli, *Eur. J. Biochem.* **1**, 170 (1967).
[4] R. H. Villet and K. Dalziel, *Biochem. J.* **115**, 639 (1969).

logical and physicochemical analysis, and the results of these studies are summarized here.

Assay Method

The method of determination of enzymic activity is based on the spectrophotometric measurement of NADPH formed during the oxidative decarboxylation of the substrate. Assays were conducted at room temperature[13,16] or at 43°.[14] The standard incubation mixture (1.0 ml) consisted of 6-phosphogluconic acid and NADP (0.3 mM each) in 0.1 M Tris-HCl buffer, pH 8.0. The reaction was initiated by the addition of the enzyme (usually 5–10 μl containing 0.1–3 μg of pure enzyme) and followed by the rate of NADPH formation at 340 nm using a thermostatted spectrophotometer. The reaction was linear with time for 1–3 min, and the reaction rate was a linear function of the protein concentration within the ranges employed. Activities were calculated from initial slopes.

An enzyme unit is defined as the amount of enzyme that catalyzes the reduction of 1 mol of NADP per minute. Specific activity is expressed as units per milligram of protein. Protein concentrations were estimated from measurements of absorbance at 280 and 260 nm, using bovine serum albumin as a standard.[18] The concentration of the homogeneous enzyme was evaluated on the basis of $A_{280\ nm}^{0.1\%} = 1.05$.

Purification Procedure

The following purification procedure is a modification of that previously described.[14] All steps are performed at 0–5°.

[5] D. Procsal and D. Holten, *Biochemistry* **11**, 1310 (1972).

[6] M. Silverberg and K. Dalziel, *Eur. J. Biochem.* **38**, 229 (1973).

[7] W. A. Scott, and T. Abramsky, *J. Biol. Chem.* **248**, 3535 (1973).

[8] W. A. Scott and T. Abramsky, *J. Biol. Chem.* **248**, 3542 (1973).

[9] B. M. F. Pearse and M. A. Rosemeyer, *Eur. J. Biochem.* **42**, 213 (1974).

[10] B. M. F. Pearse and M. A. Rosemeyer, *Eur. J. Biochem.* **42**, 225 (1974).

[11] D. Cottreu, P. Boivin, A. Kahn, A. Milani, and J. Marie, *Biochimie* **57**, 325 (1975).

[12] R. B. Bridges, M. P. Palumbo, and C. L. Wittemberger, *J. Biol. Chem.* **250**, 6093 (1975).

[13] F. M. Veronese, E. Boccù, and A. Fontana, *Biochemistry* **15**, 4026 (1976).

[14] F. M. Veronese, E. Boccù, A. Fontana, C. A. Benassi, and E. Scoffone, *Biochim. Biophys. Acta* **334**, 31 (1974).

[15] F. M. Veronese, C. Grandi, E. Boccù, and A. Fontana, *in* "Enzymes and Proteins from Thermophilic Microorganisms (H. Zuber, ed.), pp. 147–155. Birkhaeuser, Basel, 1976.

[16] B. M. F. Pearse and I. J. Harris, *FEBS Lett.* **38**, 49 (1973).

[17] F. Quadri and P. D. G. Dean, *Biochem. J.* **191**, 53 (1980).

[18] E. Layéne, this series, Vol. 3, p. 447.

Preparation of Cell-Free Extract. Large cultures of *B. stearother-mophilus* NCA 1503 are grown at 60° at the Microbiological Research Establishment (Porton Down, Wiltshire, England), as described.[19,20] The cells are stored at −20°. One kilogram of cell paste is suspended in 2 liters of 0.05 *M* Tris-HCl buffer, pH 7.5, containing EDTA and 2-mercaptoethanol each at 1 m*M* concentration (buffer A), then ruptured by three passages through a Manton–Gaulin Model 15M-8TBA French press homogenizer at an operating pressure of 8000 psi. The suspension is stirred for 30 min and then clarified by centrifugation at 25,000 *g* for 30 min using a Sorvall Model R5 centrifuge.

Step 1. Ammonium Sulfate Fractionation. Solid ammonium sulfate (18.9 g/100 ml of solution) is added to the supernatant fraction in step 1 (crude extract) to 0.48 saturation, the pH of the solution being maintained at 7.5 with 1 *N* NaOH. The solution is stirred for 30 min, and the precipitated proteins are removed by centrifugation at 25,000 *g*. The supernatant fraction is brought to 0.75 saturation with ammonium sulfate (30 g/100 ml of solution) and stirred for 30 min. The protein precipitate is collected by centrifugation, dissolved in 300 ml of buffer A, and dialyzed against several changes of the same buffer.

Step 2. DEAE-Cellulose Column Chromatography. The dialyzed solution is applied to a DEAE-cellulose DE-52, Whatman, Springfield Mill, Maidstone, Kent, England) column (5 × 60 cm) equilibrated with buffer A. The column is washed with 1 liter of buffer A at a flow rate of 30 ml/hr, and then with buffer A containing 0.15 *M* KCl until the absorbance at 280 nm of the effluent drops below 0.1 (about 2 liters). Elution is carried out with a linear gradient of buffer A containing 0.15–0.8 *M* KCl (total volume 5 liters). Fractions of 8 ml are collected and analyzed for protein content, 6-PGDH activity, and salt concentration.

Step 3. Hydroxyapatite Column Chromatography. The fractions showing 6-PGDH activity from step 2 are combined and directly applied on a hydroxyapatite (BioGel HTP, Bio-Rad, Richmond, California) column (2.5 × 45 cm) equilibrated with 3 m*M* potassium phosphate buffer, pH 6.8, containing EDTA and 2-mercaptoethanol at 1 m*M* concentration each (buffer B). The column is washed with this buffer until no more protein is eluted, then a linear gradient is established between 3 m*M* (500 ml) and 0.2 *M* (500 ml) buffer B. Fractions of 5 ml are collected at a flow rate of 10 ml/hr. Fractions having 6-PGDH activity are pooled and dialyzed against 0.1 *M* sodium acetate buffer, pH 6.2, containing EDTA and 2-mercaptoethanol at 1 m*M* concentration each (buffer C).

[19] K. Sargeant, D. N. East, A. R. Whitaker, and R. Elsworth, *J. Gen. Microbiol.* **65**, iii (1971).
[20] A. Atkinson, G. T. Banks, C. J. Bruton, M. J. Comer, R. Jakes, A. Kamalogaran, A. R. Whitaker, and G. P. Winter, *J. Appl. Biochem.* **1**, 247 (1979).

TABLE I
PURIFICATION OF 6-PHOSPHOGLUCONATE DEHYDROGENASE FROM
Bacillus stearothermophilus

Step and fraction	Total protein (mg)	Total activity[a] (units)	Specific activity[a] (units/mg of protein)	Yield (%)	Purification factor
Crude extract	83,000	2500	0.03	—	—
1. Ammonium sulfate	29,000	2300	0.08	95	2.7
2. DEAE-Cellulose	2800	2100	0.75	88	25
3. Hydroxyapatite	84	1265	15	53	176
4. Phosphocellulose	46	1050	23	44	500
5. BioGel	30	840	27	35	930

[a] Activity was determined at 22° using the assay method described in the text.

Step 4. Phosphocellulose Column Chromatography. The dialyzed enzyme solution from step 3 is applied to a phosphocellulose (P-11, Reeve Angel, Clifton, New Jersey) column (1.5 × 15 cm) equilibrated with buffer C. The column is washed with this buffer, then elution is carried out with a linear gradient of buffer A of increasing concentration from 3 mM to 0.2 M (total volume 400 ml). Fractions of 3 ml are collected at a flow rate of 10 ml/hr. The fractions with 6-PGDH activity are combined and then dialyzed against a saturated solution of ammonium sulfate.

Step 5. Gel Filtration with BioGel A-0.5m. The protein precipitate from step 4 is collected after centrifugation, dissolved in 0.1 M Tris-HCl buffer, pH 7.2, containing EDTA and 2-mercaptoethanol at 1 mM concentration each, and applied to a BioGel A-0.5 m (Bio-Rad, Richmond, California) column (2.5 × 100 cm) equilibrated with the same buffer. Fractions of 1.5 ml are collected at a flow rate of 10 ml/hr. The fractions with 6-PGDH activity are pooled and the enzyme is precipitated after dialysis against saturated ammonium sulfate solution. The enzyme is stored at 4° as a saturated ammonium sulfate suspension.

The method, summarized in Table I, results in a 930-fold overall purification with about 35% recovery of 6-PGDH activity.

Comments. The thermophilic 6-PGDH has been isolated from the same strain (NCA 1503) of *B. stearothermophilus* by Pearse and Harris.[16] The purification procedure involved initial steps on hydroxyapatite and DEAE-Sephadex for the large-scale purification of a few other enzymes, besides 6-PGDH.[20] The enzyme solution obtained after these steps was then subjected to ion-exchange chromatography on a CM-Sephadex column equilibrated at pH 5.6, resulting in a 10-fold purification. The active fractions resulting from this step were concentrated to a protein concentration of 20 mg/ml in acetate buffer, pH 6.0, containing 1 mM EDTA, 0.1% 2-mercaptoethanol, and 10 μM NADP. Microcrystals of the enzyme formed on standing at 0–4°.

More recently, the thermophilic 6-PGDH was purified by affinity chromatography using immobilized triazine dyes (Cibacron Blue F3G-A and Procion Red HE-3B) as chromatographic ligands.[17] A crude enzyme extract was prepared from *B. stearothermophilus* NCA 1503 paste by disruption of the cells ultrasonically and centrifugation. Fractionation of the supernatant was carried out by ammonium sulfate (35–75% saturation). The precipitate was redissolved, dialyzed, and applied to a column prepared with Procion Red covalently bound to Sepharose. The enzyme was eluted with a linear gradient of KCl (0 to 1 M). The active fractions were pooled, dialyzed and then applied to a Cibacron Blue-Sepharose. The enzyme was eluted with 1 mM NADP and finally purified by gel filtration on a Sephadex G-200 column. The purified enzyme was shown to be homogeneous on polyacrylamide gel electrophoresis.

Properties

Purity. The purified enzyme was shown to be homogeneous on alkaline polyacrylamide gels at pH 8.5.[14] The single protein band obtained by staining the gels with a 0.5% solution of Amido Black in 7% acetic acid corresponded to that obtained when the gels were stained for 6-PGDH activity.[14] For this purpose, the gels were immersed in 5 ml of 0.3 M Tris-HCl buffer, pH 7.6, containing 1.3 mg of NADP, 0.15 mg of phenazine methosulfate, 1.5 mg of nitro blue tetrazolium, and 2.8 mg of 6-phosphogluconic acid and then warmed at 60° for 1–2 min. Brown-blue bands developed in the gels where the enzyme was located. Purified enzyme preparations[14,16,17] gave single protein bands also on disc gel electrophoresis in the presence of sodium dodecyl sulfate (SDS).[21,22] A specific activity of 27 (this report) or 87 units/mg[14] at 22° or 43°, respectively, was obtained from several preparations. The specific activity of the crystalline enzyme when assayed under different experimental conditions from those described here (see above), at 25° in Tris buffer, pH 9.0, was found to be 20 units/mg.[16]

Kinetic Parameters. The enzyme is active in the pH range of 6 to 9 and exhibits maximum of activity around pH 8 both at 25° and 43°.[14] The pH optimum is similar to that found for the *Candida utilis* enzyme,[23] and about 1 unit lower than that observed with the rat liver enzyme.[24] The enzyme

[21] K. Weber and M. Osborn, *J. Biol. Chem.* **244**, 4406 (1969).
[22] U. K. Laemmli, *Nature (London)* **227**, 680 (1970).
[23] M. Rippa, M. Signorini, and S. Pontremoli, *Ital. J. Biochem.* **19**, 361 (1970).
[24] G. E. Glock and P. McLean, *Biochem. J.* **55**, 400 (1953).

follows Michaelis–Menten kinetics in the temperature range of 25° to 60°, and the apparent K_m values for NADP and 6-phosphogluconate at 43° were found to be $2.5 \times 10^{-5} M$ and $2.0 \times 10^{-5} M$, respectively.

Inhibitors and Activators. The *B. stearothermophilus* 6-PGDH shares with the enzyme from several other sources the sensitivity to reagents that react with —SH groups of proteins. The native enzyme was inactivated upon incubation in presence of iodoacetamide,[16] *p*-chloromercuribenzoate,[14] Hg^{2+} ions,[14] 5,5′-dithiobis-(2-nitrobenzoic acid) (Ellman's reagent),[14] or 7-chloro-4-nitrobenzo-2-oxa-1,3-diazole (NBD-Cl).[25] The enzyme was, however, partially protected from inactivation by its substrate, 6-phosphogluconate, or its coenzyme, NADP,[14,16,25] implying that these reagents react in the vicinity of the active site. It has been found that complete inactivation of the enzyme is achieved after modification of two cysteine residues per mole of enzyme, i.e., one residue per subunit.[16,25] The enzyme inactivated by the above —SH reagents, with the exception of iodoacetamide, recovers fully its catalytic activity upon incubation with excess 2-mercaptoethanol.[14,25]

The enzyme is activated by $MgCl_2$ at low concentrations, whereas at high concentrations it is inhibited.[14] The observed effect is specific for Mg^{2+} ions, since neither KCl or NaCl or sodium acetate show any effect on the enzymic activity of the enzyme. This is at variance with the effects of ions observed with the *Candida utilis* enzyme.[1] It is of interest to note that the activating effect of Mg^{2+} ions on the *B. stearothermophilus* 6-PGDH was about four times greater at 55° than at 30°.[14]

Coenzyme Binding. The binding of NADPH to the enzyme at 25°, evaluated by measuring both the tryptophan protein fluorescence quenching and the increase of NADPH fluorescence, indicated two identical binding sites with a dissociation constant of $0.34 \times 10^{-6} M$.[26]

Physical Properties and Amino Acid Composition. A molecular weight (M_r) of about 100,000 was calculated from sucrose density gradient centrifugation[14] or equilibrium sedimentation experiments.[16] The native enzyme, at pH 8, sediments as a single boundary with an $s_{20,w}$ value of 5.8 S.[16] Denaturation with SDS in the presence of a reducing agent[21,22] and polyacrylamide gel electrophoresis gave a single sharp band with a mobility corresponding to M_r 50,000.[14,16] Thus, the native enzyme exists as a dimer composed of subunits having the same molecular weight and hence resembles its counterparts from mesophilic sources, including human erythrocytes,[9,10] *Candida utilis*,[3] and *E. coli*.[13] The amino acid composition of 6-PGDH from *B. stearothermophilus* is shown in Table II. The enzyme crystallized from an ammonium sulfate, solution as distorted hexagonal

[25] A. Fontana, L. Mantovanelli, E. Boccù, and F. M. Veronese, *Int. J. Peptide Protein Res.* **9**, 329 (1977).
[26] F. M. Veronese, E. Boccù, and A. Fontana, unpublished results.

TABLE II

AMINO ACID COMPOSITION OF 6-PHOSPHOGLUCONATE DEHYDROGENASE
FROM *Bacillus stearothermophilus* [a]

Amino acid	No. of residues[b]	Amino acid	No. of residues[b]
Aspartic acid	39	Methionine	10
Threonine	18	Isoleucine	33
Serine	17	Leucine	38
Glutamic acid	53	Tyrosine	19
Proline	18	Phenylalanine	18
Glycine	40	Tryptophan[d]	5
Alanine	48	Lysine	33
Cysteine[c]	3	Histidine	10
Valine	28	Arginine	21

[a] From Veronese *et al.*[13]
[b] Moles of amino acid residue per mole of enzyme subunit (M_r 50,000).
[c] Determined by titration of the enzyme with 5,5'-dithiobis-(2-nitrobenzoic) acid [Ellman's reagent: G. L. Ellman, *Arch. Biochem. Biophys.* **80**, 70 (1969)] in 8 *M* urea.
[d] Determined spectrophotometrically by the method of H. Edelhoch [*Biochemistry* **6**, 1948 (1967)].

prisms with trigonal symmetry, space group $P3_131$ (or $P3_221$) and unit cell dimensions of $a = b = 115$ Å, $c = 156$ Å.[16]

Stability. The 6-PGDH from *B. stearothermophilus* retains full activity for several months when stored at 4° as an ammonium sulfate suspension or in acetate buffer, pH 6.0, in the presence of a crystal of thymol to prevent bacterial growth.[16] In addition, the enzyme can be lyophilized without loss of activity.[14] As expected from its thermophilic origin,[27–29] the enzyme is stable at relatively high temperatures. The enzyme retains full activity after heating for 15 min at 60° (the growth temperature of the organism), whereas it is denatured at higher temperatures.[13–16] Figure 1 compares the effect of heating for 15 min at different temperatures on the activity of *B. stearothermophilus* and *E. coli* 6-PGDH. It is seen that the mesophilic and thermophilic enzyme are inactivated in the temperature range of 50–60° and 65–75°, respectively.

Far-ultraviolet circular dichroism (CD) and fluorescence emission measurements at different temperatures of both thermophilic and mesophilic 6-PGDH[13,15] (Fig. 2) show a cooperative transition at 60–75°

[27] R. Singleton, Jr., and R. E. Amelunxen, *Bacteriol. Rev.* **37**, 320 (1973).
[28] R. E. Amelunxen and A. L. Murdock, *in* "Microbial Life in Extreme Environments" (D. J. Kushner, ed.), p. 250. Academic Press, New York, 1978.
[29] H. Zuber, "Enzymes and Proteins from Thermophilic Microorganisms. Birkhaeuser, Basel, 1976.

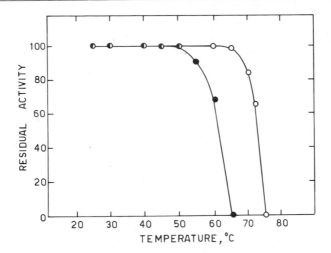

FIG. 1. Effect of heat on the activity of the *Bacillus stearothermophilus* (○) and *Escherichia coli* (●) 6-phosphogluconate dehydrogenase. Solutions of the enzyme (0.1 mg/ml) in 0.1 M potassium phosphate buffer, pH 7.2, containing 1 mM EDTA and 1 mM dithiothreitol were incubated in closed vials for 15 min at the indicated temperatures. The vials were cooled in ice, opened, and then assayed for activity. Reproduced from Veronese *et al.*[13]

and at 45–60° with the thermophilic and mesophilic enzyme, respectively. The "melting" temperatures observed were 67° and 51° for the *B. stearothermophilus* and *E. coli* enzyme, respectively. Thus, the thermal stability of enzymic activity correlates with that of the secondary structure of the protein molecules. Analogous results were obtained by measuring the effect of the temperature on the fluorescence properties of the two enzymes.

The *B. stearothermophilus* 6-PGDH shows higher stability than the mesophilic enzyme toward other protein denaturants, such as urea,[13,15] organic solvents,[26] and proteolytic enzymes.[26] Considering that the three-dimensional structure of the *B. stearothermophilus* 6-PGDH, as determined by CD and fluorescence measurements, appears to be highly similar to that of the corresponding mesophilic enzyme from *E. coli*,[13] it seems that thermophilic proteins do not possess particular structural features responsible for their enhanced stability; on the other hand, stability could arise from very subtle structural features.

If the effect of temperature on the activity of *B. stearothermophilus* 6-PGDH is reported in the form of an Arrhenius plot, two straight lines are obtained, with a point of discontinuity near 43°.[14,26] The activation energy is 11,150 cal/mol below and 15,500 cal/mol above the point of discontinuity, respectively.[14] On the contrary, a similar plot for the mesophilic

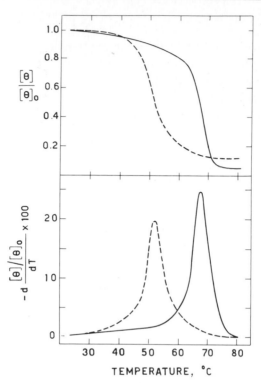

Fig. 2. The "melting profile" of *Bacillus stearothermophilus* (——) and *Escherichia coli* (---) 6-phosphogluconate dehydrogenase as determined by circular dichroism measurements. *Top:* The $[\theta]/[\theta]_0$ ratio measured at 219 nm is reported as a function of temperature, where $[\theta]_0$ is the ellipticity value at 22°. *Bottom:* The derivative curve is reported as a function of temperature. The protein concentration was 0.03 mg/ml in 0.1 M potassium phosphate buffer, pH 6.8, 1 mM EDTA, and 1 mM 2-mercaptoethanol. Reproduced from Veronese *et al.*[15]

enzyme from *E. coli* was linear.[13] A common trend in the literature is to explain nonlinear Arrhenius plots in terms of temperature-dependent conformational transitions, with two or more interconvertible conformers of the enzyme possessing different kinetic properties.[30–33] However, a study of the temperature dependence of the spectroscopic properties (far-ultraviolet CD and fluorescence emission) did not give indications of changes in conformation near 43°.[13] Similar negative results were obtained

[30] V. Massey, B. Curti, and H. Ganther, *J. Biol. Chem.* **241**, 2347 (1966).
[31] A. Orengo and G. F. Saunders, *Biochemistry* **11**, 1761 (1972).
[32] M. Dixon and E. C. Webb, "The Enzymes," 2nd ed., pp. 159–166. Academic Press, New York, 1964.
[33] M. H. Han, *J. Theor. Biol.* **32**, 543 (1972).

in studying the fluorescence properties vs temperature of the extrinsic fluorophore 1-nitrobenzo-2-oxa-1,3-diazole (NBD) covalently bound to an —SH group near the active site of the enzyme as a probe of its microenvironment.[25] These results could indicate that the broken Arrhenius plot is not due to conformational changes of the *B. stearothermophilus* enzyme near 43°, but that the break can be explained in terms of changes in rate-limiting steps of different activation energies.[33]

Acknowledgments

This research was supported in part by the Italian National Council of Research (C.N.R.), special project on Fine Chemicals.

[50] 6-Phospho-D-gluconate Dehydrogenase from *Gluconobacter suboxydans*

By OSAO ADACHI and MINORU AMEYAMA

6-Phospho-D-gluconate + NADP $\rightleftarrows$ D-ribulose 5-phosphate + NADPH + CO_2

6-Phospho-D-gluconate dehydrogenase is a key enzyme of pentose phosphate pathway in carbohydrate metabolism. The enzyme from acetic acid bacteria has been known to be different in coenzyme specificity when compared with the enzyme from other sources.[1] The enzyme from *Gluconobacter* species reduces NADP and NAD at the same rate, whereas the enzyme from yeast and mammals requires NADP specifically. Of special interest for the enzyme from *Gluconobacter* species is the existence of an active pentose cycle that makes reasonable the finding that the Krebs cycle appears to be absent.

Assay Method

Principle. 6-Phospho-D-gluconate dehydrogenase (EC 1.1.1.44) is measured spectrophotometrically by following the rate of NADPH formation at 340 nm.

Reagents

NADP, 1.5 mM, in distilled water; store at $-20°$ until use
6-Phospho-D-gluconate, trisodium salt, 10 mM, in distilled water

---

[1] V. H. Cheldelin, *in* "Metabolic Pathways in Microorganisms," pp. 1–29 (1960) Wiley, New York, 1960.

Tris-acetate buffer, 0.3 M, pH 6.5

Enzyme, dissolved in 0.01 M potassium phosphate, pH 6.0, containing 1 mM 2-mercaptoethanol

Procedure. The enzyme activity is measured by recording the increase of absorbance at 340 nm caused by NADPH formation in a spectrophotometer thermostatted with circulating water (25°) or placed in a thermostatted room (25°). Prior to the assay, the buffer solution and distilled water are warmed to 25°. The complete reaction mixture contains 0.1 ml of NADP, 0.15 ml of 6-phospho-D-gluconate, 0.1 ml of Tris-acetate buffer, and enzyme solution in a total volume of 3.0 ml. The reaction is initiated by the addition of enzyme solution.

Definition of Unit and Specific Activity. One unit of activity forms 1 μmol of NADPH per minute under the above assay conditions. Specific activity (units per milligram of protein) is based on the spectrophotometric protein estimation by measuring the absorbance at 280 nm. An $E_{\text{cm}}^{1\%}$ value of 10.0 was obtained with the crystalline enzyme on the basis of absorbance and dry weight determinations.

Source of Enzyme

Microorganism and Culture. Gluconobacter suboxydans IFO 12528 can be obtained from the Institute for Fermentation, Osaka (17-85, Juso-honmachi 2-chome, Yodogawa-ku, Osaka 532, Japan). The stock culture is maintained on a potato–glycerol slant (see this volume [24]).

Culture medium for *G. suboxydans* consists of sodium D-gluconate (20 g), D-glucose (5 g), glycerol (3 g), yeast extract (3 g), polypeptone (3 g), and 200 ml of potato extract in 1 liter of tap water. The pH of the medium is 6.5 when all these ingredients are mixed. *Gluconobacter suboxydans* is inoculated into 100 ml of the medium in a 500-ml shake flask, and the cultivation is carried out at 30° for 24 hr with reciprocal shaking. In the case of large-scale cultures, 1.5 liters of inoculum (15 shaking flasks) are transferred into 30 liters of the medium in a 50-liter jar fermentor. In this case, the potato extract is usually omitted from the medium and cultivation is performed at 30° for 24 hr under vigorous aeration at 30 liters of air per minute with 500 rpm agitation.

Purification Procedure[2]

All operations are carried out at 0–5° unless otherwise stated. Potassium phosphate buffer, pH 6.0, containing 1 mM 2-mercaptoethanol is used. The following procedure was performed with 300 g of wet cells.

---

[2] O. Adachi, K. Osada, K. Matsushita, E. Shinagawa, and M. Ameyama, *Agric. Biol. Chem.* **46**, 391 (1982).

Step 1. Preparation of Cell-Free Extract. The cell paste is suspended in 0.01 M buffer and passed twice through a French press at 1000 kg/cm^2. The resulting cell homogenate is centrifuged at 100,000 g for 90 min, and the yellowish clear supernatant (860 ml) is obtained. To the supernatant, 5% protamine sulfate solution is added dropwise to about 2.5% of total protein in the cell-free extract and stirred for 30 min. The precipitate is removed by centrifugation at 12,000 g for 30 min, then ammonium-sulfate is added to 3.9 M (51.5 g/100 ml) and stirred overnight. The precipitate is collected by centrifugation and dissolved in a minimal volume of 0.005 M buffer and dialyzed thoroughly against the same buffer.

Step 2. Affinity Chromatography on Blue-Dextran Sepharose. The dialyzed enzyme (270 ml) is applied to a Blue-Dextran Sepharose 4B column (5 × 30 cm) that has been equilibrated with 0.005 M buffer. The column is treated with the same buffer. After elution of a large amount of nonadsorbable protein, the enzyme elutes from the column as a sharp peak. This indicates that the enzyme has a weak affinity to the Blue-Dextran under the above conditions.

Step 3. DEAE-Sephadex Column Chromatography. Pooled enzyme solution (160 ml) is applied to a DEAE-Sephadex A-50 column (5 × 30 cm) that has been equilibrated with 0.01 M buffer. Elution of the enzyme is made with 0.2 M and 0.3 M buffer stepwise. The enzyme protein is collected by ammonium sulfate precipitation at 3.9 M, and the precipitate is dissolved in 0.002 M buffer. The enzyme solution is dialyzed against 0.002 M buffer.

Step 4. Hydroxyapatite Column Chromatography. The dialyzed enzyme solution (155 ml) is passed through a hydroxyapatite column (3 × 5 cm) that has been equilibrated with 0.002 M buffer. About two-thirds of the total protein is passed through the column, and some impurities are adsorbed onto the column. 6-Phospho-D-gluconate dehydrogenase is found in the passed fraction. To the combined enzyme fraction, potassium phosphate buffer is added to 0.1 M.

Step 5. DEAE-Sephadex Column Chromatography. The enzyme solution (310 ml) is applied to a DEAE-Sephadex A-50 column (2 × 10 cm) which has been equilibrated with 0.1 M buffer. After adsorption of the enzyme onto the column, elution is made by a linear gradient formed between 200 ml each of 0.1 M buffer and 0.4 M buffer. Six-milliliter fractions are collected, and the peak fraction of the enzyme activity elutes at about 0.2 M potassium phosphate.

Step 6. Crystallization. The enzyme is precipitated by ammonium sulfate to 3.9 M, and the precipitate is dissolved in a minimal volume of 0.1 M buffer. Insoluble materials are removed by centrifugation at 12,000 g for 20 min. A few grains of solid ammonium sulfate are added to the superna-

PURIFICATION OF 6-PHOSPHO-D-GLUCONATE DEHYDROGENASE FROM
Gluconobacter suboxydans

Fraction	Total protein (mg)	Total activity (units)	Specific activity (units/mg of protein)	Yield (%)
Cell-free extract	47,300	6450	0.14	100
Blue-Dextran Sepharose 4B	1,594	5420	3.40	84
DEAE-Sephadex A-50 (I)	337	4500	13.35	70
Hydroxyapatite	173	4350	25.20	67
DEAE-Sephadex A-50 (II)	132	3720	27.25	58
Crystalline enzyme	93	3600	38.70	56

tant, and the enzyme solution is stoppered and stored in a refrigerator. Crystals in the form of fine needles appear after the solution has stood for 2 days.

A typical purification of 6-phospho-D-gluconate dehydrogenase is summarized in the table. The enzyme is purified about 280-fold with an overall yield of 56%.

Properties[2]

Homogeneity. The crystalline enzyme gives a symmetrical sedimentation peak on analytical ultracentrifugation with an apparent sedimentation constant of 7.0 S. The crystalline enzyme also shows a single protein band on conventional polyacrylamide gel electrophoresis, and enzyme activity is detected in the same protein band.

Molecular Properties. The absorption maximum is found at 280 nm, and the $E_{cm}^{1\%}$ value is estimated to be 10.8. The partial specific volume of the enzyme is calculated to be 0.73 on the basis of its amino acid composition. The molecular weight of the enzyme is 175,000 by gel filtration on a Sephadex G-200 column. Sodium dodecyl sulfate gel electrophoresis gives a single protein band having a molecular weight of 43,000. Thus, the enzyme is composed of four identical subunits whereas the enzyme from other sources has been reported to be composed of two monomers (M_r approximately 50,000 each).[3]

Titration of the sulfhydryl group with 5,5'-dinitrobis(2-nitrobenzoic) acid (DTNB) reveals 2 mol of free sulfhydryl groups per mole of enzyme. An additional alternative 2 moles of sulfhydryl group appear on denatura-

[3] See this series, Vol. 41, Chapters 48–52.

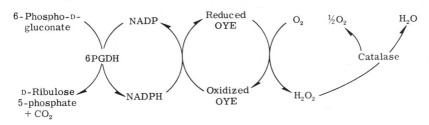

SCHEME 1. 6 PGDH, 6-phospho-D-gluconate dehydrogenase; OYE, old yellow enzyme.

tion with sodium dodecyl sulfate. The isoelectric point is estimated to be 4.0 by isoelectrofocusing.

Catalytic Properties. The enzyme is strictly specific to 6-phospho-D-gluconate, and the optimum pH is 6.5. The dehydrogenases from other sources[3] show pH optima at 7.5 to 8.0. The enzyme catalyzes oxidative decarboxylation of 6-phospho-D-gluconate with NAD as well as NADP, and apparent Michaelis constants for NADP and NAD are 0.1 mM. The apparent Michaelis constant for 6-phospho-D-gluconate is 0.07 mM and is independent of the source of coenzyme. Maximum specific activity is 55 units per milligram of protein with both NADP and NAD. The enzyme from other sources has been reported to be specific to NADP.[1,3] The enzyme reaction is not stimulated appreciably by divalent ions, and activity is inhibited competitively by NADPH, NADH, ATP, and D-fructose 1,6-diphosphate.

Cyclic Regeneration of NADP in Gluconobacter. A flavoprotein diaphorase similar to old yellow enzyme of yeast has been found in cytoplasm of *Gluconobacter*.[4] Hence NADPH generated by the NADP-dependent dehydrogenases in *Gluconobacter* species such as D-glucose dehydrogenase (see this volume [26]), aldehyde·dehydrogenase,[5] D-glucose-6-phosphate dehydrogenase,[4] and 6-phospho-D-gluconate dehydrogenase can be reoxidized to NADP by the old yellow enzyme as shown in Scheme 1. By this means, formation of pentose phosphate proceeds smoothly in *Gluconobacter* species, without inhibition by NADPH.

[4] O. Adachi, K. Matsushita, E. Shinagawa, and M. Ameyama, *J. Biochem.* (*Tokyo*) **86**, 699 (1979).

[5] O. Adachi, K. Matsushita, E. Shinagawa, and M. Ameyama, *Agric. Biol. Chem.* **44**, 155 (1980).

[51] sn-Glycerol-3-phosphate Dehydrogenase (Soluble) from Drosophila melanogaster

By David W. Niesel, Glenn C. Bewley, Chi-Yu Lee, and Frank B. Armstrong

$$\underset{sn\text{-Glycerol 3-phosphate}}{\overset{\displaystyle CH_2OH}{\underset{\displaystyle CH_2OH-OP_3^{2-}}{HO-C-H}}} + NAD^+ \rightleftharpoons \underset{\text{Dihydroxyacetone phosphate}}{\overset{\displaystyle CH_2OH}{\underset{\displaystyle CH_2-OPO_3^{2-}}{C=O}}} + NADH + H^+$$

Soluble glycerol-3-phosphate dehydrogenase (sn-glycerol-3-phosphate : NAD$^+$ oxidoreductase, EC 1.1.1.8) in *Drosophila melanogaster* represents a family of three distinct isozymes designated GPDH-1, -2, and -3, in order of their decreasing mobility toward the anode during starch gel electrophoresis.[1,2] GPDH-1, the predominant soluble activity in the adult organism, is located primarily in the thoracic flight muscle, where it participates in the glycerol phosphate cycle, which supplies energy in the form of reducing equivalents for flight metabolism.[3,4] GPDH-3 is associated with the fat body in both larvae and adults and provides glycerol 3-phosphate as a precursor for lipid biosynthesis.[5] It has also been suggested that both GPDH-1 and GPDH-3 function in maintaining cytosolic NAD$^+$: NADH ratios.[6] The two major soluble forms perform different metabolic roles, display tissue specificity,[7] and possess distinct temporal expression.[8] A minor isozyme, GPDH-2, occurs only in adult tissues that express the two major isozymes and is regarded as a heterodimer composed of a GPDH-1 and a GPDH-3 subunit. Genetic studies indicate that the multiple forms of glycerol-3-phosphate dehydrogenase in *Drosophila* are the product of the same structural gene mapped to the left arm of chromosome II and arise by an epigenetic mechanism.[6]

[1] E. H. Grell, *Science* **158**, 1319 (1967).

[2] D. A. Wright and C. R. Shaw, *Biochem. Genet.* **3**, 343 (1969).

[3] B. Sacktor and A. Dick, *J. Biol. Chem.* **237**, 3259 (1962).

[4] S. J. O'Brien and R. J. MacIntyre, *Genetics* **71**, 127 (1972).

[5] L. I. Gilbert, *Adv. Insect Physiol.* **4**, 69 (1967).

[6] G. C. Bewley and S. Miller, in "Isozymes: Current Topics in Biological and Medical Research" (M. C. Rattazzi, J. G. Scandalios, and G. S. Whitt, eds.), Vol. 3, p. 23. Liss, New York, 1979.

[7] M. C. Rechsteiner, *J. Insect Physiol.* **16**, 1179 (1970).

[8] G. C. Bewley, *Dev. Genet.* **2**, 113 (1981).

Assay Method

Principle. Glycerol-3-phosphate dehydrogenase activity is determined spectrophotometrically at 340 nm using glycerol 3-phosphate or dihydroxyacetone phosphate as substrate and measuring the rate of NAD^+ reduction or NADH oxidation, respectively. The assay utilizing glycerol 3-phosphate is more convenient and is, therefore, used routinely.

The glycerol 3-phosphate assay contains 2.5 mM glycerol 3-phosphate, 1.0 mM NAD^+, and 0.1 M glycine buffer, pH 10.2. For the dihydroxyacetone phosphate reaction, the assay mixture contains 0.75 mM dihydroxyacetone phosphate, 0.1 mM NADH, 50 mM dithiothreitol (DTT), 0.5 mg of bovine serum albumin, and 50 mM Tris-maleate buffer, pH 6.8, for GPDH-1 and pH 7.4 for GPDH-3. Both reactions are initiated by the addition of 1–100 μl of enzyme, and the total volume of each assay mixture is 1.0 ml. One unit of enzyme activity is defined as 1 μmol of NAD^+ reduced per minute, using a molar absorbance of 6.22×10^3 for NADH at 340 nm.

Drosophila Culture

Adult *Drosophila* of a Samarkand wild-type strain,[9] homozygous for the fast electrophoretic allele of soluble glycerol-3-phosphate dehydrogenase, were cultured in half-pint bottles at 25° on standard cornmeal–molasses–yeast extract medium containing propionic acid and Tegosept-M as mold inhibitors.[10] Flies were collected between 2 and 5 days postmergence and frozen until used.

Larvae of the same strain were cultured by collecting eggs over a 12-hr period and spreading 1.5 to 2 ml of eggs on a surface of banana medium containing yeast extract–agar–bananas–Karo syrup, and Tegosept-M.[10] Larvae were harvested as late third instars approximately 120 hr posthatching and were washed from the food with a stream of water, cleaned by repeated rinsing and decanting, and frozen at $-20°$ until used.

Purification Protocols for GPDH-1 and GPDH-3

Unless stated otherwise, all steps were performed at 4°.

Extraction. Frozen adult flies were homogenized in a Waring blender at 300 mg of flies per milliliter of 10 mM sodium phosphate buffer, pH 7.5, containing 1 mM DTT, 1 mM EDTA, and 0.1 mM phenylmethylsulfonyl

[9] D. L. Lindsley and E. H. Grell, Genetic Variations of *Drosophila melanogaster*. *Carnegie Inst. Washington Publ.* **627** (1968).

[10] M. Ashburner and V. N. Thompson, Jr., *in* "The Genetics and Biology of *Drosophila*" (M. Ashburner and T. R. F. Wright, eds.), Vol. 2A, p. 1. Academic Press, New York, 1978.

TABLE I
PURIFICATION OF GPDH-1 FROM *Drosophila melanogaster*

Fraction	Total units	Total protein (mg)	Specific activity (units/mg of protein)	Purification (fold)	Yield (%)
Crude extract	2150	5243	0.41	—	—
Post heat	1526	2312	0.66	1.6	71
Post (NH$_4$)$_2$SO$_4$	1333	635	2.1	5.2	62
DEAE peak	924.5	14.3	64.8	158	43
ATP peak	752.5	4.2	179.8	438.5	35

fluoride. Frozen larvae were homogenized at 500 mg of larvae per milliliter of the same buffer. Crude debris was removed by first passing each extract through buffer-saturated glass wool and then centrifuging the filtrates at 23,000 g for 30 min. The supernatant fractions were decanted and used for the purification of GPDH-1 and GPDH-3.

Summaries of the purification protocols are presented in Tables I and II.

Heat Step for GPDH-1. The supernatant fluid from adult extracts containing primarily GPDH-1 activity was incubated at 50° for 25 min to denature the GPDH-3 component.[11] The heat-treated extract was cooled on ice and then centrifuged, as described above, to remove denatured protein.

Ammonium Sulfate Fractionation. Both adult and larval preparations were fractionated by the slow addition of solid ammonium sulfate [(NH$_4$)$_2$SO$_4$] at 4°. GPDH-1 activity from adult preparations precipitated between 1.85 M and 2.87 M (NH$_4$)$_2$SO$_4$, and GPDH-3 activity from larval preparations precipitated between 2.05 M and 2.67 M (NH$_4$)$_2$SO$_4$. Pellets from both fractionations were obtained by centrifugation at 23,000 g for 30 min, resuspended in one-tenth the original volume with homogenization buffer, and dialyzed against two changes of a 500-fold (v/v) excess of homogenization buffer for 10 hr.

DEAE-Column Chromatography. A dialyzed sample (500–2000 mg) from an (NH$_4$)$_2$SO$_4$ sulfate pellet was applied directly to a column (2.5 × 40 cm) containing a bed volume of 160–210 ml of Whatman DE-52 equilibrated with homogenization buffer. The sample was washed into the column with three column volumes of homogenization buffer. The column was then developed with a 500-ml linear NaCl gradient (0 to 0.25 M), and

[11] D. W. Niesel, G. C. Bewley, S. G. Miller, F. B. Armstrong, and C.-Y. Lee, *J. Biol. Chem.* **255**, 4073 (1980).

TABLE II
PURIFICATION OF GPDH-3 FROM *Drosophila melanogaster*

Fraction	Total units	Total protein (mg)	Specific activity (units/mg protein)	Purification (fold)	Yield (%)
Crude extract	1112	5947	0.187	—	—
Post $(NH_4)_2SO_4$	767	1669	0.46	2.5	69
DEAE peak	634	33.5	18.9	101.1	57
ATP peak	425	2.7	156	834	38

6-ml fractions were collected. Glycerol-3-phosphate dehydrogenase activity appeared coincident with the first A_{280} material to elute in the gradient. Fractions with GPDH activity were pooled, concentrated using an Amicon unit fitted with a PM-10 membrane, and dialyzed against two changes of a 500-fold excess of 10 mM sodium phosphate buffer, pH 6.5, containing 1 mM EDTA and 1 mM DTT (equilibration buffer) for 6 hr.

ATP-Sepharose 4B Chromatography. The concentrated, dialyzed post-DEAE sample was applied to a column (1.5 × 30 cm) containing a bed volume of 50–70 ml of 8-(6-aminohexyl)amino-ATP Sepharose 4B that had been thoroughly washed with equilibration buffer. The column was then washed with three column volumes of equilibration buffer, and GPDH-1 was eluted with a linear NADH gradient (0 to 0.2 mM). For GPDH-3 preparations, contaminating malate dehydrogenase activity was first eluted utilizing a reduced NAD^+-oxaloacetate adduct. The adduct was prepared according to the procedure described by Kaplan et al.[12] for the preparation of a reduced NAD^+-pyruvate adduct. The reduced adduct was synthesized using 200 mg each of NAD^+ and oxaloacetate and, prior to its chromatographic use, was diluted to 0.2 A/ml at 340 nm with equilibration buffer. Malate dehydrogenase activity elutes within the first two column volumes. GPDH-3 is then eluted with a linear NADH gradient (0 to 0.2 mM). Peak fractions of GPDH activity were pooled and concentrated to 1–3 mg of protein per milliliter by ultrafiltration, using the Amicon unit described above. The behavior of the glycerol-3-phosphate dehydrogenase isozymes on this matrix is consistent with the *in vitro* kinetics of ATP as a weak inhibitor (see Table III). Under identical conditions, 8-(6-aminohexyl)amino-AMP-Sepharose does not bind either isozyme. Preparations of both isozymes, after affinity chromatography, are more than

[12] N. O. Kaplan, J. Everse, J. E. Dixon, F. E. Stolzenbach, C.-Y. Lee, C. L. T. Lee, S. S. Taylor, and K. Mosbach, *Proc. Natl. Acad. Sci. U. S. A.* **71**, 3450 (1974).

TABLE III
PHYSICOCHEMICAL AND KINETIC PARAMETERS

Parameter	GPDH-1	GPDH-3
Thermal stability (50°)[11,14]	Stable	Labile
Electrophoretic mobility, pH 7[11,14]	Anodal	Less anodal
Native isoelectric point[11]	pH 5.2	pH 5.6
Native molecular weight[14] (gel filtration)	66,000	66,000
Subunit molecular weight[11] (SDS-gel electrophoresis)	32,000	32,000
pH optimum (DHAP → α-GP)[6]	6.8	7.4
K_m dihydroxyacetone phosphate[6] (mM)	0.52	0.027
K_m NADH[6] (mM)	0.078	0.012
K_m glycerol 3-phosphate[6] (mM)	0.59	0.29
K_m NAD[+6] (mM)	0.38	0.30
K_i ATP[6] (mM)	7.26	1.15

95% pure, as judged by electrophoretic criteria under both native and denaturing conditions.

Preparation and Maintenance of Affinity Matrix

The affinity matrix, 8-(6-aminohexyl)amino-ATP-Sepharose 4B, can be prepared by the facile procedure described in this volume [61]. Ligand density is typically estimated as follows. From the prepared gel, 1 ml is extensively washed with water to remove uncoupled ligand and allowed to dry on filter paper. From this material, 100 mg are resuspended in 3 ml of 50% glycerol, and an ultraviolet scan is performed from 230 to 360 nm using uncoupled Sepharose 4B as a blank (1 g of Sepharose = 1.5 ml gel). Ligand density is estimated, utilizing $A_{278 \text{ nm}}^{\text{Ade}} = 17,700\ M^{-1}\ \text{cm}^{-1}$ for the 8-substituted adenyl derivative,[13] by the following expression:

$$\text{Ligand density/ml wet gel} = \frac{\Delta A_{278 \text{ nm}} \times 10 \times 3\ \text{ml}}{1.5 \times 17.7}$$

A ligand density of 1.2–1.7 mol of ATP per milliliter of Sepharose has proved to be effective for the purification of the GPDH isozymes.

After use, the affinity column is regenerated by washing with one-half column volume of 6 M urea followed by one column volume of 2 M NaCl and finally two column volumes of equilibration buffer. After regeneration, the column is unpacked, suspended in 50% glycerol, and stored at

---

[13] I. P. Trayer, H. R. Trayer, D. A. P. Small, and R. C. Bottomley, *Biochem. J.* **139,** 609 (1974).

−20°. This treatment provides a high-capacity matrix for the purification of both isozymes from 5–10 separate preparations.

Properties

Stability of Preparations. GPDH-1 activity is stable for over a month at 5° in 10 mM sodium phosphate buffer, pH 6.5, containing 1 mM EDTA, 1 mM DTT, and 10 mM NADH. GPDH-3 activity is intrinsically more labile and loses about 50% activity when stored under the same conditions for a month (see Table III).[6,11,14]

Immunological Characterization. Both molecular forms exhibit homologous precipitation when tested by Ouchterlony double-diffusion and immunoactivation experiments with antibodies elicited from either purified GPDH-1 or GPDH-3.[11] The two enzymes are, therefore, antigenically very similar, if not identical.

Structural Characterization. Amino acid composition and peptide mapping data on tryptic digests of GPDH-1 and -3 reveal that the primary structures of the two isozymes are very similar.[11] These results corroborate the existing genetic data that predict a single structural gene for the GPDH isozymes.

[14] G. C. Bewley, J. M. Rawls, Jr., and J. C. Lucchessi, *J. Insect Physiol.* **20**, 153 (1974).

[52] Glyceraldehyde-3-phosphate Dehydrogenase from Human Tissues

By FRITZ HEINZ and BARBARA FREIMÜLLER

D-Glyceraldehyde 3-phosphate + NAD$^+$ + P$_i$ $\rightleftarrows$
  3-phospho-D-glyceroyl phosphate + NADH + H$^+$

The isolation method described herein for glyceraldehyde-3-phosphate dehydrogenase (GAPDH) from human liver has been used by us without change for isolation of GAPDH from human brain, heart muscle, skeletal muscle, kidney, lung, and erythrocytes in crystalline homogeneous form. The method is applicable also for the isolation of GAPDH from animal tissues.

Assay Method

The assay for glyceraldehyde-3-phosphate dehydrogenase is derived from the method of Beisenherz et al.[1] The substrate for the so-called "back reaction," 3-phospho-D-glycerol phosphate, is produced by the ATP-dependent phosphorylation of D-glycerate 3-phosphate catalyzed by phosphoglycerate kinase. Then 3-phospho-D-glycerol phosphate is reduced by NADH to D-glyceraldehyde 3-phosphate by glyceraldehyde-3-phosphate dehydrogenase.

Reagents

Triethanolamine chloride buffer, 0.05 M, pH 7.5
NADH (14 mM), 10 mg of disodium salt per milliliter
MgSO$_4$, 0.5 M
ATP (16.5 mM), 10 mg of disodium salt · 3 H$_2$O per milliliter
Glycerate 3-phosphate (93 mM), 50 mg of tricyclohexylammonium salt · 3 H$_2$O per milliliter
L-cysteine (114 mM), 20 mg of cysteine hydrochloride per milliliter, prepared daily
3-Phosphoglycerate kinase from yeast (10 mg/ml, 450 units/mg), crystalline suspension in 3.2 M ammonium sulfate (Boehringer Mannheim)

Enzyme Dilution. If necessary, the enzyme is diluted in 0.05 M triethanolamine, pH 7.5, 0.01 M EDTA, and 0.05 M mercaptoethanol to 50–400 units/ml.

Procedure. The following reagents are added to a cuvette with a 10-mm light path: 2.0 ml of buffer, 0.05 ml of NADH, 0.04 ml of MgSO$_4$, 0.2 ml of ATP, 0.1 ml of glycerate 3-phosphate, 0.1 ml of L-cysteine, 0.005 ml of phosphoglycerate kinase, and water to a total volume of 3.0 ml. The reaction is started by the addition of 0.01–0.05 ml of enzyme solution at 25°. The change in absorbance is recorded at 365 nm, 334 nm, or 340 nm.

Definition of Unit and Specific Activity. One unit is defined as the amount of enzyme that catalyzes the reduction of 1 μmol of 3-phospho-D-glycerol phosphate to D-glyceraldehyde 3-phosphate per minute. The units can be calculated from the change in optical density per minute using the following absorbance coefficients: NADH (Hg): 334 nm = 6.176 × 10^2; 340 nm = 6.317 × 10^2 and 365 nm = 3.441 × 10^2 (liter mol^{-1} mm^{-1}).[2] Specific activity is based on a spectrophotometric determination of protein concentration according to Lowry et al.[3]

[1] G. Beisenherz, H. J. Boltze, T. Bücher, R. Czok, K. H. Garbade, E. Meyer-Arendt, and G. Pfleiderer, *Z. Naturforsch.* **8b**, 555 (1953).
[2] H. Netheler, in "Grundlagen der enzymatischen Analyse" (H. U. Bergmeyer and K. Gawehn, eds.), p. 145. Verlag Chemie, Weinheim, 1977.

Purification Procedure

Human tissues (e.g., liver, kidney, heart and skeletal muscle, lung, and brain) were obtained 24–48 hr postmortem; blood, connective tissues, and fat were removed, then the samples were frozen and stored up to 3 months at −20°. Blood was obtained from the blood bank.

All purification steps are performed at 4°.

Step 1. Extraction. Tissues: 500 g frozen tissue are thawed at 5° overnight. Small portions are homogenized in a blender with 3 volumes (w/v) of triethanolamine buffer 0.005 *M*, 0.01 *M* EDTA, and 0.02 *M* mercaptoethanol, pH 7.5 (buffer I). The suspension is stirred for 30 min and centrifuged for 45 min at 10,000 g. To remove fat and small particles, the supernatant is filtered through a layer of quartz wool.

Erythrocytes: 1 liter of packed erythrocytes is washed three times with the same volume of isotonic NaCl solution and lysed by the addition of 5 liters of buffer I. Insoluble material is removed by centrifugation.

Step 2. Ammonium Sulfate Fractionation. To 100 ml of extract, 39 g of powdered ammonium sulfate are added during 30 min; stirring is continued for 1 hr, and then the suspension is centrifuged at 23,000 g for 30 min. To 100 ml of the clear red supernatant, 22.7 g of ammonium sulfate are added and the pH is adjusted to 8.6 (as measured directly with pH indicator paper) by the addition of 5 *N* ammonium hydroxide solution. After stirring for another 30 min, the suspension is filtered overnight in large funnels through filter paper (Schleicher & Schüll, No. 1573, Dassel, GFR).

Step 3. Chromatography on DEAE-Sephadex G-50. At this step it is necessary to keep the volume as small as possible. Therefore the precipitate is scraped from the filter paper and directly packed into dialysis tubing. The precipitate is dialyzed for approximately 40 hr against 100 volumes of buffer, the buffer being changed 3 or 4 times. The precipitate dissolves at the beginning of this procedure. At the end of the dialysis, undissolved particles are removed by centrifugation. If the volume of the solution exceeds 70 ml, it is concentrated by dialysis against 30% polyethylene glycol in buffer I.

The protein solution is carefully layered on a DEAE-Sephadex G-50 column (5 × 72 cm) equilibrated with buffer I and eluted with the same buffer; 15-ml fractions are collected, and the absorbance is measured at 280 nm. The first protein peak eluted contains the enzyme. The GAPDH-containing fractions are collected, and the proteins are precipitated by the addition of 66.2 g of ammonium sulfate per 100 ml at pH 8.6 adjusted with

---

[3] O. H. Lowry, N. J. Rosebrough, A. L. Farr, and R. J. Randall, *J. Biol. Chem.* **193**, 265 (1951).

PURIFICATION OF GLYCERALDEHYDE-3-PHOSPHATE DEHYDROGENASE FROM HUMAN LIVER

Step	Volume (ml)	Specific activity (units/mg protein)	Protein (mg)	Purification (fold)	Yield (%)
1. Extraction	1,370	0.99	18,500	1	100
2. Ammonium sulfate fractionation	65	2.69	3,900	2.7	58
3. Chromatography on DEAE-Sephadex	156	43.5	186	44	44
4. Molecular sieve chromatography	46	51.0	81.6	52	20
5. Crystallization	54	68.1	62.0	69	23

5 N ammonium hydroxide solution. After 30 min the precipitate is collected by centrifugation, dissolved in buffer I, and dialyzed overnight against buffer I.

Step 4. Molecular Sieve Chromatography. The dialyzed solution is applied to an Ultrogel AcA-34 (LKB Stockholm) column (3.1 × 135 cm) equilibrated with buffer I. The enzyme is eluted with the same buffer, and GAPDH-containing fractions are collected.

Step 5. Crystallization. To 100 ml of eluate, 47.2 g of powdered ammonium sulfate are added, and the precipitate is removed by centrifugation. The pH of the supernatant is raised to 8.6 with 5 N ammonium hydroxide solution, and ammonium sulfate is slowly added until a silky luster appears. The suspension is kept at 4° for 10 days. Small needles appear; later these change into a rhombohedral form in the case of the heart and liver enzyme. The crystal suspension can be stored at 5° for more than 3 months without loss of activity.

A typical purification is summarized in the table.

Properties

Homogeneity. The purified enzyme prepared from the different sources were homogeneous when examined by SDS–polyacrylamide gel electrophoresis. In zone electrophoresis at pH 8.6 and 4.9, using cellogel strips, the samples migrate as one zone with identical mobility. In the presence of 6 M urea identical mobility of the subunits was observed for all samples from human tissues.

Molecular Weight. As calculated from sedimentation equilibrium runs, a molecular weight of 142,000 could be detected for glyceraldehyde-3-phosphate dehydrogenase from human tissues.

pH Optimum. With glyceraldehyde 3-phosphate as substrate and in the presence of arsenate, the pH optimum was found to be 8.0–8.3. With 3-phospho-D-glyceroylphosphate, the pH optimum was 7.2–7.3.

K_m and K_i Values. At pH 7.0, K_m values of 10 to 21 × $10^{-6} M$ and at pH 8.2 of 10 to 32 × $10^{-6} M$ were obtained for glyceraldehyde 3-phosphate. The corresponding data for NAD$^+$ were 55 to 210 × $10^{-6} M$ at pH 7.0 and 13 to 98 × $10^{-6} M$ at pH 8.2. NADH is a powerful competitive inhibitor showing K_i values of 2 to 18 × $10^{-6} M$ and 1 to 11.7 × $10^{-6} M$ at pH 7.0 and 8.2, respectively.[4]

For the substrate, 3-phospho-D-glyceroyl phosphate at pH 7.0 and 8.2, the K_m values were 4 to 21 × $10^{-6} M$ and 7 to 14 × $10^{-6} M$; the corresponding data for NADH were 5 to 14 × $10^{-6} M$ and 3 to 9 × $10^{-6} M$. NAD acts as a weak competitive inhibitor with K_i of 171 to 430 × $10^{-6} M$ at pH 7.0 and 102 to 284 × $10^{-6} M$ at pH 8.2.[4]

Immunological Properties. Sheep antibodies against glyceraldehyde-3-phosphate dehydrogenase from skeletal muscle and liver were prepared. In immunodiffusion experiments according to Ouchterlony, the two antibodies react similarly with the preparations from different tissues. In kinetic experiments the two antibodies inactivate completely all enzymes from human tissues.

[4] G. Dinkel and C. Dinkel, Dissertation, Medizinische Hochschule Hannover, 1976.

[53] Glyceraldehyde-3-phosphate Dehydrogenase from Rabbit Muscle

By R. M. Scheek *and* E. C. Slater

D-Glyceraldehyde 3-phosphate + NAD$^+$ + P$_i$ ⇌ 1,3-diphosphoglycerate + NADH + H$^+$

Its ready availability has made rabbit muscle one of the most frequently used sources for glyceraldehyde-3-phosphate dehydrogenase [D-glyceraldehyde-3-phosphate : NAD$^+$ oxidoreductase (phosphorylating), EC 1.2.1.12]. Classical isolation procedures rest upon the high solubility of the NAD$^+$-containing enzyme in $(NH_4)_2SO_4$ solutions[1]; several recrystallization steps from concentrated $(NH_4)_2SO_4$ solutions at pH values above 8.0 complete these procedures, but poorly understood differences were found, most significant in the NAD-binding properties, between the enzyme obtained in this way and that from other muscle sources

[1] R. E. Amelunxen and D. O. Carr, this series, Vol. 41, p. 264.

(e.g., sturgeon muscle[2]). These differences largely disappear when the isolation is carried out as described below.[3]

Assay

The assay method of Krebs has been presented in detail elsewhere in this series.[1,4] We shall describe here another frequently used assay, first published by Ferdinand,[5] that is routinely used in our laboratory.

Reagents

DL-Glyceraldehyde-3-phosphate (DL-GAP) in water, prepared from the water-insoluble barium salt of the diethylacetal (Boehringer) by treatment of the suspension in water with a cation-exchange resin (Dowex 50W-X8, H^+ form) to dissolve the material and heating on a boiling water bath for 2–3 min to restore the aldehyde function. The concentration of D-GAP is determined enzymically as described in this series[1,4] and adjusted to 10 mM.

NAD$^+$ (10 mM) in water, made up with commercially available NAD$^+$; its concentration is determined with ethanol and alcohol dehydrogenase (EC 1.1.1.1) as described elsewhere.[6]

Assay buffer: EDTA, 0.2 mM, Na$_2$HPO$_4$, 50 mM, and triethanolamine, 40 mM; the pH is adjusted to 8.9.

Procedure. A typical assay mixture contains 2.0 ml of the assay buffer, 0.2 ml of 10 mM NAD$^+$ solution, and 0.1 ml of the enzyme solution (containing about 1 μg of enzyme). The reaction is started by the addition of 0.2 ml of 10 mM D-GAP solution and monitored continuously during the earliest 30 sec of the reaction at 340 nm in a 1-cm spectrophotometer cell, thermostated at 25°. The extinction coefficient of NADH at this wavelength ($6.22 \times 10^3 M^{-1}$ cm^{-1}) is used to arrive at the specific activity, expressed in micromoles of NADH produced per minute per milligram of protein.

Isolation

Extraction. A rabbit is killed by a blow on the head, and the skeletal muscles (typically about 300 g) are rapidly removed from the back and

---

[2] N. Kelemen, N. Kellershohn, and F. Seydoux, *Eur. J. Biochem.* **57**, 69 (1975).

[3] R. M. Scheek and E. C. Slater, *Biochim. Biophys. Acta* **526**, 13 (1978).

[4] E. G. Krebs, this series, Vol. 1, p. 407.

[5] W. Ferdinand, *Biochem. J.* **92**, 578 (1964).

[6] M. M. Ciotti and N. O. Kaplan, this series, Vol. 3, p. 891.

legs. These are minced at 5° into a solution (about 120 ml per 100 g of muscle) containing 30 mM KOH, 5 mM EDTA, and 1 mM dithiothreitol (DTT). (The use of a Waring blender in this step was described by Amelunxen and Carr.[1]) All subsequent steps are carried out between 0 and 5°. After stirring for 15–20 min the extract is separated by centrifugation (20 min at 30,000 g) and the mince is reextracted with an equal volume of fresh medium.

Ammonium Sulfate Fractionation. To the combined extracts, the pH of which is about 7, solid $(NH_4)_2SO_4$ (Merck, Suprapur) is added in three stages: 30 g per 100 ml of extract at the first, 11.7 g per 100 ml of the supernatant solution at the second, and 4 g per 100 ml of the remaining solution at the third. The pH is kept between 7.0 and 7.5 with a 35% (w/w) NH_3 solution. After each of the first two additions, the suspension is allowed to stand for 30 min and the precipitated protein is removed by centrifugation (60 min at 30,000 g). After the last addition, the pH is brought to 7.8–8.0 with the NH_3 solution. After standing overnight the suspension is centrifuged (60 min at 30,000 g), and the pellet is dissolved in about 50 ml of a buffer (pH 6.4) containing 5 mM morpholinopropane sulfonate (MOPS), 1 mM EDTA, and 1 mM DTT.

Chromatography.[7] The deep-red clear solution is desalted by passing through a column (length 40 cm, diameter 4 cm) of Sephadex G-50 (fine grade), equilibrated with the same MOPS–EDTA–DTT buffer (pH 6.4). (This step can be replaced by extensive dialysis, which, however, is more time-consuming.) The filtrate is transferred to a column of CM-Sephadex (C-25, K+ form) of the same dimensions. After washing with the same buffer until the first protein appears in the eluate, the enzyme is eluted with a buffer (pH 7.5) containing 5 mM MOPS, 1 mM EDTA, 1 mM DTT, and 0.1 M $(NH_4)_2SO_4$. Pure enzyme appears in an asymmetrical peak, showing a sharp rise and some tailing. The combined fractions (about 100 ml containing 7–10 mg enzyme per milliliter at pH 6.5) are made 1 mM in NAD^+ and brought to pH 7.0 with the NH_3 solution. Then 50 g of solid $(NH_4)_2SO_4$ are added per 100 ml of the yellow holoenzyme solution. The enzyme crystallizes within a few days. From 100 g of muscle, about 300 mg of enzyme are obtained.

Storage. The enzyme is stable for several months when stored in this suspension at 5°. Occasional refreshment of the medium (by centrifugation and resuspension in fresh medium) is advisable when storage over long periods is necessary. Freezing in liquid nitrogen always results in partial loss of activity.

---

[7] W. Bloch, R. A. MacQuarrie, and S. A. Bernhard, *J. Biol. Chem.* **246**, 780 (1971).

Experimental Procedures

Preparation of Holoenzyme. This is carried out at 17°.[3] The holoenzyme is collected by centrifugation of the suspension in $(NH_4)_2SO_4$, dissolved in a small volume of the buffer to be used in the experiments and passed through a column of Sephadex G-50, equilibrated with the same buffer.

Preparation of Apoenzyme. This is carried out at 17°.[3] To prepare NAD^+-free enzyme, the suspension in $(NH_4)_2SO_4$ is centrifuged, the pellet is dissolved in the MOPS–EDTA–DTT buffer (pH 6.4) and the solution is passed through a Sephadex G-50 column, equilibrated with this buffer. The filtrate is applied to a CM-Sephadex (C-25, K^+ form) column of the same dimensions, equilibrated with the same buffer, and the NAD^+ is eluted with this buffer. When the elution is complete, as judged by the absorbance at 260 nm, the apoenzyme is eluted with the MOPS–EDTA–DTT buffer (pH 7.5) containing 0.1 M $(NH_4)_2SO_4$. The NAD^+ content is determined with ethanol and alcohol dehydrogenase on the neutralized supernatant after precipitating the protein with 4% trichloroacetic acid; the enzyme contains no detectable NAD^+ when treated in this manner.

Preparation of Tetra(3-phosphoglyceroyl) Enzyme.[8] Fully acylated enzyme is prepared by the addition of an 8- to 12-fold molar excess of 1,3-diphosphoglycerate (prepared and purified as described by Furfine and Velick[9]) to the holoenzyme. Acylation can be followed spectrophotometrically at 360 nm and was found to be complete. The lifetime of this enzyme species is limited by the hydrolysis of the 3-phosphoglyceroyl groups, covalently attached to Cys-149, which occurs with a pseudo first-order rate constant of 0.18 min^{-1} at 20° in the MOPS–EDTA–DTT buffer (pH 7.0). Hydrolysis can be slowed down by decreasing temperature and/or increasing ionic strength: at 10° and with 0.1 M $(NH_4)_2SO_4$ present in the same buffer it occurs with a rate constant of 0.02 min^{-1}; the enzyme remains fully acylated, however, as long as 1,3-diphosphoglycerate is still present in the medium (unpublished observations[10]).

Properties

The specific activity of the enzyme prepared as described is about 185 units per milligram of protein. Protein concentrations are determined spectrophotometrically at 280 nm using 1.00 ml mg^{-1} cm^{-1} and 0.83 ml

[8] R. M. Scheek, J. A. Berden, R. Hooghiemstra, and E. C. Slater, *Biochim. Biophys. Acta* **569**, 124 (1979).

[9] C. S. Furfine and S. F. Velick, *J. Biol. Chem.* **240**, 844 (1965).

[10] R. Hooghiemstra and R. M. Scheek (1979).

mg^{-1} cm^{-1} for the extinction coefficients of the holo- and apoenzyme, respectively. The enzyme migrates as a single band in polyacrylamide gel electrophoresis and has a molecular weight of 145,000. It contains 3.8–4.0 highly reactive cysteine residues, as determined with dithionitroben-zoate,[11] and can bind 4.0 molecules of NAD^+ or NADH per enzyme molecule.

The most striking property of the enzyme isolated and purified in the way described above, when compared with the classical preparations used in various laboratories, is the two orders of magnitude increase in binding strength of both oxidized and reduced NAD molecules at high saturation of the binding sites. Binding was measured at 16° in the MOPS–EDTA–DTT buffer (pH 7.6) containing 0.1 M $(NH_4)_2SO_4$; for NAD^+ microscopic dissociation constants of 0.08 and 0.18 μM were found at low and high saturation, respectively, whereas for NADH these constants are 0.5 at low saturation and 1.0 μM at high saturation.[8]

When ³H-labeled NAD^+ is bound to the enzyme, it exchanges rapidly and completely, even at 5°, with unbound NAD^+ present in the eluting buffer during a gel filtration on Sephadex G-50; any other bound nucleotide (ADPR, NADH, ATP) can be easily replaced by NAD^+ in this manner.

The absorbance at 360 nm, which appears upon NAD^+ binding (Racker band, $\epsilon = 1.00$ mM^{-1} cm^{-1}) is linearly related to the saturation of the four binding sites. Alkylation (e.g., with iodoacetate) or acylation (e.g., with the product 1,3-diphosphoglycerate) of the holoenzyme results in complete disappearance of this broad charge-transfer band and a sharp decrease of the NAD^+-binding strength. The tetra(3-phosphoglyceroyl) enzyme binds NAD^+ with a K_d (about 65 μM) consistent with the K_i value of NAD^+ for inhibition of the back reaction, measured under the same conditions.[8]

[11] F. Seydoux, S. Bernhard, O. Pfenninger, M. Payne, and O. P. Malhotra, *Biochemistry* 12, 4290 (1973).

[54] Glyceraldehyde-3-phosphate Dehydrogenase from Pig Liver[1-3]

By SHAWKY M. DAGHER[4] and WILLIAM C. DEAL, JR.

Glyceraldehyde 3-Phosphate + NAD^+ + P_i ⇌ 1,3-diphosphoglycerate + NADH + H^+

In contrast to the muscle and heart enzymes, liver glyceraldehyde-3-phosphate dehydrogenase (GAPD) operates in the backward direction when conditions require gluconeogenesis. With the rabbit muscle enzyme, the backward reaction is strongly inhibited by NAD,[5] whereas the liver enzyme is only weakly inhibited by NAD.[2] Other preparations of liver GAPD have used ethanol precipitation at subzero temperatures.[5] The procedure described here makes use of the fact that the pig liver enzyme is highly soluble in metal ion and high salt concentrations; a major step is initial precipitation of large amounts of unwanted proteins with $ZnCl_2$.

Assay Methods

The reaction sequence above is the physiological reaction; in the laboratory assay for the enzyme, we replace P_i by arsenate. The product then is 1-arseno-3-phosphoglycerate, which spontaneously and rapidly decomposes to 3-phosphoglycerate, thereby driving the reaction far to the right.

The reaction velocity may show a decrease with time unless precautions are taken to avoid this: (*a*) product inhibition by NADH; (*b*) product inhibition by 1,3-diphosphoglycerate; and (*c*) deterioration of glyceraldehyde-3-P with time. For these reasons, accurate data should be obtained in the first few seconds of the assay, before NADH has accumulated or glyceraldehyde-3-P has deteriorated; also, glyceraldehyde-3-P should be added to the assay last. The use of arsenate in the assay, in place of phosphate, avoids the inhibition by 1,3-diphosphoglycerate.

---

[1] EC 1.2.1.12, D-glyceraldehyde-3-phosphate: NAD^+ oxidoreductase(phosphorylating).

[2] Parts of this chapter were adapted, with permission, from S. M. Dagher and W. C. Deal, Jr., *Arch. Biochem. Biophys.* **179**, 643 (1977).

[3] This is paper 21 in a series entitled Metabolic Control and Structure of Glycolytic Enzymes. Paper 20 by C. S. Johnson and William C. Deal, Jr., *J. Biol. Chem.* **257**, 913 (1982).

This work was supported in part by grants from NIH (GM-11170) and the Michigan Agricultural Experiment Station (Hatch 1273, Publication No. 9916).

[4] Present address: Faculty of Agricultural Sciences, American University of Beirut, Beirut, Lebanon.

[5] C. M. Smith and S. F. Velick, *J. Biol. Chem.* **247**, 273 (1972).

Reagents

Tris-HCl, 0.1 M, pH 8.5 (60 mM)
KCl, 2.0 M (100 mM)
Disodium arsenate, 0.2 M (10 mM)
Cysteine-HCl, 0.06 M (7.5 mM)
NAD, 0.01 M, H$^+$, (1 mM) or 0.18 M, pH 6.0 (18 mM)
Enzyme, 50 μg/ml (1.25 μg/ml)
DL-Glyceraldehyde 3-phosphate (H$^+$), 10 mM (0.5 mM)

The final pH of the assay is 7.8; the high pH of the stock buffer compensates for the acidity of several of the added components. A several months supply of stock Tris–arsenate–KCl is made by mixing the appropriate volumes of these reagents. An additional assay stock, cysteine–Tris–arsenate–KCl is stable for about 6 hr maximum, so it is prepared fresh daily when numerous assays must be run. An additional stock assay mix, containing NAD$^+$, along with all assay components except enzyme and substrate, is stable for about 2 hr, so it is used only when assays must be run in rapid succession. The protocol described below is for addition of individual components.

Our 0.4 ml assay contains 0.24 ml of Tris-HCl, 0.02 ml of KCl, 0.02 ml of disodium arsenate, 0.05 ml of cysteine (H$^+$), 0.04 ml of NAD$^+$, 10 μl of enzyme, and 0.02 ml of DL-glyceraldehyde-3-P (H$^+$). All components except enzyme and substrate are added and mixed, then enzyme is added and mixed, and the blank is run. The assay is then begun by adding glyceraldehyde-3-P. If NADH product inhibition is a problem for routine assays, then a "high NAD$^+$ assay" is recommended. This involves an 18-fold higher concentration of NAD$^+$, and a 2-fold higher concentration of glyceraldehyde-3-P. To achieve this, the high stock NAD$^+$ solution is used, 0.04 ml of glyceraldehyde-3-P is used instead of 0.02 ml, and the buffer is reduced from 0.24 ml to 0.22 ml.

NAD$^+$ is prepared at a concentration of 0.2 M. It is then diluted to 0.01 M for the "low NAD" (standard) assay, divided up into several tubes and frozen. For the "high NAD$^+$" assay, the 0.2 M NAD is neutralized to pH 6.0 with 2 M NaOH, adjusted to 0.18 M final concentration, and frozen.

GAPD activity in the "back" reaction, i.e., with NADH and 1,3-diphosphoglycerate as substrates, is determined by the method of Rossner.[6] The reaction is started by addition of the enzyme. This allows the coupling system of phosphoglycerate kinase and 3-phosphoglyceric acid to equilibrate prior to initial velocity measurements.

Unit. An enzyme unit is defined as the amount of enzyme oxidizing or reducing 1 μmol of substrate per minute at 25°. The specific activity is expressed as enzyme units per milligram of protein.

[6] G. Rossner, *Arch. Klin. Exp. Dermatol.* **222,** 383 (1965).

Substrate. The concentration of the substrate, GAP, and the coenzyme, NAD, are measured spectrophotometrically using an end-point enzymic analysis. The assay for GAP contains the following: 50 mM Tris, pH 8.5; 18 mM NAD; 10 mM sodium arsenate; 50 mM KCl, 50 μg/ml of rabbit muscle GAPD, and 0.19 ml of the GAP solution to be tested. The total volume is 0.4 ml. Matched cuvettes are used. The blank contains all the above ingredients except that GAP is replaced with an equal volume of water. After the spectrophotometer has been zeroed with the blank, the reaction is started by addition of GAP to the second cuvette. The progress of the reaction is monitored with a recorder. Upon completion, the increase in absorbance is calculated from the initial and final readings on the absorbance indicator. For determination of NAD concentration, the procedure is identical except that NAD replaces GAP as the variable component. For each determination, at least two aliquots of different concentrations of the component to be tested are used. The concentration of GAP or NAD is calculated according to the following equation:

$$\text{concentration } (\mu\text{mol/ml}) = \frac{A_{340\,\text{nm}} \times 0.1608 \times \text{total assay volume}}{\text{volume of unknown added to assay}}$$

Purification Procedure[2]

The purification procedure described here is for 300 g of pig liver. However, it has been successfully scaled up to samples of whole liver (approximately 1200 g). All steps are carried out at room temperature unless indicated otherwise.

Step 1. Homogenization in ZnCl$_2$ Solutions. A 300-g sample of freshly thawed liver is homogenized in a Waring blender with 300 ml of a cold solution of 10 mM ZnCl$_2$ containing 25 mM 2-mercaptoethanol. This "crude homogenate" is then centrifuged for 15 min at 27,300 g. The supernatant is collected and made 50 mM in 2-mercaptoethanol and 50 mM in EDTA to remove the excess ZnCl$_2$. Other studies have shown that the ZnCl$_2$ step is as effective with the *crude* homogenate as with the supernatant of the *centrifuged crude* homogenate, so it is not necessary to centrifuge before the ZnCl$_2$ step.

Step 2. Ammonium Sulfate Fractionation. The supernatant solution obtained in step 1 is chilled in an ice bath. Solid ammonium sulfate is added slowly with constant stirring. The precipitate that forms between 2.67 M and 3.9 M is collected by centrifugation at 27,000 g for 30 min and dissolved in 20 ml of 10 mM imidazole buffer, pH 7.5, containing 25 mM 2-mercaptoethanol. The clear red solution is then desalted into deionized water on a column of Sephadex G-25 (6 × 96 cm) equilibrated with deionized water. The salt-free eluate, which has an average conductivity of 45 μmho as measured by a Radiometer conductivity meter, is made 50 mM in 2-mercaptoethanol and concentrated to 20 ml in an Amicon TCF 10 ultrafiltration cell using a PM-30 membrane.

Step 3. Chromatography on DEAE-Sephadex. The salt-free concentrated enzyme solution is adjusted to pH 7.3 with 0.1 N NH$_4$OH and centrifuged at 23,500 g for 10 min. The dark brown supernatant is applied to a DEAE-Sephadex A-50 column (2.5 × 45 cm) previously equilibrated with 10 mM imidazole buffer, pH 7.3, containing 25 mM 2-mercaptoethanol. The column is eluted with the same buffer, and the enzyme appears with the solvent front in a clear, colorless fraction. Recovery of the enzyme activity in this step is complete, and a fourfold purification is obtained.

Step 4. Chromatography on SE-Sephadex. The DEAE eluates are concentrated to 2 ml under N$_2$ pressure in an Amicon Model 52 ultrafiltration cell employing a PM-30 membrane. The concentrated enzyme solution is made 5 mM in EDTA and 0.2 mM in NAD and adjusted to pH 6.6 with 0.1 N acetic acid. After removal of the white precipitate by ultracentrifugation at 23,500 g for 10 min, the supernatant (1–2 ml) is applied to an SE-Sephadex C-25 column (1.5 × 30 cm) previously equilibrated with 10 mM imidazole buffer, pH 6.6, containing 5 mM EDTA and 25 mM 2-mercaptoethanol. The column is then washed with 150 ml (one column volume) of the equilibrating buffer, and the enzyme is eluted, in the same buffer, with a linear gradient of 0 to 0.3 M KCl (150 ml in each reservoir). Essentially pure enzyme emerges from the column in the third protein peak.

Step 5. Crystallization. The tubes containing the third protein peak from the SE-Sephadex column are pooled and concentrated by ultrafiltration as described in step 4. Minor impurities are removed by ammonium sulfate crystallization of the enzyme, according to Jakoby,[7] in solutions also containing 5 mM 2-mercaptoethanol and 0.2 mM NAD. The stock saturated (4°) ammonium sulfate solution is also adjusted to pH 8.4 with NH$_4$OH; by convention, this pH determination is made on a sixfold diluted sample at 0°. Crystals in the form of thin plates appear overnight. The highest yields are obtained at 3.2 M ammonium sulfate.

Detailed results of a typical purification of GAPD from pig liver are given in Table I. The major steps include precipitation of some undesired protein with 10 mM ZnCl$_2$, 2.67–3.9 M ammonium sulfate fractionation, desalting, concentration, negative absorption on DEAE-Sephadex, concentration, positive absorption on SE-Sephadex, elution with KCl, and,

[7] W. B. Jakoby, this series, Vol. 22, p. 248.

TABLE I

PURIFICATION OF GLYCERALDEHYDE-3-PHOSPHATE DEHYDROGENASE (GAPD) FROM PIG LIVER[a]

Fraction	Volume (ml)	Total protein (mg)	Specific activity (units/mg protein)	Purification (fold)	Cumul recov (%)
Supernatant[b] of crude homogenate	352	55,552	0.51	—	10(
Supernatant of zinc chloride treatment	274	15,892	1.1	2	8:
Ammonium sulfate precipitate	32	704	22	40	6'
DEAE-Sephadex A-50	71	142	69	130	6:
SE-25 Sephadex	59	26	132	260	2.
Crystals	4	18	150	300	1(

[a] Details are given in the text under Purification Procedure. Enzyme activity was measured at 25° data are based on 300 g of whole liver.

[b] A 300-g sample of freshly thawed liver was homogenized in a Waring blender with 300 ml of deionized distilled water. This "crude homogenate" was then centrifuged for 30 min at 27,300 g supernatant was collected and made 25 mM in 2-mercaptoethanol.

finally, crystallization with ammonium sulfate. The enzyme crystallized twice from ammonium sulfate solution was considered pure because further crystallizations failed to increase the specific activity, and all physical tests indicated homogeneous protein; a single symmetric peak was observed in sedimentation velocity experiments and a single band on SDS gel electrophoresis.

Fresh, quick-frozen liver (frozen in Dry Ice within 10 min of sacrifice of the animal) did not fractionate as well as livers aged in ice 2–3 hr after the sacrifice of the animal before being frozen to −20°. However, the total number of units of enzyme of the unaged liver samples was only slightly less than that of aged samples. The main difficulty encountered with very fresh or quick-frozen livers was the failure to obtain clear supernatants in the relatively low-speed centrifugation steps used in this procedure.

Properties

Absorption Spectrum. The ultraviolet absorption properties of the enzyme suggest the presence of a prosthetic group, probably 2 mol or more of tightly bound NAD per mole of apoenzyme, as estimated from the data of Fox and Danliker[8]; the $A_{280\,nm} : A_{260\,nm}$ ratio was found to be 1.19. In this respect, pig liver GAPD is similar to the rabbit muscle and rabbit liver enzymes, which crystallize with 2–3 mol of bound NAD,[8] but different

[8] J. B. Fox and W. B. Dandliker, *J. Biol. Chem.* **221**, 1005 (1956).

from the yeast enzyme, which crystallizes with no detectable NAD unless very special steps are taken.

Subunit Structure and Physical Properties. Sodium dodecyl sulfate–polyacrylamide gel electrophoresis experiments showed a single protein band, indicating that the enzyme is pure and probably consists of similar, if not identical, subunits. A subunit molecular weight of 38,000 was calculated. The enzyme sediments as a single symmetrical peak throughout sedimentation velocity experiments, and its sedimentation coefficient is practically independent of protein concentration, yielding a value of $s_{20,w}^0 = 7.8$ S. From a graph of sedimentation coefficients as a function of molecular weight for a series of proteins known to be globular in shape, it is estimated that this value corresponds to a molecular weight of 145,000. Sedimentation equilibrium experiments yield a molecular weight value of 1.48×10^5, assuming a value of 0.74 ml/g for the partial specific volume.[5] The isoelectric point of the enzyme is pH 8.8. The enzyme is dissociated into subunits by ATP[2] and can be reversibly dissociated by urea.[2]

Catalytic Properties. The optimum pH range for maximum catalytic activity is pH 8.4–8.8, with half-maximum activity at pH 7.4 and 9.6. The decrease in enzyme activity at the higher pH region is not as sharp as that

TABLE II

DETERMINATION OF KINETIC CONSTANTS FOR PIG LIVER GLYCERALDEHYDE-3-PHOSPHATE DEHYDROGENASE (GAPD)[a]

Reaction	Substrate or inhibitor	K_m rabbit muscle GAPD[b] (μM)	K_m or K_i^c pig liver GAPD (μM)	V_{max} pig liver GAPD[d] (μmol/min)	Amount of protein per assay[d] (μg)
+ NAD + As	GAP	90	300 ± 30	153 ± 7	1.4
+ NAD + As	NAD	13	6 ± 2	287 ± 12	0.7
i-PGA[e] + NADH	NADH	3.3	23	50	0.4
i-PGA[e] + NADH	NAD(I)	—	850[f] or 870[g]	—	0.4

All assays were carried out at 25°. In the forward direction, the concentration of substrates was varied up to 1.1 m*M* for GAP and 240 m*M* for NAD. In the backward direction, NADH concentration was varied up to 150 μM. *V* decreases with decreased enzyme concentration.

Data for rabbit muscle GAPD are from S. F. Velick and C. Furfine *in* "The Enzymes" (P. D. Boyer, H. Lardy, and K. Myrbäck, eds.), 2nd ed., Vol. 7, p. 243, Academic Press, New York, 1963.

K_i for NADH inhibition of oxidative phosphorylation was not determined because the Lineweaver–Burk plots were curved.

The V_{max} values shown have been converted to activity in a solution of 1 mg of protein per liter.

Produced from 3-PGA using excess phosphoglycerate kinase.

From a Dixon plot.

Computer analysis of Lineweaver–Burk data.

reported by Bondi et al.[9] for the crude and purified enzymes from rabbit muscle, liver, and heart. Kinetic constants for pig liver GAPD for both the forward and the reverse reaction are shown in Table II. Of special interest is the large value for K_i (NAD) for the back reaction; the value of 860 μM is much larger than the value of 100 μM reported for the muscle enzyme under almost identical conditions,[5] but not as high as the values of 1000–2000 reported[5] for the rabbit liver enzyme (determined as half-maximum inhibition under saturating conditions).

[9] E. Bondi, J. Watkins, and M. E. Kirtley, *Biochim. Biophys. Acta* **185**, 305 (1969).

[55] Glyceraldehyde-3-phosphate Dehydrogenase (Glycolytic Form) from Spinach Leaves

By M. L. Speranza and G. Ferri

D-Glyceraldehyde 3-phosphate + NAD$^+$ + P$_i$ $\rightleftharpoons$
$$1,3\text{-diphosphoglycerate} + NADH + H^+ \quad (1)$$
D-Glyceraldehyde 3-phosphate + NAD$^+$ $\xrightarrow{\text{HAsO}_4^-}$
$$3\text{-phosphoglycerate} + NADH + H^+ \quad (2)$$

Two different glyceraldehyde-3-phosphate dehydrogenases exist in green leaves. One (EC 1.2.1.13), localized in the chloroplast, is active with either NADP$^+$ or NAD$^+$, and is thought to function in the photosynthetic production of hexoses. The other (EC 1.2.1.12), located in the cytoplasm, requires NAD$^+$ for activity and is associated with the glycolytic process.

A procedure for the purification of the latter enzyme is given below.

Assay Method

Principle. The enzyme is assayed spectrophotometrically at 25° by measuring the rate of increase in absorbance at 340 nm due to the reduction of NAD$^+$ in the presence of arsenate [reaction (2)].

Reagents

DL-Glyceraldehyde 3-phosphate solution, prepared from the barium salt of the diethyl acetal and assayed for the D-isomer content according to the specifications of the manufacturer (Sigma Chemical Co., St. Louis, Missouri). The standardized solution is diluted to 30 mM with respect to the D-isomer.

NAD⁺, 20 mM

Sodium pyrophosphate buffer, 45 mM, pH 8.4, containing 5 mM EDTA

Sodium arsenate, 0.1 M

Procedure. The reaction cuvette contains, in a final volume of 1.5 ml: 1 ml of pyrophosphate buffer (containing EDTA), 0.1 ml of arsenate, 0.05 ml of NAD⁺, 0.05 ml of D-glyceraldehyde 3-phosphate. The reaction is started by the addition of 5–10 μl of enzyme solution.

Definition of Unit and Specific Activity. One enzyme unit produces 1 μmol of NADH per minute, under the above conditions. Specific activity is expressed as units of enzyme activity per milligram of protein. Protein is determined by a modification of the biuret method.[1]

Purification Procedure

Unless otherwise stated, the following general conditions are used. All operations are carried out at 0–4°. Ammonium sulfate fractionations are made by the slow addition, under stirring, of solid salt.

Centrifugations are carried out at 14,000 g for 30 min. Ultrafiltrations are performed by an Amicon cell (Amicon N.V., Oostrhout-N.B., Holland), equipped with PM-30 membrane.

Fresh spinach purchased locally is used as starting material.

Step 1. Extraction. About 4 kg of spinach leaves, cleaned of stems and midribs, are thoroughly washed in tap water and rinsed in distilled water. Lots of 150 g of tissue are homogenized for 30 sec at low speed and for 60 sec at high speed in a domestic blender with 300 ml of 0.1 M potassium phosphate buffer, pH 9, containing 30 mM EDTA. The homogenates are combined and filtered through several layers of muslin.

Step 2. Heat Fractionation. The filtered solution is heated, in 2–3-liter portions, to 60° in a boiling water bath, maintained at 60° for 3 min, and cooled on ice. The heated solution is centrifuged, and the sediment is discarded.

Step 3. Ammonium Sulfate Fractionation. To the supernatant from step 2, 39% (w/v) of ammonium sulfate is added. After 30 min of additional stirring, the precipitate is removed by centrifugation. To the resulting supernatant, 22% (w/v) of ammonium sulfate is added; after standing overnight, the precipitate is collected by centrifugation (to reduce the volume, the upper clear part of the supernatant can be easily removed by suction). The precipitate is dissolved in a minimal volume of 20 mM

[1] R. F. Itzhaki and D. M. Gill, *Anal. Biochem.* **9**, 401 (1964).

Purification of Glyceraldehyde-3-phosphate Dehydrogenase
(Glycolytic) from Spinach Leaves

Fraction	Protein (mg)	Total units[a]	Specific activity (units/mg protein)
Ammonium sulfate precipitate	3850	3700	0.96
Acetone	780	3000	3.80
Sephadex G-100	270	1750	6.50
DEAE-Sephadex	17	742	43.60
NAD-Sepharose	9.2	740	80.40

[a] Up to Sephadex G-100 fraction, some residual photosynthetic enzyme contributes to the observed activity.

potassium phosphate buffer, pH 7.5, containing 1 mM EDTA and 1 mM 2-mercaptoethanol. The same buffer is used in all of the following steps, except the last one.

Step 4. Acetone fractionation. The above solution is placed in an ice–salt bath, and 0.8 volume of acetone, previously cooled at $-20°$, is added rapidly with stirring. The precipitate is collected by centrifugation ($20,000\,g$, 10 min) and suspended in about 40 ml of buffer. The undissolved protein is removed by centrifugation.

Step 5. Sephadex G-100 Chromatography. The solution obtained above is chromatographed on a 2000-ml column of Sephadex G-100 (5 × 100 cm) at a flow rate of 24 ml/hr. The enzyme activity is collected after about 700 ml of eluent have passed through the column.

Step 6. DEAE-Sephadex Chromatography. The collected active fractions from step 5 are applied to a 180-ml column of DEAE-Sephadex A-50 (2.4 × 40 cm). Elution is performed through a 500-ml linear gradient of KCl from 0 to 0.2 M, in phosphate buffer. Fractions of 10 ml are collected at 30-min intervals. The enzyme emerges from the column after about 350 ml of the gradient. The active fractions are pooled and concentrated by ultrafiltration to a protein content of about 2 mg/ml.

Step 7. Affinity Chromatography. The final enzyme purification is achieved by affinity chromatography on a column (1 × 2 cm) of NAD–hexane–Sepharose (P-L Biochemicals, Milwaukee, Wisconsin) equilibrated with 45 mM sodium pyrophosphate buffer, pH 8.5, containing 5 mM EDTA and 0.5 M sodium chloride.

The enzyme solution from DEAE-Sephadex is subjected to affinity chromatography in portions 2.5–3 ml. After washing with 16 ml of the equilibrating buffer, the enzyme is eluted with 5 ml of 10 mM NAD$^+$. Fractions of 2 ml are collected at 6-min intervals. The active fractions are concentrated by ultrafiltration to a protein content of about 10 mg/ml and stored at −20°.

A summary of a typical purification is given in the table.

Properties

The activity of the purified enzyme declines steadily at −20°; about 30% of the original activity remains after 6 months.

The native enzyme has a molecular weight of 150,000 and a subunit molecular weight of 37,000.[2]

Half-maximum velocity is attained with 0.24 mM D-glyceraldehyde 3-phosphate. The turnover number is 3000 per subunit.[3]

The enzyme shows optimum activity in the pH interval 8.9 to 9.1.

Each subunit has the following amino acid composition[2]: Asp-38, Thr-25, Ser-22, Glu-25, Pro-13, Gly-33, Ala-29, Cys-6, Val-36, Met-5, Ile-18, Leu-22, Tyr-8, Phe-15, Lys-28, His-8, Arg-12, Trp-4.

The N-terminal amino acid is alanine.

[2] M. L. Speranza and C. Gozzer, *Biochim. Biophys. Acta* **522**, 32 (1978).
[3] M. L. Speranza, M. Bolognesi, and G. Ferri, *Ital. J. Biochem.* **29**, 113 (1980).

[56] Glyceraldehyde-3-phosphate Dehydrogenase from Pea Seeds

By RONALD G. DUGGLEBY and DAVID T. DENNIS

Glyceraldehyde 3-phosphate + NAD⁺ + P$_i$ ⇌ 1,3-diphosphoglycerate + NADH + H⁺

D-Glyceraldehyde 3-phosphate : NAD⁺ oxidoreductase (EC 1.2.1.12; G3PDH) catalyzes the only oxidative reaction in glycolysis. This enzyme has been studied from a wide variety of species including several higher plants.[1–8] In addition to G3PDH, many plant tissues contain a second

[1] R. G. Duggleby and D. T. Dennis, *J. Biol. Chem.* **249**, 162 (1974).
[2] R. G. Duggleby and D. T. Dennis, *J. Biol. Chem.* **249**, 167 (1974).
[3] R. G. Duggleby and D. T. Dennis, *J. Biol. Chem.* **249**, 175 (1974).
[4] R. E. McGowan and M. Gibbs, *Plant Physiol.* **54**, 312 (1974).
[5] R. Cerff, *Eur. J. Biochem.* **94**, 243 (1979).
[6] O. P. Malhotra, Srinivasan, and D. K. Srivastava, *Biochim. Biophys. Acta* **526**, 1 (1978).
[7] O. P. Malhotra, D. K. Srivastava, and Srinivasan, *Arch. Biochem. Biophys.* **197**, 302 (1979).
[8] P. Pupillo and R. Faggiani, *Arch. Biochem. Biophys.* **194**, 581 (1979).

glyceraldehyde-3-phosphate dehydrogenase (EC 1.2.1.13) that can utilize NAD$^+$ or NADP$^+$ as a substrate.[9] In selecting a plant tissue from which to isolate G3PDH, it is advantageous to avoid those that contain EC 1.2.1.13, and pea seeds fulfill this criterion. They are also readily available in substantial quantities, and purification of G3PDH from this tissue is relatively straightforward. Several procedures have been reported for purifying the pea seed enzyme,[1,4,6] all of which are similar. The procedure described here is the simplest of the three, since it employs no chromatographic steps. It is so simple that Foster[10] has reported that it can be performed by undergraduate students.

Assay Method

The activity of the enzyme is measured by monitoring the increase in absorbance at 340 nm due to the production of NADH. Accurate determinations of initial rates require selection of conditions to avoid nonlinear assays that can result from the instability of glyceraldehyde 3-phosphate (G3P) and inhibition by accumulation of products.

We initially reported[2] that G3P was unstable at alkaline pH and was converted to unknown substances that were not substrates for the enzyme, an observation that has been confirmed.[11,12] This instability has now been demonstrated to result from a reaction of G3P with the Tris buffer used in the assay. Segal and Boyer[13] had in fact reached the same conclusion many years earlier. For this reason, we cannot recommend our original assay buffer,[2] and the buffer of choice is triethanolamine, in which G3P is stable.[11,12] Byers and Koshland[14] reported that G3P is also stable in pyrophosphate buffer, but this would not be suitable for detailed kinetic work because it usually is contaminated with substantial and variable amounts of phosphate.

Product inhibition by accumulation of 1,3-diphosphoglycerate (1,3-diPGA) contributes to the observed nonlinearity of the assay. For this reason, arsenate is often substituted for phosphate in the assay on the assumption that 1-arseno-3-PGA is rapidly hydrolyzed, thereby preventing any inhibition. Teipel and Koshland[15] have questioned this assumption and have suggested that 1-arseno-3-PGA may accumulate in sufficient

[9] G. R. Yonuschot, B. J. Ortwerth, and O. J. Koeppe, *J. Biol. Chem.* **245**, 4193 (1970).
[10] J. M. Foster, *Bioscience* **29**, 539 (1979).
[11] V. L. Crow and C. L. Wittenberger, *J. Biol. Chem.* **254**, 1134 (1979).
[12] W. O. Weischet and K. Kirschner, *Eur. J. Biochem.* **65**, 365 (1976).
[13] H. L. Segal and P. D. Boyer, *J. Biol. Chem.* **204**, 265 (1953).
[14] L. D. Byers and D. E. Koshland, *Biochemistry* **14**, 3661 (1975).
[15] J. Teipel and D. E. Koshland, *Biochim. Biophys. Acta* **198**, 183 (1970).

amounts to cause significant inhibition. In contrast, Byers et al.[16] have indicated that nonlinear progress curves result from preferential utilization of contaminating phosphate in the assay. Irrespective of the merits of these conflicting claims, the nonlinearity of the assay of the pea seed enzyme is more pronounced when arsenate is substituted for phosphate. Product inhibition due to 1,3-diPGA accumulation can be relieved by adding 0.5 mM ADP, 0.2 mM Mg^{2+}, and 1.7 units of 3-PGA kinase (EC 2.7.2.3) per milliliter to the assay. The addition of coupling enzymes to remove inhibitory products is not widely used, although Wang and Alaupovic[17] have also used PGA kinase to remove 1,3-diPGA from the assay of G3PDH from erythrocyte membranes, and Scopes[18] has used the technique in a study of 3-PGA kinase.

Another factor that must be considered in the assay of G3PDH is that all three substrates, but especially G3P, are inhibitory at higher concentrations. Fortunately, NAD^+ is the least inhibitory substrate, and by increasing its concentration the inhibition by G3P can be reduced. Thus, it is possible to find conditions that give almost maximum activity of the pea seed enzyme: 1 mM DL-G3P, 1 mM NAD^+, and 10 mM phosphate, at an assay pH of 8.8.

Purification Procedure

The purification procedure is based on that described by Hageman and Arnon.[19] An ethanol precipitation step was introduced into their procedure and this step gives a great increase in specific activity.

Reagents. The buffer used throughout the purification contains 1.5 mM EDTA in 10 mM K_2HPO_4/KH_2PO_4, pH 7.2, at 25°. Except as noted below, this buffer is kept and used at 0–4°. A yield of 15–20 mg of homogeneous enzyme can be obtained from 80 g of peas and requires 2.5 liters of the above buffer, 2 liters of acetone at −20°, and 110 ml of 95% ethanol at −20°.

Peas. Dried soup peas, available at low cost in most supermarkets, are a reliable source of enzyme. These seeds are fully viable and can be grown to the seedling stage.[20] In addition to their low cost and availability, soup peas have the additional advantage of being free of fungicides that may contain mercury. This is an important consideration, since G3PDH has a cysteine residue at the active site and is sensitive to mercury salts.

[16] L. D. Byers, H. S. She, and A. Alayoff, *Biochemistry* 18, 2471 (1979).
[17] C.-S. Wang and P. Alaupovic, *Arch. Biochem. Biophys.* 205, 136 (1980).
[18] R. K. Scopes, *Eur. J. Biochem.* 85, 503 (1978).
[19] R. H. Hageman and D. I. Arnon, *Arch. Biochem. Biophys.* 55, 162 (1955).
[20] R. G. Duggleby and D. T. Dennis, *J. Biol. Chem.* 245, 3745 (1970).

Acetone Powder. Peas (80 g) are rinsed with distilled water and soaked overnight in distilled water at 0–4°. The peas are drained, blotted dry, then homogenized with 700 ml of acetone at −20° for 1 min in a Waring blender. The slurry is filtered on a Büchner funnel until the filter cake cracks, and this is then homogenized with a further 700 ml of acetone at −20°. The resulting homogenate is filtered as before and washed with three lots of 150 ml of cold acetone. The filter cake is broken up using a spatula and dried on the funnel under continuous suction with occasional stirring until the smell of acetone is not detectable. Finally, it is crushed to a powder and dried overnight under vacuum over P_2O_5 at 0–4°. Starting with 80 g of dried peas, approximately 73 g of acetone powder are obtained, which can be stored for many months at −20° with no loss of G3PDH activity. Malhotra *et al.*[6] have eliminated this step of the purification and prepare a crude extract directly from the imbibed seeds. We have not attempted to reproduce this procedure, but it appears to have no adverse effects on the success of the subsequent steps.

Crude Extract. Acetone powder is stirred with buffer (6 ml per gram of powder) for 15 min at 0° then clarified by centrifugation for 15 min at $10,000 g$. The supernatant is filtered through glass wool while the pellet is resuspended in buffer (5 ml per gram of acetone powder). After stirring, centrifugation, and filtration of the supernatant as before, the filtrates are combined to give approximately 700 ml of crude extract from 80 g of dried peas. All volumes reported below are based on this quantity of peas, but they may be scaled up or down as appropriate.

Heat Treatment. The reproducibility of a heat treatment depends critically on the volume of the enzyme solution, since it takes longer to heat larger volumes to a given temperature. For this reason a programmed heating regime, which works equally well on large or small volumes, was devised in which the temperature increase is achieved by immersing a beaker containing the enzyme in an electrically heated water bath. The temperature of the enzyme solution is monitored while the rate of temperature increase of the water bath is adjusted using a variable transformer so that the enzyme solution follows a chosen heating curve. For pea seed G3PDH, the crude extract is heated to 55° in 18 min while following a curve that starts at 2° and approaches 60° exponentially. On reaching 55° the extract is cooled in an ice bucket. The rate of cooling is not critical, provided that a temperature below 30° is achieved within 5 min. Any volume of crude extract may be treated in this way, but batches of 300–400 ml are more convenient. Although the description of this heat treatment appears to be complex, it is not in fact difficult to perform. The precipitate that forms during this treatment is removed with the first ammonium sulfate precipitation. The heat treatment gives only a small purification, but it greatly increases the stability of the enzyme. In contrast, the crude extract loses 90% of its activity over 10 days at 0–4°.

First Ammonium Sulfate Fractionation. Solid ammonium sulfate (390 g/liter) is slowly added to the heat-treated enzyme with continuous stirring, the pH being maintained at 7.2 by the addition of 3.5 N NH_4OH. The solution is stirred for 30 min, and the precipitate is removed by centrifugation for 15 min at 10,000 g. The enzyme is precipitated by the addition of a further 240 g of solid ammonium sulfate per liter of supernatant, the pH again being maintained at 7.2. The suspension is allowed to stand for 4–6 hr at 0–4° when the precipitate is collected by centrifugation (10,000 g for 30 min). The pellet is dissolved in 60 ml of buffer and dialyzed for 1.5 hr against 600 ml of buffer and then overnight against 600 ml of fresh buffer. It is convenient to bring several batches of enzyme to this stage of purification and then combine them for the last three steps of the procedure.

Second Ammonium Sulfate Fractionation. The pH of the dialyzed enzyme solution is adjusted to 7.8 with 3.5 N NH_4OH and solid ammonium sulfate (480 g/liter) is added with continuous stirring, the pH being maintained at 7.8. After stirring for a further 30 min, inactive protein is removed by centrifugation (34,000 g for 10 min) and the enzyme is precipitated by the addition of 87 g of solid ammonium sulfate per liter of supernatant, the pH being maintained at 7.8. Stirring is continued for 30 min, and the precipitate is collected by centrifugation for 10 min at 30,000 g. The pellet is dissolved in 60 ml of buffer.

Ethanol Fractionation. Inactive protein is precipitated by the slow addition, with rapid stirring, of 700 ml of 95% ethanol (at −20°) per liter of enzyme solution. After 10 min, the precipitate is removed by centrifugation for 5 min at 34,000 g and a further 600 ml of ethanol per liter of supernatant is added. Very vigorous stirring is essential at this stage to prevent the precipitated enzyme from coagulating into an intractable gum. The enzyme is collected by centrifugation (34,000 g for 5 min) and dissolved in 10 ml of buffer. McGowan and Gibbs,[4] who also developed a purification of pea seed G3PDH, use an acetone precipitation of the enzyme that appears to be just as effective as our ethanol fractionation.

Acid Precipitation. The enzyme solution is adjusted to pH 5.1 with 1 N HCl, the precipitate is removed immediately by centrifugation (34,000 g for 5 min), and the supernatant is adjusted to pH 7.2 with 3.5 N NH_4OH.

Yield. A representative purification obtained using this procedure is summarized in the table. The recoveries and specific activities shown are fairly typical except that a specific activity closer to 3 units/mg after the first ammonium sulfate fractionation is more usual. The specific activity of the final product is generally within the range 200–230 units/mg and has never exceeded 240 units/mg. A yield of 15–20 mg of enzyme from 80 g of

PURIFICATION[a] OF PEA SEED GLYCERALDEHYDE-3-PHOSPHATE DEHYDROGENASE

Fraction	Volume (ml)	Activity (units)	Protein (mg)	Recovery (%)	Specific activity (units/mg protein)	Purificat (fold)
Crude	702	11,500	11,910	100	0.97	1.0
Heated[b]	691	11,470	7,600	100	1.5	1.5
First $(NH_4)_2SO_4$	111[c]	7,920	4,240	69	1.9	2.0
Second $(NH_4)_2SO_4$	59	6,870	950	60	7.2	7.4
Ethanol	10.4	5,820	37.0	51	157	162
Acid	10.2	3,480	17.2	33	223	230

[a] Starting from 80 g of dried peas.
[b] Based on a sample from which the precipitate had been removed by centrifugation prior to as for protein and enzymic activity.
[c] Volume after dialysis.

peas can be obtained routinely. In 1 week it is possible to isolate 70 mg of G3PDH from two purifications, each of 160 g of peas. The two preparations are combined after the first ammonium sulfate fractionation.

Purity and Stability. The purified enzyme migrates as a single protein band in SDS gel electrophoresis, and the bulk of the material sediments as a single symmetrical boundary in the ultracentrifuge. A minor component seen using the latter technique probably represents a dissociated form of this tetrameric enzyme. The enzyme can be stored in buffer at 0–4°, as it is obtained from the final acid precipitation step of the purification. There is no substantial loss of activity during several months of storage under these conditions.

Properties

Structure. The mobility in SDS gels indicates a subunit molecular weight of 36,000–37,000, and sedimentation velocity experiments give an $s_{20,w}$ of 7.4×10^{-13} sec, consistent with a molecular weight of 145,000. These data suggest that pea seed G3PDH, like that from animal sources, is tetrameric. Amino acid analysis indicates a similar composition to the enzyme from a wide range of species.[1] McGowan and Gibbs[4] reported an isoelectric point at pH 5.1 that is surprisingly low, as the amino acid composition shows that there are 51 basic residues per subunit and 63 Asx + Glx. If these data are correct, there must be relatively few amides in the enzyme.

Tightly Bound NAD^+. The purification procedure described above yields an enzyme that contains 0.5–0.7 mol of NAD^+ per mole of subunit. This can be increased to 1 mol per mole by precipitation of the enzyme from a 1 mM solution of NAD^+ by ammonium sulfate. Complete removal of NAD$^+$ is difficult, and treatment with charcoal, the standard procedure[21] for removing NAD$^+$ from G3PDH, removes only part of the NAD$^+$ from the pea seed enzyme. Charcoal treatment can be made effective if bound NAD$^+$ is first converted to NADH by the addition of 0.1 mM G3P. After removal of the charcoal containing the bound NADH by centrifugation, the NAD$^+$-free enzyme can be precipitated by ammonium sulfate. This procedure yields a partially acylated enzyme (0.4 mol per mole of subunit), but this will deacylate spontaneously if redissolved in buffer and left for 24 hr at 0°. The NAD$^+$-free enzyme is labile in solution but may be stored for several weeks, without loss of activity, as a suspension in 3.5 M ammonium sulfate. Malhotra *et al.*[6] reported that chromatography of pea seed G3PDH on DEAE-cellulose removes bound NAD$^+$.

Kinetics. The enzyme shows maximal activity in the pH range 8.5–9.0. It is absolutely specific for NAD$^+$; NADP$^+$ will not sustitute for NAD$^+$ and will not inhibit, indicating that it is unable to bind to the enzyme. Some synthetic NAD$^+$ analogs (such as thionicotinamide adenine dinucleotide) are reduced by the enzyme, but none is as effective as NAD$^+$. Each of the physiological substrates show some degree of substrate inhibition, and this property is most pronounced with G3P. At concentrations of NAD$^+$ and phosphate in the vicinity of their respective Michaelis constants, the K_i for substrate inhibition by G3P is only four times greater than the K_m for G3P. As a result, there is a well-defined optimum in the G3P saturation curve at approximately twice the K_m.

At substrate concentrations below the inhibitory range, the enzyme obeys the rate equation:

$$v = V/[1 + K_a/A + K_b/B + K_c/C + (K_{ia}K_b)/(AB)$$

in which A, B and C represent the concentrations of NAD$^+$, phosphate, and G3P, respectively. Values for the kinetic constants are $K_a = 0.13$ mM, $K_{ia} = 0.99$ mM, $K_b = 0.40$ mM and $K_c = 0.28$ mM. Inhibition by NADH is competitive with NAD$^+$ and with G3P, but it is noncompetitive with phosphate. These and other data are consistent with the mechanism shown below.

It is now accepted that this mechanism, or a variant of it, is generally applicable to G3PDH from plant and nonplant sources.[22-24]

[21] S. F. Velick, J. E. Hayes, and J. Harting, *J. Biol. Chem.* **203,** 527 (1953).

[22] P. J. Harrigan and D. R. Trentham, *Biochem. J.* **135,** 695 (1973).

[23] F. J. Seydoux, N. Kelemen, N. Kellershohn, and C. Roucous, *Eur. J. Biochem.* **64,** 481 (1976).

[24] J.-C. Meunier and K. Dalziel, *Eur. J. Biochem.* **82,** 483 (1978).

[57] Glyceraldehyde-3-phosphate Dehydrogenase from Yeast

By Larry D. Byers

D-Glyceraldehyde 3-phosphate + NAD⁺ + P_i ⇌
$$1,3\text{-diphosphoglycerate} + \text{NADH} + \text{H}^+ \quad (1)$$

D-Glyceraldehyde 3-phosphate + NAD⁺ $\xrightarrow{\text{HAsO}_4^{-2}}$
$$3\text{-phosphoglycerate} + \text{NADH} + \text{H}^+ \quad (2)$$

Assay Method

Principle. The equilibrium constant of D-glyceraldehyde-3-phosphate dehydrogenase (EC 1.2.1.12) for the physiological reaction, Eq. (1), is $5.6 \times 10^{-8} M$ (25°, $\mu = 0.6 M$).[1] Although the phosphoric anhydride product is thermodynamically labile ($\Delta G^{\circ\prime} = -11.8$ kcal/mol for hydrolysis, pH 6.9, 25°),[2] it is kinetically stable ($t_{1/2} \sim 1$ hr for hydrolysis, $5.5 < \text{pH} < 8.4$, 25°).[1] As a result of the small equilibrium constant ($K'_{eq} = 0.6$ at pH 7) and product inhibition, assay procedures based on the physiological reaction (either NAD⁺ reduction in the "forward" reaction or NADH oxidation in the "back" reaction) are inconvenient.[3] The most commonly used procedures for monitoring the reversible oxidative phosphorylation reaction are the Ferdinand assay[4] for the "forward" reaction and the coupled assay of Kirschner and Voigt[5,6] for the "back" reaction (Table I). Other assay conditions, which involve *in situ* generation of substrates and/or removal of products, have been described.[7–10]

The "arsenolysis" reaction, Eq. (2), is nearly irreversible. This reaction, first used by Warburg and Christian,[11] has been described by Ve-

[1] L. D. Byers, H. S. She, and A. Alayoff, *Biochemistry* **18**, 2471 (1979).

[2] M. R. Atkinson and R. K. Morton, in "Comparative Biochemistry" (M. Florkin and H. S. Mason, eds.), Vol. 2, p. 1. Academic Press, New York, 1960.

[3] At initial substrate concentrations of 1 mM NAD⁺, 1 mM D-glyceraldehyde 3-phosphate, and 0.1 M P_i, for example, the equilibrium concentrations of 1,3-diphosphoglycerate and NADH will be 0.2 mM at pH 7 and 0.6 mM at pH 8.5. The K_m of 1,3-diphosphoglycerate is smaller than that of D-glyceraldehyde 3-phosphate by a factor of ~ 10.

[4] W. Ferdinand, *Biochem. J.* **92**, 578 (1964).

[5] K. Kirschner and B. Voigt, *Hoppe Seyler's Z. Physiol. Chem.* **349**, 632 (1968).

[6] The addition of 50 μM NAD⁺ to the assay mixture abolishes the lag phase in NADH disappearance.

[7] G. Rossner, *Arch. Klin. Exp. Dermatol.* **222**, 383 (1965).

[8] D. R. Trentham, C. H. McMurray, and C. I. Pogson, *Biochem. J.* **114**, 19 (1969).

[9] J. J. Aragón and A. Sols, *Biochem. Biophys. Res. Commun.* **82**, 1098 (1978).

[10] J. Ovádi and T. Keleti, *Eur. J. Biochem.* **85**, 157 (1978).

[11] O. Warburg and W. Christian, *Biochem. Z.* **303**, 40 (1939).

lick[12] and by Krebs.[13] The arsenolysis reaction provides a particularly convenient and efficacious assay for glyceraldehyde-3-phosphate dehydrogenase activity, since the immediate product, 1-arseno-3-phosphoglycerate, is rapidly hydrolyzed ($t_{1/2}$ < 2.5 sec, pH = 7, 25°).[1] The traditional arsenolysis assay[11-13] is carried out at low levels of substrates. The reaction conditions are given in Table I. Since the assay is nonlinear (owing to the non-zero-order conditions), the amount of NADH produced in the 30-sec interval between the first 15 sec and 45 sec after initiation of the reaction is generally used to calculate the "initial" velocity. The assay procedure described here is based on the method of Stallcup et al.[14] The use of higher substrate concentrations[15] improves the linearity and enhances the sensitivity of the traditional "standard" assay.

The production of NADH is monitored spectrophotometrically (ϵ_{340} = 6.3 × 10^3 M^{-1} cm^{-1}, ϵ_{365} = 3.4 × 10^3 M^{-1} cm^{-1}).[16] Alternatively, the reaction can be monitored by (a) fluorometric detection of NADH, either directly (λ_{ex} = 340 nm, λ_{em} = 460 nm) or indirectly[20]; (b) bioluminescent detection of NADH using the luciferase coupling reaction[21]; and (c) continuous potentiometric titration by means of a pH-stat.[22]

Reagents

DL-Glyceraldehyde 3-phosphate (G3P) is prepared by treatment of the barium salt of the diethyl acetal with a cation exchange resin. A 20 mM solution of D-G3P is prepared as follows: 12 g of Dowex 50 X4-200R (or 50-X8) is washed with boiling deionized water and added to 16 ml of hot water; 800 mg of the barium salt of the acetal is added to the resin and mixed thoroughly. The slurry is incubated

---

[12] S. F. Velick, this series, Vol. 1, p. 401 (1955).

[13] E. G. Krebs, this series, Vol. 1, p. 407 (1955).

[14] W. B. Stallcup, S. C. Mockrin, and E. D. Koshland, Jr., *J. Biol. Chem.* **247**, 6277 (1972).

[15] See, for example, W. C. Deal, Jr., *Biochemistry* **8**, 2795 (1969).

[16] R. B. McComb, L. W. Bond, R. W. Burnett, R. C. Keech, and G. N. Bowers, Jr., *Clin. Chem.* **22**, 141 (1976).[17]

[17] The molar extinction coefficient in use since 1948 (ϵ_{340} = 6.22 × 10^3)[18] has been redetermined enzymatically with glucose-6-phosphate dehydrogenase from *L. mesenteroides*[16] and with glutamate and lactate dehydrogenases.[19] The "revised" value (pH 7.8, 25°) is ~2% higher than the value generally used.[18]

[18] B. L. Horecker and A. Kornberg, *J. Biol. Chem.* **175**, 385 (1948).

[19] J. Ziegenhorn, M. Senn, and T. Bücher, *Clin. Chem.* **22**, 150 (1976).

[20] See, for example, G. G. Guilbault, this series, Vol. 41, p. 53 (1975).

[21] W. Cantarow and B. D. Stollar, *Anal. Biochem.* **71**, 333 (1976).

[22] This method takes advantage of the reaction catalyzed by the enzyme in the presence of molybdate, which is analogous to the "arsenolysis" reaction.[1] The substitution of molybdate (pK_{a2} = 4) for arsenate (pK_{a2} = 7) reduces the buffering capacity of the assay mixture in the pH range where the enzyme is most active.

in a boiling water bath for 2.5 min and stirred intermittently. The resin is removed by filtration, with suction, into an ice bath and washed with 4 ml of water. The pH of the filtrate is adjusted to ~4 with 0.1 M NaOH. The concentration of D-G3P is determined enzymically by oxidation in Bicine buffer (pH 8.5) containing 1 mM NAD$^+$, 25 mM arsenate, and sufficient G3P dehydrogenase (~35 μg) to completely oxidize the D-enantiomer in ~3 min. The substrate solution is diluted with an appropriate amount of water (usually ~5 ml) to yield a 20 mM solution of D-G3P. The substrate is stable for several months when stored at $-20°$.

NAD$^+$ is dissolved in deionized water. The pH is adjusted to ~4, and the solution is diluted to a final concentration of 10 mM. The NAD$^+$ solution is standardized by enzymic reduction and by absorbance at 260 nm ($\epsilon = 1.73 \times 10^4 \, M^{-1} \, cm^{-1}$).[23] The solution is stored frozen.

Na$_2$HAsO$_4$ is dissolved in deionized water to a concentration of 0.5 M. The pH is adjusted to 8.5 with HCl.

Bicine [N,N-bis(2-hydroxyethyl)glycine] buffer, 0.05 M, is prepared with deionized water (1 M in sodium acetate and 1 mM in EDTA). The pH is adjusted to 8.5 with NaOH.

Procedure. An assay mixture (1 ml) containing 1 mM D-G3P, 1 mM NAD$^+$, 25 mM arsenate, 0.04 M Bicine, 0.8 M acetate, and 0.8 mM EDTA (pH 8.5, $\mu \approx 0.87 \, M$) is prepared by first adding 10 μl of the enzyme solution (~0.02 mg of G3P dehydrogenase per milliliter), 0.1 ml NAD$^+$, and 0.05 ml arsenate to 0.8 ml of buffer in a 1-cm light path cuvette. The mixture is incubated at 25° for ~5 min, and the reaction is initiated by the addition of 50 μl of G3P. The aldehyde is added last, since it is not very stable under the assay conditions. (The amount of G3P available for enzymic oxidation decreases with a half-life of ~10 hr at pH 8.5. Adduct formation between NAD$^+$ and G3P is a much slower process.)

Under these assay conditions the rate of production of NADH is linearly dependent on enzyme concentration (0.05–5 μg/ml). The rate of increase in the absorbance at 340 nm is "constant" (less than a 5% decrease) up to an absorbance change of 1 absorbance unit provided that the assay mixture is relatively free of contaminating inorganic phosphate.[24]

---

[23] E. Haid, cited in "Methods of Enzymatic Analysis" (H. U. Bergmeyer, ed.), 2nd English ed., Vol. 1, p. 546. Academic Press, New York, 1974.

[24] The nonlinearity in the "arsenolysis" assay (with substrate levels ≥0.5 mM) has been attributed to phosphate contamination of G3P.[1] Levels of contaminating phosphate that are less than 25% of the D-G3P can generally be tolerated. The G3P can be separated from inorganic phosphate by ion-exchange chromatography.[25]

[25] J. G. Belasco, J. M. Herlihy, and J. R. Knowles, *Biochemistry* **17**, 2971 (1978).

TABLE I
SPECIFIC ACTIVITY OF YEAST GLYCERALDEHYDE-3-PHOSPHATE DEHYDROGENASE
UNDER VARIOUS ASSAY CONDITIONS

Assay conditions (25°)		Units/mg protein
ïnand[4]		
2 M triethanolamine, pH 8.9	0.84 mM D-G3P, 0.84 mM NAD$^+$, 42 mM P$_i$	400
oux et al.[28]		
1 M imidazole, 0.1 M KCl, pH 7.0	0.45 mM D-G3P, 0.45 mM NAD$^+$, 50 mM P$_i$	190
s[13]		
26 M PP$_i$, pH 8.5	0.25 mM D-G3P, 0.25 mM NAD$^+$, 5.7 mM As$_i$	120
up et al.[14]		
26 M PP$_i$, pH 8.5	1.0 mM D-G3P, 1.0 mM NAD$^+$, 10 mM As$_i$	155
4 M Bicine, pH 8.5	1.0 mM D-G3P, 1.0 mM NAD$^+$, 10 mM As$_i$	65
4 M Bicine, 0.8 M NaOAc, pH 8.5	1.0 mM D-G3P, 1.0 mM NAD$^+$, 10 mM As$_i$	235
4 M Bicine, 0.8 M NaOAc, pH 8.5	1.0 mM D-G3P, 1.0 mM NAD$^+$, 25 mM As$_i$	470
oux et al.[28]		
1 M imidazole, 0.1 M KCl, pH 7.0	0.08 mM 1,3-diphosphoglycerate, 0.08 mM NADH	390
hner and Voigt[5]		
M triethanolamine, pH 7.6	7 mM phosphoglycerate, 0.35 mM ATP, 3.3 mM MgSO$_4$, ~300 units of phosphoglycerate kinase per milliliter, 0.15 mM NADH	220

Definition of Specific Activity. The specific activity is the number of enzyme activity units (= number of micromoles of NADH produced per minute at 25°) in 1 mg of protein. The relationship between the specific activity under the conditions specified in this assay and that under the conditions of some of the more commonly used alternative procedures is given in Table I. Under a given set of conditions [i.e., substrate concentrations, ionic strength (maintained by a specific salt), temperature, and pH], the specific activity has generally been found to be fairly independent of the particular buffer used in the assay. Some buffers, however, should be avoided: borate forms complexes with NAD[26] and Tris [tris(hydroxymethyl)aminomethane, THAM] forms imines with G3P and other aldehydes.[27] The specific activity of the enzyme is anomalously high in the presence of triethanolamine.[4,28]

A reliable determination of the specific activity in crude extracts is difficult, presumably as a consequence of the presence of triosephosphate isomerase and glycerol phosphate dehydrogenase. Although these inter-

[26] K. W. Smith and S. L. Johnson, *Biochemistry* **15**, 560 (1976).

[27] J. W. Ogilvie and S. C. Whitaker, *Biochim. Biophys. Acta* **445**, 525 (1976).

[28] F. Seydoux, S. Bernhard, O. Pfenninger, M. Payne, and O. P. Malhotra, *Biochemistry* **12**, 4290 (1973).

fering reactions are not as substantial in the crude extract from yeast as they are in crude extracts from muscle[29] or bacteria,[30] the linearity of the assay can be improved by the addition of 0.1 mM phosphoglycolate (an inhibitor of triosephosphate isomerase) to the assay mixture.

Purification Procedure

Glyceraldehyde-3-phosphate dehydrogenase is present in high concentrations in most organisms.[31] The enzyme was shown by Krebs to be the most abundant protein in bakers' yeast, composing 5% of the dry cell weight and nearly 20% of the total soluble protein.[32] The uncommonly high level of the enzyme facilitates its purification from the crude extracts of yeast. The main difficulty in the isolation of the enzyme from this rich source is the general problem of cell disruption. In the original purification procedure of Warburg and Christian,[11] as well as in the traditional procedure of Krebs,[13] the method of solubilization is autolysis. Although this method has been used in the purification of G3P dehydrogenase for over 30 years,[33] it does have some disadvantages. The long incubation periods required for autolysis and the abundance of proteolytic enzymes in yeast cells result in varying degrees of degradation of G3P dehydrogenase.[14] Some other, more rapid, solubilization methods that have been used in purification procedures for yeast G3P dehydrogenase are grinding (either in a mortar using an alumina abrasive[14,34] or in a blender with glass beads[35]), sonication,[36] and shearing with a French press.[37,38] The scheme described here involves lysis of the cells with a Manton–Gaulin homogenizer.[39] The purification procedure is essentially that of Stallcup et al.[14] with an affinity chromatography step in place of the time-consuming recrystallizations.[40] This procedure consistently yields stable preparations

---

[29] For example, from *Agkistrodon piscivorus* (water moccasin); S. Shames and L. D. Byers, unpublished observations.

[30] *Bacillus stearothermophilus;* S. McCaul and L. D. Byers, unpublished observations.

[31] J. I. Harris and M. Waters, *in* "The Enzymes" (P. D. Boyer, ed.), 3rd ed., Vol. 13, p. 1. Academic Press, New York, 1976.

[32] (a) E. G. Krebs, *J. Biol. Chem.* **200**, 471 (1953). (b) E. G. Krebs, G. W. Rafter, and J. M. Junge, *J. Biol. Chem.* **200**, 479 (1953).

[33] P. J. G. Butler and G. M. T. Jones, *Biochem. J.* **118**, 375 (1970).

[34] L. S. Gennis, *Proc. Natl. Acad. Sci. U.S.A.* **73**, 3928 (1976).

[35] For example, "Bead-Beater" cell disruptor (Biospec Products, Bartlesville, Oklahoma, or Techmar Co., Cincinnati, Ohio); 0.3–0.5 mm glass beads.

[36] A. F. Chaffotte, C. Roucous, and F. Seydoux, *Eur. J. Biochem.* **78**, 309 (1977).

[37] M. J. Holland and E. W. Westhead, *Biochemistry* **12**, 2264 (1973).

[38] A. G. Brownlee, D. R. Phillips, and G. M. Polya, *Eur. J. Biochem.* **109**, 39 (1980).

[39] S. C. Mockrin, L. D. Byers, and D. E. Koshland, Jr., *Biochemistry* **14**, 5428 (1975).

[40] L. D. Byers, *Arch. Biochem. Biophys.* **186**, 335 (1978).

of G3P dehydrogenase of high specific activity. The source of the enzyme is *Saccharomyces cerevisiae.* The procedure is based on 500 g of cells (pressed wet cake of commercial bakers' yeast[41]) and can easily be scaled up or down. The other materials needed are 100 ml of Cibacron Blue F3GA Sepharose 6B,[43] 1.2 liters of DEAE-Sephadex A-50 (45 g in 0.01 M Tris-HCl, pH 8.3), 1.5 liters of Sephadex G-50, and 50 ml of 1 M Tris base. Buffers are prepared from (10 times concentrated) stock solutions: 500 ml of 1 M Tris-HCl, 10 mM EDTA, pH, 9.0; and 2 liters of 0.1 M Tris-HCl, 10 mM EDTA, pH 8.3. The buffers used in all of the fractionation steps contain 1 mM EDTA and 1 mM dithiothreitol (or dithioerythritol). All steps, up through the ion-exchange chromatography, are carried out at ~4°. Buffer pH measurements are made with a glass electrode at room temperature.

Step 1. Extraction. The yeast cake (500 g) is crumbled into 750 ml of precooled buffer (pH 9, 0.1 M Tris-HCl, 1 mM, EDTA, 1 mM dithio-threitol), stirred until suspension is complete, and passed twice through a precooled Manton–Gaulin homogenizer at 9000 psi. The suspension is immediately adjusted to pH ~8.3 and centrifuged at 7500 g (= 9000 rpm in a GSA rotor) for 30 min.

Step 2. 50% Ammonium Sulfate Fractionation. Solid ammonium sulfate 313 g/liter, is slowly added, with constant stirring. The addition takes ~15 min, and stirring is continued for an additional 45 min. The pH (pH paper) is kept between 8 and 8.5 by small additions of concentrated NH_4OH. The suspension is centrifuged at 7500 g for 45 min, and the pellet is discarded.

Step 3. 80% Ammonium Sulfate Fractionation. Powdered ammonium sulfate, 250 g per liter of original supernatant (step 1), is added to the clear yellow supernatant from step 2. The stirring is continued for 1–2 hr. The suspension is centrifuged at 9000 g (= 10,000 rpm) for 1 hr. The turbid supernatant is discarded, and the pellet is dissolved in a minimal amount of 0.1 M Tris/HCl, pH 9 buffer (~150 ml). The pH (pH paper) is maintained between 8 and 8.5 with 1 M Tris base.

Step 4. Desalting by Gel Filtration. The enzyme solution is applied to a 5 × 90 cm Sephadex G-50 column equilibrated with 0.01 M Tris-HCl, pH 8.3. The yellow-pink protein fraction separates from some yellow material.

Step 5. DEAE-Sephadex Chromatography. The desalted protein solu-

[41] Red Star Yeast Co. (strain F1, grown on a limiting carbon source and oxygen). Holland and Holland[42] have shown that the levels of the glycolytic enzymes in yeast cells can be influenced by growth conditions.

[42] M. J. Holland and J. P. Holland, *Biochemistry* **17**, 4900 (1978).

[43] H. J. Böhme, G. Kopperschläger, J. Schulz, and E. Hofmann, *J. Chromatogr.* **69**, 209 (1972).

tion is adsorbed onto a 5 × 60 cm DEAE-Sephadex A-50 column that has been equilibrated with the same buffer used in step 4 (0.01 M Tris-HCl, pH 8.3). After addition of the protein is complete, the column is washed with ~1 liter of buffer (0.01 M Tris-HCl, pH 8.3). The enzyme is eluted with a linear gradient by allowing 1.5 liters of buffer (0.01 M Tris, pH 8.3) containing NaCl (0.15 M) to run into a mixing chamber containing 1.5 liters of buffer (0.01 M Tris, pH 8.3); 10-ml fractions are collected at a rate of ~50 ml/hr. Three peaks of activity are resolved. The first fraction (~10% of the total activity) generally begins to appear at an elution volume ~800 ml, and the second fraction (~30% of the total activity) at an elution volume ~1100 ml. The peak of activity containing the most acidic isozyme generally appears between an elution volume of 1500 and 2000 ml. This fraction is pooled and concentrated by ammonium sulfate precipitation; 725 g of solid ammonium sulfate per liter is slowly added, with constant stirring, to the pooled fractions at room temperature. Gentle stirring is continued for ~1 hr, and the suspension is allowed to stand an additional 1–2 hr. The sample is centrifuged at 13,000 g (10,000 rpm is a SS-34 rotor) for 1 hr. The pellet is dissolved in a minimal amount of 0.1 M Tris-HCl, pH 9 buffer (~70 ml), and the solution is desalted on a 4 × 45 cm column of Sephadex G-50 equilibrated with 0.01 M Tris-HCl, pH 7.5.

Step 6. Affinity Chromatography on Cibacron Blue F3G-A Sepharose 6B. The protein solution is applied to a 3 × 14 cm column of affinity medium equilibrated with 0.01 M Tris-HCl, pH 7.5. The column is washed with 300 ml (~3 column volumes) of the same buffer at room temperature to remove unadsorbed protein. The enzyme is eluted by applying a 0.05 M Tris-HCl buffer, pH 8.6, containing 0.2 M $(NH_4)_2SO_4$.[44] The enzyme begins to elute after ~70 ml, and activity is contained in the subsequent 80–110 ml. The enzyme is precipitated by adding 70 g of ammonium sulfate per 100 ml, and the pH is maintained between 7.5 and 8 with NH_4OH. The suspension is allowed to stand overnight at 4°, and the precipitate is collected by centrifugation at 13,000 g for 2 hr. The enzyme is stored (4°) as the ammonium sulfate precipitate. Before use a sample of the precipitate is dissolved in a minimal amount of buffer containing 10 mM dithiothreitol, allowed to stand at room temperature for ~30 min, and desalted on a small Sephadex G-25 column.

[44] The enzyme can also be eluted with 0.05 M Tris-HCl, pH 8.6, containing either 0.5 M NaCl or 10 mM NAD$^+$.[45] After the enzyme is eluted, the column is washed with 3 column volumes of 1 M NaCl and 4 column volumes of deionized water. The affinity agent is stored at 4° in 0.02% NaN_3. The agent has been used nine times over a 3-year period with no loss in effectiveness and can also be stored dried.[45]

[45] R. L. Easterday and I. M. Easterday, *in* "Immobilized Biochemicals and Affinity Chromatography" (B. R. Dunlap, ed.), Vol. 2, p. 123. Plenum, New York, 1974.

TABLE II
PURIFICATION OF YEAST GLYCERALDEHYDE-3-PHOSPHATE DEHYDROGENASE

Fraction	Volume (ml)	Protein[a] (mg)	Total units[b]	Specific activity (units/mg protein)	Recovery (%)
Crude extract (500 g bakers' yeast)	900	55,000	165,000	3	100
50% Ammonium sulfate supernatant	940	16,600	162,000	9.7	98
80% Ammonium sulfate precipitate	320	6,800	145,000	21	88
DEAE-Sephadex pool	400	1,160	74,000	64	45
Blue-Sepharose eluate	105	465	72,000	155	44

[a] The protein concentration was estimated from the $A_{280}:A_{260}$ absorbance ratio [E. Layne, this series, Vol. 3, p. 447 (1953)] and by the turbidimetric procedure of T. Bücher [*Biochem. Biophys. Acta* **1**, 292 (1947)] with G3P dehydrogenase as a standard: c (mg/ml) $= 0.27 \Delta A_{340} = 0.66 \Delta A_{500}$.

[b] Assay conditions are those of Stallcup *et al.*[14] (see Table I).

The purification procedure is summarized in Table II. Other affinity agents have been used in the purification of the yeast enzyme. Gennis[34] describes a rapid procedure using NAD⁺-azobenzamidopropyl-Sepharose 4B, and Chaffotte *et al.*[36] describe a procedure using N^6-(6-aminohexyl)-AMP-Sepharose 4B.

Properties

The enzyme is homogeneous as judged by the following criteria[39]: (*a*) isoelectric focusing (pI = 6.1); (*b*) SDS-gel electrophoresis ($M_r \sim 35,000$), (*c*) native-gel electrophoresis ($\mu = -4 \times 10^{-5}$ cm²/volt-sec at pH = 7.25, $\mu = 0.1\,M$)[13]; (*d*) sedimentation analysis ($s^\circ_{20,w} = 7.70$ S; $D^\circ_{20,w} = 5.3$ Fick units); (*e*) gel filtration ($M_r \sim 145,000$, $f/f_0 \sim 1.2$), (*f*) amino-terminal analysis (Val); (*g*) titration of active-site sulfhydryl group with 5,5′-dithiobis(2-nitrobenzoate), 4,4′-bis(dimethylamino)diphenylcarbinol, and p-nitrophenyl acetate (3.9 ± 0.2 groups per tetramer); and (*h*) rapid transient production of NADH in the absence of acyl acceptors (4.0 ± 0.1 per tetramer). The $A_{280}:A_{260}$ ratio is 2.10 (±0.02) indicating the absence of any bound NAD⁺.[46]

Molecular Weight and Other Physical Properties. The apoenzyme is a

[46] K. Kirschner, E. Gallego, I. Schuster, and D. Goodall, *J. Mol. Biol.* **58**, 29 (1971).

monodisperse tetramer consisting of four identical polypeptide chains.[47] The enzyme is rapidly inactivated below pH 4.5 and above pH 11 as a result of dissociation.[47,48] At intermediate pH values, however, the enzyme is quite stable. The half-life for spontaneous inactivation of a 2 μM solution of the enzyme (10 mM dithiothreitol, 1 mM EDTA, 0.3 M NaCl) is ~2 days at pH 8.5 and 27° and ~3 hr at pH 7.5 and 50°. The stability is enhanced by increasing the enzyme concentration.[39] The dissociation, and subsequent inactivation, is inhibited by D_2O and by NAD^+ and is promoted by high ionic strength[47] and by ATP and low temperature.[49]

The amino acid sequence has been determined for yeast G3P dehydrogenase.[50,51] The molecular weight of the tetramer, based on the amino acid sequence,[51] is 142,604. Each subunit contains two free sulfhydryl groups (Cys-149 and Cys-153). There is no striking difference in the amino acid composition of the yeast enzyme compared to that of the enzyme from either *B. stearothermophilus* (54% sequence homology) or lobster (67% sequence homology). Of the 68 amino acid residues that are within 5 Å of another subunit in the lobster enzyme[52] 82% are identical to those in the sequence of the yeast enzyme. Of the 18 residues that interact with bound NAD^+ in either the lobster muscle enzyme[52] or the *B. stearothermophilus* enzyme,[53] 17 are identical with those of the yeast enzyme. Yeast G3P dehydrogenase readily forms hybrids with the enzymes from a variety of sources.[54–56] The extinction coefficient of the apoenzyme (0.05 M P_i, 7 ≤ pH ≤ 8, 20°) is $A_{280}^{0.1\%} = 0.894$ cm^2/mg.[46] Based on a molecular weight of 142,604, this corresponds to a molar absorptivity of $\epsilon_{280} = 1.28 \times 10^5 M^{-1}$ cm^{-1}.

NAD$^+$ Binding. The yeast enzyme differs from the enzymes isolated from a number of other sources in that it can readily be prepared free of NAD^+ and the apoenzyme is stable. The binding of NAD^+ to the yeast enzyme is relatively weak. At pH = 7.5 NAD^+ binding is characterized by a dissociation constant of $7 \times 10^{-5} M$ with no evidence of cooperativity.[34,39,57] At pH 8.5, NAD^+ binding to the enzyme isolated from Fleisch-

[47] R. Jaenicke, D. Schmid, and S. Knof, *Biochemistry* **7**, 919 (1968).

[48] Y. Shibata and M. J. Kronman, *Arch. Biochem. Biophys.* **118**, 410 (1967).

[49] S. M. Constantinides and W. C. Deal, Jr., *J. Biol. Chem.* **244**, 5695 (1969).

[50] G. M. T. Jones and J. I. Harris, *FEBS Lett.* **22**, 185 (1972).

[51] S. Bayne, B. Martin, and I. Svendsen, *Carlsberg Res. Commun.* **45**, 195 (1980).

[52] D. Moras, K. W. Olsen, M. N. Sabesan, M. Buehner, G. C. Ford, and M. G. Rossmann, *J. Biol. Chem.* **250**, 9137 (1975).

[53] J. E. Walker, A. F. Carne, M. J. Runswick, J. Bridgen, and J. I. Harris, *Eur. J. Biochem.* **108**, 549 (1980).

[54] W. B. Stallcup and D. E. Koshland, Jr., *J. Mol. Biol.* **80**, 41 (1973).

[55] K. Suzuki and J. I. Harris, *J. Biochem.* **77**, 587 (1975).

[56] H. H. Osborne and M. R. Hollaway, *Biochem. J.* **157**, 255 (1976).

[57] C. W. Niekamp, J. M. Sturtevant, and S. F. Velick, *Biochemistry* **16**, 436 (1977).

mann's yeast is characterized by positive cooperativity[34,57] and binding to the enzyme isolated from Red Star yeast is characterized by mixed (positive and negative) cooperativity.[39,58] The enzyme has been shown to bind a variety of NAD^+ analogs.[31,59-61]

Catalytic Properties. Yeast G3P dehydrogenase is similar to the enzyme from a variety of other sources with respect to the nature of the "nonphysiological" reactions that it catalyzes. These reactions (e.g., acyl phosphatase and esterase activities) have been reviewed by Harris and Waters.[31] The pH optimum of the phosphorolysis and arsenolysis reactions [Eqs. (1) and (2)] is ~8.5. The enzyme catalyzes the oxidation of a variety of aldehydes, but at rates (k_{cat}/K_m) that are at least 10^4 times smaller than that for D-glyceraldehyde 3-phosphate.[40,62] The 3-acetylpyridine analog of NAD^+ can substitute for NAD^+ in the enzyme-catalyzed oxidation of these aldehydes.[63] NAD^+ facilitates the transfer of the phosphoglyceroyl group from the active-site sulfhydryl residue of Cys-149 to inorganic phosphate.[64] In addition to arsenate, molybdate and a variety of phosphonates can serve as acyl acceptors in reactions that are analogous to the physiological oxidative phosphorylation reaction.[1] At pH 8.5 (25°) $k_{cat} = 10^3$ sec^{-1} and the K_m values are 1.5 mM for P_i, 0.6 mM for D-G3P, and 0.1 mM for NAD^+.

[58] L. D. Byers and M. Kahn, *Fed. Proc., Fed. Am. Soc. Exp. Biol.* **39**, 1853 (1980).
[59] A. Stockell, *J. Biol. Chem.* **234**, 1293 (1959).
[60] P. L. Luisi, A. Baici, F. J. Bonner, and A. A. Aboderin, *Biochemistry* 14, 362 (1975).
[61] A. G. Brownlee and G. M. Polya, *Eur. J. Biochem.* 109, 51 (1980).
[62] M. S. Kanchuger, P.-M. Leong, and L. D. Byers, *Biochemistry* 18, 4373 (1979).
[63] N. O. Kaplan, M. M. Ciotti, and F. E. Stolzenbach, *J. Biol. Chem.* **221**, 833 (1956).
[64] L. D. Byers and D. E. Koshland, Jr., *Biochemistry* 14, 3661 (1975).

[58] Glyceraldehyde-3-phosphate Dehydrogenase from *Thermus thermophilus*

By Tairo Oshima, Shinobu C. Fujita, and Kazutomo Imahori

$$D\text{-Glyceraldehyde 3-phosphate} + NAD^+ + P_i \rightleftharpoons 1,3\text{-diphosphoglycerate} + NADH + H^+$$

Assay Method

Principle. Glyceraldehyde-3-phosphate dehydrogenase (EC 1.2.1.12) from *Thermus thermophilus*[1] is assayed spectrophotometrically by the

[1] S. C. Fujita, T. Oshima, and K. Imahori, *Eur. J. Biochem.* **64**, 57 (1976).

method of Krebs[2] with modifications. In the assay, arsenate is used instead of P_i, since the product arseno-3-phosphoglycerate was reported to decompose readily to 3-phosphoglycerate and make the reaction irreversible.

Reagents

Tris-HCl buffer, 0.5 M, pH 8.3

NH_4Cl, 1.0 M

Na_2HAsO_4, 0.17 M

2-mercaptoethanol, 0.1 M (fresh)

DL-Glyceraldehyde 3-phosphate, 0.02 M, prepared from the diethyl acetal according to the Sigma Technical Bulletin.[3]

Assay mixture: In 21 ml of H_2O, dissolve 215 mg NAD, add 12 ml of Tris-HCl buffer, 6 ml of NH_4Cl, 5 ml of Na_2HAsO_4, and 6 ml of 2-mercaptoethanol. This assay mixture is prepared fresh and kept at room temperature, but unused portions may be stored in a refrigerator for several days.

Procedures. Into a microcuvette (2 mm wide, 22 mm high, 10 mm light path), 250 μl of the assay mixture and 20 μl of enzyme solution are measured. The cuvette is then placed in a spectrophotometer with the cuvette chamber thermostatted at 30°. After 5 min of preincubation, 30 μl of glyceraldehyde 3-phosphate are added with rapid mixing, and the change in absorbance at 340 nm is recorded. The reaction rate is calculated from the initial portion of the curve. One unit of activity is defined as 1 μmol of NAD reduced per min,[1] with 6.22 × 10^3 as the molar absorption coefficient of NADH. Protein concentration in purified fractions is determined spectrophotometrically using absorption coefficient at 280 nm, $A_{280}^{0.1\%}$ = 0.727 for apoenzyme and 0.807 for holoenzyme.[1] During the purification, the protein concentration is calculated by the method of Kalckar.[4]

Purification Procedure

Bacterial Culture. Thermus thermophilus[5] strain HB8 was deposited in the American Type Culture Collection (ATCC 27634). The cells can be grown in a nutrient broth (0.5% peptone, 0.3% yeast extract, 0.1% glucose, and 0.2% NaCl, pH 7–7.2) or in a synthetic medium.[6] The synthetic medium consists of 20 g of sucrose, 20 g of monosodium glutamate, 2 g of

[2] E. G. Krebs, this series, Vol. 1, p. 407.

[3] Sigma Technical Bulletin No. 10, Sigma Chemical Co., St. Louis, Missouri.

[4] H. M. Kalckar, *J. Biol. Chem.* **167**, 461 (1947).

[5] T. Oshima and K. Imahori, *Int. J. Syst. Bacteriol.* **24**, 102 (1974).

[6] T. Oshima and M. Baba, *Biochem. Biophys. Res. Commun.* **103**, 156 (1981).

NaCl, 0.25 g of KH_2PO_4, 0.5 g of K_2HPO_4, 0.5 g of $(NH_4)_2SO_4$, 0.125 g of $MgCl_2 \cdot 6 H_2O$, 0.025 g of $CaCl_2 \cdot 2 H_2O$, 6 mg of $FeSO_4 \cdot 7 H_2O$, 0.8 mg of $CoCl_2 \cdot 6 H_2O$, 20 μg of $NiCl_2 \cdot 6 H_2O$, 1.2 mg of $Na_2MoO_4 \cdot 2 H_2O$, 0.1 mg of $VOSO_4$, 0.5 mg $MnCl_2 \cdot 4 H_2O$, 60 μg of $ZnSO_4 \cdot 7 H_2O$, 15 μg of $CuSO_4 \cdot 5 H_2O$, 100 μg of biotin, and 1 mg of thiamine per 1000 ml, pH 7–7.2. The growth temperature is 75°, the aeration rate is 20 liters of air per minute per 20 liters of culture broth, and the cells are harvested at the late log phase. Under these conditions, the generation time is 70–90 min, and the yield is about 8–10 g of wet cells per liter of either nutrient or synthetic medium.

Step 1. Frozen cells (500 g) are homogeneously suspended in 1 liter of 50 mM Tris-HCl buffer, pH 7.5, containing 1 mM EDTA, 50 mM 2-mercaptoethanol, 14 mM magnesium acetate, and 0.14 M KCl. The suspension in 50-μl aliquots is subjected to sonic disruption for 6 min each. The cell debris is removed by centrifugation at 11,000 g for 40 min, and the supernatant is further clarified by ultracentrifugation at 75,000 g for 240 min to yield a clear, brown supernatant.

Step 2. To each 100 ml of supernatant from step 1, 37.3 g of ammonium sulfate are added; after at least 1 hour, the precipitate is removed by centrifugation at 11,000 g for 10 min. Further 8 g of ammonium sulfate per initial 100 ml are added. The resulting precipitate is collected by ultracentrifugation at 56,000 g for 20 min. The pellet is taken up in a smallest possible volume of 50 mM Tris-HCl buffer, pH 7.5, containing 50 mM 2-mercaptoethanol or 0.2 mM EDTA. (This buffer is referred to below as the standard buffer).

Step 3. The dissolved pellet from step 2 is desalted by using a column of Sephadex G-25 or dialysis, and then added to a column (5 × 15 cm) of DEAE-Sephadex A-50 equilibrated with the standard buffer. The proteins adsorbed are eluted with 1 liter of the standard buffer containing a linear gradient of NaCl from 0 to 0.25 M. The flow rate is controlled at approximately 76 ml/hr. The enzyme elutes at about 0.15 M NaCl, and the fractions containing the enzymic activity are pooled.

Step 4. To the pooled fractions from step 3 is added $(NH_4)_2SO_4$ (39.8 g/100 ml). After 1 hour the precipitate is removed by centrifugation, and further 7.8 g of $(NH_4)_2SO_4$ are added per initial 100 ml. The flocculent precipitate is collected by ultracentrifugation at 56,000 g for 20 min. The pellet is taken up in 20 ml of 5 mM phosphate buffer, pH 7.0, containing 50 mM 2-mercaptoethanol, and the solution is passed through a calcium phosphate gel[7] column (3 × 15 cm). The elution is performed with a linear gradient of phosphate buffer from 5 mM to 300 mM in a total volume of

---

[7] Calcium phosphate gel was prepared according to M. Koike and M. Hamada, this series, Vol. 22, p. 339.

PURIFICATION OF GLYCERALDEHYDE-3-PHOSPHATE DEHYDROGENASE
FROM *Thermus thermophilus*[a]

Fraction	Activity[b] (units)	Protein (mg)	Specific activity (units/mg protein)
Extract	774	7332	0.11
Ammonium sulfate precipitated fraction	419	736	0.57
DEAE-Sephadex	220	302	0.73
After step 4	190	42	4.5
First crystals	150	29	5.2
Third crystals	19	3.5	5.4

[a] The data are taken from Fujita *et al.*[1]
[b] The activity was assayed in the absence of NH_4Cl.

400 ml, pH 7.0, in the presence of 50 mM 2-mercaptotheanol, the flow rate being controlled at 30 ml/hr. The enzyme elutes at around 0.235 M phosphate.

Step 5. To the pooled fractions from step 4 is added 47.6 g of $(NH_4)_2SO_4$/100 ml, and the resulting precipitate is again collected by centrifugation. The pellet is taken up in the smallest possible volume of the standard buffer. Either $(NH_4)_2SO_4$-saturated standard buffer or finely ground solid $(NH_4)_2SO_4$ is added in minute quantities until the slightest opacity is noted. This will turn into the "silky shimmer" characteristic of microcrystals standing overnight. The microcrystals are collected by centrifugation and taken up in a small volume of the standard buffer. The crystallization is repeated. The crystals of holoenzyme are readily obtainable in the presence of added NAD (1 mM).

The course of purification is summarized in the table.

Apoenzyme. The enzyme solution (10 mg of protein in 1.4 ml of 50 mM Tris-HCl buffer, pH 7.5) is mixed with 100 mg of acid-treated charcoal. After stirring for 60 min, the mixture is filtered through a Millipore filter, and the filtrate is used as apoenzyme. The apoenzyme preparation may contain no more than 0.02 NAD per subunit.

Notes on the Purification Procedures

1. The enzyme is rapidly inactivated by a trace amount of heavy-metal ion(s) and is unstable in tap water. 2-Mercaptoethanol (50 mM) or 1 mM

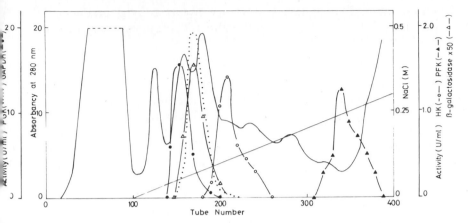

FIG. 1. DEAE-cellulose DE-32 chromatography of glyceraldehyde-3-phosphate dehydrogenase and other enzymes of *Thermus thermophilus*. Precipitate formed 16.4–51.6 g of ammonium sulfate per 100 ml of crude extract (prepared from 2 kg of wet cells), was applied on a column (6 × 32 cm), and eluted with 8 liters of 50 mM Tris-HCl buffer, pH 7.5, containing 50 mM 2-mercaptoethanol and a linear gradient of 0 to 0.5 M NaCl (20 ml/tube). Solid curve, A_{280}, solid straight line, NaCl concentration; –●–, glyceraldehyde-3-phosphate dehydrogenase; –△–, β-galactosidase; –○–, hexokinase; ·······, phosphoglycerate kinase; –▲–, phosphofructokinase.

EDTA prevents the inactivation.[8] In the presence of 50 mM 2-mercaptoethanol, the enzyme is so stable that all purification procedures can be carried out at room temperature.

2. In step 3, the enzyme was sometimes eluted at lower salt concentrations than 0.15 M NaCl,[8] especially when the precipitate formed at lower ammonium sulfate concentration was combined in step 2 (see below). The reason is not clear.

3. Other enzymes could also be purified from the extract described in step 1. When other enzymes are also to be fractionated, 16.4 g of $(NH_4)_2SO_4$ per 100 ml are added to the extract from step 1. After removal of the precipitate, 35.2 g of $(NH_4)_2SO_4$ per 100 ml are added to the supernatant. The precipitate formed is dissolved in an appropriate volume of the standard buffer and applied on a DEAE-cellulose (DE-32) column. An example of the elution profile[8] is shown in Fig. 1. Fractions containing the enzyme activity are collected and subjected to the purification procedures listed in step 4 and thereafter.

4. When the purity of the preparation after step 4 is insufficient, the collected fractions should be dialyzed and applied on a DEAE-Sephadex

[8] T. Oshima, unpublished data.

A-50 column. The column is developed with a linear gradient of NaCl from 0 to 0.4 M in the standard buffer. The enzyme will be eluted at around 0.15 M NaCl.

Properties[1]

Stability. The most conspicuous feature of glyceraldehyde-3-phosphate dehydrogenase from *T. thermophilus* is its stability to heat[9] and chemical denaturants. The enzyme was remarkably stable to heat, being slowly inactivated at 90°. The enzyme was only 40% inactivated in 8 M urea, and was fully active in 1% sodium dodecyl sulfate. The enzyme was activated by the addition of some organic solvents. The activation was about 5-fold at an ethanol concentration of 30% (v/v). Other alcohols and acetone also activated the enzyme. However, at higher concentrations, inactivation due to denaturation was observed. The enzyme can be stored at least for 2 years at −20° without significant loss of the activity.

Catalytic Properties. The thermophile dehydrogenase is activated by the presence of NH_4Cl, CsCl, or KCl. The optimum concentration is 0.1 M NH_4Cl, and under these conditions the activation was 25-fold. Ammonium sulfate and LiCl also activate the enzyme, but to a lesser degree; the enzyme activity was increased about 10-fold in the presence of 0.045 M $(NH_4)_2SO_4$. Na_2SO_4 and NaCl activated the enzyme up to 5-fold.

Other enzymic properties of *T. thermophilus* glyceraldehyde-3-phosphate dehydrogenase are similar to those from mesophilic and moderately thermophilic organisms.[10,11] The K_m for glyceraldehyde 3-phosphate is 0.3 mM, and that for NAD is 0.1 mM (30°, pH 8.3). Specific activity is 55 μmol/min per milligram of enzyme at 30°, pH 8.3, in the presence of 90 mM NH_4Cl; the optimum pH is 8.3. The enzyme presumably contains an essential sulfhydryl group at the active site. Monoiodoacetate and sodium tetrathionate are potent inhibitors. Monoiodoacetamide is also an inhibitor, but less potent than monoiodoacetate. $HgCl_2$, $CdCl_2$, and $CuCl_2$ are also strong inhibitors.

Molecular Properties. The enzyme is a tetramer of identical subunits with a molecular weight of 33,500. The $\lambda_{max} = 278$ nm, $\lambda_{min} = 251$ nm for the apoenzyme; and $\lambda_{max} = 274$ and 268.5 nm, $\lambda_{min} = 249$ nm, for the holoenzyme. The ratio of $A_{280} : A_{260} = 1.89$ for the apoenzyme and 1.05 for the holoenzyme.

[9] S. C. Fujita and K. Imahori, *in* "Peptides, Polypeptides and Proteins" (E. R. Blout, F. A. Bovey, M. Goodman, and N. Lotan, eds.), p. 217. Wiley, New York, 1974.

[10] S. F. Velick and C. Furfine, *in* "The Enzymes" (P. D. Boyer, H. Lardy, and K. Myrbäck, eds.), Vol. 7, p. 243. Academic Press, New York, 1963.

[11] J. I. Harris and M. Waters, *in* "The Enzymes" (P. D. Boyer ed.), 3rd ed., Vol. 13, p. 1. Academic Press, New York, 1976.

[59] Hydroxypyruvate Reductase (D-Glycerate Dehydrogenase) from *Pseudomonas*

By LEONARD D. KOHN and JOHN M. UTTING

Assay Method

Principle. The reaction may be measured spectrophotometrically by utilizing the decreases in absorption at 340 nm due to the conversion of NADH to NAD in the presence of hydroxypyruvate.

Reagents

Potassium phosphate, M, pH 6.6
Lithium hydroxypyruvate, 0.01 M, freshly prepared each hour by dissolving 12.8 mg in 10 ml of water
NADH, 2 mM, or NADPH, 1 mM, adjusted to pH 8.0 with KOH

Procedure. The reagents are added in the order indicated and brought to a final volume of 1.0 ml: phosphate buffer, 100 μl; hydroxypyruvate, 400 μl; appropriate amounts of enzyme; and reduced pyridine nucleotide, 100 μl. The reaction is conveniently measured at 340 nm with a spectrophotometer equipped with an automatic cuvette changer. The reaction is linear at 25° with respect to time, and protein concentration when absorbance changes of less than 0.5 are recorded. Since limiting quantities of reduced pyridine nucleotide are used, care must be exercised to include the specified concentration of that reagent in order to maintain reproducibility of the assay.

Definition of Unit. A unit of activity is defined as that amount of enzyme required to convert 1 μmol of NADH to NAD$^+$ per minute using the above-noted procedure. Specific activity is defined as the number of enzyme units per milligram of protein. Protein is measured by the method of Lowry *et al.*,[1] using bovine serum albumin as a standard.

---

[1] O. H. Lowry, N. J. Rosebrough, A. L. Farr, and R. J. Randall, *J. Biol. Chem.* **193**, 265 (1951).

Purification Procedure

Growth. The enzyme is obtained from a strain of *Pseudomonas acidovorans* (ATCC 17455), originally isolated as *Pseudomonas* A by U. Bachrach[2] from an enrichment culture containing uric acid. Maximal induction of enzyme is achieved by growing the bacteria on a medium containing the following (per liter): K_2HPO_4, 1.1 g; KH_2PO_4, 0.48 g; NH_4NO_3, 1 g; and $MgSO_4 \cdot 7 H_2O$, 0.25 g. The magnesium salt is added last, just prior to autoclaving; after the medium returns to room temperature, sterile filtered yeast extract, 0.1 g/liter, and sodium glyoxylate, 3.0 g/liter, are added. A fresh agar slant culture is used to inoculate 20 ml of the medium, which, after 24 hr of growth, serves as an inoculum for 1 liter of medium in a 6-liter flask. After 24 hr of vigorous aeration on a reciprocating shaker, the contents of the flask are transferred to a 5-gallon carboy containing 5 liters of medium that is vigorously aerated for 18 hr at room temperature. Cells are harvested in the cold with a Sharples centrifuge, washed once with 0.02 M potassium phosphate at pH 7.2, and stored at $-15°$. Enzyme activity remains constant for at least 1 year under these conditions.

Purification

Except where specifically indicated, all procedures are carried out at approximately $0°$.

Step 1. Cell Extraction. Frozen cells are suspended in 0.02 M potassium phosphate, pH 7.2, containing 3 mM 2-mercaptoethanol and 100 μg of phenylmethylsulfonyl fluoride per milliliter. After being thawed and diluted to a wet weight concentration of 20 g of cells to 100 ml of suspension, the cells are disrupted by either sonication or pressure. Sonication is for 16 min in a 10-kc Raytheon sonic oscillator cooled with water at $2°$. Pressure disruption is achieved by two passes in a Gaulin homogenizer (Manton–Gaulin Manufacturing Co., Everett, Massachusetts); cell extracts are chilled to $2°$ between passes and are collected in containers packed in ice. In each case, the sediment obtained after 30 min of centrifugation at 18,000 g and $5°$ is discarded.

Step 2. Ammonium Sulfate Precipitation. With constant stirring, ammonium sulfate, 24.3 g/100 ml, is slowly added to the supernatant from step 1, and the resultant solution is allowed to equilibrate for at least 1 hr. The precipitate obtained by this addition is discarded after centrifugation at 18,000 g for 10 min, and an additional 20.5 g of ammonium sulfate per 100 ml are slowly added with constant stirring. After 1 hr of equilibration,

[2] U. Bachrach, *J. Gen. Microbiol.* **17**, 1 (1957).

the precipitate is removed by centrifugation, solubilized in a minimal volume of 0.04 M Tris-chloride, pH 7.4, containing 25% (v/v) glycerol, 100 μg of phenylmethylsulfonyl fluoride per milliliter, and 3 mM 2-mercaptoethanol, and finally dialyzed against two successive 6-liter batches of the same buffer to achieve a final conductivity of 0.6 mmho.

Step 3. DEAE-Cellulose Chromatography. The dialyzed material is applied to a column (5 cm × 77 cm) of microgranular DEAE-cellulose (Whatman DE-52), which has previously been equilibrated with 0.04 M Tris-chloride at pH 7.4, containing 25% (v/v) glycerol and 3 mM 2-mercaptoethanol (conductivity, 0.45 mmho). The column is washed with approximately 750 ml of the equilibrating buffer and eluted with a linear gradient of 1500 ml of equilibrating buffer and 1500 ml of equilibrating buffer made 0.2 M in sodium chloride. A flow rate of 1 ml/min is maintained, and fractions of 20 ml are collected. The main enzymic activity is contained in fractions 103 to 149 collected between 1600 and 2600 ml after the initiation of the gradient.

Step 4. Hydroxyapatite Chromatography. A column of hydroxyapatite (Bio-Rad Laboratories, Rockville Center, New York), 5 cm × 42 cm, is prepared and equilibrated with 0.04 M Tris-chloride, pH 7.4, containing 25% (v/v) glycerol and 3 mM 2-mercaptoethanol. The fractions containing the main peak of enzyme activity from the DEAE-cellulose column are pooled and dialyzed against this equilibrating buffer to a final conductivity of 0.6 mmho. The pooled material is pumped onto the hydroxyapatite column, which is then washed with 400 ml of equilibrating buffer. Elution is with a linear gradient of 2 liters of 0.07 M potassium phosphate, pH 7.5, containing 25% (v/v) glycerol and 3 mM 2-mercaptoethanol into 2 liters of the Tris-chloride equilibrating system; fractions of 15 ml are collected at an elution rate of 1 ml/min. The induced enzyme is collected in fractions 118 to 170; a constitutive enzyme[3] appears in fractions 103 to 108. The fractions containing induced enzyme are pooled and concentrated in an Amicon Model 402 pressure cell over a UM-10 membrane.

Step 5. Gel Filtration Chromatography. A Sephadex G-150 column (5 cm × 87 cm) is prepared in the cold and equilibrated with 0.04 M Tris-chloride, pH 7.4, containing 25% (v/v) glycerol, 3 mM 2-mercaptoethanol, and 0.1 M sodium chloride. The concentrated pool from step 4 is applied to the column and eluted with equilibrating buffer at a flow rate of 30 ml/hr. Fractions of 15 ml are collected. Fractions from the first half of the enzyme peak have a constant specific activity of 57 units/mg; those toward the second half are contaminated with a small amount of protein, <5%, having about twice the extinction coefficient at 280 nm as the induced hydroxypyruvate reductase. The pooled concentrate of the latter fractions

[3] L. D. Kohn and W. B. Jakoby, this series, Vol. 9, p. 229.

FRACTIONATION OF HYDROXYPYRUVATE REDUCTASE FROM *Pseudomonas*

Step	Total volume (ml)	Total protein (mg)	Total activity (units)	Specific activity (units/mg protein)	Recov (%
Extraction	1250	18,375	61,294	3.3	
Ammonium sulfate, 40–70%, precipitation	460	6,486	56,206	8.7	92
DEAE-cellulose chromatography	970	2,716	47,174	17.3	77
Hydroxyapatite chromatography	795	789	32,154	40.8	52.
Gel filtration (G-150) chromatography	250	351	20,155	57.4	32.

are recycled through the gel filtration procedure to ensure complete elimination of the contaminating protein. Fractions from the first half of the
initial gel filtration procedure and from the first two-thirds of the second,
are pooled and stored in 0° in 0.04 M Tris-chloride, pH 7.4, containing
25% (v/v) glycerol, 0.1 M sodium chloride, and 3 mM 2-mercaptoethanol.
A typical fractionation is summarized in the table.

Properties

Although the reduction of hydroxypyruvate by the enzyme appears to
be analogous to that catalyzed by lactic dehydrogenase and glyoxylic acid
reductase, the hydroxypyruvate reductase differs in its specificity for hydroxypyruvate. Thus the following compounds do not serve as substrate:
glyoxal, glycoaldehyde, glyoxylate, glyceraldehyde, glyceraldehyde
3-phosphate, dihydroxyacetone, pyruvate, mercaptopyruvate, bromo-
pyruvate, α-ketobutyrate, β-ketobutyrate, oxaloacetate, oxaloglycolate,
α-ketoglutarate, and β-ketoglutarate. Solutions of tartronic semialdehyde
have no activity beyond that accounted for by the contaminating presence
of hydroxypyruvate.

At NADH or NADPH concentrations of 0.15 mM, a K_m of 8 mM is
obtained for hydroxypyruvate. At a hydroxypyruvate concentration of 10
mM, the K_m values for NADH and NADPH are 20 μM and 100 μM,
respectively. The appropriate kinetic constants obtained by the method of
Alberty[4] are presented in Eq. (1).

$$\frac{V_M}{v} = 1 + \frac{8.3 \times 10^{-3} M}{[\text{hydroxypyruvate}]} + \frac{2 \times 10^{-5} M}{[\text{NADH}]}$$
$$+ \frac{1.0 \times 10^{-7} M}{[\text{hydroxypyruvate}][\text{NADH}]} \quad (1)$$

[4] R. A. Alberty, *J. Am. Chem. Soc.* **75**, 1932 (1953).

The protein yields one band on disc gel electrophoresis and is homogeneous in the ultracentrifuge with an $s_{25,w}$ of 6.34 after extrapolation to zero protein concentration. The diffusion constant is 0.697×10^{-6} cm^2 sec^{-1}, and the partial specific volume is 0.73 ml/g; the latter is an estimate based on amino acid analysis. From these data the molecular weight and frictional coefficient may be calculated as 84,500 and 1.19, respectively. Meniscus depletion sedimentation equilibrium experiments yield a molecular weight of $85,000 \pm 6\%$. The enzyme is composed of two apparently identical subunits.[5]

The original pH for hydroxypyruvate reduction is at 6.5 with half of maximal activity at pH 5.5 and 8.0. The enzyme is anion-inhibited (NO$_3$ > Cl > SO$_4$) when hydroxypyruvate is the substrate; anions have no effect when D-glycerate is the substrate. At pH 8.5 in 0.2 M Tris-chloride, the reaction is experimentally reversible in the presence of D-glycerate and either NAD or NADP.

The enzyme is stable for 1 year when stored at 0°. Dilute solutions of protein (<0.01 mg/ml) in storage buffer without glycerol appear to be stable for only 6 hr; with glycerol, stability is for several days.

[5] J. M. Utting and L. D. Kohn, *J. Biol. Chem.* **250**, 5233 (1975).

[60] L-Lactate Dehydrogenase Isozymes from Sweet Potato Roots

By KAZUKO ÔBA and IKUZO URITANI

$$\text{L-Lactate} + \text{NAD}^+ \rightleftharpoons \text{pyruvate} + \text{NADH} + \text{H}^+$$

Sweet potato (*Ipomoea batatas* Lam) contains two isozymes of L-lactate dehydrogenase [L-lactate : NAD$^+$ oxidoreductase, EC 1.1.1.27]. The isozymes were separated by a DE-52 cellulose column chromatography.[1]

Assay Method

The routine assay is performed spectrophotometrically at 25° by measuring the decrease in absorbance at 340 nm of NADH oxidation coupled with pyruvate reduction to L-lactate. An alternative method for measuring the oxidation of lactate is used for kinetic analysis.

[1] K. Ôba, S. Murakami, and I. Uritani, *J. Biochem. (Tokyo)* **81**, 1193 (1977).

METHODS IN ENZYMOLOGY, VOL. 89

Reagents

For reduction of pyruvate

N-2-Hydroxyethylpiperazine-N-2-ethanesulfonic   acid   (HEPES)–
KOH buffer, 0.25 M, pH 7.3 at 25°

NADH, 5.9 mM, prepared fresh daily

Pyruvate, sodium salt, 0.05 M, stored at −20°

For oxidation of L-lactate

Tris-maleate buffer, 0.5 M, pH 8.2 at 25°

or

N,N-Bis(2-hydroxyethyl)glycine (Bicine)–KOH buffer, 0.5 M, pH
8.8 at 25°

L-Lactate, sodium salt, 1.0 M, stored at −20°

NAD$^+$, 0.16 M, stored at −20°, adjusted to pH 7 before use

Procedure for Reduction of Pyruvate. In a reference quartz cell (0.7 ml, 1-cm light path), are placed 0.08 ml of buffer, 0.02 ml of pyruvate, 0.005 ml of enzyme, and water to make the final volume 0.4 ml. In a test quartz cell (0.7 ml, 1-cm light path) are placed 0.08 ml of buffer, 0.01 ml of NADH, 0.005 ml of enzyme, and water to a total volume of 0.38 ml. When the decrease in absorbance at 340 nm ceases, the reaction is initiated by the addition of 0.02 ml of pyruvate. Since crude extracts and some preparations from the early fractionation steps contain low but significant levels of NADH oxidase, a correction must be made for this activity.

Procedure for Oxidation of L-Lactate. In a quartz cell (0.7 ml, 1-cm light path), are placed 0.04 ml of buffer, 0.01 ml of NAD$^+$, 0.01 ml of enzymes, and water to make the final volume of 0.38 ml. When the absorbance at 340 nm becomes constant the reaction is initiated by the addition of 0.02 ml of L-lactate.

Definition of Enzyme Unit and Specific Activity. One unit of enzyme activity is defined as the pyruvate-dependent oxidation of 1.0 nmol of NADH per minute, which is equivalent to a decrease in absorbance at 340 nm of 0.016 per minute under the above conditions. Specific activity is defined as enzyme units per milligram of protein. Protein content is calculated by multiplying the nitrogen content by 6.25. Nitrogen content is determined with Nessler's reagent[2] after digestion of the trichloroacetic acid-insoluble fraction with 10 N H_2SO_4 containing 0.002% $CuSeO_3$.

Purification Procedure

Sampling of Sweet Potato Root Tissues. Sweet potato (*Ipomoea batatas* Lam, cv. Norin No. 1) roots are harvested in the autumn and stored at 10–14° until use. Roots are dipped in a solution of 0.1% sodium

[2] M. J. Johnson, *J. Biol. Chem.* **137**, 575 (1941).

hypochlorite for 20 min for sterilization, washed with water for 20 min, and sliced (2 cm thick). The slices are inoculated on the cut surfaces with a spore suspension of *Ceratocystis fimbriata* (Ell. et Halst) (1×10^7 spores/ml). After incubation for 45 hr at 29°, slices 1.0–1.5 mm thick are taken from the noninfected tissue adjacent to the infected region. This material is designated as diseased tissue. A sample is also prepared from the central part of parenchymatous tissue of fresh roots and designated as healthy tissue.

Preparation of Endoconidial Suspension of Ceratocystis fimbriata. Laboratories of Plant Pathology of universities in countries where the sweet potato is cultivated have ascospores of *Ceratocystis fimbriata* from sweet potato. Ascospores are used to inoculate 100 ml of potato extract medium containing 1% sucrose. The inoculated medium is shaken on a shaker at 140 strokes per minute at 30°. After 2–3 days, the medium is passed through two layers of Miracloth to remove mycelia and centrifuged to collect endoconidia, which are then washed twice with water and finally suspended in water to 1×10^7 spores/ml.

Crude Extracts. All the steps below are carried out at 0–4°. Healthy or diseased tissue (200 g) is mixed with 600 ml of chilled (4°) 20 mM potassium phosphate buffer (pH 7.5) containing 0.7 M mannitol, 1 mM EDTA, 6 g of potassium isoascorbate, and 40 g of Polyclar AT (GAF Corp., New York), and homogenized twice (20 sec and then 30 sec) in a blender (Sakuma Co. Ltd) at maximum speed. The homogenate is squeezed through 4 layers of cotton gauze and centrifuged at 20,000 g for 30 min at 4°.

Gel Filtration, Sephadex G-25. The supernatant solution is passed through a Sephadex G-25 column (8×70 cm) preequilibrated with the above medium not containing isoascorbate to remove polyphenol compounds.

Ammonium Sulfate Precipitation. To the protein fraction obtained from the effluent of a Sephadex G-25 column is added solid ammonium sulfate (53.2 g/100 ml) at pH 7.0. The precipitate obtained by centrifugation at 20,000 g for 30 min is dissolved in 75 ml of 20 mM potassium phosphate buffer (pH 7.0) containing 1 mM DTT, and the solution is desalted by passage through a Sephadex G-25 column (2×31 cm) preequilibrated with 20 mM potassium phosphate buffer (pH 7.0) containing 1 mM DTT.

First DE-52 Cellulose Column Chromatography, pH 7.5. The enzyme solution is diluted twofold with 1 mM DTT and applied to a DE-52 cellulose column (4.5×12 cm) preequilibrated with 5 mM potassium phosphate buffer (pH 7.5) containing 10% (w/v) sucrose and 1 mM DTT. The column is washed with 200 ml of the equilibration buffer, and the enzymes are then eluted with a linear gradient of 0 to 0.4 M KCl in the same buffer (800 ml) at a flow rate of 1 ml/min. Two enzyme fractions are located in the concentration range of 0.04–0.06 M KCl and 0.07–0.09 M KCl, and they are designated as isozymes I and II, respectively. The activity of isozyme I is higher than that of isozyme II in a healthy tissue preparation, but the activity of isozyme II is higher than that of isozyme I in a diseased tissue preparation. Therefore, isozyme I is purified from a healthy tissue preparation and isozyme II from a diseased tissue preparation.

Purification of Isozyme I from Healthy Tissue Preparation

Solid ammonium sulfate (49.4 g/100 ml) is added at pH 7.0 to the first active protein fraction (116 ml) (containing isozyme I) eluted from a DE-52 cellulose column. The precipitated protein is dissolved in 8 ml of the equilibration buffer, and the solution is passed through a Sephadex G-25 column (2.0 × 20 cm).

Gel Filtration, Sephadex G-200. The effluent (15 ml) is concentrated to 2.8 ml using a Diaflo ultrafiltration membrane (PM-30). Then, 0.84 ml of the enzyme solution is applied to the bottom of a Sephadex G-200 column (1.4 × 90 cm) preequilibrated with 5 mM potassium phosphate buffer (pH 7.5) containing 10% (w/v) sucrose and 1 mM DTT, and 50 ml of the same buffer is passed upward through the column at a flow rate of 1.5 ml/hr. The active protein fractions are pooled and concentrated to 1.2 ml as described above, and the concentrates are stored at −20°.

Purification of Isozyme II from Diseased Tissue Preparation

Solid ammonium sulfate (49.4 g/100 ml) is added at pH 7.0 to the active fraction (195 ml, containing isozymes I and II) obtained by the first DE-52 cellulose column chromatography. The precipitated protein fraction is dissolved in 5 mM potassium phosphate buffer (pH 7.5) containing 10% (w/v) sucrose and 1 mM DTT, and ammonium sulfate is removed from the preparation by passage through a Sephadex G-25 column.

Second DE-52 Cellulose Column Chromatography, pH 7.5. The enzyme solution (18.5 ml) is applied to a second DE-52 cellulose column (2.5 × 6.5 cm) preequilibrated with 5 mM potassium phosphate buffer (pH 7.5) containing 10% (w/v) sucrose and 1 mM DTT. The column is washed with 100 ml of the equilibration buffer and the enzymes are then eluted with a linear gradient of 0 to 0.3 M KCl in the same buffer (300 ml) at a flow rate of 30 ml per hour. The active protein fractions containing mainly isozyme II, are collected and concentrated as above.

Third DE-52 Cellulose Column Chromatography, pH 6.5. Two milliliters of the enzyme solution are applied to a third DE-52 cellulose column (1.3 × 10 cm) preequilibrated with 5 mM potassium phosphate buffer (pH 6.5) containing 10% (w/v) sucrose and 1 mM DTT. The column is washed

TABLE I
SEPARATION AND PURIFICATION OF L-LACTATE DEHYDROGENASE ISOZYMES FROM
HEALTHY AND DISEASED TISSUES

Step	Isozyme	Total activity (units)	Total protein (mg)	Specific activity (units/mg of protein)
Isozyme I from healthy tissue				
Sephadex G-25	I + II	8,800	238	37.1
Ammonium sulfate	I + II	7,520	98.4	76.5
1st DE-52 cellulose	I	2,560	6.31	406
Sephadex G-200	I	868	1.22	714
Isozyme II from diseased tissue				
Sephadex G-25	I + II	12,800	228	56.4
Ammonium sulfate	I + II	10,900	171	63.7
1st DE-52 cellulose	I + II	8,870	44.8	198
2nd DE-52 cellulose	II	4,810	11.8	408
3rd DE-52 cellulose	II	2,830	3.10	913
Sephadex G-200	II	1,300	0.694	1870

with 50 ml of the equilibration buffer, and the enzymes are then eluted with a linear gradient of 0 to 0.25 M KCl in the same buffer (160 ml) at a flow rate of 20 ml/hr. The fractions containing isozyme II with the highest specific activity are pooled and concentrated to 1.5 ml.

Gel Filtration, Sephadex G-200. The concentrated isozyme II (1.4 ml) is applied to the bottom of a Sephadex G-200 column (1.4 × 90 cm) preequilibrated with 5 mM potassium phosphate buffer (pH 7.5) containing 10% (w/v) sucrose and 1 mM DTT, and 50 ml of the same buffer is passed upward through the column at a flow rate of 2 ml/hr. Active protein fractions are collected and concentrated to 1 ml using a Diaflo ultramembrane (PM-30), and the concentrate is stored at $-20°$. A plot of enzyme activity as a function of the volume of eluate from a Sephadex G-200 column shows a single symmetrical peak of isozyme I or II. A summary of the purification of L-lactate dehydrogenase isozymes is given in Table I.

Properties

Purity. Isozymes I and II obtained after chromatography of Sephadex G-200 give several bands on polyacrylamide gel electrophoresis.[3] The activities of isozymes I and II are located in minor bands by staining the

[3] J. B. Davis, *Ann. N.Y. Acad. Sci.* **121**, 404 (1964).

TABLE II

KINETIC CONSTANTS OF L-LACTATE DEHYDROGENASE ISOZYMES FROM SWEET POTATO

Isozyme	pH^a	Varied substrate	Constant substrate	K_m (apparent) for varied substrate (mM)	k_i (apparent) for varied subs (mM)
I	7.3	Pyruvate	NADH (0.148 mM)	0.34	13.5
	6.5	Pyruvate	NADH (0.148 mM)	1.5	12.5
	7.3	NADH	Pyruvate (2.5 mM)	0.0125	—
	8.8	L-Lactate	NAD$^+$ (5.2 mM)	22	—
II	7.3	Pyruvate	NADH (0.148 mM)	0.42	15.0
	6.5	Pyruvate	NADH (0.148 mM)	0.50	14.0
	7.3	NADH	Pyruvate (2.5 mM)	0.0125	—
	8.2	L-Lactate	NAD$^+$ (4.0 mM)	17	—
	8.2	DL-Lactate	NAD$^+$ (4.0 mM)	33	—
	8.2	NAD$^+$	L-Lactate (50 mM)	0.95	—

^{a} The buffers used are 50 mM histidine-HCl buffer (pH 6.5), 50 mM HEPES–KOH buffer (pH 7.3), 50 Tris–maleate buffer (pH 8.2), and 50 mM Bicine–KOH buffer (pH 8.8).

gel for 1 hr in the dark at 30° in 5 ml of 50 mM Bicine–KOH buffer (pH 8.4) containing 10 mM potassium lactate, 32 mM NAD$^+$, 0.4 mg of phenazine methosulfate, and 1.6 mg of p-nitro blue tetrazolium. Relative mobilities of isozymes I and II in the 7% polyacrylamide gel (pH 8.9) are 0.35 and 0.37, respectively.

Molecular Weight. The molecular weights of isozymes I and II are determined by polyacrylamide gel electrophoresis according to Hedrick and Smith.[4] Straight lines are obtained in plots of log R_m of the active enzyme band against gel concentration. The molecular weights of both isozymes were determined to be 150,000. It seems likely that the sweet potato L-lactate dehydrogenase isozymes may be charge isomers, not size isomers. It is not yet clear whether the isozymes are composed of subunits.

Stability. Little or no loss of activity of isozyme I or II occurred on storage at −20° for several months. Dithiothreitol and sucrose are necessary to retain enzymic activity during the purification process.

pH Optimum. The optimum pH for the reduction of pyruvate by NADH is pH 7.3 with isozymes I and II. The pH optimum for oxidation of L-lactate with NAD$^+$ is pH 7.8 to 8.8 for both isozymes I and II.

Kinetic Properties. Pyruvate inhibits the enzyme activity at concentrations higher than 2.5 mM. At pH 7.3, the inhibitor constants (K_i) of pyruvate for isozymes I and II are 13.5 and 15.0 mM, respectively (Table II), and extrapolation of the double-reciprocal plots indicates that the

[4] J. L. Hedrick and A. J. Smith, *Arch. Biochem. Biophys.* **126**, 155 (1968).

apparent Michaelis constants (K_m) of pyruvate for isozymes I and II are 0.34 and 0.42 mM, respectively. When the pH is reduced to 6.5, the K_m value of pyruvate increases markedly in the case of isozyme I, but only slightly in the case of isozyme II. The enzyme apparently shows normal Michaelis–Menten kinetics, and the K_m values for L-lactate and coenzymes (NADH and NAD$^+$) are as shown in Table II. The finding that the K_m value of L-lactate is half of that of DL-lactate in the case of isozyme II (Table II) indicates that isozyme II is L-lactate dehydrogenase. Although the K_m value of DL-lactate in the case of isozyme I is not determined, it appears that isozyme I is also L-lactate dehydrogenase, since the value for L-lactate is almost the same as that of isozyme II.

Inhibitors and Regulation. A number of mononucleotides, coenzymes, organic acids, ipomeamarone (a major furanoterpene produced in diseased sweet potato root tissue) and its precursor were tested for effects on the two isozymes of L-lactate dehydrogenase.[1] Isozymes I and II are inhibited by various mononucleotides, phosphoenolpyruvate and NAD$^+$, but not by the organic acids, acetyl-CoA, or ipomeamarone. A plot of inhibition as a function of concentration of ATP yielded a sigmoidal inhibition curve; isozyme I is more sensitive to inhibition by adenine nucleotide than isozyme II. The inhibitions caused by the nucleotides are reduced by the addition of MgCl$_2$ and are overcome by amounts of MgCl$_2$ larger than the stoichiometric amount. The observation that inhibition by CTP or GTP is much less than that by ATP indicates that ATP functions as a specific inhibitor. The reduction of pyruvate by NADH is inhibited by NAD$^+$ competitively with respect to NADH.

[61] Lactate Dehydrogenase Isozymes from Mouse

By Chi-Yu Lee, James H. Yuan, and Erwin Goldberg

$$\text{L-Lactate} + \text{NAD}^+ \rightleftharpoons \text{pyruvate} + \text{NADH} + \text{H}^+$$

Three homotetrameric isozymes of lactate dehydrogenase (EC 1.1.1.27; L-Lactate : NAD$^+$ oxidoreductase) designated as A$_4$, B$_4$, and C$_4$ or X are found in mice.[1-3] Each isozyme is expressed by a separate

[1] C. L. Markert, J. B. Shaklee, and G. S. Whitt, *Science* **189**, 102 (1975).

[2] E. Goldberg, *in* "Isozymes" (M. C. Rattazzi, J. G. Scandalios, and G. S. Whitt, eds.), Vol. 1, pp. 79–124. Liss, New York, 1977.

[3] S.-M. T. Chang, C.-Y. Lee, and S. Li, *Biochem. Genet.* **17**, 715 (1979).

METHODS IN ENZYMOLOGY, VOL. 89

structural gene.[2] In addition, heterotetramers of lactate dehydrogenase, such as A_3B, A_2B_2, and AB_3, are also present in different mouse tissues.[2] Lactate dehydrogenase C_4 or lactate dehydrogenase X is found only in mature testis and spermatozoa.[2] All the lactate dehydrogenase isozymes require NAD^+ or NADH as the coenzyme and catalyze the same chemical reactions.[2] The purification of these enzymes can simply be achieved by "general ligand affinity chromatography"[4] in combination with one more steps of ion-exchange chromatography to separate the multiple forms.

General ligand affinity chromatography is based on the principle that a single immobilized ligand is able to adsorb a group or groups of enzymes that utilize the same coenzyme, inhibitors, or activators.[4] The biospecific elution of enzymes can usually be achieved by an eluent containing (a) a competitive inhibitor specific to a particular enzyme; (b) a combination of coenzyme and substrate or inhibitor that form a specific ternary complex with the enzyme of interest; or (c) a concentration gradient of coenzymes.

In the last 5 years, several coenzyme-dependent enzymes from mouse and *Drosophila melanogaster* have been purified in our laboratory by general ligand affinity chromatography[4] (see also this volume [43,51,75,94] and Vol. 90 [3,21]). This methodology has the potential to purify as much as 30% of the 2000 known enzymes that depend on one of the following coenzymes for activity: NAD^+, $NADP^+$, ATP, and coenzyme A. Owing to the conservation of the biospecificity of enzymes, there is little variation as to the general protocols for using affinity chromatography for the purification of the same enzymes among different species.

Assay Methods

Activity of lactate dehydrogenase isozymes is determined at 25° spectrophotometrically by following the changes in absorbance at 340 nm. The reaction mixture in a total volume of 1 ml contains 0.1 M Tris-HCl, pH 8.0, 1.0 mM pyruvate, 0.15 mM NADH and a suitable amount of enzyme to produce a decrease in absorbance of 0.05 to 0.1 per minute at 340 nm. In testicular extracts, lactate dehydrogenase X can be distinguished from other lactate dehydrogenase isozymes by using 22 mM α-ketoglutarate as the substrate instead of pyruvate. Other lactate dehydrogenase isozymes are virtually inactive with α-ketoglutarate as the substrate. One unit of enzyme activity is defined as the amount of enzyme that catalyzes the formation of 1 μmol of NAD^+ and lactate per minute under the described conditions.

[4] C.-Y. Lee and A. F.-T. Chen, *in* "The Pyridine Nucleotide Coenzymes" (J. Everse, B. M. Anderson, and K.-S. You, eds.), Chapter 6, pp. 189–224. Academic Press, New York, 1982.

Affinity Columns

Since lactate dehydrogenase is an NAD^+-dependent enzyme, derivatives of NAD^+, AMP, or ATP immobilized on Sepharose can be employed as general ligands for the purification of this enzyme from numerous sources.[4,5] These derivatives are either coenzymes or inhibitors of lactate dehydrogenase and many other dehydrogenases.[4-6] Among them are 8-(6-aminohexyl)amino-AMP, N^6-(6-aminohexyl)-AMP, 8-(6-amino-hexyl)amino-ATP, 8-(6-aminohexyl)amino-NAD^+, and N^6-[(6-amino-hexyl)carbamoylmethyl]-NAD^+.[7-9] Procedures for the preparation of these ligands and affinity gels have been described[7-10] and reviewed.[4-6] An easy procedure for the preparation of 8-(6-aminohexyl)amino-AMP and -ATP has been developed and allows a large-scale preparation of these affinity columns.[4]

AMP or ATP (20 g) is dissolved in water, and the pH is adjusted to 4.5. The solution is diluted to 100 ml with 1 M NaOAc, pH 4.5. Bromine (5 ml, 15 g) is then added dropwise with vigorous stirring, causing the pH to drop to 3.6. The pH is readjusted to 4.5 by addition of 2 N NaOH. The extent of bromination of AMP or ATP is followed spectrophotometrically by measuring the absorbance ratio, $A_{260} : A_{259}$. The initial ratio, 0.957, shifts to 1.03 after 15 min of reaction (8-Br-AMP or 8-Br-ATP shows $A_{263} : A_{259} = 1.05$). After 20 min, no further change in absorbance ratio is observed. The solution is transferred to a separatory funnel and extracted three times with 100 ml of CCl_4. Solid $NaHSO_3$ is added to bleach the color to a pale yellow. Three volumes of cold ethanol (chilled to $-20°$) are added with rapid stirring. After 2 hr at $-40°$, the supernatant is decanted off. Centrifugation of the lower layer is followed by dissolving the combined oily and solid materials in H_2O (100 ml) to which 1,6-diaminohexane (50 g) is added, the incubating at room temperature (pH 11.5). The progress of the reaction is monitored spectrophotometrically by UV absorbance. After 24 hr, the absorbance maximum shifts to 274 ± 1 nm and $A_{278} : A_{263}$ changes from 0.74 to between 1.1 and 1.25 depending on the amount of unbrominated AMP or ATP present. No further change in $A_{278} : A_{263}$ or λ_{max} is observed after 24 hr. (The product, 8-(6-aminohexyl)amino-AMP or -ATP shows $\lambda_{max} = 278$ nm and $A_{278} : A_{263} = 1.4$.) The major product of the

[5] C.-Y. Lee, and N. O. Kaplan, *J. Macromol. Sci., Chem.* **A-10**, 15 (1976).

[6] K. Mosbach, *Adv. Enzymol.* **46**, 205 (1978).

[7] H. Guilford, P.-O. Larsson, and K. Mosbach, *Chem. Scr.* **2**, 165 (1972).

[8] C.-Y. Lee, D. Lappi, B. Wermuth, J. Everse, and N. O. Kaplan, *Arch. Biochem. Biophys.* **163**, 561 (1974).

[9] M. Lindberg, P.-O. Larsson, and K. Mosbach, *Eur. J. Biochem.* **40**, 187 (1973).

[10] C.-Y. Lee, L. H. Lazarus, D. S. Kabakoff, M. B. Larver, P. J. Russell, and N. O. Kaplan, *Arch. Biochem. Biophys.* **178**, 8 (1977).

reaction, 8-(6-aminohexyl)amino-AMP or -ATP is precipitated with 10 volumes of cold ethanol (−20°). The trace of 1,6-diaminohexane in the resulting product is removed by gradual addition of fresh Dowex 50-X8 cation-exchange resin (H⁺ form) until the pH drops from 11.0 to 6.5. The resulting solution is then lyophilized to complete dryness.

Thin-layer chromatography of the product is performed on cellulose plates with a fluorescent indicator (Eastman Kodak). The solvent system isobutyric acid–NH_4OH–H_2O, 66 : 1 : 33, is used. The chromatograms are visualized by UV light and ninhydrin. A major UV- and ninhydrin-positive spot ($\simeq$90% in intensity) is observed corresponding to that of 8-(6-aminohexyl)amino-AMP or -ATP. A minor UV positive spot corresponds to that of AMP or ATP. No 1,6-diaminohexane is found. With respect to the prepared affinity ligands, the overall yield of the pinkish product is about 60% in either case.[11]

These AMP or ATP derivatives are readily coupled to Sepharose by cyanogen bromide activation.[12] Typically, 100 ml of wet-packed Sepharose-4B is suspended in 0.5 M sodium carbonate buffer (100 ml, pH 11.0, 4°) in a 1000-ml beaker. Cyanogen bromide ($\simeq$20 g) dissolved in 40 ml of acetylnitrile is added dropwise to the Sepharose solution over a period of 5 min. The mixture is gently stirred under the hood with an occasional addition of ice chips to maintain the low temperature and a constant adjustment of the solution pH with 4 N NaOH to 11.0 ± 0.2. About 10 min after the addition of cyanogen bromide, the gel is filtered on a Büchner funnel and washed with distilled water at 0° for 2 min until the pH is neutral.

Immediately after activation, the packed gel (100 ml) is added to 50 ml of 8-(6-aminohexyl)amino-AMP or -ATP (ligand concentration: 20–50 mg/ml in H_2O with pH adjusted to 9.5–10.0 with NaOH). The Sepharose–ligand mixture is gently stirred or shaken at 4° for 12–24 hr. After coupling, the liganded Sepharose solution is filtered through a funnel. The first filtrate can be saved, lyophilized, and reused for the next coupling to freshly activated Sepharose. The affinity gel is then extensively washed with 1 liter of 2 M NaCl and then distilled water. The affinity gel can be stored at −20° in 50% glycerol.

Cyanogen bromide-activated Sepharose-4B is available commercially as the freeze-dried powder containing lactose or dextran as the stabilizer. These additives are washed away when the gel is reconstituted by washing

---

[11] For the preparation of 8-(6-aminohexyl)amino-ATP, we take advantage of the fact that ATP derivatives are stable under the mild alkaline conditions (pH 11–11.5). Little degradation of the triphosphate group of ATP derivatives is observed after 24 hr at room temperature in the presence of 50% 1,6-diaminohexane.

[12] R. Axén, J. Porath, and S. Ernbäck, *Nature (London)* **214**, 1302 (1967).

first with 1 mM HCl and then with distilled water at 4°. The washed gel (15 g of powder is equivalent to 50 ml of swollen gel) is then mixed with the ligand solution for coupling as described above.

The ligand density can be estimated according to the procedure described in this volume.[13] After each use, the affinity gel can be regenerated by washing with a solution of 6 M urea and 2 M NaCl. Typically, a ligand density of 1.2–1.7 μmol per milliliter of AMP- or ATP-Sepharose is considered effective for routine enzyme purification.

The affinity ligands and gels described in this chapter are commercially available from one of the following companies: P-L Biochemicals Inc., Milwaukee, Wisconsin; Sigma Chemical Company, St. Louis, Missouri; Pharmacia Fine Chemicals, Piscataway, New Jersey.

Preparation of Reduced NAD$^+$-Pyruvate Adduct. Reduced NAD$^+$-pyruvate adduct was used for biospecific elution of lactate dehydrogenases from the affinity columns,[8] since it is a specific inhibitor of lactate dehydrogenase from many sources.[14] The adduct is routinely prepared by mixing NAD$^+$ (100 mg) and pyruvate (100 mg) in 2 ml of distilled water. The pH of the solution is adjusted and maintained at 11.5 by a dropwise addition of 1 N NaOH at room temperature for 20 min. The mixture is then diluted to give an absorbance of 2 at 340 nm in 10 mM phosphate, pH 6.5, before being used as such. In this preparation, the absorbance ratio, $A_{260} : A_{340}$ is about 2.8.

Purification of Lactate Dehydrogenase X from Mouse Testis[15]

The following purification procedure was employed to obtain 1 g of lactate dehydrohydrogenase X from testes of 15,000 mice[14] (Jackson Laboratory, Bar Harbor, Maine).

Frozen testes (about 140 g) from C57BL/6J mice (Jackson Laboratory) are homogenized in 300 ml of 20 mM Tris-HCl, pH 7.4 and centrifuged at 27,000g for 20 min at 4°. The supernatant is collected by passage through a double-layered cheesecloth, and the pellet or lipid is discarded. The supernatant is heated to 60° for 15 min in a 65° water bath. The precipitated proteins are removed by centrifugation (27,000g for 20 min). The pH of the clear supernatant is then adjusted to 6.5 by proper additions of 0.5 M KH$_2$PO$_4$. The enzyme solution is loaded on an 8-(6-aminohexyl)amino-AMP-Sepharose column (2.5 × 20 cm) that has been preequilibrated with 10 mM potassium phosphate, pH 6.5. No sig-

[13] D. W. Niesel, G. C. Bewley, C.-Y. Lee, and F. B. Armstrong, this volume [51].

[14] C.-Y. Lee, B. Pegoraro, J. Topping, and J. H. Yuan, *Mol. Cell. Biochem.* 18, 49 (1977).

[15] Unless otherwise specified, all the phosphate buffers used in enzyme purification are the potassium salt.

PURIFICATION OF LACTATE DEHYDROGENASE X (C₄) FROM MOUSE TESTIS

Step	Total protein (mg)	Total activity (units/ml)	Specific activity (units/mg protein)
Crude extract	11,584	5792	0.5
Heat treatment (60°)	2,900	3500	1.2
8-(6-Aminohexyl) amino-AMP-Sepharose column	25[a]	2000	80

[a] Protein was determined by fluorescamine assays of P. Böhlen, S. Stein, W. Dairman, and S. Udenfriend [*Arch. Biochem. Biophys.* **155**, 213 (1973)], using bovine serum albumin as the protein standard.

nificant activity of lactate dehydrogenase is found in the eluent. The affinity column is then washed with 100 ml of 10 mM potassium phosphate, pH 6.5, followed by another 300 ml of 50 mM potassium phosphate of the same pH. Little enzyme activity is detected in the eluent of the column wash. Lactate dehydrogenase is then eluted biospecifically with 150 ml of potassium phosphate, pH 6.5 containing reduced NAD$^+$-pyruvate adduct (A_{340} = 2.0). Lactate dehydrogenase X appears immediately after the column volume with less than 1% contamination of other lactate dehydrogenase isozymes.

A typical purification of this enzyme is summarized in the table.

This scheme has also been used in a single purification with as much as 480 g of testes from random-bred Swiss–Webster retired breeders.[2]

A similar procedure can be used to purify lactate dehydrogenase X from testes of rat, hamster, and guinea pig. However, some modifications are required to obtain homogeneous enzyme. In these rodent species other lactate dehydrogenase isozymes are not completely denatured by heating. Therefore, one more purification step utilizing DEAE-Sephadex ion exchange chromatography is applied to separate lacate dehydrogenase X from other residual lactate dehydrogenase isozymes. After heat treatment and the affinity column step, the recovered enzyme solution is concentrated and dialyzed against 10 mM Tris-HCl, pH 8.0 (1 liter with two changes, 10 hr each). It is then loaded on a DEAE-Sephadex A-50 ion-exchange column (about 10 ml of resin per milligram of protein) that has been equilibrated with 10 mM Tris-HCl, pH 8.0. Lactate dehydrogenase X is readily eluted with a 0 to 50 mM NaCl gradient (300 ml by 300 ml for a 100-ml column). Fractions with homogeneous lactate dehydrogenase X are determined by separate assays using pyruvate or α-ketoglutarate as the substrate. Those with the highest activity ratio between

α-ketoglutarate and pyruvate are pooled; lactate dehydrogenase X usually appears at the beginning of the gradient.

During routine enzyme preparation by this procedure, we noticed that N^6-(6-aminohexyl)-AMP-Sepharose is better both in affinity and capacity than 8-(6-aminohexyl)amino-AMP-Sepharose for purification of lactate dehydrogenase X from rat testis.[16]

Purification of Lactate Dehydrogenase A$_4$ and Lactate Dehydrogenase B$_4$

As with lactate dehydrogenase X, lactate dehydrogenase A$_4$ and B$_4$ can be purified from mouse muscle and kidney, respectively, by a procedure almost identical to that described above,[3] except that the heat treatment is eliminated. Typically, frozen mouse tissues (muscle or kidney) are homogenized in 10 mM phosphate, pH 6.5 (1 : 3, w/v) containing 1 mM α-thioglycerol. After centrifugation at 27,000 g for 20 min, the supernatant is loaded on an 8-(6-aminohexyl)amino-AMP-Sepharose column equilibrated with the same buffer. Since this affinity column has a relatively high capacity for mouse lactate dehydrogenase, about 100 units of enzyme can be quantitatively adsorbed to 1 ml of affinity gel. After loading, the column is washed first with two column volumes of 10 mM phosphate, pH 6.5, followed by 50 mM phosphate, pH 6.5, until the 280 nm absorbance decreases to the baseline. The adsorbed lactate dehydrogenase is readily eluted biospecifically with reduced NAD$^+$-pyruvate adduct ($A_{340} = 2.0$) in 10 mM phosphate, pH 6.5. The recovery of lactate dehydrogenase from the affinity column step ranges from 80 to 95%.

After affinity chromatography, lactate dehydrogenase purified from mouse muscle is composed of lactate dehydrogenase A$_4$ ($\approx$80%) and other heterotetramers. On the other hand, only about 20% of lactate dehydrogenase purified from mouse kidney is lactate dehydrogenase B$_4$. Separation of the multiple forms of lactate dehydrogenase is simply achieved by DEAE-Sephadex ion-exchange chromatography as demonstrated in Fig. 1 for the purification of lactate dehydrogenase B$_4$ from mouse kidney. The specific activity of homogeneous lactate dehydrogenases A$_4$ and B$_4$ is about 410 and 320 units/mg, respectively, under the standard assay conditions.

Properties

By the described procedure, the homotetrameric lactate dehydrogenase isozymes from mouse are purified to homogeneity as judged by

[16] S.-M. T. Chang, C.-Y. Lee, and S. Li, *Int. J. Biochem.* **11,** 1 (1980).

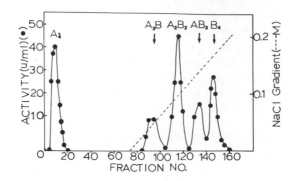

FIG. 1. Separation of mouse lactate dehydrogenase A_4 and B_4 isozymes by DEAE-Sepharose chromatography. Lactate dehydrogenase A_4 and B_4 as well as their hybrids from mouse kidney are adsorbed on a DEAE-Sepharose column (5 × 25 cm) that is equilibrated and eluted with 10 mM Tris-HCl, pH 8.0, containing 1 mM dithiothreitol. After isozyme A_4 is eluted, a linear NaCl gradient (0 to 0.3 M; 2 liters by 2 liters) is used to separate the different isozymes (20 ml per fraction).

acrylamide gel electrophoresis with or without sodium dodecyl sulfate.[3,13] This procedure yielded gram quantities of mouse lactate dehydrogenases A_4 and C_4 for total amino acid sequencing.[17]

Biochemical properties of mouse lactate dehydrogenases A_4 and B_4 are quite similar to those of the corresponding chicken lactate dehydrogenase isozymes (M_4 and H_4). General properties of lactate dehydrogenase X (or C_4) have been studied and reviewed by Goldberg.[18] The isoelectric points were shown to be 7.0, 5.3, and 7.5, respectively for lactate dehydrogenase A_4, B_4, and C_4 from mouse. Antisera raised in rabbits against each isozyme revealed no mutual cross-reactivity.[19]

Structural relatedness among the three mouse lactate dehydrogenase isozymes have been studied by amino acid composition, peptide mapping, and partial amino acid sequencing. The preliminary results suggest that lactate dehydrogenases A_4 and B_4 are more closely related to each other than to lactate dehydrogenase C_4.[3] The amino acid sequence of the loop peptide region of the mouse C subunit seems to be markedly different from all known A and B sequences of lactate dehydrogenase from many species.[3,17]

[17] Y.-C. E. Pan, S. Huang, J. P. Marciniszyn, C.-Y. Lee, and S. Li, *Hoppe-Seyler's Z. Physiol. Chem.* **361**, 795 (1980).
[18] E. Goldberg, this series, Vol. 41 [70].
[19] R. S. Holmes and R. K. Scopes, *Eur. J. Biochem.* **43**, 167 (1974).

[62] L(+)-Lactate Dehydrogenase from *Homarus americanus*

By R. D. EICHNER

$$\text{Pyruvate} + \text{NADH} + \text{H}^+ \rightleftarrows \text{L-lactate} + \text{NAD}^+$$

Assay Method

Principle. The activity of lactate dehydrogenase (EC 1.1.1.27) from the tail muscles of *Homarus americanus* is measured by monitoring spectrophotometrically the decrease (reduction of pyruvate) or the increase (lactate oxidation) in absorbance at 340 nm.

Reagents

For reduction of pyruvate
Pyruvate, sodium salt, 10 mM
NADH, 1.7 mM
Sodium phosphate buffer, 0.2 M, pH 7.6
For oxidation of lactate
L-Lactate, sodium salt, 0.5 M
NAD$^+$, 14 mM
Sodium phosphate buffer, 0.5 M, pH 7.6
Pyruvate, lactate, and NAD$^+$ solutions were stored at $-20°$. NADH solutions were prepared daily.

Procedure for Reduction of Pyruvate. Each 3-ml cuvette contains 3 μmol of pyruvate, 0.5 μmol of NADH, and 300 μmol of sodium phosphate, pH 7.6. The reaction is initiated by adding an amount of lactate dehydrogenase sufficient to produce a change of 0.1–0.2 absorbance per minute. Measurements are continued for at least 2 min; resultant activities are quantified on linear portions of these time-course progress curves.

Procedure for Oxidation of Lactate. Each 3-ml cuvette contains 200 μmol of L-lactate, 4.2 μmol of NAD$^+$, and 600 μmol of sodium phosphate, pH 7.6. Reactions are initiated by addition of lactate dehydrogenase, and measurements of changes in optical density are made for at least 5 min. Time-course progress curves are nonlinear in solutions of intermediate ionic strength (where $\mu > 0.4$)[1,2]; initial velocities are defined as the rate 30 sec after addition of enzyme.

[1] H. D. Kaloustian, F. E. Stolzenbach, J. Everse, and N. O. Kaplan, *J. Biol. Chem.* **244**, 2902 (1969).
[2] R. D. Eichner and N. O. Kaplan, *Arch. Biochem. Biophys.* **191**, 666 (1978).

Units. One unit of enzyme activity oxidizes NADH or reduces NAD^+ at a rate of 0.48 μmol per minute. Specific activity is the activity per milligram of protein.

Purification[1,3]

The tails from live adult East Coast lobsters, weighing between 0.5 and 1.5 kg, are severed and placed on ice for immediate purification or stored at $-20°$ for later use. No appreciable loss in activity occurs upon cold storage. Seasonal variations were not observed in LDH activities. All subsequent steps are performed at 4°, and all preparative centrifugations require 13,000 g for 20 min.

Crude Extract. Fifteen to twenty tail muscles are carefully separated from both the exoskeleton and gastrointestinal tract; 1500 g of muscle are homogenized in 2 volumes of buffer (Tris, 10 mM–EDTA, 1 mM; pH 7.6) using a commercial Waring blender. The extract is stirred for 2 hr and then centrifuged. The pellet contains no LDH activity and is discarded.

Ammonium Sulfate Precipitation and Fractionation. The above supernatant is made 1.89 M in ammonium sulfate and allowed to stir an additional hour, whereupon it is centrifuged; again, the enzyme activity is recovered in the supernatant.

Solid ammonium sulfate is then added slowly to the supernatant to 2.6 M concentration. Stirring for an additional hour followed by centrifugation concentrates essentially all the enzymic activity in the pellet.

The 2.65 M pellet is back-extracted with ammonium sulfate solutions (2.65 M, 2.50 M, 2.35 M, 2.20 M, 2.09 M, and 1.82 M saturation). The back-extraction is performed beginning with the most concentrated ammonium sulfate solution. All resulting solutions of 2.35 M saturation and higher contain many proteins but little or no LDH activity; these solutions are discarded. The remaining solutions (1.82–2.20 M concentration) contain 90% of the total activity; these solutions are pooled and then exhaustively dialyzed (v/v, 1 : 20, three changes of buffer).

DEAE Column Chromatography. The dialyzed solution is used to charge a DEAE-11 column (4.6 cm × 75 cm) preequilibrated with buffer (Tris, 10 mM, pH 7.6). The column is then washed with 0.75 column volumes of buffer, and a linear NaCl gradient in the above buffer (0–0.3 M, 3 column volumes) is employed to elute the enzyme (flow rate = 120 ml/hr). Starch gel electrophoresis[4] of the collected fractions indicates a clear separation of the LDH isoenzymes. Only fractions containing the

---

[3] R. D. Eichner and N. O. Kaplan, *Arch. Biochem. Biophys.* **181**, 490 (1977).
[4] I. H. Fine and L. Costello, this series, Vol. 6, p. 958.

PURIFICATION OF L(+)-LACTATE DEHYDROGENASE FROM LOBSTER TAIL

Step	Enzyme units	Total protein (g)	Specific activity (μmol/min/mg protein)	Purification (fold)	Yield (%)
...de extract	800,000	200	1.9	1.0	100
...monium sulfate, 1.89 M	900,000	36	12	6.25	112
...k extraction, after dialysis	800,000	4.5	86	45	100
...AE chromatography	540,000	0.54	480	250	67
...electric focusing	—	—	480	250	—

single slow-moving isoenzyme are subjected to further purification and chemical and kinetic analysis.

The enzyme is then precipitated with solid ammonium sulfate (2.5 M). It can be stored for several months at 4° with little loss in activity. Isoelectric focusing[5] is performed on an aliquot of the suspension that has been exhaustively dialyzed against 1% glycine.

A typical purification is summarized in the table.

Properties

Purity. This procedure yields a 250-fold purification with 60% recovery. Starch gel electrophoresis indicates one protein staining band corresponding to enzymic activity. Sedimentation velocity profiles are symmetrical. Ouchterlony double-diffusion plating with prepared antisera reveals single precipitant lines.

Kinetic Properties.[2,6] The pH optimum for both reduction of pyruvate and oxidation of lactate is 7.5 (phosphate buffer). The K_Ds for NAD$^+$ and NADH are $1.6 \times 10^{-4} M$ and $7.1 \times 10^{-6} M$, respectively; both coenzymes bind to the enzyme in an open conformation as determined by proton nuclear magnetic resonance spectroscopy.[7] The apparent K_ms for pyruvate and NADH are $2.4 \times 10^{-4} M$ and $6.1 \times 10^{-5} M$, respectively. However, at 11°, a more physiological temperature, the value for pyruvate is $5.0 \times 10^{-5} M$, and marked substrate inhibition occurs (90% at 10 mM pyruvate). The enzyme forms an abortive ternary complex with NAD$^+$ and pyruvate. These latter properties are indicative of those of a heart-

[5] M. B. Hayes and D. Wellner, *J. Biol. Chem.* **244**, 6636 (1969).
[6] R. D. Eichner and N. O. Kaplan, *Arch. Biochem. Biophys.* **181**, 501 (1977).
[7] C.-Y. Lee, R. D. Eichner, and N. O. Kaplan, *Proc. Natl. Acad. Sci. U.S.A.* **70**, 1593 (1973).

type LDH. The turnover number is 1.4×10^5 mole of NADH oxidized per minute per mole of enzyme.

The kinetics of lactate oxidation exhibit sigmoidal time course progress curves in solutions of intermediate ionic strength. The apparent K_ms for lactate and NAD^+ in solutions of low ionic strength ($0.1\,M$ Tris) are 1.1 M and $1.1 \times 10^{-3}\,M$, respectively; in intermediate ionic strength solutions ($0.2\,M$ sodium phosphate), $0.052\,M$ and $2.8 \times 10^{-4}\,M$, respectively; in high ionic strength ($0.1\,M$ Tris, $1.1\,M$ ammonium sulfate), $0.028\,M$ and $1.5 \times 10^{-4}\,M$, respectively. The Hill coefficient is 1.8 ($0.2\,M$ sodium phosphate): the apparent K_m for lactate at $11°$ is $0.0067\,M$ ($0.2\,M$ sodium phosphate).

Physical Properties.[3] The pI of the enzyme is 5.2. The enzyme is stoichiometrically inactivated by thiol reagents (pHMB and DTNB) indicating one essential sulfhydral group per subunit. The extinction coefficient at 280 nm is $2.0 \times 10^5\,M^{-1}\,cm^{-1}$.

The values for $s_{20,w}$ and molecular weight (as measured by analytical ultracentrifugal techniques) are affected by ionic strength. In low ionic strength solutions the $s_{20,w}$ and molecular weight are 7.3 and 145,000, respectively; in high ionic strength solutions, 3.8 and 75,000, respectively. The enzyme probably exists as a tetramer in low ionic strength solutions and a dimer in high ionic strength solutions. The intermediate case can be represented by an equilibrium between tetramer and dimer, which in turn can be modulated by the NADH mole-fraction, which results in sigmoidal time-course progress curves.

[63] L-Lactate Dehydrogenase,[1] FDP-Activated, from *Streptococcus cremoris*

By Alan J. Hillier and G. Richard Jago

$$\text{Pyruvate} + \text{NADH} + \text{H}^+ \leftrightarrow \text{L-lactate} + \text{NAD}^+$$

Assay Method

Principle. L-Lactate dehydrogenase activity is measured spectrophotometrically at 340 nm by following either the oxidation of NADH in the presence of pyruvate and FDP or the reduction of NAD in the presence of L-lactate and FDP.

---

[1] L-Lactate : NAD^+ oxidoreductase, EC 1.1.1.27.

METHODS IN ENZYMOLOGY, VOL. 89

Reagents

For reduction of pyruvate
Pyruvate, sodium salt, 0.1 M
NADH, 4 mM
Fructose 1,6-diphosphate (FDP), 30 mM
Triethanolamine-hydrochloride, 0.1 M, pH 6.5, at 25°
For oxidation of L-lactate
L-Lactate, sodium salt, 0.5 M
NAD, 10 mM
Fructose 1,6-diphosphate (FDP), 30 mM
Triethanolamine-hydrochloride, 0.1 M, pH 7.5, at 25°

Procedure. The purified enzyme is diluted before assay in 0.1 M triethanolamine-hydrochloride, pH 6.0, to obtain a concentration of 30–300 μg/ml.

For reduction of pyruvate, use per 3.0 ml of reaction mixture contained in a silica cell of 1-cm light path: 0.2 ml of sodium pyruvate, 0.1 ml of NADH, 0.1 ml of FDP, 0.1 ml of enzyme solution, and 2.5 ml of triethanolamine-hydrochloride, pH 6.5. The decrease of absorbance at 340 nm is recorded at 25°, and the maximal (initial) rate is used for calculations.

For oxidation of L-lactate, use per 3.0 ml of reaction mixture contained in a silica cell of 1-cm light path: 0.5 ml of sodium L-lactate, 0.2 ml of NAD, 0.1 ml of FDP, 0.2 ml of enzyme solution, and 2.0 ml of triethanolamine-hydrochloride, pH 7.5. The increase of absorbance at 340 nm is recorded at 25°, and the maximal (initial) rate is used for calculations.

Units. One unit of lactate dehydrogenase activity is defined as the amount of enzyme that catalyzes the pyruvate-dependent oxidation of 1.0 μmol of NADH per minute or the lactate-dependent reduction of 1.0 μmol of NAD per minute at 25°. It has been also defined as that amount of enzyme causing a change in the absorbance at 340 nm of 0.1 per minute (this corresonds to the oxidation or reduction of 0.048 μmol of coenzyme per minute).[2] Specific activity is defined as units of enzyme activity per milligram of protein.

Purification Procedure

The FDP-activated L-lactate dehydrogenase from *S. cremoris* has been studied in greater detail in strain US3 than in any other strain of this

[2] G. R. Jago, L. W. Nichol, K. O'Dea, and W. H. Sawyer, *Biochim. Biophys. Acta* **250**, 271 (1971).

species.[2–6] The organism is obtainable from the CSIRO Division of Food Research, Dairy Research Laboratory, Highett, Victoria 3190 (Australia). Purification of the L-lactate dehydrogenase in strain US3 is achieved by the following procedures, which yield 5–10 mg of pure enzyme from 60 g of bacterial cells grown in 5 liters of broth.[2]

Step 1. Growth of Organism. The organism is grown in broth containing tryptone, 30 g; yeast extract, 10 g; lactose, 30 g; KH_2PO_4, 5 g; beef extract, 2 g; and water to 1 liter. The broth is inoculated (0.5% v/v) and incubated at 30° until the cells reach early stationary phase ($\approx$16 hr). The pH of the broth is maintained at 6.3 throughout the incubation by the addition of 10 M NaOH. The cells are harvested by centrifugation, washed in 0.154 M NaCl, and resuspended in 0.01 M sodium phosphate buffer, pH 7.0 (1 g wet weight of cells per 4 ml of buffer).

Step 2. Preparation of Cell-Free Extracts. Cell-free extracts are best prepared by extruding the washed cells through a French pressure cell at 154 MPa. The supernatant obtained after centrifugation at 35,000 g for 30 min is dialyzed overnight at 4° against 0.01 M sodium phosphate buffer, pH 7.0. All subsequent steps in the purification procedure are carried out at 4°.

Step 3. Fractionation with Streptomycin Sulfate. A 15% solution (w/v) of streptomycin sulfate is added dropwise to the dialyzed cell-free extract with constant stirring until no further precipitation of the nucleic acid component is obtained. The precipitate is removed by centrifugation at 35,000 g and discarded.

Step 4. First Fractionation with Ammonium Sulfate. Solid ammonium sulfate is added to the supernatant (24.3 g/100 ml). After stirring for 15 min, the precipitate is removed by centrifugation at 35,000 g and discarded. The ammonium sulfate concentration is further increased by the addition, with constant stirring, of 13.2 g of solid ammonium sulfate per 100 ml of supernatant. After centrifugation at 35,000 g, the precipitate is retained and the supernatant is discarded. The precipitate is dissolved in 0.05 M sodium phosphate, pH 7.0, and dialyzed overnight against the same buffer.

Step 5. Chromatography on DEAE-Cellulose. The ammonium sulfate fraction is applied to a DEAE-cellulose (Whatman DE-11) column (1.3 × 120 cm) equilibrated in 0.05 M sodium phosphate buffer (pH 7.0). The enzyme is eluted with a linear NaCl gradient (0 to 0.5 M) in the same buffer at a flow rate of 0.6 ml/min. Enzymically active fractions are pooled

[3] M. K. Dynon, G. R. Jago, and B. E. Davidson, *Eur. J. Biochem.* **30**, 348 (1972).
[4] H. A. Jonas, R. F. Anders, and G. R. Jago, *J. Bacteriol.* **111**, 397 (1972).
[5] L. G. Streader, Ph.D. Thesis, University of Melbourne, 1975.
[6] L. G. Crossley, G. R. Jago, and B. E. Davidson, *Biochim. Biophys. Acta* **581**, 342 (1979).

and concentrated by ultrafiltration in an Amicon Diaflo apparatus containing a PM-10 membrane.

Step 6. Chromatography on DEAE-Sephadex. The concentrated enzyme solution from step 5 is dialyzed against 0.05 M sodium phosphate in 0.1 M NaCl (pH 7.0) and applied to a DEAE-Sephadex A-50 column (1.3 × 25 cm) previously equilibrated with the same buffer. The enzyme is eluted with a linear NaCl gradient (0.1 to 0.5 M) in the sodium phosphate buffer at a flow rate of 0.6 ml/min. Enzymically active fractions are pooled and concentrated by ultrafiltration and dialyzed against 0.05 M sodium phosphate (pH 7.0).

Step 7. Second Fractionation with Ammonium Sulfate. The concentrated enzyme from step 6 is precipitated with ammonium sulfate (47.2 g/100 ml of 0.05 M sodium phosphate, pH 7.0) and extracted sequentially with a series of ammonium sulfate solutions (39, 35.1, 31.3, 27.7, 24.3, 20.9, and 17.6 g of 0.05 M sodium phosphate per 100 ml, pH 7.0) as described by Jakoby.[7] The supernatant obtained after each extraction is retained, and the precipitate is reextracted with the next solution of ammonium sulfate. A fraction is obtained by this procedure that has high specific activity and exhibits only a single protein band after polyacrylamide gel electrophoresis.

Larger quantities of bacterial cell-free extract can be processed (giving approximately 50 mg of pure enzyme from 250 g of bacterial cells grown in 20 liters of broth) by using DE-52 cellulose in step 5 and replacing steps 6 and 7 with chromatography on hydroxyapatite and Sepharosyl-ε-aminohexanoyl–NAD, respectively, as described by Streader.[5]

Properties

Stability. The stability of the enzyme is dependent on both the type and pH of the buffer in which it is stored.[4] Cell-free extracts prepared in acidic to neutral buffers (pH 6.0–7.0) and stored at 5° retain more lactate dehydrogenase activity than those prepared in alkaline buffers (pH 7.5–8.5). More activity is retained in alkaline phosphate buffers than in alkaline triethanolamine buffers, but the loss of activity in alkaline triethanolamine buffers can be reduced by adding 10 mM sodium phosphate or 1 mM FDP. The purified enzyme obtained after chromatography on DEAE-cellulose is stable for several months in 50 mM sodium phosphate buffer, pH 7.0, at −20°.

Lactate dehydrogenase activity is destroyed by incubating the enzyme at 50° in 0.01 M sodium phosphate buffer, pH 7.0. Protection against heat

[7] W. B. Jakoby, *Anal. Biochem.* **26**, 295 (1968).

inactivation is provided by 1 mM FDP, 1 mM ATP, 1 mM glucose 6-phosphate, and increased levels of phosphate.[4]

Activators and Inhibitors. The activity of the lactate dehydrogenase from *S. cremoris* is markedly stimulated by the allosteric activators FDP and its C-4 epimer, tagatose 1,6-diphosphate.[4,8] In the absence of FDP, a sharp peak of low activity for pyruvate reduction and lactate oxidation is observed at approximately pH 8.0 in both triethanolamine-hydrochloride and phosphate buffers, but the enzyme is virtually inactive at pH values below 7.0. The addition of FDP to the reaction mixture activates the enzyme approximately 500-fold and changes the shape of the pH curve for pyruvate reduction to a broad plateau with maximal activity between pH 7.0 and 5.0. The pH optimum for lactate oxidation is displaced to about pH 7.5 in the presence of FDP, but there is no appreciable change in the shape of the pH curve. The concentration of FDP required to give half-maximal activation of lactate dehydrogenase activity (pyruvate reduction) in triethanolamine-hydrochloride buffer is approximately 8×10^{-6} M at pH 6.0, 1×10^{-5} M at pH 7.0, and 1×10^{-4} M at pH 8.0.[4] Glucose 1,6-diphosphate is the only other known activator of *S. cremoris* lactate dehydrogenase, but the concentration of this compound required to achieve half-maximal activation of the enzyme is 200 times greater than that of FDP or tagatose 1,6-diphosphate.[8] The Michaelis constants for pyruvate and NADH in the presence and in the absence of FDP are: pyruvate, 1.15 mM; NADH, 4.4×10^{-2} M (in 0.06 M triethanolamine-hydrochloride, pH 6.0, and 1 mM FDP); pyruvate, 7.7 mM; NADH, 0.4 mM (in 0.06 M triethanolamine-hydrochloride, pH 8.0, and no FDP).[4] The kinetic properties of the enzyme are consistent with a Theorell–Chance mechanism in which NADH or NAD binds to the enzyme before pyruvate or lactate, respectively.[5]

The activity of the *S. cremoris* lactate dehydrogenase is inhibited by ATP (competitive inhibition with respect to NADH), oxamate (competitive inhibition with respect to pyruvate), and the sulfhydryl reagent *p*-chloromercuribenzoate. The enzyme is not inhibited by the sulfhydryl reagents iodoacetamide, iodoacetate, *N*-ethylmaleimide, and 5,5-dithiobis(2-nitrobenzoic acid).

The activation of lactate dehydrogenase activity by FDP is inhibited by phosphate and by treatment of the enzyme with pyridoxal phosphate followed by reduction with sodium borohydride.[3–5]

Physical and Chemical Properties. The molecular weight of the enzyme in phosphate or acetate buffers (pH 5.0–8.0) or in triethanolamine-hydrochloride buffer (pH 6.0–7.0) is 140,000 (ultracentrifuge studies). At pH values ≥ 7.8 in triethanolamine-hydrochloride buffer, the enzyme

[8] T. D. Thomas, *Biochem. Biophys. Res. Commun.* **63**, 1035 (1975).

undergoes a slow unfolding and dissociation that is seen in sedimentation velocity patterns as two incompletely resolved forms of the enzyme. At pH 9.0 in triethanolamine-hydrochloride buffer, the protein exists as a species of molecular weight 70,000 that is enzymically inactive. The substrate NADH and the activator FDP partially stabilize the 140,000 molecular weight form of the enzyme against dissociation in triethanolamine-hydrochloride buffer at pH 8.0.[2] The active enzyme consists of four identical subunits of molecular weight 35,000. At pH 6.0, each subunit has the ability to bind 1 molecule of FDP (with a mixture of positive and negative cooperativity) and 1 molecule of NADH (noncooperative binding).[3,5]

The extinction coefficient of the pure enzyme in dilute phosphate buffer is $E_{1cm}^{1\%} = 11.3$ at 280 nm at pH 7.0.[5]

The amino acid composition of the enzyme and the amino acid sequence of the 20 N-terminal amino acids and that of a 53-residue tryptic polypeptide containing the single cysteine residue, have all been determined. The isoelectric point of the enzyme is 4.3.[6]

[64] Lactate-Oxaloacetate Transhydrogenase from Veillonella alcalescens

By S. H. GEORGE ALLEN

L-Lactate + oxaloacetate ⇌ pyruvate + L-malate

Thus far, the lactate–oxaloacetate transhydrogenase or malate–lactate transhydrogenase (EC 1.1.99.7) has been found in only one genus, *Veillonella*. The transhydrogenase was originally isolated from *Micrococcus lactilyticus*,[1,2] which was subsequently renamed *Veillonella alcalescens*. This bacterium has been isolated from the rumen of sheep[3] as well as from the oral cavity of man.[4] The transhydrogenase is also present in *Veillonella parvula*.[5] The *Veillonella* are obligate anaerobes and are found in environments rich in lactic and other organic acids produced by other microorganisms. Rogosa and co-workers[6] have shown that the *Veillonella* lack

[1] M. I. Dolin, E. F. Phares, and M. V. Long, *Biochem. Biophys. Res. Commun.* **21**, 303 (1965).
[2] S. H. G. Allen, *J. Biol. Chem.* **241**, 5266 (1966).
[3] A. T. Johns, *J. Gen. Microbiol.* **5**, 317 (1951).
[4] R. J. Gibbons and J. van Houte, *Annu. Rev. Microbiol.* **29**, 19 (1975).
[5] S. K. C. Ng and L. R. Hamilton, *J. Bacteriol.* **105**, 999 (1971).
[6] M. Rogosa, M. I. Krichevsky, and F. S. Bishop, *J. Bacteriol.* **90**, 164 (1965).

METHODS IN ENZYMOLOGY, VOL. 89

hexokinase and, in fact, cannot metabolize any exogenous carbohydrate. Furthermore, these bacteria do not contain lactate dehydrogenase activity. Instead the lactate-oxaloacetate transhydrogenase functions to oxidize lactate to pyruvate probably by coupling with an L-malate dehydrogenase present in *Veillonella*.

$$\text{L-Lactate + oxaloacetate} \rightleftharpoons \text{pyruvate + L-malate}$$
$$\underline{\text{L-Malate + NAD}^+ \rightleftharpoons \text{oxaloacetate + NADH H}^+}$$
$$\text{Net: L-Lactate + NAD}^+ \rightleftharpoons \text{pyruvate + NADH H}^+$$

Thus, lactate entering the cell would be converted to pyruvate which is metabolized either to acetic or to propionic acids via pathways similar to these described for the propionic acid bacteria.[6]

Assay Method

Principle. The transhydrogenase is readily reversible, and thus enzyme activity can be assayed in either direction.[1,2] The specific activity of the enzyme is 50% higher with L-malate and pyruvate as substrates than it is with L-lactate and oxaloacetate. Routinely, because of the higher activity and the relative instability of oxaloacetate solutions, the assay is performed with L-malate and pyruvate as substrates. Two assay procedures are used. The *direct assay* depends upon the increase in absorbance at 258 nm due to oxaloacetate formation. A molar extinction coefficient for oxaloacetate was determined experimentally to be $E_{258} = 8.4 \times 10^2 \ M^{-1}$ cm^{-1}. Initial rates with this assay are linear with time and enzyme concentration. The reverse reaction, i.e., the disappearance of oxaloacetate, can also be measured in this type of assay. The *indirect assay,* employing NADH oxidation, which was approximately 7 times more sensitive than the direct assay, measures the formation of oxaloacetate from L-malate and pyruvate by coupling the transhydrogenase with the malate dehydrogenase.

$$\text{L-Malate + pyruvate} \xrightleftharpoons{\text{transhydrogenase}} \text{oxaloacetate + L-lactate}$$
$$\underline{\text{Oxaloacetate + NADH + H}^+ \xrightleftharpoons{\text{malate dehydrogenase}} \text{L-malate + NAD}^+}$$
$$\text{Net: Pyruvate + NADH + H}^+ \rightleftharpoons \text{L-lactate + NAD}^+$$

Initial rates with this assay were also linear with time and enzyme concentration. The reverse reaction, i.e., the appearance of pyruvate from oxaloacetate and L-lactate, can also be measured by this type of assay, except that lactate dehydrogenase rather than malate dehydrogenase is used. With all these assays, one unit of enzyme is defined as that amount catalyzing the oxidation of 1 μmol of either L-malate or L-lactate per minute.

Direct Assay

Reagents

Tris-HCl buffer, 0.5 M, pH 7.8
Sodium pyruvate, 0.1 M (Sigma Chemical Company, St. Louis, Missouri)
Tris-L-malate, 0.2 M (Sigma)

Procedure. The reagents listed above are added to a 0.5-ml quartz spectrophotometric cell (1-cm light path) in the following order and amounts: Tris-HCl buffer, 0.05 ml; sodium pyruvate, 0.10 ml; Tris-L-malate, 0.05 ml; distilled water, 0.04 ml; lactate–oxaloacetate transhydrogenase, 0.01 ml of an appropriate dilution.

Usually larger but proportional volumes of the reagents are combined to form a mixture that could be added as a single volume of 0.20 ml. All reactants except the transhydrogenase are kept at room temperature or in a bath at the same temperature as the spectrophotometric cell chamber (usually 25°) prior to adding the enzyme that initiates the reaction. The cell contents are mixed well, and the reaction at 258 nm is measured. Since pyruvate itself absorbs some light at this wavelength, a cuvette with all the reagents except the enzyme can be used as a reagent blank.

In measuring the rate in the reverse direction, 0.05 ml of 0.2 M DL-lithium lactate (Sigma Chemical Company) and 0.01 ml of 10 mM Tris-oxaloacetate, pH 6.5, are substituted for pyruvate and malate. All other conditions are the same as described above. Oxaloacetic acid is adjusted to pH 6.5 with Tris base using bromocresol green indicator, and this solution is made fresh daily.

Indirect Assay

Reagents

Tris-HCl buffer, 0.5 M, pH 7.8
NADH, 4 mM (Sigma Chemical Company)
Sodium pyruvate, 0.1 M (Sigma Chemical Company)
Tris-L-malate, 0.2 M (Sigma Chemical Company)
Malate dehydrogenase (Boehringer-Mannheim); dilution: 0.01 ml contains 0.1 unit (dilution of the commercial preparation is made in 1% bovine serum albumin)

Procedure. The reagents listed above are added to a 0.50-ml cuvette (1-cm light path) in the following order and amounts: Tris-HCl buffer, 0.05 ml; NADH, 0.01 ml; sodium pyruvate, 0.08 ml; Tris-L-malate, 0.05 ml; malate dehydrogenase, 0.01 ml; distilled water, 0.03 ml; lactate–oxaloacetate transhydrogenase, 0.01 ml of an appropriate dilution.

Usually, larger but proportional volumes of the reagents are combined to form a mixture that can be added as a single volume of 0.20 ml.

The enzyme is added to initiate the reaction, which is measured at 340 nm. In the reverse direction, 0.05 ml of 0.2 M DL-lithium lactate (Sigma Chemical Company), 0.01 ml of 10 mM Tris-oxaloacetate, pH 6.5 (Sigma Chemical Company), and 0.01 ml of lactate dehydrogenase (Boehringer-Mannheim) containing 0.1 unit are substituted for the pyruvate, malate, and malate dehydrogenase. All other conditions are the same as described above. Since oxaloacetate tends to decarboxylate at a slow rate, the rate of pyruvate formation should be measured in a cuvette containing all the reagents except transhydrogenase, at each level of oxaloacetate used.

Purification Procedure

Growth of Cells. Veillonella alcalescens, ATCC 12641 or 12642, is grown on thioglycolate medium without dextrose (Baltimore Biological Laboratories) supplemented with 0.4% yeast extract, 1.7% sodium lactate, and 0.1 mg% resazurin. Cultures are grown at 37° under strict anaerobic conditions in an atmosphere of 100% CO_2 in serum bottles as described by Miller and Wolin.[7] Large-scale cultivation is carried out in 25-liter carboys containing 15 liters of a medium consisting of 1% yeast extract, 1% tryptone, and 2% sodium lactate as described by Delwiche et al.[8] In this preparation, the culture medium contains 6 μCi per liter of [7-^{14}C]nicotinamide (New England Nuclear), which has a specific radioactivity of 43 μCi/nmol. With freshly autoclaved medium and a 5% active culture as inoculum, no further anaerobic precautions are needed to obtain vigorous growth at 30°. In both media, the presence of lactate is required in order to obtain maximal transhydrogenase production. After 3 days of cultivation the cells are harvested with a Sharples centrifuge. Approximately 50 g wet weight of cells are obtained per 15 liters of medium.

Crude Extract. A 30% wet weight to volume suspension of cells is made in 0.2 M potassium phosphate buffer, pH 7.0. Pyrex beads (120 μM) equal to the weight of cells (204 g) are added to this suspension, which is then placed in an Eppenbach mill (Gifford-Wood Co., Hudson, New York), and the cells are ruptured at 0–5°. For smaller preparations, cells can be ruptured by sonication at 0–5°. The cells and cell debris are removed by centrifugation at 20,000 g for 20 min. A clear brown extract is obtained as previously described.[9] Specific activities of 20–80 μmol/

[7] T. L. Miller and M. J. Wolin, *Appl. Microbiol.* **27**, 985 (1974).
[8] E. A. Delwiche, E. F. Phares, and S. F. Carson, *J. Bacteriol.* **71**, 598 (1956).
[9] S. H. G. Allen and J. H. Patil, *J. Biol. Chem.* **247**, 909 (1972).

minute per milligram of protein are usually found in crude extracts of cells. Since large dilutions of these extracts must be made in order to assay the enzyme, direct spectrophotometric assays are possible without interference with other reactions. The extract contains 11.5 g of protein, as measured with the biuret test, and approximately 10^6 units of transhydrogenase, which is assayed with malate and pyruvate as substrates and coupled to malate dehydrogenase as previously described.[10]

Ion-Exchange Chromatography

Purification to homogeneity requires only a two-step procedure, using DEAE-cellulose batch treatment followed by QAE-Sephadex chromatography as described by Allen and Patil.[9]

DEAE-Cellulose. The crude extract (560 ml) is dialyzed against 10 liters of 0.05 M phosphate buffer, pH 7.5, for 15 hr, with three changes of buffer. Dilute the dialyzed extract to 3000 ml with cold distilled water. Add 1000 ml of packed DEAE-cellulose (Schleicher and Schuell, type 40, 0.84 meg/g) that have been washed and equilibrated with 0.005 M potassium phosphate, pH 7.5. Stir at 4° for 20 min, remove a 1.0-ml sample, centrifuge, and test the clear supernatant for transhydrogenase activity. If more than 5% of the total activity remains in the supernatant, add more packed DEAE-cellulose and cold water. When the enzyme is absorbed, filter on a Büchner funnel with Whatman No. 4 paper. Resuspend the DEAE-cellulose in 5 liters of 0.005 M potassium phosphate buffer, pH 7.4. Stir at 4° for 20 min, then filter and discard the colloidal, yellow filtrate. Resuspend the DEAE-cellulose in 5 liters of 0.20 M potassium phosphate buffer, and stir at 4° for at least 3 hr. Filter and assay the clear yellow solution for transhydrogenase activity. Table I shows that 74% of the starting activity is recovered in this eluate and that the specific activity of the enzyme is twice that in the crude. Solid ammonium sulfate (Schwarz-Mann) is added to a final concentration of 3.7 M, and the precipitate is collected by centrifugation at 16,000 g for 45 min. This precipitate can be stored at −20° without loss of activity.

QAE-Sephadex. The ammonium sulfate precipitate is dissolved in 200 ml of 0.05 M Tris-HCl, pH 7.5, and dialyzed against 4 liters of 0.05 M potassium phosphate, pH 7.4, for 15 hr with three changes of buffer. The preparation is absorbed onto a QAE-Sephadex (Pharmacia) column (5 × 88 cm) that has been equilibrated with 0.05 M potassium phosphate pH 7.4. Wash the absorbed protein with 1 liter of the same buffer, and then elute with a linear gradient of 0.05 to 0.40 M potassium phosphate, pH 7.4. Ten-milliliter samples are collected at a flow rate of 0.2 ml/min. A single major protein peak, which contains the transhydrogenase, is eluted.

[10] S. H. G. Allen, this series, Vol. 13, p. 265.

TABLE I
PREPARATION OF ^{14}C-LABELED TRANSHYDROGENASE[a]

Step	Total protein (mg)	Specific activity (units/mg protein)	Total units (μmol/min)	Radio-activity (cpm/mg)	Recovery of units (%)
Crude extract dialyzed	11,536	79	1,011,344	811	100
DEAE-cellulose, 0.2 m eluate	4,504	166	747,662	1276	74
QAE-Sephadex chromatography fractions 327–354	1,110	368	405,000	3920	40

[a] Units refer to the micromoles of product formed per minute.

Specific activities and radioactive activities are constant across the center of the peak. Fractions with the highest activity are pooled (Table I) and the protein is precipitated at 3.7 M ammonium sulfate. Centrifuge at 24,000 g for 45 min, suspend in 0.05 M Tris-HCl, pH 7.4, and store at $-20°$.

Figure 1 shows a typical elution profile in which DEAE-Sephadex has been used instead of QAE-Sephadex. The profiles obtained from either resin are similar; however, the specific activity of the enzyme obtained from the QAE-Sephadex columns is approximately 2 times higher. Thus, QAE-Sephadex is the resin of choice.

Properties

Properties of the Pure Enzyme. The colorless protein was homogeneous as measured in the analytical ultracentrifuge[11] and by disc gel electrophoresis. It had a maximum specific activity of approximately 400 units/mg. During storage at $-20°$ the specific activity fell to about 200, but the preparation appeared to be stable at this activity level indefinitely, when stored at $-20°$. An absorbance coefficient at 280 nm, based on dry weight of the pure protein was 1.27 cm^2 mg^{-1}. An $s_{20,w}$ corrected for protein concentration was found to be 5.26 S, and the diffusion coefficient, $D_{20,w}$ was 6.57 Fick units.[11] Calculation of the molecular weight based on the Svedberg equation and using a partial specific volume of 0.726 mg/g was found to be 71,000 $\pm$ 3600. A frictional ratio was 1.19, which corresponds to a prolate ellipsoid with an axial ratio of approximately 4. The molecular weight was also determined with the high speed (miniscus-

[11] S. H. G. Allen, *Eur. J. Biochem.* **35**, 338 (1973).

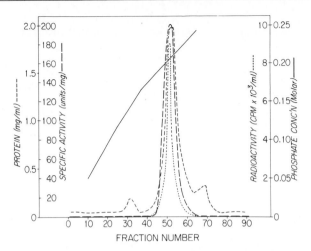

FIG. 1. The chromatographic purification of lactate-oxaloacetate transhydrogenase. After desalting on a Sephadex G-25 column (2.5 × 18 cm), 132 mg of protein (as measured by absorbance at 280 nm) containing 15,850 units of transhydrogenase were absorbed onto a DEAE-Sephadex (A-50) column (2 × 50 cm) that had been equilibrated with 0.005 M potassium phosphate buffer, pH 7.0. The approximate flow rate was 0.10 ml/min; 7.5-ml fractions were collected. The protein was eluted with a phosphate buffer (pH 7.0) gradient of 0.05 to 0.4 M, with 500 ml in the mixing bottle (——). Protein (- - -) was determined by 280 nm absorbance and specific activity (– – –) using the spectrophotometric assay coupled with malate dehydrogenase and NADH. Radioactivity (.......) was measured with a Nuclear Mark I scintillation spectrometer (Nuclear Chicago). Ninety-six milligrams of protein containing 7700 units of transhydrogenase activity were recovered from the column.

depletion) sedimentation-equilibrium method at four different protein concentrations. A value of 69,700 ± 2500 was obtained from these results.[11]

Amino acid analysis is given in Table II; the partial specific volume was calculated from the amino acid content as described by Cohn and Edsall[12] to be 0.726 mg/g.

Molecular Weight of Subunits. Sedimentation-diffusion analysis in a 5 M guanidine-HCl and 5 mM 2-mercaptoethanol revealed a single sedimenting species. A value of 1.87 S was obtained when the $s_{20,w}$ was extrapolated to zero protein concentration. Diffusion experiments revealed $D_{20,w}$ of 4.13 ± 10 Fick units. Assuming a partial specific volume of 0.726 mg/g, the molecular weight was calculated to be 40,000 ± 1000. The frictional ratio was calculated at 2.28, which corresponds to a prolate ellipsoid having an axial ratio of approximately 28. Molecular weight values obtained by sucrose gradient centrifugation and SDS–polyacrylamide gel

[12] E. J. Cohn and J. T. Edsall, "Proteins, Amino Acids and Peptides," p. 374. Van Nostrand-Reinhold, Princeton, New Jersey, 1943.

TABLE II
AMINO ACID COMPOSITION OF LACTATE-OXALOACETATE
TRANSHYDROGENASE

Amino acid residue	Number of residues (moles/70,000 g of protein)	
	15 Hours	23 Hours
Lysine	47	47
Histidine	17	16
Arginine	14	14
Aspartic acid	62	49
Threonine	28	27
Serine	33	30
Glutamic acid	62	59
Proline	28	26
Glycine	69	67
Alanine	54	53
Half-cystine	12	12
Valine	32	33
Methionine	25	22
Isoleucine	30	31
Leucine	38	38
Tyrosine	18	18
Phenylalanine	27	25
Tryptophan	7	8

electrophoresis yielded molecular weight values of 30,000 and 43,000, respectively.[9] These results also agree with the minimum molecular weight values of between 30,000 and 40,000 obtained by Dolin et al.,[1] which were based on the total pyridine nucleotide content. Thus, the enzyme is composed of two identical subunits.

Nature of NAD/NADH Binding. The pyridine nucleotide is very firmly bound to the protein and cannot be removed by charcoal absorption or by ammonium sulfate fractionation. It appears that the removal of the NAD requires profound changes in the structure of the protein. Treatment with 35% perchloric acid (final concentration) at room temperature and heating at 100° at pH 10 both release the prosthetic group. Concentrations of urea of 3.5 M and higher, 0.1% SDS, and 5 M guanidine HCl also remove the bound pyridine nucleotide. Treatment of the transhydrogenase with crude NADase (Sigma Chemical Company) had been reported[9] to remove the [14C]nicotinamide from the transhydrogenase. Subsequent experiments revealed that protease activity in this NADase preparation caused breakdown of the protein and subsequent liberation of the prosthetic group.

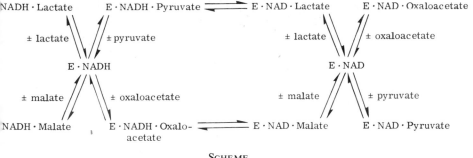

SCHEME

Purified NADase[13] did not cause release of nicotinamide or loss of transhydrogenase[14] activity. NADH can be removed by succinylation or citraconylation of the transhydrogenase.[11] In all cases studied, removal of the prosthetic group is accompanied by dissociation of the transhydrogenase into subunits and complete loss of enzymic activity. Reassociation of the subunits in the presence of either the isolated, purified prosthetic group or authentic NADH has not been possible.

Mechanism of Action of the Transhydrogenase. The transhydrogenase has been shown to catalyze a "Ping-Pong" type of reaction mechanism.[9] Evidence for this is based on the fact that the two half-reactions:

$$\text{L-Lactate} + E \cdot NAD^+ \rightleftharpoons \text{pyruvate} + E \cdot NAD + H^+$$
$$\text{L-Malate} + E \cdot NAD \rightleftharpoons \text{oxaloacetate} + E \cdot NAD + H^+$$

can be carried out separately and stoichiometrically. Second, kinetic analysis at low substrate concentrations, yields the parallel lines on Lineweaver–Burk plots that are diagnostic of the "Ping-Pong" mechanism.[9] At higher substrate concentrations, inhibition is observed, and the full mechanism is more complicated. All substrates form abortive complexes with the enzyme[15] and evidence has been presented to support the participation of ternary complexes in the mechanism. The significance of these is not clear. A simplified mechanism can be written as shown in the reaction scheme.

A puzzling aspect of the mechanism is the apparent ability of the enzyme to maintain about 40% of the bound pyridine nucleotide prosthetic groups in the reduced state, i.e., as NADH. Treatment of the native enzyme with pyruvate or oxaloacetate quenches the NADH fluorescence

[13] F. J. Fehrenbach, *Eur. J. Biochem.* **18**, 94 (1971). The NAD⁺ glycohydrolase (EC 3.2.2.5) was a kind gift from Dr. Fehrenbach, University of Freiberg, Federal Republic of Germany.

[14] S. H. G. Allen, unpublished data, 1972.

[15] M. I. Dolin, *J. Biol. Chem.* **244**, 5273 (1969).

and yields either lactate or malate, which can be recovered stoichiometrically after either gel filtration or dialysis to remove the enzyme. Upon standing at 4°, the enzyme-bound NAD slowly becomes reduced[1,2] until about 40% of the total prosthetic group are as NADH. Furthermore, the native enzyme has been shown to contain covalently linked pyruvate.[9] The exact nature of the attachment of the pyruvate to the protein as well as its role, if any, in the mechanism are unknown. Treatment of the native transhydrogenase with either hydroxylamine or sodium borohydride have little effect on the enzymic activity. Yet reduction with [3H]sodium borohydride results in incorporation of radioactivity into the protein, which after repeated lyophilization and acid hydrolysis can be identified chromatography as [3H]lactate.[9] Dolin[15] has also shown the presence of tightly bound pyruvate using the lactate dehydrogenase assay. The possibility exists that these molecules act as a reservoir of reducing equilalents that interact with the bound NAD, keeping a constant NAD:NADH ratio in the native enzyme. This may be necessary for keeping the transhydrogenase in the proper configuration or important in allowing the enzyme to react more readily with keto acid substrates. The amount of bound pyruvate present on each enzyme molecule has not been determined, and the role of this bound keto acid has not been ascertained.

[65] Pyruvate Dehydrogenase Complex from Bovine Kidney and Heart

By Flora H. Pettit and Lester J. Reed

$$\text{Pyruvate} + \text{CoA} + \text{NAD}^+ \rightarrow \text{acetyl-CoA} + \text{CO}_2 + \text{NADH} + \text{H}^+$$

In eukaryotic cells the pyruvate dehydrogenase complex is located in mitochondria, within the inner membrane-matrix compartment. The complex consists of three catalytic components: pyruvate dehydrogenase (E_1) (EC 1.2.4.1), dihydrolipoyl transacetylase (E_2) (EC 2.3.1.12), and dihydrolipoyl dehydrogenase (E_3) (EC 1.6.4.3). These three enzymes, acting in sequence, catalyze the above overall reaction.

Assay Method

Principle. Activity of this multienzyme complex is assayed spectrophotometrically by measurement of NADH production. The method is essentially that of Linn *et al.*[1] The assay is used routinely at all levels of purity of the enzyme complex beginning with mitochondrial extracts prepared by the freeze–thaw procedure described below. However, the assay is less satisfactory with crude extracts that contain lactate dehydrogenase or NADH oxidase. With these preparations $^{14}CO_2$ production from [1-^{14}C]pyruvate is measured,[2] or the overall reaction above is coupled with the arylamine acetyltransferase-catalyzed reaction, and acetylation of an appropriate arylamine is measured spectrophotometrically.[3]

Reagents

Potassium phosphate buffer, 1 M, pH 8.0
$MgCl_2$, 0.1 M
NAD^+, 0.25 M
Thiamin pyrophosphate, 0.02 M
CoA, 1.3 mM, and dithiothreitol, 3.2 mM, prepared before use
Sodium pyruvate (Sigma type II), 0.1 M
Enzyme complex, diluted in 0.05 M sodium 2-(N-morpholino)propane sulfonate (MOPS), pH 7.0, 1 mM $MgCl_2$, 0.3% bovine serum albumin

Procedure. The reaction mixture contains, in micromoles: potassium phosphate (pH 8.0), 50; NAD^+, 2.5; thiamin pyrophosphate, 0.2; CoA, 0.13; dithiothreitol, 0.32; $MgCl_2$, 1; sodium pyruvate, 2; and 0.008–0.06 unit of enzyme complex in a total volume of 1.0 ml. The pH of the reaction mixture is 7.4. After equilibration of the cuvette to 30°, the reaction is started by addition of enzyme complex, and the increase in absorbance at 340 nm is followed with a recording spectrophotometer. An increase in absorbance of 0.05–0.4 per minute during the initial phase of the reaction is a linear function of enzyme concentration.

Units. Units are expressed as micromoles of NADH produced per minute at 30°; and specific activities, as units per milligram of protein. Protein is determined by the biuret method[4] with crystalline bovine serum albumin as standard.

Purification of Pyruvate Dehydrogenase Complex from Bovine Kidney

The α subunit of pyruvate dehydrogenase (E_1) and the subunit of dihydrolipoyl transacetylase (E_2) are very sensitive to proteolysis.[5]

[1] T. C. Linn, J. W. Pelley, F. H. Pettit, F. Hucho, D. D. Randall, and L. J. Reed, *Arch. Biochem. Biophys.* **148**, 327 (1972).
[2] S. C. Dennis, M. DeBuysere, R. Scholz, and M. S. Olson, *J. Biol. Chem.* **253**, 2229 (1978).
[3] H. G. Coore, R. M. Denton, B. R. Martin, and P. J. Randle, *Biochem. J.* **125**, 115 (1971).
[4] A. G. Gornall, C. J. Bardawill, and M. M. David, *J. Biol. Chem.* **17**, 751 (1949).
[5] D. M. Bleile, M. L. Hackert, F. H. Pettit, and L. J. Reed, *J. Biol. Chem.* **256**, 514 (1981).

TABLE I
PURIFICATION OF PYRUVATE DEHYDROGENASE COMPLEX FROM BOVINE KIDNEY[a]

Step	Volume (ml)	Protein (mg)	Specific activity[b]	Recovery (%)
1. Mitochondrial extract	4560	53,860	0.17	—
2. Incubation with $MgCl_2$	4600	53,860	0.23	100
3. First PEG[c] fractionation	300	3,760	2.7	83
4. First ultracentrifugation	84	2,490	3.9	79
5. Second PEG fractionation	30	490	18.3	74
6. Second ultracentrifugation	10	345	23.0	65

[a] From about 45 pounds of kidney.
[b] Micromoles of NADH produced per minutes per milligram of protein.
[c] PEG, polyethylene glycol.

Therefore, care must be taken to remove or inhibit endogenous proteases. In the procedure described below, a lysozome-enriched fraction is carefully removed during preparation of mitochondria, and the mitochondrial fraction is washed extensively to remove proteases released from ruptured lysozomes. In addition, the serine protease inactivator phenylmethanesulfonyl fluoride, the thiol protease inactivator benzyloxycarbonyl-Phe-Ala diazomethyl ketone[6] (Z-Phe-AlaCHN$_2$), and rabbit serum (a source of protease inhibitors, particularly α_2-macroglobulin) are added in the early steps of the purification procedure. The protease inhibitor leupeptin has also been used.[7] Digitonin has been used to rupture lysozomes, followed by extensive washing of the mitochondrial fraction.[8]

The purification procedure is modified from that of Linn et al.[1] and is summarized in Table I. Unless specified otherwise, all operations are performed at 4°. This procedure results in highly purified preparations of the pyruvate dehydrogenase complex and the α-ketoglutarate dehydrogenase complex as well as partially purified preparations of the branched-chain α-keto acid dehydrogenase complex, pyruvate dehydrogenase phosphatase, and pyruvate dehydrogenase kinase.

Preparation of Tissue

Fifteen to eighteen pounds of bovine kidneys are collected and chilled immediately after slaughter. Slices (0.5–1 cm) of cortical tissue are removed and soaked in deionized water for 1 hr. The tissue slices are passed

[6] H. Watanabe, G. D. J. Green, and E. Shaw, *Biochem. Biophys. Res. Commun.* **89**, 1354 (1979).
[7] G.-B. Kresze and L. Steber, *Eur. J. Biochem.* **95**, 569 (1979).
[8] F. Machicao and O. H. Wieland, *Hoppe-Seyler's Z. Physiol. Chem.* **361**, 1093 (1980).

through an electric meat grinder, and the ground meat is suspended in an equal volume of 0.25 M sucrose (Schwarz-Mann, enzyme grade) containing 10 mM potassium phosphate, pH 7.6, and 0.1 mM EDTA. 2-Mercaptoethanol is added to give a final concentration of 0.01 M, and the suspension (~8 liters) is stored overnight (~16 hr).

Preparation and Washing of Mitochondria

The pH of the suspension is adjusted to 6.8 with NaOH, and the suspension is passed through a continuous-flow homogenizer[9] operated at a speed of about 1000 rpm. The homogenate is diluted with an equal volume of the sucrose–phosphate–EDTA solution, and the pH is readjusted to 6.8. The homogenate is centrifuged in 1-liter bottles at 2000 g for 10 min. The supernatant fluid is strained through eight layers of cheesecloth. The mitochondrial fraction is collected by centrifugation at 22,000 g (maximal centrifugal force) for 15 min in the type 15 rotor of a Beckman Model L3-40 centrifuge. The mitochondrial paste is resuspended in the sucrose–phosphate–EDTA solution (total volume, 6 liters) with the aid of the continuous-flow homogenizer equipped with a Delrin pestle and operated at a speed of about 400 rpm. The pH is adjusted to 7.8, and the suspension is centrifuged at 22,000 g for 15 min. The lysozome-enriched fluffy layer is carefully removed from the mitochondrial pellet. The mitochondrial paste is washed again with the sucrose-phosphate-EDTA solution, care being taken to remove any fluffy layer that is visible after centrifugation. The mitochondrial paste is resuspended in 6 liters of 0.014 M 2-mercaptoethanol, the pH is adjusted to 6.8, and the suspension is centrifuged at 22,000 g for 20 min. The mitochondrial paste is washed twice with 6-liter portions of 0.02 M potassium phosphate, pH 6.5; the paste is collected by centrifugation for 10 min.[10] The yield of washed mitochondrial paste is 800–1000 g (wet weight). It is resuspended with a minimal volume (about 150 ml) of 0.02 M potassium phosphate, pH 6.5. One-thousandth volume of 0.1 M phenylmethanesulfonyl fluoride (Sigma) and 0.001 volume of 0.01 M Z-Phe-AlaCHN$_2$[6] in methanol are added. Five hundred-milliliter portions of the suspension are shell-frozen in 4-liter flasks in a −70° (solid CO_2–isopropanol) cooling bath. The frozen preparation is stored at −20°. It is stable for at least 1 month under these conditions, but is usually processed within 1–2 weeks.

[9] D. M. Ziegler and F. H. Pettit, *Biochemistry* **5**, 2932 (1966). A Waring blender is usually used for large-scale homogenization of tissue.[8] However, use of the continuous-flow homogenizer results in a significantly higher yield of intact mitochondria.

[10] The last three washes remove proteases and cause swelling of the mitochondria. These washes are essential for subsequent release of the pyruvate dehydrogenase complex from mitochondria by freezing and thawing.

Purification Procedure

Step 1. Preparation of Mitochondrial Extract. Three batches of frozen mitochondria (from 45–50 pounds of kidney) are thawed under running tap water. The suspension is diluted with 0.02 M potassium phosphate, pH 6.5, to a final volume containing about 0.5 g of mitochondrial paste per milliliter. Ten milliliters of 5 M NaCl, 10 ml of rabbit serum (Pel-Freez Biologicals, Rogers, Arkansas, type 2), and 1 ml of 0.01 M Z-Phe-AlaCHN$_2$ are added, with stirring, for every liter of suspension. The mixture is adjusted to pH 6.4 with 10% acetic acid and centrifuged at 30,000 g for 30 min in the J-14 rotor of a Beckman Model J2-21 centrifuge. The supernatant fluid is decanted carefully; the residue is discarded.

Step 2. Incubation with MgCl$_2$ and Isoelectric Precipitation of Branched-Chain α-Keto Acid Dehydrogenase Complex. Two-tenths milliliter of 0.1 M thiamin pyrophosphate, 10 ml of rabbit serum, and 10 ml of 1 M MgCl$_2$ are added for every liter of mitochondrial extract, and the mixture is stirred for 20 min with a magnetic stirrer.[11] The pH is carefully lowered to 6.1 by dropwise addition, with stirring, of 10% acetic acid. The mixture is stirred for 20 min, and the precipitate is collected by centrifugation at 30,000 g for 10 min. The pellet contains the branched-chain α-keto acid dehydrogenase complex, which is purified further as described elsewhere.[12]

Step 3. First Polyethylene Glycol Fractionation. To the supernatant fluid from step 2 is added dropwise, with stirring, 0.06 volume of 50% (w/w) polyethylene glycol (molecular weight 6000, J. T. Baker Co., Phillipsburg, Pennsylvania). The suspension is stirred for 10 min, then the precipitate is collected by centrifugation for 10 min. This precipitate contains both the pyruvate dehydrogenase complex and the α-ketoglutarate dehydrogenase complex. It is resuspended, by means of a large glass homogenizer equipped with a motor-driven Teflon pestle, in 300 ml of buffer A (0.05 M MOPS, pH 7.0; 2 mM dithiothreitol; 1 mM MgCl$_2$; 0.02 mM thiamin pyrophosphate; 0.01 volume of rabbit serum; 0.01 mM Z-Phe-AlaCHN$_2$). The suspension is stored overnight in a refrigerator.

Step 4. First Ultracentrifugation-Separation of Pyruvate Dehydrogenase Phosphatase. The suspension from step 3 is centrifuged at 40,000 g for 20 min; the precipitate is discarded. To the clear supernatant fluid is added 0.01 volume of 0.2 M ethylene glycol bis(2-aminoethyl)-N,N'-tetraacetate (EGTA), pH 7.0. After 20 min the solution is centrifuged at 105,000 g for 3.5 hr in a Beckman type 30 rotor. The supernatant fluid contains pyru-

---

[11] The mitochondrial extract contains pyruvate dehydrogenase phosphatase. In the presence of this enzyme and 10 mM Mg^{2+}, any phosphorylated, inactive pyruvate dehydrogenase complex is converted to the dephosphorylated, active form.

[12] F. H. Pettit, S. J. Yeaman, and L. J. Reed, *Proc. Natl. Acad. Sci. U. S. A.* **75**, 4881 (1978).

vate dehydrogenase phosphatase.[13] The amber pellet contains both the pyruvate and α-ketoglutarate dehydrogenase complexes. It is resuspended, with the aid of a glass–Teflon homogenizer, in 80–100 ml of buffer A, and the suspension is stored overnight in a refrigerator. The mixture is clarified by centrifugation at 40,000 g for 20 min; the precipitate is discarded.

Step 5. Second Polyethylene Glycol Fractionation-Separation of Pyruvate and α-Ketoglutarate Dehydrogenase Complexes. All procedures in this step are carried out at 20–25°. The supernatant fluid from step 4 is diluted with buffer A to a protein concentration of 5 mg/ml. The pH is adjusted to 6.2, and the solution is clarified, if necessary, by centrifugation at 30,000 g for 10 min. To the clear solution is added dropwise, with stirring, 0.02 volume of 50% polyethylene glycol. The preparation is stirred for 10 min, then the precipitate is collected by centrifugation for 10 min. Most of the α-ketoglutarate dehydrogenase complex is recovered in this precipitate.[1] Addition of the polyethylene glycol to the supernatant fluid is continued, in 0.02 volume increments, followed by stirring and centrifugation, until the α-ketoglutarate dehydrogenase complex activity[15] of the supernatant fluid is reduced to less than 1% of the pyruvate dehydrogenase complex activity. The final supernatant solution is made 10 mM with respect to MgCl$_2$ by addition, with stirring, of 0.01 volume of 1 M MgCl$_2$. The mixture is stirred for 10 min, then the precipitate is collected by centrifugation for 10 min. This precipitate contains the pyruvate dehydrogenase complex. It is dissolved, with the aid of a glass–Teflon homogenizer, in 30–40 ml of cold buffer B (0.05 M potasssium phosphate, pH 7.0; 1 mM MgCl$_2$; 0.02 mM thiamin pyrophosphate; 0.1 mM EDTA, pH 7.0; 0.4 mM dithiothreitol; 1 mM NAD$^+$).

Step 6. Second Ultracentrifugation. The clear yellow solution from step 5 is diluted with 0.05 M phosphate buffer, pH 7.0, to a protein concentration of 5 mg/ml, and 0.01 volume of 0.2 M EGTA is added. After 20 min the solution is centrifuged at 144,000 g for 2.25 hr. The bright yellow gelatinous pellets are carefully suspended in about 8 ml of buffer B and allowed to dissolve overnight in a refrigerator. The solution is clarified, if necessary, by centrifugation at 25,000 g for 15 min.

---

[13] In the presence of Ca^{2+}, pyruvate dehydrogenase phosphatase binds to the dihydrolipoyl transacetylase (E$_2$) component of the complex.[14] In the presence of the Ca^{2+} chelating agent EGTA, the phosphatase dissociates from the complex and can be separated from the latter by ultracentrifugation.

[14] F. H. Pettit, T. E. Roche, and L. J. Reed, *Biochem. Biophys. Res. Commun.* **49**, 563 (1972).

[15] Activity of the α-ketoglutarate dehydrogenase complex is determined in the same manner as that of the pyruvate dehydrogenase complex except that α-ketoglutarate is substituted for pyruvate in the assay mixture.

The solution is divided into small aliquots and stored at $-50°$. The enzyme complex is stable for several months under these conditions, but it is sensitive to freezing and thawing.

Properties of Complex from Bovine Kidney

Homogeneity. The enzyme complex obtained from step 6 is at least 90% pure as judged by analytical ultracentrifugation and sodium dodecyl sulfate (SDS)–polyacrylamide gel electrophoresis.[16] The gel pattern shows four major bands corresponding to dihydrolipoyl transacetylase (E_2), dihydrolipoyl dehydrogenase (E_3), and the α and β subunits of pyruvate dehydrogenase (E_1). Two minor bands are usually present, with molecular weights of about 50,000 and 30,000, which apparently correspond to proteolysis products of E_2 and $E_1\alpha$, respectively.

Molecular Properties.[1,17] The kidney pyruvate dehydrogenase complex has a molecular weight by sedimentation equilibrium of about 7,000,000[18] and a sedimentation coefficient ($s_{20,w}^{\circ}$) of 70 S. Polyacrylamide gel electrophoresis in SDS[19] gives subunit molecular weights of 74,000 (E_2), 53,000 (E_3), 41,000 ($E_1\alpha$), and 36,000 ($E_1\beta$). By sedimentation equilibrium analysis the molecular weight of the E_2 subunit is about 52,000. The E_2 subunit shows anomalous migration in SDS–polyacrylamide gel electrophoresis.[5] Highly purified E_1 has a molecular weight of about 154,000 and possesses the subunit composition $\alpha_2\beta_2$. E_2 has a molecular weight of about 3,100,000 and consists of 60 apparently identical subunits arranged into a pentagonal dodecahedron.[20] E_3 has a molecular weight of about 110,000 and contains two apparently identical polypeptide chains and two molecules of flavin adenine dinucleotide. The kidney pyruvate dehydrogenase complex contains 20 E_1 tetramers and about 6 E_3 dimers that are attached to the E_2 core by noncovalent bonds. The complex also contains a small amount of tightly bound pyruvate dehydrogenase kinase, which is purified further as described elsewhere.[21] The isolated complex binds about 10 additional E_1 tetramers, but no additional E_3 dimers.[18]

Catalytic Properties. The kidney pyruvate dehydrogenase complex is specific for NAD^+ and pyruvate. The complex shows little activity, if any, with α-ketoglutarate, α-ketoisovalerate, α-ketoisocaproate, or α-keto-

[16] U. K. Laemmli, *Nature (London)* 227, 680 (1970).
[17] C. R. Barrera, G. Namihira, L. Hamilton, P. Munk, M. H. Eley, T. C. Linn, and L. J. Reed, *Arch. Biochem. Biophys.* 148, 343 (1972).
[18] L. Hamilton, P. Munk, and L. J. Reed, unpublished results.
[19] K. Weber and M. Osborn, *J. Biol. Chem.* 244, 4406 (1969).
[20] L. J. Reed and R. M. Oliver, *Brookhaven Symp. Biol.* 21, 397 (1968).
[21] F. H. Pettit, S. J. Yeaman, and L. J. Reed, Vol. 90 [30].

TABLE II
PURIFICATION OF PYRUVATE DEHYDROGENASE COMPLEX FROM BOVINE HEART[a]

Step	Volume (ml)	Protein (mg)	Specific activity[b]	Recovery (%)
1. Mitochondrial extract	2930	11,120	0.5[c]	—
2. First PEG[d] fractionation	287	2,690	2.4	100
3. First ultracentrifugation	75	1,630	4.1	104
4. Second PEG fractionation	35	352	15.8	86
5. Second ultracentrifugation	10	289	18.7	84

[a] From about 80 pounds of heart.
[b] Micromoles of NADH produced per minute per milligram of protein.
[c] After incubation with $MgCl_2$. This is a minimum value because of interference by lactate dehydrogenase and NADH oxidase. NaCN (1 mM) was included in the assay mixture to inhibit NADH oxidase.
[d] PEG, polyethylene glycol.

β-methylvalerate. With α-ketobutyrate the rate of oxidation is about 45% of that observed with pyruvate. Apparent K_m values are 42, 20, and 78 μM for pyruvate, CoA, and NAD$^+$, respectively.[22] The activity of the complex is inhibited by the products of the reaction, acetyl-CoA and NADH, which are competitive with respect to CoA and NAD$^+$, respectively.

Purification of Pyruvate Dehydrogenase Complex from Bovine Heart

In contrast to bovine kidney, bovine heart is relatively free of endogenous proteases. Therefore, it is not necessary to take precautions to remove or inhibit proteases during isolation of the pyruvate dehydrogenase complex from bovine heart. However, because of the presence of myofibrils in heart tissue, it is more difficult to obtain good yields of mitochondria from heart than from kidney tissue. It is also more difficult to rupture heart mitochondria by freezing and thawing in dilute phosphate buffer. Accordingly, the procedures used in the preparation and extraction of kidney mitochondria are modified to overcome these difficulties. Nonionic detergents such as Triton X-100 or Tween-80 have also been used to extract the pyruvate dehydrogenase complex from heart tissue.[23]

The purification procedure is modified from that of Linn et al.[1] and is summarized in Table II. Unless specified otherwise, all operations are performed at 4°.

[22] C. S. Tsai, M. W. Burgett, and L. J. Reed, J. Biol. Chem. 248, 8348 (1973).
[23] C. J. Stanley and R. N. Perham, Biochem. J. 191, 147 (1980).

Preparation of Tissue

Twenty to twenty-five pounds of bovine heart are collected and chilled immediately after slaughter. After 24 hr the hearts are trimmed to remove fat and connective tissue, and the heart tissue is passed through an electric meat grinder. The ground meat is suspended in one-third volume of 0.25 M sucrose containing 10 mM potassium phosphate, pH 7.6, and 0.1 mM EDTA. The pH is adjusted to 7.6 with NaOH, and the suspension (~ 12 liters) is stored overnight.

Preparation and Washing of Mitochondria

The pH is adjusted to 6.8, and the suspension is passed through a continuous-flow homogenizer[9] operated at a speed of about 2000 rpm. The pH is readjusted to 6.8, and the homogenate is centrifuged in 1-liter bottles at 2000 g for 10 min. To recover trapped mitochondria, the pellets are resuspended in an equal volume of the sucrose–phosphate–EDTA solution and processed like the original suspension. The supernatant fluids are combined and strained through eight layers of cheesecloth. The mitochondrial fraction is collected by centrifugation at 22,000 g for 17 min. The mitochondrial paste is resuspended in 4 liters of deionized water with the aid of the continuous-flow homogenizer. The pH is adjusted to 6.8, and the suspension is centrifuged at 22,000 g for 20 min. The mitochondrial paste is washed twice with 4-liter portions of 0.02 M potassium phosphate, pH 6.5; the paste is collected by centrifugation for 10 min. The yield of washed mitochondrial paste is 500–800 g (wet weight). It is resuspended with a minimal volume (about 150 ml) of 0.02 M potassium phosphate, pH 6.5. The homogenate is shell-frozen and stored at $-20°$ for as long as 1 month.

Purification Procedure

Step 1. Preparation of Mitochondrial Extract. Four batches of frozen mitochondria (from 80–90 pounds of heart) are thawed under running tap water. Ten milliliters of 5 M NaCl are added for every liter of suspension, and the pH is adjusted to 6.4. The suspension is passed through a Manton–Gaulin Laboratory Homogenizer (Gaulin Corp.) operated at 4000 psi. The pH is adjusted to 6.2, and the mixture is centrifuged at 30,000 g for 30 min.[24] The supernatant fluid is decanted carefully and warmed to 23°. One-hundredth volume of 1 M MgCl$_2$ is added, and the mixture is stirred for 20 min with a magnetic stirrer.[11]

Step 2. First Polyethylene Glycol Fractionation. To the mitochondrial extract from step 1 is added dropwise, with stirring, 0.1 volume of 50%

[24] The pellets are saved for isolation of pyruvate dehydrogenase phosphatase.[25]

(w/w) polyethylene glycol. The suspension is stirred for 10 min at 23°, then the precipitate is collected by centrifugation for 10 min. This precipitate contains both the pyruvate dehydrogenase complex and the α-ketoglutarate dehydrogenase complex. The copious brown pellet is resuspended, by means of a large glass homogenizer equipped with a motor-driven Teflon pestle, in 300 ml of cold buffer C (0.05 M MOPS, pH 7.0; 1 mM MgCl$_2$; 2 mM dithiothreitol; 0.02 mM thiamin pyrophosphate). The suspension is stored overnight in a refrigerator.

Step 3. First Ultracentrifugation-Separation of Pyruvate Dehydrogenase Phosphatase. The suspension from step 2 is centrifuged at 40,000 g for 20 min; the precipitate is discarded. To the clear amber supernatant fluid is added 0.01 volume of 0.2 M EGTA, pH 7.0. After 20 min the solution is centrifuged at 105,000 g for 3.5 hr in a Beckman type 30 rotor. The supernatant fluid contains pyruvate dehydrogenase phosphatase,[13] which is purified further as described elsewhere.[25] The amber pellet contains both the pyruvate and α-ketoglutarate dehydrogenase complexes. It is resuspended, with the aid of a glass–Teflon homogenizer, in 80–100 ml of buffer C, and the suspension is stored overnight in a refrigerator. The mixture is clarified by centrifugation at 40,000 g for 20 min; the precipitate is discarded.

Step 4. Second Polyethylene Glycol Fractionation-Separation of α-Ketoglutarate Dehydrogenase Complex. The clear yellow supernatant fluid from step 3 is diluted with buffer C to a protein concentration of 5 mg/ml and warmed to 23°. One-tenth volume of 0.1 M EDTA is added, and the pH is adjusted to 6.3. The solution is clarified, if necessary, by centrifugation at 30,000 g for 10 min at 20°. To the clear solution is added dropwise, with stirring, 0.04 volume of 50% polyethylene glycol. The preparation is stirred for 10 min, then the precipitate is collected by centrifugation for 10 min. The pellet contains most of the α-ketoglutarate dehydrogenase complex.[1] Addition of 50% polyethylene glycol to the supernatant fluid is continued, in 0.02-volume increments, followed by stirring and centrifugation, until the α-ketoglutarate dehydrogenase complex activity[15] of the supernatant fluid is reduced to less than 2% of the pyruvate dehydrogenase complex activity. Usually, 0.06 volume of 50% polyethylene glycol is required. The final supernatant fluid is made 10 mM with respect to MgCl$_2$ by addition, with stirring, of 0.01 volume·of 1 M MgCl$_2$, and another 0.02 volume of 50% polyethylene glycol is added. The mixture is stirred for 10 min, then the precipitate is collected by centrifugation for 10 min. This precipitate contains the pyruvate dehydrogenase complex. It is dissolved, with the aid of a glass–Teflon homogenizer, in 30–40 ml of cold buffer B.

[25] F. H. Pettit, W. M. Teague, and L. J. Reed, Vol. 90 [67].

Step 5. Second Ultracentrifugation. The clear yellow solution from step 4 is diluted with 0.05 M potassium phosphate buffer, pH 7.0, to a protein concentration of 5 mg/ml, and 0.01 volume of 0.2 M EGTA is added. After 20 min the solution is centrifuged at 144,000 g for 2.25 hr. The bright yellow gelatinous pellets are carefully suspended in about 8 ml of buffer B and allowed to dissolve overnight in a refrigerator. The solution is clarified, if necessary, by centrifugation at 25,000 g for 15 min. The solution is divided into small aliquots and stored at $-50°$. The enzyme complex is stable for several months under these conditions, but it is sensitive to freezing and thawing.

Properties of Complex from Bovine Heart[1,17]

The enzyme complex obtained from step 5 is about 95% pure as judged by analytical ultracentrifugation and SDS–polyacrylamide gel electrophoresis.[16] The gel pattern shows four major bands corresponding to dihydrolipoyl transacetylase (E_2), dihydrolipoyl dehydrogenase (E_3), and pyruvate dehydrogenase ($E_1\alpha$ and $E_1\beta$). The bovine heart pyruvate dehydrogenase complex has a molecular weight by sedimentation equilibrium of about 8,500,000[18] and a sedimentation coefficient ($s_{20,w}^\circ$) of about 90 S. The subunit molecular weights, macromolecular organization, and catalytic properties of the bovine heart pyruvate dehydrogenase complex are very similar, if not identical, to those of the complex from bovine kidney. The heart pyruvate dehydrogenase complex contains about 30 E_1 tetramers ($\alpha_2\beta_2$) and about 6 E_3 dimers that are attached to the E_2 core by noncovalent bonds. The complex also contains a small amount of pyruvate dehydrogenase kinase that is tightly bound to E_2.

[66] Pyruvate Dehydrogenase Complex from *Neurospora*

By Roy W. Harding, Dina F. Caroline, and Robert P. Wagner

$$\text{Pyruvate} + \text{CoA} + \text{NAD}^+ \rightarrow \text{acetyl CoA} + \text{CO}_2 + \text{NADH} + \text{H}^+$$

The pyruvate dehydrogenase complex (PDC), which catalyzes the above reaction, has been isolated from a number of sources including *Escherichia coli*[1] and other bacteria,[2,3] pigeon muscle,[4,5] mammalian tis-

---

[1] L. J. Reed and C. R. Willms, this series, Vol. 9, p. 247.
[2] R. Lüderitz and J.-H. Klemme, *Z. Naturforsch. C: Biosci.* **32C**, 351 (1977).

METHODS IN ENZYMOLOGY, VOL. 89

ISBN 0-12-181989-2

sues,[6,7] *Neurospora crassa,*[8] and higher plants.[9-11] The conversion of pyruvate to acetyl-CoA is catalyzed by three enzymes associated in an enzyme complex[12,13]: pyruvate dehydrogenase (EC 1.2.4.1), dihydrolipoyl transacetylase (EC 2.3.1.12), and dihydrolipoyl dehydrogenase (EC 1.6.4.3). A procedure for the purification and characterization of the *Neurospora* pyruvate dehydrogenase complex has been previously published[8] and is described in detail below.

Assay Method

Principle. The pyruvate dehydrogenase complex activity is measured by monitoring NADH formation spectrophotometrically at 340 nm essentially as previously described.[14]

Reagents

Solution A: thiamine pyrophosphate (TPP), 0.3 mM; NAD$^+$, 3.4 mM; potassium phosphate buffer (pH 8.0), 71.5 mM

Solution B: CoA, 1.3 mM; cysteine · HCl, 26 mM

Potassium phosphate buffer (pH 7.0), 20 mM

Sodium pyruvate, 100 mM

Enzyme: Purified enzyme is diluted with potassium phosphate buffer (pH 7.0), 20 mM; dithiothreitol, 2 mM; TPP, 0.5 mM; and MgSO$_4$, 0.5 mM to a final concentration of 75–2500 μg of protein per milliliter.

Procedure. Add 0.16 ml of the potassium phosphate buffer, 0.7 ml of solution A, 0.1 ml of solution B, and 0.02 ml of enzyme to a 1-cm

[3] T. W. Bresters, R. A. De Abreu, A. De Kok, J. Visser, and C. Veeger, *Eur. J. Biochem.* **59**, 335 (1975).

[4] V. Jagannathan and R. S. Schweet, *J. Biol. Chem.* **196**, 551 (1952).

[5] R. S. Schweet, B. Katchman, R. M. Bock, and V. Jagannathan, *J. Biol. Chem.* **196**, 563 (1952).

[6] T. Hayakawa, H. Muta, M. Hirashima, S. Ide, K. Okabe, and M. Koike, *Biochem. Biophys. Res. Commun.* **17**, 51 (1964).

[7] E. Ishikawa, R. M. Oliver, and L. J. Reed, *Proc. Natl. Acad. Sci. U. S. A.* **56**, 534 (1966).

[8] R. W. Harding, D. F. Caroline, and R. P. Wagner, *Arch. Biochem. Biophys.* **138**, 653 (1970).

[9] B. J. Rapp and D. D. Randall, *Plant Physiol.* **65**, 314 (1980).

[10] M. Williams and D. D. Randall, *Plant Physiol.* **64**, 1099 (1979).

[11] E. E. Reid, P. Thompson, C. R. Lyttle, and D. T. Dennis, *Plant Physiol.* **59**, 842 (1977).

[12] M. Koike, L. J. Reed, and W. R. Carroll, *J. Biol. Chem.* **238**, 30 (1963).

[13] T. Hayakawa, T. Kanzaki, T. Kitamura, Y. Fukuyoshi, Y. Sakurai, K. Koike, T. Suematsu, and M. Koike, *J. Biol. Chem.* **244**, 3660 (1969).

[14] E. R. Schwartz, L. O. Old, and L. J. Reed, *Biochem. Biophys. Res. Commun.* **31**, 495 (1968).

pathlength cuvette. The reaction is started by the addition of 0.02 ml of the sodium pyruvate solution. The resulting solution is rapidly mixed, and the increase in absorbance at 340 nm is recorded as a function of time at 25°.

Definition of Unit and Specific Activity. One unit of activity is the amount of enzyme required to produce 1 μmol of NADH per hour. The specific activity is expressed as units per milligram of protein. Protein is determined by the procedure of Lowry *et al.*[15] using bovine serum albumin as a standard.

Preparation of Cell-Free Extracts

Reagents

Washing solution: Tris-HCl, 100 mM (pH 7.8); sucrose, 100 mM
Grinding solution: sucrose, 250 mM; bovine serum albumin, 0.15%; TPP, 0.043 mM; MnSO$_4$, 0.36 mM; cysteine · HCl, 5.3 mM
Sonication solution: Tris-HCl, 100 mM, pH 8.0; sucrose, 100 mM

Procedure. The procedure used to prepare mitochondrial pellets is essentially that of Caroline *et al.*[16] The wild-type *Neurospora crassa* strain used [Em(LSDT1969)] is available from the Fungal Genetics Stock Center, Humboldt State University Foundation, Arcata, California 95521 as FGSC No. 2460. Mycelial cultures are grown for 18 hr on a rotating shaker at 29° in Vogel's medium (type N)[17,18] supplemented with 2% sucrose. The mycelia are harvested on a Büchner funnel, washed with Tris-sucrose solution, and ground with a mortar and pestle in 1.5 ml of grinding solution per gram fresh weight plus 1.5 g of acid-washed sand per gram fresh weight.

The homogenate is centrifuged twice at 1500 g for 15 min, and the supernatant is centrifuged at 12,000 g for 30 min. The resulting mitochondrial pellet is washed with grinding solution and centrifuged for 15 min at 39,000 g. The washed mitochondrial pellet is suspended in sonication solution (12 ml per 50 g fresh weight of mycelia). Sonication of 3.5-ml aliquots of the enzyme solution in 10-ml beakers is carried out with a Branson Sonifier, Model LS75 (Branson Sonic Power Co., Danbury, Connecticut) for 20 sec with a setting of 4 amperes. This suspension is centrifuged at 39,000 g for 15 min. The resulting sonication supernatant is used in subsequent purification steps.

[15] O. H. Lowry, N. J. Rosebrough, A. L. Farr, and R. J. Randall, *J. Biol. Chem.* **193**, 265 (1951).

[16] D. F. Caroline, R. W. Harding, H. Kuwana, T. Satyanarayana, and R. P. Wagner, *Genetics* **62**, 487 (1969).

[17] H. J. Vogel, *Am. Natur.* **98**, 435 (1964).

[18] R. H. Davis and F. J. De Serres, this series, Vol. 17A, p. 79.

Neurospora PDC is now known to be localized on the outer mitochondrial membrane.[19] The detergent Lubrol-WX can be used in place of sonication to liberate PDC from the mitochondria.[19] It has not been determined whether the use of this detergent would require modification of the purification procedure described below.

Purification Procedure

Reagents

Protamine sulfate solution, 2%
Potassium phosphate buffer, 20 mM (pH 7.0); dithiothreitol, 2 mM; TPP, 0.5 mM; MgSO$_4$, 0.5 mM
RNA solution, 1% (Sigma; *Torula* yeast RNA) in potassium phosphate buffer, 20 mM (pH 7.0)

Procedure. A 0.061 volume of 2% protamine sulfate solution is added dropwise to the sonication supernatant. The mixture is stirred for 15 min and centrifuged for 15 min at 39,000 g. The activity remains in the supernatant. An additional 0.0045 volume of 2% protamine sulfate is added to the supernatant, and the stirring and centrifugation procedure is repeated. The resulting supernatant and pellet are assayed, and if the enzyme activity remains in the supernatant, additional 0.0045 volumes of 2% protamine sulfate are added until the activity is precipitated. The total amount of 2% protamine sulfate required is generally 0.061–0.079 volume. The pellet containing the activity is suspended in potassium phosphate buffer, 20 mM (pH 7.0), supplemented as described above.

In order to precipitate protamine, RNA solution is added to give a final concentration of 0.1% RNA. The mixture is centrifuged at 39,000 g for 15 min, and the resulting supernatant is centrifuged for 1 hr at 35,000 rpm in a Spinco Model L ultracentrifuge, No. 50 rotor. The pellet is suspended in potassium phosphate buffer, 20 mM (pH 7.0), supplemented as described above, and diluted to a concentration of 1–3 mg of protein per milliliter. This fraction is stored at −20°. The results of a typical purification are presented in the table. An alternative procedure for purification of *Neurospora* PDC has also been published.[20]

Comments on the Purification Procedure. α-Ketoglutarate dehydrogenase complex activity is not detected at any stage of the purification of the *Neurospora* pyruvate dehydrogenase complex. NADH oxidase activity, which is present in the sonication supernatant, is removed by the protamine fractionation.

[19] E. Song, J. Briggs, and J. B. Courtright, *Biochim. Biophys. Acta* **544**, 453 (1978).
[20] O. H. Wieland, U. Hartmann, and E. A. Siess, *FEBS Lett.* **27**, 240 (1972).

PURIFICATION OF PYRUVATE DEHYDROGENASE COMPLEX FROM *Neurospora*[a]

Fraction	Volume (ml)	Protein (mg)	Specific activity (μmol NADH/ mg protein hr^{-1})	Recovery (%)
Sonication supernatant	40	477	5	100
Protamine sulfate precipitate	11	34	59	87
RNA supernatant	12	26	65	74
35,000 rpm pellet	2	4	225	39

[a] Modified after Harding *et al.*[8]

Properties

Physical Properties. Only one peak is observed ($s_{20,w}$ of 85 S) in sedimentation studies of *Neurospora* PDC in a Beckman Model E analytical centrifuge. The diameter of the complex is 300–350 Å as calculated from electron micrographs.

Substrate Specificity. Both sodium pyruvate and sodium α-ketobutyrate are substrates for the reduction of NAD$^+$ by *Neurospora* PDC. The K_ms for these substrates are 2.6×10^{-4} M and 8.2×10^{-4} M, respectively. There is no activity when α-ketoglutarate is tested as a substrate.

Characterization of the Purified PDC. The enzyme activity is linear over the range 1.5–50.0 μg of protein per assay. The pH optimum is between 8 and 9. The PDC activity is completely dependent on CoA and NAD$^+$. The K_m for CoA is 1.1×10^{-5} M and the K_m for NAD$^+$ is 1.5×10^{-4} M. The activity of the enzyme is reduced by 91% when TPP is omitted from the grinding solution, the purification buffer, and the assay mixture. There is no demonstrable requirement for Mg^{2+} by *Neurospora* PDC.

Inhibitors of PDC Activity. NADH and acetyl-CoA are competitive inhibitors of *Neurospora* PDC activity. The K_i for NADH is 3.4×10^{-5} M, and the K_i for acetyl-CoA is 1.8×10^{-5} M.

Adenosine triphosphate at 5 mM inhibits enzyme activity by 93%. Inhibition of *Neurospora* PDC by ATP is due to phosphorylation of the complex by an associated kinase.[20] Regulation of PDC isolated from other organisms by phosphorylation and dephosphorylation has also been demonstrated.[21–23]

[21] T. C. Linn, F. H. Pettit, and L. J. Reed, *Proc. Natl. Acad. Sci. U. S. A.* **62**, 234 (1969).

[22] T. C. Linn, F. H. Pettit, F. Hucho, and L. J. Reed, *Proc. Natl. Acad. Sci. U. S. A.* **64**, 227 (1969).

[23] For a recent review of enzyme cascade systems see P. B. Chock, S. G. Rhee, and E. R. Stadtman, *Annu. Rev. Biochem.* **49**, 813 (1980).

Acknowledgments

This investigation was supported in part by Grant GM-12323 from the National Institutes of Health and a grant from the Robert A. Welch Foundation, Houston, Texas, and was carried out in the Department of Zoology, University of Texas, Austin, Texas.

[67] Pyruvate Dehydrogenase Complex from *Escherichia coli*

By JAAP VISSER and MARIJKE STRATING

$$\text{Pyruvate} + \text{CoA-SH} + \text{NAD}^+ \rightarrow \text{acetyl-S-CoA} + \text{CO}_2 + \text{NADH} + \text{H}^+ \tag{1}$$

The pyruvate dehydrogenase complex of *E. coli*, catalyzing reaction (1), consists of multiple copies of pyruvate decarboxylase (E_1; EC 1.2.4.1) and lipoamide dehydrogenase (E_3; EC 1.6.4.3), which bind independently to dihydrolipoamide acetyltransferase (E_2; EC 2.3.1.12).[1,2] The latter enzyme forms the structural core of the complex, which appears in electron micrographs as a cube and consists of 24 identical subunits arranged with octahedral (432) symmetry.[3,4] The E_1 and E_2 structural genes, *aceE* and *aceF*, constitute an operon at 2.6 min on the *E. coli* linkage map, closely linked with *lpd*, the structural gene for E_3.[5] Mutants in all three components of the complex are known.[6,7] Direct analysis,[8] enzymic acetylation,[9,10] and pyruvate dependent inactivation of the complex by N-ethyl[2,3-^{14}C]maleimide[11] indicate the presence of two functional lipoyl moieties per E_2 chain. Average intermolecular distances measured between the catalytic sites[10,12] are inconsistent with the "swinging arm" model.[2] Evidence ac-

[1] L. J. Reed and R. M. Oliver, *Brookhaven Symp. Biol.* **21**, 397 (1968).

[2] L. J. Reed, *Acc. Chem. Res.* **7**, 40 (1974).

[3] D. J. DeRosier, R. M. Oliver, and L. J. Reed, *Proc. Natl. Acad. Sci. U. S. A.* **68**, 1135 (1971).

[4] C. C. Fuller, L. J. Reed, R. M. Oliver, and M. L. Hackert, *Biochem. Biophys. Res. Commun.* **90**, 431 (1979).

[5] J. R. Guest, *Adv. Neurol.* **21**, 219 (1978).

[6] U. Henning, J. Dietrich, K. N. Murray, and G. Deppe, *in* "Molecular Genetics" (H. G. Wittmann and H. Schuster, eds.), p. 223. Springer-Verlag, Berlin, 1968.

[7] J. R. Guest and I. F. Creaghan, *J. Gen. Microbiol.* **81**, 237 (1974).

[8] M. Koike, L. J. Reed, and W. R. Carroll, *J. Biol. Chem.* **238**, 30 (1963).

[9] D. C. Speckard, B. H. Ikeda, S. S. Wong, and P. A. Frey, *Biochem. Biophys. Res. Commun.* **77**, 708 (1977).

[10] G. Shepherd and G. G. Hammes, *Biochemistry* **15**, 311 (1976).

[11] M. J. Danson and R. N. Perham, *Biochem. J.* **159**, 677 (1976).

[12] O. A. Moe, Jr., D. A. Lerner, and G. G. Hammes, *Biochemistry* **12**, 2552 (1974).

cumulates that acyl group and electron pair transfer occur between multiple lipoic acid residues.[13–15] The molecular weights (M_r) of the component enzymes are: E_1, 96,000–100,000; E_2, 65,000–89,000; and E_3, 54,000–56,000.[16–18] The E_2 component becomes easily degraded. Some fragments are enzymically active,[17–19] and different functional domains can be identified.[20] The subunit stoichiometry of the complex is still a matter of debate. Eley *et al.*[16] and others[9,21] report a chain ratio ($E_1 : E_2 : E_3$) of 2 : 2 : 1, whereas on the basis of radioamidination experiments a disproportionate chain ratio is found of 1.3–1.8 : 1.0 : 0.8–1.0.[22] Reed and Willms[23] described a very general method to purify and resolve the pyruvate dehydrogenase complex. The purification involves fractionation with protamine sulfate, addition of yeast ribonucleic acid, ultracentrifugation, and isoelectric precipitation. A further improvement is obtained by introducing several precipitation steps between pH 5.7 and pH 4.9 and by introducing calcium phosphate gel cellulose chromatography.[24] Another method is more rapid and uses BioGel A-50 m followed by calcium phosphate gel cellulose chromatography.[25] Two other principles for purifying the complex from *E. coli* using affinity chromatography have since been described.[26,27]

Assay Method

The various principles for determining the overall reaction and the partial reactions catalyzed by the pyruvate dehydrogenase complex can

[13] D. L. Bates, M. J. Danson, G. Hale, E. A. Hooper, and R. N. Perham, *Nature (London)* **268**, 313 (1977).

[14] J. H. Collins and L. J. Reed, *Proc. Natl. Acad. Sci. U. S. A.* **74**, 4223 (1977).

[15] K. J. Angelides and G. G. Hammes, *Proc. Natl. Acad. Sci. U. S. A.* **75**, 4877 (1978).

[16] M. H. Eley, G. Namihira, L. Hamilton, P. Munk, and L. J. Reed, *Arch. Biochem. Biophys.* **152**, 655 (1972).

[17] O. Vogel, *Biochem. Biophys. Res. Commun.* **74**, 1235 (1977).

[18] T. Gebhardt, D. Mecke, and H. Bisswanger, *Biochem. Biophys. Res. Commun.* **84**, 508 (1978).

[19] G. Hale and R. N. Perham, *Eur. J. Biochem.* **94**, 119 (1979).

[20] D. M. Bleile, P. Munk, R. M. Oliver, and L. J. Reed, *Proc. Natl. Acad. Sci. U. S. A.* **76**, 4385 (1979).

[21] K. J. Angelides, S. K. Akiyama, and G. G. Hammes, *Proc. Natl. Acad. Sci. U. S. A.* **76**, 3279 (1979).

[22] M. J. Danson, G. Hale, P. Johnson, R. N. Perham, J. Smith, and P. Spragg, *J. Mol. Biol.* **129**, 603 (1979).

[23] L. J. Reed and C. R. Willms, this series Vol. 9 [50].

[24] D. C. Speckhard and P. A. Frey, *Biochem. Biophys. Res. Commun.* **62**, 614 (1975).

[25] O. Vogel, H. Beikirch, H. Müller, and U. Henning, *Eur. J. Biochem.* **20**, 169 (1971).

[26] J. Visser, M. Strating, and W. van Dongen, *Biochim. Biophys. Acta* **524**, 37 (1978).

[27] J. Visser, W. van Dongen, and M. Strating, *FEBS Lett.* **85**, 81 (1978).

be found in a previous volume.[23] The procedure for the overall reaction followed here at 25° is described in this volume.[28]

Other Assay Methods. A continuous assay[29] for the pyruvate decarboxylase (E_1) is based on the decrease in absorbance at 420 nm due to the reduction of ferricyanide ($\epsilon = 1030\ M^{-1}\ cm^{-1}$) according to reaction (2).

$$CH_3COCOOH + 2\ Fe(CN)_6^{-3} + H_2O \rightarrow CH_3COOH + CO_2 + 2\ Fe(CN)_6^{4-} + 2\ H^+ \quad (2)$$

A more sensitive variant of this procedure is based on the release of $^{14}CO_2$ in the reaction.[30]

Purification Procedure

Reagents

Potassium phosphate buffer, 1 M, pH 7.0
Magnesium chloride, 1 M
EDTA, 0.1 M, pH 7.0
2-Mercaptoethanol
Dithiothreitol (DTT)
Phenylmethylsulfonyl fluoride (PMSF), 50 mM in 95% ethanol
Thiamine pyrophosphate (TPP)
DNase
RNase
1,4-Butanediol-diglycidyl ether
Ethanolamine, 0.1 M or 0.2 M in sodium bicarbonate buffer, 0.2 M,
    pH 9.5–10
BioGel A-50m
Sepharose-2B
Ammonium sulfate
Sodium chloride
Potassium chloride
Sodium hydroxide, 0.6 M
Sodium borohydride
Glycerol

Preparation of Ethanol-Sepharose 2B. The procedure is based on the method described by Sundberg and Porath.[31] Suction-dried Sepharose 2B (50 g) is washed with water on a Büchner funnel and then mixed with 50 ml of 1,4-butanediol-diglycidyl ether and 50 ml of 0.6 M NaOH containing 2

[28] J. Visser, H. Kester, K. Jeyaseelan, and R. Topp, this volume [68].
[29] E. R. Schwartz, L. O. Old, and L. J. Reed, *Biochem. Biophys. Res. Commun.* **31**, 495 (1968).
[30] G. B. Kresze, *Anal. Biochem.* **98**, 85 (1979).
[31] L. Sundberg and J. Porath, *J. Chromatogr.* **90**, 87 (1974).

mg of $NaBH_4$ per milliliter. The reaction mixture is gently rotated overnight (use, e.g., a Büchi Rotavapor) at 25°. The activated Sepharose is carefully washed on a Büchner funnel with large volumes of deionized water (25–50 liters). The activated matrix is then suspended in 200 ml of sodium bicarbonate buffer 0.2 M, containing 0.1 M or 0.2 M ethanolamine (pH 9.5–10 final pH) and allowed to react for 24 hr at room temperature.[27] A modification that leads to a more constant degree of substitution requires coupling of ethanolamine in 0.01 M NaOH at 40° for 2 hr.

Growth of Microorganisms and Preparation of Crude Extract. A constitutive strain, K1-1 LR8-13,[25] is grown aerobically at 37° in complete medium containing per liter Bacto peptone (Difco), 10 g; yeast extract (Difco), 5 g; sodium chloride, 5 g; glucose, 1 g; thymine, 50 mg, and adjusted to pH 7.0 with 1.0 M NaOH. The bacterial cells are harvested by centrifugation and stored at −60°. The crude extract is prepared by breaking the frozen cells (60–120 g) with an X-press and suspending the material in extraction buffer (5 ml/g of cell paste). The extraction buffer contains: potassium phosphate buffer (50 mM, pH 7.0), EDTA (0.1 mM), $MgCl_2$ (3 mM), PMSF (50 μM), 2-mercaptoethanol (5 mM), DNase (5 μg/ml), RNase (10 μg/ml), and TPP (100 μM). Extraction is continued overnight at 4°, and the extract is then sedimented at 20,000 g for 20 min at 0° to separate the cell debris and the supernatant.

Affinity Chromatography. The clear supernatant is dialyzed twice against a 10-fold excess of buffer which is identical to the extraction buffer except for the absence of DNase, RNase, and TPP. The extract, which has a specific activity of 0.3–0.6 unit/mg, in various experiments, is then loaded on a column filled with ethanol-Sepharose 2B (2.5 × 30–35 cm) equilibrated with the same buffer as that used for dialysis. A single column is loaded with extract equivalent to 60 g of cells. The column is eluted and washed at a flow rate of 40 ml/hr until the absorbance at 280 nm has a value at least below 0.9. The enzyme is then eluted with 800 ml of a linear gradient of potassium or sodium chloride (0 to 200 mM) in the same buffer. A somewhat steeper gradient can also be applied successfully. An elution profile using a gradient from zero to 300 mM NaCl is shown in Fig. 1. The fractions containing the pyruvate dehydrogenase complex are yellow and opalescent. The active fractions are pooled, and the enzyme complex is precipitated by ammonium sulfate (43 g/100 ml). The precipitate is collected by centrifugation at 21,000 g for 15 min and resuspended in buffer, the composition of which depends on which step is chosen next. The recovery of the combined affinity chromatography and ammonium sulfate precipitation step has repeatedly been found to be approximately 70%.

BioGel A-50 Chromatography. The enzyme pellets are resuspended in potassium phosphate buffer (pH 7.0, 0.1 M), which contains glycerol (5%

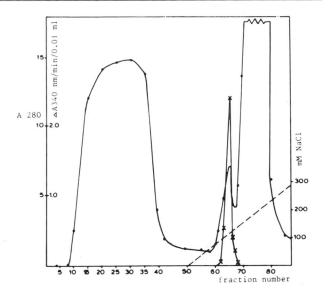

FIG. 1. Affinity chromatography of pyruvate dehydrogenase complex on ethanol–Sepharose 2B (column dimensions 2.5 × 30 cm). Crude extract is prepared from 60 g of bacteria. Fractions of 15 ml are collected. A_{280} (●——●) and enzyme activity (×——×) are shown. From Visser *et al.*[27]

v/v), DTT (4 mM), MgCl$_2$ (3 mM), and TPP (2 mM), to give a protein concentration of approximately 30–40 mg/ml, and the concentrated enzyme is immediately dialyzed. A column is filled with BioGel A-50m (2.5 × 90 cm bed) and equilibrated with the same buffer. A 2-ml sample of enzyme is applied (60 mg of protein) and elution is carried out with the same buffer using a pump delivering 90 ml/hr. Fractions are assayed for pyruvate dehydrogenase complex activity and analyzed by SDS–polyacrylamide gel electrophoresis on 10% gels. The intact complex elutes first and separates from complex molecules containing transacetylase breakdown products. The enzyme is precipitated with ammonium sulfate (43 g/100 ml). The pellets are resuspended in a small volume of 30 mM potassium phosphate buffer (pH 7.0) containing EDTA (2.5 mM) and 2-mercaptoethanol (5 mM) and dialyzed against the same buffer. The concentrated enzyme is stored at a concentration of 20–35 mg/ml at −20°. The yield of intact complex in this step is approximately 50%, and the specific activity is about 18 units. The enzyme is pure as indicated by SDS–polyacrylamide gel electrophoresis (Fig. 2).

Calcium Phosphate Gel–Cellulose Chromatography. Alternatively the enzyme pellets obtained in the affinity chromatography step are resus-

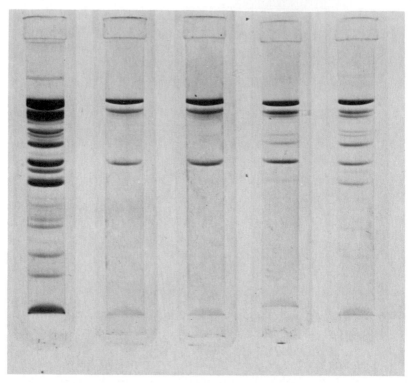

FIG. 2. Sodium dodecyl sulfate–polyacrylamide disc gel electrophoresis patterns of different stages of the pyruvate dehydrogenase complex purification. From left to right: enzyme pool after ethanol–Sepharose 2B and $(NH_4)_2SO_4$ precipitation; 3rd, 7th, 10th, and 14th active fractions obtained by BioGel chromatography. From Visser et al.[27]

pended in 50 mM potassium phosphate buffer, pH 7.0, containing EDTA (1 mM), β-mercaptoethanol (5 mM), $MgCl_2$ (3 mM), and TPP (100 μM) and dialyzed twice against 50 mM potassium phosphate buffer, pH 7.0. Calcium phosphate gel adsorbed on cellulose is used to prepare at room temperature a 3.2 × 18 cm column.[28] After equilibration of the column at 4° with potassium phosphate buffer, pH 7.0, several batches of enzyme obtained by the previous affinity step are adsorbed at a protein concentration of 10 mg/ml using a flow rate of 25 ml/hr. Subsequently, the column is washed with 100 mM potassium phosphate buffer, pH 7.5 containing 1% (w/v) $(NH_4)_2SO_4$ until the absorbance at 280 nm, which is monitored continuously, reaches zero. The enzyme complex is then released by eluting with the same buffer containing 4% $(NH_4)_2SO_4$. Initially fractions of low specific activity appear (8% of the activity applied to the column), followed by enzyme that is pure on the basis of SDS–polyacrylamide gel electrophoresis. The total recovery in this step is approximately 80%. The precipitated enzyme is dialyzed against potassium phosphate buffer (pH 7.0, 30 mM) containing EDTA (sodium salt, 2.5 mM). The enzyme is stored at $-20°$ at a protein concentration of at least 25 mg/ml. Storage at 4° in the presence of azide (0.02% w/v) has also been reported.[19]

A typical purification is summarized in the table.

Comments. Slight leakage of the complex is observed under the loading conditions used for affinity chromatography (2.5 ml bed volume per gram of cells; 50 mM phosphate buffer). At 30 mM phosphate no leakage is observed. As the specific activity of the pooled enzyme is lower, a procedure using 50 mM is preferred. The *E. coli* strain K1-1 LR8-13 overproduces the pyruvate dehydrogenase complex and is relatively low in α-ketoglutarate dehydrogenase complex activity. In other strains, when both complexes are present in noticeable amounts the α-ketoglutarate complex is almost completely separated in the affinity chromatography step. The enzyme elutes in the gradient behind another unidentified yellow protein fraction but before the pyruvate dehydrogenase complex. The purification procedure followed here does not involve an isoelectric precipitation at low pH, which is usually required to separate both complexes.

Proteolysis of E_2 is a general problem in purifying the pyruvate dehydrogenase complex. PMSF and EDTA effectively limit degradation in the

PURIFICATION OF PYRUVATE DEHYDROGENASE COMPLEX FROM *Escherichia coli*[a]

Step	Volume (ml)	Protein (mg)	Total activity (units)	Specific activity[b] (units/mg protein)	Recovery (%)
Preparation of crude extract[c]	300	5098	2600	0.51	100
Ethanol-Sepharose 2B (NH$_4$)$_2$SO$_4$ precipitation	5.7	265	1802	6.8	69[d]
BioGel A-50m[e], (NH$_4$)$_2$SO$_4$ precipitation	0.5	12	216	18	37
Calcium phosphate gel–cellulose, (NH$_4$)$_2$SO$_4$ precipitation[f]	1.6	54	1258	23.3	48

[a] Sixty grams of *E. coli* K1-1 LR8-13.
[b] Specific activities are determined at 25°.
[c] Results are the average of five separate purifications.
[d] Recovery of the affinity chromatography step is generally better than 75%.
[e] The experiment is based on 60 mg of protein from preceding step.
[f] The actual experiment was performed on six times larger scale.

extraction buffer. The presence of Mg^{2+} improves the affinity chromatography step. In the final preparation a high concentration of EDTA (2–2.5 mM) in 20–30 mM phosphate buffer protects the complex well.

Behavior of Component Enzymes and Complex Mutants on Ethanol-Sepharose 2B

Preparation of Subunits. Principles for the resolution of the complex have been described by Reed and Willms.[23] E_1 dissociates at alkaline pH, e.g., in 20 mM ethanolamine–25 mM potassium phosphate buffer, pH 9.5–10. E_1 is either selectively eluted from the complex, which is previously adsorbed to calcium phosphate gel–cellulose[20,23] or separated by gel filtration on Sepharose 2B.[16,22] E_3 can be prepared from the complex or from the E_2E_3 subcomplex and dissociates in the presence of 4–5 M urea. E_3 remains selectively adsorbed to calcium phosphate gel.[20,23] Gel filtration of the subcomplex E_2E_3 on BioGel A-5m in the presence of 6 M guanidine-HCl has been used as an alternative.[22]

Affinity Chromatography. E_3, applied to ethanol–Sepharose 2B in 20 mM phosphate buffer, pH 7.0, which contains 2-mercaptoethanol (5 mM) and EDTA (1 mM), does not bind. Contaminating partial complexes do bind and elute with a pulse of 0.3 M KCl in buffer. E_1 itself has low affinity for ethanol-Sepharose 2B; the major part is eluted between 50 and 75 mM potassium phosphate.[32] Contaminating E_2 subcomplexes behave like the intact complex and require a higher ionic strength for elution. Pyruvate dehydrogenase complex binding to ethanol-Sepharose 2B thus depends on the E_2 component enzyme. This allows another principle to be applied to dissociate the complex in an adsorbed state. With calcium phosphate gel E_3 binds most strongly, whereas in this case E_2 remains bound. Increasing the concentration of urea first releases E_1 at 6 M urea from the complex followed by E_1 and some E_3 with 7 M urea.

Mutants of the Pyruvate Dehydrogenase Complex. The analysis of several *E. coli* strains mutated in one of the complex components confirms that E_2 is involved in the binding to ethanol-Sepharose 2B.[33] Chromatography of a crude extract of a deletion strain lacking both E_1 and E_2 like $K\Delta$-15[34] results in frontal elution of uncomplexed E_3. From strain *lpd8*, which is immunologically inactive toward antibodies against E_3,[35] a complex is isolated that behaves on ethanol–Sepharose 2B as the wild type.

---

[32] The ligand concentration on the matrix influences the phosphate molarity required for elution. Slight leakage is continuously observed during washing.

[33] J. Visser, M. Strating, and J. R. Guest, unpublished data, 1979.

[34] D. Langley and J. R. Guest, *J. Gen. Microbiol.* **106**, 103 (1978).

[35] J. R. Guest and I. T. Creaghan, *J. Gen. Microbiol.* **81**, 237 (1974).

The purified complex lacks a flavin spectrum, but it still contains a small residual amount of E_3 protein ($\pm 15\%$). The complexes of other *lpd* mutants analyzed contain amounts of E_3 corresponding to the wild type. In *lpd7*, which maps rather close to *lpd8*, also no flavin spectrum is found in the final preparation. In *lpd5* and *lpd9* the flavin becomes fully reduced by NADH; in *lpd5* the spectral changes due to reduction by reduced lipoamide are small, in contrast to those in *lpd9* where the half-reduced state is formed.[33]

In *Pseudomonas aeruginosa*[36] and in *Aspergillus nidulans*,[37] pyruvate dehydrogenase complex mutants have also been successfully analyzed by the affinity chromatography described.

[36] K. Jeyaseelan, J. R. Guest, and J. Visser, *J. Gen. Microbiol.* **120**, 393 (1980).
[37] J. Visser, M. Strating, C. J. Bos, and C. F. Roberts, submitted for publication, 1981.

[68] Pyruvate Dehydrogenase Complex from *Bacillus*

By Jaap Visser, Harry Kester, Kandiah Jeyaseelan, and Randy Topp

The pyruvate dehydrogenase complex is a multienzyme complex catalyzing the oxidative decarboxylation of pyruvate to acetyl CoA:

$$\text{Pyruvate} + NAD^+ + CoA{-}SH \rightarrow \text{acetyl-S-CoA} + CO_2 + NADH + H^+$$

Among bacteria, the pyruvate dehydrogenase complex of *Escherichia coli* has been the most thoroughly investigated.[1,2] It consists of multiple copies of three enzymic components: pyruvate decarboxylase (E_1; EC 1.2.4.1), dihydrolipoamide acetyltransferase (E_2; EC 2.3.1.12), and lipoamide dehydrogenase (E_3; EC 1.6.4.3).

The structural core of the assembled complex is formed by enzyme E_2, which appears in electron micrographs as a cube and has octahedral (432) symmetry.[1,3] This is consistent with 24 polypeptide E_2 chains in the intact complex. Each transacetylase (E_2) chain contains two covalently bound lipoyl moieties.[4,5] The molecular weights of the three different polypeptide

[1] L. J. Reed, *Acc. Chem. Res.* **7**, 40 (1974).
[2] D. L. Bates, M. J. Danson, G. Hale, E. A. Hooper, and R. N. Perham, *Nature (London)* **268**, 313 (1977).
[3] D. J. De Rosier, R. M. Oliver, and L. J. Reed, *Proc. Natl. Acad. Sci. U. S. A.* **68**, 1135 (1971).
[4] M. J. Danson and R. N. Perham, *Biochem. J.* **159**, 677 (1976).
[5] J. H. Collins and L. J. Reed, *Proc. Natl. Acad. Sci. U. S. A.* **74**, 4223 (1977).

chains that have been reported are for E_1: M_r 96,000–100,000; for E_2: M_r 65,000–89,000; and for E_3: M_r 54,000–56,000.[1,6–8] Estimates of the complex molecular weight vary between 4.6×10^6 and 6.1×10^6.[6,9]

The pyruvate dehydrogenase complex from mammalian sources shows substantial differences from that of *E. coli*. A molecular weight of at least 7.6×10^6 is found.[10] The transacetylase core appears, moreover, as a pentagonal dodecahedron[1,11] consisting of 60 polypeptide chains with a molecular weight M_2 of 52,000 and each containing one lipoyl moiety.[12]

Enzyme E_1 consists of two nonidentical subunits $E_1\alpha$ and $E_1\beta$ with subunit molecular weights of 41,000 and 36,000, respectively.[12] In the native protein these subunits appear as dimers ($\alpha_2\beta_2$). The complex contains 10–12 flavoprotein E_3 chains and in addition up to five copies of two regulatory enzymes, a relatively strongly bound kinase and a loosely bound phosphatase,[13] that act through (de-)phosphorylation of the $E_1\alpha$ component.

The pyruvate dehydrogenase complex of the genus *Bacillus* is of interest for several reasons. The structural organization of the bacillar complex[14,15] differs substantially from that of *E. coli* and of other prokaryotes (*Azotobacter vinelandii*,[16] *Pseudomonas aeruginosa* PAO[17]) but resembles the eukaryote counterpart (fungi, mammals). The subunit molecular weights are almost identical to those found in eukaryotes, including two nonidentical E_1 subunits. Moreover, *Bacillus* spp. are gram-positive and the enzymes of some species are thermostable whereas the other prokaryotes studied thus far are gram-negative. These observations are reminiscent of differences in molecular size of citrate synthase and succinate thiokinase, which appear as small enzymes both in gram-positive

[6] M. H. Eley, G. Namihira, L. Hamilton, P. Munk, and L. J. Reed, *Arch. Biochem. Biophys.* **152**, 655 (1972).

[7] O. Vogel, *Biochem. Biophys. Res. Commun.* **74**, 1235 (1977).

[8] T. Gebhardt, D. Mecke, and H. Bisswanger, *Biochem. Biophys. Res. Commun.* **84**, 508 (1978).

[9] M. J. Danson, G. Hale, P. Johnson, R. N. Perham, J. Smith, and P. Spragg, *J. Mol. Biol.* **129**, 603 (1979).

[10] T. Hayakawa, T. Kanzaki, T. Kitamura, Y. Fukuyoshi, Y. Sakurai, K. Koike, T. Suematsu, and M. Koike, *J. Biol. Chem.* **244**, 3660 (1969).

[11] M. Koike and K. Koike, *Adv. Biophys.* **9**, 187 (1976).

[12] C. R. Barrera, G. Namihira, L. Hamilton, P. Munk, M. H. Eley, T. C. Linn, and L. J. Reed, *Arch. Biochem. Biophys.* **148**, 343 (1972).

[13] T. C. Linn, J. W. Pelley, F. H. Pettit, F. Hucho, D. D. Randell, and L. J. Reed, *Arch. Biochem. Biophys.* **148**, 327 (1972).

[14] C. E. Henderson, R. N. Perham, and J. T. Finch, *Cell* **17**, 85 (1979).

[15] J. Visser, H. Kester, and A. Huigen, *FEMS Microbiol. Lett.* **9**, 227 (1980).

[16] T. W. Bresters, R. A. de Abreu, A. de Kok, J. Visser, and C. Veeger, *Eur. J. Biochem.* **59**, 335 (1975).

[17] K. Jeyaseelan, J. R. Guest, and J. Visser, *J. Gen. Microbiol.* **120**, 393 (1980).

bacteria and in eukaryotes and as large enzymes in gram-negative bacteria.[18] Thus far no data are available as to whether in the case of the pyruvate dehydrogenase complex this feature is characteristic of *Bacillus* or of gram-positive species in general.[19]

The purification of the enzyme complexes of *B. subtilis* and *B. stearothermophilus* using affinity chromatography and some properties of the complexes are described.

Assay Method

The enzyme complex can be assayed spectrophotometrically by following continuously the keto acid-dependent reduction of either 3-acetylpyridine NAD^+ or NAD^+ at 366 nm (3-acetylpyridine NADH) or 340 nm (NADH). NAD^+ is preferred in the assay as it is cheaper than 3-acetylpyridine NAD^+. In crude extracts 3-acetylpyridine NAD^+ is preferred due to interference of NADH oxidase in the NAD^+ assay. The various principles of determining the overall reaction and the partial reactions catalyzed by the 2-oxoacid dehydrogenase complexes are described in previous volumes.[20,21]

Definition of Unit and Specific Activity. One unit is the amount of enzyme required to produce 1 μmol of NADH per minute under the conditions specified. Specific activity is expressed as units per milligram of protein. Protein is determined by the method of Lowry *et al.*[22]

Reagents

Potassium phosphate buffer, 1 *M*, pH 7.0
Sodium pyruvate, 20 m*M* (freshly prepared, to be kept on ice)
Magnesium chloride, 1 *M*
Thiamine pyrophosphate (TPP)
Coenzyme A (CoA)
Dithiothreitol (DTT)
NAD^+
3-Acetylpyridine NAD^+ (APNAD$^+$)

Procedure. Prepare a stock solution (10 ml) of the reaction mixture by mixing DTT (16 mg), CoA (4 mg), NAD^+ (20 mg), TPP (7.4 mg), $MgCl_2$

[18] P. D. J. Weitzman, *in* "Soc. Appl. Bacteriol. Symp. Ser. n8" (M. Goodfellow and R. G. Board, eds.), p. 107. Academic Press, New York, 1980.

[19] A partially purified complex of *Brevibacterium flavum*, another gram-positive species, has been found to be inhibited by antibodies raised against the *E. coli* complex, but not by antibodies raised against the *B. stearothermophilus* complex.

[20] L. J. Reed and C. R. Willms, this series, Vol. 9 [50].

[21] L. J. Reed and B. B. Mukherjee, this series, Vol. 13 [12].

[22] O. H. Lowry, N. J. Rosebrough, A. L. Farr, and R. J. Randall, *J. Biol. Chem.* **193**, 265 (1951).

(0.08 ml), potassium phosphate buffer (2.4 ml), and distilled water (7.6 ml). This solution can be stored at −20° and used for up to 2 weeks.

Pipette the reaction mixture (0.25 ml) into a cuvette (1 ml). Add sodium pyruvate (0.25 ml) and distilled water (0.4 ml) and incubate for 5 min at either 25° (*B. subtilis*) or 55° (*B. stearothermophilus*) in a water bath. Start the reaction by adding 0.1 ml of an appropriately diluted enzyme sample and measure the change in absorbance at 340 nm preferably in a dual-wavelength mode at 340–380 nm (e.g., Aminco-Chance DW2 spectrophotometer) with the cuvette housing maintained at the same temperature as the water bath.

Purification Procedure

Reagents

Potassium phosphate buffer, 1 M, pH 7.0
Magnesium chloride, 1 M
EDTA, 0.1 M, pH 7.0
2-Mercaptoethanol
Phenylmethylsulfonyl fluoride (PMSF), 50 mM, in 95% ethanol
Dithiothreitol (DTT)
Thiamine pyrophosphate (TPP)
DNase
RNase
Ammonium sulfate
Sodium chloride
Cellulose (Whatman CF-11, 10% w/v in distilled water)
Calcium phosphate gel[23]

Growth of Organisms and Preparation of Cell-free Extracts. Bacillus subtilis 168 is grown aerobically at 37° in a medium containing (per liter): 25 g of nutrient broth-2 (Difco), 4 g of glucose, and 2 g of sodium pyruvate. *Bacillus stearothermophilus* is grown aerobically in batches at 60° in a medium based on that described by Sargeant *et al.*[24] Both cultures are harvested in late logarithmic phase.

For direct enzyme assays the organisms can be grown as 250-ml batches in 1-liter bottles with rotary shaking (200 rpm). Large batch cultures of *B. subtilis* and *B. stearothermophilus* are obtained as a frozen cell paste from Diosynth B.V. (Oss, The Netherlands) and from the Public Health Laboratory Service CAMR (Porton, Salisbury, U.K.), respectively.

---

[23] S. M. Swingle and A. Tiselius, *Biochem. J.* **48**, 171 (1951).
[24] K. Sargeant, D. N. East, A. R. Whitaker, and R. Elseworth, *J. Gen. Microbiol.* **65**, iii (1971).

The bacterial cells are harvested by centrifugation under refrigeration and stored at $-60°$ until needed.

The frozen bacterial cells (70–100 g) are broken with an X-press and suspended in extraction buffer (5 ml/g of cell paste) and left overnight with gentle stirring at 4°. The extraction buffer contains (final concentration): potassium phosphate buffer, 10 mM, pH 7.0; EDTA, 1 mM; MgCl$_2$, 3 mM; PMSF, 50 μM; 2-mercaptoethanol, 5 mM; deoxyribonuclease I, 5 μg/ml; ribonuclease, 10 μg/ml; and TPP, 100 μM. The extract is sedimented at 20,000 g for 15 min at 0° to separate the cell debris, and the supernatant containing the pyruvate dehydrogenase complex is carefully removed.

Affinity Chromatography (Bacillus subtilis) and Concentration. The clear supernatant is dialyzed three times, each time against a 10-fold volume of extraction buffer that contains all components except DNase, RNase, and thiamine pyrophosphate. The extract is then loaded on a column filled with ethanol–Sepharose 2B (2.6 × 30 cm bed) equilibrated with buffer of the same composition used for dialysis. The preparation of the affinity matrix is described in this volume.[25] The column is eluted at a flow rate of 30–40 ml/hr and fractions of 10 ml are collected. When all unbound proteins are removed, the enzyme is eluted at the same flow rate with a linear gradient of sodium chloride (0 to 300 mM) in the same buffer (800 ml total volume). Enzyme is eluted around 120 mM sodium chloride. Fractions containing pyruvate dehydrogenase complex activity are collected and pooled (approximately 150 ml). All steps are performed at 4–6°. The diluted enzyme is concentrated at 0° by ammonium sulfate precipitation, adding 43 g/100 ml. The precipitate is collected at 4° by centrifugation for 20 min at 21,000 g. The enzyme pellet is resuspended in a small volume (2–4 ml) of 50 mM potassium phosphate buffer, pH 7.0, containing EDTA (0.5 mM), 2-mercaptoethanol (5 mM), magnesium chloride (3 mM), and TPP (50 μM). Ammonium sulfate is removed by dialyzing twice against a 200-fold volume of the same buffer.

Calcium Phosphate Gel–Cellulose Chromatography. To 50 ml of the cellulose suspension 5.4 ml of calcium phosphate gel[23] is added together with 7 ml of distilled water; the mixture is then briefly stirred on a magnetic stirrer. After the suspension settles, the supernatant is discarded to remove fine particles. Washing is repeated with 30 mM potassium phosphate buffer, pH 7.0, containing EDTA (0.5 mM). A slurry of gel cellulose is added to a column, and a bed is made by gravity packing to give a column of approximately 1.2 × 4 cm. Larger-size columns are poured by progressively opening the outlet. The enzyme solution is dialyzed against the same buffer used to wash the column and then is loaded on the col-

[25] J. Visser and M. Strating, this volume [67].

TABLE I
PURIFICATION OF PYRUVATE DEHYDROGENASE COMPLEX FROM *Bacillus subtilis* [a,b]

Step	Volume (ml)	Protein (mg)	Specific activity (units/mg protein)	Recovery (%)
Crude extract	425	10625	0.003[c]	100
Ethanol-Sepharose 2B,	98	25	0.88	76
(NH$_4$)$_2$SO$_4$ precipitation	2.5	23	1.0	81
Calcium phosphate gel cellulose, (NH$_4$)$_2$SO$_4$ precipitation	0.85	2.0	5.9	41

[a] From Visser *et al.*[15]
[b] Seventy grams of cell paste.
[c] Determined with APNAD$^+$ as cofactor; with purified enzyme, reaction rates with APNAD$^+$ and NAD$^+$ are not significantly different under the assay conditions used.

umn. The column is subsequently washed with (*a*) 50 mM potassium phosphate buffer, pH 7.0 containing 0.5 mM EDTA; (*b*) the same solution to which 1% (w/v) ammonium sulfate is added; and (*c*) the same solution to which 5% (w/v) ammonium sulfate is added. Washing is continued in each case until the absorbance of the eluate at 280 nm, which is continuously monitored, reaches the baseline. The enzyme is finally recovered by eluting the column with 100 mM potassium phosphate buffer, pH 7.0, containing EDTA (0.5 mM) and at least 5% (w/v) ammonium sulfate. Active fractions are pooled, concentrated and dialyzed as described above. The final enzyme preparation (approximately 1 ml) is stored at −20° for at least 2 weeks.

The results of a purification are summarized in Table I.

Alternative Procedure (*B. stearothermophilus*)

The recovery of the pyruvate dehydrogenase complex on ethanol–Sepharose 2B starting with crude extract is generally high, 70–80% in the case of *B. subtilis* and approximately 90% in the case of the dehydrogenase from thermophilic sources, e.g., *B. stearothermophilus*, *B. caldotenax*, and *B. caldolyticus*.[26] Nevertheless, an alternative method has been developed that is more rapid and removes the bulk of the protein before the affinity chromatography step.

[26] We find 15–30 units per gram of cell paste in different commercial batches of *B. stearothermophilus* with specific activities varying between 0.1 and 0.8 unit per milligram of protein. From the data of Henderson *et al.*,[14] 2.4 units per gram of cells can be calculated when assayed at 30°. This corresponds to 17 units if assayed at 55°.

Ultracentrifugation. The crude extract is prepared similarly and the cell debris removed as usual, starting from 100 g of cell paste homogenized in 300 ml of extraction buffer. Then the complex is sedimented by centrifugation at 140,000 g for 4.5 hr at 4°. The supernatant fluid is carefully removed from the gelatinous grayish yellow precipitates, which are dissolved overnight in extraction buffer (no DNase and RNase added). After appropriate dilution (100–150 ml), the enzyme solution is dialyzed for a period of 2 hr and, if necessary, clarified by centrifugation for 15 min at 20,000 g before the affinity chromatography step is applied. The enzyme complex is again recovered by high speed centrifugation of the pooled active fractions. The combined steps described up to this point lead generally to a 50- to 60-fold purification and an excellent recovery (at least 90%).[27]

Hydroxyapatite Chromatography. Calcium phosphate gel cellulose chromatography can be replaced by using commercially available hydroxyapatite (Bio-Rad HTP, BNA grade). Protein (approximately 5 mg/ml bed volume) is adsorbed in 50 mM potassium phosphate buffer, pH 7.0 and impurities are removed by stepwise elution with (a) 100 mM phosphate buffer; (b) the same buffer containing 1% (w/v) ammonium sulfate; and (c) the buffer containing 2% (w/v) ammonium sulfate. The enzyme is finally eluted under the same conditions used for calcium phosphate gel cellulose chromatography. Enzyme recovery (approximately 50%) is of the same order of magnitude as obtained with calcium phosphate gel suspended on cellulose.

Comments. Both procedures are reproducible. The yield in the affinity chromatography step depends on the bacterial species but has never been less than 70% if column overloading is prevented. The second procedure deserves preference when large-scale purifications are intended.

Properties

Physical Constants and Subunit Compositions. Henderson *et al.*[14] reported a sedimentation coefficient ($s_{20,w}$) of 73 S for the pyruvate dehydrogenase complex of *B. stearothermophilus* and estimated the molecular weight to be 9.1 × 10^6. The enzyme complex of *B. subtilis* has a value of 61.7 S and an estimated weight of 7.1 × 10^6.

SDS–polyacrylamide gel electrophoresis according to Laemmli[28] using a 7% gel with a 3% stacking gel leads to an identical pattern of four

---

[27] During step c some leakage of E$_3$ activity is observed, both in the case of *B. subtilis* and of *B. stearothermophilus*. This problem has been solved in the latter case by avoiding 5% ammonium sulfate in combination with 50 mM phosphate and using 2% ammonium sulfate with 100 mM phosphate instead. For the *B. subtilis* complex this is not tested yet. J. Visser, unpublished results (1980).

[28] K. K. Laemmli, *Nature (London)* **227**, 681 (1970).

TABLE II

KINETIC CONSTANTS OF BACTERIAL PYRUVATE DEHYDROGENASE COMPLEXES

	Bacillus subtilis[a]	Bacillus stearothermophilus[b]	Escherichia coli[c]	Azotobacter vinelandii[d]	Pseudomonas fluorescens[e]	Pseudomonas aeruginosa[f]
K_m pyruvate (mM)	0.2	0.37	0.2–0.6	S 0.5 = 1.9[g]	0.6	1.3
K_m NAD$^+$ (μM)	10	30	10	100	ND[h]	110
K_m CoA (μM) or	—	2.9	—	14	ND	1.3
S 0.5 CoA (μM)[g]	0.3–1.7	—	2	16	ND	—
K_i NADH (μM)	120	125	Not listed[i]	40	ND	25

[a] Visser et al.[15]

[b] Visser and Topp[29]; freeze-dried enzyme assay at 60°; other conditions as in Visser et al.[15]

[c] Bisswanger and Henning[30]; Bisswanger.[31]

[d] Bresters et al.[32]

[e] Visser.[33]

[f] Jeyaseelan.[34]

[g] S 0.5: substrate concentration is given at half saturation due to nonlinear kinetics.

[h] ND, not determined.

[i] Could not be determined owing to strong inhibition.[35]

protein bands with the complexes purified from *B. subtilis, B. stearothermophilus, B. caldotenax,* and *B. caldolyticus.* The molecular weights are 58,000–59,000, 53,000, 40,000–41,000, and 37,000. Both in *B. subtilis* and in *B. stearothermophilus* the 58,000 molecular weight component has been identified as the dihydrolipoamide acetyltransferase (E_2) and the 54,000 component as the lipoamide dehydrogenase (E_3).[14,15] The pyruvate decarboxylase (E_1) activity resides in the 37,000 molecular weight protein and in complexes of the two low molecular weight components.[15] From the flavin content of the complex of *B. stearothermophilus,* we calculate a number of lipoamide dehydrogenase dimers of 29–32 per assumed molecular weight of 9.6×10^6. Taken together with the pentagonal dodecahedron proposed for the E_2 core,[14] this suggests the presence of 60 copies of each of the E_1 components.

Temperature and pH Optima. Temperature optima for the various pyruvate dehydrogenase complexes under standard assay conditions are[29] *B. caldolyticus,* 70° (broad optimum); *B. caldotenax,* 64°; *B. stearothermophilus,* 72°; *B. subtilis,* 48°. The pH optima for the thermophilic complexes determined at 55° are[29] *B. caldolyticus,* pH 7.0–7.4 (broad optimum); *B. caldotenax,* pH 7.5; *B. stearothermophilus,* pH 7.4. For the *B. subtilis* enzyme, a relatively broad optimum around pH 7.6 exists.[15]

Kinetic Constants. In view of the different physiology and oxygen requirements of the bacteria in which the pyruvate dehydrogenase complex has been studied thus far, a comparison of kinetic parameters is of interest. In Table II the available data are compiled.[15,29–35] For experimental details, the original literature should be consulted.

Immunological Relationship. Antibodies raised in rabbits against the pyruvate dehydrogenase complex of *B. stearothermophilus* also precipitate with the complexes of other *Bacillus* species but not with the complex of *E. coli.* Similarly, antibodies against the *E. coli* complex do not form precipitates with the bacillar enzymes.[15]

[29] J. Visser and R. J. Topp, unpublished results, 1979.
[30] H. Bisswanger and U. Henning, *Eur. J. Biochem.* **24,** 376 (1971).
[31] H. Bisswanger, *Eur. J. Biochem.* **48,** 377 (1974).
[32] T. W. Bresters, A. de Kok, and C. Veeger, *Eur. J. Biochem.* **59,** 347 (1975).
[33] J. Visser, unpublished data, 1979.
[34] K. Jeyaseelan, Ph.D. Thesis, University of Sheffield, U.K., 1980.
[35] R. G. Hansen and U. Henning, *Biochim. Biophys. Acta* **122,** 355 (1966).

[69] Pyruvate Dehydrogenase Complex from Broccoli and Cauliflower[1]

By DOUGLAS D. RANDALL

Pyruvate + NAD$^+$ + coenzyme A → acetyl-CoA + O$_2$ + NADH + H$^+$

The pyruvate dehydrogenase complex is most likely composed of three enzymes in a multienzyme complex similar to the mammalian and *E. coli* enzyme complex.[2] The three enzymes are pyruvate dehydrogenase (EC 1.2.4.1), lipoate transacetylase (EC 2.3.1.12), and dihydrolipoyl dehydrogenase (EC 1.6.4.3).[2]

Assay Method

Principle. The spectrophotometric assay is based on the reduction of NAD$^+$ and the resultant increase of absorbance at 340 nm.

Reagents

Glycylglycine, 0.125 M + morpholinopropanesulfonic acid, 0.125 M buffer, pH 8.1
MgCl$_2$, 2.0 mM
Thiamine pyrophosphate, 4.0 mM
Potassium pyruvate, 50 mM (fresh)
Coenzyme A, 1.2 mM, containing 25 mM cysteine-HCl (prepared fresh)
NAD$^+$, 48 mM

Procedure. Assays are carried out in 1.2-ml cuvettes with a 10-mm light path maintained at 27°. Into the cuvette are placed 0.6 ml of buffer, 0.05 ml of thiamine pyrophosphate (TPP), 0.05 ml of MgCl$_2$, 0.05 ml of NADH, 0.1 ml of coenzyme A, enzyme, and water to a volume of 0.98 ml. The reaction is initiated with the addition of 0.02 ml of pyruvate, and the rate of increase of absorbance at 340 nm is measured. With crude enzyme, it is necessary to determine the rate of NADH oxidase activity using 0.2 mM NADH in the assay mixture instead of NAD$^+$ and leaving out the pyruvate. To reduce pipetting steps, the Mg^{2+}, TPP, and NAD$^+$ can be combined with buffer so that an aliquot of this mixture will provide the appropriate final concentration of reaction components.

[1] This work was supported by National Science Foundation Grant PCM-77-11390 and the Missouri Agricultural Experiment Station.

[2] L. J. Reed, *Acc. Chem. Res.* **7**, 40 (1974).

Units. One unit of enzyme activity is defined as the formation of 1 μmol of NADH per minute. Specific activity is expressed as units of enzyme activity per milligram of protein. Protein concentration is determined by the Lowry method[3] or using 280 and 260 nm absorbance with the Warburg and Christian[4] extinction values.

Purification of Cauliflower Pyruvate Dehydrogenase Complex[5]

Isolation of Cauliflower Mitochondria.[5] The selection of cauliflower at the market is critical. The heads must be firm and tight, not in the flowering or "loose" stage. The heads are chilled, and the outer 2–4 mm are shaved with razor blades or scalpels onto aluminum foil over ice. About 24–30 cauliflower heads will yield 1200 g of floral buds. The buds are homogenized in 100 g lots in two volumes of extraction medium (0.4 M sucrose, 50 mM potassium phosphate, and 5 mM EGTA, pH 7.7) for 30 sec with a Polytron or Waring blender at a medium speed. The homogenate is filtered through six layers of cheesecloth, pooled, and centrifuged at 400 g for 10 min. The supernatant is centrifuged at 16,000 g for 20 min. The pellet is washed by resuspending in 20 mM potassium phosphate and 20 mM 2-mercaptoethanol, pH 7.0, using a Potter–Elvehjem Teflon homogenizer. The resuspended mitochondria are centrifuged at 14,500 g for 20 min. The pellet is redissolved in a *minimal* amount of 20 mM potassium phosphate, pH 7.0 (thick but runny paste), and 1 mg of dithiothreitol (DTT) is added per milliliter of resuspension. The suspension is divided and immediately shell frozen in two 250-ml Erlenmeyer flasks in a solid CO_2–isopropanol bath and stored at $-20°$.

Purification Procedures[5]

Unless designated otherwise, all steps are performed at ice bath temperatures.

Step 1. Extraction. Shell-frozen mitochondria are thawed and diluted 2.5-fold with 25 mM potassium phosphate and 3.2 mM dithiothreitol buffer, pH 6.5, and gently homogenized or stirred for 15 min. The suspension is centrifuged at 30,000 g for 30 min, and the supernatant is designated the mitochondrial extract.

Step 2. Ultracentrifugation. The enzyme is pelleted by centrifuging for 3 hr at 200,000 g max (45,000 rpm, 60 Ti rotor). The resulting amber pellet is gently resuspended in 10–15 ml buffer (25 mM MOPS, 3.2 mM DTT,

[3] O. H. Lowry, N. J. Rosebrough, A. L. Farr, and R. J. Randall, *J. Biol. Chem.* **193**, 265 (1951).
[4] O. Warburg and W. Christian, *Biochem. Z.* **310**, 384 (1942).
[5] D. D. Randall, P. M. Rubin, and M. Fenko, *Biochim. Biophys. Acta* **485**, 336 (1977).

pH 7.0). After 2 hr to redissolve the complex, the suspension is clarified by centrifuging at 27,000 g for 20 min.

Step 3. Polyethylene Glycol Fractionation. Fraction II is warmed to 20°, and 10 μmol of $MgCl_2$ are added per milliliter. The solution is centrifuged at 27,000 g for 10 min to remove any precipitate, and the supernatant is fractionated by slowly added one-tenth volume of 50% (w/v) solution of polyethylene glycol 6000 (PEG). The solution is equilibrated for 15 min and centrifuged for 20 min at 27,000 g. The pellet, which contains the enzyme, is resuspended in 5 ml of 25 mM MOPS, 3.2 mM DTT buffer, pH 7.0, and centrifuged to remove insoluble material. The supernatant is designated the PEG fraction.

Step 4. Glycerol Gradient. The PEG fraction (III) is further purified on 37-ml 10 to 14% (v/v) glycerol gradients prepared in 25 mM MOPS, pH 7.0, 3.2 mM DTT, 0.1 mM TPP, and 1 mM NAD$^+$. A 1-ml sample is layered on the gradient, and the gradient is centrifuged in an SW-27 rotor at 25,000 rpm for 9 hr. The gradient is fractionated in 1-ml fractions, and peak fractions are pooled and precipitated by addition of 0.1 volume of 50% PEG in the presence of 10 mM Mg^{2+}. The precipitated complex is collected by centrifugation and redissolved in a minimal volume of 25 mM MOPS, pH 7.0, containing 3.2 mM DTT and 0.1 mM TPP. The enzyme may be stored for several days at 4° or quick frozen in a Dry Ice–isopropanol bath and stored at −20° for several weeks under N$_2$ in sealed tubes.

Properties of the Cauliflower Pyruvate Dehydrogenase Complex[5]

The enzyme has a molecular weight of several million as indicated by its sedimentation coefficient of 58.7 S. However, the dihydrolipoyl dehydrogenase component tends to dissociate. The complex exhibits four bands on dissociating gel electrophoresis in a manner similar to that of other eukaryotic pyruvate dehydrogenase complexes. The complex is sensitive to dilution, and every effort must be made to maximize protein concentration during purification and storage. Stability during storage also requires sulfhydryl protectants. The optimum pH for activity is between pH 7.5 and 8.0.

Cofactors and Specificity.[5] The enzyme has an absolute requirement for divalent cation, Mg^{2+}, thiamine pyrophosphate, and coenzyme A for activity. The specificity of the enzyme is absolute for NAD$^+$. Pyruvate is the primary substrate with α-ketobutyrate oxidized at 18% of the pyruvate rate, and no activity is observed with α-ketoglutarate.

Kinetic Properties.[5] The complex exhibits multisite Ping-Pong kinetics with apparent Michaelis constants of 207, 125, and 7 μM for pyruvate, NAD$^+$, and coenzyme A, respectively.

Regulation and Inhibition.[5] The purified complex is quite sensitive to the $NAD:NAD^+$ ratio. The K_i for NADH is 34 μM, and that for acetyl-CoA is 13 μM. Both of these products are competitive inhibitors with their respective substrates and noncompetitive with pyruvate. Other metabolites are generally without effect, only glyoxylate and hydroxypyruvate exhibiting very mild noncompetitive inhibition. In crude mitochondrial extracts, the enzyme can be phosphorylated by ATP with concomitant deactivation; however, this process cannot be reversed. The phosphorylation phenomenon does not occur with purified preparations of the complex.

Purification of the Broccoli Pyruvate Dehydrogenase Complex[6]

Isolation of Broccoli Mitochondria. Broccoli with firm, tight heads (nonflowering) are purchased at local markets. The floral buds are shaved from the heads and held at 4°. About 8–10 cases are required for 800–1000 g of buds. The chilled buds are homogenized for 40 sec in 2.5 volumes of extraction medium (0.5 M sucrose, 0.1 M sodium phosphate, pH 7.8, 0.1% bovine serum albumin, 5 mM EGTA, and 13.5 mM 2-mercaptoethanol) using a Polytron homogenizer, filtered through eight layers of cheesecloth, and centrifuged at 400 g for 15 min. The supernatant is centrifuged at 14,500 g for 30 min to pellet the mitochondria. The pellet is resuspended in 250 ml of 25 mM sodium phosphate, pH 7.0, containing 13.5 mM 2-mercaptoethanol, and recentrifuged as above. This wash step is repeated a second time, and the pellet is resuspended in a *minimal* volume of 25 mM sodium phosphate, pH 7.0. Dithiothreitol (1 mg/ml) is added, and the mitochondria are shell frozen in 250-ml Erlenmeyer flasks in a Dry Ice–propanol bath. The frozen mitochondria yield active enzyme for about 1 month. The crude mitochondria are dark green owing to chloroplast fragments that cosediment with the mitochondria.

Purification Procedure

All procedures are performed at 0–4° unless otherwise noted. The enzyme is kept as concentrated as possible owing to inactivation at dilute protein concentrations.

Step 1. Extraction. Frozen mitochondria are thawed and diluted with 5 volumes of ice cold phosphate buffer, pH 6.9, containing 6 mM dithiothreitol, 1 mM NAD^+, and 0.1 mM thiamine pyrophosphate. The diluted extract is gently homogenized at a low speed with a Polytron homogenizer for 2 min (or stirred for 20 min) and then centrifuged for 20 min at 12,000 g.

[6] P. M. Rubin and D. D. Randall, *Arch. Biochem. Biophys.* **178**, 342 (1977).

The supernatant is centrifuged again at 40,000 g for 30 min, and the resulting supernatant is designated the mitochondrial extract (fraction I).

Step 2. Protamine Sulfate Treatment. Protamine sulfate is added to the mitochondrial extract by additions of 0.001% volumes of 2% protamine sulfate followed by centrifugation at 40,000 g for 10 min. This process is repeated until a slight loss of enzyme activity is detected (0.004–0.015%). At this point, further addition of protamine sulfate will precipitate the enzyme and result in considerable loss of enzyme activity. The protamine sulfate should remove most of the chlorophyll.

Step 3. Ultracentrifugation. The protamine sulfate-treated enzyme (fraction II) is stabilized by addition of 5 mg of crystalline bovine serum albumin per milliliter of enzyme solution. The solution is also made 0.02% with Triton X-100. The enzyme is pelleted by centrifuging at 45,000 rpm (204,000 g, max) in a 60 Ti rotor for 3 hr. The amber pellet is gently dissolved in 25 mM phosphate, pH 6.9, containing 1 mM NAD$^+$, 0.1 mM thiamin pyrophosphate, and 6 mM dithiothreitol. After equilibrating for 2 hr, the redissolved enzyme is clarified by centrifuging for 30 min at 30,000 g, and the supernatant is designated the ultracentrifuge fraction (fraction III).

Step 4. Polyethylene Glycol Fractionation. The ultracentrifuge fraction is warmed to 20° and made to 10 mM MgCl$_2$. A 50% (w/v) PEG solution is added dropwise to a final concentration of 5% with respect to PEG. The solution is equilibrated for 15 min, and the precipitate is collected by a 10-min 27,000 g centrifugation. The pellet containing the complex is redissolved in 2.4 ml of 25 mM phosphate, 0.1 mM thiamine pyrophosphate, 1 mM EDTA, and 6 mM dithiothreitol, pH 6.9. The enzyme is clarified by centrifugation and stored on ice or further purified. Purification to this point must be accomplished as rapidly as possible for maximum yield and stability.

Step 5. Glycerol Gradient. The pyruvate dehydrogenase complex is layered on 38-ml 10 to 40% glycerol gradients made in the phosphate, thiamin pyrophosphate, NAD$^+$, and dithiothreitol buffer. The gradients are developed using an SW-27 rotor at 25,000 rpm for 8 hr. The active fractions are pooled and precipitated with PEG as in step 4. The enzyme is stored at 4° for short periods.

Typical purifications from cauliflower and broccoli are summarized in the table.

Properties of the Enzyme from Broccoli[6]

The broccoli pyruvate dehydrogenase complex is also very large, with $s_{20,w}$ of 59.3. The purified enzyme is reasonably stable at 4° for up to 2 weeks. However, freezing inactivates the enzyme. Sulfhydryl protectants

PURIFICATION OF PYRUVATE DEHYDROGENASE COMPLEX FROM
BROCCOLI AND CAULIFLOWER

Step and fraction	Total volume (ml)	Activity (units)[a]	Protein (mg)	Specific activity[b]	Recovery (%)
Cauliflower					
1. Crude extract	70	112	1600	0.07	100
Centrifuged extract	180	93	664	0.14	82
2. Ultracentrifugation	19	65	106	0.61	58
3. Polyethylene glycol	6	40.2	21	3.21	36
4. Glycerol gradient	2	15.6	2.8	5.4	14
Broccoli[c]					
1. Crude extract	45	50.6	2308	0.02	100
Centrifuged extract	225	65.2	392	0.17	129
2. Potassium sulfate	225	65.0	348	0.19	129
3. Ultracentrifugation	14.5	46.6	37.5	1.2	92
4. Polyethylene glycol	4.3	40.1	15.7	2.5	79
5. Glycerol gradient	2.0	6.3	1.0	6.3	12.

[a] Micromoles of NADH found per minute
[b] Micromoles of NADH found per minute per milligram of protein.
[c] By permission of Rubin and Randall.[6]

are necessary for stability, and the enzyme partially loses the dihydrolipoyl dehydrogenase component during purification. The enzyme has optimum activity at pH 8–8.2.

Cofactors and Specificity.[6] The complex required a divalent cation (Mg^{2+}, Mn^{2+}, or Ca^{2+}), thiamine pyrophosphate, coenzyme A, and NAD^+. $NADP^+$ will not substitute for NAD^+. Pyruvate is the preferred substrate with α-ketobutyrate oxidation at 32% of pyruvate and no activity toward α-ketoglutarate.

Kinetic Properties.[7] The apparent Michaelis constants for the three substrates are 250, 5.4, and 110 μM for pyruvate, coenzyme A, and NAD^+, respectively.

Regulation and Inhibition.[7] NADH and acetyl-CoA are competitive inhibitors with their respective substrates. The complex is most sensitive to the NADH : NAD^+ since the K_i NADH is 13 μM vs K_m NAD^+ of 110 μM. The K_i acetyl-CoA is 19 μM. Acetyl-CoA and NADH are noncompetitive with pyruvate. Hydroxypyruvate is competitive with pyruvate, and glyoxylate is noncompetitive with pyruvate. Other metabolites or analogs are without effect. There is ATP-dependent phosphorylation and deactivation in crude preparations; however, this is lost with purified preparations.

[7] P. M. Rubin, W. L. Zahler, and D. D. Randall, *Arch. Biochem. Biophys.* **188,** 70 (1978).

The magnesium-thiamine pyrophosphate exhibits a K_d of 33.8 μM, and its rapid dissociation may be involved in regulation.

Pyruvate Dehydrogenase Complex in Other Plant Tissues

The complex has been isolated from mitochondria from pea epicotyls,[8] castor bean endosperm,[9,10] spinach leaves,[11] and pea leaves.[12] The complex in leaf tissues undergoes covalent modification by phosphorylation (deactivation) and dephosphorylation (activation).[11,12] The deactivation is inhibited by pyruvate, ADP, and dichloroacetate, whereas fluoride inhibits the activation.[12] The first nonmitochondrial location of the complex in eukaryotes is found in chloroplasts[13] and developing endosperm plastids,[8,9] where it probably functions to provide two carbon units for fatty acid biosynthesis.[13]

[8] E. E. Reid, P. Thompson, C. R. Lyttle, and D. T. Dennis, *Plant Physiol.* **59**, 842 (1977).
[9] E. E. Reid, C. R. Lyttle, D. T. Canvin, and D. T. Dennis, *Biochem. Biophys. Res. Commun.* **62**, 42 (1975).
[10] B. J. Rapp and D. D. Randall, *Plant Physiol.* **65**, 314 (1980).
[11] K. P. Rao and D. D. Randall, *Arch. Biochem. Biophys.* **200**, 461 (1980).
[12] D. D. Randall, M. Williams, and B. J. Rapp, *Arch. Biochem. Biophys.* **207**, 437 (1981).
[13] M. Williams and D. D. Randall, *Plant Physiol.* **64**, 1099 (1979).

[70] Pyruvate Dehydrogenase Complex from Pigeon Breast Muscle

By Shuichi Furuta and Takashi Hashimoto

$$\text{Pyruvate} + \text{CoA} + \text{NAD}^+ \rightarrow \text{acetyl-CoA} + \text{CO}_2 + \text{NADH} + \text{H}^+$$

Assay Methods

Principle. Pyruvate dehydrogenase complex activity is determined spectrophotometrically by following the formation of acetyl-CoA or NADH.[1] When the enzyme preparation contains the lactate dehydrogenase activity, the activity of the enzyme complex is determined by measuring the formation of acetyl-CoA by coupling with the arylamine acetyltransferase reaction.

[1] S. Furuta, Y. Shindo, and T. Hashimoto, *J. Biochem.* (*Tokyo*) **81**, 1839 (1977).

Reagents

Potassium phosphate buffer, 50 mM, pH 7.5
Sodium pyruvate, 0.2 M
CoA, 10 mM
MgCl$_2$, 0.1 M
Thiamine pyrophosphate, 0.1 M
2-Mercaptoethanol, 2 M
NAD$^+$, 10 mM
p-Nitroaniline, 4 mM
Lactate dehydrogenase (pig heart; Boehringer), 5 mg/ml
Lipoamide dehydrogenase (pig heart; Boehringer), 5 mg/ml
Arylamine acetyltransferase prepared from pigeon liver.[2]

Preparation of Arylamine Transferase. Livers can be kept at $-20°$ for 6 months. Frozen livers (80 g, approximately 8 g per adult bird) are homogenized with 240 ml of ice-cold 1 mM EDTA, pH 7.5 in a Waring blender for 1 min at top speed. The homogenate is centrifuged at 8000 g for 20 min. When the supernatant (210 ml) starts to freeze in the cooling mixture (NaCl–ice), 168 ml of cold acetone are added slowly with stirring. The temperature is kept at 0 to $-1°$ during the addition of half volume of acetone and at $-3°$ during the addition of the remaining acetone. The precipitate is discarded by centrifugation at $-5°$. To the supernatant 420 ml of cold acetone are added slowly with stirring to keep below $-5°$. The precipitate is collected by centrifugation at $-5°$ and dissolved in 2 ml of 10 mM potassium phosphate buffer, pH 8.0. The fraction is dialyzed against 500 ml of the same buffer for a day with three changes. An equal volume of glycerol is added to the dialyzed enzyme preparation. The preparation is stable at $-20°$ for 2 months. From 80 g of liver, about 2 units of the enzyme are obtained. The activity of arylamine acetyltransferase is assayed in a reaction mixture containing the following components in a final volume of 1 ml: 100 mM Tris-HCl, pH 8.0, 50 mM 2-mercaptoethanol, 0.5 mM p-nitroacetanilide, 50 μM aniline, 10 mM EDTA, pH 8.0, and the enzyme. The reaction is carried out at 25°. The increase in the absorbance at 405 nm is recorded. The molar absorption coefficient of p-nitroaniline at 405 nm is 12,000.

Procedure: Method A. The reaction mixture contains 25 mM potassium phosphate buffer, pH 8.0, 20 mM sodium pyruvate, 0.1 mM CoA, 1 mM MgCl$_2$, 1 mM thiamine pyrophosphate, 2 mM 2-mercaptoethanol, 0.5 mM NAD$^+$, 1 mM p-nitroaniline, 25 μg of lactate dehydrogenase, 10 milliunits of arylamine acetyltransferase, 25 μg of lipoamide dehydrogenase, and the enzyme fraction (up to 3 milliunits) in a final volume of 1 ml. The

[2] O. H. Wieland, C. Patzelt, and G. Löffler, *Eur. J. Biochem.* **26**, 426 (1972).

reaction is started by addition of the enzyme complex and carried out at 25°. The acetylation of p-nitroaniline is followed by the decrease in the absorbance at 405 nm. The molar absorption coefficient of p-nitroaniline at 405 nm is 12,000.

Procedure: Method B. The activity of the purified enzyme complex is determined by measuring the formation of NADH at 340 nm. The reaction mixture contains 25 mM potassium phosphate buffer, pH 7.5, 2 mM sodium pyruvate, 0.1 mM CoA, 1 mM MgCl$_2$, 1 mM thiamine pyrophosphate, 2 mM 2-mercaptoethanol, 0.5 mM NAD$^+$, 25 μg of lipoamide dehydrogenase, and the enzyme fraction in a final volume of 1 ml. The reaction is carried out at 25°.

Units. One unit of pyruvate dehydrogenase complex activity is defined as the amount that catalyzes the acetylation of 1 μmol of p-nitroaniline or the formation of 1 μmol of NADH per minute under the assay conditions described. Essentially identical activities are measured by both methods. Specific activity is expressed as units per milligram of protein. Protein is determined by the method of Lowry *et al.*[3] using bovine serum albumin as a standard.

Comments. Lipoamide dehydrogenase (EC 1.6.4.3), one of the component enzymes of the pyruvate dehydrogenase complex, is not firmly bound in the complex. Therefore, the addition of lipoamide dehydrogenase to the reaction mixture is necessary to get the maximal activity of the enzyme complex. Lactate dehydrogenase activity, which interferes with the assay by Method B, is not removed during the early stages of the purification procedure. The complete removal of the lactate dehydrogenase is attained at the final step. To monitor the activity of the enzyme complex during the purification after the first acid precipitation at pH 5.75 at step 4, which is described below, Method B can be used to obtain a rough estimate of the activity by following the initial rate of NADH formation (up to 30 sec).

Purification Procedure

All operations are carried out at 4°. All potassium phosphate buffers employed contain 1 mM EDTA and 5 mM 2-mercaptoethanol.

Step 1. Washing of Muscles. Pigeons are purchased from a local bird shop. Breast muscles (200 g, approximately 60 g per adult bird) are minced with a meat grinder and suspended in 1000 ml of ice-cold 1 mM EDTA, pH 7.5, with gentle stirring. After standing for 10 min, the super-

[3] O. H. Lowry, N. J. Rosebrough, A. L. Farr, and R. J. Randall, *J. Biol. Chem.* **193**, 265 (1951).

natant is decanted. The residue is squeezed through 2–4 sheets of gauze. The squeezed residue is further washed three times with 1000 ml of 1 mM EDTA, pH 7.5, as described above. The washed muscle pulp is frozen at $-20°$ for up to 3 months.

Step 2. Acid Precipitation at pH 5.4. The washed frozen muscle pulp is thawed and homogenized in a Waring blender at top speed for 2 min with 800 ml of 10 mM potassium phosphate buffer, pH 7.5. The homogenate is centrifuged at 3000 g for 10 min. The supernatant (extract) is adjusted to pH 5.4 with addition of 1 M acetic acid with stirring and then centrifuged at 8000 g for 10 min. The precipitate is dissolved in 50 ml of 50 mM potassium phosphate buffer, pH 7.5 (pH 5.4 fraction). All of the precipitate collected by centrifugation is easily dispersed in the buffer by means of a glass homogenizer with a Teflon pestle.

Step 3. Polyethylene Glycol Fractionation. The fraction is adjusted to pH 6.6 with addition of 1 M acetic acid and 0.1 volume of 25% (w/v) polyethylene glycol (M_r 7800–9000) solution is added with stirring. After 15 min, the suspension is centrifuged at 8000 g for 10 min. To the supernatant is added 0.1 volume of 25% polyethylene glycol and 0.01 volume of 1 M MgCl$_2$ (each based on the volume prior to addition of polyethylene glycol), and the mixture is stirred for 15 min. The suspension is centrifuged at 8000 g for 10 min. The precipitate is dispersed in 20 ml of 50 mM potassium phosphate buffer, pH 7.5. The fraction is adjusted to pH 6.6 with 1 M acetic acid and treated with 0.12 volume of 25% polyethylene glycol. The suspension is centrifuged at 8000 g for 10 min, and 0.08 volume of 25% polyethylene glycol and 0.01 volume of 1 M MgCl$_2$ are added to the supernatant. The precipitate is collected by centrifugation at 8000 g for 10 min and dissolved in 5 ml of 50 mM potassium phosphate buffer, pH 7.0 (polyethylene glycol fraction). Most of the 2-oxoglutarate dehydrogenase complex activity[4] is precipitated at a lower concentration of polyethylene glycol at this step. The complete removal of this activity is accomplished by the next acid precipitation. 2-Oxoglutarate dehydrogenase complex can be purified from the fraction precipitated by 0.1 volume of 25% polyethylene glycol.[1]

Step 4. Acid Precipitation and Ultracentrifugation. The pH of the fraction is carefully adjusted to pH 5.75 with gentle stirring by dropwise addition of 1 M acetic acid. The sediment is removed by centrifugation at 8000 g for 5 min. This procedure is important to precipitate the 2-oxoglutarate dehydrogenase complex completely. However, the pyruvate dehydrogenase complex starts to precipitate at pH 5.75. Activities of

---

[4] The activity of the 2-oxoglutarate dehydrogenase complex is determined by Method B except that 2 mM potassium 2-oxoglutarate is substituted for sodium pyruvate in a reaction mixture.

PURIFICATION OF PYRUVATE DEHYDROGENASE COMPLEX FROM
PIGEON BREAST MUSCLE

Fraction	Volume (ml)	Total activity (units)	Protein (mg)	Specific activity (units/mg of protein)
Extract	600	156	4240	0.037
pH 5.4 precipitate	56	165	1770	0.093
Polyethylene glycol	5	55	86	0.64
First ultracentrifugation	2	44	25	1.76
Second ultracentrifugation	2	40	17.6	2.27

the two enzyme complexes should be assayed to follow the recovery of the pyruvate dehydrogenase complex and the 2-oxoglutarate dehydrogenase complex. When the recovery of the pyruvate dehydrogenase complex in the supernatant is low, the sediment and the supernatant are combined and homogenized. The suspension is adjusted to pH 7.0, by addition of 0.1 M KOH and treated again at a slightly higher pH than 5.75.

The pyruvate dehydrogenase complex is precipitated by adjusting the supernatant to pH 5.4. The precipitate is collected by centrifugation and dissolved in 2 ml of 50 mM potassium phosphate buffer, pH 7.0. This fraction is centrifuged at 144,000 g for 60 min. The precipitate collected after ultracentrifugation is transparent and yellow. The brown material usually seen at the bottom of the precipitate is an impurity, and it is largely insoluble in the following treatment. The precipitate is dispersed in 2 ml of 50 mM potassium phosphate buffer, pH 7.0, and centrifuged at 8000 g for 10 min to remove insoluble materials (first ultracentrifugal fraction). The fraction is treated again as described above. The enzyme complex collected by ultracentrifugation is dispersed in 2 ml of 20 mM potassium phosphate buffer, pH 7.0. Insoluble material is discarded by centrifugation at 8000 g for 10 min (second ultracentrifugal fraction).

The preparation is pale yellow and can be stored in the frozen state for a month without significant loss in activity.

A summary of the purification procedure is given in the table. The purity of the final preparation is checked by sodium dodecyl sulfate polyacrylamide gel electrophoresis.

Properties

Purity and Structure. Analytical ultracentrifuge shows one protein peak with the sedimentation coefficients of $s_{20,w}^0 = 77$ S.[1] The electropherogram of the enzyme complex in sodium dodecyl sulfate (SDS)–polyacrylamide gel gives four bands which from top to bottom correspond to lipoate acetyltransferase (EC 2.3.1.12), lipoamide dehydrogenase, α- and β-subunit of pyruvate dehydrogenase (EC 1.2.4.1).[1]

Low Content of Lipoamide Dehydrogenase in the Enzyme Complex. Preparation. The purified preparation of the enzyme complex contains a lower amount of lipoamide dehydrogenase than does the pig heart enzyme complex[5] with respect to its activity and densitometric assay of the component enzymes after SDS polyacrylamide gel electrophoresis.[1] The activity of the purified preparation of pigeon breast muscle pyruvate dehydrogenase complex is increased about 5-fold by addition of lipoamide dehydrogenase to the reaction mixture.[1] All attempts to obtain the enzyme complex with a higher content of lipoamide dehydrogenase by the reconstitution procedure have failed.

pH. The optimum pH of the activity of the enzyme complex is 7.7.

Effects of Thiamine Pyrophosphate and Mg^{2+}. The activity of purified enzyme complex is completely dependent on the presence of thiamine pyrophosphate. The concentration of thiamine pyrophosphate giving half-maximal activity is lowered in the presence of Mg^{2+}, but not Ca^{2+}. In the presence of 1 mM $MgCl_2$, the value for thiamine pyrophosphate is less than 10 μM.

Effects of Substrates. The Michaelis constant of each substrate is determined at saturating fixed levels of the other two substrates as follows: pyruvate, 17 μM; CoA, 27 μM; NAD$^+$, 200 μM

Inhibitors. Acetyl-CoA (K_i = 18 μM) is competitive with respect to CoA, and NADH (K_i = 30 μM) is competitive with NAD$^+$. Kynurenate (K_i = 0.15 mM) and xanthurenate (K_i = 0.2 mM) are inhibitory. They inhibit lipoamide dehydrogenase, but not the other two component enzymes.

Phosphorylation and Dephosphorylation. The enzyme complex is converted into a phosphorylated (inactive) form in the presence of MgATP^{2-} by kinase, which tightly bound to the purified enzyme complex.[6] The phosphorylated (inactive) enzyme complex is reactivated with liberation of P$_i$ in the presence of both Mg^{2+} and Ca^{2+} by the phosphatase, which is purified from pigeon breast muscle according to the method of Siess and Wieland.[7]

Resolution of the Enzyme Complex. The three component enzymes of the enzyme complex are prepared from the purified preparation of the complex.[8] The complex is first resolved into lipoate acetyltransferase–pyruvate dehydrogenase subcomplex and lipoamide dehydrogenase by

[5] R. H. Cooper, P. J. Randle, and R. M. Denton, *Biochem. J.* **143**, 625 (1974).

[6] S. Furuta, Y. Shindo, and T. Hashimoto, unpublished observation, 1977.

[7] E. A. Siess and O. H. Wieland, *Eur. J. Biochem.* **26**, 96 (1972).

[8] S. Furuta, *J. Biochem.* (*Tokyo*) **86**, 183 (1979).

the urea treatment as described by Hayakawa *et al.*[9] Resolution of lipoate acetyltransferase–pyruvate dehydrogenase subcomplex is performed according to the method of Linn *et al.*[10]

[9] T. Hayakawa, T. Kanzaki, T. Kitamura, Y. Fukuyoshi, Y. Sakurai, K. Koike, T. Suematsu, and M. Koike, *J. Biol. Chem.* **244**, 3660 (1969).
[10] T. C. Linn, J. W. Pelley, F. H. Pettit, F. Hucho, D. D. Randall, and L. J. Reed, *Arch. Biochem. Biophys.* **148**, 327 (1972).

[71] Pyruvate Dehydrogenase Complex from *Hansenula*

By T. HARADA and T. HIRABAYASHI

Assay Method

Principle. The multienzyme complexes, collectively called pyruvate dehydrogenase complexes (PDC) are resolved into three component enzymes, i.e., pyruvate dehydrogenase (EC 1.2.4.1), dihydrolipoyl transacetylase (EC 2.3.1.12), and dihydrolipoyl dehydrogenase (EC 1.6.4.3). These component enzymes catalyze a series of five reactions that constitute the following overall reaction.[1]

$$\text{Pyruvate} + \text{CoA} + \text{NAD}^+ \rightarrow \text{acetyl-S-CoA} + CO_2 + \text{NADH} + H^+$$

The amount of NADH formed is proportional to the concentration of the multienzyme complex. This chapter describes the purification and properties of the enzyme complex from *Hansenula miso* IFO 0146.[2]

Reagents

Tris-HCl buffer, 0.3 M, pH 8.0
$MgCl_2$, 0.01 M
CoA, 8 mM
L-Cysteine, 0.2 M
NAD$^+$, 10 mM
TPP, 10 mM
Sodium pyruvate, 0.1 M

Procedure. The overall oxidation of pyruvate is determined essentially as described by Schwartz *et al.*[3] The reaction mixture contains 0.5 ml of

[1] T. Hayakawa, T. Kanzaki, T. Kitamura, Y. Fukuyoshi, Y. Sakurai, K. Koike, T. Suematsu, and M. Koike, *J. Biol. Chem.* **244**, 3660 (1969).
[2] T. Hirabayashi and T. Harada, *J. Biochem.* (*Tokyo*) **71**, 797 (1972).
[3] E. R. Schwartz, O. O. Lynn, and L. J. Reed, *Biochem. Biophys. Res. Commun.* **31**, 495 (1968).

Tris-HCl buffer, 0.05 ml of $MgCl_2$, 0.01 ml of CoA, 0.025 ml of L-cysteine, 0.03 ml of NAD^+, 0.05 ml of TPP, and 0.05 ml of sodium pyruvate in a final volume of 2 ml. The reaction is initiated at 25° by addition of the enzyme solution, and the rate of formation of NADH is followed at 340 nm using a Hitachi Model 124 spectrophotometer in conjunction with a Hitachi Model QPD 34 recorder.

Definition of Unit and Specific Activity. One unit of PDC is defined as the amount of enzyme that catalyzes the oxidation of 1 μmol of pyruvate per minute. Specific activity is expressed in terms of units per milligram of protein. Protein is determined by the biuret method, using crystalline bovine serum albumin as standard.[4]

Growth of Organism. The enzyme is obtained from *Hansenula miso* IFO (Institute for Fermentation, Osaka) 0146.[5] This species is a synonym of *Hansenula anomala.* The organism is grown in synthetic media as described below.

The medium contains 2 g of glucose, 400 mg of $(NH_4)_2HPO_4$, 100 mg of KH_2PO_4, 50 mg of $MgSO_4 \cdot 7 H_2O$, 1 mg each of NaCl, $MnCl_2 \cdot 4 H_2O$, and $FeCl_3 \cdot 6 H_2O$, and distilled water to a final volume of 100 ml. The medium is adjusted to pH 7.2 with 1 N HCl. A 500-ml conical flask containing 95 ml of the medium is inoculated with 5 ml of a seed culture that was grown in the same medium. The inoculated mixture is shaken reciprocally at 120 strokes per minute with an amplitude of 7 cm at 30° for 2 days. After cultivation, the cells are harvested by centrifugation at 10,000 g for 15 min and washed twice with 0.02 M potassium phosphate buffer, pH 7.0, at 5°. The pyruvate oxidizing activity of cell-free extract from this organism is about 20 times and 3 times higher than those from bakers' yeast and *Candida albicans* IFO 0583, respectively.

Purification Procedure

All operations are carried out at 0–5°, and solutions are prepared with deionized water. The amount of wet cells used for each purification experiment is 500 g; this is obtained from about 200 liters of culture broth.

Step 1. Preparation of Cell-Free Extract. The cells washed twice with potassium phosphate buffer are disrupted in a Braun cell homogenizer Model MSK as follows. Ten grams of cells are suspended in 10 ml of cold 0.02 M potassium phosphate buffer (pH 7.0) and cooled. The suspension is transferred into a cold 75-ml flask containing 40 g of glass beads. The chamber and flask are cooled for a few seconds with gaseous CO_2 and then

---

[4] E. Layne, this series, Vol. 3, p. 450.
[5] This organism is obtainable from the American Type Culture Collection as strain ATCC 20211.

agitated for 2 min under cooling with CO_2 at 4000 cycles per minute. The fluid is separated from the beads by decantation and centrifuged for 15 min at 2000 g. Triton X-100 is added to the turbid supernatant (final concentration, 2%), and the mixture is stirred for 1 hr in an ice bath. Then the cell-free extract is obtained by centrifugation for 20 min at 20,000 g.

Step 2. Protamine Sulfate Fractionation. The cell-free extract is diluted with 0.02 M potassium phosphate buffer (pH 7.0) to a protein concentration of approximately 10 mg/ml. The diluted extract is adjusted to pH 6.0 with 1 N acetic acid, and 0.009 volume of 2% protamine sulfate solution (pH 5.0) is added dropwise with stirring. The mixture is stirred for 15 min and then centrifuged for 20 min at 15,000 g. This first precipitate contains most of the α-ketoglutarate dehydrogenase complex (KGDC)[6] activity. To the supernatant fluid, 0.0025 volume of 2% protamine sulfate solution (pH 5.0) is added. The mixture is stirred for 15 min, and the precipitate is collected by centrifugation. The activity is eluted from this second precipitate with 50 ml of 0.2 M potassium phosphate buffer (pH 7.0), and the eluate is dialyzed against 0.1 M potassium phosphate buffer (pH 7.0) overnight and centrifuged at 25,000 g for 20 min to remove the precipitate.

Step 3. First Ultracentrifugation. The supernatant fluid obtained in step 2, which is almost free from KGDC, is diluted with 0.1 M potassium phosphate buffer (pH 7.0) to a protein concentration of 5 mg/ml. The diluted supernatant is centrifuged for 2 hr at 144,000 g. The supernatant is discarded. The resultant yellow pellet is suspended in 0.2 M potassium phosphate buffer (pH 7.0) using a Potter–Elvehjem homogenizer, and the homogenate is centrifuged for 20 min at 25,000 g. The precipitate is discarded.

Step 4. Ammonium Sulfate Fractionation. To the yellow supernatant from step 3, solid ammonium sulfate is added with stirring. The mixture is first brought to 30% saturation (162 g/liter at 0°), stirred for 1 hr, and centrifuged at 25,000 g for 20 min. The supernatant fluid is brought to 40% saturation (224 g/liter at 0°) by addition of solid ammonium sulfate, stirred for 1 hr, and centrifuged as before. The precipitate is resuspended in a minimal volume of 0.2 M potassium phosphate buffer (pH 7.0).

Step 5. Second Ultracentrifugation. The solution from step 4 is diluted with 0.2 M potassium phosphate buffer (pH 7.0) to a protein concentration of 5 mg/ml. The diluted solution is centrifuged for 2 hr at 144,000 g. The supernatant is discarded. The pellet is dissolved in 0.2 M potassium phosphate buffer (pH 7.0). Then the solution is centrifuged for 20 min at 25,000 g and the precipitate is discarded. The solution is opalescent with bright yellow fluorescence.

[6] T. Hirabayashi and T. Harada, Biochem. Biophys. Res. Commun. **45**, 1369 (1971).

PURIFICATION OF PYRUVATE DEHYDROGENASE COMPLEX

Step	Total protein (mg)	Total activity (units)	Specific activity (units/mg protein)	Recovery (%)
Crude extract	7040	—	—	—
Eluate of protamine precipitate	762	305	0.4	100
144,000 g, pellet	78	140	1.8	46
Ammonium sulfate precipitate (0.3–0.4 sat.)	29	101	3.5	33
144,000 g pellet	7.2	53	7.4	17

A summary of the data obtained in a typical purification experiment is presented in the table.

Properties

Catalytic Activity. The pH optimum of the overall reaction is approximately 8.2 in the presence of Tris-HCl buffer. This enzyme shows specific activities with pyruvate as substrate of 7.42 units/mg. The specific activities with α-ketobutyrate and α-ketovalerate are 4.41 and 1.01 units/mg, respectively. The PDC activity is completely dependent on CoA, NAD$^+$, and TPP. Omission of Mg^{2+} causes 70% loss of activity. The K_m value for sodium pyruvate is $1.6 \times 10^{-4} M$.

α-Acetolactate and Acetoin Formation. Samples of purified PDC have an ability to form α-acetolactate and acetoin from pyruvate and acetaldehyde. In addition to α-acetolactate, some acetoin is also detected when pyruvate is used as substrate. With acetaldehyde as substrate, a small amount of acetoin is formed, but no α-acetolactate. When both pyruvate and acetaldehyde are added to the medium, the amount of α-acetolactate formed is much less than when pyruvate alone is used, but the amount of acetoin formed is much larger.

Molecular Properties. The sedimentation constant $s^{\circ}_{20,w}$ is 50.7 S. The absorption maximum of ultraviolet spectrum is at 276 nm. It shows a broad shoulder from 400 to 450 nm, with a minor peak at 455 nm. From the spectrum, the complex seems to have a flavin component.

[72] Alcohol Oxidase from *Candida boidinii*

By HERMANN SAHM, HORST SCHÜTTE, and MARIA-REGINA KULA

$$RCH_2OH + O_2 \rightarrow RCHO + H_2O_2$$

Alcohol oxidase or alcohol:oxygen oxidoreductase (EC 1.1.3.13) catalyzes the oxidation of lower primary alcohols to the corresponding aldehydes and H_2O_2. This enzyme was originally isolated by fractional precipitation with polyethylene glycol from the mycelium of an unidentified basidiomycete.[1,2] This enzyme has been detected also in a number of methanol-utilizing yeasts.[3] There can be little doubt that this alcohol oxidase is a key enzyme in the metabolism of methanol, as the enzyme is catalyzing the oxidation of methanol to formaldehyde. The purification and properties of the enzyme, isolated from the yeasts *Kloeckera* sp., *Candida* N-16, *Candida boidinii*, and *Hansenula polymorpha*, have been described.[4–7]

Assay Method

Principle. The enzyme activity is most conveniently measured by determining the color produced from H_2O_2 in a coupled reaction utilizing peroxidase and 2,2'-azino-bis(3-ethylbenzthiazoline 6-sulfonate) (ABTS).[8] The reaction rate is measured by the increase in absorbance at 420 nm ($\epsilon = 43.2$ cm^2/μmol). Alcohol oxidase can also be assayed by determining the rate of oxygen consumption in an oxygen electrode cell (Yellow Springs Instrument Co.)[9] or by measuring the formaldehyde formed as described by Tani *et al.*[4]

Reagents

Potassium phosphate buffer, 0.15 M, adjusted to pH 7.5 (A)
Methanol, 0.4 M

[1] F. W. Janssen and H. W. Ruelius, *Biochim. Biophys. Acta* **151**, 330 (1968).
[2] F. W. Janssen, R. M. Kerwin, and H. W. Ruelius, this seris, Vol. 41, p. 364.
[3] N. Kato, Y. Tani, and K. Ogata, *Agric. Biol. Chem.* **38**, 675 (1974).
[4] Y. Tani, T. Miya, H. Nishikawa, and K. Ogata, *Agric. Biol. Chem.* **36**, 68 (1972).
[5] T. Fujii and K. Tonomura, *Agric. Biol. Chem.* **36**, 2297 (1972).
[6] H. Sahm and F. Wagner, *Eur. J. Biochem.* **36**, 250 (1973).
[7] N. Kato, Y. Omori, Y. Tani, and K. Ogata, *Eur. J. Biochem.* **64**, 341 (1976).
[8] A. W. Wahlefeld, *Quad. Sclavo Diagn. Clin. Lab.* **7**, 232 (1971).
[9] H. Sahm, *Arch. Microbiol.* **105**, 179 (1975).

2.2′-Azinobis(3-ethylbenzthiazoline 6-sulfonate) (ABTS), 6 mM in potassium phosphate buffer (A)

Horseradish peroxidase (EC 1.11.1.7), 10 units per milliliter of potassium phosphate buffer (A)

HCl, 4.0 N

Procedure. In a test tube are added 1.0 ml of ABTS, 1 ml of peroxidase, and 0.8 ml of methanol. The solution is aerated with bubbling air for 1 min, and then the reaction is immediately initiated by addition of 0.2 ml of appropriately diluted enzyme solution. After incubation for 30 min at 30° the reaction is stopped by the addition of 0.2 ml of 4 N HCl. The color produced is measured in a spectrophotometer at 420 nm. The reading is corrected for the blank which is obtained with enzyme and all reagents present except methanol.

Definition of Enzyme Unit and Specific Activity. One unit of enzyme activity is defined as the amount of enzyme producing 1 μmol of H_2O_2 per minute under these assay conditions. Specific activity is defined as the number of units per milligram of protein.

Production of Alcohol Oxidase

Cultures of *Candida boidinii* (ATCC 32 195) may be obtained from the American Type Culture Collection, Rockville, Maryland, and maintained on malt agar slants.[10] The organism is subcultured every month onto fresh slants, grown at 30°, and stored at 4°. Cells are grown in 10-liter or 80-liter fermentors under forced aeration (1 liter of air per culture volume per minute) and stirring (800 rpm) at 30° in a growth medium consisting of 1 g of KH_2PO_4, 2 g of K_2HPO_4, 2 g of $(NH_4)_2SO_4$, 2 g of NH_4NO_3, 1 g of $Na_2HPO_4 \cdot 2 H_2O$, 0.2 g of KCl, 0.2 g of $MgSO_4 \cdot 7 H_2O$, 0.5 mg of H_3BO_3, 0.04 mg of $CuSO_4 \cdot 5 H_2O$, 0.1 mg of KI, 0.2 mg of $FeCl_3 \cdot 6 H_2O$, 0.4 mg of $MnSO_4 \cdot H_2O$, 0.4 mg of $ZnSO_4 \cdot 7 H_2O$, 0.2 mg of ammonium heptamolybdate, 0.1 mg of biotin, 1 mg of thiamine, and 10 ml of methanol dissolved in 1000 ml of distilled water. After sterilizing this medium 100 ml of an inoculum grown on a complex medium (4 g of yeast extract, 10 g of malt extract, and 4 g of glucose in 1000 ml of distilled water) is transferred aseptically into the fermentor. The pH of the medium is maintained at 5.0 as growth proceeds by addition of NaOH. The yeast cells are harvested by centrifugation at the end of the exponential growth phase at a cell density of approximately 3 g dry weight per liter and stored at $-25°$. It is of crucial importance that the yeast be grown on methanol as a carbon and energy source, since the formation of the alcohol oxidase is correlated with growth on methanol.

[10] H. Sahm and F. Wagner, *Arch. Mikrobiol.* **84**, 29 (1972).

Purification Procedure

All operations are carried out in the cold room (4°).

Step 1. Preparation of Crude Extract. Frozen *Candida boidinii* cells (4 kg, wet weight) are thawed at 4° and suspended in 10 m*M* potassium phosphate buffer (pH 7.5) using a Waring blender. The cells are disrupted in a glass bead mill (Dyno-Mill, Bachofen, Basel) at a rotational speed of 2000 rpm, glass beads of a diameter of 0.25–0.50 mm and a flow rate of 5 liter/hr; 85% of the free volume of the grinding chamber is filled with glass beads. The cell suspension is pumped twice through the disintegrator in a single-pass mode. This procedure is sufficient to break the cells, disintegrate the microbodies, and solubilize the alcohol oxidase. The cellular debris are removed by centrifugation at 25,000 g for 1 hr using a Sorvall centrifuge RC-2B and the GSA-Rotor.

Step 2. Streptomycin Sulfate Precipitation. A 10% solution of streptomycin sulfate (in water) is added to the crude extract with constant stirring to give a final concentration of 1%. The pH is checked and adjusted to 7.5 if necessary. The resulting precipitate is removed by centrifugation at 25,000 g for 90 min.

Step 3. DEAE-Cellulose Column Chromatography. A column (15 × 70 cm) is packed with DEAE-cellulose equilibrated with 10 m*M* potassium phosphate buffer (pH 7.5). The enzyme solution obtained in step 2 is applied to the column. After washing the column with 20 liters of equilibration buffer, followed by 20 liters of 50 m*M* and 20 liters of 100 m*M* potassium phosphate buffer, pH 7.5, the enzyme is eluted with a linear gradient between 25 liters of 100 m*M* potassium phosphate buffer (pH 7.5) and 25 liters of the same buffer containing 500 m*M* sodium chloride at a flow rate of 1400 ml/hr. Fractions of 700 ml are collected and tested for enzyme activity. Alcohol oxidase is eluted between 250 m*M* and 350 m*M* sodium chloride. The peak fractions are combined and concentrated by ultrafiltration through an Amicon hollow-fiber cartridge.

PURIFICATION OF ALCOHOL OXIDASE FROM *Candida boidinii*

Fraction	Volume (ml)	Total protein (g)	Total activity (units)	Specific activity (units/mg protein)	Recovery (%)	Purification (fold)
Crude extract	5750	215	95,000	0.44	100	1
Streptomycin sulfate	5500	209	94,000	0.45	99	1.02
DEAE-cellulose	718	21.2	69,000	3.25	73	7.39
Gel filtration (Sephadex G-200)	280	5.3	58,000	10.94	61	24.86

Step 4. Sephadex G-200 Gel Filtration. The concentrated enzyme solution from step 3 is applied to a column (21.5 × 80 cm) packed with Sephadex G-200 and equilibrated with 30 mM potassium phosphate buffer (pH 7.5). Elution of the enzyme is achieved with the same buffer at a flow rate of 400 ml/hr. Fractions of 100 ml are collected and assayed for alcohol oxidase. After it has been concentrated by ultrafiltration, the enzyme is finally dialyzed against 20 mM potassium phosphate buffer (pH 7.5) and stored at 4° or longer at −20°.

A summary of the purification procedure is given in the table.

Properties

Molecular Weight and Subunit Structure. The molecular weight of the alcohol oxidase as determined by the sedimentation equilibrium method is about 600,000. Polyacrylamide gel electrophoresis in the presence of sodium dodecyl sulfate indicated that this enzyme is most probably an octamer composed of eight probably identical subunits (molecular weight 74,000), which are noncovalently associated.[6]

Prosthetic Group. The native enzyme has an absorption spectrum with maxima at 460 nm, 372 nm, and 290 nm. This resembles the absorption spectrum of a flavoprotein. When the enzyme solution is boiled, FAD in the supernatant fluid can be detected by chromatography. The FAD content of the enzyme was calculated as 8 mol per mole of native enzyme.

Specificity. Studies of substrate specificity have shown that the enzyme is not specific for methanol, although it is a key enzyme in the methanol metabolism of this yeast. The alcohol oxidase catalyzes also the oxidation of different short-chain primary aliphatic alcohols such as ethanol, 1-propanol, 1-butanol, 2-propen-1-ol, 2-buten-1-ol, 2-chloroethanol, and also formaldehyde; but it is inactive toward alcohols of alkyl-chain lengths longer than C_5, secondary or tertiary alcohols, and aromatic alcohols. However, the K_m values indicate that methanol is the best substrate for this enzyme; the affinity of alcohol oxidase for alcohols decreases with increasing length of the alkyl chain.

Effect of pH and Temperature. The enzyme has a broad pH optimum from 7.5 to 9.5; below pH 7 the activity falls off rapidly. The alcohol oxidase is stable from pH 7 to 10. The activity of the enzyme is optimal at 30°.

Inhibitors. The enzyme activity is completely inhibited by the sulfhydryl reagent p-chloromercuribenzoic acid (1 mM). Furthermore, the activity is inhibited to 95% by Cu^{2+} at a concentration of 1 mM. No metal ion was found which could stimulate the enzyme activity.

Localization. Cells of *Candida boidinii* grown on methanol contained a cluster of microbodies varying in size from 0.4 to 1.0 μm in diameter, containing crystalloid inclusions.[11] After isolating these microbodies it could be demonstrated that the alcohol oxidase and catalase are the main enzymes present in these microbodies.[12]

Distribution. It seems probable that alcohol oxidase is a common mechanism for the oxidation of methanol in methanol-utilizing yeasts[3,13] and in lignin-degrading basidiomycetes.[2] An alcohol oxidase has been purified also from the brown rot fungus *Poria contigua.*[14] This enzyme is similar to the alcohol oxidases from methanol-utilizing yeasts and is also located in microbodies.[15]

[11] H. Sahm, R. Roggenkamp, F. Wagner, and W. Hinkelmann, *J. Gen. Microbiol.* **88**, 218 (1975).
[12] R. Roggenkamp, H. Sahm, W. Hinkelmann, and F. Wagner, *Eur. J. Biochem.* **59**, 231 (1975).
[13] H. Sahm, *Adv. Biochem. Eng.* **6**, 77 (1977).
[14] S. Bringer, B. Sprey, and H. Sahm, *Eur. J. Biochem.* **101**, 563 (1979).
[15] S. Bringer, H. P. Bochem, B. Sprey, and H. Sahm, *Zentralbl. Bakteriol. Hyg. C* **1**, 193 (1980).

[73] Alcohol Dehydrogenase from Horse Liver, Steroid-Active SS Isozyme

By REGINA PIETRUSZKO

Horse liver alcohol dehydrogenase (EC 1.1.1.1) was first crystallized by Bonnichsen and Wassén in 1948.[1] The purification and crystallization procedure described previously in this series[2] ensured its commercial availability and extensive investigation of its properties. Despite observations of Dalziel[3] and McKinley-McKee and Moss[4] suggesting that preparations of horse liver alcohol dehydrogenase may not be homogeneous, the enzyme was treated during the 1960s and the early 1970s as a single homogeneous component long after definite proof of its heterogeneity was provided.[5,6] This may have been due to the fact that even though the

[1] R. K. Bonnichsen and A. M. Wassén, *Arch. Biochem. Biophys.* **18**, 361 (1948).
[2] R. K. Bonnichsen and N. G. Brink, this series, Vol. 1, p. 495.
[3] K. Dalziel, *Acta Chem. Scand.* **12**, 459 (1958).
[4] J. S. McKinley-McKee and D. W. Moss, *Biochem. J.* **96**, 583 (1965).
[5] R. Pietruszko, A. F. Clark, J. Graves, and H. J. Ringold, *Biochem. Biophys. Res. Commun.* **23**, 526 (1966).
[6] H. Theorell, S. Taniguchi, Å. Åkeson, and L. Skursky, *Biochem. Biophys. Res. Commun.* **24**, 603 (1966).

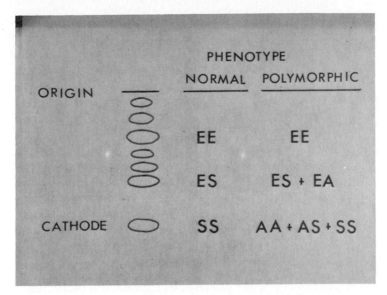

FIG. 1. Schematic diagram of a starch gel electrophoretic pattern representing zones of alcohol dehydrogenase activity and their composition in normal and polymorphic horse livers. The subunit composition of unmarked zones is not known.

classical purification procedure[1,2] produces an isozymically heterogeneous preparation, the EE isozyme, which is totally inactive with steroids, constitutes major part (60–80%) of the preparation.[5] The relative instability and high solubility of the S subunit accounts for its virtual absence from classical preparations of horse liver alcohol dehydrogenase.

The schematic representation of alcohol dehydrogenase isozymes present in horse liver, as visualized by starch gel electrophoresis of crude liver homogenates, is shown in Fig. 1, where nomenclature in terms of subunit composition for the known species is given. The identity of unnamed bands is not known; it has been suggested that they may be conformational variants or deamidation products of the respective parent isozymes.

Alcohol dehydrogenase heterogeneity results from distinct gene loci, located longitudinally on chromosomes, coding for distinct subunits that then dimerize randomly to form isozymes.[7,8] Mutations occasionally occur in the gene loci, thus introducing allelic genes and further heterogeneity due to polymorphic forms. In the horse liver one polymorphic form is known (A subunit) whose homodimer, AA, and the AS heterodimer

[7] M. Smith, D. A. Hopkinson, and H. Harris, *Ann. Hum. Genet.* **34**, 251, (1971).
[8] R. Pietruszko and H. Theorell, *Arch. Biochem. Biophys.* **131**, 288 (1969).

migrate on electrophoresis similarly to the SS isozyme.[9] The catalytic site located on the A subunit has no steroid activity. Thus, there are two steroid-active isozymes present in all horse livers: the ES heterodimer and the SS homodimer; the third steroid-active form, AS, occurs only in some horse livers. The purification described here is concerned with the SS homodimer.[10] Since no method is at present available to distinguish SS isozyme from a mixture of AA, AS, and SS isozymes prior to partial purification via chromatography on DEAE-cellulose, purification of SS isozyme from more than one horse liver should be avoided.

The SS isozyme (MW = 80,000) consists of two S subunits (MW = 40,000)[11] and its extinction coefficient (E_{280} 1 mg/ml solution, 1-cm light path) is 0.46[10] like that of the classical preparations of horse liver alcohol dehydrogenase.[1,2] The S subunit differs from the E subunit in six amino acids in the total sequence.[11] The SS isozyme forms a tight and highly fluorescent binary complex with NADH ($K_D = 0.09 \mu M$,[12]) whose fluorescence intensity is further increased and the dissociation constant further decreased by isobutyramide. This complex is useful for fluorometric titration of the active sites on the enzyme. Some catalytic properties of the SS isozyme with ethanol, acetaldehyde, cyclohexanone, and some steroid substrates are listed in Table I. SS isozyme can also utilize NADP(H) as coenzyme.[13]

Enzyme Assays for Activity Ratio Determinations

SS isozyme is detected during purification procedures by determining the ratio of activity with acetaldehyde (or cyclohexanone) to activity with 3-oxo-5β-androstan-17β-ol which will be referred to as acetaldehyde : steroid or cyclohexanone : steroid activity ratio. The assays used for the determination of these activities contain 1.2 mM acetaldehyde or 0.114 mM 3-oxo-5β-androstan-17β-ol in 0.1 M sodium phosphate buffer, pH 7.0, and 0.17 mM NADH. The 3-oxo-5β-androstan-17β-ol is added to the assay mixture from a 10 mg/ml solution in dioxane (0.010 ml; final volume of 3 ml).

Acetaldehyde is freshly distilled prior to each use, cyclohexanone, (MCB Manufacturing Chemists) is redistilled once and stored at 4°, 3-oxo-5β-androstan-17β-ol (Sigma Chemical Co., St. Louis Missouri) is

[9] R. Pietruszko and C. N. Ryzewski, *Biochem. J.* **153**, 249 (1976).
[10] C. N. Ryzewski and R. Pietruszko, *Arch. Biochem. Biophys.* **183**, 73 (1977).
[11] H. Jörnvall, *Eur. J. Biochem.* **16**, 41 (1970).
[12] H. Theorell, Å. Åkeson, B. Liszka-Kopeć, and C. de Zalenski, *Arch. Biochem. Biophys.* **139**, 241 (1970).
[13] R. Pietruszko, *Biochem. Biophys. Res. Commun.* **54**, 1046 (1973).

TABLE I
SOME CATALYTIC PROPERTIES OF SS ISOZYME

Substrate	pH	NAD (mM)	NADH (mM)	K_m (mM)	V^a	Reference[b]
Ethanol	6.1	0.5	—	118	1.0	(1)
Ethanol	7.0	0.5	—	42	1.2	(1)
Ethanol	7.6	0.5	—	22	1.6	(1)
Ethanol	8.8	0.5	—	16	2.0	(1)
Ethanol	10.0	0.5	—	7	3.6	(1)
Acetaldehyde	7.0	—	0.17	6	32	(2)
Cyclohexanone	7.0	—	0.17	14	5.9	(2)
3-Oxo-5β-androstan-17β-ol	7.0	—	0.17	0.03	2.05	(2)
3β-Hydroxy-5β-androstan-17-one	9.5	0.5	—	0.003	1.1	(3)
3β-Hydroxy-5β-cholanoate	10.0	0.5	—	<0.002	0.42	(4)
3β-Hydroxy-5α-cholanoate	10.0	0.5	—	<0.0004	0.23	(4)

[a] V = turnover number per active site per second.
[b] Key to references: (1) C. N. Ryzewski and R. Pietruszko, *Biochemistry* **21**, 4843 (1980); (2) C. N. Ryzewski and R. Pietruszko, *Arch. Biochem. Biophys.* **183**, 73 (1977); (3) R. Pietruszko and C. N. Ryzewski, *Biochem. J.* **153**, 249 (1976); (4) T. Cronholm, C. Larsen, J. Sjövall, H. Theorell, and Å. Åkeson, *Acta Chem. Scand.* **B29**, 571 (1975).

used as supplied. Dioxane is double distilled as described by Vogel[14] and stored at $-10°$. The specific activities of SS isozyme at 25° in the above conditions are 4.2 IU/mg with 1.2 mM acetaldehyde; 4.2 IU/mg with 12.3 mM cyclohexanone; and 3.0 IU/mg with 0.114 mM 3-oxo-5β-androstan-17β-ol.

Gel Electrophoresis

To identify the SS isozyme by electrophoresis, two kinds of gels are required: starch gels and ionagar gels. Starch gels are made by adding 11.5 g of starch (Otto Hiller Co., Madison, Wisconsin) per 100 ml of 0.025 M Tris-HCl buffer, pH 8.5. Electrophoresis is done on horizontal starch gels (vertical gels give about the same resolution) at 4° overnight (15–20 hr) at 200 V. In the electrode vessels 0.3 M Tris-HCl buffer, pH 8.5, is used.

Ionagar (from Wilson Diagnostics, Glenwood, Illinois) is added at 1.5% w/v to 5 mM phosphate buffer, pH 7.0, and dissolved by heating. About 3.5 ml of the hot ionagar solution is pipetted evenly onto clean microscope slides (25 mm × 75 mm) and left to cool. Slots are made just before sample application by dipping the edge of 3-4 mm strip of dry

[14] A. I. Vogel, "A Textbook of Practical Organic Chemistry," p. 177. Longman Group, London, 1956.

TABLE II

PROTEIN AND ACTIVITY PROFILE OF THE PURIFICATION OF SS ISOENZYME FROM
1 KG OF HORSE LIVER

Step	Total protein (mg)	Total activity[a] (IU/min)		Specific activity[b] (IU/min/mg)		Activity ratio	Purification (fold)
		Cyclo[c]	Steroid[c]	Cyclo[c]	Steroid[c]		
1	82,100	37,760	1940	0.46	0.024	19.2	1.0
2	31,580	28,390	1122	0.90	0.035	25.7	1.5
3	400	178	139	0.44	0.35	1.3	14.6
4	37	105	73	2.8	1.95	1.4	81.0
5	12.5	53	40	4.2	3.2	1.3	133.0

[a] All purification procedures were performed at 4°.
[b] Assay procedures are described in detail under Enzyme Assays.
[c] Cyclo, cyclohexanone (12.3 mM); steroid, 3-oxo-5β-androstan-17β-ol (114 μM).

Whatman No. 3 filter paper into the gel. Electrophoresis is performed in an apparatus built to the specifications of Feinstein.[15] The electrode gels consist of 1.5% bacteriological agar (Matheson, Coleman and Bell) in 5 mM sodium phosphate buffer, pH 7.0, which is also used as the well buffer. Light petroleum (bp 30–60°) or hexane is used in the center compartment as a coolant. All electrophoretic runs are performed in a cold room at 100 V for 45 min. To detect the isozymes, NAD (15 mg, Boehringer Mannheim Co.); nitro blue tetrazolium (10 mg, Sigma); phenazine methosulfate (1 mg, Sigma) are dissolved successively in 25 ml of 25 mM Tris-HCl buffer, pH 8.6. For ethanol substrate staining, ethanol is added to give a concentration of ca. 15 mM; for steroid substrate staining, 3β-hydroxy-5β-androstan-17-one (50 μM final concentration), is added in double-distilled dioxane (80 μl of dioxane per 25 ml of staining solution). When a combined ethanol–steroid stain is required, 3β-hydroxy-5β-androstan-17-one is dissolved in 95% ethanol and 0.1 ml of this solution is used.

Purification Procedure

The purification procedure used is summarized in Table II. Since proper identification of the SS isozyme cannot be done prior to the DEAE chromatography step, it is advisable to run a pilot-scale purification up to this step. This could be especially valuable with large livers, part of which could be left for subsequent purifications.

[15] A. Feinstein, in "Chromatographic and Electrophoretic Techniques" (I. Smith, ed.), 2nd ed., Vol. 2, p. 194. Wiley (Interscience), New York 1968.

Identification of a Suitable Liver. Small samples (1–20 g) of horse livers (obtained from local slaughterhouse) are homogenized in 5 mM sodium phosphate buffer, pH 7.5 (1 ml/g liver), preferably with a good homogenizer employing rapid tissue rupture techniques or sonication and extracted at 4° for 30 min. The homogenate is then centrifuged at 10,000 g for 30 min. The supernatants are electrophoresed on starch gel overnight, and two mirror images (obtained after slicing of starch gel) are stained with ethanol activity stain and steroid activity stain. The pattern obtained with ethanol stain is like that diagrammed in Fig. 1. The livers selected for purification are those showing heaviest staining of the fastest cathodal band with steroid stain.

Step 1. Extraction. Fresh horse livers are frozen immediately and stored at −70° prior to use. All preparative procedures are performed at 4°. Thawed liver (1 kg) is cut into small pieces and minced in a meat grinder, then suspended in 5 mM sodium phosphate buffer, pH 7.5 (1 liter per kilogram of liver) with constant stirring for 4 hr. The mixture is then strained through cheesecloth.

Step 2. Ammonium Sulfate Fractionation. Ammonium sulfate (enzyme grade, Schwarz-Mann) is dissolved in the filtrate, and the slurry is allowed to stand for 60 min before removal of precipitated proteins by centrifugation at 10,000 g for 45 min. Solid ammonium sulfate, 210 g/liter, is then dissolved in the supernatant, and the slurry is allowed to stand for 60 min before recovery of the precipitate by centrifugation at 10,000 g for 45 min.

Step 3. DEAE-Cellulose Chromatography. The final ammonium sulfate precipitate is dissolved in a minimum volume of 20 mM Tris-HCl buffer, pH 9.15 (conductivity, 0.235 mmho/cm at 25°), and dialyzed for 72 hr against at least four changes of 16 times its volume of the same Tris-HCl buffer. The red dialyzate is centrifuged at 10,000 g for 30 min before loading on a column (400 × 3.5 cm packed gel dimensions) of DEAE-cellulose (Whatman DE-11) preequilibrated with 20 mM Tris-HCl buffer. This preequilibration should be done with care, employing both pH and conductivity meters to ensure complete equilibration. The fastest migrating isozyme band (Fig. 1) is separated from all other alcohol dehydrogenase isozymes by this chromatography. If this step is done correctly, subsequent chromatography is solely concerned with removal of extraneous protein. The column is eluted with the same buffer, and fractions (30–40 ml) are monitored either (*a*) by absorbance at 280 nm using an ultraviolet monitor; (*b*) by measuring enzyme activity with acetaldehyde or cylcohexanone and 3-oxo-5β-androstan-17β-ol under the conditions described above and determining their ratios (1.4 for SS isozyme and ca 20 for a mixture of AA, AS, and SS isozymes); or (*c*) by starch gel and ionagar gel electrophoresis. Starch gel electrophoresis should show only a single band (see Fig. 1); if SS isozyme is obtained, ionagar electrophoresis should also show a single band.

If the activity ratio is higher than 1.4, and no contamination with ES and other isozymes is apparent on electrophoresis on starch, and if multiple banding is visualized on ionagar, the purification has to be discontinued and another purification started using a different horse liver.[15a]

Step 4. Affinity Chromatography. The DEAE eluate containing SS isozyme as identified by starch and agar gel electrophoresis is applied on the column containing 5 g (dry weight) of 5'-AMP Sepharose 4B (Pharmacia Fine Chemicals) prewashed with the Tris-HCl buffer used in the DEAE chromatography step. A column of this size readily binds 50 mg of enzyme. The column is washed with 100 ml of the same Tris-HCl buffer followed by a wash with 200 ml of 0.1 M sodium phosphate buffer, pH 7.0.[16] SS isozyme is then eluted with the same buffer containing 0.2 mM NAD and 1.5 mM sodium cholate.

Step 5. CM-Cellulose Chromatography. Combined fractions containing SS isozyme are dialyzed against four changes of 50 times their volume of 5 mM sodium phosphate buffer, pH 6.9, and then applied on a column (1.5 × 40 cm packed gel dimensions) of CM-cellulose (CM-Cellex, Bio-Rad Laboratories) preequilibrated with the same buffer. The column elution procedure is similar to that previously reported[17] except that the column is washed with 10 mM sodium phosphate buffer, pH 6.9 (conductivity, 1.0 mmho/cm at 4°) followed by a gradient of 10 mM sodium phosphate from pH 6.9–7.5, and finally with 10 mM phosphate buffer, pH 7.5, until no activity with cyclohexanone is detected in the fractions. SS isozyme is eluted with the ionic strength gradient (1.0–2.2 mmho/cm) of sodium phosphate at pH 7.5. The enzyme is concentrated by ultrafiltration using a PM-10 membrane (Amicon Corp., Lexington, Massachusetts) and stored in solution in 0.1 M sodium phosphate, pH 7.0, containing 20% v/v glycerol at 4°. When chromatography on CM-cellulose precedes the 5'-AMP Sepharose step, homogeneous SS isozyme is also obtained, but this reversal of steps necessitates additional steps to remove NAD and cholate used during the elution of the affinity gel.

Acknowledgment

Supported by Research Scientist Development Award AA 00046 and a Grant AA 00186 from NIAAA.

[15a] Our preliminary experiments indicate that the frequency of occurrence of the A subunit is ca. 10%.

[16] L. Andersson, H. Jörnvall, Å. Åkeson, and K. Mosbach, Biochim. Biophys. Acta 364, 1 (1974).

[17] U. M. Lutstorf, P. M. Schurch, and J. P. von Wartburg, Eur. J. Biochem. 17, 497 (1970).

[74] Alcohol Dehydrogenase from Horse Liver by Affinity Chromatography

By LARS ANDERSSON and KLAUS MOSBACH

$$RCH_2OH + NAD^+ \rightleftharpoons RCHO + NADH + H^+$$

Alcohol dehydrogenase (alcohol : NAD$^+$ oxidoreductase, EC 1.1.1.1) from horse liver is one of the best studied enzymes.[1] It is a dimeric enzyme (M_r 80,000) containing two zinc ions per enzyme subunit; one zinc is required for catalysis, and the other may have a structural function. The enzyme has three isozymic forms, usually named EE, ES, and SS (normally SS is found in smaller amounts than EE and ES in the liver). The primary structures of the two subunits E and S are very similar, and a difference of only six amino acid residues has been found by peptide mapping.[2] Subfractions of each isozyme are also known[3] as well as a possible polymorphic form of the SS isozyme.[4] Alcohol dehydrogenase reversibly catalyzes the oxidation of a variety of different alcohols by the coenzyme NAD$^+$. In addition, it is known that its isozymes differ in substrate specificity; all three isozymic forms are active with ethanol as a substrate, but only isozymes with an S subunit are active with certain 3β-hydroxysteroids (3α-hydroxysteroids are strong inhibitors of the enzyme). The complete amino acid sequence of the EE isozyme is known as well as its three-dimensional structure determined to a resolution of 2.4 Å by X-ray crystallography. The mechanism of action of the liver enzyme has also been studied in great detail.[1,5]

Horse liver alcohol dehydrogenase has been purified with conventional methods, involving purification steps such as fractionation with ammonium sulfate, ethanol–chloroform treatment, ion-exchange chromatography, and crystallization with ethanol.[6] Since these methods are time consuming and often result in low yields, and since there is great interest for the enzyme, there is need for rapid and efficient methods to isolate alcohol dehydrogenase. In this chapter we describe affinity methods,

[1] C.-I. Brändén, H. Jörnvall, H. Eklund, and B. Furugren, in "The Enzymes" (P. D. Boyer, ed.), 3rd ed., Vol. 11, pp. 103–190. Academic Press, New York, 1975.

[2] H. Jörnvall, Eur. J. Biochem. 16, 41 (1970).

[3] U. M. Lutstorf and J. P. von Wartburg, FEBS Lett. 5, 202 (1969).

[4] R. Pietruszko and C. N. Ryzewski, Biochem. J. 153, 249 (1976).

[5] J. P. Klinman, in "Critical Reviews in Biochemistry" (G. D. Fasman, ed.), Vol. 10, pp. 39–78. CRC Press, Boca Raton, Florida, 1981.

[6] R. K. Bonnichsen and N. G. Brink, this series, Vol. 1, p. 495.

based on the general ligand concept,[7] that are useful for purification of horse liver alcohol dehydrogenase on columns containing an immobilized 5'-AMP analog. We also show that use of magnetic affinity adsorbents obtained after treatment of affinity gels with magnetic ferrofluids could be a valuable tool for quick recovery of the enzyme from crude or viscous extracts.

Assay Methods

Alcohol dehydrogenase activity toward ethanol is measured according to the method of Dalziel.[8] Alcohol dehydrogenase activity toward 5β-dihydrotestosterone is measured at pH 7 in 0.03 M sodium phosphate by following the oxidation of NADH as a decrease in absorbance at 340 nm.[9]

Protein is determined according to the method of Lowry *et al.*[10] Bovine serum albumin is used as a standard, and corrections are made for interference from coenzyme and glutathione in the protein determinations. The specific activity of alcohol dehydrogenase is expressed as units per milligram of enzyme; 1 unit converts 1 μmol of substrate per minute under the assay conditions described.

Purification of Alcohol Dehydrogenase by Affinity Chromatography on 5'-AMP-Sepharose

Preparation of 5'-AMP-Sepharose

Sepharose 4B (10 ml) is suspended in water (5 ml) and activated with CNBr (500 mg), which is dissolved in water (5 ml).[11] After activation for 6–8 min with pH maintained at 11 with 4 M NaOH, the gel is washed with cold 0.1 M NaHCO$_3$, pH 9, and then added to the carbonate buffer (5 ml) containing 200 mg (450 μmol) of N^6-(6-aminohexyl)-5'-AMP.[12] [The 5'-AMP analog, N^6-(6-aminohexyl)adenosine 5'-monophosphate, is commercially available from Sigma Chemical Co., St. Louis, Missouri.] The 5'-AMP analog is allowed to couple overnight at 4° while the suspension is gently agitated on a head-over-end table. After coupling, the gel is washed on a glass filter with, successively, the carbonate buffer, water, 0.5 M sodium chloride and water. The amount of bound ligand is determined from the phosphate content, analyzed after total combustion of freeze-

[7] K. Mosbach, *Adv. Enzymol.* **46**, 205 (1978).
[8] K. Dalziel, *Acta Chem. Scand.* **11**, 397 (1957).
[9] R. Pietruszko and H. Theorell, *Arch. Biochem. Biophys.* **131**, 288 (1969).
[10] O. H. Lowry, N. J. Rosebrough, A. L. Farr, and R. J. Randall, *J. Biol. Chem.* **193**, 265 (1951).
[11] R. Axén, J. Porath, and S. Ernback, *Nature (London)* **214**, 1302 (1967).
[12] H. Guilford, P.-O. Larsson, and K. Mosbach, *Chem. Scr.* **2**, 165 (1972).

dried gel.[13] 5′-AMP-Sepharose used in our studies contained about 6 μmol of bound ligand per milliliter of settled gel. 5′-AMP-Sepharose 4B containing about 2 μmol of bound ligand per milliliter of settled gel is commercially available from Pharmacia Fine Chemicals (Piscataway, New Jersey) and Sigma Chemical Co. (St. Louis, Missouri).

One-Step Purification of Alcohol Dehydrogenase from a Horse Liver Crude Extract

Pyrazole, a specific and potent inhibitor of alcohol dehydrogenase, is known to form a very strong abortive ternary complex with the enzyme in the presence of NAD^+ (K_{EPyr,NAD^+} = 0.1 μM at pH 7).[14] This property of alcohol dehydrogenase to form a ternary complex with NAD^+ and pyrazole has been exploited to desorb affinity-bound alcohol dehydrogenase specifically from the Sepharose-bound general ligand N^6-(6-aminohexyl)-5′-AMP.[15]

Crude Extract Preparation. A frozen horse liver, or parts of it, is thawed, ground in a meat grinder, and extracted at 4° overnight with two volumes of distilled water. After filtration through cheesecloth, centrifugation (23,000 g, 4 hr), and filtration through glass wool, the crude extract is applied to the affinity column.

Affinity Chromatography on 5′-AMP-Sepharose. All chromatographic procedures are performed at 4°. Crude extract (2 ml), containing about 40 mg of protein per milliliter, is applied to a column (1.5 × 0.5 cm) packed with 3 ml of 5′-AMP-Sepharose, preequilibrated with 0.1 M sodium phosphate, pH 7.5, 1 mM in glutathione. The column is then washed, successively, with the buffer (15 ml), 0.3 M NaCl (30 ml), and 0.1 mM NAD^+ (35 ml) in the same buffer; the latter two solutions are applied to remove non-biospecifically bound proteins and weakly bound nucleotide-dependent proteins, respectively. When using 5′-AMP-Sepharose of a lower level of ligand substitution (e.g., 1–2 μmol per milliliter of settled gel), washing with solutions containing high salt concentration should be avoided, since in the presence of many competing enzymes (in the crude extract) showing affinity for the immobilized ligand there is a risk that affinity-bound alcohol dehydrogenase is desorbed from the column by the salt treatment. Alcohol dehydrogenase is eluted specifically from the affinity column by applying 20 ml of the buffer containing 0.1 mM NAD^+ and 1 mM pyrazole to the affinity column (Fig. 1). Prior to enzyme assay, the enzyme solution is dialyzed against the buffer.

[13] G. R. Bartlett, *J. Biol. Chem.* **234**, 466 (1958).
[14] H. Theorell and T. Yonetani, *Biochem. Z.* **338**, 537 (1963).
[15] L. Andersson, H. Jörnvall, Å. Åkeson, and K. Mosbach, *Biochim. Biophys. Acta* **364**, 1 (1974).

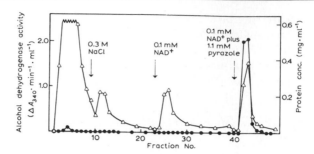

FIG. 1. Purification of alcohol dehydrogenase from a horse liver crude extract on a column containing 5'-AMP-Sepharose; elution was effected by application of a solution containing 0.1 mM NAD$^+$ and 1 mM pyrazole to the affinity column. Washing of the column with a high concentration of salt should not be carried out when affinity gels with a lower ligand concentration (1–2 μmol per milliliter of settled gel) are used. ●——●, Alcohol (ethanol) dehydrogenase activity; △——△, protein concentration.

In this one-step purification of horse liver alcohol dehydrogenase on 5'-AMP-Sepharose a homogeneous enzyme is obtained, as judged by sodium dodecyl sulfate–gel electrophoresis.[16] The enzyme is purified 22 times and the yield is 36%, corresponding to 1.3 mg of alcohol dehydrogenase. (Problems associated with scaling up of the method are not anticipated.) Moreover, elution effected by ternary complex formation involving enzyme, NAD$^+$, and pyrazole has also been successfully employed in purifications of chicken liver[17] and rat liver alcohol dehydrogenase[18] on 5'-AMP-Sepharose.

In the above experiments pyrazole was used as a specific eluent for alcohol dehydrogenase. Alternatively, it can be immobilized to Sepharose and used as a specific adsorbent (as opposed to the general ligand approach described above) for the enzyme. It was found that alcohol dehydrogenase is strongly adsorbed on immobilized 4-[3-(N^6-aminocaproyl)-aminopropyl]pyrazole in the presence of NAD$^+$.[19] In this specific ligand approach, the enzyme was eluted by adding ethanol, propanol, or butanol to the NAD$^+$-containing buffer to form a second strong productive ternary complex competing (in the solution) with the first one. This "double-ternary complex affinity chromatography" has been employed after pre-purification on DEAE-cellulose to isolate and purify alcohol dehydrogenase from human, horse, and rabbit livers. With both procedures pure enzyme is obtained, containing the three isozyme forms found in the liver extract.

[16] K. Weber and M. Osborn, *J. Biol. Chem.* **244**, 4406 (1969).
[17] H. von Bahr-Lindström, L. Andersson, K. Mosbach, and H. Jörnvall, *FEBS Lett.* **89**, 293 (1978).
[18] R. J. S. Duncan, J. E. Kline, and L. Sokoloff, *Biochem. J.* **153**, 561 (1976).
[19] L. G. Lange and B. L. Vallee, *Biochemistry* **15**, 4681 (1976).

Preparative Purification of Homogeneous Steroid-Active Isozyme of Horse Liver Alcohol Dehydrogenase

The main isozymes of horse liver alcohol dehydrogenase have been purified with conventional methods: for example, EE as by Taniguchi *et al.*[20]; ES as by Theorell *et al.*[21]; and SS as by Lutstorf.[22] Separation of a mixture of the two homogeneous isozymes, EE and SS, on a column of Sepharose-bound 5'-AMP has also proved to be possible.[15] Specific elution of affinity-adsorbed alcohol dehydrogenase from the column was effected by ternary complex formation of enzyme with NAD^+ and the steroid inhibitor cholic acid ($3\alpha,7\alpha,12\alpha$-trihydroxy-5β-cholanoic acid); cholic acid is preferentially bound to the isozyme with the highest steroid activity. Based on this successful isozyme separation on 5'-AMP-Sepharose, a method has been developed for preparative purification of the SS form from a horse liver crude extract, prepurified on CM-cellulose, by affinity chromatography on 5'-AMP-Sepharose.[23] Binding of the SS isozyme to the cellulose was necessary in order to lower the amount of enzyme other than alcohol dehydrogenase and to reduce the concentration in the crude extract of EE and ES competing with the SS isozyme for the immobilized ligand.

Pretreatment of Liver Crude Extract on CM-Cellulose and Ammonium Sulfate Precipitation. Frozen horse liver (260 g) is thawed and extracted overnight with two volumes of 10 mM sodium phosphate, pH 7.5, under continuous stirring. After filtration and centrifugation as described above, the extract is dialyzed for 20 hr with one change against a 10-fold volume of 10 mM sodium phosphate, pH 7.5. Preswollen CM-cellulose (200 g, equilibrated with 10 mM sodium phosphate, pH 7.5) is added to this solution. The mixture is stirred for 2 hr and then centrifuged (10,000 g, 10 min). The supernatant is discarded, and the cellulose is washed with equilibrium buffer. After centrifugation the exchanger is stirred continuously with 500 ml of 0.1 M sodium phosphate, pH 7.5, for 2 hr. The mixture is centrifuged (10,000 g, 10 min) and the supernatant is collected. The exchanger is washed once with 0.1 M sodium phosphate, pH 7.5, and the supernatants are combined. Protein is then precipitated by addition of solid ammonium sulfate (47 g per 100 ml of solution) to the enzyme solution (during ammonium sulfate addition the pH is kept at 7–8 by addition of ammonia). After centrifugation the precipitated protein is dis-

[20] S. Taniguchi, H. Theorell, and Å. Åkeson, *Acta Chem. Scand.* **21**, 1903 (1967).
[21] H. Theorell, S. Taniguchi, Å. Åkeson, and L. Skurský, *Biochem. Biophys. Res. Commun.* **5**, 603 (1966).
[22] U. M. Lutstorf, P. M. Schürch, and J.-P. von Wartburg, *Eur. J. Biochem.* **17**, 497 (1970).
[23] L. Andersson, H. Jörnvall, and K. Mosbach, *Anal. Biochem.* **69**, 401 (1975).

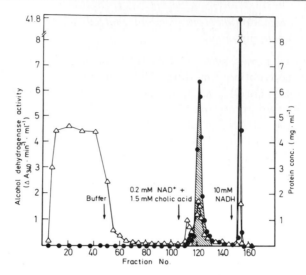

FIG. 2. Affinity chromatography of the SS isozyme of horse liver alcohol dehydrogenase from a CM-cellulose-pretreated horse liver crude extract on a column packed with 5'-AMP-Sepharose; elution was effected by application of a solution containing 0.2 mM NAD$^+$ and 1.5 mM cholic acid to the affinity column. ●——●, Alcohol (ethanol) dehydrogenase activity; △——△, protein concentration.

solved in 0.1 M sodium phosphate, pH 7.5, 1 mM in glutathione, and dialyzed overnight against a 10-fold volume of the same buffer. The dialyzed enzyme is centrifuged (39,000 g, 1 hr) and then subjected to affinity chromatography on 5'-AMP-Sepharose.

Affinity Chromatography on 5'-AMP-Sepharose. All affinity chromatographic procedures are performed at 4°. The affinity column (1.5 × 15 cm) is packed with 22 ml of 5'-AMP-Sepharose and equilibrated with 0.1 M sodium phosphate, pH 7.5, 1 mM in glutathione. The CM-cellulose-treated extract (150 ml) is applied to the affinity column, and the column is then washed with about 10 volumes of the buffer. When the eluate is free of protein, elution of alcohol dehydrogenase is effected by application of buffer (150 ml) containing 0.2 mM NAD$^+$ and 1.5 mM cholic acid to the affinity column, and fractions containing alcohol dehydrogenase activity are pooled (55 ml). This affinity step yields the SS isozyme, which is slightly contaminated by other proteins and isozymes as judged by sodium dodecyl sulfate–gel electrophoresis[16] and agarose-gel electrophoresis[15] (Figs. 2 and 3). It is worth mentioning that enzyme, affinity-bound to commercially available 5'-AMP-Sepharose, is desorbed from the affinity column at a lower NAD$^+$ concentration (0.1 mM) in the elution buffers; the lower immobilized ligand concentration of the commercial preparation accounts for this change in binding strength.

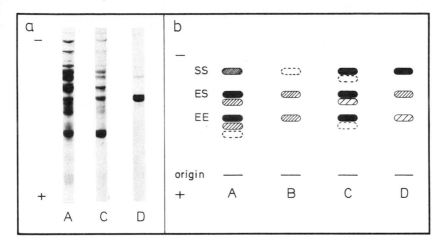

FIG. 3. (a) Sodium dodecyl sulfate–gel electrophoresis. (b) Agarose-gel electrophoresis. A, Horse liver crude extract; B, protein not adsorbed onto CM-cellulose; C, protein applied to the affinity column after the pretreatment step; D, protein from pooled fractions of hatched peak in Fig. 2. Agarose-gel electrophoresis was carried out in 0.05 M Tris-HCl, pH 8.5, and followed by staining for enzymic activity.[15] Hatched areas indicate less intense bands, and dotted outlines indicate hardly visible bands. Forms just below the main isozymes are their subfractions.

The partially purified SS isozyme is rechromatographed on the same affinity column after the column has been regenerated with 10 mM NADH to desorb remaining affinity-bound NAD$^+$-dependent proteins (see Fig. 2) and reequilibrated with buffer. However, prior to affinity chromatography the enzyme solution is filtered through a Sephadex G-25 column to remove NAD$^+$ and cholic acid from the buffer (dilution of the enzyme solution to lower the concentration of NAD$^+$ and cholic acid in the buffer is also possible). Before elution, affinity-bound SS isozyme is washed with buffer solution (270 ml) containing 0.2 mM NAD$^+$ (0.1 mM NAD$^+$ when using commercially available 5′-AMP-Sepharose) to desorb NAD$^+$-dependent proteins other than alcohol dehydrogenase. (Such a prewash of the column with NAD$^+$ in the first affinity step would have resulted in appreciable loss of alcohol dehydrogenase activity due to its weak binding in the presence of many other competing enzymes.) The SS isozyme is eluted, as in the first affinity step, with buffer solution (295 ml) containing 0.2 mM NAD$^+$ and 1.5 mM cholic acid. After dialysis against 0.1 M sodium phosphate, pH 7.5, the enzyme is precipitated by slowly adding solid ammonium sulfate (47 g per 100 ml of buffer) to the enzyme solution while keeping the pH at 7–8 with ammonia. The precipitated enzyme is stored at 4°.

After the second affinity step a completely pure SS isozyme is obtained, as judged by agarose-gel electrophoresis[15] and sodium dodecyl sulfate–gel electrophoresis.[16] The results from the purification are shown in Table I. Noteworthy is the enzyme yield of 18% determined using 5β-dihydrotestosterone as a substrate. The true yield, however, is higher, considering the fact that the ES form present in the crude extract is also active toward this steroid substrate. It deserves mentioning that similarly specific elution of affinity-bound alcohol dehydrogenase from 5'-AMP-Sepharose with NAD^+ and cholic acid was reported later as a purification step for isolation of steroid-active isozyme from so-called S-type horse livers.[24] Alternatively, the EE isozyme from commercially available horse liver alcohol dehydrogenase preparations, containing less steroid-active isozymes (2–5%[1]) than the preparations obtained by the affinity procedure given here, has been obtained by prior elution of affinity-bound steroid-active isozymes using only cholic acid and a low concentration of NAD^+. The EE preparation obtained subsequently at a higher NAD^+ concentration had very little steroid activity and has been used for hybridization studies of horse liver alcohol dehydrogenase on solid phase.[25]

Application of Magnetic Affinity Adsorbents for Purification of Horse Liver Alcohol Dehydrogenase

Magnetic polymers have been successfully applied in various areas of biochemistry, such as immunochemistry[26] and enzyme technology.[27] The inherent advantage in the use of magnetic affinity adsorbents is, in particular, the quick and easy recovery of the adsorbents from especially viscous solutions by applying a magnetic field, obviating, for example, centrifugation and filtration procedures. We have developed a simple procedure for preparation of magnetic affinity adsorbents from polymers that already have been substituted with ligands.[28] According to the method, these affinity adsorbents are "postmagnetized" by their treatment with a magnetic fluid, ferrofluid, which can be characterized as a fluid consisting of ultramicroscopic ferrite particles kept in a colloidal suspension. The ferrite particles (100 Å in diameter) are stabilized by a coating, approximately 25 Å thick, that prevents their sticking together in a magnetic field. Ferrofluids containing ferrite particles are available with either water or hydrocarbon as the carrier from Ferrofluidics Corporation (40 Simon Street, Nashua, New Hampshire, or Ferrox Ltd, Blackhorse House 11 West Way, Botley, Oxford, England).

[24] C. N. Ryzewski and R. Pietruszko, *Arch. Biochem. Biophys.* **183**, 73 (1977).
[25] L. Andersson and K. Mosbach, *Eur. J. Biochem.* **94**, 557 (1979).
[26] J.-L. Guesdo and S. Avrameas, *Immunochemistry* **14**, 443 (1977).
[27] P.-J. Halling and P. Dunnill, *Enzyme Microb. Technol.* **2**, 2 (1980).
[28] K. Mosbach and L. Andersson, *Nature (London)* **270**, 259 (1977).

TABLE I

PURIFICATION OF THE SS ISOZYME OF HORSE LIVER ALCOHOL DEHYDROGENASE BY
AFFINITY CHROMATOGRAPHY ON A SEPHAROSE-BOUND 5'-AMP ANALOG[a]

	Enzyme activity		Total protein (mg)	Specific activity (units/mg protein)	Puri- fication (fold)	Yield (%)
Step	Units/ml	Total units				
Crude extract	0.67	321	23000	0.014	1	100
CM-cellulose	0.13	115	1110	0.10	7.1	36
Ammonium sulfate precipitation	0.67	108	930	0.12	8.6	34
First affinity chromatography	1.0	71	40	1.8	129	22
Second affinity chromatography	0.84	58	29	2.0	143	18

[a] The figures for specific activity, purification, and yield apply to alcohol dehydrogenase activity tested with 5β-dihydrotestosterone as a substrate.

Preparation of Magnetic 5'-AMP-Sepharose

One gram of moist 5'-AMP-Sepharose 4B (Pharmacia), preswollen in 0.1 M sodium phosphate, pH 7.5, and prewashed with water, is packed in a column. 6.5 ml of ferrofluid (base, H_2O; magnetic saturation, 200 g; trademark, A05) are pumped through the column and cycled for 4 hr at a flow rate of 50 ml/hr (room temperature). After this treatment the gel is washed with 1 liter of water, on a glass filter, and incubated overnight at 4° with 100 mg of bovine serum albumin in 0.1 M Tris-HCl, pH 7.6, 5 mM in EDTA and 1 mM in 2-mercaptoethanol. The gel is subsequently washed with 200 ml of 1 M NaCl and 200 ml of the Tris buffer. The resulting dark brown gel beads are now magnetic and ready to use. Posttreatment of the magnetic affinity gels with serum albumin is necessary in order to avoid inactivation of enzyme molecules in the subsequent affinity experiments.

One-Step Batch Purification of Alcohol Dehydrogenase from a Horse Liver Crude Extract on Magnetic 5'-AMP-Sepharose

A horse liver crude extract containing about 45 mg of protein per milliliter is prepared as described above. About 4 ml of the extract are applied to 1 g of moist, magnetic 5'-AMP-Sepharose in a test tube. After incubation (60 min) on a head-over-end table, the gel is settled by applying on the outside of the test tube a permanent magnet with a pull of approximately 5 kg. The solution is decanted, and 10 ml of 0.1 M sodium phosphate, pH 7.5, are added. The mixture is incubated for 15 min, and the "supernatant" is removed. This procedure is repeated three times. The

TABLE II
PURIFICATION OF HORSE LIVER ALCOHOL DEHYDROGENASE BY BATCH AFFINITY
CHROMATOGRAPHY ON MAGNETIC 5′-AMP-SEPHAROSE IN ONE STEP[a]

Step	Specific activity (units/mg protein)	Purification (fold)	Capacity (mg of enzyme/ g of moist gel)
Crude extract	0.08	1	—
Magnetic 5′-AMP-Sepharose	1.70	21	0.7
5′-AMP-Sepharose	1.98	25	1.5

[a] The lower set of values refers to a parallel experiment using nonmagnetic 5′-AMP-Sepharose under otherwise identical batch conditions. Enzyme activity was measured using ethanol as a substrate.

magnetic affinity gel is then washed with, successively, 4 ml of 0.1 mM NAD$^+$ in the phosphate buffer (15 min) and three times with buffer only (3 × 10 ml). Affinity-bound alcohol dehydrogenase is specifically eluted from the affinity adsorbent on incubation of the gel for 30 min with buffer solution (3 ml) containing 0.1 mM NAD$^+$ and 5 mM pyrazole. All these batch affinity chromatography steps as well as the subsequent dialysis of eluted enzyme against the phosphate buffer are carried out at 4°. Table II shows that an alcohol dehydrogenase preparation is obtained with a specific activity comparable with that found for enzyme isolated with nonmagnetic 5′-AMP-Sepharose under batch conditions (although the binding capacity of the former preparations is lower). The magnetic preparations can be used repeatedly, and adsorption of enzyme from crude noncentrifuged preparations is also possible.[28]

Concluding Remarks

In this work, methods have been described to purify horse liver alcohol dehydrogenase by general ligand affinity chromatography on 5′-AMP-Sepharose. Alternative affinity systems for purification and studies of alcohol dehydrogenase have also been reported. Noteworthy is the use of Blue Sepharose.[29] In this case, the blue dye (commercially known as Cibacron Blue F3G-A) is assumed to bind the enzyme by interacting specifically with the AMP binding domain of the enzyme structure.[30] In another study, a Sepharose-bound NADH analog was used to make monoalkylated alcohol dehydrogenase on solid phase.[31] The method in-

[29] S. K. Roy and A. H. Nishikawa, *Biotechnol. Bioeng.* **21**, 775 (1979).
[30] S. T. Thompson, K. H. Cass, and E. Stellwagen, *Proc. Natl. Acad. Sci. U.S.A.* **72**, 669 (1975).
[31] L. Andersson and K. Mosbach, *Eur. J. Biochem.* **94**, 565 (1979).

volves carboxymethylation of an active-site cysteine (Cys-46) in the "free," not affinity-bound, subunit while the enzyme is strongly affinity-bound to the immobilized NADH analog. Finally, it is tempting to speculate that a combination of affinity chromatography and high-performance liquid chromatography (HPLC) in the future will provide very rapid and efficient methods for isolation of enzymes. Such experiments have already been attempted, and this novel affinity technique, named high-performance liquid affinity chromatography (HPLAC), has proved to be useful for very quick separation of alcohol dehydrogenase and lactate dehydrogenase, on an analytical scale, on silica derivatized with the 5'-AMP analog used here.[32]

---

[32] S. Ohlson, L. Hansson, P.-O. Larsson, and K. Mosbach, *FEBS Lett.* **93**, 5 (1978).

[75] Alcohol Dehydrogenase from *Drosophila melanogaster*

By Chi-Yu Lee

$$\text{Alcohol} + \text{NAD}^+ \rightleftharpoons \text{aldehyde(ketone)} + \text{NADH} + \text{H}^+$$

Alcohol dehydrogenase (EC 1.1.1.1; alcohol: NAD$^+$ oxidoreductase) has been well studied from a number of species.[1-3] Among the gene–enzyme systems in *Drosophila melanogaster*, alcohol dehydrogenase is the most extensively studied.[4,5] The gene coding for alcohol dehydrogenase in *D. melanogaster* has been cloned, sequenced, and partially characterized.[6] Biochemistry of alcohol dehydrogenase from *D. melanogaster* has been reported by many workers.[3,7-10] However, the purification of this enzyme was mainly based on multistep conventional procedures.[3,7,11] In view of its importance as an enzyme marker for studies in biochemical genetics and

---

[1] R. K. Bonnichsen and A. M. Wassén, *Arch. Biochem.* **18**, 361 (1948).

[2] L. Andersson, H. Jörnvall, and K. Mosbach, *Anal. Biochem.* **69**, 401 (1975).

[3] W. Sofer and H. Ursprung, *J. Biol. Chem.* **243**, 3110 (1968).

[4] T. H. Day, P. C. Hillier, and B. Clarke, *Biochem. Genet.* **11**, 141 (1974).

[5] C. L. Vigue and F. M. Johnson, *Biochem. Genet.* **9**, 213 (1973).

[6] D. Goldberg, *Proc. Natl. Acad. Sci. U.S.A.* **77**, 5794 (1980).

[7] K. B. Jacobson, J. B. Murphy, and F. C. Hartman, *J. Biol. Chem.* **245**, 1075 (1970).

[8] K. B. Jacobson, J. B. Murphy, J. A. Knopp, and J. R. Ortiz, *Arch. Biochem. Biophys.* **149**, 22 (1972).

[9] J. A. Knopp and K. B. Jacobson, *Arch. Biochem. Biophys.* **149**, 36 (1972).

[10] K. B. Jacobson and P. Pfuderer, *J. Biol. Chem.* **245**, 3938 (1970).

[11] J. I. Elliott, and J. A. Knopp, this series, Vol. 41, p. 374.

environmental mutagenesis,[12] facile affinity chromatographic procedures have been developed for large-scale enzyme purification from frozen adult *Drosophila*.[12,13]

Assay Procedure

Alcohol dehydrogenase activity is assayed spectrophotometrically in 1 ml of assay mixture containing 0.1 M Tris-base (pH 9.8), 1 mM NAD$^+$, and 5% (w/v) 2-propanol at 25°. The fly enzyme exhibits a maximum specific activity and is stable under these conditions. The activity is followed by an increase in absorbance at 340 nm upon the addition of a suitable amount of enzyme. One unit of enzyme catalyzes the formation of 1 μmol of NADH per minute under these conditions.

Adult Drosophila Culture

Adult *Drosophila* (strain Samarkand or *cn, bw, ri, e*) are cultured and collected according to the procedure described in this volume.[14]

Affinity Columns

Since alcohol dehydrogenase from *Drosophila melanogaster* is an NAD$^+$-dependent enzyme, 8-(6-aminohexyl)amino-AMP- or ATP-Sepharose columns[12,13] can be employed as the main step in enzyme purification. In either case, the adsorbed or retarded enzyme can be eluted biospecifically with NAD$^+$ and pyrazole which are known to form a specific abortive ternary complex with this enzyme.[15] The preparation, maintenance, and commercial sources of these two affinity gels are described in the chapter on lactate dehydrogenase isozymes.[16] The estimation of ligand density of these affinity gels is described in this volume.[14]

Purification Procedures

Two different affinity chromatographic procedures are described depending on the types of affinity columns used as the major step of enzyme purification. Unless otherwise specified, all the operations are performed at 4°. The phosphate buffers used are potassium salts.

[12] C.-Y. Lee, D. Charles, and D. Bronson, *J. Biol. Chem.* **254**, 6375 (1979).
[13] A. Leigh-Brown and C.-Y. Lee, *Biochem. J.* **179**, 479 (1979).
[14] D. W. Niesel, G. C. Bewley, C.-Y. Lee, and F. B. Armstrong, this volume [51].
[15] H. Theorell and T. Yonetani, *Biochem. Z.* **338**, 537 (1963).
[16] C.-Y. Lee, J. H. Yuan, and E. Goldberg, this volume [61].

Procedure I

Step 1. Preparation of Drosophila Extract. Crude homogenates are prepared by homogenizing frozen flies (about 5 g) in approximately 5 volumes (v/w) of 10 mM phosphate, pH 6.5, containing 1 mM dithiothreitol, using a Polytron homogenizer (Brinkmann Instrument Co., Westbury, New York). The homogenate is centrifuged at 27,000 g for 20 min, and the supernatant is filtered through glass wool. Ammonium sulfate (28.2 g/100 ml) is added to the clear supernatant with gentle stirring. The slurry is centrifuged at 27,000 g for 20 min, and the precipitate is discarded. More ammonium sulfate (32 g/100 ml of supernatant) is added to the supernatant, followed by centrifugation as above. The precipitate is dissolved in 2–3 ml of homogenization buffer and dialyzed overnight against 1 liter of the same buffer.

Step 2. 8-(6-Aminohexyl)amino-AMP-Sepharose Affinity Chromatography. After dialysis, the extract is concentrated to about 1 ml by ultrafiltration. The concentrated extract is then applied to the top of an 8-(6-aminohexyl)amino-AMP-Sepharose column (1.5 × 28 cm) which has been equilibrated with 10 mM phosphate, pH 6.5. Any dilution of the applied sample should be avoided. The column is then washed with buffer to elute the main protein peak. Fractions of about 2 ml are collected and assayed for alcohol dehydrogenase activity. When the minor peak of activity (≤10%) is detected, elution of the retarded enzyme is initiated by the addition of 0.2 mM NAD$^+$ and 5 mM pyrazole in one column volume of phosphate buffer. Most of the enzyme activity is recovered in two fractions (about 4 ml). The purity of the pooled enzyme is about 70% as judged by sodium dodecyl sulfate–acrylamide gel electrophoresis.

Step 3. DEAE-Sepharose Column Chromatography. The partially purified enzyme preparation is passed directly over a DEAE-Sepharose column (1 × 10 cm) preequilibrated with 10 mM Tris-HCl, pH 8.5. After loading, the column is washed with two column volumes of the same buffer; alcohol dehydrogenase activity is then eluted with a NaCl gradient (0 to 0.1 M, 30 ml by 30 ml) in the same buffer. The eluted homogeneous enzyme has a specific activity of 34.7 units/mg with protein concentration determined by the Lowry method[17] using bovine serum albumin as the standard.

Procedure II

The 8-(6-aminohexyl)amino-ATP-Sepharose column exhibits a much better affinity for *Drosophila* alcohol dehydrogenase than the correspond-

[17] O. H. Lowry, N. J. Rosebrough, A. L. Farr, and R. J. Randall, *J. Biol. Chem.* **193**, 265 (1951).

ing AMP column. The enzyme can be quantitatively adsorbed and eluted from the ATP affinity column with NAD$^+$ and pyrazole. Under proper conditions, the homogeneous enzyme can be obtained from this affinity column step with high recovery.[12]

Details for preparation of the crude extract from 10 g of frozen adult *Drosophila* are essentially the same as those described in step 1 of Procedure 1. The dialyzed extract is loaded on an 8-(6-aminohexyl)amino-ATP-Sepharose column (2.5 × 30 cm) preequilibrated with 10 mM phosphate, pH 6.0, containing 1 mM dithiothreitol and 1 mM EDTA (buffer A). Less than 1% of the enzyme activity appears in the eluent. After washing with 1 liter of buffer A, alcohol dehydrogenase is eluted biospecifically with 0.1 mM NAD$^+$ and 10 mM pyrazole in 150 ml of buffer A. The appearance of alcohol dehydrogenase activity coincides with that of NAD$^+$ and pyrazole in the eluent. In peak fractions (about 60 ml), the eluted enzyme is homogeneous as judged by sodium dodecyl sulfate–acrylamide gel electrophoresis. Based on the fluorescamine method[18] of protein determination (bovine serum albumin as the standard), the homogeneous enzyme has a specific activity of 55 units/mg.

Typical purifications of *Drosophila* alcohol dehydrogenase by these two different procedures are summarized in Table I and Table II, respectively.

Comments on Enzyme Purification

Alcohol dehydrogenase activity is not retained when a crude fly extract is passed over an 8-(6-aminohexyl)amino-AMP-Sepharose column. Washing with loading buffer is sufficient to elute all the enzyme activity. The activity is eluted as two peaks under these conditions, the first of which is detected at the tail of the main protein fractions. The second major activity peak is retarded by the column. When two different volumes of crude homogenate containing the same amount of enzyme activity and total protein were passed over one 40 ml of AMP-Sepharose column,[13] both samples were eluted with the buffer alone. Retardation of alcohol dehydrogenase activity is increased when the sample volume is decreased. This phenomenon has been used as the basis for the purification of *Drosophila* alcohol dehydrogenase on the 8-(6-aminohexyl)amino-AMP-Sepharose column.[19,20]

[18] P. Bohlen, S. Stein, W. Dairman, and S. Udenfriend, *Arch. Biochem. Biophys.* **155**, 213 (1973).

[19] P. Cuatrecasas, *Adv. Enzymol.* **36**, 29 (1972).

[20] C.-Y. Lee, A. J. Leigh-Brown, C. H. Langley, and D. Charles, *J. Solid-Phase Biochem.* **2**, 213 (1978).

TABLE I

PURIFICATION OF *Drosophila* ALCOHOL DEHYDROGENASE BY PROCEDURE I

Step	Total activity (units)	Total protein[a] (units)	Specific activity (units/mg protein)	Purification (fold)	Yield (%)
Crude extract	380	280	1.35	—	—
Ammonium sulfate fractionation	97.9	92.8	1.05	1	25.7
8-(6-Aminohexyl)amino- AMP-Sepharose affinity chromatography	90.0	2.86	31.5	23.3	23.6
DEAE-Sepharose chromatography	60	1.73	34.7	25.7	16

[a] Protein determinations are done by the Lowry method.[17]

TABLE II

PURIFICATION OF *Drosophila* ALCOHOL DEHYDROGENASE BY PROCEDURE II

Step	Total protein[a] (mg)	Total activity (mg)	Specific activity (units/mg protein)	Yield (%)
Crude homogenate	1200	560	0.47	—
Ammonium sulfate fractionation	250	410	1.6	73
8-(6-Aminohexyl)amino- ATP-Sepharose chromatography	6.4	351	55	63

[a] Based on fluorescamine protein determinations.[18]

Drosophila alcohol dehydrogenase exhibits a relatively good affinity for the 8-(6-aminohexyl)amino-ATP-Sepharose column. The enzyme can be readily purified to homogeneity simply by adsorption and biospecific elution,[21] in a manner similar to that of lactate dehydrogenases purified by AMP-affinity chromatography.[22] NAD^+ and pyrazole are able to form an abortive ternary complex specifically with alcohol dehydrogenase.[15] The apparent K_i of pyrazole for *Drosophila* alcohol dehydrogenase is on the order of 1 μM. The use of this potent inhibitor together with NAD^+

[21] C.-Y. Lee, and A. F.-T. Chen, *in* "The Pyridine Nucleotide Coenzymes" (J. Everse, B. M. Anderson, and K.-S. You, eds.), chapter 6, pp. 189–224. Academic Press, New York, 1982.

[22] C.-Y. Lee, D. Lappi, B. Wermuth, J. Everse, and N. O. Kaplan, *Arch. Biochem. Biophys.* **163**, 561 (1974).

permits elution of all retarded or adsorbed alcohol dehydrogenase activity in one or two fractions; this results in substantial enzyme purification in either procedure. After the elution of alcohol dehydrogenase from ATP-Sepharose affinity column with NAD^+ and pyrazole, the native enzyme can be obtained by extensive dialysis with the desired buffer.

Both procedure I and procedure II yield homogeneous preparations of alcohol dehydrogenase from *Drosophila melanogaster* as judged by sodium dodecyl sulfate–acrylamide gel electrophoresis, even though the final specific activity was reported differently, based on different methods of protein determination.[12,13]

[76] Alcohol Dehydrogenase from Acetic Acid Bacteria, Membrane-Bound

By MINORU AMEYAMA and OSAO ADACHI

Primary alcohol + acceptor → aldehyde + reduced acceptor

Alcohol dehydrogenase of acetic acid bacteria acts on a wide range of primary alcohols except for methanol. The enzyme has role as vinegar producer by coupling with aldehyde dehydrogenase (see this volume [82]). The enzyme is localized on the outer surface of cytoplasmic membrane, and oxidation of substrate is linked to its respiratory chain. The enzyme differs from alcohol dehydrogenase (EC 1.1.99.8) of methanol-utilizing bacteria, which without exception requires ammonia for full activity.[1]

Assay Method

Principle. The reaction rate is estimated (a) by spectrophotometry in the presence of 2,6-dichlorophenolindophenol and phenazine methosulfate; (b) by colorimetry in the presence of potassium ferricyanide; (c) by polarography with an oxygen electrode; or (d) by manometry in a conventional Warburg apparatus. The assay method with potassium ferricyanide is employed because of its simplicity for routine assay. This method is described by Wood *et al.*[2] in the assay of D-gluconate dehydrogenase. Some modifications are described below.

[1] C. Anthony and L. J. Zatman, *Biochem. J.* **104**, 953 (1967).
[2] W. A. Wood, R. A. Fetting, and B. C. Hertlein, this series, Vol. 5, p. 287.

METHODS IN ENZYMOLOGY, VOL. 89

Reagents

Potassium ferricyanide, 0.1 M, in distilled water

Ethyl alcohol, 1 M, in distilled water

10% Triton X-100 in distilled water

McIlvaine buffer, pH 4.0 and 5.5. These buffer solutions are prepared by mixing 0.1 M citric acid and 0.2 M disodium phosphate.

Ferric sulfate–Dupanol reagent containing 5 g of $Fe_2(SO_4)_3 \cdot n$ H_2O, 3 g of Dupanol (sodium lauryl sulfate), 95 ml of 85% phosphoric acid, and distilled water to 1 liter

Procedure. The reaction mixture contains 0.1 ml of potassium ferricyanide, 0.5 ml of McIlvaine buffer, enzyme solution, 0.1 ml of Triton X-100, and 0.1 ml of ethyl alcohol in a total volume of 1.0 ml. After preincubation for 5 min at 25°, the reaction is initiated by the addition of potassium ferricyanide and stopped by adding 0.5 ml of ferric sulfate–Dupanol reagent. Then, 3.5 ml of water are added to the reaction mixture. The resulting Prussian blue color is measured at 660 nm after standing for 20 min at 25°. For the assay with the enzyme from *Acetobacter*, McIlvaine buffer, pH 4.0, is to be used, whereas McIlvaine buffer, pH 5.5, is recommended for the assay with the enzyme from *Gluconobacter*.

Definition of Unit and Specific Activity. One unit of enzyme activity catalyzes the oxidation of 1 μmol of ethyl alcohol per minute under the conditions described above: 4.0 absorbance units equal 1 μmol of ethyl alcohol oxidized. Specific activity (units per milligram of protein) is based on the protein estimation by Lowry *et al.*[3] with bovine serum albumin as the standard. A modified method described by Dulley and Grieve[4] is employed for the samples that contain detergent. In this method, sodium lauryl sulfate is added to the alkaline solution.

Source of Enzyme

Microorganisms and Cultures. Gluconobacter suboxydans IFO 12528 and *Acetobacter aceti* IFO 3284 can be obtained from the Institute for Fermentation, Osaka (17-85, Juso-honmachi 2-chome, Yodogawa-ku, Osaka 532, Japan). Stock cultures are maintained on a potato–glycerol slant (see this volume [24]).

Culture medium for *G. suboxydans* consists of sodium D-gluconate (20 g), D-glucose (5 g), glycerol (3 g), yeast extract (3 g), polypeptone (2 g) and 200 ml of potato extract in 1 liter of tap water. The pH of the medium is 6.5. *Gluconobacter suboxydans* is inoculated into 100 ml of the medium in a

---

[3] O. H. Lowry, N. J. Rosebrough, A. L. Farr, and R. J. Randall, *J. Biol. Chem.* **193**, 265 (1951).

[4] J. R. Dulley and P. A. Grieve, *Anal. Biochem.* **64**, 136 (1975).

500-ml shake flask, and the cultivation is carried out at 30° for 24 hr with reciprocal shaking. In the case of large-scale culture, 1.5 liters of inoculum (15 shake flasks) are transferred into 30 liters of the medium in 50-liter jar fermentor. In this case, the potato extract is usually omitted from the medium and cultivation is performed at 30° for 24 hr under vigorous aeration at 30 liters of air per minute with 500 rpm agitation.

The culture medium for A. aceti contains glycerol (20 g), yeast extract (1 g), NH_4Cl (2 g), 500 mg each of KH_2PO_4 and K_2HPO_4, 200 mg of $MgSO_4 \cdot 7 H_2O$, and 10 mg each of $FeSO_4 \cdot 7 H_2O$, NaCl, and $MnSO_4 \cdot 7 H_2O$ in 1 liter of tap water.

The procedures for shake flask culture and jar fermentor are essentially the same as described above for G. suboxydans. The cells are harvested at the late exponential phase by continuous flow centrifugation.

Purification Procedure

All operations are carried out at 0–5° unless otherwise stated. The buffer used throughout is potassium phosphate buffer, pH 6.0, with Triton X-100 (0.1%) and 2-mercaptoethanol (1 mM). The following example uses 350 g of wet cells for G. suboxydans and 100 g for A. aceti.

Preparation from G. suboxydans [5]

Step 1. Preparation of Membrane Fraction. Cells of G. suboxydans are suspended in 0.01 M buffer and passed twice through a French press (American Instrument Co.) at 1000 kg/cm². Intact cells are removed by centrifugation at 5000 g for 10 min. The resulting cell homogenate is centrifuged at 68,000 g for 90 min. The sedimented membrane fraction can be stored at −20° for over 6 months without appreciable loss of activity.

Step 2. Solubilization of Enzyme. The membrane fraction is suspended in 0.01 M buffer, pH 6.0, and the protein concentration is adjusted to 30 mg/ml. To the suspension, 10% Triton X-100 and 2-mercaptoethanol are added to a final concentration of 1% and 1 mM, respectively. The suspension is gently stirred for 3 hr at 0° and centrifuged at 68,000 g for 60 min. A rose-red supernatant is obtained as the solubilized enzyme.

Step 3. DEAE-Sephadex Column Chromatography (I). To the solubilized enzyme solution (560 ml), polyethylene glycol 6000 is added to 20% to precipitate the enzyme. After 30 min of stirring in an ice bath, the enzyme solution is centrifuged at 12,000 g for 20 min. The precipitate is suspended in a small volume of 0.01 M buffer, and the thick suspension is dialyzed overnight against 0.002 M buffer containing 0.1% Triton X-100. The dialyzed solution (120 ml) is applied to a DEAE-Sephadex A-50 col-

[5] O. Adachi, K. Tayama, E. Shinagawa, K. Matsushita, and M. Ameyama, *Agric. Biol. Chem.* **42**, 2045 (1978).

TABLE I
PURIFICATION OF ALCOHOL DEHYDROGENASE OF *Gluconobacter suboxydans*

Fraction	Total protein (mg)	Total activity (units)	Specific activity (units/mg protein)	Yield (%)
Homogenate	32,920	131,700	4	100
Solubilized fraction	5,660	108,200	19	82
DEAE-Sephadex A-50 (I)	2,080	104,000	50	79
DEAE-Sephadex A-50 (II)	550	71,400	130	54
Hydroxyapatite	290	56,500	195	43

umn (5 × 30 cm) which has been equilibrated with 0.002 M buffer containing 0.1% Triton X-100. After washing with 500 ml of the same buffer, the enzyme is eluted with 0.1 M buffer containing 1% Triton X-100. Enzyme solution in dialyzing tubing is concentrated by dehydration by embedding the enzyme in dry polyethylene glycol 6000. The concentrated fraction is then dialyzed thoroughly against 0.002 M buffer containing 0.1% Triton X-100. The insoluble material is removed by centrifugation at 12,000 g for 20 min.

Step 4. DEAE-Sephadex Column Chromatography (II). After adsorption of the enzyme solution into a column of DEAE-Sephadex A-50 (1.5 × 20 cm), the column is first treated with 0.015 M buffer containing 0.05% Triton X-100 and elution of the enzyme is performed by a linear gradient elution composed of 0.015 M and 0.1 M phosphate buffer. Each buffer reservoir contains 500 ml and 0.05% Triton X-100 is added throughout this step. This second step of DEAE-Sephadex A-50 column chromatography is convenient to bring about in removing an impurity which is co-chromatographed with alcohol dehydrogenase in the next step. Pooled enzyme fraction is dialyzed against 0.01 M buffer containing 0.05% Triton X-100 overnight.

Step 5. Hydroxyapatite Column Chromatography. The dialyzed enzyme is applied to a hydroxyapatite column (5 × 5 cm) that has been equilibrated with 0.01 M buffer containing 0.1% Triton X-100. The column is then treated stepwise with 0.02 M, 0.05 M, and 0.1 M buffer containing 0.1% Triton X-100. Much of the activity is eluted with 0.1 M buffer. The peak fractions, which show a constant level of specific activity of over 175 units per milligram protein, are collected. Pooled solution is concentrated in a dialysis bag with solid polyethylene glycol 6000 as above. The dehydrated fraction with a heavy rose-red precipitate is dialyzed thoroughly against 0.05 M buffer containing 0.1% Triton X-100, and insoluble material is removed by centrifugation at 12,000 g for 20 min.

The results of a typical purification are given in Table I. The purified alcohol dehydrogenase, prepared as above, usually has a specific activity of about 190 units per milligram of protein under the standard assay conditions.

Preparation from A. aceti[6]

The procedures for preparation of the membrane fraction (Step 1) and solubilization of the enzyme (Step 2) with Triton X-100 are similar to those for purification of the enzyme from G. *suboxydans* as described above.

Step 3. *DEAE-Sephadex Column Chromatography.* The solubilized and dialyzed enzyme solution is applied to a DEAE-Sephadex A-50 column (4.5 × 30 cm) that has been equilibrated with 0.002 M buffer containing 0.1% Triton X-100. The column is washed with 1.5 liters of the same buffer to remove nonadsorbable materials. The enzyme is eluted with 0.05 M buffer containing 0.1% Triton X-100 and polyethylene glycol 6000 is added to 20% to the pooled fractions (900 ml) to precipitate the enzyme. The precipitate collected by centrifugation at 12,000 g for 20 min is dissolved in a minimum volume of 0.01 M buffer, containing 0.1% Triton X-100 and dialyzed thoroughly against the same buffer. The dialyzed enzyme is applied to the second column of DEAE-Sephadex A-50 (3.5 × 20 cm), which has been equilibrated with the buffer used for dialysis. After washing with 0.015 M buffer, elution is made by a linear gradient elution formed by 500 ml of 0.015 M buffer and 500 ml of 0.06 M buffer. Triton X-100 is supplemented to 0.1% to both buffer solutions. Fractions (15 ml) are collected, and the enzyme is eluted at 300 to 600 ml of eluate. Judging from the elution pattern, the enzyme is eluted at about 0.035 M potassium phosphate, pH 6.0. Pooled enzyme solution (300 ml) is concentrated by a ultrafiltration (Toyo UP-50) to about 75 ml and dialyzed against 0.002 M buffer. Insoluble materials are removed by centrifugation.

Step 4. *Hydroxyapatite Column Chromatography.* The dialyzed enzyme from the preceding step is applied to a column of hydroxyapatite (2.5 × 8 cm) that has been equilibrated with 0.002 M buffer. The column is washed thoroughly with 0.1 M buffer. Elution is effected by a linear gradient using 450 ml of 0.1 M buffer and 450 ml of 0.2 M buffer, and 10-ml fractions are collected. Most of the activity is eluted at 0.12 to 0.14 M of potassium phosphate, pH 6.0. The pooled fractions (300 ml) are dialyzed against 3 liters of 0.1% Triton X-100 to reduce phosphate to about 0.05 M. The dialyzed solution is applied to the second column of hydroxyapatite (2.5 × 4 cm), which has been equilibrated with 0.05 M buffer. The column is washed with 200 ml of 0.08 M buffer, then the enzyme is eluted with 0.12 M buffer. Fractions of specific activity over 120 are pooled and concentrated by the addition of polyethylene glycol 6000 to 20%. The precipitate is dissolved in a minimum volume of 0.05 M buffer containing 0.1%

[6] O. Adachi, E. Miyagawa, E. Shinagawa, K. Matsushita, and M. Ameyama, *Agric. Biol. Chem.* **42**, 233 (1978).

TABLE II
PURIFICATION OF ALCOHOL DEHYDROGENASE OF *Acetobacter aceti*

Fraction	Total protein (mg)	Total activity (units)	Specific activity (units/mg protein)	Yield (%)
Cell homogenate	7510	19,520	2.6	100
Solubilized fraction	3765	19,950	5.3	100
DEAE-Sephadex A-50 (I)	742	17,080	23.0	88
DEAE-Sephadex A-50 (II)	282	15,240	54.0	78
Hydroxyapatite (I)	129	14,250	110.5	73
Hydroxyapatite (II)	57	10,750	188.6	55

Triton X-100 and dialyzed thoroughly against the same buffer. Insoluble materials are removed by centrifugation at 12,000 g for 20 min.

The results of a typical purification are summarized in Table II. The purified enzyme from A. *aceti* usually has a specific activity of about 180 units per milligram of protein under the standard assay conditions.

Crystallization. The purified enzyme (50 mg of protein in 0.01 M potassium phosphate, pH 6.0, containing 0.1% Triton X-100) is applied to a hydroxyapatite column (1 × 2 cm) that has been equilibrated with 0.01 M buffer. After adsorption of the enzyme, the column is washed extensively with 0.01 M buffer to remove Triton X-100 until the absorbance at 280 nm of the eluate is 0.005 or below. The enzyme is eluted with 0.15 M potassium phosphate, pH 6.0, containing 1 mM 2-mercaptoethanol and no Triton X-100, and the solution containing 10–20 mg of protein per milliliter is retained. Through these treatments alcohol dehydrogenase is converted from monomer to dimer. Crystallization is induced by adding ammonium sulfate or polyethylene glycol 6000. The enzyme (dimer) crystallizes at 1.79 M of ammonium sulfate (23.7 g/100 ml). A huge rose-red prism occurs after standing for several days at 4°. Alternatively, the enzyme was crystallized after a few days by gradual addition of polyethylene glycol 6000 to about 12.5%. The crystals are rose-red rods. The enzyme crystallizes as a dimer in both cases.

Properties[5,6]

Homogeneity. The sedimentation pattern of the enzyme shows a single peak with an apparent sedimentation constant of 5.9 S in the presence of Triton X-100. When Triton X-100 is removed, the enzyme also shows a symmetrical sedimentation peak having an apparent sedimentation constant of 9.8 S, indicating that the enzyme is fully converted to the dimer form. Gel filtration also shows a symmetrical elution peak of protein coincident with enzyme activity and an apparent molecular weight of about 150,000.

Absorption Spectra. The purified enzyme from both genera is rose-red and shows a typical cytochrome *c*-like absorption spectrum, since the enzyme is purified as an enzyme complex composed of dehydrogenase protein, cytochrome *c*, and one or two small subunits. Absorption maxima at wavelengths of 553 nm, 522 nm, and 417 nm are observed with the reduced enzyme, and a sole peak at 409 nm is seen with the oxidized enzyme. The heme in the enzyme preparation is almost completely reduced, suggesting that the hemoprotein is quite stable and poorly autooxidizable.

Subunit Composition. The purified enzyme from *G. suboxydans* comprises three components that dissociate in sodium dodecyl sulfate gel electrophoresis to gene components with apparent molecular weights of 85,000, 49,000, and 14,400; the sum of these components is about 150,000. When the purified enzyme is subjected to conventional polyacrylamide gel electrophoresis at pH 8.3 and run at pH 9.4, three protein bands are also stained. When an unstained gel is viewed under fluorescent light, intense fluorescence is observed only with the slowest moving protein band (R_f = 0.14). Activity is detected in this band when the gel is incubated in the medium containing phenazine methosulfate, nitro blue tetrazolium, and ethanol at pH 6.0. When an unstained gel is scanned at 550 nm, a middle protein band (R_f = 0.62) is seen; it is cytochrome *c*. The function of the smallest component (R_f = 0.80) remains unclear.

The purified enzyme from *A. aceti* dissociates into four components having an apparent molecular weight of 63,000 (dehydrogenase), 44,000 (cytochrome *c*), 29,000, and 13,500 and the sum of these gives 149,000 of total molecular weight. The function of the two smaller subunits is unknown.

Heme Content. The purified enzyme from *G. suboxydans* contains 3 mol of heme *c* per mole of enzyme complex when the heme content is calculated from a pyridine hemochrome spectrum. The midpoints of oxidation-reduction potential of the cytochrome is determined to be +340 mV and +260 mV at pH 7.0 (25°) from the results of titration with ferrocyanide in the presence of ferricyanide. Two of three moles of heme *c* are carried by the cytochrome component of molecular weight 49,000. The remaining mole of heme *c* is bound to the dehydrogenase component having a molecular weight of 85,000.

Catalytic Properties. Alcohol dehydrogenase can be assayed *in vitro* in the presence of any one of the following dyes as an electron acceptor: potassium ferricyanide, phenazine methosulfate, nitro blue tetrazolium, or 2,6-dichlorophenolindophenol. NAD, NADP, and oxygen are completely inactive as electron acceptors. Ammonia is not required for enzyme activity, whereas the enzyme from methanol-utilizing bacteria requires ammonia for full activity.[1] Aliphatic primary alcohols having a carbon chain length of 6 or less, except methanol, are rapidly oxidized. Secondary and tertiary alcohols and aromatic alcohols are not oxidized. The apparent Michaelis constant for ethanol is $1.6 \times 10^{-3}\ M$ with both enzyme preparations. Optimum pH of ethanol oxidation and stability against various pH levels are somewhat different: the optimum pH at 6.0 is found with the enzyme from *G. suboxydans* and at 4.0 from *A. aceti*.

Stability. The enzyme at pH 6.0 from both genera can be stored at 4° for several weeks without appreciable loss of enzyme activity when 0.1% Triton X-100 is present. The enzyme from *G. suboxydans* retains full activity from pH 5 to 8. On the other hand, the enzyme form *A. aceti* is stable from pH 4 to 8. The presence of a detergent such as Triton X-100 facilitates enzyme stability, and removal of the detergent causes gradual inactivation. Addition of glycerol, sucrose, or polyethylene glycol stabilizes activity.

Prosthetic Group. The absorption spectrum at 25° shows maxima at 264 nm, 330 nm, and 406 nm. Fluorescence emission at 25° is at 480 nm with excitation at 370 nm. Two excitation peaks are observed, at 370 nm and 260 nm, with emission at 480 nm. These characteristics show the presence of pyrroloquinoline quinone (PQQ)[7] as the prosthetic group; PQQ has been found as the prosthetic group of methanol dehydrogenase of methylotroph and D-glucose dehydrogenase of *Acinetobacter*.[8,9] It is also detected from D-glucose dehydrogenase of *Pseudomonas* (see this volume [24]) and *Gluconobacter*[10] and aldehyde dehydrogenase of acetic acid bacteria (see this volume [82]).

[7] S. A. Salisbury, H. S. Forrest, W. B. T. Crure, and O. Kennard, *Nature (London)* **280**, 843 (1979).
[8] J. A. Duine, J. Frank, Jr., and J. Westerling, *Biochim. Biophys. Acta* **524**, 277 (1978).
[9] J. A. Duine, J. Frank, Jr., and J. K. Van Zeeland, *FEBS Lett.* **108**, 443 (1979).
[10] M. Ameyama, E. Shinagawa, K. Matsushita, and O. Adachi, *Agric. Biol. Chem.* **45**, 851 (1981).

[77] Covalent Enzyme–Coenzyme Complexes of Liver Alcohol Dehydrogenase and NAD

By MATS-OLLE MÅNSSON, PER-OLOF LARSSON, and KLAUS MOSBACH

The advent of immobilized enzymes in the early 1970s raised considerable expectations of an extended use of enzymes both in well-known fields and in exciting new areas. Yet, so far, only processes based on relatively simple hydrolytic reactions have been utilized to an appreciable extent in practice. The more complicated reactions, e.g., those leading to the syn-

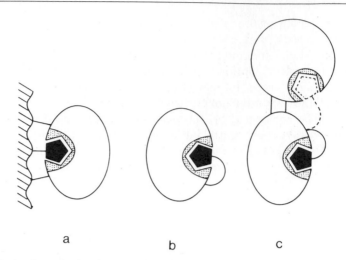

a   b   c

FIG. 1. A schematic drawing of (a) enzyme–coenzyme–agarose complex, (b) enzyme–coenzyme complex, and (c) hypothetical enzyme–coenzyme–enzyme complex.

thesis of compounds instead of hydrolysis, usually require coenzymes, such as NAD and ATP. The high cost and stoichiometric need of these coenzymes have hitherto precluded the use of such enzymes in enzyme reactors. The obstacle is that the coenzymes must be efficiently regenerated and, to allow their use in flow-through systems, also properly retained. These facts have been discussed in many reports on immobilized coenzymes[1] and membrane reactors.[2] Various ways of covalently coupling the coenzyme to both soluble and insoluble supports are available. Thus, when the coenzyme has been attached to a higher molecular weight support it is much easier to retain it, e.g., within an enzyme reactor. The coenzyme can be regenerated, i.e., converted back to its original state, both enzymically and nonenzymically. Regeneration by the nonenzymic methods, however, is less specific, leading to deterioration of the coenzyme, and this limits its usefulness at the present state of development.[3]

To illustrate how the two problems of retention and regeneration can be solved, we have devised two methods for the preparation of complexes between alcohol dehydrogenase and nicotinamide adenine dinucleotide. The first method[4] consists of the immobilization of the reduced NAD analog, N^6-[N-(6-aminohexyl)carbamoylmethyl]NADH and the enzyme liver alcohol dehydrogenase to a support such as activated agarose. This results in the formation of a complex as depicted in Fig. 1a. This method,

[1] K. Mosbach, Adv. Enzymol. **46**, 205 (1978).
[2] C. Wandrey, R. Wichmann, A. F. Bückmann, and M. R. Kula, in "Enzyme Engineering" (H. H. Weetall and G. P. Royer, eds.), Vol. 5, pp. 453–456. Plenum, New York, 1980.
[3] R. D. Schmid, Process Biochem. **14**, 2 (1979).
[4] S. Gestrelius, M.-O. Månsson, and K. Mosbach, Eur. J. Biochem. **57**, 529 (1975).

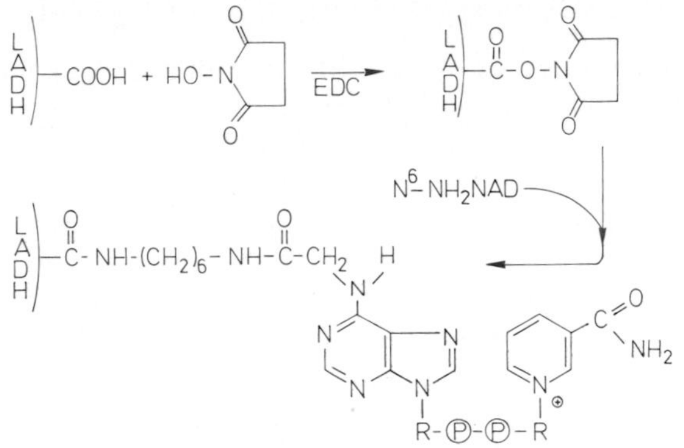

FIG. 2. Reactions involved in the preparation of the enzyme–coenzyme complex. An enzyme carboxyl group is activated with carbodiimide and N-hydroxysuccinimide. The active ester is then allowed to react with the NAD analog to form the enzyme–coenzyme complex. LADH, liver alcohol dehydrogenase. Reproduced, with permission, from Månsson et al.[5]

however, gives a complex that has limitations as far as regeneration is concerned. The second method[5,6] provides more alternatives for regeneration. It is similar to the first method in that the coenzyme is more or less permanently fixed at or near the active site of the enzyme. The difference is that here the solid support has been replaced by a covalent bond between the enzyme and the coenzyme (Fig. 1b).

Synthesis of Covalent Enzyme– Coenzyme Complexes

The NAD analog, N^6-[N-(6-aminohexyl)carbamoylmethyl]NAD (Fig. 2) was synthesized as described elsewhere.[7] The analog is also available from Sigma Chemical Co. (St. Louis, Missouri). Alcohol dehydrogenase from horse liver (Boehringer, Mannheim, West Germany) was purified further[8] in the following way. One milliliter of enzyme suspension (10 mg/ml) was centrifuged at 3000 g for 10 min, and the supernatant was discarded. The enzyme crystals were then dissolved in 1 ml of 0.1 M sodium pyrophosphate buffer, pH 8.7, containing 10 mM dithioerythritol, after which the enzyme was transferred to the desired buffer by gel filtration on a Sephadex G-50 (Pharmacia Fine Chemicals, Uppsala, Sweden) column (1 × 20 cm).

[5] M.-O. Månsson, P.-O. Larsson, and K. Mosbach, Eur. J. Biochem. **86**, 455 (1978).
[6] M.-O. Månsson, P.-O. Larsson, and K. Mosbach, FEBS Lett. **98**, 309 (1979).
[7] M. Lindberg, P.-O. Larsson, and K. Mosbach, Eur. J. Biochem. **40**, 187 (1973).
[8] M. F. Dunn and J. S. Hutchinson, Biochemistry **12**, 4882 (1973).

Liver Alcohol Dehydrogenase–NAD(H)–Agarose Complex

Liver alcohol dehydrogenase and the reduced coenzyme N^6-[N-(6-aminohexyl)carbamoylmethyl]NADH were simultaneously coupled to Sepharose 4B. First, the Sepharose was activated with 75 mg of cyanogen bromide[9] per gram of wet Sepharose. To this end, 4 M NaOH was added dropwise to the stirred suspension of Sepharose in water (2 ml of water per gram of wet Sepharose) in order to keep the pH constant at 11 for 8 min. The activated Sepharose was washed with 0.1 M NaHCO$_3$ and then added to the coupling solution, which consisted of a mixture of the enzyme and the coenzyme in 0.1 M NaHCO$_3$. After coupling had been allowed to proceed overnight at 4°, the Sepharose beads were washed in large volumes of cold 0.1 M NaHCO$_3$, 0.5 M NaCl, and distilled water. The reduced form of the coenzyme was used because it forms a stronger complex with the enzyme than does the oxidized form, the dissociation constants being 0.46 μM and 28 μM, respectively, at pH 8.0.[10]

The desired ratio of agarose bound enzyme to coenzyme can be achieved by varying the ratio between enzyme and coenzyme in the coupling solution (see the table, p. 465). Four milligrams of liver alcohol dehydrogenase (0.1 μmol of subunit) and 2.5 μmol of NADH analog, for example, will give a ratio of 10 NADH per enzyme subunit.

Soluble Liver Alcohol Dehydrogenase–NAD Complex

All the coupling (Fig. 2) and separation steps were carried out at 4°. To 5 mg of liver alcohol dehydrogenase in 1 ml of 0.1 M triethanolamine, pH 7.5, the coenzyme analog in its oxidized form, N^6-[N-(6-aminohexyl)-carbamoylmethyl]NAD, was added to a final concentration of 5 mM. The coupling reagents 1-ethyl-3-(3-dimethylaminopropyl)carbodiimide and N-hydroxysuccinimide were added in a 250-fold excess compared to the concentration of enzyme subunit. After 20 hr of coupling time, the reaction solution was made 0.1 M with respect to glycine in order to consume remaining carbodiimide molecules and to react with activated carboxyl groups. The coupling solution was then dialyzed against 2 × 3 liters of 0.05 M NaHCO$_3$, pH 7.5, before the final separation of unreacted coenzyme from the complex on a Sephacryl S-200 (1 × 95 cm) gel filtration column equilibrated with 0.05 M NaHCO$_3$, pH 7.5. The liver alcohol dehydrogenase–NAD complex appeared in the fraction corresponding to that of native alcohol dehydrogenase.

The protein containing fractions were collected and could be lyophilized. The lyophilized protein was stored at $-20°$.

[9] R. Axén, J. Porath, and S. Ernback, Nature (London) 214, 1302 (1967).
[10] H. Theorell and A. D. Wiener, Arch. Biochem. Biophys. 83, 291 (1959).

Composition of the Complexes

The amounts of coimmobilized enzyme and of coenzyme were determined with protein[11] and phosphate[12] analysis after lyophilization of the Sepharose beads. As expected, the ratio of coenzyme to enzyme varied with the ratio in the coupling solution.

The concentration of coenzyme and enzyme in the soluble complex was determined spectrophotometrically with the use of the following equations, which are based on the molar absorption coefficients at 290 nm and 266 nm for the NAD analog and the enzyme to which had been coupled 6-aminohexanol instead of NAD:

$$13,200 \text{ [enzyme]} + 1890 \text{ [NAD]} = A_{290}$$
$$15,800 \text{ [enzyme]} + 21,500 \text{ [NAD]} = A_{266}$$

This spectrophotometric procedure was used as a convenient routine procedure and its validity was checked by several other procedures. The concentration of subunit was thus determined by the Lowry method,[13] and the concentration of coenzyme in the complex was determined from its phosphate content. In some cases a coenzyme labeled with ^{14}C was used, and then its concentration could be determined from the radioactivity. The different complexes are also characterized by the activities measured with and without addition of exogenous coenzyme (0.24 mM). The activity ratio obtained is an indicator of the number of the active sites occupied by an active coenzyme molecule.[4]

Assay of Enzyme–Coenzyme Complexes

According to the method of synthesis, a catalytic unit consists of one coenzyme molecule and one enzyme subunit. Thus, if there is no regeneration of the coenzyme, the reaction will stop after it has been used only once. Hence, the assay of the enzyme should include both regeneration of the coenzyme and measurement of product formation. Different regeneration–assay methods can be used, both enzymic (described below) and nonenzymic.[14]

Coupled Substrate Regeneration Assay

This is one of the methods used for the soluble enzyme–coenzyme complex, and it is the only satisfactory method for the enzyme–

[11] P. H. Spackman, W. H. Stein, and S. Moore, *Anal. Chem.* **30**, 1190 (1958).
[12] G. R. Bartlett, *J. Biol. Chem.* **234**, 466 (1958).
[13] O. H. Lowry, N. J. Rosebrough, A. L. Farr, and R. J. Randall, *J. Biol. Chem.* **193**, 265 (1951).
[14] A. Torstensson, G. Johansson, M.-O. Månsson, P.-O. Larsson, and K. Mosbach, *Anal. Lett.* **13** (B10), 837 (1980).

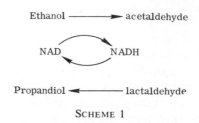

Ethanol ⟶ acetaldehyde

NAD ⟶ NADH

Propandiol ⟵ lactaldehyde

Scheme 1

coenzyme–Sepharose complex.[4] Since the Sepharose together with the enzyme restricts the accessibility of the NAD analog, a second enzyme cannot interact with the coenzyme (Fig. 1a).

Coupled substrate regeneration means that one substrate, e.g., ethanol, reduces the coenzyme and the other substrate, lactaldehyde, oxidizes the coenzyme (Scheme 1). Both reactions are thus carried out by the same enzyme, liver alcohol dehydrogenase.

The experimental setup is shown in Fig. 3. The reaction vessel contains, in a total volume of 5.2 ml, 50 mM ethanol, 7 mM lactaldehyde,[15] 0.2 M potassium phosphate buffer, pH 8.0, and the appropriate amount of complex.

The reaction is driven by removal of one of the products, the volatile acetaldehyde, by bubbling nitrogen gas (15 liters/hr) through the reaction vessel. The acetaldehyde is then trapped in a second vessel containing 10 ml of 12 mM semicarbazide in 0.1 M sodium phosphate buffer, pH 7.2. The increase in absorbance at 224 nm due to formation of acetaldehyde semicarbazone is continuously measured by pumping the solution at a rate of 50 ml/hr through a flow cuvette placed in a spectrophotometer. Both the reaction vessel and the vessel containing semicarbazide are kept at 30° in a thermostatically controlled water bath. The activity of the complex is also measured in the presence of an excess of exogenous NAD in order to measure the activity when all the active sites are occupied by an NAD molecule. The concentration of free NAD is then 0.24 mM.

The molar extinction coefficient at 224 nm for acetaldehyde semicarbazone is $15,500 \ M^{-1} \ cm^{-1}$,[15] and from this value the recovery of acetaldehyde as semicarbazone was calculated to be 70%, which is in accordance with earlier findings.[16]

An alternative spectrophotometric coupled substrate assay described in the literature is based on the couples ethanol–acetaldehyde and benzyl alcohol–benzaldehyde.[17]

[15] M. P. Schulman, N. K. Gupta, A. Omachi, G. Hoffman, and W. E. Marshall, *Anal. Biochem.* **60**, 302 (1974).

[16] M. Ramart-Lucas and J. Klein, *Bull. Soc. Chim. Fr.* 454 (1949).

[17] C. W. Fuller, J. R. Rubin, and H. J. Bright, *Eur. J. Biochem.* **103**, 421 (1980).

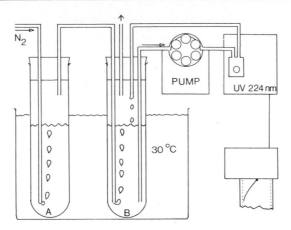

FIG. 3. Experimental setup for determining enzyme activity in the coupled substrate assay. A, the reaction vessel; B, the semicarbazide vessel.

Coupled Enzyme Regeneration Assay

This method is used for the soluble enzyme–coenzyme complex only (Fig. 1b). The first reaction of Scheme 2 is catalyzed by liver alcohol dehydrogenase, and the oxidation of NADH is catalyzed by lactate dehydrogenase or malate dehydrogenase. The auxiliary enzyme can thus interact with the coenzyme covalently bound to liver alcohol dehydrogenase.

One way of measuring the cycling rate is to follow the product formation (lactate).[18] A more convenient way is to use a fluorometric assay to measure the amount of reduced coenzyme. When excited at 340 nm, NADH (present in the active site of liver alcohol dehydrogenase) emits light with a maximum at 425 nm. This fact was used to monitor the amount of reduced coenzyme in a direct fluorometric assay. Thus, when ethanol is added to a solution of enzyme–coenzyme complex and lactate dehydrogenase, the fluorescence increases to a constant level due to NADH formation. When pyruvate is added, the bound NADH is oxidized and the fluorescence decreases rapidly (Fig. 4). When pyruvate is consumed, the fluorescence increases again until all the coenzyme molecules are reduced. With knowledge of the amount of pyruvate added, the number of coenzyme cycles can be calculated. The system may also be used as a sensitive assay for pyruvate.

The assay mixture consisted of 50 mM NaHCO$_3$, pH 7.5, 1.2 μM enzyme–coenzyme complex (based on enzyme subunit concentration),

[18] O. H. Lowry and J. V. Passoneau, "A Flexible System of Enzymatic Analysis," p. 196. Academic Press, New York, 1972.

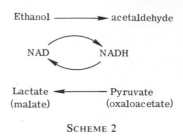

SCHEME 2

30 mM ethanol, 14 μM lactate dehydrogenase (malate dehydrogenase), and 25 nmol of pyruvate (oxaloacetate) in a total volume of 1.1 ml.

General Properties of Covalent Liver Alcohol Dehydrogenase– NAD(H) Complexes

The coenzyme retention strategy emphasized here, i.e., the permanent fixation of the coenzyme to the enzyme in the fashion of a prosthetic group, leads to interesting consequences with respect to catalytic efficiency, inhibition properties, thermal stability as well as areas of application. This is briefly discussed below together with the molecular organization of the complexes.

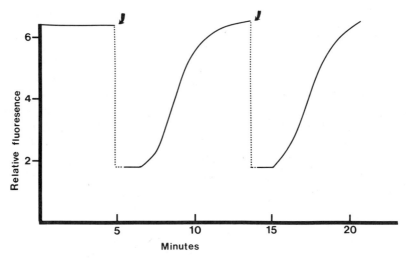

FIG. 4. Enzymic cyling of NAD covalently bound to liver alcohol dehydrogenase. The principal components of the assay mixture are alcohol dehydrogenase–NAD complex, lactate dehydrogenase, ethanol, and pyruvate. Additions of pyruvate are indicated by the arrows. Reproduced, with permission, from Månsson et al.[6]

.OPERTIES OF LIVER ALCOHOL DEHYDROGENASE–NADH–AGAROSE COMPLEX (I, II) AND LIVER
ALCOHOL DEHYDROGENASE–NAD COMPLEX (III)

eparation	NAD/subunit	Activity (μmol min^{-1} mg protein^{-1})		Internal activity relative to maximum (%)	NAD/NADH cycles (hr^{-1})
		Internal	Maximum		
I	1	0.8	3.9	20	1900
II	10	5.7	17.3	33	1400
III	1.6	4.2	10.6	40	10,000 (300[a])

[a] The activity was measured with the coupled enzyme assay. Other values are from experiments where the activity was measured with the coupled substrate assay.

Kinetic Properties. A selection of data of the properties of enzyme–coenzyme complexes is given in the table. Preparations I and II are both (insoluble) liver alcohol dehydrogenase–NAD(H)–agarose complexes, but differ in NAD : subunit ratios. The cycling rates are approximately the same for the two preparations. The internal activity (i.e., activity in the absence of exogenous NAD) and the maximum activity (i.e., activity in the presence of exogenous NAD), on the other hand, are both much higher for the preparation with a high NAD : subunit ratio, indicating that the excess of NAD analogs protects the enzyme during the coupling to Sepharose and/or that the excess increases the probability of obtaining Sepharose-bound coenzymes, which are properly located in relation to the enzyme.

Preparation III, the soluble liver alcohol dehydrogenase–NAD complex, has a considerably higher cycling rate, 10,000 cycles per hour. Careful investigation of the properties of the complex revealed that only every fourth active site carried a covalently attached and properly functioning coenzyme.[5] If this is taken into account, one might expect an ideal preparation to exhibit the impressive cycling rate of 40,000 hr^{-1} or more. The cycling rate, 300 hr^{-1}, given in parentheses in Table I, refers to recycling by a second enzyme, lactate dehydrogenase, and pyruvate. The remarkable difference in cycling rates reflects the difference between conditions under which the two systems operate. In the substrate couple regeneration assay the coenzyme that is covalently fixed via its spacer to the region of the active site remains constantly positioned within the site throughout the catalytic cycle. This means that the rate-limiting step in dehydrogenase catalysis, the dissociation of coenzyme,[19] is overruled and very

[19] C. I. Brändén, H. Jörnvall, H. Eklund, and B. Furugren, *in* "The Enzymes" (P. D. Boyer, ed.), 3rd ed., Vol. 11, pp. 103–190. Academic Press, New York, (1975).

high conversion rates are possible. In the coupled enzyme system, the coenzyme cannot constantly remain in the active site of liver alcohol dehydrogenase. After completion of the first half-cycle the coenzyme has first to dissociate and swing out of the active-site region into the solution and then await a proper collision with the second enzyme, lactate dehydrogenase. There then remains another slow dissociation step (from lactate dehydrogenase) followed by swinging back and reassociation with the liver alcohol dehydrogenase for completion of the cycle.

Obviously, this latter system must be comparatively very slow. A way of increasing its efficiency would be to fix the second enzyme permanently to alcohol dehydrogenase and in such a way that the alcohol dehydrogenase-bound coenzyme could readily alternate between the active sites of the two enzymes (Fig. 1c). The bonus of proximity and correct orientation of such a configuration should at least principally lead to effective self-contained catalysts. Initial efforts in this direction have, however, so far been met with only limited success.[6]

Closer analysis of the kinetic properties would require investigation of homogeneous preparations of the covalent enzyme–coenzyme complex, i.e., where all active sites are occupied with a functioning coenzyme. The easiest way of achieving this would probably be to fractionate the present preparations by a combination of techniques, including affinity chromatography.

The permanent fixation of an NAD molecule to the active site of liver alcohol dehydrogenase also results in a preparation characterized by a constant and very high local concentration of coenzyme in the active site. Regardless of the dilution, the enzyme will experience the same coenzyme concentration. This was indeed verified in a set of experiments where the activity of the complex was determined for a range of dilutions and compared with a reference system based on native alcohol dehydrogenase and NAD. The native enzyme system was shown to obey a rate equation of the second order: $v = k$ [enzyme]^2 whereas the covalent complex followed a rate equation of the first order[5] $v = k$ [enzyme].

Also inhibition studies of the soluble covalent enzyme–coenzyme complex demonstrated the high local coenzyme concentration in the active site. For example, when the covalent enzyme–coenzyme complex with a formal coenzyme concentration of 0.01 μM was treated with the competitive inhibitor AMP (10 mM), the internal activity dropped by only 50%. In the reference system with the same concentration of native enzyme, but with a higher concentration of NAD, 0.24 mM, the activity decreased by 80%. This means that it was more difficult to inhibit active sites provided with a covalently bound coenzyme than it was to inhibit an ordinary active site even though the gross coenzyme concentration of the latter system was more than 20,000 times higher.

Stability. Stability is always an important property when using enzymes for practical purposes. The permanent attachment of a coenzyme to the very vicinity of the active site could be expected to influence the stability positively. This was also found to be the case, in particular for the Sepharose based enzyme–coenzyme complex. For example, at 50° free enzyme lost 95% of its activity in 30 min, whereas the intrinsic activity of the Sepharose-bound complex decreased only by 10%. At the same time the maximum activity (i.e., in the presence of exogenous NAD) dropped by 65%, thus clearly indicating that those sites that have a coenzyme attached in their vicinity are much better protected than those lacking the coenzyme.

Molecular Organization. The properties of the soluble enzyme–coenzyme complex raise questions about its molecular organization, including the exact point of coenzyme attachment. The amino acids that are possible points of attachment for the active NAD analog are Asp-223, Asp-273, Asp-227, and Glu-366. Asp-273 would give an enzyme–coenzyme complex where the interactions between the active site and coenzyme are the same as in a normal binary complex. Coupling to the other amino acid residues, however, would require that the normal binary complex was somewhat distorted with respect to binding of the adenine part of the coenzyme (C. I. Brändén, personal communication).

A way to direct the coupling to the proper carboxyl group, leading to an active enzyme–coenzyme complex, would then be to make sure that the coenzyme is situated at the active site when the coupling takes place, i.e., in accordance with the principles of affinity labeling.

Such an experiment was carried out by forming a ternary complex between stoichiometric amounts of enzyme, the NAD-analog, and excess pyrazole.[20] The coupling agents were added when all the sites were occupied by an NAD–analog. However, these experiments did not result in any coupling at all. The same negative result was obtained when an alternative ternary complex was used, enzyme–NADH–analog and isobutyramide.[21]

These results might suggest that coupling takes place only when the coenzyme to be coupled is absent from the active site. This view is supported by the results of another experiment, where equimolar concentrations of NAD-analog and unmodified NADH were present during coupling. The much lower dissociation constant of NADH (about 150 times[10]) ensures that all the active sites are occupied by a nonreactive NADH molecule. The result of such a coupling was similar to that obtained in the absence of NADH.

[20] H. Theorell and T. Yonetani, *Biochem. Z.* **338,** 537 (1963).
[21] L. Andersson, P.-O. Larsson, and K. Mosbach, *FEBS Lett.* **88,** 167 (1978).

Another interpretation of the absence of any coupling when the NAD-analog was in the active site as a ternary complex is that the ternary complex forming molecule sterically hinders the coupling reaction. This interpretation is strengthened by the result of a coupling reaction where NAD-analog was used in excess at the same time as 0.1 M pyrazole was present. This experiment gave a very low coupling yield indicating that the pyrazole was involved.

As mentioned earlier, Asp-223 is a possible point of attachment for an active NAD-analog. This amino acid residue normally also forms a hydrogen bond with the adenine ribose of the coenzyme.[19] This means that when NAD is in the active site, the Asp-223 is blocked and no coupling can take place. The coupling can occur only when the active site is unoccupied and when pyrazole or another ternary complex forming molecule is present, the coenzyme is bound very firmly and the active site is always occupied, thus making it impossible for the coupling reaction to take place. The binary complexes between liver alcohol dehydrogenase and NAD or NADH are not so strong as the ternary complex, and in these cases the active site is sometimes vacant and the coupling can occur.

This hypothesis is compatible with the results of the two other interpretations and also includes coupling to an amino acid residue known to be a possible point of attachment.

If the active NAD-analog is covalently bound to Asp-223, the coenzyme will no longer have any hydrogen bond to Asp-223. The normal binding has therefore been distorted, and the interesting consequence is that the binding role of the adenosine part of the coenzyme has partly been replaced by the covalent bond between the enzyme and the coenzyme.

Conclusions

The broad substrate specificity of liver alcohol dehydrogenase for $C{=}O \rightleftarrows CH{-}OH$ redox reactions makes it a versatile enzyme for use in organic reactions.[22]

The complex of liver alcohol dehydrogenase–NADH and agarose is stable and easy to prepare, and might thus be useful in the production of various ketones, aldehydes, and alcohols.

The soluble complex of liver alcohol dehydrogenase and NAD may also function in enzyme reactions with other NAD-dependent dehydrogenases where both enzymes will share a common coenzyme, though it is covalently bound to liver alcohol dehydrogenase.

---

[22] J. B. Jones, this series, Vol. 44, p. 831.

[78] Aldehyde Dehydrogenase[1] from Bakers' Yeast

By NANAYA TAMAKI and TAKAO HAMA

$$RCHO + NAD(P)^+ \xrightarrow{K^+} RCOOH + NAD(P)H + H^+$$

Assay Method

Principle. Aldehyde dehydrogenase activity is assayed at 25° by measuring the rate of formation of NADH at 340 nm.

Reagents

Tris-HCl buffer, 0.2 M, pH 8.3, including bovine serum albumin (0.2 mg/ml)

KCl, 1 M

Pyrazole, 0.1 M

2-Mercaptoethanol, 1 M, prepared freshly each day and stored in ice

Propionaldehyde, 0.1 M, prepared freshly each day and stored in ice

Procedure. The above reagents and water are added to a final volume of 2.5 ml: 1.25 ml of Tris-HCl buffer, 0.1 ml of KCl, 0.02 ml of pyrazole, 0.02 ml of 2-mercaptoethanol, 0.02 ml of NAD, an appropriate amount of enzyme, and 0.02 ml of propionaldehyde. After several minutes for temperature equilibration in the 10-mm light path cuvette in a thermostattable sample compartment, the reaction is initiated by the addition of the aldehyde solution and is followed by measuring the change in absorbance at 340 nm with a Hitachi 200-10 recording spectrophotometer. The yeast aldehyde dehydrogenase may be diluted in 50 mM potassium phosphate buffer, pH 6.8, containing 10 mM 2-mercaptoethanol and 1 mM EDTA, prior to assay for enzyme activity. Pyrazole is added as an inhibitor for alcohol dehydrogenase.

Definition of Unit and Specific Activity. One unit of enzyme is defined as the amount that catalyzes the conversion of 1 μmol of aldehyde to acid per minute at 25°. Specific activity is expressed in units per milligram of protein. Protein concentration is measured by the biuret method[2] with albumin as protein standard.

Purification Procedure[2a]

The yeast aldehyde dehydrogenase was originally studied by Black[3,4] and then purified and crystallized by Steinman and Jakoby.[5] However,

---

[1] Aldehyde : NAD(P)$^+$ oxidoreductase, EC 1.2.1.5.

[2] G. Beisenherz, H. J. Boltze, T. Bücher, R. Czok, K.-H. Garbade, E. Meyer-Arendt, and G. Pfleiderer, *Z. Naturforsch.* **8**, 555 (1953).

[2a] N. Tamaki, M. Nakamura, K. Kimura, and T. Hama, *J. Biochem.* (*Tokyo*) **82**, 73 (1977).

Clark and Jakoby[6] reported that the proteolytic products from the native enzyme were found in the process of the purification of aldehyde dehydrogenase from yeast. The preparation method in the presence of PMSF[7] as a serine esterase inhibitor diminishes the amount of proteolytic degradation products.[6,8] Diisopropyl fluorophosphate is desirable in terms of preventing proteolysis,[6,8] but is too dangerous for use during homogenization.[8]

From these findings it appears necessary to develop a new mild procedure for the isolation of sufficient amounts of the native yeast aldehyde dehydrogenase. The purification procedure, described below and carried out under mild conditions in the presence of PMSF, utilizes protamine sulfate and ammonium sulfate treatments and a chromatographic technique. No treatment with acids, heat, or organic solvents is used during the purification procedure. Therefore, during the purification procedure, three different proteins are not observed as reported by Steinman and Jakoby[5] and Clark and Jakoby[6] in the DEAE chromatography step.

Fresh bakers' yeast (a commercially available product from Oriental Kobo Ltd., Japan) is washed twice with 3 volumes of cold distilled water and stored at $-25°$ before use. All subsequent steps are performed at about $4°$.

Step 1. Preparation of Crude Extract. Frozen bakers' yeast is thawed in a 2-liter beaker at room temperature, suspended in 1 liter of buffer A (100 mM potassium phosphate, pH 7.8, containing 2 mM 2-mercaptoethanol, 1 mM EDTA, 2 mM propionic acid, and 0.6 mM PMSF), and then homogenized with glass beads in a DYNO-Laboratory Mill, type KDL (Willy A. Bachofen, Maschinenfabrik, Basel, Switzerland), at 0° for 30 min. The homogenate is adjusted to pH 7.0 with 1 N KOH and allowed to stand for 30 min, then centrifuged at 27,000 g for 20 min.

Step 2. Protamine Sulfate Precipitation. A freshly prepared 1.5% (w/v) protamine sulfate solution, which is adjusted to pH 6.3 with 1 N KOH, is slowly added with stirring to the supernatant to give a final ratio of 1 mg of protamine sulfate to 12 mg of protein. After standing for 30 min, the resulting precipitate is removed by centrifugation and discarded.

Step 3. First Ammonium Sulfate Fractionation. Solid ammonium sulfate is added (243 g/liter) slowly with gentle stirring to the supernatant from

---

[3] S. Black, *Arch. Biochem. Biophys.* **34**, 86 (1951).

[4] S. Black, see this series, Vol. 1 [81].

[5] C. R. Steinman and W. B. Jakoby, *J. Biol. Chem.* **242**, 5019 (1967).

[6] J. F. Clark and W. B. Jakoby, *J. Biol. Chem.* **245**, 6065 (1970).

[7] The following abbreviations are used: PMSF, phenylmethylsulfonyl fluoride; SDS, sodium dodecyl sulfate.

[8] S. L. Bradbury, J. F. Clark, C. R. Steinman, and W. B. Jakoby, see this series, Vol. 41 [77].

step 2. After standing for 30 min, the precipitate is removed by centrifugation for 20 min. Additional solid ammonium sulfate is then added (132 g per liter of supernatant solution), and the precipitate is collected as above. The precipitate is dissolved in 500 ml of buffer B (50 mM potassium phosphate, pH 6.8, containing 2 mM 2-mercaptoethanol, 1 mM EDTA, 2 mM propionic acid, and 0.6 mM PMSF).

Step 4. Second Ammonium Sulfate Fractionation. Solid ammonium sulfate is added (263 g/liter) to the enzyme solution from step 3. The resulting suspension is allowed to stand for 30 min, then centrifuged for 30 min to remove the precipitate. To the supernatant, a further 77 g of ammonium sulfate per liter is added gradually, and the mixture is stirred for 30 min. The precipitate is collected by centrifugation at 27,000 g for 20 min and dissolved in 50 ml of buffer B.

Step 5. CM-Sepharose Chromatography. The enzyme solution is desalted by passage through a Sephadex G-25 column (8 × 25 cm, 150 ml/hr) equilibrated with buffer C (10 mM potassium phosphate, pH 6.8; 0.25%, v/v, 2-mercaptoethanol; 1 mM EDTA; 2 mM propionic acid; and 0.6 mM PMSF). The protein fraction containing aldehyde dehydrogenase is directly applied to a CM-Sepharose CL-6B column (4 × 40 cm, 120 ml/hr), equilibrated with buffer C. Aldehyde dehydrogenase passed through the column with buffer C, and alcohol dehydrogenase remained on the column.

Step 6. DEAE-Sepharose Chromatography. The fraction containing aldehyde dehydrogenase from the CM-Sepharose column is also applied directly to a DEAE-Sepharose CL-6B column (4.6 × 25 cm, 75 ml/hr), which is washed with buffer C. Then the enzyme is eluted with 600 ml of the same buffer containing KCl in a continuous gradient of 0 to 500 mM. One peak of aldehyde dehydrogenase appears between 100 and 200 mM KCl. To the active fraction, solid ammonium sulfate is added (561 g/liter). The precipitate is collected by centrifugation, and the sediment is dissolved in 10 ml of buffer B.

Step 7. Blue Sepharose Chromatography. The enzyme solution is then applied to a Blue Sepharose CL-6B column (2.8 × 35 cm, 35 ml/hr), equilibrated with buffer C. Blue Sepharose CL-6B is purchased from Pharmacia or prepared by binding Cibacron Blue F3G-A to Sepharose CL-6B according to Rinderknecht et al.[9] After absorption of the enzyme solution onto the column, it is washed with buffer C containing 0.1 M KCl, and then the enzyme is eluted with the same buffer containing 1 M KCl. To the active fraction, solid ammonium sulfate is added (561 g/liter). The precipitate is dissolved in 3 ml of buffer B.

Step 8. Gel Filtration. The enzyme preparation obtained by the Blue

[9] H. Rinderknecht, P. Wilding, and B. J. Harverback, *Experientia* **23**, 805 (1967).

PURIFICATION OF ALDEHYDE DEHYDROGENASE FROM BAKERS' YEAST

Steps	Total activity (units)	Total protein (mg)	Specific activity (units/mg protein)	Purification (fold)
Extract	23,810	91,922	0.26	1.0
Protamine treatment	27,755	36,644	0.78	2.9
1st $(NH_4)_2SO_4$ fractionation	23,882	20,400	1.17	4.5
2nd $(NH_4)_2SO_4$ fractionation	12,507	5,000	2.50	9.6
CM chromatography	7,234	2,793	2.59	10
DEAE chromatography	4,038	909	4.44	17
Blue Sepharose chromatography	1,897	155	12.2	47
Ultrogel chromatography	1,728	84	20.6	79

Sepharose procedure is applied to an Ultrogel AcA-34 column (1.6 × 150 cm, 25 ml/hr), equilibrated with buffer B containing 1 M ammonium sulfate. Elution is carried out with the same buffer. Contents of fractions that contain enzymic activity are pooled, and ammonium sulfate is added (390 g/liter). The suspension of aldehyde dehydrogenase is centrifuged, and the precipitate is dissolved in 2 ml of buffer B.

A summary of the isolation procedure from 1 kg of bakers' yeast is shown in the table.

Properties

Purity. The purified enzyme migrates as a single component when subjected to polyacrylamide electrophoresis in the absence and in the presence of SDS and does not show any sign of proteolytically degraded forms as shown by Clark and Jakoby[6] and Bostian and Betts.[10]

Stability. The enzyme preparation can be stored in buffer B (10 mg/ml) at −25° for at least 3 months without loss of activity. When the enzyme suspension is stored in 3.2 M ammonium sulfate at 4°, a 50% decrease in the specific activity is observed within 10 days. When incubated at 0° for 5 hr, the enzyme preparation (0.1 mg/ml) in 0.1 M Tris-HCl buffer loses about 20% of its specific activity at pH 7.3 and 50% at pH 8.3. This loss of activity is completely prevented in the presence of 1 mM propionic acid or 10 mM 2-mercaptoethanol.

Molecular Properties. In order to test whether the molecular weight of the enzyme preparation had changed compared with that of the crude yeast extract in step 1, gel filtration on Sephadex G-200 is carried out with the crude extract. The enzymic activity is eluted as one single peak at the same position as that observed with the highly purified preparation. The

[10] K. A. Bostian and G. F. Betts, *Biochem. J.* **173,** 773 (1978).

molecular weight by sucrose density gradient centrifugation is 207,000 ± 13,000. Cross-linking patterns obtained with yeast aldehyde dehydrogenase after treatment with a series of diimidoesters of increasing chain lengths with different reaction times result in the appearance of tetramers as the largest cross-linked product of the enzyme subunits.[11] The molecular weights of its monomer, dimer, trimer, and tetramer are 57,000, 114,000, 171,000, and 228,000, respectively, as estimated from their mobilities on SDS-electrophoresis. Monomers are probably assembled in tetramers in a heterologous square arrangement.

Ultraviolet Light Absorption Spectra. At pH 7.0, the purified enzyme solution has an absorption maximum at 279 nm ($A_{0.1\%}$ = 0.94) and a minimum at 252 nm ($A_{0.1\%}$ = 0.35). The ratio $A_{max} : A_{min}$ = 2.69 ($A_{280} : A_{260}$) = 1.92.

pH Optimum. The pH optimum for propionaldehyde is 9.3. Different pH profiles are not obtained when the buffers and ligand conditions are changed. Moreover, the same pH optimum (pH 9.3) is found when NADP is used as an oxidizing agent.

SUBSTRATE SPECIFICITY. Evidence from alternative substrate analysis and product-inhibition studies as well as NAD-enzyme binding studies in the absence of aldehyde, indicate an ordered sequential mechanism where $NAD(P)^+$ is the first substrate bound and the carboxylic product dissociates last.[10,12]

The enzyme is active in catalyzing the oxidation of a wide variety of aliphatic and aromatic aldehydes including formaldehyde, acetaldehyde, propionaldehyde, *n*-butylaldehyde, isobutylaldehyde, *n*-valeraldehyde, caproaldehyde, benzaldehyde, glycolaldehyde, D-glyceraldehyde, malonic semialdehyde, and succinic semialdehyde. It is inactive on *o*-, *m*-, and *p*-hydroxybenzaldehydes, D-glucuronolactone, and glyoxal. NADP is active for the enzyme as well as NAD. The enzyme does not catalyze the reduction of acetic, propionic, and malonic acids using NADH as a cofactor.

Other Properties. An investigation of the temperature dependence of the maximal velocity yields linear Arrhenius plots between 6° and 32°. The activation energy is 18.3 kcal/mol at pH 8.3 in the presence of propionaldehyde and NAD.[2a]

[11] N. Tamaki, K. Kimura, and T. Hama, *J. Biochem.* (*Tokyo*) **83**, 821 (1978).
[12] K. A. Bostian and G. F. Betts, *Biochem. J.* **173**, 787 (1978).

[79] Isozymes of Aldehyde Dehydrogenase from Horse Liver

By JOHN H. ECKFELDT and TAKASHI YONETANI

Aldehyde + NAD → acid + NADH

Two isozymes of aldehyde dehydrogenase (EC 1.2.1.3) are found in horse liver,[1,2] one cytosolic (F1) and one mitrochondrial (F2).[3] Both enzymes can oxidize a wide variety of aliphatic and aromatic aldehydes and show strong, but not absolute, specificity for NAD over NADP. While both isozymes can oxidize acetaldehyde to acetate, we have suggested that the cytosolic (F1) isozyme is more important quantitatively for ethanol metabolism based on its extreme sensitivity to disulfiram (Antabuse) inhibition and its lower K_m for NAD.[1]

Assay Method

Principle. Aldehyde dehydrogenase activity is determined by monitoring the rate of NADH formation spectrophotometrically or fluorimetrically. To facilitate comparison with previous work, we used the assay system of Feldman and Weiner[2] during purification.

Reagent

Sodium pyrophosphate, 0.1 M, pH 9.0, containing 1.45 mM NAD and
        69 μM propionaldehyde
Procedure. Prepare the above reaction mixture fresh daily. Add 5–50 μl of enzyme to enough assay reagent to bring the volume to 3.0 ml in a 1-cm cuvette in a 25° thermostatted spectrophotometer. One unit of aldehyde dehydrogenase activity will catalyze the formation of 1 μmol of NADH per minute under these conditions.

Kinetic Studies of Purified Enzymes. Studies with the purified enzymes were done in $\mu = 0.1$ sodium phosphate, pH 7.0, at 25°. Aldehyde, NAD, and NADP kinetic constants were determined by preincubating the enzyme with NAD or NADP, adding aldehyde dissolved in distilled water to initiate the reaction, and monitoring NADH production.

[1] J. Eckfeldt, L. Mope, K. Takio, and T. Yonetani, *J. Biol. Chem.* **251**, 236 (1976).
[2] R. I. Feldman and H. Weiner, *J. Biol. Chem.* **247**, 260 (1972).
[3] J. Eckfeldt and T. Yonetani, *Arch. Biochem. Biophys.* **131**, 371 (1976).

TABLE I
PURIFICATION OF TWO ALDEHYDE DEHYDROGENASE ISOZYMES FROM HORSE LIVER

F1/F2 mixture	Total protein[a] (g)	Total activity[b] (μmol NADH/min)	Specific activity (μmol NADH/min/g)
Crude extract	60	1400	23
35% Ammonium sulfate	38	1100	29
55% Ammonium sulfate	24	800	33
CM-cellulose chromatography	12	500	42
F1 Isozyme			
DEAE-cellulose chromatography			
imidazole buffer	0.61[c]	100	160
Gel filtration chromatography	0.20[c]	74	370
F2 Isozyme			
DEAE-cellulose chromatography			
imidazole buffer	133[d]	200	150
Gel filtration chromatography	0.23[d]	131	570
DEAE-cellulose chromatography			
phosphate buffer	0.10[d]	80	800

[a] By the Lowry method except as noted below.
[b] In the pH 9 sodium pyrophosphate buffer system with propionaldehyde substrate.
[c] Using E_{280} = 0.95 mg ml^{-1} cm^{-1} for the F1 isoenzyme.
[d] Using E_{280} = 1.05 mg ml^{-1} cm^{-1} for the F2 isoenzyme.

Purification Procedure

Obtain fresh horse livers periodically, and within 2 hr after removal perfuse the liver with approximately 8 liters of cold 0.15 M NaCl to remove excess blood. Remove the liver capsule and large blood vessels, divide into approximately 800-g lots, and freeze for later use. Perform all subsequent procedures below 5°, using nitrogen-saturated buffers and a nitrogen atmosphere where possible.

Extraction. When beginning a preparation, partially thaw an 800-g lot of liver and mince in a meat grinder. To the minced liver, add 1 liter of 36 mM mercaptoethanol containing 2 mM EDTA and homogenize the mixture for 3 min using a Torax Tissumizer. Remove the insoluble debris by centrifugation at 13,000 g for 60 min.

Ammonium Sulfate Fractionation. To this extract, add, per liter, 2 ml of mercaptoethanol and 209 g of solid ammonium sulfate. After dissolution of the ammonium sulfate, allow the slurry to stand for 60 min and then centrifuge at 24,000 g for 60 min. Discard the precipitate. To the supernatant, add 129 g of solid ammonium sulfate and 1 ml mercaptoethanol per liter. After dissolution of the ammonium sulfate, allow the slurry to stand 90 min and then centrifuge at 24,000 g for 30 min. Discard the supernatant and suspend the pellet in approximately 400 ml of 10 mM sodium phosphate, pH 6.3, containing 2.5 ml of mercaptoethanol per liter and 1 mM EDTA. Dialyze this material overnight against several 10-liter changes of the suspending buffer.

CM-Cellulose Chromatography. Recentrifuge the dialyzed material at 24,000 g for 30 min to remove insoluble material and apply to a CM-cellulose column (4 × 15 cm) equilibrated with 10 mM sodium phosphate, pH 6.3, containing 2.5 ml of mercaptoethanol per liter and 1 mM EDTA. Elute the materials adsorbed with the same buffer. The eluate immediately following the void volume contains the aldehyde dehydrogenase activity. Dialyze this eluate overnight against several 10-liter changes of 5 mM imidazole hydrochloridé buffer, pH 7.2, containing 2.5 ml of mercaptoethanol per liter.

Separation of F1 and F2 Isozymes Using DEAE-Cellulose Chromatography. After dialysis, apply the sample to a DEAE-cellulose column (5 × 20 cm) equilibrated with the same imidazole buffer. All the aldehyde dehydrogenase activity is firmly bound under these conditions. First, remove several extraneous proteins by washing the column with 250 ml of the equilibration buffer. Then elute the aldehyde dehydrogenase activity with a linear salt gradient prepared with 1 liter of 5 mM imidazole hydrochloride, pH 7.2, containing 2.5 ml of mercaptoethanol per liter and with 1 liter of 5 mM imidazole hydrochloride, pH 7.2, containing 2.5 ml of mercaptoethanol per liter and 200 mM sodium chloride. This gradient resolves the two isozymes as shown in Fig. 1.

Final Purification of F1 Isozyme. Prior to gel filtration chromatography, combine the fractions containing the F1 isozyme from the above column and concentrate the pooled solution on a small DEAE-cellulose column with the above two imidazole buffers. Apply the concentrated sample to a BioGel A-1.5m column (2.5 × 90 cm) equilibrated with 10 mM sodium phosphate buffer, pH 5.5, containing 2.5 ml mercaptoethanol per liter and 1 mM EDTA. Elute with the same buffer. Collect and combine the fractions that contained the F1 isozyme activity and store under nitrogen at 4° for further use.

Final Purification of F2 Isozyme. Combine the fractions from the DEAE-cellulose column containing the F2 isozyme and concentrate on DEAE-cellulose with imidazole buffer as above. Apply the sample to the same BioGel A-1.5m column, this time equilibrated with 10 mM sodium phosphate, pH 5.8, containing 2.5 ml of mercaptoethanol per liter and 1 mM EDTA. Elute with the same buffer. Combine the fractions containing the F2 isozyme and apply to a DEAE-cellulose column (2.5 × 18 cm)

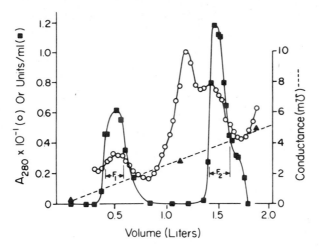

FIG. 1. Elution profile of aldehyde dehydrogenase isozymes from a DEAE-cellulose column with a salt gradient in imidazole hydrochloride as described in the text. The 280 nm absorption (O) and aldehyde dehydrogenase (■) activity are shown.

equilibrated with the same pH 5.8 phosphate buffer. Elute the bound F2 isozyme with a linear salt gradient prepared with 300 ml of 10 mM sodium phosphate, pH 5.8, and 300 ml of 200 mM sodium phosphate, pH 5.8, both containing 2.5 ml of mercaptoethanol per liter and 1 mM EDTA. Combine the fractions containing the F2 isozyme and store at 4° under nitrogen for further use.

Properties

Native Protein and Subunit Molecular Weights. Multiporosity polyacrylamide gel electrophoresis[4] gave molecular weight estimates for the F1 and F2 isozymes of 230,000 and 240,000. The results of sodium dodecyl sulfate–polyacrylamide gel electrophoresis gave subunit molecular weight estimates of 52,000 and 53,000.

Absorption Spectra. The absorption spectra of both isozymes from 250 to 600 nm showed only a single peak at 280 nm with no absorption in the visible region. The extinction coefficients at 280 nm in sodium phosphate, pH 7.0, were calculated to be 0.95 and 1.05 ml mg^{-1} cm^{-1} for the F1 and F2 isozymes, respectively. The A_{280}/A_{260} was between 1.90 and 1.95 for both the isozymes. The lack of visible spectra and the high A_{280}/A_{260} indicates that the isozymes are not flavoproteins, nor do they contain bound nucleotide as does glyceraldehyde-3-phosphate dehydrogenase.

[4] J. L. Hedrick and A. J. Smith, *Arch. Biochem. Biophys.* **126,** 155 (1968).

Amino Acid Composition. The two isozymes have somewhat similar amino acid compositions.[1] Differences in ionizable amino acids agree with the electrophoretic and chromatographic behavior. The cysteine contents as measured by the mercury orange method of Sakai[5] were calculated to be 1.1 and 0.8 free sulfhydryl groups per 10^4 g of protein for the F1 and F2 isozymes, respectively.

Carbohydrate and Lipid Content. Because recoveries on amino acid analysis were slightly lower than expected (81% for F1 and 82% for F2),[1] tests for the presence of lipid and carbohydrate were performed. An approximately 10% higher 280-nm extinction in 6 M guanidine hydrochloride was observed after chloroform–methanol extraction of the distilled water-dialyzed, lyophilized proteins compared to the unextracted isozymes. These results suggest that some lipid may be present in the purified samples, although possible changes in water content caused by the extraction process must be considered. No carbohydrate was detected in the purified isozymes by the phenol sulfuric acid method of Hirs.[6]

Aldehyde Dehydrogenase Activity. As shown in Table II, both isozymes show rather broad aldehyde specificity. The K_m values reported in Table II were derived from Lineweaver–Burk plots, the aldehyde concentration in the solutions being based either on the volume of pure aldehyde added or, for the low K_m aldehydes, on NADH production after allowing the reaction to progress to completion. However, in aqueous solutions aldehydes exist as a mixture of the free aldehyde, RCHO, the hydrate, $RCH(OH)_2$, the exact ratio being determined by the gem-diol equilibrium constant.[7] It is likely that only one form is the true substrate for the dehydrogenase,[8-10] and the large range in K_m values for different aldehydes may be due in part to the widely varying fractions existing as free aldehyde.

Both isozymes show a relative, but not absolute, specificity for NAD over NADP as coenzyme. In sodium phosphate, pH 7.0, with 1 mM acetaldehyde as substrate, the K_m values for NAD and NADP were 3 μM and 700 μM, respectively, for the F1 isozyme and 30 μM and approximately 10 mM, respectively, for the F2 isozyme. For each isozyme, the V_{max} values with the two coenzymes were about the same.

Inhibition Studies. Chloral hydrate, p-chloromercuribenzoate, and disulfiram (tetraethylthiuram disulfide, Antabuse) were all found to be inhib-

[5] H. Sakai, *Anal. Biochem.* **26**, 269 (1968).
[6] C. H. W. Hirs, this series, Vol. 11, p. 411.
[7] R. P. Bell, *Adv. Phys. Org. Chem.* **4**, 1 (1966).
[8] F. H. Bodley and A. H. Blair, *Can. J. Biochem.* **49**, 1 (1971).
[9] D. R. Trentham, C. H. McMurray, and C. I. Pogson, *Biochem. J.* **114**, 19 (1969).
[10] D. Gregory, P. A. Goodman, and J. E. Meany, *Biochemistry* **11**, 4472 (1972).

TABLE II

RELATIVE MAXIMAL VELOCITIES AND K_m VALUES OF TWO ISOZYMES FOR VARIOUS ALDEHYDES[a]

Aldehyde	R group	F1 isozyme Relative V_{max}[b]	K_m (μM)	F2 isozyme Relative V_{max}[b]	K_m (μM)
Aliphatic aldehydes					
Isovaleraldehyde	$(CH_3)_2CH\ CH_2-$	1.1	0.5	0.6	<0.1
Valeraldehyde	$CH_3(CH_2)_3-$	1.1	0.1		
Propionaldehyde	CH_3CH_2-	1.2	5	0.7	<0.1
Acetaldehyde	CH_3-	1.0	70	1.0	0.2
Phenylacetaldehyde	$(C_6H_5)CH_2-$	1.9	4	1.4	0.3
Formaldehyde	$H-$	0.8	940	1.6	270
Glycolaldehyde	$HOCH_2-$	0.9	130	3.2	50
Chloroacetaldehyde	$ClCH_2-$	0.7	30	3.5	10
Conjugated aldehydes					
Benzaldehyde	C_6H_5-	0.6	<0.1	0.11	<0.1
Cinnamaldehyde	$(C_6H_5)CH{=}CH-$	0.5	<0.1	0.10	<0.1

[a] Assay system was sodium phosphate buffer, $\mu = 0.1$, pH 7.0, containing 1 mM NAD.
[b] The V_{max} values are relative to acetaldehyde as substrate for each isozyme. With 1 mM NAD and 1 mM acetaldehyde in this pH 7.0 buffer, the specific activities of the purified F1 and F2 isozymes are approximately 0.13 and 0.35 μmol of NADH min^{-1} mg^{-1}, respectively.

itors of both isozymes. Using approximately 10 μg of protein per milliliter in the pH 7 phosphate buffer with 1 mM acetaldehyde and 1 mM NAD as substrates, the concentration of chloral hydrate resulting in 50% inhibition was approximately 0.25 mM and 0.4 mM for the F1 and F2 isozymes, respectively. Preincubation of dilute solutions of either isozyme with 1 μM p-chloromercuribenzoate without protective reducing agents gave complete inhibition. With disulfiram, a physiologically important inhibitor, the behavior of the F1 and F2 isozymes is markedly different. In the absence of protective reducing agents, approximately 30 μM disulfiram gave 50% inhibition of the F2 isozyme in the above reaction mixture, while 1 μM disulfiram completely inhibited the F1 isozyme. Further studies on the F1 isozyme showed disulfiram to be an essentially stoichiometric inhibitor, requiring a single disulfiram per monomer for inhibition.[1] The inhibition by disulfiram under these conditions is complete within a minute; and addition of a large excess of mercaptoethanol (0.4 M) leads to more than 90% reactivation of the inhibited enzyme. Details of the F1 and F2 isozyme reaction mechanism are presented elsewhere.[11,12]

[11] J. Eckfeldt and T. Yonetani, Arch. Biochem. Biophys. 173, 273 (1976).
[12] R. I. Feldman and H. Weiner, J. Biol. Chem. 247, 267 (1972).

[80] Aldehyde Dehydrogenase from *Proteus vulgaris*[1]

By SHOJI SASAKI and YASUTAKE SUGAWARA

$$\text{R-CHO} + \text{NADP}^+ \xrightarrow{\text{NH}_4^+} \text{R-COOH} + \text{NADPH} + \text{H}^+$$

Assay Method

Principle. The activity of aldehyde dehydrogenase (aldehyde : NADP$^+$ oxidoreductase, EC 1.2.1.4) is monitored by measuring the production of NADPH spectrophotometrically at 340 nm with isovaleraldehyde as substrate.[2] Isovaleraldehyde can be replaced with propionaldehyde or acetaldehyde, both which are also good substrates for this enzyme.

Procedures. The reaction mixture in a silica cuvette consists of the following components in a final volume of 2 ml: Tris-HCl buffer,[3] pH 8.7, 20 mM; ammonium sulfate, 15 mM; NADP$^+$, 0.12 mM; an appropriate amount of enzyme. After preincubation of the mixture at 30° for a few minutes, the reaction is started by addition of isovaleraldehyde, 0.2 mM, and the absorption change is recorded.

Definition of Units. One unit of activity is defined as the amount of enzyme that reduces 1 μmol of NADP$^+$ per minute. Specific activity is defined as units of enzyme activity per milligram of protein. Protein is determined by the methods of Lowry *et al.*[4] and of Warburg and Christian.[5]

Purification Procedure

Growth and Cell Extracts. Proteus vulgaris, which is a strain carried for a long time in our laboratory, is cultured at 37° in four 3-liter flasks each of which contains 1 liter of liquid medium consisting of 1% each of peptone and meat extract, pH 7.0. After 15 hr of vigorous shaking, the cells are harvested by continuous centrifugation and washed three times with

---

[1] Y. Sugawara and S. Sasaki, *Biochim. Biophys. Acta* **480**, 343 (1977).

[2] This enzyme was found during a study on L-leucine metabolism, and we used exclusively isovaleraldehyde as substrate.

[3] Phosphate buffer is preferable to Tris because of the possibility of binding between aldehydes and Tris.

[4] O. H. Lowry, N. J. Rosebrough, A. L. Farr, and R. J. Randall, *J. Biol. Chem.* **193**, 265 (1951).

[5] O. Warburg and W. Christian, *Biochem. Z.* **310**, 384 (1941). See also this series, Vol. 3 (73).

deionized water by centrifugation at about 4000 g for 15 min. About 8 g of cell paste are obtained per flask.

All procedures described below are performed at 0° to 4° unless otherwise stated. Washed cells are suspended in an appropriate amount of 20 mM potassium phosphate buffer, pH 7.0, to give about 8 g/40 ml. Aliquots of 40 ml of the cellular suspension are treated by sonic oscillation at 20 kc for 8 min at 0° to 2° so as to avoid thermal inactivation. The supernatant from similar aliquots obtained by centrifugation at 15,000 g for 30 min are combined.

Step 1. Protamine Sulfate Fraction. Protamine sulfate (1% aqueous solution) is dropped into the sonic extract with stirring to about 0.1% concentration. It is necessary to choose a suitable amount of protamine sulfate so as to obtain the best purification while avoiding a marked loss of enzyme yield, since excess protamine sulfate diminishes the yield. After stirring for about 30 min, the extract is centrifuged at 15,000 g for 30 min, and the precipitate is discarded. For convenience during the next fractionation, the volume of the protamine sulfate fraction is adjusted to 120 ml with 20 mM potassium phosphate buffer, pH 7.0.

Step 2. Acetone Fraction. One-half volume (60 ml) of acetone chilled at −20° is added slowly to the protamine sulfate fraction with stirring. After more about 30 min of stirring, the mixture is centrifuged and the precipitate is discarded. Then 120 ml of cold acetone is added to the supernatant as above. After stirring for about 30 min, the precipitate formed between 33% and 60% of acetone is collected and dissolved in about 100 ml of 20 mM potassium phosphate buffer, pH 7.0. After gentle stirring for 30 min, insoluble materials are removed by centrifugation. The supernatant is adjusted to 120 ml by addition of the same buffer.

Step 3. Ammonium Sulfate Fraction. The acetone fraction is fractionated with powdered ammonium sulfate. Ammonium sulfate, 18.2 g, is slowly added to 120 ml of the acetone fraction with stirring. After 30 min, the precipitate is removed by centrifugation at 15,000 g for 30 min. An additional 36.5 g of ammonium sulfate are added to the supernatant, which has been adjusted to 120 ml as before. The resultant precipitate is collected by centrifugation, dissolved in about 25 ml of 20 mM potassium phosphate buffer, pH 7.0, and dialyzed overnight against 1 liter of 20 mM potassium phosphate buffer, pH 6.1, with several changes of buffer.

Step 4. Hydroxyapatite Column Chromatography. Prior to be chromatography, the ammonium sulfate fraction is clarified by centrifugation at 15,000 g for 30 min. The supernatant is applied to a hydroxyapatite column (2.4 × 12 cm) previously equilibrated with buffer A consisting of 20 mM potassium phosphate buffer, 1 mM EDTA, and 3 mM dithiothreitol, pH 6.1. After washing with more than 200 ml of buffer A, the

TABLE I

PURIFICATION OF ALDEHYDE DEHYDROGENASE FROM
Proteus vulgaris[a]

Fraction	Specific activity (units/mg protein)	Yield (%)	Purity
Sonic extract	0.51	100.0	1
Protamine sulfate	1.01	95.3	2
Acetone	1.82	62.5	3.6
Ammonium sulfate	2.46	59.3	4.8
Hydroxyapatite	12.0	48.7	23.5
DEAE-cellulose	14.5	46.5	28.4
Sephadex G-200	14.6	34.9	28.6

[a] From Sugawara and Sasaki.[1]

enzyme is eluted successively with 0.1 M, 0.111 M, and 0.125 M potassium phosphate buffer containing 1 mM EDTA and 3 mM dithiothreitol, pH 6.1. Aldehyde dehydrogenase activity is found exclusively in 0.111 M eluate. Active fractions are combined and dialyzed against 20 volumes of buffer A with stirring for 18 hr.

Step 5. DEAE-Cellulose Column Chromatography. About 50 ml of the dialyzed enzyme are loaded onto a DEAE-cellulose column (1.4 × 10 cm) equilibrated with buffer A. After washing with about 300 ml of buffer A, elution is performed with 150 ml of 0.12 M KCl in buffer A. Active fractions are combined and concentrated to about 2 ml using a collodion bag.

Step 6. Sephadex G-200 Gel Filtration. The concentrated enzyme solution is loaded onto a Sephadex G-200 column (2.4 × 36 cm) equilibrated with buffer A, and eluted with the same buffer at a flow rate of 2.5 ml/hr. The enzyme eluting between about 100 ml and 120 ml of effluent is purified to a homogeneous state as judged by polyacrylamide gel electrophoresis and ultracentrifugal analysis. The purified enzyme preparation differs from intermediate fractions in its stability: when it is stored at 0° to 4° in 20 mM potassium phosphate buffer containing 1 mM EDTA and 8 mM dithiothreitol, pH 6.1, the enzyme activity loss is only slight after a month or so, and the loss is about 50% even after 4 months.

A summary of the purification is shown in Table I.

Crystallization. About 20 ml of the purified enzyme solution is concentrated to about 1 ml. Powdered ammonium sulfate is added to the solution to about 1.2 M. Thenceforth, powdered ammonium sulfate is gradually added with great care until cloudiness is visible. Then a small amount of distilled water or 20 mM potassium phosphate buffer, pH 6.1, is added until the cloudiness just disappears. The solution is kept at room tempera-

TABLE II
SUBSTRATE SPECIFICITY OF ALDEHYDE DEHYDROGENASE[a]

Substrate	Concentration (M)	Relative activity $(\%)$	K_m (M)
Propionaldehyde	1.0×10^{-4}	181	1.7×10^{-5}
Acetaldehyde	1.2×10^{-3}	131	4.0×10^{-5}
Isovaleraldehyde	2.0×10^{-4}	100	3.0×10^{-5}
Benzaldehyde	6.4×10^{-5}	6	—

[a] Sugawara and Sasaki.[1]

ture over a period of one week and crystals are formed if the crystallization is successful. However, the catalytic activity is lost after crystallization as reported for yeast aldehyde dehydrogenase by Steinman and Jakoby.[6]

Properties

The reaction absolutely depends on NH_4^+ ion. Rb^+ and K^+ also activate, but the effect is about half that of NH_4^+; Cs^+, Li^+, and Na^+ have little or no effect.

The pH optimum is about 9.0, when the activity is assayed with 50 mM potassium chloride instead of ammonium sulfate, which cannot function as cofactor at higher pH values.

This enzyme is not isovaleraldehyde-specific aldehyde dehydrogenase, as can be seen in Table II. Propionaldehyde and acetaldehyde are better substrates for the enzyme. Since the enzyme is inhibited by all three aldehydes in higher concentrations, apparent K_m values for the respective aldehydes have to be determined in the range of lower concentrations (Table II).

The aldehyde dehydrogenase is specific for $NADP^+$, which cannot be replaced with NAD^+. CoA is not required for the enzyme reaction. K_m for $NADP^+$ is 2.1×10^{-5} M.

The molecular weight of the purified enzyme is calculated to be about 130,000 by the gel filtration method.[7]

[6] C. R. Steinman and W. B. Jakoby, *J. Biol. Chem.* **242**, 5019 (1967).
[7] P. Andrews, *Biochem. J.* **96**, 595 (1965).

[81] Aldehyde Dehydrogenases from *Pseudomonas aeruginosa*

By J. P. VANDECASTEELE and L. GUERRILLOT

$$CH_3(CH_2)_nCHO + NAD(P^+) + H_2O \rightarrow CH_3(CH_2)_nCOOH + NAD(P)H + H^+$$

Several specific aldehyde dehydrogenases, such as α-ketoglutarate semialdehyde dehydrogenase[1] and succinate semialdehyde dehydrogenase,[2] have been described in *Pseudomonas*. In *Pseudomonas aeruginosa* an aldehyde dehydrogenase induced by growth on ethanol has also been studied.[3] The two enzymes described below, which are noninducible soluble enzymes from *P. aeruginosa* and utilize a wide range of aliphatic aldehydes, have been studied by Guerrillot and Vandecasteele.[4] Another different aldehyde dehydrogenase with a wide specificity has been reported in *Pseudomonas fluorescens*.[5]

NAD$^+$-Linked Aldehyde Dehydrogenase[4] (EC 1.2.1.3)

Assay Method

Principle. The reduction of NAD$^+$ is measured at 30° with a recording spectrophotometer at 340 nm.

Reagents

Butanal (from Fluka) 0.21 mM dissolved in potassium pyrosphate buffer 54 mM, pH 8.6. Because of the limited solubility of aldehydes, butanal or any other aldehyde assayed is dissolved in the buffer at a concentration close to the final assay concentration. In these conditions, aldehydes are present as true solutions in the buffer to avoid the artifacts resulting from the use of emulsions, as reported in the case of long-chain alcohols.[6]
NAD$^+$, 30 mM

[1] E. Adams and G. Rosso, *J. Biol. Chem.* **242**, 1802 (1967).
[2] D. M. Callewaert, M. S. Rosemblatt, K. Suzuki, and T. T. Tchen, *J. Biol. Chem.* **248**, 6009 (1973).
[3] R. G. von Tigerström and W. E. Razzel, *J. Biol. Chem.* **243**, 6495 (1968).
[4] L. Guerrillot and J. P. Vandecasteele, *Eur. J. Biochem.* **81**, 185 (1977).
[5] W. B. Jakoby, *J. Biol. Chem.* **232**, 89 (1958).
[6] J. P. Tassin and J. P. Vandecasteele, *C. R. Hebd. Seances Acad. Sci. Ser. D* **272**, 1024 (1971).

Procedure. To the cuvettes (10 mm light path) are added 2.8 ml of the butanal solution in buffer, 0.1 ml of NAD^+, water, and enzyme to a final volume of 3 ml. (Final concentrations are 2 mM for butanal, 50 mM for potassium pyrophosphate and 1 mM for NAD^+.) The reaction is initiated by addition of enzyme.

Unit. Enzymic activities are expressed in enzyme units. One unit is defined as the amount that transforms 1 μmol of substrate per minute.

Protein Assay. A modified Folin method is used to avoid interfering reactions between reagent and SH groups that are present in elution buffers. Typically, 50–150 μg of protein in 0.5 ml are precipitated by 2 ml of 25% trichloroacetic acid. This preparation is kept for 15 min at 0° before centrifugation for 15 min at 10,000 g at 2°. The pellet is resuspended in 0.5 ml of NaOH, and then the assay is performed by the method of Lowry *et al.*[7]

Purification Procedure

Step 1. Growth of the Microorganism. The strain used, *Pseudomonas aeruginosa* 196 Aa, isolated by Traxler and Bernard,[8] is grown in sterile conditions in a 12-liter fermentor (Magnaferm, from New Brunswick). The salts medium used[9] contains per liter: NH_4Cl, 2 g; KH_2PO_4, 4 g; $Na_2HPO_4 \cdot 12 H_2O$, 6 g; $Mg SO_4 \cdot 7 H_2O$, 0.3 g; $FeSO_4 \cdot 7 H_2O$, 10 mg; $ZnSO_4 \cdot 7 H_2O$, 0.1 mg; $CuSO_4 \cdot 5 H_2O$, 0.1 mg; H_3BO_3, 40 μg; $MnSO_4 \cdot H_2O$, 22 μg; $Mo_7O_{24}(NH_4)_6 \cdot 4 H_2O$, 153 μg. This medium is adjusted to pH 7.0 by 5 N KOH and sterilized 25 min at 120°. The carbon source is 2% (w/v) glucose (final concentration). It is added to the sterile salts medium as a concentrated (40% w/v) solution that has been sterilized separately (20 min at 105°). The culture is performed at 30° with stirring (1000 rpm) and forced aeration (0.5 liter liter^{-1} min^{-1}). The pH is regulated at 7.0 by 4 N NH_4OH. Cells are harvested in the exponential phase of growth (about 9 hr after inoculation) by centrifugation, washed with 10 mM phosphate buffer (pH 7.1), and kept at −30° until used.

Step 2. Preparation of Extracts. Cells (80 g wet weight) are suspended by portions of 10 g in 200 ml of potassium phosphate buffer (pH 7.2) containing 7 mM 2-mercaptoethanol and 1 mM dithiothreitol; 1.6 mg of DNase (Sigma, DHC) and 8 ml of 0.2 M magnesium sulfate are added. The cell suspensions, immersed in an ice bath, are disintegrated by sonication

[7] O. H. Lowry, N. J. Rosebrough, A. L. Farr, and R. J. Randall, *J. Biol. Chem.* **193**, 265 (1951).

[8] R. W. Traxler and J. M. Bernard, *Int. Biodeterior. Bull.* **5**, 21 (1969).

[9] D. S. Robinson, *Antonie Van Leeuwenhoek J. Microbiol. Serol.* **30**, 303 (1964).

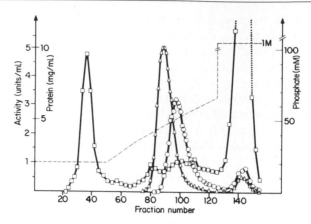

FIG. 1. Separation of the NAD$^+$-dependent and NADP$^+$-dependent aldehyde dehydrogenases by DEAE-cellulose chromatography. A 38 mm × 370 mm column is used. The elution rate is 95 ml hr^{-1}, and fractions of 10 ml are collected. △——△, NADP$^+$-dependent aldehyde dehydrogenase activity; ○——○, NAD$^+$-dependent aldehyde dehydrogenase activity; □——□, protein (mg × ml^{-1}); ----, concentration of elution phosphate buffer (pH 7.8). Protein concentration in fraction 140 is 22 mg ml^{-1}. Adapted from Guerrillot and Vandecasteele,[4] with permission.

(six periods of 15–20 sec, to prevent a temperature increase above 12°). Then the pooled preparations are centrifuged at 3° for 60 min at 100,000 g.

Step 3. Protamine Sulfate Treatment. A 2% protamine sulfate (7.2 ml) solution is added dropwise to the extract (138 ml), which is stirred for 30 min and then centrifuged for 20 min at 48,000 g. From this step on, *all enzyme preparations and chromatography or dialysis buffers* (unless otherwise stated) contain 16% glycerol and either 7 mM 2-mercaptoethanol or 1 mM dithiothreitol (final concentrations) to maintain sufficient stability (see Properties). The protamine sulfate supernatant is dialyzed against 20 mM phosphate buffer pH 7.3.

Step 4. DEAE-Cellulose Chromatography. The dialyzed extract (178 ml) adjusted to pH 7.8 is adsorbed on the top of a DEAE-cellulose column previously equilibrated with 20 mM phosphate buffer (pH 7.8) and eluted with a concentration gradient of the phosphate buffer (pH 7.8) as shown in the elution diagram (Fig. 1). This gradient is obtained with a programmable gradient pump system [Dialagrad from Instrument Specialties Co., Lincoln, Nebraska (Isco)]. A slightly convex gradient with similar initial and final concentrations can also be used and is obtained with a classical two-vessels device, presented earlier in this series.[10] As shown in Fig. 1, separation of the two aldehyde dehydrogenase takes place on this column.

[10] E. A. Peterson and H. A. Sober, this series, Vol. 5 [1].

TABLE I

PURIFICATION OF NAD-DEPENDENT ALDEHYDE DEHYDROGENASE[a,b]

Preparation	Volume (ml)	Protein (mg/ml)	Total activity (units)	Specific activity (units/mg protein)	Yield (%)	Purification (fold)
Supernatant, 100,000 g	138	44	830	0.136	—	—
Protamine supernatant	142	44.8	850	0.134	—	—
Dialyzed supernatant	178	36	710	0.112	100	1
DEAE-cellulose eluate (fractions 95–120)	254	1.9	280	0.57	40	5.1
Hydroxyapatite eluate	232	0.18	140	3.4	20	30
Sephadex G-200 eluate (173–240 ml)	67	0.099	100	15.2	14	135
Fraction (198–203 ml) of Sephadex G-200	5	0.11	11	20	—	178

[a] Adapted from Guerrillot and Vandecasteele,[4] with permission.
[b] Activity determinations are performed in 50 mM $K_3HP_2O_7$ (pH 8.6) with 0.2 mM butanal.

In spite of various attempts, however, complete separation has not been achieved in this step. From this step on, purification of the two enzymes is carried out independently. Fractions 95–120 are pooled together and then dialyzed against 20 mM phosphate buffer (pH 6.8) (Table I).

Step 5. Hydroxyapatite Chromatography. The preparation is concentrated to a volume of 39.5 ml by ultrafiltration (Amicon, UM-20E membrane) and subjected to chromatography on a hydroxyapatite column (25 × 340 nm) previously equilibrated with 20 mM phosphate buffer (pH 6.8). Elution, performed at a flow rate of 30 ml/hr, is started with the same buffer for 250 ml, then continued with a concentration gradient of the phosphate buffer (pH 6.8), obtained by the methods described in step 4, the phosphate concentration increasing from 20 mM at 250 ml to 70 mM at 1000 ml of elution volume.

Step 6. Sephadex G-200 Chromatography. The preparation is then adjusted to 100 mM potassium phosphate (pH 7.1). The Sephadex column (25 × 780 mm) is equilibrated and eluted (flow rate 8 ml/hr) with the same phosphate buffer.

The purest preparation (eluate between 198 and 203 ml) has a specific activity of 20 units per milligram of protein. Electrophoresis of this preparation on polyacrylamide gel allows a rough estimation of enzyme purity of about 20%.

Typical results of the various purification steps are summarized in Table I.

Properties

Molecular Weight. A molecular weight of 225,000 ± 15,000 has been determined by gel filtration on Sephadex G-200.

Apparent Kinetic Constants. Aliphatic aldehydes in a range from C_2 to at least C_{10} are used by this enzyme. Kinetic constants for these aldehydes have been determined in 50 mM potassium pyrophosphate (pH 8.6) in the presence of 1 mM NAD$^+$. K_m values decrease with increasing aldehyde chain length from 165 μM for ethanal, 45 μM for butanal, 9.5 μM for hexanal, and 1.3 μM for decanal, whereas variations of V are small. In the same buffer, an apparent K_m value for NAD$^+$ of 44 μM is found in the presence of 200 μM butanal.

Influence of pH, Buffer, and Ions on Activity. Small variations of K_m values for aldehydes with pH and buffer composition are observed. In Tris buffer, however, a much higher K_m value is found, possibly because of the formation of imine bonds between Tris and aldehyde. V values for aldehydes somewhat increase with pH. Potassium and ammonium ions have a slight activation effect on the enzyme.

Enzyme Stability. Glycerol and reducing agents have a clear stabilization effect on enzyme activity. Dithiothreitol or 2-mercaptoethanol can also partially reactivate the enzyme that has been inactivated by storage in the absence of reducing agents. Alkaline pH values also inactivate the enzyme. At pH 7.4 the enzyme is stable, but at pH 8.6 (in 50 mM potassium pyrophosphate) 90% activity is lost at 30° in 10 min. In the presence of 1 mM NAD$^+$, however, no inactivation takes place under the same conditions. As a consequence, the enzyme is routinely stored and handled in 50 mM potassium phosphate, pH 7.4, containing 16% glycerol and 1 mM dithiothreitol. At −8° under these conditions the purified enzyme loses 25% activity in 6 months.

Regulation of Enzyme Synthesis. Very similar enzyme activities are found in high-speed supernatants of extracts of cells grown on glucose, succinate, malonate, n-heptane, or n-hexadecane. The enzyme is soluble; very little activity is observed in the particulate fraction of the extracts except when hydrocarbons are used as carbon sources. Growth on hydrocarbon induces the synthesis of a specific membrane-bound NAD$^+$-dependent aldehyde dehydrogenase.

Kinetic Studies. Detailed steady-state kinetic studies including bireactant initial velocity and product inhibition experiments have been performed.[11] The results indicate the existence of a sequential mechanism.

---

[11] L. Guerrillot, "Contribution à l'étude du métabolisme des hydrocarbures chez les microorganismes. Etude des aldéhyde déshydrogénases de *Pseudomonas aeruginosa*." Doctoral Thesis No. 2013, University of Paris-Sud-Orsay, 1978.

TABLE II
PURIFICATION OF NADP-DEPENDENT ALDEHYDE DEHYDROGENASE[a,b]

Preparation	Volume (ml)	Protein (mg/ml)	Total activity (units)	Specific activity (units/mg protein)	Yield (%)	Purification (fold)
Supernatant, 100,000 *g*	138	44	850	0.14	—	—
Protamine supernatant	142	44.8	830	0.13	—	—
Dialyzed supernatant	178	36	650.	0.10	100	1
DEAE-cellulose eluate (fractions 80–94)	148	1.5	310	1.31	48	12.8
Hydroxyapatite eluate	170	0.065	180	17.0	28	166
Sephadex G-200 eluate (175–235 ml)	60	0.08	156	32.0	23.5	314
Fraction (200–205 ml) of Sephadex G-200	5	0.09	16.7	37.0		363

[a] Adapted from Guerrillot and Vandecasteele,[4] with permission.
[b] Activity determinations are performed in 50 mM $K_3HP_2O_7$ (pH 8.6) with 4 mM pentanal.

The product inhibition pattern suggests an ordered pathway, but its complexity precludes a simple interpretation.

NADP$^+$-Dependent Aldehyde Dehydrogenase[4] (EC 1.2.1.4)

Assay Method

Assay of enzyme activity is performed as described for the NAD$^+$-dependent enzyme except that the reagents used are 4.3 mM pentanal (from Fluka) dissolved in 54 mM potassium pyrophosphate buffer, pH 8.6, and 15 mM NADP$^+$. The final concentrations in the assay are 4 mM for pentanal, 50 mM for potassium pyrophosphate, and 0.5 mM for NADP$^+$.

Purification Procedure

Steps 1–4 which are the same as for the NAD$^+$-dependent enzyme have been described above. Fractions 80–94 of the chromatography on DEAE-cellulose (Fig. 1), which contain the NADP$^+$-dependent activity, are further purified by chromatography on hydroxyapatite (step 5) and on Sephadex G-200 (step 6) using the same conditions as for the NAD$^+$-dependent enzyme. The results of the complete purification procedure are given in Table II. The purest fraction of the Sephadex G-200 eluate (200–

205 ml) is 90–95% pure from gel electrophoresis determinations. It contains no NAD^+-dependent activity.

Properties

Molecular Weight. A molecular weight of 215,000 ± 15,000 has been obtained by gel filtration on Sephadex G-200.

Apparent Kinetic Constants. Kinetic constants have been determined in 50 mM potassium pyrophosphate, pH 8.6, for aldehyde chain lengths ranging from C_2 to C_{14}, in the presence of 0.5 mM $NADP^+$. K_m values decrease with increasing aldehyde chain lengths from 35,000 μM for ethanal to 435 μM for pentanal, 250 μM for hexanal and 15 μM for tetradecanal. V values present two maxima for pentanal and for decanal. In the same buffer, an apparent K_m value for $NADP^+$ of 65 μM is found in the presence of 4 mM pentanal.

Influence of pH, Buffer, Ions on Activity. K_m and V values for aldehydes increase with pH in phosphate, pyrophosphate, or glycine–KOH buffers. Tris-HCl cannot be used, as very low activities are observed in this buffer. Strong enhancement of enzyme activity by alkaline ions, in particular by potassium ions, are observed.

Enzyme Stability. The protective effect of glycerol as well as the stabilization and reactivation by thiols are also observed for this enzyme. Inactivation at alkaline pH values, however, does not take place in this case. The $NADP^+$-dependent enzyme is handled and stored in the same medium as the NAD^+-dependent enzyme. At $-8°$, there is a 12% loss in activity in 6 months under these conditions.

Regulation of Enzyme Synthesis. The $NADP^+$-dependent enzyme is soluble. Cell-free extracts from cells grown on the carbon sources mentioned above contain quite similar activities. Extracts from ethanol-grown cells, however, contain an additional soluble $NADP^+$-dependent aldehyde dehydrogenase that has not been characterized in detail.

Kinetic Studies. The same studies have been performed as for the NAD^+-dependent enzyme,[11] and the results, although different in some respects, support the same general conclusions.

[82] Aldehyde Dehydrogenase from Acetic Acid Bacteria, Membrane-Bound

By MINORU AMEYAMA and OSAO ADACHI

R-CHO + acceptor → R-COOH + reduced acceptor

Aldehyde dehydrogenase of acetic acid bacteria acts on a wide range of aliphatic aldehydes except for formaldehyde. Aldehydes having a carbon chain length of 2–4 are oxidized most rapidly by the enzyme from genera of both *Acetobacter* and *Gluconobacter*. The enzyme is localized on the outer surface of cytoplasmic membrane of the organisms, and oxidation of aldehyde is linked to its respiratory chain. This enzyme acts as vinegar producer in acetic acid bacteria by coupling with alcohol dehydrogenase (see this volume [76]). During alcohol oxidation, no aldehyde liberation is observed, indicating that alcohol dehydrogenase and aldehyde dehydrogenase function as multienzyme complexes to produce acetic acid.

Assay Method

Principle. The rate of aldehyde oxidation is estimated (*a*) by spectrophotometry in the presence of 2,6-dichlorophenolindophenol and phenazine methosulfate; (*b*) by colorimetry in the presence of potassium ferricyanide; (*c*) by polarography with an oxygen electrode; or (*d*) by manometry in a conventional Warburg apparatus. The assay method with potassium ferricyanide is employed here because of its simplicity for routine assay; it has been described by Wood *et al.*[1] for the assay of D-gluconate dehydrogenase. Some modifications are described below.

Reagents

Potassium ferricyanide, 0.1 *M*, in distilled water
Acetaldehyde, 1 *M*, in distilled water. Redistillation of commercial product and storage in a refrigerator is recommended.
McIlvaine buffer, pH 4.0. This buffer solution is prepared by mixing 0.1 *M* citric acid and 0.2 *M* disodium phosphate.
Ferric sulfate–Dupanol reagent containing 5 g of $Fe_2(SO_4)_3 \cdot n\ H_2O$, 3 g of Dupanol (sodium lauryl sulfate), 95 ml of 85% phosphoric acid, and distilled water to 1 liter
Triton X-100, 10% solution

[1] W. A. Wood, R. A. Fetting, and B. C. Hertlein, this series, Vol. 5, p. 287.

METHODS IN ENZYMOLOGY, VOL. 89

Procedure. The reaction mixture contains 0.1 ml of potassium ferricyanide, 0.5 ml of McIlvaine buffer (pH 4.0), 0.1 ml of 10% Triton X-100, enzyme solution, and 0.1 ml of acetaldehyde in a total volume of 1.0 ml. Aldehyde dehydrogenase is dissolved in 0.01 M sodium acetate buffer (pH 5.3), containing 0.5% Triton X-100, 25 mM benzaldehyde, and 10% sucrose. After preincubation for 5 min at 25°, the reaction is initiated by the addition of potassium ferricyanide and stopped by adding 0.5 ml of ferric sulfate–Dupanol reagent. Then, 3.5 ml of water are added to the reaction mixture. The resulting Prussian blue color is measured at 660 nm after the mixture has stood at 25° for 20 min.

Definition of Unit and Specific Activity. One unit of enzyme activity catalyzes the oxidation of 1 μmol of acetaldehyde per minute under the conditions described above: 4.0 absorbance units equal 1 μmol of acetaldehyde oxidized. Specific activity (units per milligram of protein) is based on the protein estimation by Lowry *et al.*[2] with bovine serum albumin as the standard. A modified method described by Dulley and Grieve[3] is employed for samples that contain detergent. In this method, sodium lauryl sulfate is added to the alkaline solution.

Source of Enzyme

Microorganisms and Cultures. Gluconobacter suboxydans IFO 12528 and *Acetobacter aceti* IFO 3284 can be obtained from the Institute for Fermentation, Osaka (17-85, Juso-honmachi 2-chome, Yodogawa-ku, Osaka 532, Japan). Stock cultures are maintained on potato-glycerol slants (see this volume [24]).

Culture media for *G. suboxydans* and *A. aceti* are the same as those used for study of alcohol dehydrogenase of acetic acid bacteria (see this volume [76]).

Purification Procedure

All operations are carried out at 0–5°, unless otherwise stated. Sodium acetate buffer, pH 5.3, is used for enzyme purification from *G. suboxydans.* Potassium phosphate buffer is used for the purification from *A. aceti.*

[2] O. H. Lowry, N. J. Rosebrough, A. L. Farr, and R. J. Randall, *J. Biol. Chem.* **193**, 265 (1951).
[3] J. R. Dulley and P. A. Grieve, *Anal. Biochem.* **64**, 136 (1975).

Preparation from G. suboxydans [4]

Step 1. Preparation of Membrane Fraction. Cells of *G. suboxydans* (160 g wet weight) harvested from 25 liters of a 24-hr culture are suspended in 0.01 M buffer, pH 5.3. The cell suspension (420 ml) is passed twice through a French press at 1000 kg/cm^2 and subjected to sonication for 60 sec at 20 kHz. Cell debris is removed by centrifugation at 5000 g for 10 min. The cell-free extract is centrifuged at 68,000 g for 90 min. The resulting precipitate is designated as the membrane fraction and suspended in 0.01 M buffer (220 ml) for further purification.

Step 2. Solubilization of Aldehyde Dehydrogenase. Aldehyde dehydrogenase is solubilized from the membrane after alcohol dehydrogenase has been solubilized. To the membrane fraction, 10% Brij 58 and 10% Tween 80 are added to a final concentration of 2% and 1%, respectively. After stirring overnight, the membrane fraction is centrifuged at 68,000 g for 90 min. Aldehyde dehydrogenase is not detected in the red supernatant, which contains alcohol dehydrogenase and other proteins. Thus, the residual precipitate containing aldehyde dehydrogenase is free of alcohol dehydrogenase. To the precipitate are added 250 ml of 0.01 M buffer containing 1.5% Triton X-100, 50 mM benzaldehyde, and 10% sucrose. The mixture is homogenized with a glass homogenizer and subjected to sonication for 60 sec at 20 kHz. The homogenate is stirred for 48 hr and centrifuged at 68,000 g for 90 min; the reddish supernatant contains the solubilized aldehyde dehydrogenase.

Step 3. DEAE-Cellulose Column Chromatography. The solubilized enzyme (245 ml) is applid to a column of DEAE-cellulose (4 × 17 cm) that has been equilibrated with 0.01 M buffer containing 0.5% Triton X-100, 25 mM benzaldehyde, and 10% sucrose. Under these conditions, only protein impurities are adsorbed. The column is washed with an additional 25 ml of the same buffer.

Step 4. CM-Cellulose Column Chromatography. The pooled effluents (250 ml) are passed through a CM-cellulose column (3 × 15 cm) that has been equilibrated with 0.01 M buffer containing 0.5% Triton X-100, 25 mM benzaldehyde, and 10% sucrose; the column is washed with 150 ml of the same buffer. After the column has been washed with 450 ml of 0.06 M buffer containing the same additions as above, elution of the enzyme is effected with 0.075 M buffer. Specific activity of the peak fraction shows a constant specific activity, and fractions with a specific activity of over 420 units per milligram of protein are collected. The pooled fractions (75 ml)

[4] O. Adachi, K. Tayama, E. Shinagawa, K. Matsushita, and M. Ameyama, *Agric. Biol. Chem.* **44**, 503 (1980).

TABLE I
PURIFICATION OF ALDEHYDE DEHYDROGENASE FROM
Gluconobacter suboxydans

Fraction	Total protein (mg)	Total activity (units)	Specific activity (units/mg protein)	Yield (%)
Cell homogenate	15,200	26,000	1.7	100
Membrane fraction	6,200	21,100	3.4	80
Solubilized fraction	558	20,600	37.0	78
DEAE-Cellulose	263	20,110	76.5	76
CM-Cellulose	14	6,060	430.0	23

are concentrated by ultrafiltration (Toyo UP-50) or dehydration with sucrose powder. The enzyme is purified about 250-fold with a yield of 23% (Table I).

Preparation from A. aceti [5]

The preparation of the membrane fraction is essentially the same as described above (step 1). The following purification is made with 150 g of wet cells of *A. aceti* harvested from 30 liters of culture broth.

Step 2. Solubilization of Aldehyde Dehydrogenase. The membrane fraction is suspended in 0.005 M potassium phosphate, pH 7.0, and the protein concentration is adjusted to about 30 mg/ml. Triton X-100 and cetylpyridinium chloride (CPC) are added to 5% and 2%, respectively, to the membrane suspension; solubilization of the enzyme is performed by stirring the suspension for 2 hr. Benzaldehyde is also added to 10 mM as a stabilizing agent. Solubilized enzyme is obtained by centrifugation at 100,000 g for 90 min. Polyethylene glycol 6000 is added to 20% to the supernatant solution, and the precipitate is centrifuged at 12,000 g for 30 min. The precipitate is dissolved in a minimal volume of 0.005 M buffer, pH 7.0, containing Triton X-100 (5%), CPC (2%), and benzaldehyde (10 mM), and insoluble materials are removed by centrifugation.

Step 3. DEAE-Cellulose Column Chromatography. The enzyme is applied to a DEAE-cellulose column (5 × 35 cm) that has been equilibrated with 0.005 M buffer, pH 7.0, containing 5% Triton X-100, 2% CPC, and 10 mM benzaldehyde. The column is washed with the same buffer, and the activity is eluted with 0.02 M buffer, pH 6.0, containing the same additives as above. Polyethylene glycol 6000 is added to 20% to the eluate

[5] M. Ameyama, K. Osada, E. Shinagawa, K. Matsushita, and O. Adachi, *Agric. Biol. Chem.* **45**, 1889 (1981).

TABLE II
PURIFICATION OF ALDEHYDE DEHYDROGENASE FROM *Acetobacter aceti*

Fraction	Total protein (mg)	Total activity (units)	Specific activity (units/mg protein)	Yield (%)
Cell homogenate	13,860	37,800	2.7	100
Membrane fraction	9,720	34,100	3.5	90
Solubilized fraction	1,525	30,000	19.6	79
PEG[a] precipitate	1,120	28,200	25.0	75
DEAE-cellulose (I)	82	10,500	128.0	28
DEAE-cellulose (II)	13	5,800	450.0	15
Hydroxyapatite	4	2,500	654.0	7

[a] Polyethylene glycol 6000.

(180 ml), the precipitate redissolved in a minimal volume of 0.005 M buffer, pH 6.0, containing the additives; the insoluble materials are removed by centrifugation. The enzyme solution is applied to the second DEAE-cellulose column (2 × 35 cm), which has been equilibrated with 0.005 M buffer as above. The activity is eluted with a linear gradient elution formed between 0.005 M and 0.02 M buffer, pH 6.0, and the peak fraction of the enzyme appears at about 0.013 M potassium phosphate buffer.

Step 4. Hydroxyapatite Column Chromatography. The enzyme solution is diluted to 0.01 M buffer with a solution containing 5% Triton X-100, 2% CPC, and 10 mM benzaldehyde and is applied to a hydroxyapatite column (1 × 2 cm) that has been equilibrated with 0.01 M buffer, pH 6.0, containing the same additives. After thorough washing with the same buffer, the enzyme is eluted with 0.03 M buffer, pH 6.0. Polyethylene glycol 6000 is added to 20% to the pooled fraction (300 ml), and the precipitate is treated as above. The precipitate or concentrated enzyme solution can be stored at −20° or below without appreciable loss of activity.

A typical purification from *A. aceti* is summarized in Table II.

Properties[4,5]

Homogeneity. Purified enzyme preparations from both genera are homogeneous by a conventional polyacrylamide gel electrophoresis at pH 8.3 (7.3% polyacrylamide). The single protein band with enzyme activity appears at an R_f value of 0.22.

Absorption Spectra. Both enzymes are a rose-red and show cytochrome c-like absorption spectra with absorption maxima at 551 nm, 523 nm, and 418 nm (reduced enzyme) and a sole peak at 410 nm (oxidized enzyme).

Subunit Composition. The enzyme from *G. suboxydans* comprises two components, which are dissociated by sodium dodecyl sulfate gel electrophoresis. One component, of molecular weight (M_r) 86,000, is the dehydrogenase protein having intense fluorescence. The second component (M_r 55,000) is cytochrome *c*. The sum of these two components gives 140,000 total molecular weight. The enzyme from *A. aceti* dissociates, in sodium dodecyl sulfate gel electrophoresis, into three components of molecular weights 78,000, 45,000, and 14,000 (a total molecular weight of 137,000). The dehydrogenase is the largest component; the second subunit (M_r 45,000) is the cytochrome component; the smallest component has no known function.

Heme Content. The purified enzyme from *G. suboxydans* contains 1 mol of heme *c* per 140,000 g of protein (1 mol of enzyme) when calculated from the pyridine hemechrome spectrum. The midpoint redox potential of the cytochrome is determined to be $+135$ mV at pH 7.0 (25°) from titration with ferrous ammonium sulfate in the presence of ferric ammonium sulfate and EDTA anaerobically.

Stability. The purified enzyme from both genera can be stored at 4° for several weeks without appreciable loss of enzyme activity when Triton X-100 and benzaldehyde are added to 1% and 10 mM, respectively. Both preparations are most stable at around pH 4.0. Sucrose or glycerol in the enzyme solution facilitates the enzyme stability. Removal of detergent causes aggregation and rapid loss of activity.

Catalytic Properties. Aldehyde dehydrogenase from both genera can be assayed *in vitro* in the presence of any one of the following dyes: potassium ferricyanide, phenazine methosulfate, nitro blue tetrazolium, or 2,6-dichlorophenolindophenol. NAD, NADP, or oxygen are completely inactive as electron acceptors. Aliphatic aldehydes with a straight carbon chain length of 6 or less (except formaldehyde) are rapidly oxidized by the enzyme from *G. suboxydans*. Benzaldehyde added during purification and subsequent storage is also oxidized to about 5% of the rate for acetaldehyde. Substrate specificities of the two enzymes are quite similar. The apparent Michaelis constant for acetaldehyde is 3.3 mM (*G. suboxydans*) and 2.9 mM (*A. aceti*). The pH optima of aldehyde oxidation with the enzyme from *G. suboxydans* and *A. aceti* are 4.0 and 5.0, respectively.

Prosthetic Group. The absorption spectrum of the extracted prosthetic group shows maxima at 268 nm and 298 nm. The fluorescence emission peak is at 450 nm with excitation at 370 nm, and two excitation peaks at 360 nm and 290 nm are observed with emission at 450 nm. These spectral characteristics are quite similar to those obtained with methanol dehydrogenase of methanol-utilizing bacteria[6] and D-glucose dehydrogenase of *Acinetobacter*.[7] It is quite probable that aldehyde dehydrogenase contains pyrroloquinoline quinone[8] as the prosthetic group. D-Glucose dehydrogenase of *Pseudomonas* (this volume [24]) and alcohol dehydrogenase of acetic acid bacteria (this volume [76]) also have pyrroloquinoline quinone as a prosthetic group.

[6] J. A. Duine, J. Frank, Jr., and J. Westerling, *Biochim. Biophys. Acta* **524**, 277 (1978).
[7] J. A. Duine, J. Frank, Jr., and J. K. Van Zeeland, *FEBS Lett.* **108**, 443 (1979).
[8] S. A. Salisbury, H. S. Forrest, W. B. T. Crure, and O. Kennard, *Nature (London)* **280**, 843 (1979).

[83] Aldehyde Dehydrogenase from Bovine Liver

By ANDREA WESTERHAUSEN, WOLFGANG LEICHT, and FRITZ HEINZ

$$RCHO + NAD^+ + H_2O \rightarrow RCOOH + NADH + H^+$$

An isolation method for aldehyde dehydrogenase from bovine liver is described in this chapter.

Assay Method

Principle. The enzyme catalyzes the conversion of DL-glyceraldehyde to the corresponding acid in the presence of its coenzyme, NAD^+. NADH formation is followed spectrophotometrically.

Reagents

Sodium pyrophosphate buffer, 0.87 M, pH 8.5
NAD^+, 0.65 mM
DL-Glyceraldehyde, 9.65 mM

Procedure. Assay mixture (1.1 ml), containing 0.5–2 enzyme units, is incubated at 25° in a semimicro glass cuvette with a light path of 10 mm. The reaction is initiated by the addition of 50 μl of DL-glyceraldehyde solution. The increase of absorbance at 366 nm is recorded.

Unit. One unit is defined as the amount of enzyme producing 1 μmol of NADH in 1 min under the given assay conditions.

Purification Procedure

All operations are carried out at 4°. Protein concentrations are determined by the method of Bradford.[1]

Step 1. Preparation of the Cell-Free Extract. Frozen bovine liver (0.5 kg; −20°) is cut into small pieces and homogenized in a blender with three parts (v/w) of a solution containing 5 mM triethanolamine, 10 mM EDTA, and 20 mM mercaptoethanol; pH 8.0 (buffer A). After centrifugation (15 min at 13,000 g), the supernatant is filtered through glass wool.

Step 2. Fractionation by Alcohol Precipitation. Cold ethanol (0.7 volume; −20°) is added dropwise to one volume of the supernatant; the solution is stirred gently, and its temperature is kept between 0° and −3° using a cold finger. After centrifugation (15 min, 13,000 g) another 0.4 volume (relative to the cell-free extract) of cold ethanol is slowly added, the temperature being held between −5° and −8°. The suspension is centrifuged (20 min, 13,000 g), and the precipitate is serially extracted three times with 3 mM triethanolamine–200 mM mercaptoethanol, pH 8.0 (buffer B). The extracts are combined for further purification.

Step 3. CM-Cellulose Treatment. The extract is diluted by adding 0.5 volume of cold distilled water. To evaluate the optimum amount of CM-cellulose to be used for the batch treatment, the following procedure is used: An aliquot of the enzyme solution is adjusted to pH 5 with acetic acid. Starting with 1 g wet weight, increasing amounts of CM-cellulose (equilibrated with 50 mM acetate buffer, pH 5) are added successively. The suspension is stirred for 5 min and filtered in a Büchner funnel; the specific enzyme activity of the filtrate is determined. The amount of CM-cellulose giving an optimum yield and specific activity is scaled up for the removal of contaminating material from the bulk preparation. The solution is then immediately adjusted to pH 8 with 3 M ammonium hydroxide solution.

Step 4. Ammonium Sulfate Fractionation. Powdered ammonium sulfate (31.3 g/100 ml) is added to the solution with stirring; the precipitate is removed by centrifugation (15 min; 15,000 g), then 17.6 g/100 ml of ammonium sulfate are added to the supernatant and stirring is continued for 1 hr. The precipitated protein is collected by centrifugation (15 min; 15,000 g), dissolved in buffer B, and dialyzed for 15 hr against three 2-liter portions of buffer B.

Step 5. Ion-Exchange Chromatography on DEAE-Cellulose. The dialyzed enzyme solution (15 ml) is applied to a DEAE-cellulose column (2.5 × 45 cm), washed with buffer B, and eluted with a linear 1.4-liter gradient of 0 to 0.1 M sodium chloride in buffer B. The enzymically active fractions are pooled and concentrated by ammonium sulfate precipitation

[1] M. M. Bradford, *Anal. Biochem.* **72**, 248 (1976).

TABLE I
PURIFICATION OF ALDEHYDE DEHYDROGENASE

Step	Specific activity (units/mg)	Protein[a] (mg)	Purification (fold)	Yield (%)
1. Preparation of cell-free extract	0.013	23,000	1	100
2. Ethanol fractionation	0.108	2,035	8.3	74
3. CM-cellulose treatment	0.14	1,200	11	56
4. Ammonium sulfate fractionation	0.6	282	46	57
5. DEAE-cellulose chromatography	0.84	37	65	10.4
6. Molecular sieve chromatography	1.4	19	100	8.3

[a] Protein concentrations are determined by the method of Bradford.[1]

(51.6 g/100 ml). The precipitate is sedimented (20 min; 15,000 g) and can be stored at $-20°$.

Step 6. Molecular Sieve Chromatography. The precipitated protein is dissolved in 2 ml of buffer B, dialyzed overnight against 600 ml of buffer B, and subjected to molecular sieve chromatography on an ACA-34 Ultrogel (LKB, Sweden) column (1.5 × 85 cm). Enzymically active fractions are pooled, concentrated by ammonium sulfate precipitation (51.6 g/100 ml), and stored at $-20°$.

A typical purification is summarized in Table I.

Properties

Stability. The enzyme is stable for at least 6 weeks when frozen at $-20°$ as an ammonium sulfate precipitate at either step 5 or 6. At 4°, aldehyde dehydrogenase solutions in buffer B retained 90% of their activity for 20 days in the presence of 1% Ampholine pH 3–10 (LKB, Sweden). Without the Ampholine, 30% of the activity was lost within 5 days.[2]

Homogeneity. The enzyme appears to be homogeneous when subjected to electrophoresis in 5% polyacrylamide gels according to Ornstein and Davis[3] and in sodium dodecyl sulfate disc gels.[4] Isoelectric focusing in an Ampholine gradient (pH 3 to 10) shows a single band containing all of

[2] A. Westerhausen, Diplomarbeit, University of Hannover, Hannover, Germany, 1981.
[3] L. Ornstein and B. Davis, *Ann. N. Y. Acad. Sci.* **121**, 404.
[4] K. Weber and M. Osborn, *J. Biol. Chem.* **244**, 4406 (1969).

TABLE II
MICHAELIS CONSTANTS FOR ALDEHYDE
DEHYDROGENASE[a]

Substrate	K_m (μM)	Relative reactivities
DL-Glyceraldehyde	170	1
D-Glyceraldehyde	170	—
L-Glyceraldehyde	170	—
Ethanal	130	1.4
Propanal	120	1.2
Butanal	97	1.0
Pentanal	100	1.0
Hexanal	80	0.8
Heptanal	77	0.8
Octanal	66	0.8
Nonanal	60	0.8
Decanal	55	0.8
NAD$^+$	47	—

[a] The constants are determined fluorometrically in an assay mixture containing 0.1 M sodium pyrophosphate, pH 8, 1.0 mM NAD$^+$, and 0.02 enzyme unit.[2]

the enzymic activity. In sedimentation velocity experiments in an analytical ultracentrifuge, one symetrical peak is observed.

Physical Properties.[5] A molecular weight of 220,000 ± 10,000 can be determined by equilibrium experiments in an analytical ultracentrifuge using a partial specific volume of 0.734 cm^3/g, calculated from the amino acid composition. Analysis by sodium dodecyl sulfate disc electrophoresis indicates that the molecular weight of the subunits is 55,000, thus demonstrating a tetrameric enzyme structure. Charcoal-treated aldehyde dehydrogenase has an absorption coefficient of 1.03 cm^2 mg^{-1} at 280 nm based on enzyme dry weight. The $s_{20,w}$ value is 8,[6] and the isoelectric point is 5.

Chemical Composition.[5] The amino acid composition shows the high content of cysteine (32 mol per mole of enzyme). The amino-terminal amino acid is alanine, and the carboxyl-terminal sequence is Glx-Ala.

Enzymic Properties. The enzyme is activated by divalent cations such as Mg^{2+}, Mn^{2+} and Ca^{2+}; 0.2 mM Mg^{2+} enhances the activity twofold. Aldehyde dehydrogenase readily catalyzes the oxidation of DL- as well as D- and L-glyceraldehyde and straight-chain saturated aldehydes. The apparent K_m values are given in Table II.

[5] W. Leicht, F. Heinz, and B. Freimüller, *Eur. J. Biochem.* **83**, 189 (1978).
[6] W. Leicht, Diplomarbeit, University of Hannover, Hannover, Germany, 1974.

[84] Aldehyde Reductase (L-Hexonate : NADP Dehydrogenase) from Pig Kidney

By T. G. FLYNN, J. A. CROMLISH, and W. S. DAVIDSON

D-Glucuronate + NADP + H$^+$ $\rightleftharpoons$ L-gulonate + NADP$^+$

D-Glyceraldehyde + NADPH + H$^+$ $\rightleftharpoons$ Glycerol + NADP$^+$

Pyridine-3-aldehyde + NADPH + H$^+$ $\rightleftharpoons$ pyridine-3-carbinol + NADP$^+$

p-Nitrobenzaldehyde + NADPH + H$^+$ $\rightleftharpoons$ p-nitrobenzyl alcohol + NADP$^+$

Aldehyde reductase (L-hexonate : NADP dehydrogenase, EC 1.1.1.19) from pig kidney is a monomeric oxidoreductase of broad substrate specificity.[1,2] The enzyme catalyzes the reduction of a wide range of aliphatic and aromatic aldehydes and in this and other respects is very similar to another monomeric NADPH-linked oxidoreductase, aldose reductase (alditol : NADP$^+$ 1-oxidoreductase, EC.1.1.1.21). These enzymes, however, are not identical as shown by lack of immunological reactivity between antisera to aldose reductase and aldehyde reductase from the same species.[3]

Assay

Procedure. Aldehyde reductase is assayed spectrophotometrically by determining the decrease in absorbance of NADPH at 340 nm. The reaction is carried out in quartz cuvettes in a double-beam recording spectrophotometer at 25°. Reaction mixtures contain a final concentration of 100 mM sodium phosphate buffer, pH 7.0, 160 μM NADPH, 5 mM pyridine-3-aldehyde, and 0.1 ml of enzyme solution in a total volume of 3.0 ml. Reactions are initiated by the addition of enzyme. Blank cuvettes contain all reagents at the same concentration, but with pyridine-3-aldehyde omitted.

Enzyme Units and Specific Activity. One unit of enzyme activity is defined as the amount of enzyme that causes the oxidation of 1 μmol of NADPH per minute at 25°. Specific activity is expressed in units of activity per milligram of protein. Protein is determined by the Bio-Rad protein assay,[4,5] using bovine γ-globulin as standard.

[1] W. F. Bosron and R. L. Prairie, *J. Biol. Chem.* **247**, 4485 (1972).

[2] T. G. Flynn, J. Shires, and D. J. Walton, *J. Biol. Chem.* **250**, 2933 (1975).

[3] K. H. Gabbay and E. S. Cathcart, *Diabetes* **23**, 460 (1974).

[4] Bio-Rad Technical Bulletin 1051, April 1977.

[5] M. M. Bradford, *Anal. Chem.* **72**, 248 (1976).

Purification Procedure

The procedure used is essentially that of Davidson and Flynn[6] with minor modifications. All procedures are carried out at 0–4°.

Buffers

Buffer A: 10 mM sodium phosphate buffer, pH 7.0, containing 5 mM 2-mercaptoethanol

Buffer B: 10 mM sodium phosphate buffer, pH 7.0, containing 5 mM 2-mercapatoethanol and 1 M NaCl

Buffer C: 5 mM sodium phosphate buffer, pH 7.4, containing 5 mM 2-mercaptoethanol

Step 1. Pig kidneys fresh from the slaughterhouse or thawed after storage at $-20°$ are cut into small pieces after fat and the outer capsule have been removed. It is convenient to start the procedure with tissue from 5–7 kidneys (about 500–800 g). The tissue is homogenized in two volumes of buffer A in a Waring blender for 3–4 min. The homogenate is centrifuged at 13,200 g for 90 min; the supernatant is filtered through glass wool and retained.

Step 2. Solid $(NH_4)_2SO_4$ (23.0 g/100 ml) is added with stirring over a period of 45 min. After stirring a further 15 min, the mixture is centrifuged and the precipitate is discarded. The supernatant is collected and filtered through glass wool. To this supernatant solid $(NH_4)_2SO_4$ (14.2 g/100 ml) is added with stirring over a period of 45 min and further stirred for 15 min. The mixture is centrifuged and the supernatant is discarded. The precipitate is dissolved in a minimal volume (about 140 ml) of buffer A.

Step 3. The dissolved precipitate from step 2 is applied to a column of Sephadex G-100 (10 × 90 cm) previously equilibrated with buffer A. The column is eluted with the same buffer and fractions (about 15 ml per fraction) are collected. Fractions with enzyme activity are pooled.

Step 4. Blue Dextran-Sepharose 4B prepared by the method of Ryan and Vestling[7] is washed several times in buffer A and allowed to stand in this buffer for 12 hr. The pooled fractions from step 3 are added to a slurry of Blue Dextran-Sepharose 4B in buffer A with stirring, and the mixture is stirred for a further 12 hr. The mixture is allowed to settle, then the supernatant is removed by aspiration and discarded. The remaining slurry of Blue Dextran-Sepharose 4B is poured into a glass column (2.5 × 60 cm) and washed with 1 liter of buffer A. The column is then eluted with a linear salt gradient (750 ml of buffer A + 750 ml of buffer B). Fractions (approximately 5 ml) containing enzyme activity are pooled and concentrated to 20 ml by ultrafiltration.[8]

[6] W. S. Davidson and T. G. Flynn, *Biochem. J.* **177**, 595 (1979).

[7] L. D. Ryan and C. S. Vestling, *Arch. Biochem. Biophys.* **160**, 279 (1974).

[8] PM-10 membrane, Amicon Corp., Lexington, Massachusetts.

TABLE I

PURIFICATION OF PIG KIDNEY ALDEHYDE REDUCTASE[a]

Fraction	Volume (ml)	Total protein (mg)	Total activity (units)	Specific activity (units/mg protein)	Recovery (%)	Purification (fold)
Supernatant from crude extract	1010	28,600	977	0.034	100	1.0
Sephadex G-100	1150	6,330	779	0.123	80	3.6
Blue Dextran-Sepharose 4B	430	258	583	2.230	60	66.0
Sephadex G-100	225	124	588	4.900	60	138.0
DEAE-cellulose (DE-52)	128	51	278	5.450	28	267.0

[a] From Davidson and Flynn.[6]

TABLE II
AMINO ACID COMPOSITION OF PIG KIDNEY
ALDEHYDE REDUCTASE[a]

Amino acids	Residues
Lysine	18
Histidine	6
Arginine	17
Aspartic acid and asparagine	28
Threonine	14
Serine	13
Glutamic acid and glutamine	37
Proline	20
Glycine	20
Alanine	28
Valine	24
Methionine[b]	3
Isoleucine	14
Leucine	32
Tyrosine	8
Phenylalanine	8
Cysteine[c]	5
Tryptophan[d]	4
Amide ammonia[e]	11
Total[f]	299

[a] Values are calculated assuming a molecular weight of 33,000.
[b] Estimated as sulfone in performic acid-oxidized enzyme.
[c] Estimated as cysteic acid and as carboxymethyl derivative.
[d] Estimated spectrophotometrically by the method of T. W. Goodwin and R. A. Morton, *Biochem. J.* **40**, 628 (1946).
[e] Estimated by microdiffusion after hydrolysis with 1 N sulfuric acid by the method of C. H. W. Hirs, W. H. Stein, and S. Moore, *J. Biol. Chem.* **211**, 941 (1954).
[f] Excluding amide ammonia.

Step 5. The concentrated pool of enzyme activity from step 4 is applied to a Sephadex G-100 column (5 × 85 cm) previously equilibrated with buffer A, and the column is eluted with the same buffer. Fractions (5 ml) containing enzyme activity are pooled and concentrated by ultrafiltration.

Step 6. The concentrated enzyme solution is applied to a column (2.5 × 30 cm) of DEAE-cellulose (DE-52) previously equilibrated with buffer C. The column is eluted with buffer C. Aldehyde reductase activity is asso-

TABLE III
K_m VALUES FOR SUBSTRATES OF
ALDEHYDE REDUCTASE

Substrate	K_m (mM)
D-Glyceraldehyde	4.8[a]
D-Glucoronate	8.4[b]
Pyridine-3-aldehyde	2.6[b]
4-Nitrobenzaldehyde	0.35[b]

[a] Taken from Davidson and Flynn.[6]
[b] Taken from Bosron and Prairie.[1]

ciated with the major protein peak. Active fractions are pooled, dialyzed exhaustively (48 hr with 4 × 2 liters of water), and lyophilized.

A typical purification is summarized in Table I.

Properties

Purity. When subjected to polyacrylamide gel electrophoresis at pH 8.8, the purified enzyme appears as a single protein staining band. Sometimes two or three very faint additional protein bands appear. These may be removed by dialyzing the enzyme solution exhaustively against distilled water followed by chromatography on DEAE-Sephacel using a linear salt gradient (1000 ml of buffer C + 1000 ml of 200 mM phosphate buffer, pH 7.4, 5 mM 2-mercaptoethanol).

Molecular Weight. On polyacrylamide gels in sodium dodecyl sulfate, aldehyde reductase migrates as a single band; by reference to appropriate molecular weight standards a molecular weight of 36,700 is calculated. The molecular weight obtained from gel filtration on Sephadex G-100 is 33,000. Aldehyde reductase is therefore monomeric.

Amino Acid Composition. The amino acid composition of pig kidney aldehyde reductase is shown in Table II. It is unusual only in the proline content, which is relatively high (20 residues per 300 residues).

Coenzyme Specificity and Binding. Aldehyde reductase uses NADPH exclusively as coenzyme. The dissociation constant, K_D, for binding of NADPH, as determined by fluorescence spectroscopy is 25 μM.[2] The K_m for NADPH is 9.6 μM.[6]

Substrate Specificity and Kinetic Constants. The relative rates of reduction of different aldehydes by aldehyde reductases are (values in parentheses): 4-nitrobenzaldehyde (290), pyridine-3-aldehyde (320), D-glyceraldehyde (100), D-glucuronate (54), D-erythrose (6), and D-arabinose (0). K_m values for various substrates are shown in Table III.

Stereochemistry of Reduction. In the reduction of D-glyceraldehyde catalyzed by pig kidney aldehyde reductase the *pro-4R* "A" hydrogen is transferred from NADPH. In addition, the *pro-4R* "A" hydrogen attacks the *re* face of the carbonyl group of D-glyceraldehyde.

Kinetic Mechanism. Initial-velocity analysis and product-inhibition data shown that pig kidney aldehyde reductase follows an Ordered BiBi reaction mechanism in which NADPH binds first.

Inhibition. Phenobarbital is a potent inhibitor of pig kidney aldehyde reductase, inhibiting both substrate and coenzyme noncompetitively ($K_i =$ 80 μM and 67 μM, respectively). The enzyme is also inhibited by high concentrations of p-chloromercuribenzoate, and the inhibition can be partially reversed by 2-mercaptoethanol. Pig kidney aldehyde reductase is also inhibited by arginine-specific reagents, e.g., 2,3-butanedione and phenylglyoxal, and the loss in activity may be attributed to a single essential arginine.[9] The enzyme is also inhibited by pyridoxal 5'-phosphate,[10] diethylpyrocarbonate,[10] and by iodoacetic acid,[1] hydroxylamine,[1] and sodium fluoride.[1]

[9] W. S. Davidson and T. G. Flynn, *J. Biol. Chem.* **254**, 3724 (1979).
[10] W. S. Davidson and T. G. Flynn, *Fed. Proc., Fed. Am. Soc. Exp. Biol.* **37**, 1509 (1978).

[85] Aldehyde Reductase from Human Tissues

By JEAN-PIERRE VON WARTBURG and BENDICHT WERMUTH

$$\text{Aldehyde} + \text{NADPH} + \text{H}^+ \rightarrow \text{alcohol} + \text{NADP}^+$$
$$\text{D-Glucuronate} + \text{NADPH} + \text{H}^+ \rightarrow \text{L-gulonate} + \text{NADP}^+$$
$$\text{Mevaldate} + \text{NADPH} + \text{H}^+ \rightarrow \text{mevalonate} + \text{NADP}^+$$
$$\text{L-Lactaldehyde} + \text{NADPH} + \text{H}^+ \rightarrow \text{1,2-propanediol} + \text{NADP}^+$$

Aldehyde reductase (EC 1.1.1.2; alcohol : NADP$^+$ oxidoreductase) is a ubiquitous enzyme in mammalian tissues, catalyzing the reduction of aromatic and aliphatic aldehydes. A comparison of the properties of aldehyde reductase with other carbonyl reductases shows that it is identical with glucuronate reductase (EC 1.1.1.19; L-gulonate : NADP$^+$ oxidoreductase, or L-gulonate dehydrogenase, or L-hexonate dehydrogenase), with mevaldate reductase (EC 1.1.1.33; mevalonate : NADP$^+$ oxidoreductase), and with lactaldehyde reductase (EC 1.1.1.55; 1,2-

propanediol : NADP$^+$ oxidoreductase).[1,2] On the other hand, aldehyde reductase is different from aldose reductase,[3-5] although they share some structural properties and overlapping substrate specificity. Based on specific activities, aldehyde reductase has also been termed daunorubicin reductase[6] and succinic semialdehyde reductase.[7] The latter term, however, should be reserved for an enzyme that is even more specific for succinic semialdehyde and differs in structure.[3,8,9]

Assay Method

Principle. Aldehyde reductase activity is measured spectrophotometrically by monitoring the oxidation of NADPH at 340 nm as a function of time. The presence of other NADPH oxidizing enzymes may lead to nonspecific blank reactions in the absence of any aldehyde substrate. Especially with crude enzyme preparations, the rate of the blank must be recorded before addition of the substrate. Depending on the substrate used, aldose reductase and alcohol dehydrogenase may interfere. The interference with alcohol dehydrogenase can be excluded by the addition of pyrazole. An estimation of the nonspecific contribution from other enzymes may also be obtained if aldehyde reductase is inhibited by the addition of 1.0 mM phenobarbitone or diphenylhydantoin. Furthermore, it is possible to distinguish between aldehyde and aldose reductase on the basis of activity measurements with substrates preferred by the one enzyme or by the other.

Reagents

Sodium phosphate buffer, 0.1 M, pH 7.0
NADPH, 1.6 mM, prepared daily in deionized water and kept on ice
p-Nitrobenzaldehyde, 5 mM in assay buffer
Sodium D-glucuronate, 0.1 M, in assay buffer
Inhibitors
    Sodium phenobarbitone, 0.01 M in deionized water
    Diphenylhydantoin, 0.01 M, dissolved in 0.01 M NaOH

[1] J. P. von Wartburg and B. Wermuth, *in* "Enzymatic Basis of Detoxication" (W. B. Jakoby, ed.), Vol. 1, p. 249. Academic Press, New York, 1980.
[2] R. L. Felsted and N. R. Bachur, *Drug Metab. Rev.* **11**, 1 (1980).
[3] P. L. Hoffman, B. Wermuth, and J. P. von Wartburg, *J. Neurochem.* **35**, 354 (1980).
[4] M. M. O'Brien and P. J. Schofield, *Biochem. J.* **187**, 21 (1980).
[5] B. Wermuth and J. P. von Wartburg, this volume [30].
[6] R. I. Felsted, M. Gee, and N. R. Bachur, *J. Biol. Chem.* **249**, 3672 (1974).
[7] C. Cash, M. Maître, and P. Mandel, *C. R. Hebd. Seances Acad. Sci. Ser. D* **286**, 1829 (1978).
[8] B. Tabakoff and J. P. von Wartburg, *Biochem. Biophys. Res. Commun.* **63**, 957 (1975).
[9] C. Cash, M. Maître, and P. Mandel, *J. Neurochem.* **33**, 1169 (1979).

Aldehyde substrates and inhibitors not readily water soluble can be dissolved in 12.5% methanol, giving a final methanol concentration of 1.25% in the assay mixture, which does not interfere with enzyme activity. Special procedures exist for the preparation of succinic semialdehyde from L-glutamic acid[10] or biogenic aldehydes from biogenic amines.[11]

Procedure. The standard reaction mixture contains 0.1 ml of sodium glucuronate, 0.05 ml of NADPH, 0.005–0.1 ml of enzyme, and sodium phosphate buffer to a total volume of 1.0 ml. Reactions are initiated by addition of substrate. If necessary the activity must be corrected for blank reactions without substrate. The reverse reaction is assayed at pH 8.8 in 0.1 M sodium pyrophosphate buffer containing 1 mM NADP$^+$ and 0.1 M benzyl alcohol or 20 mM L-gulonate. Because of the unfavorable equilibrium of the reaction, it is essential to use a scale expander.

Units. One unit of enzyme is defined as the amount that causes the oxidation of 1 μmol of NADPH per minute. Specific activity is defined as number of units per milligram of protein. Protein concentration of crude preparations is estimated by measuring the absorption at 280 and 260 nm,[12] and of pure preparations from the absorption at 280 nm using an extinction coefficient of 54,300 M^{-1} cm^{-1}.

Purification Procedure

The method of purification is a modification of the procedure described by Wermuth *et al.*[13] Human livers that appeared normal were obtained from legal autopsies, causes of death being traffic accidents, homicides, or cardiovascular diseases. The organs were frozen 6–20 hr after death and stored at $-20°$. To avoid bacterial growth, all buffers contain 1 mM NaN$_3$ (which has no effect on the catalytic activity of aldehyde reductase), and the entire purification is carried out at 4°.

Step 1. Extraction. One liver (1–1.5 kg) is homogenized with twice the volume (v/w) of 50 mM Na$_2$HPO$_4$ in a Waring blender for 3 min. This concentration of secondary phosphate buffer is needed in order to obtain a final pH of the extract of approximately 7. The homogenate is centrifuged for 1 hr at 15,000 g. The lipid layer is carefully removed before collection of the supernatant. The extract is dialyzed for 2 days twice against 15 volumes of 5 mM sodium phosphate buffer, pH 7.0.

Step 2. DEAE-Cellulose Chromatography. Adsorption of aldehyde reductase to DE-52 cellulose is carried out using a batchwise procedure.

[10] R. A. Anderson, R. F. Ritzmann, and B. Tabakoff, *J. Neurochem.* **28**, 633 (1977).
[11] B. Tabakoff, R. A. Anderson, and S. G. A. Alivisatos, *Mol. Pharmacol.* **9**, 428 (1973).
[12] E. Layne, this series, Vol. 3, p. 447.
[13] B. Wermuth, J. D. B. Münch, and J. P. von Wartburg, *J. Biol. Chem.* **252**, 3821 (1977).

TABLE I
PURIFICATION OF ALDEHYDE REDUCTASE FROM HUMAN LIVER

Fraction	Volume (ml)	Protein (mg)	Total activity (units)	Specific activity (units/mg protein)	Yield (%)
1. Extract	1970	190,000	500	0.0026	100
2. DEAE-cellulose	560	4400	343	0.078	69
3. Sephadex G-100	154	169	304	1.8	61
4. Cibacron Blue–Sepharose	350	31.7	254	8.0	51
5. Hydroxyapatite	370	15.0	160	10.7	32
6. DEAE-Sepharose	105	11.5	136	11.8	27

Half a gram of wet cellulose per gram of original liver is added to the dialyzed extract and kept suspended by light stirring for 1 hr. The cellulose is then sedimented for 5 min at 2000 g. The supernatant is discarded and the cellulose is washed three times with 4 volumes of 5 mM sodium phosphate buffer, pH 7, before it is packed into a column (40 × 7 cm). The absorbed proteins are eluted with six column volumns of a linear gradient of 5 to 100 mM sodium phosphate buffer, pH 7.0. Fractions with aldehyde reductase activity are pooled and concentrated in an Amicon ultrafiltration apparatus to obtain approximately 100 mg of protein per milliliter.

Step 3. Gel Filtration. The concentrated aldehyde reductase solution is applied to a Sephadex G-100 column (190 × 7 cm). Elution is performed with 5 mM sodium phosphate buffer, pH 6.5. Fractions with enzyme activity are pooled and used directly for the next step.

Step 4. Cibacron Blue-Sepharose. Cibacron Blue-Sepharose is prepared as described in this volume.[5] The pooled enzyme is applied to a column of Cibacron Blue-Sepharose (7 × 30 cm) equilibrated with 5 mM sodium phosphate buffer, pH 6.5. The column is washed with 700 ml of the same buffer before elution of the enzyme with 100 mM sodium phosphate buffer, pH 7.0 containing 250 mM ammonium sulfate. Fractions containing aldehyde reductase are pooled and dialyzed against 10 mM sodium phosphate buffer, pH 7.0.

Step 5. Hydroxyapatite. Hydroxyapatite, 1 g/5 mg of protein is packed into a column (30 × 3.5 cm) and equilibrated with 10 mM sodium phosphate, pH 7.0. The applied enzyme is not adsorbed to the hydroxyapatite. Low yields are obtained when fresh hydroxyapatite is used. This can be avoided by washing the hydroxyapatite with bovine serum albumin before applying the aldehyde reductase.

Step 6. DEAE-Sepharose. The enzyme from the hydroxyapatite chromatography is applied to DEAE-Sepharose equilibrated with 10 mM

TABLE II

PHYSICOCHEMICAL PROPERTIES OF HUMAN LIVER
ALDEHYDE REDUCTASE

Molecular weight[a]	36,200
Stokes' radius	2.65 nm
Sedimentation coefficient	2.9 S
Diffusion constant	7.5×10^{-7} cm^2 sec^{-1}
Frictional ratio	1.18
Isoelectric point	pH 5.3
Extinction coefficient	54,300 M^{-1} cm^{-1}
α helicity[b]	12%
Partial specific volume	0.74 cm^3/g

[a] Average from determinations by sodium dodecyl
sulfate–polyacrylamide gel electrophoresis, gel
filtration, and ultracentrifugation.
[b] Average of values computed from the ellipticities
at 208 nm and 222 nm.

sodium phosphate buffer, pH 7.0 and packed in a column (20 × 2.4 cm).
Aldehyde reductase is eluted with six column volumes of a linear gradient
of 10 to 100 mM sodium phosphate, pH 7.0. Fractions containing aldehyde reductase activity are pooled, concentrated in an Amicon ultrafiltration apparatus to approximately 1 mg of protein per milliliter, and stored
at 4°.

A typical purification is summarized in Table I.

Alternative purification procedures have been described for human
brain.[4,14]

Properties

Purity and Stability. The described purification procedure yields a homogeneous enzyme that is stable for several weeks when stored in 10 mM
phosphate buffer, pH 7.0, at 4°. The enzyme purified from brain still
contains some contaminating proteins.[4,14]

Physicochemical Properties. The physicochemical properties of aldehyde reductase from human liver are listed in Table II. One molecule of
NADPH binds to the enzyme, causing a red shift of the absorption maximum of the coenzyme. The amino acid composition of the enzyme is given
in Table III and compared with aldose reductase from human brain.[5]

pH Optimum. For aldehyde reduction with NADPH the enzyme has an
optimum activity at pH 6.6. In the reverse reaction the optimum is at pH
9–9.5.

[14] M. M. Ris and J. P. von Wartburg, *Eur. J. Biochem.* **37,** 69 (1973).

TABLE III

AMINO ACID COMPOSITION OF HUMAN ALDEHYDE AND ALDOSE REDUCTASE

Amino acid	Aldehyde reductase[a]	Aldose reductase[b]
Asparagine or aspartic acid	28	29
Threonine	12	14
Serine	16	15
Glutamine or glutamic acid	35	33
Proline	27	22
Glycine	21	17
Alanine	29	19
Valine	21	23
Methionine	4	5
Isoleucine	14	16
Leucine	35	31
Tyrosine	13	9
Phenylalanine	8	11
Histidine	9	11
Lysine	20	24
Arginine	15	14
Cysteine	6	9
Tryptophan	7	ND[c]

[a] From Wermuth et al.[13]

[b] See B. Wermuth and J. P. von Wartburg, this volume [30].

[c] ND, not determined.

Substrate Specificity, Michaelis Constants and Catalytic Constants, Stereospecificity. Aldehyde reductase catalyzes the reduction of a number of aromatic and medium chain length aliphatic aldehydes and ketones in the presence of NADPH (see Table IV). With NADH as the coenzyme, the activity is less than 5% of that observed with NADPH. Aldehyde reductase also reduces biogenic aldehydes, preferentially aldehydes derived from β-hydroxylated amines.[15] Further, possibly physiological, reactions comprise the reduction of the 17-aldol side chain of isocorticosteroids[16] and of succinic semialdehyde to γ-hydroxybutyric acid.[3] Xenobiotic substances, such as the anthracycline antibiotic daunorubicin, are also reduced. The reduction of glucuronolactone represents an activity that is also attributed to glucuronolactone reductase (EC 1.1.1.20, L-gulono-γ-lactone : NADP⁺ oxidoreductase). The rate of the reverse reaction is less than 2% of the rate observed with p-nitrobenzaldehyde.

Kinetic Mechanism. Product inhibition studies indicate a sequential ordered Bi Bi mechanism in which NADPH binds to the enzyme before

[15] B. Wermuth and J. D. B. Münch, *Biochem. Pharmacol.* **28**, 1431 (1979).

[16] V. Lippman and C. Monder, *J. Biol. Chem.* **253**, 2126 (1978).

TABLE IV

K_m AND K_{cat} VALUES FOR VARIOUS SUBSTRATES OF HUMAN LIVER
ALDEHYDE REDUCTASE[a]

Substrate	K_m (mM)	K_{cat} (sec^{-1})
p-Nitrobenzaldehyde	0.15	9.5
p-Carboxybenzaldehyde	0.025	7.3
m-Nitrobenzaldehyde	2.4	4.7
p-Chlorobenzaldehyde	0.94	1.6
Benzaldehyde	3.4	0.47
Methylglyoxal	1.3	7.3
DL-Glyceraldehyde	4.0	3.0
Butyraldehyde	1.3	0.67
Propionaldehyde	2.9	0.46
Camphorquinone	1.4	1.9
D-glucuronate	3.2	9.0
D-Xylose	260	1.0
D-Glucose	390	0.08
4-Hydroxyphenylglycolaldehyde	0.056	11.6
3-Methoxy-4-hydroxyphenylglycolaldehyde	0.043	5.6
Succinic semialdehyde	0.086	6.1
α-Isocortisol	0.20	1.4
α-11-Deoxyisocorticosterone	0.003	12.6
Daunorubicin	0.12	2.5
L-Gulonate	5.0	0.05
NADPH	0.002	—
NADP$^+$	0.006	—

[a] Results were compiled from references cited in text footnotes 3, 4,
13, and 15.

D-glucuronate, and L-gulonate dissociates before NADP$^+$. The hydrogen transfer occurs from the pro-4R position on the dihydronicotinamide ring of the coenzyme to the re face of the carbonyl carbon atom of the substrate.

Inhibitors. The enzyme is inactivated upon modification of more than one sulfhydryl group by p-mercuribenzoate and other thiol reagents[4,13] or by the amidination of amino groups with bifunctional imido esters.[17] These inactivations are prevented by NADPH. Anticonvulsive drugs, such as barbiturates, succinimides, and hydantoins, are potent inhibitors of the enzyme from liver[13] and brain.[18]

[17] B. Wermuth, J. D. B. Münch, J. Hajdu, and J. P. von Wartburg, *Biochim. Biophys. Acta* **566**, 237 (1979).
[18] M. M. Ris, R. A. Deitrich, and J. P. von Wartburg, *Biochem. Pharmacol.* **24**, 1865 (1975).

Immunological Properties. Antibodies induced to human liver aldehyde reductase prepared in rabbits cross-react with aldehyde reductase from other human tissues, such as kidney and brain. There is no cross reaction with aldose reductase.

Sources

Aldehyde reductase has been found in the following human tissues: liver,[13,19,20] kidney,[19] brain,[9,14,21] erythrocytes,[22,23] leukocytes and platelets.[23]

[19] W. F. Bosron and R. L. Prairie, *Arch. Biochem. Biophys.* **154**, 166 (1973).
[20] N. K. Ahmed, R. L. Felsted, and N. R. Bachur, *Biochem. Pharmacol.* **27**, 2713 (1978).
[21] V. G. Erwin, *Biochem. Pharmacol.* **23**, *Suppl. 1*, 110 (1974).
[22] E. Beutler and E. Guinto, *J. Clin. Invest.* **53**, 1258 (1974).
[23] B. Wermuth, J. D. B. Münch, and J. P. von Wartburg, *Experientia* **35**, 1288 (1979).

[86] 2-Ketoaldehyde Dehydrogenase from Rat Liver

By David L. Vander Jagt

$$CH_3\overset{O}{\underset{\|}{C}}\text{—CHO} + NADP^+ \rightarrow CH_3\text{—}\overset{O}{\underset{\|}{C}}\text{—COOH} + NADPH + H^+$$

Considerable work has been reported on the oxidative metabolism of methylglyoxal which is converted by the glyoxalase system to lactic acid in an intramolecular redox reaction or is oxidized by 2-ketoaldehyde dehydrogenase (2-oxoaldehyde : NAD(P)$^+$ oxidoreductase, EC 1.2.1.23) to pyruvic acid. The dehydrogenase from sheep liver[1,2] can use either NAD$^+$ or NADP$^+$ and is not specific for methylglyoxal. The dehydrogenase from rat liver[3] appears to be specific both for NADP$^+$ and for methylglyoxal. Both dehydrogenases require an amine with a vicinal hydroxyl group for activity. The activating amine *in vivo* is not known. Here we describe a scheme for the purification of 2-oxoaldehyde dehydrogenase from rat liver and an assay procedure that uses L-serine methyl ester as the activating amine.

[1] C. Monder, *J. Biol. Chem.* **242**, 4603 (1967).
[2] J. Dunkerton and S. P. James, *Biochem. J.* **149**, 609 (1975).
[3] D. L. Vander Jagt and L. M. Davison, *Biochim. Biophys. Acta* **484**, 260 (1977).

Reagent

Methylglyoxal (Aldrich), purified by distillation at atmospheric pressure. The distillate was passed through Amberlite CG-400 (HCO_3^- form) on a sintered-glass funnel to remove lactic acid.

Enzyme Assay

2-Ketoaldehyde dehydrogenase was assayed in 0.1 M sodium pyrophosphate buffer, pH 9, containing 0.5 mM NADP$^+$, 5 mM methylglyoxal, and 50 mM L-serine methyl ester. Background activity at 340 nm, 25°, was monitored for 1 min prior to addition of the enzyme sample. One unit of activity corresponds to the production of 1 μmol of NADPH per minute.

Purification Procedure

Materials. CM-Sephadex C-50 and DEAE-Sephadex A-50 were prepared according to the manufacturer's instructions. Blue Dextran affinity material was prepared according to the procedure of Ryan and Vestling[4] except that commercial cyanogen bromide-activated Sepharose 4B (Pharmacia) was used. Proteins were determined by the procedure of Bradford.[5]

Step 1. Livers from 24 male Sprague–Dawley rats, 150–175 g, which had been killed by decapitation, were cut into small pieces and homogenized in a glass tissue homogenizer with a close-fitting Teflon pestle, in 0.04 M potassium phosphate buffer, pH 7.2, containing 0.1 M sucrose, 0.05 M KCl, 0.03 M EDTA, and 0.01 M dithioerythritol in the ratio 3 ml of buffer per gram of liver. The homogenate was centrifuged at 10,000 g for 30 min, and the supernatant was then centrifuged at 330,000 g for 15 min, giving 270 ml of supernatant with a total activity of 393 units.

Step 2. Solid ammonium sulfate, 66 g, was added slowly at 0° to the 270 ml of supernatant. The resulting solution was centrifuged at 10,000 g for 15 min. The supernatant was treated with 83 g of ammonium sulfate, and the solution was centrifuged at 10,000 g for 15 min. Most of the 2-ketoaldehyde dehydrogenase activity was in the pellet which was dissolved in 0.007 M phosphate buffer, pH 7, giving 52 ml of solution; total activity was 332 units.

Step 3. After concentration of the sample by ultrafiltration, a portion (110 units) was placed on a DEAE-Sephadex A-50 column (2.6 × 28 cm) equilibrated at 4° with 0.05 M Tris, pH 8. Protein was eluted with this buffer using a linear salt gradient from 0 to 0.15 M KCl over an elution

[4] L. D. Ryan and C. S. Vestling, *Arch. Biochem. Biophys.* **160**, 279 (1974).
[5] M. M. Bradford, *Anal. Biochem.* **72**, 248 (1976).

PURIFICATION OF 2-KETOALDEHYDE DEHYDROGENASE FROM RAT LIVER

Step and fraction	Volume (ml)	Units	Protein (mg/ml)	Specific activity (units/mg protein)	Yield (%)	Overall purification (fold)
1. 330,000 g supernatant fraction	270.0	393.0	24.6	0.059	100	1.0
2. $(NH_4)_2SO_4$ fraction, 40–80%	52.0	332.0	75.1	0.085	84	1.4
3. DEAE-Sephadex A-50, pH 8, concentrate	7.8	41.0	5.0	1.05	44	18.0
4. Blue Dextran affinity column concentrate	7.2	31.0	0.81	5.39	100	91.0
5. CM-Sephadex C-50, pH 6, concentrate	3.0	7.4	0.34	7.25	89	123.0

volume of 300 ml. A single peak of activity was collected in a volume of 66 ml, which had 45 units (44%) of activity. The sample was concentrated to 7.8 ml by ultrafiltration; 41 units of activity remained.

Step 4. A sample of concentrate from step 3 (31 units) was placed on a Blue Dextran affinity column (1.5 × 26 cm) equilibrated with 0.007 M phosphate buffer, pH 7. Proteins were eluted with this buffer until the effluent appeared to be free of protein. The column then was developed with a linear KCl gradient from 0.02 to 1.5 M over an 80-ml volume. 2-Ketoaldehyde dehydrogenase activity appeared as a single peak, which was collected in a volume of 17.5 ml and concentrated to 7.2 ml with 100% recovery in this step.

Step 5. A sample of concentrate from step 4 (8.7 units) was placed on a CM-Sephadex C-50 column (0.9 × 25 cm) equilibrated with 0.03 M phosphate buffer, pH 6. Protein was eluted using a linear KCl gradient from 0.1 to 0.4 M over a 30-ml volume. Activity appeared as a single peak, which was collected in a volume of 18 ml containing 7.9 units (89%) of activity. Concentration of the sample to 3 ml left 7.4 units of activity. This sample was stored in 50% glycerol at −20°. Activity slowly decreased, but the rate of decrease was slowed by addition of NADP[+].

The purification scheme is summarized in the table.

Comments

1. The use of DL-2-amino-1-propanol in place of L-serine methyl ester gave equally effective activation of 2-ketoaldehyde dehydrogenase from rat liver. Tris, however, is not an effective activator of this enzyme, although it is a good activator of the sheep liver enzyme.[1]

2. In the absence of an activating amine, no activity is detectable.

[87] Diacetyl Reductase

By ROBERTO MARTÍN SARMIENTO and JUSTINO BURGOS

$$\text{Diacetyl} + \text{NADH(NADPH)} \rightleftarrows \text{acetoin} + \text{NAD}^+ \text{(NADP}^+)$$

Diacetyl reductase (EC 1.1.1.5) is included in the list of the IUPAC–IUB Enzyme Commission under the name acetoin : NAD$^+$ oxidoreductase on the basis of work with a partially purified preparation from *Staphylococcus aureus*, which catalyzes the reduction of diacetyl coupled to NADH oxidation.[1] However, there is insufficient evidence for the existence of diacetyl reductases that are specifically linked to NADH. Several similar enzymes have been isolated from microorganisms and animal tissues, but, in all cases in which sufficient specificity data are available, diacetyl reductase is either NADPH-dependent or able to accept both NADH and NADPH.[2-7]

Assay Method

Diacetyl reductase can be conveniently assayed by measuring the decrease in absorbance at 340 nm caused by NADH oxidation during diacetyl reduction to acetoin. The assay is performed at 25° in 0.1 M Na$_2$-K phosphate buffer (pH 6.1). The test solution contains the enzyme preparation, NADH (0.6 μmol) and diacetyl (30 μmol) in a total volume of 3 ml. Crude extracts have NADH-oxidase activity; to correct for this, assays must be performed using a control without diacetyl. Beef liver diacetyl reductase has a K_m^{diacetyl} lower than that from other sources and shows activation by excess of substrate[8,9]; concentrations of 4 mM diacetyl in the reaction mixture are recommended for this enzyme.

One unit is defined as the amount of enzyme that oxidizes 1 μmol of coenzyme per minute under the specified conditions.

[1] H. J. Strecker and I. Harary, *J. Biol. Chem.* **211**, 263 (1954).
[2] A. L. Branen and T. W. Keenan, *Can. J. Microbiol.* **16**, 947 (1970).
[3] M. A. Gabriel, H. Jabara, and U. A. S. Al-Khalidi, *Biochem. J.* **124**, 793 (1971).
[4] J. Burgos and R. Martín, *Biochim. Biophys. Acta* **268**, 261 (1972).
[5] P. Silber, H. Chung, P. Gargiulo, and H. Schulz, *J. Bacteriol.* **118**, 919 (1974).
[6] V. Díez, J. Burgos, and R. Martín, *Biochim. Biophys. Acta* **350**, 253 (1974).
[7] P. López Lorenzo, R. Martín, L. Herrero, and J. Burgos, *An. Fac. Vet. León Univ. Oviedo* **21**, 391 (1975).
[8] R. Martín and J. Burgos, *Biochim. Biophys. Acta* **289**, 13 (1972).
[9] F. Provecho, R. Martín, and J. Burgos, unpublished work, 1980.

METHODS IN ENZYMOLOGY, VOL. 89

Disc Gel Electrophoresis

Disc gel electrophoresis is performed at 0–4° and pH 7.8 in 0.1 M Na_2-K phosphate buffer, applying 8 mA/tube (0.6 cm internal diameter). Gels are prepared with a 7.7% total acrylamide (bisacrylamide versus acrylamide, 2.7%) in 0.05 M Na_2-K phosphate buffer (pH 7.8). Protein can be stained by usual procedures. For activity staining, gels are incubated in 12 mM diacetyl, 1.5 mM NADH in 0.5 M Na_2-K phosphate buffer (pH 6.1) for 15–20 min. At the end of this period, the incubation medium is discarded. Then the gels are maintained dry for 30 min to exhaust the reduced pyridine nucleotide in the enzyme position. After this, the acrylamide cylinders are transferred to a freshly prepared solution containing 25 mg of nitro blue tetrazolium and 12.5 mg of phenazine methosulfate per 100 ml of 0.5 M Tris–boric buffer (pH 9). Diacetyl reductase bands appear as colorless zones on a dark purple background.

Purification Procedures

At least three different enzymes described as diacetyl reductases have been isolated: enzymes from beef liver, pigeon liver, and *Escherichia coli*.

Enzyme from Beef Liver

The original purification procedure has been modified to obtain electrophoretically pure preparations.[2,8]

Step 1. Extraction. Fresh beef liver is decapsulated and homogenized in a blade-type homogenizer with 2 volumes of acetone at −15° in the cold room. Another 8 volumes of acetone at −15° are added to the suspension, which is then stirred for 15 min and filtered through a porous filter paper under vacuum. The cake obtained is homogenized by the same procedure with 10 volumes more of cold acetone and stirred and filtered as before. This second cake is dried, first by suction and then by hand pressing between several layers of filter paper. Diacetyl reductase activity in the resulting cake is stable for at least 3 months at −18°.

Step 2. Acetone Precipitation. The actone cake is extracted with 4 volumes of glass-distilled water at 0–4° with stirring for 30 min. The aqueous extract is cleared by centrifugation in a refrigerated centrifuge (2–4°) at 15,000 g for 10 min. Then, 0.95 volume of acetone is added dropwise with continuous stirring, and the precipitate is removed by centrifugation as before. After filtering the supernatant through two layers of filter paper, another 0.35 volume of cold acetone is added. The pellet, obtained by centrifugation as before, is resuspended in a minimum volume of water and freeze-dried to remove acetone traces with the water.

TABLE I
PURIFICATION OF DIACETYL REDUCTASE FROM 1 KG OF BEEF LIVER

Step	Total protein (mg)	Total activity (units)	Specific activity (units/mg protein)	Purifi- cation (fold)[a]	Yie (9
1. Aqueous extraction of acetone cake	36,000	200	0.002	2–2.5	1(
2. Actone (0.95–1.3 volumes) precipitation	500	100	0.2	100	!
3. DEAE-cellulose chromatography	100	80	0.8	400	
4. Electrofocusing	2.5	40	16	8000	

[a] Relative to a water extract obtained by homogenizing liver with 10 volumes of glass-disti⬛ water in a blade-type homogenizer at 0–2°.

Step 3. DEAE Cellulose Chromatography. The lyophylized powder is suspended (50 mg/ml) in 0.25 M sucrose in 0.025 M Na$_2$-K phosphate buffer (pH 7.5). The suspension, cleared by centrifugation (40,000 g for 10 min or equivalent), is chromatographed on a column of DEAE-22 cellulose by elution with the buffered sucrose solution used to dissolve the enzyme. A DEAE bed of 2.5 × 20 cm and a fraction volume of 5 ml are suitable for 15 ml of sample; in these conditions, the enzyme activity is recovered between fractions 15 and 24.

Step 4. Electrofocusing. From the DEAE chromatography step, those tubes containing maximal enzyme activity are electrofocused in the pH range 4–8 in a Svensson and Vesterberg column (LKB Produktur, Stockholm, Sweden) at 0–2° for 70 hr at no more than 4 W. A 440-ml column is appropriate for the material recovered from a 2.5 × 20 cm DEAE-cellulose column. About 90% of total diacetyl reductase activity is eluted in a single peak with a pI of 6.95. The ampholites are removed by passing fractions twice (with a freeze-drying concentration stage in between) through short and wide columns (2.5 × 25 cm) of Sephadex G-25 Coarse equilibrated and eluted with 0.25 M sucrose in 0.05 M phosphate buffer (pH 7.5). The central tubes of this band are electrophoretically pure. Sucrose and a slightly alkaline buffer are needed to avoid loss of activity.

A minor peak is collected at around pH 6.2. Although able to reduce diacetyl, the enzyme species of this peak is most likely a L-glycol dehydrogenase (see this volume [88]).

Table I summarizes a typical purification.

Enzyme from Pigeon Liver

The method proposed by Díez *et al.* is recommended.[6] All operations are performed at 0–4°.

Step 1. Aqueous Extraction. After slaughter, the liver is quickly removed, chilled in ice (made with distilled water), decapsulated, and cut into small pieces. It is then homogenized in 4 volumes of ice-cold distilled water in either a blade-type or a Potter–Elvehjem homogenizer. The homogenate is filtered through four layers of cheesecloth, and the filtrate is centrifuged, if possible at 100,000 g for 100 min, or at least at 20,000 g for 1 hr (or equivalent).

Step 2. Acetone Precipitation. Acetone (1.2 volumes) at $-18°$ is added to the supernatant with continuous stirring. The precipitate obtained after standing for 10 min is removed by centrifugation (15,000 g for 10 min). Then, 1.3 volumes more of acetone at $-18°$ are added to the supernatant as before. This second precipitate is collected by centrifugation (15,000 g for 10 min), dissolved in a minimum volume of distilled water, and freeze-dried. At this stage, the preparation can be kept for several months at $-18°$ with high retention of activity.

Step 3. Sephadex Gel Filtration. The acetone powder is extracted with 10 volumes of 1 M Na_2-K phosphate buffer (pH 6.1) and cleared by centrifugation at 105,000 g for 15 min (or equivalent). The supernatant is then chromatographed on a column of Sephadex G-100 equilibrated and eluted with the phosphate buffer (if the gel bed is 2.5 × 40 cm, the peak of activity is obtained at an elution volume of around 100 ml). Tubes with the highest activity are pooled and freeze-dried.

The purity of the preparations obtained by this method has been estimated to be about 40% on the basis of polyacrylamide gel electrophoresis.[6] The procedure is summarized in Table II.

Enzyme from Escherichia coli

The following procedure, developed by Silber *et al.*[5] can be used. Enzyme activity must be monitored through ethyl acetoacetate reduction by a method similar to that described for diacetyl reductase assay, except that the buffer was 0.1 M potassium phosphate (pH 7.0), the coenzyme was 0.5 μmol of NADPH, and the substrate was 75 μmol of ethyl acetoacetate.[5]

Organism. *Escherichia coli* B ATCC 11303 can be grown at $37°$ in a medium containing 10 g of dextrose, 10 g of yeast extract, and 5 g of casein hydrolyzate per liter of water. Cells can also be purchased as frozen paste from Grain Processing Corp., Muscatine, Iowa. The purification procedure here described uses this frozen paste as starting material. All operations must be performed at $0–4°$.

Step 1. Extraction. Escherichia coli cells (1000 g) are suspended in 1 liter of 0.05 M potassium phosphate buffer (pH 7.0). The suspension is treated

TABLE II
PURIFICATION OF DIACETYL REDUCTASE FROM 200 G OF PIGEON LIVER

Step	Total protein (mg)	Total activity (units)	Specific activity (units/mg protein)	Purification (fold)[a]	Y: (°)
1. Aqueous extraction	9000	900	0.1	3.5	
2. Acetone (1.2–2.5 volumes) precipitation	300	250	0.8	25	
3. Sephadex G-100 filtration	6	70	12	400	

[a] Relative to a homogenate obtained by homogenizing liver with 4 volumes of cold distilled wate a Potter-Elvehjem homogenizer.

with a Branson sonifier for 10 min. The crude sonicate is centrifuged at 16,000 g for 30 min, and the pellet is discarded.

Step 2. Ammonium Sulfate Precipitation. Solid ammonium sulfate, 25.8 g per 100 ml of supernatant, is added. After stirring for 30 min, the suspension is cleared by centrifugation (16,000 g for 30 min), and the pellet is discarded. To the supernatant another 19.9 g of solid ammonium sulfate per 100 ml of new supernatant are added with stirring, as before. The precipitate is collected by centrifugation as above, suspended in a minimum volume of 0.01 M potassium phosphate (pH 7.0), and dialyzed overnight with three changes against the same buffer.

Step 3. First DEAE-Cellulose Chromatography. The dialyzed solution is applied to a DEAE-cellulose column (6.6 × 40 cm) equilibrated with the buffer used for dialysis and the column is washed with the same buffer until no absorption at 280 nm is detected in the eluate. The column is then developed with 0.1 M NaCl in 0.01 M potassium phosphate (pH 7.0) until all reductase activity is eluted. Active fractions are bulked, and the enzyme is precipitated by addition of solid ammonium sulfate (60.3 g/100 ml) as described, collected by centrifugation, suspended in the phosphate buffer, and dialyzed overnight against several changes of buffer.

Step 4. Second DEAE-Cellulose Chromatography. The protein solution is chromatographed on a DEAE-cellulose column (4 × 45 cm) and eluted with a linear salt gradient made from 1 liter of 0.01 M potassium phosphate (pH 7.0) and 1 liter of 0.3 M NaCl in the same buffer. The fractions containing active enzyme are pooled and the protein is precipitated with ammonium sulfate (60.3 g/100 ml), collected by centrifugation, and redissolved in a minimum volume of the potassium phosphate buffer.

Step 5. Sephadex Gel Filtration. The enzyme solution is chromatographed on a Sephadex G-100 column (5 × 45 cm) using 0.01 M phosphate

TABLE III
PURIFICATION OF DIACETYL REDUCTASE FROM 1 KG OF FROZEN *Escherichia coli* CELLS

Step	Total protein (mg)	Total activity (units)[a]	Specific activity (units/mg protein)	Purification (fold)	Yield (%)
Extraction, crude homogenate	72,000	220	0.003	—	100
Ammonium sulfate precipitation	24,000	170	0.007	2.3	77
First DEAE-cellulose chromatography	3000	71	0.024	8	32
Second DEAE-cellulose chromatography	350	27.5	0.078	26	12.5
Sephadex G-100 filtration	80	46	0.58	193	20.9
DEAE-Sephadex chromatography	1.4	3.3	2.45	800	1.5

[a] Enzyme activity was determined using ethyl acetoacetate as substrate. Data are taken from Silber *et al.*[5]

buffer (pH 7.0) as eluent. Fractions with highest activity are pooled and solid ammonium sulfate is added as in step 4. The precipitated protein is collected by centrifugation and suspended in the phosphate buffer.

Step 6. DEAE-Sephadex Chromatography. After dialysis overnight against 0.01 M Tris-HCl buffer (pH 7.0), the solution is chromatographed on a DEAE-Sephadex A-50 column (2.5 × 30 cm) and eluted with a linear salt gradient made from 1 liter of 0.01 M Tris-HCl (pH 7.0) and 1 liter of 0.3 M NaCl in the same buffer. Again fractions with the highest activity are pooled, and the protein is precipitated as before with solid ammonium sulfate, collected by centrifugation, and stored in this form at $-20°$.

Polyacrylamide gel electrophoresis of the preparations obtained by this procedure, summarized in Table III, indicates that diacetyl reductase, although of high specific activity, is still contaminated with noticeable amounts of inactive protein.[5]

Properties

Diacetyl reductase activity is preferentially distributed in beef liver between the soluble and mitochondrial fractions.[10] This enzyme has an optimum pH of 6.1, an activation energy of 14.4 kcal/mol and shows a very high affinity for diacetyl ($K_m = 40 \ \mu M$) and NADH ($K_m = 0.1 \ mM$).[4,8] The reaction follows a Theorell–Chance mechanism, the coenzyme being the leading substrate. The reaction is reversibly inhibited by ethanol ($K_i =$

[10] R. Martín and J. Burgos, *Biochim. Biophys. Acta* **212**, 356 (1970).

$0.6\,M$) and acetate ($K_i = 62$ mM).[8,11,12] Preparations obtained as described here are not specific for NADH, using NADPH with similar efficiency, but show a fairly good specificity for diacetyl: they do not accept monoketones, keto acids, nonvicinal diketones, glyceraldehyde, glyoxal, and methylglyoxal, although they do reduce 2,3-pentanedione and β-keto acid esters.[4,9] The molecular weight has been estimated as around 76,000.[4,9] Activity is allosterically controlled, being activated by a moderate excess of substrate and inhibited at higher concentrations of substrate.[9]

Pigeon liver diacetyl reductase is located entirely in the soluble fraction.[10] It follows the same reaction mechanism as that from beef liver and shows similar activation energy, optimum pH, coenzyme affinity, and specificity.[6,13] It differs from the mammalian enzyme in molecular weight (110,000) and affinity for diacetyl, which is much lower ($K_m^{\text{diacetyl}} = 3.1$ mM).[6,13] No substrate other than diacetyl is known for this enzyme, but no comprehensive specificity tests have yet been performed. A study of the effects of pH on the kinetic parameters of the reaction have revealed that an ionizing group with a pK around 7, active in the protonated form, participates in the interaction of the enzyme with NADH and NAD; a second group with a pK 8.4, also active in the protonated form, takes part in the binding of diacetyl to E-NADH; and a third group of pK 4.7–5, active in the unprotonated form, is involved in both the dissociation of the complex E-NAD and attachment of diacetyl to E-NADH.[14]

The enzyme from *Escherichia coli* clearly differs from those of animal origin in molecular weight (10,000). It has a similar optimum pH, shows a K_m^{diacetyl} of 4.4 mM and is inhibited by 50 mM diacetyl. This reductase is NADPH-dependent; it also accepts NADH, but very poorly ($K_m^{\text{NADPH}} = 20$ μM, $K_m^{\text{NADH}} = 460$ μM; V_{max} is 3.5-fold higher with NADPH). Besides diacetyl, it also uses keto acid esters; its activity against other α-dicarbonyls has not been tested.[5]

Other Dehydrogenases Able to Reduce Diacetyl

There is a wide variety of enzymes that reduce keto groups. Some could occasionally interfere with diacetyl reductase assays in crude extracts. Bryn *et al.* have purified from *Aerobacter aerogenes* an NADH-

---

[11] L. Herrero, R. Martín, J. Burgos, and P. López Lorenzo, *An. Fac. Vet. León Univ. Oviedo* **20**, 405 (1974).

[12] R. Martín, J. Burgos, P. López Lorenzo, and L. Herrero, *An. Fac. Vet. León Univ. Oviedo* **21**, 399 (1975).

[13] J. Burgos, R. Martín, and V. Díez, *Biochim. Biophys. Acta* **364**, 9 (1974).

[14] R. Martín, V. Díez, and J. Burgos, *Biochim. Biophys. Acta* **429**, 293 (1976).

dependent dehydrogenase, named by them diacetyl(acetoin) reductase, which, besides 2,3-pentanedione, acetylethylcarbinol, and acetoin, accepted diacetyl.[15] An L-glycol dehydrogenase has been described (see this volume [88]) which reduces diacetyl and all kinds of α-dicarbonyls to α-hydroxycarbonyls, and these to L(+)-glycols.[16] Furthermore, diacetyl is accepted, although poorly, as substrate by some alcohol dehydrogenases.[17]

[15] K. Bryn, O. Hetland, and F. C. Størmer, *Eur. J. Biochem.* **18**, 116 (1971).
[16] A. Bernardo, J. Burgos, and R. Martín, *Biochim. Biophys. Acta* **659**, 189 (1981).
[17] E. Juni and G. Heym, *J. Bacteriol.* **74**, 757 (1957).

[88] L-Glycol Dehydrogenase from Hen Muscle

By JUSTINO BURGOS and ROBERTO MARTÍN SARMIENTO

L-Glycol dehydrogenase (proposed systematic name, L(+)-glycol : NAD(P) oxidoreductase, EC 1.1.1 . . .) is an enzyme purified from hen muscle and characterized in the authors' laboratory, which reversibly reduces uncharged vicinal dicarbonyls and α-hydroxycarbonyls to L(+)-glycols.[1-5]

$$R_1—CO—CO—R_2 + NAD(P)H \leftrightarrow R_1—CO—CHOH—R_2 + NAD(P)^+$$
$$R_1—CO—CHOH—R_2 + NAD(P)H \leftrightarrow R_1—CHOH—CHOH—R_2 + NAD(P)^+$$

Assay Method

L-Glycol dehydrogenase activity can be conveniently measured by determining the decrease in absorbance at 340 nm due to NADPH oxidation in the following system: pH 7 sodium-potassium phosphate buffer, 300 μmol; NADPH, 0.6 μmol; acetoin, 12 μmol; total volume, 3 ml; assay temperature, 25°. Acetoin must be washed three times with a large volume

[1] F. Robla, J. Burgos, and R. Martín, *An. Fac. Vet. León Univ. Oviedo* **18**, 743 (1972).
[2] A. Bernardo, "Purificación, propiedades y múltiples formas de la butilénglicol deshidrogenasa de músculo de gallina y su caracterización como un nuevo enzima," pp. 149–190. Ph.D. Thesis, University of Oviedo, Oviedo, Spain, 1977.
[3] A. Bernardo, F. Robla, J. Burgos, and R. Martín, *An. Fac. Vet. León Univ. Oviedo* **25**, 273 (1979).
[4] J. Burgos, R. Martín, and A. Bernardo, *An. Fac. Vet. León Univ. Oviedo* **25**, 285 (1979).
[5] A. Bernardo, J. Burgos, and R. Martín, *Biochim. Biophys. Acta* **659**, 189 (1981).

(50 ml/g) of peroxide-free anhydrous ether to remove traces of diacetyl, as recommended by Westerfeld[6]; after washing it can be stored at $-18°$ for at least 6 months.

Units are defined as the amount of enzyme that reduces 1 μmol of acetoin per minute under these assay conditions.

Electrophoresis

Disc gel electrophoresis in 7.5% polyacrylamide gels (bisacrylamide versus acrylamide, 1 : 37.5) at $0–4°$ in 0.1 M sodium–potassium phosphate buffer (pH 7.8) is recommended; 8 mA per tube of 6 mm i.d. must be applied. Protein can be stained by usual methods. Enzyme activity is easily stained, following the method described for diacetyl reductase in this volume [87] except for the molarity of NADH (1 mM instead of 1.5 mM) and the pH of the phosphate buffer (7.5 instead of 6.1).

Purification Procedure[5]

All operations must be performed in the cold room.

Step 1. Aqueous Extraction. Leg muscles from culled laying hens, free of surface fat and connective tissue, are homogenized with 5 volumes of cold distilled water using a blade-type homogenizer. The homogenate is then centrifuged for 10 min at 1500 g, and the pellet is discarded.

Step 2. Calcium Phosphate Gel Adsorption. Calcium phosphate gel is added to the crude extract until 10–15% of the enzyme activity is bound (usually about 0.5 volume of gel containing 2.8 g dry weight/100 ml), which results in the adsorption of about 80% of the protein. The gel is removed by centrifugation for 10 min at 1500 g. The supernatant is collected and freeze-dried.

Step 3. Sephadex G-100 Gel Filtration. Eight grams of the lyophylyzed powder are suspended in 20 ml of 3 mM sodium-potassium phosphate buffer (pH 7). The suspension is cleared by centrifugation (20,000 g for 30 min or equivalent) and chromatographed on a column (5 × 50 cm) of Sephadex G-100 equilibrated with the phosphate buffer. $E_{280\,nm}$ and L-glycol dehydrogenase activity are monitored in the fractions, and those of the highest specific activities are bulked and freeze-dried.

Step 4. Sephadex G-75 Superfine Gel Filtration. The enriched material from the step 3 is redissolved in 2.5 ml of water (about 60 mg of protein per milliliter) and sieved through a bed of 2.5 × 40 cm of Sephadex G-75 Superfine using 3 mM sodium–potassium phosphate buffer pH 7 as eluent.

[6] W. W. Westerfeld, *J. Biol. Chem.* **161**, 495 (1945).

The fractions with the highest specific activity are bulked, lyophylyzed, and stored at $-18°$ until step 5 is initiated.

Step 5. Hydroxyapatite Chromatography. Material from 3 or 4 G-75 gel filtrations (step 4) is redissolved by adding water to the original volume and loaded onto a column of hydroxyapatite (Bio-Rad Laboratories, Richmond, California) of 1.5×15 cm. The column is eluted with a discontinuous gradient of pH 7 sodium–potassium phosphate consisting first of 3 bed volumes of 3 mM buffer; second, 4 bed volumes of 10 mM; and finally 6 bed volumes of 30 mM. The enzyme is recovered in the 30 mM eluate.

Step 6. Electrofocusing. The fractions with the highest activity are pooled and electrofocused in the pH range 4–8 in a Svensson and Vesterberg column (LKB Produktur, Stockholm, Sweden) refrigerated with ice-cold water. After 70 hr of focusing at no more than 4 W, the column is unloaded and activity is measured in the collected fractions. Several enzyme forms of L-glycol dehydrogenase are detected, the pIs of the three major ones being 4.8, 6.2, and 7.2. The relative proportions between these forms vary widely in different batches, but the total recovered activity is usually distributed as follows: pI 7.2 : pI 6.2 : pI 4.8 = 10 : 2 : 1.3. The best fractions of the bands corresponding to these three major enzyme species are collected. If necessary, they can be freed of sucrose and ampholites by two consecutive filtrations, with a freeze-drying stage in between, through short and wide columns (2.5×25 cm) of Sephadex G-25 (coarse). These preparations can be stored at $-18°$ either frozen or previously freeze-dried.

The central tubes of the bands of pI 7.2 and 6.2 are electrophoretically pure. Those of the pI 4.8 band seem to be contaminated with some inactive protein.

This purification method results (see the table) in an apparent enrichment of 4000, 3000, and 500 times the original specific activity for the enzyme species of pI 7.2, 6.2, and 4.8, respectively. However, these factors are probably an underestimate, since they have been calculated with respect to total original activity, which is due to all the enzymic forms, not to that of each individual species. Were the distribution of the three purified forms at the end of the purification process the same as in the original tissue, the enrichment factors would be 5300 for the pI 7.2 species, 20,000 for that of pI 6.2, and 5000 for the pI 4.8 form.

Properties

Preparations up to stage 4 of the purification procedure are very stable in both water and pH 7 phosphate buffers of low molarity at $0-4°$, but they are easily inactivated in high-molarity buffers.[3,5] At any stage of purifica-

PURIFICATION OF L-GLYCOL DEHYDROGENASE FROM HEN MUSCLE[a]

Step	Total protein (mg)	Total activity (units)	Specific activity (units/mg protein)	Purification (fold)	Yield (%)
1. Aqueous extraction	40,000	30	0.0007	—	—
2. Calcium phosphate gel (unadsorbed)	13,000	25	0.002	3	90
3. Sephadex G-100 filtration	300	20	0.06	80	65
4. Sephadex G-75 Superfine filtration	100	15	0.15	180	50
5. Hydroxyapatite chromatography	30	13	0.4	500	45
6. Electrofocusing					
pI 7.2 form	3	8	2.9	4000	25
pI 6.2 form	1	2	2.2	3000	5
pI 4.8 form	3	1	0.4	500	3

[a] From 1 kg of muscle (about 10 hens).

tion the enzyme can be stored for several months at $-18°$, either frozen or freeze-dried, with high retention of activity.

The three forms of the enzyme show a pH profile of activity with a plateau from pH 5 to around pH 6.6 and then descending at higher pH values.[3,5] Although this enzyme functions well with NADH, it must operate *in vivo* essentially linked to NADPH, in view of the much higher affinity for this coenzyme ($K_m^{NADPH}/K_m^{NADH} = 30$–45 in the different forms).[4,5,7] Diacetyl, pentane-2,3-dione, glyoxal, and methylglyoxal among the vicinal dicarbonyls, and glyceraldehyde among α-hydroxycarbonyls, are the best substrates.[2,4–5,7] The reduction of these compounds catalyzed by the major enzyme form (pI 7.2), the only form so far investigated, is inhibited by acetone ($K_i = 196$ m$M \pm 33\%$) and follows an Ordered Bi-Bi mechanism in which the coenzyme is the first substrate to bind to the enzyme.[7]

The molecular weight of L-glycol dehydrogenase has been calculated as 28,000.[5]

[7] J. González Prieto, "Estudios cinéticos de la L-glicol deshidrogenasa de músculo de gallina," pp. 64–201, Ph.D. Thesis Univ. León, Spain, 1980.

[89] Formaldehyde Dehydrogenase from *Candida boidinii*

By HORST SCHÜTTE, MARIA-REGINA KULA, and HERMANN SAHM

$$\text{HCHO + glutathione (GSH)} \xrightarrow{\text{spontaneous}} \underset{\overset{|}{\text{OH}}}{\text{H}_2\text{C}}\text{—SG}$$

$$\underset{\overset{|}{\text{OH}}}{\text{H}_2\text{C}}\text{—SG + NAD} \rightleftharpoons \underset{\overset{\parallel}{\text{O}}}{\text{HC}}\text{—SG + NADH}_2$$

Studies have shown that in methanol-utilizing yeasts, the methanol is successively oxidized to carbon dioxide by the inducibly formed enzymes: alcohol oxidase, formaldehyde dehydrogenase, and formate dehydrogenase.[1] In all methanol-utilizing yeasts studied so far, an NAD-linked and glutathione-dependent formaldehyde dehydrogenase (EC 1.2.1.1) has been found.[2] It has been assumed that the hemimercaptal spontaneously formed between formaldehyde and glutathione is the true substrate of formaldehyde dehydrogenase.[3,4] Using highly purified formaldehyde dehydrogenases from *Candida boidinii* and *Hansenula polymorpha,* it was demonstrated that the product of the oxidation of formaldehyde is *S*-formylglutathione.[5,6] Furthermore, only *S*-formylglutathione, but not formate, is able to support the reverse reaction. *S*-formylglutathione is hydrolyzed to formate and glutathione in *Candida boidinii* by a special enzyme, *S*-formylglutathione hydrolase.[7]

Assay Method

Principle. Formaldehyde dehydrogenase is measured spectrophotometrically by following the rate of NADH formation at 340 nm in the presence of saturating amounts of formaldehyde, glutathione, and NAD.

Reagents

Sodium phosphate buffer, 0.1 M adjusted to pH 8.0
Formaldehyde, 30 mM

[1] H. Sahm, *Adv. Biochem. Eng.* **6**, 77 (1977).
[2] N. Kato, Y. Tani, and K. Ogata, *Agric. Biol. Chem.* **38**, 675 (1974).
[3] Z. B. Rose and E. Racker, *J. Biol. Chem.* **237**, 3279 (1962).
[4] L. Uotila and M. Koivusalo, *J. Biol. Chem.* **249**, 7653 (1974).
[5] H. Schütte, J. Flossdorf, H. Sahm, and M. R. Kula, *Eur. J. Biochem.* **62**, 151 (1976).
[6] J. P. van Dijken, G. J. Oostra-Demkes, R. Otto, and W. Harder, *Arch. Microbiol.* **111**, 77 (1976).
[7] I. Neben, H. Sahm, and M. R. Kula, *Biochim. Biophys. Acta* **614**, 81 (1980).

Glutathione, 60 mM
NAD, 30 mM

Procedure. The assay is carried out at 30° in a recording spectrophotometer. The reaction mixture consists of 1.0 ml of buffer, 0.1 ml of glutathione, 0.1 ml of NAD, limiting amounts of enzyme, and distilled water to give a total volume of 2.9 ml. The reaction is carried out in a 3.0-ml quartz cell with a 1.0-cm light path. After sufficient warming of the solution, the reaction is initiated by the addition of 0.1 ml of formaldehyde and thorough mixing. The rate of absorbance change at 340 nm is followed for at least 2 min, and activities are calculated by using $\epsilon = 6.22$ cm^2/μmol for NADH at 340 nm.

Definition of Enzyme Unit and Specific Activity. One unit of enzyme activity is defined as the amount of enzyme necessary to oxidize 1 μmol of formaldehyde or to reduce 1 μmol of NAD per minute under the conditions of the assay. Specific activity is expressed as units per milligram of protein.

Production of Formaldehyde Dehydrogenase

For enzyme preparations *Candida boidinii* (ATCC 32 195) is grown in fermentors on a minimal medium with methanol as the sole carbon and energy source as described earlier.[8] Cells are harvested by centrifugation at the end of the exponential growth phase and stored frozen until used. Since the formaldehyde dehydrogenase is induced, it is important that the yeast be grown on methanol.

Purification Procedure

All operations are carried out at 4° and with 0.2% (v/v) 2-mercaptoethanol in all buffers to maintain the enzyme activity.

Step 1. Preparation of Crude Extract. Frozen cells (1000 g wet weight) are softened overnight at 4° and suspended with 3000 ml of 10 mM potassium phosphate buffer (pH 7.5) using a Waring blender. The pH is checked in the suspension and adjusted to 7.5 if necessary with 10% ammonia. The cells are disrupted in a glass bead mill (Dyno-Mill, Bachofen, Basel) at a rotational speed of 2000 rpm, and with glass beads of a diameter of 0.25–0.50 mm; 85% of the free volume of the grinding chamber is filled with glass beads, and the cell suspension is pumped twice through the disintegrator with a flow rate of 5 liter/hr. The pH of the crude extract is maintained at pH 7.5. Cell debris are removed by centrifugation at 25,000 g for 1 hr using a Sorvall centrifuge RC-2B and the GSA-rotor.

---

[8] H. Sahm, H. Schütte, and M. R. Kula, this volume [72].

Step 2. Streptomycin Sulfate Precipitation. To the crude extract obtained in step 1, a 10% solution of streptomycin sulfate in water is added with constant stirring to give a final concentration of 1% streptomycin sulfate. The pH is checked and adjusted to 7.5 by adding ammonia if necessary. The resulting precipitate is removed by centrifugation at 25,000 g for 90 min.

Step 3. DEAE-Cellulose Chromatography. A glass column (10×90 cm) is packed with DEAE-cellulose (Whatman DE-52) under slight pressure and equilibrated with 10 mM potassium phosphate buffer (pH 7.5). The clear supernatant from step 2 is directly applied to the column. The column is washed with 10 liters of equilibration buffer. Elution is carried out by increasing the concentration of the potassium phosphate buffer stepwise to 50 mM, 100 mM, 100 mM + 100 mM sodium chloride. The flow rate of the column is maintained at 500 ml/hr, and 500-ml fractions are collected. Formaldehyde dehydrogenase is eluted with the 100 mM potassium phosphate buffer. Active fractions are combined and concentrated by ultrafiltration (Amicon hollow fiber type H1P10). The concentrated enzyme solution is dialyzed against 10 mM potassium phosphate buffer (pH 7.5).

Step 4. Second DEAE-Cellulose Column Chromatography. The dialyzed enzyme solution obtained in step 3 is applied to a second DEAE-cellulose column (5×50 cm) equilibrated with 10 mM potassium phosphate buffer (pH 7.5). The column is washed with two column volumes of equilibration buffer at a flow rate of 60 ml/hr. Elution of formaldehyde dehydrogenase is carried out by a linear gradient between 2 liters of 50 mM potassium phosphate buffer, pH 7.5, and 2 liters of 50 mM potassium phosphate buffer, pH 7.5, containing 100 mM sodium chloride. Fractions of 15 ml are collected and analyzed for enzyme activity. The active fractions are combined and concentrated with an Amicon column eluate concentrator using a PM-10 membrane and dialyzed against 10 mM potassium phosphate buffer, pH 7.5.

Step 5. Hydroxyapatite Column Chromatography. A column (5×50 cm) is packed with hydroxyapatite prepared in our laboratory according to Levin[9] and equilibrated against 10 mM potassium phosphate buffer, pH 7.5. The enzyme solution is applied to the column and washed with the same buffer. The column is eluted stepwise with 10 mM, 20 mM, 50 mM, and 100 mM potassium phosphate buffer at a flow rate of 45 ml/hr. The enzyme appears in the eluate at 100 mM potassium phosphate buffer. Analytical electrophoresis is carried out across the enzyme peak, and fractions showing a single band only are combined and concentrated by

[9] Ö. Levin, this series, Vol. 5, p. 27.

PURIFICATION OF FORMALDEHYDE DEHYDROGENASE FROM *Candida boidinii*

Fraction	Volume (ml)	Protein (mg)	Total activity (units $\times$ 10^{-3})	Specific activity (units/mg protein)	Yield (%)	Purificat (fold)
Crude extract	2000	48,000	18.0	0.37	100	1
Streptomycin sulfate	2040	40,000	17.0	0.42	95	1.1
DEAE-cellulose I	2300	2,990	15.0	5.02	83	13.6
DEAE-cellulose II	460	736	15.0	20.4	83	55.1
Hydroxyapatite	440	282	13.5	47.9	75	129.5

ultrafiltration. The enzyme is finally dialyzed against 20 mM potassium phosphate buffer (pH 7.5) containing 1 mM dithioerythritol and stored at 4°. For storage at $-20°$ glycerol has to be added to a final concentration of 50% to stabilize the enzyme.

A summary of the purification procedure is given in the table.

Properties

Molecular Weight and Subunit Structure. The molecular weight of the formaldehyde dehydrogenase was determined by sedimentation diffusion equilibrium to be 80,000 $\pm$ 5000. Furthermore, from the sedimentation and diffusion coefficient a molecular weight of 81,000 $\pm$ 12,000 was calculated. Polyacrylamide gel electrophoresis in the presence of sodium dodecyl sulfate indicated that the enzyme is a dimer composed of two probably identical subunits with a molecular weight of 40,000.

Specificity. The enzyme requires glutathione for activity; other thiol compounds, such as cysteine, 2-mercaptoethanol, or thioglycolate, are not able to replace glutathione as a cofactor. The formaldehyde dehydrogenase is dependent on NAD; no reaction was observed using NADP. Formaldehyde as well as methylglyoxal can be used as substrates for the enzymes, but no activity was found toward acetaldehyde, propionaldehyde, benzaldehyde, glycolaldehyde, and glyoxal. The Michaelis constants were found to be 0.25 mM for formaldehyde, 1.2 mM for methylglyoxal, 0.13 mM for glutathione, 0.09 mM for NAD.

Effect of pH. The enzyme has a broad pH optimum from 7.5 to 9.0; below pH 7.5 the activity decreases. The maximal enzymic activity can be obtained at pH 8.5 and 34°.

Inhibitors. Since in *Candida boidinii* formaldehyde dehydrogenase is a key enzyme of the dissimilatory pathway of the methanol metabolism, a possible regulatory effect of ATP, ADP and AMP on the enzyme activity was investigated.[10] Kinetic studies indicate that these nucleoside phos-

[10] N. Kato, H. Sahm, and F. Wagner, *Biochim. Biophys. Acta* **566**, 12 (1979).

phates are competitive inhibitors with respect to S-hydroxymethyl-glutathione. The K_i values at pH 7.0 are 1.65 mM for ATP, 0.40 mM for ADP, and 4.0 mM for AMP. The activity of formaldehyde dehydrogenase may be regulated *in vivo* by these nucleoside phosphates. Furthermore, the enzyme activity is completely inhibited by the following metal ions (1 mM): Cd^{2+}, Cu^{2+}, Hg^{2+}; and by the sulfhydryl reagent p-chloromercuribenzoic acid (1 mM).

[90] Formate Dehydrogenase from *Pseudomonas oxalaticus*

By TH. HÖPNER, U. RUSCHIG, U. MÜLLER, and P. WILLNOW

$$HCO_2^- + NAD^+ \rightleftharpoons CO_2 + NADH$$
$$HCO_2^- + \text{electron acceptor} \rightleftharpoons CO_2 + \text{acceptor reduced}$$

When *Pseudomonas oxalaticus* is grown on formate as the main carbon and energy source, formate dehydrogenase is the key enzyme that generates NADH and CO_2. The CO_2 enters the ribulose diphosphate carboxylase reaction.[1] There is no other example of a soluble iron–sulfur flavoprotein within the very heterogeneous group of formate dehydrogenases.[2,3]

Assay Method

Principle. The continuous spectrophotometric assay[4] is based on formate oxidation by NAD^+ or ferricyanide. The appearance of NADH is followed at 334 or 365 nm; ferricyanide decrease is followed at 405 nm. The NADH oxidase activity of the enzyme does not interfere in the presence of 5 mM sulfide.

Reagents

Potassium phosphate buffer, 0.1 M, pH 7.0
NAD^+, 12 mM. Some preparations contain formate, which is removable by repeated lyophilization of the acidified (pH 3) solution.

[1] M. A. Blackmore and J. R. Quayle, *Biochem. J.* **107**, 705 (1968).
[2] R. K. Thauer, G. Fuchs, and K. Jungermann, *in* "Iron–Sulfur Proteins" (W. Lovenberg, ed.), Vol. 3, p. 121. Academic Press, New York, 1977.
[3] L. G. Ljungdahl, *in* "Molybdenum and Molybdenum-containing Enzymes" (M. Coughlan, ed.), p. 465. Pergamon, Oxford, 1980.
[4] U. Müller, P. Willnow, U. Ruschig, and T. Höpner, *Eur. J. Biochem.* **83**, 485 (1978).

> Sodium formate, 15 mM
> Sodium sulfide, 50 mM
> Potassium ferricyanide, 15 mM

Procedure. For the NAD$^+$ method the 1-ml cuvette (light path 10 mm) contains 0.5 ml of buffer, 0.1 ml of NAD$^+$, 0.1 ml of formate, 0.1 ml of sulfide, and water to 1.0 ml. The mixture is incubated for some minutes (oxygen removal). The reaction is started by addition of enzyme.

For the ferricyanide method, NAD$^+$ is replaced by 0.1 ml of ferricyanide solution. Sulfide and preincubation are omitted.

Definition of Unit and Specific Activity. One unit of enzyme is defined as the amount that catalyzes the formation of 1 μmol of NADH or ferrocyanide per minute at 25°. Specific activity is expressed as units per minute per milligram of protein. Protein is determined by the biuret method.[5]

Growth of *Pseudomonas oxalaticus*

Pseudomonas oxalaticus (culture No. 8642, National Collection of Industrial Bacteria, Torry Research Station, Aberdeen, Scotland) is grown aerobically in a medium of the following composition (tap water): 0.02 M NH$_4^+$, 0.07 M K$^+$, 0.155 M Na$^+$, 1 mM Mg^{2+}, 0.01 mM Fe^{2+}, 0.01 mM Ca^{2+}, 0.02 M SO$_4^{2-}$, 0.12 M HPO$_4^{2-}$, 0.10 M HCOO$^-$, 0.02 mM Cl$^-$, 0.01 mM MoO$_4^{2-}$, 5 mM pyruvate, pH 7.8. Stock culture and preculturing have been described elsewhere.[6] Preparative growth takes place under vigorous aeration in six 10-lier stainless steel flasks placed in a water bath at 30°. The flasks are fed continuously by means of a six-channel peristaltic pump from a linear gradient of formic acid constructed from 2 liters of 2.3 M HCOOH (10% v/v) and 2 liters of 5.75 M HCOOH (25%) containing 0.05 and 0.15 M pyruvic acid, respectively. By selecting the velocity of addition, the pH value of the media is maintained at 8.0–8.6. Cells are harvested at the end of the logarithmic growth after 2 days (100–150 g of wet cells) and stored at −18°.

Purification Procedure

This purification procedure[4] is based on several earlier attempts[7,8] of obtaining preparations useful for enzymic formate assays.[9,10]

[5] G. Beisenherz, H. J. Boltze, T. Bücher, R. Czok, K. H. Garbade, K. H. Meyer-Arendt, and G. Pfleiderer, *Z. Naturforsch. B: Anorg. Chem., Org. Chem., Biochem. Biophys. Biol.* **8B**, 555 (1953).
[6] T. Höpner and A. Trautwein, *Arch. Mikrobiol.* **77**, 26 (1971).
[7] P. A. Johnson and J. R. Quayle, *Biochem. J.* **93**, 281 (1964).

All steps are performed at 0° with oxygen and light excluded as far as possible. Unless otherwise stated the buffer (pH 5.6) contains 50 mM histidine hydrochloride, 5 mM sodium sulfide, and 0.5 mM EDTA. Stored enzyme solutions and all solutions for the chromatographic steps are kept in an atmosphere of argon.

Eighty grams of *P. oxalaticus* are thawed and washed twice at 0° in the growth medium in which formate and pyruvate have been replaced by chloride. This is necessary to prevent inactivation of the enzyme by formate after disintegration of the cells. The cells are suspended in 240 ml of buffer solution, and the pH is adjusted to 5.8 with 1 M HCl; 2 mg of NAD$^+$ (to remove residual formate), 0.5 mg each of ribonuclease and deoxyribonuclease, 2 mg of glucose oxidase, and 300 mg of glucose are added. The suspension (320 ml) is sonicated for 10 min at 0–10° in 80-ml portions with a Branson sonifier B-12. Cell debris is removed by centrifugation at 30,000 g for 30 min.

The pH of the extract is adjusted to 5.2 by addition of 0.5 M HCl. After centrifugation (15 min at 30,000 g) the pH is readjusted to 5.6 with a solution of 0.5 M Na$_2$S and 5 mM EDTA.

A polyethyleneimine solution, 1.5%, v/v (PEI 18, Dow Chemicals, 10% in water, pH adjusted to 5.6 by HCl) is added; after stirring for 15 min the mixture is centrifuged (10 min at 30,000 g). From the supernatant, active protein is precipitated by degassed and argonized ammonium sulfate (1.95 M) and centrifuged off as above. The supernatant is discarded, the pellet is suspended in 40 ml of buffer containing 1.33 M ammonium sulfate and recentrifuged. The new pellet is suspended in 40 ml of buffer containing 0.80 M ammonium sulfate and centrifuged. Formate dehydrogenase appears in the supernatant. The enzyme solution is desalted in an anaerobic column of Sephadex G-25 medium (2.5 × 40 cm; no Na$_2$S/EDTA).

The brown filtrate is adjusted to pH 5.2 with 1 M acetic acid, and the solution is clarified by centrifugation. To the supernatant 7% (v/v) acetone of −15° is added, the temperature being kept at −3°. After centrifugation the pellet is discarded, and 20% (v/v) acetone is added to the supernatant at −5°. The solution is centrifuged (15 min at 30,000 g), and the pellet is suspended in 20 ml of anaerobic buffer without Na$_2$S/EDTA. Insoluble material is centrifuged off; this extraction process is repeated.

[8] J. R. Quayle, this series, Vol. 9, p. 360.

[9] P. A. Johnson, M. C. Jones-Mortimer, and J. R. Quayle, *Biochim. Biophys. Acta* **89**, 351 (1964).

[10] T. Höpner and J. Knappe, *in* "Methoden der enzymatischen Analyse" (H. U. Bergmeyer, ed.), 3rd German ed., p. 1596. Verlag-Chemie, Weinheim, 1974.

To the combined brownish-yellow supernatants KCl is added up to 0.05 M. The solution is applied to an anaerobic column of DEAE-Sephadex A-25 (1.6 × 20 cm) equilibrated with 0.05 M histidine, pH 5.6–0.05 M KCl (no Na_2S/EDTA) yielding a diffuse brown zone. After washing with the equilibrating buffer, the enzyme is eluted by a linear KCl gradient (300 ml, 0.05–0.30 M, 20 ml/hr). Addition of Na_2S/EDTA enhances the loss of activity. Combined active fractions are applied to a column of hydroxyapatite (1.6 × 5 cm, BioGel HT, Bio-Rad Laboratories Richmond, California) equilibrated with anaerobic 0.05 M buffer containing 0.05 M potassium phosphate. After washing with the equilibrating buffer, formate dehydrogenase is eluted by a gradient of potassium phosphate (100 ml, 0.05–0.30 M, no Na_2S/EDTA) in the buffer.

From these combined active fractions the enzyme is precipitated by 3.5 M ammonium sulfate. After centrifugation (40 min at 38,000 g), the pellet is suspended in the buffer containing 1.55 M ammonium sulfate and stored at 0° under argon. Under these conditions the activity has a half-life of approximately 2 months.

For the final purification step, the products of four purification procedures are combined and centrifuged for 40 min at 38,000 g. The pellet is dissolved in a minimum volume of anaerobic buffer. Each 0.4 ml of the

PURIFICATION OF FORMATE DEHYDROGENASE[a]

Step	Volume (ml)	Activity units	Protein (mg)	Specific activity (units/mg protein)	Purification (fold)	Y (9
1. Crude extract	1320	5050	22200	0.227	1.0	10
2. pH 5.2 supernatant	1320	4750	16000	0.30	1.3	9
3. Polyethyleneimine	1320	4740	12900	0.37	1.6	9
4. Ammonium sulfate fractionation	173	3620	4080	0.88	3.9	7
5. Gel filtration	212	1940	3180	0.61	2.7	3
6. Acetone fractionation	106	950	1100	0.86	3.8	1
7. DEAE-Sephadex chromatography	98	416	58	7.1	31	
8. Hydroxyapatite chromatography[b]	24	230	17	14	62	
9. Sucrose density gradient (fractions 4–14)	11	110	4	28	120	
Enzyme I (fractions 4–6)	3.6	50	1.2	42	185	
Enzyme II (fractions 12–14)	2.7	16	0.4	32	140	

[a] Reproduced, with permission, from Müller *et al.*[4]

[b] Steps 1–8 are performed in four parallel procedures, each comprising a quarter of the amou given here.

solution is layered on a centrifuge tube containing a 13-ml sucrose gradient (5 to 20% w/v) in the buffer (without sulfide and EDTA). Centrifugation is performed at 0° and at 40,000 rpm for 21 hr in a Beckman SW-40 Ti rotor. Fifteen fractions are collected in Thunberg tubes under an atmosphere of argon. Two enzymically active species are obtained: formate dehydrogenase I (fractions 4–6) and II (fractions 12–14). At 0° under anaerobic conditions the fractions are stable for some weeks.

The table summarizes the results of the whole purification procedure. The overall yield in activity never exceeded 5%. The purification for formate dehydrogenase I and II is 185-fold and 140-fold, respectively. Enzymes I and II show the same multiple bands after polyacrylamide electrophoresis when stained for protein or enzyme activity. Sucrose gradient centrifugation fractions 8–11 contain a single enzymically inactive protein.

Essential conditions of the purification procedure are: exclusion of light and oxygen, absence of formate, low temperature, and rapid performance with possible interruption only after precipitation by ammonium sulfate of the eluate of the hydroxyapatite chromatography. Chemical removal of oxygen by sulfide cannot be applied to the chromatography steps, where it causes severe losses of activity.

Properties

Molecular Properties. From sucrose density gradient centrifugations in the presence of internal standards, the following values for the molecular weights were obtained: formate dehydrogenase I $s_{20,w}$ = 13.2, M_r = 315,000; formate dehydrogenase II $s_{20,w}$ = 9.1, M_r = 175,000. It is assumed that formate dehydrogenase I is a dimer of formate dehydrogenase II. A freshly prepared cell-free extract does not contain formate dehydrogenase II. Formate dehydrogenase II appears during the purification steps. In crude preparations (not after purification) I is slowly converted into II.

From assays of the sucrose density gradient centrifugation fractions, the coenzyme contents are as follows: I: 2 FMN, 18–25 Fe, 15–20 S^{2-}; II: 1 FMN, 6–10 Fe, 6–8 S^{2-}. The identity of the flavin as well as the protein : flavin ratio have been confirmed by the reactivation of the deflavo enzyme with FMN (see below).

Catalytic Properties. Formate dehydrogenation by NAD^+ is catalyzed by enzyme I with a specific activity of 42 units/mg. The pH optimum is 7.4. K_m values for formate are $1.35 \times 10^{-4}\ M$ (enzyme I) and $1.30 \times 10^{-4}\ M$ (II); for NAD, 1.05×10^{-4} (I) and 1.10×10^{-4} (II).

Carbonate reduction by NADH is catalyzed by enzyme I with a specific activity of 1.4 units/mg.[11] The pH optimum is 6.6. CO_2 (not HCO_3^-) is the active species in the equilibrium. The K_m value for CO_2 is 30–50 mM. K_m values for other electron acceptors are also similar or identical for enzymes I and II. In the presence of 1.5 mM formate, the following values have been found: ferricyanide 4.6×10^{-4} (I) : 5.0×10^{-4} (II); benzylviologen 2.5×10^{-4} : 2.4×10^{-4}; dichlorophenolindophenol 3.7×10^{-5} : 4.2×10^{-5} M. In the presence of 0.2 mM NADH the values are: ferricyanide 6.9×10^{-4} : 6.2×10^{-4}; benzyl viologen ca. 10^{-2} : 2×10^{-3}; dichlorophenolindophenol 2×10^{-4} : 2×10^{-4} M.

The enzyme acts with other electron acceptors, including methylviologen, phenazine methosulfate, methylene blue, nitro blue tetrazolium salt, FMN, FAD, riboflavin, and oxygen (product: H_2O_2).

N_3^-, NO_3^-, NADH, CN^-, p-hydroxymercuribenzoate, and Hg^{2+} at 0.25 mM each are inhibitors; SO_3^{2-} and $H_2PO_2^-$ are not. The inhibition by N_3^-, NO_3^-, and NADH is partly overcome by addition of formate or NAD. Inhibition by Hg^{2+} and p-hydroxymercuribenzoate can be reversed to 70% by addition of mercaptoethanol (0.06 M). Addition of CN^- leads to a slow irreversible inactivation. Its velocity is dependent on the concentration of the inhibitor. Formate inactivates the enzyme in the absence of an electron donor in a slow reaction, the velocity of which is dependent on the formate concentration.[12]

Preparation of the Deflavo Enzyme and Reactivation[4]

For flavin removal the enzyme is dialyzed for 24 hr against the following medium: 50 mM histidine hydrochloride, 1.2 M ammonium sulfate, 5 mM sodium sulfide, 0.5 mM EDTA, pH 5.6. To achieve an absolutely inactive preparation it was necessary to remove flavin from the dialysis equilibrium by adding charcoal to the outer medium. The removal proceeds only under a low oxidation reduction potential (approximately −0.4 mV).

The deflavo protein reincorporates FMN under aerobic conditions and regains full activity. The incorporation can be followed by fluorescence titration because the fluorescence of the incorporated flavin is quenched compared to the free flavin. While the fluorescence quench of added FMN is observed immediately, the appearance of the enzyme activity is a slow process, the velocity of which is dependent on the concentration and on the reduction state of the FMN. Half-maximal reactivation velocity is

[11] U. Ruschig, U. Müller, P. Willnow, and T. Höpner, *Eur. J. Biochem.* **70**, 325 (1976).
[12] T. Höpner and A. Trautwein, *Z. Naturforsch. B: Anorg. Chem., Org. Chem., Biochem. Biophys. Biol.* **27B**, 1075 (1972).

observed in the presence of 0.1 μM FMN or 2.5 μM FMNH$_2$. Formate dehydrogenase I incorporates 2 mol of FMN per mole, whereas dehydrogenase I incorporates 1 mol per mole.

FAD and riboflavin are also incorporated under the same stoichiometry and with the same fluorescence titration observations. The nonphysiological flavin enzymes exhibit no formate dehydrogenase activity. This does not hold for the formate and NADH oxidase activity, which are also shown by the FAD and riboflavin enzymes. The specific formate oxidase activities (in units per milligram) are: FMN-enzyme, 2; FAD-enzyme, 1.4; riboflavin enzyme, 0.6. The NADH-oxidase activities (units/mg) are: FMN–enzyme, 2; FAD–enzyme, 1; riboflavin-enzyme, 1. The affinity between the deflavo enzyme and the flavins increases in the direction of riboflavin–FAD–FMN, as has been shown by displacement experiments.

[91] Formate Dehydrogenase from *Escherichia coli*

By HARRY G. ENOCH and ROBERT L. LESTER

Formate dehydrogenase has a widespread occurrence in animals, plants, and microorganisms.[1] The enzyme of *Escherichia coli* is a membrane-associated multifunctional respiratory component[2,3] (formate : cytochrome b_1 oxidoreductase, EC 1.2.2.1). Growth of *E. coli* anaerobically in the presence of nitrate leads to the formation of high levels of formate dehydrogenase,[4] as well as cytochrome *b* and nitrate reductase.[5] The importance of these components is indicated by the fact that they may comprise 62% of the cytoplasmic membrane protein during anaerobic nitrate respiration.[6] The procedure described below has been used for the purification of formate dehydrogenase in a form suitable for use in the reconstruction of the formate–nitrate reductase activity.[6,7]

[1] H. G. Enoch, Ph.D. dissertation, University of Kentucky, Lexington, 1975.

[2] E. Itagaki, T. Fujita, and R. Sato, *J. Biochem. (Tokyo)* **52**, 131 (1962).

[3] A. W. Linnane and C. W. Wrigley, *Biochim. Biophys. Acta* **77**, 408–418 (1963).

[4] R. L. Lester and J. A. DeMoss, *J. Bacteriol.* **105**, 1006 (1971).

[5] M. K. Showe and J. A. DeMoss, *J. Bacteriol.* **95**, 1305 (1968).

[6] H. G. Enoch and R. L. Lester, *J. Biol. Chem.* **250**, 6693 (1975).

[7] H. G. Enoch and R. L. Lester, *Biochem. Biophys. Res. Commun.* **61**, 1234 (1974).

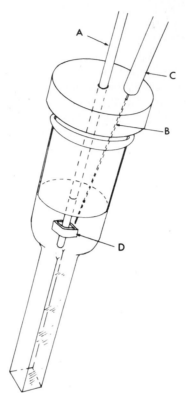

Fig. 1. Anaerobic cuvette. A, gas inlet line; B, gas exit and sample port; C, microsyringe; D, sample holder.

Assay Method

Formate dehydrogenase is measured by spectrophotometric assay of the reduction of dichlorophenolindophenol in the presence of phenazine methosulfate[4] in an anaerobic cuvette.[8]

Anaerobic Cuvette. An anaerobic cuvette can be constructed by a simple procedure using cheap materials (Fig. 1). The body of the cuvette is made by joining a length of 10 mm i.d. square tubing (Ace Glass) to a standard-taper (24/25) ground-glass joint and sealing the bottom. A plug is made from a Teflon bar tapered to give a gastight fit requiring no lubricant. A hole is drilled through the center of the plug to accommodate a length of heavy-wall polyethylene tubing (0.042 inch i.d., 0.114 inch o.d.), which serves as the gas inlet line (A, Fig. 1). The fit should be just tight enough to

[8] H. G. Enoch and R. L. Lester, *J. Bacteriol.* **110,** 1032 (1972).

allow the tubing to be moved up and down. Another hole serves as a gas exit and sample port (B, Fig. 1). This hole is drilled at an angle to intercept the sample holder about 1 cm below the Teflon plug. The hole is just large enough to accommodate the needle of a 100-μl microsyringe (Hamilton), but must allow gas to exit around the needle. A sample holder (D, Fig. 1) is made by cutting a 4-mm length of 0.25 inch o.d. Teflon tubing, flattening it as shown, and slipping it on the end of the gas inlet line.

To operate, the gas inlet line is lowered into the assay mixture, which is deoxygenated by vigorously bubbling for 2 min with purified argon.[4] The inlet line is then raised, and the sample (up to 50 μl) is added with a microsyringe. The sample may be degassed by allowing it to equilibrate for several minutes in the argon atmosphere. The reaction is initiated by plunging the gas inlet line into the solution (where mixing is rapidly accomplished by the bubbling gas) and then quickly withdrawing it. The argon should continue to blow over the surface of the solution during the assay. The cuvette can be adapted to fit many spectrophotometers, and a black cloth can usually be arranged to substitute for the sample compartment lid. As an indication of the performance of the cuvette, the autoxidation of reduced benzylviologen in this system is very slow ($\Delta A_{600} < 0.002$ per minute).

Assay Procedure. A 3-ml reaction mixture containing 75 μM dichlorophenolindophenol, 75 μg of phenazine methosulfate per milliliter, and 50 μM sodium phosphate buffer, pH 7.2 is gassed with argon at 30° and then the absorbance is followed at 600 nm. Enzyme is added and an endogenous rate is recorded. The reaction is then initiated by the addition of 15 μl of 4.8 M sodium formate. The endogenous rate is subtracted from the rate with formate. An extinction coefficient of 21 mM^{-1} cm^{-1} is used.

Bacterial Growth

Escherichia coli HfrH (thi⁻) used throughout these studies was obtained from J. A. DeMoss.[9] It is likely that other strains, including the wild type, would also be satisfactory. The bacteria for large-scale purification are grown in 45-liter batches in 12-gallon carboys, which are capped with rubber stoppers fitted with a sintered-glass sparger tube and a gas exit port. The growth medium has the following composition (per liter of final medium): part A: 9.16 g of $K_2HPO_4 \cdot 3 H_2O$, 1.0 g of $(NH_4)_2SO_4$, 0.1 g of $MgSO_4 \cdot 7 H_2O$, 2.0 g of KH_2PO_4, 0.4 g of sodium citrate $\cdot$ 2 H_2O, 0.5 mg of $(NH_4)_2Fe(SO_4)_2$, 10.0 g of $KHCO_3$, 10.0 g of KNO_3; part B: 10.0 g of D-glucose; part C: 5 mg of thiamine $\cdot$ HCl, 0.1 μmol of Na_2SeO_3, 0.1 μmol of Na_2MoO_4.

[9] J. Ruiz-Herrera, M. K. Showe, and J. A. DeMoss, *J. Bacteriol.* **97**, 1291 (1969).

The medium is sterilized by autoclaving parts A and B separately, and adding part C, which is prepared from filter-sterilized solutions. The final culture medium is gassed for 1 hr or more with 95% N_2–5% CO_2, at which point the pH should be 8.1–8.2. The slightly alkaline initial pH allows a higher growth yield to be obtained under anaerobic (acid-producing) conditions.[4] Growth is carried out at 37° with continuous gassing (N_2–CO_2) through the sparger tube; the flow rate is adjusted to provide a vigorous agitation, adequate to keep the bacteria suspended. The culture is inoculated with 1 liter of a culture which has just reached the end of exponential growth. To prevent an excessive lag period, the inoculum should be produced in the same medium and growth conditions as the larger batch. The culture is allowed to reach the end of exponential growth (an absorbance at 600 nm of 1.2–1.4 with a 1-cm path length), at which point the cells are rapidly chilled to 0–4° and then harvested by centrifugation. The cells are washed twice by suspension in NaCl–phosphate buffer (0.9 g of NaCl/100 ml, 50 mM sodium phosphate, pH 7.2), and centrifugation. The final supernatant is discarded and the cell paste is stored at $-15°$. The cells were used within a week of harvesting. The enzyme activity of frozen cells is stable for months, but such cells have not yet been used for enzyme preparation.

Purification Procedure

All steps are carried out at 0–5°. After disruption of the cells, subsequent steps are carried out in closed vessels under an atmosphere of N_2. Strict anaerobic conditions are not required—the major precautions include the use of deaerated solutions and providing an atmosphere of N_2 above the solutions at all times.

Preparation of Membrane Fraction. The cells from four 45-liter batches (about 500 g wet weight) are thawed and suspended in 500 ml of 0.1 M sodium phosphate buffer, pH 7.2 (about 35 mg of protein per milliliter). The cells are disrupted by two passages through a Manton–Gaulin homogenizer at 6000 psi. The warm stream exiting the homogenizer is immediately passed through a stainless steel coil immersed in an ice-water bath.

Solid ammonium sulfate is slowly added to the homogenate to a final concentration of 230 g/liter, and the mixture is stirred for 1 hr. The crude membranes are collected by centrifugation at 15,000 g for 2 hr. The membranes are resuspended in 500 ml of 0.1 M Tris-HCl buffer, pH 8, with the aid of a glass homogenizer with a loose-fitting Teflon pestle. The protein concentration is then measured[4] and the mixture is diluted to a concentration of 4–6 mg of protein per milliliter using the same Tris buffer (final volume 2–4 liters).

Solubilization and Salt Fractionation. Sodium deoxycholate solution (10 g of deoxycholic acid per 100 ml of distilled water, adjusted to pH 8 with NaOH) is added with stirring to give a final concentration of 1 mg of deoxycholate per milligram of protein. Ammonium sulfate solution (saturated at 2–4° in 0.1 M Tris-HCl buffer, pH 8) is slowly added to give 30% saturation (requires the addition of three-sevenths volumes of ammonium sulfate solution). The mixture is stirred for 3 hr, and the precipitate is then removed by centrifugation at 15,000 g for 3 hr. The supernatant solution is brought to 40% saturation ammonium sulfate (requires the addition of one-sixth volume of the saturated ammonium sulfate solution based on volume of the 30% saturated supernatant) and stirred for 1 hr. The mixture is centrifuged at 15,000 g for 1 hr, and the precipitate is again discarded. The supernatant solution is then brought to 50% saturation ammonium sulfate (requires the addition of one-fifth volumes of the saturated ammonium sulfate solution), and the mixture is stirred for 1 hr. The precipitate is collected by centrifugation at 15,000 g for 1 hr. The reddish-brown precipitate is dissolved in approximately 50 ml of 0.1 M Tris-HCl buffer, pH 7.2, and dialyzed against 20 volumes of the same buffer for 24 hr, and then for an additional 24 hr against 20 volumes of 1 M NaCl in 0.1 M Tris-HCl buffer, pH 7.2. The dialyzed enzyme solution is centrifuged at 15,000 g for 1 hr, and any precipitate is discarded. Triton X-100 is slowly added to the supernatant to give 0.5 g of Triton X-100 per 100 ml (requires the addition of one-nineteenth volume of concentrated detergent solution: 10 g of Triton X-100 per 100 ml of distilled water), and the mixture is stirred for 1 hr.

Agarose Gel Chromatography. Approximately 2 liters of agarose gel (BioGel A-1.5m, Bio-Rad Laboratories) is washed twice with 5 volumes of Triton–NaCl–Tris (buffer (0.5 g of Triton X-100/100 ml, 1 M NaCl, 0.1 M Tris-HCl buffer, pH 7.2), taking care to remove the fines. The gel is loaded in a glass column (5 × 92 cm) and, after settling, is washed with 2 liters of anaerobic Triton–NaCl–Tris buffer. The enzyme preparation from the preceding step (about 90 ml) is layered on top of the gel and under anaerobic Triton–NaCl–Tris buffer (the enzyme solution is more dense than the buffer). The column is developed with the Triton–NaCl–Tris buffer at a flow rate of 30 ml/hr. Fractions of 9–10 ml are collected in the presence of air. The fractions are assayed immediately for enzyme activity and protein,[10] and the appropriate fractions are pooled within 3 hr of elution. The pooled enzyme preparation is deaerated and then dialyzed against 15 volumes of anaerobic Triton–Tris buffer (0.1 g of Triton X-100 per 100 ml, 10 mM Tris-HCl buffer, pH 8) for 40 hr (buffer is changed once at 20 hr).

[10] O. H. Lowry, N. J. Rosebrough, A. L. Farr, and R. J. Randall, *J. Biol. Chem.* **193**, 265 (1951).

PURIFICATION OF FORMATE DEHYDROGENASE

Fraction	Volume (ml)	Protein (mg)	Total activity (μmol/min)	Specific activity (μmol/min per mg protein)
Cell suspension	1000	35,400	45,800	1.3
Crude membranes	2600	13,000	45,300	3.5
Deoxycholate extract	3900	5,880	33,500	5.8
Ammonium sulfate fraction	70	1,700	27,900	16.4
BioGel A-1.5m	128	492	21,600	44
DEAE-BioGel	133	80	15,700	196

Anion-Exchange Chromatography. Approximately 250 ml of DEAE-agarose (DEAE-BioGel A, Bio-Rad) is washed twice with several volumes of Triton–Tris buffer and loaded in a glass column (2.5 × 42 cm). The column is washed with 2 liters of anaerobic Triton–Tris buffer, and then the dialyzed enzyme solution (100–150 ml) is applied. The enzyme is absorbed in the top 1–2 cm of the gel as indicated by the dark brown color. After washing with 1 column volume of Triton–Tris buffer, the column is developed with a 2-liter linear gradient of NaCl (0 to 0.2 M NaCl in Triton-Tris buffer). The gravity flow rate is adjusted initially to 100 ml/hr and fractions of 9 ml are collected by drop counting. An anaerobic atmosphere is provided for the eluting enzyme by placing a minifraction collector (Gilson Medical Electronics) in a polyethylene bag that is kept inflated with a continuous stream of N_2. Homogeneous fractions are identified by polyacrylamide gel electrophoresis in the presence of 0.25% Triton X-100[6] and pooled. The enzyme may be stored for months at 4–5° under an atmosphere of N_2.

A summary of the protocol together with protein and enzyme yields is shown in the table.

Properties

Composition. The enzyme has the oxidized and reduced spectra of a b-type cytochrome. The cytochrome of formate dehydrogenase is completely reduced in the presence of formate. The heme content of the purified enzyme is 6.5 nmol per milligram of protein. Chemical analysis indicates that the enzyme also contains (per mole of heme): 0.95 mol of molybdenum, 0.95 mol of selenium, 14 mol of nonheme iron, and 13 mol of acid-labile sulfide; quinone, flavin, and phospholipid are not detected.

Molecular Size and Subunits. The molecular weight of the intact protein estimated in the presence of Triton X-100 by the method of Tanford *et al.*[11] is 590,000 ± 59,000. The protein–detergent complex has a Stokes' radius of 76 Å and a sedimentation coefficient, $s_{20,w}$ of 18.1 S. The minimal molecular weight based on heme content is 154,000, suggesting the enzyme contains 4 molecules of heme.

Formate dehydrogenases can be dissociated into three subunits in the presence of sodium dodecyl sulfate. The molecular weights of the subunits determined by polyacrylamide gel electrophoresis[12] are 110,000, 32,000, and 20,000. Electrophoresis of formate dehydrogenase obtained from cells grown in [^{75}Se]selenite indicates that selenium is covalently bound to the largest subunit.

Catalytic Characteristics. The highest specific activity of formate dehydrogenase observed was 200 μmol of dichlorophenolindophenol reduced per minute per milligram of protein, yielding a turnover number of 33,800 mol per minute per mole of heme. The K_m for formate is 0.12 mM. The enzyme also displays activity with other electron acceptors including ferricyanide, tetrazolium dyes, methylene blue, and coenzyme Q_6; there is no activity with FAD, FMN, NAD$^+$, or NADP$^+$. The activity with dichlorophenolindophenol–phenazine methosulfate is almost completely inhibited in the presence of 5 mM NaCN, but the activity with methylene blue is unaffected, indicating that the enzyme has multiple catalytic sites for interaction with artificial substrates.

Reconstitution of Formate–Nitrate Reductase. The procedure for the purification of formate dehydrogenase also yields a highly purified preparation of nitrate reductase.[6] It is possible to reconstruct the formate–nitrate reductase *in vitro* by mixing the two enzyme preparations.[7] Activity requires both enzymes, formate, nitrate, and anaerobic conditions. The reaction is stimulated 33-fold by the further addition of coenzyme Q_6. The turnover number of the reconstructed system is 30% of turnover number of nitrate reductase. The rate with vitamin K_2-30 is 34% of that with coenzyme Q_6. *n*-Heptyl hydroxyquinoline-*N*-oxide, which inhibits formate–nitrate reductase *in vivo*,[13] also inhibits the reconstructed activity.

[11] C. Tanford, Y. Nozaki, J. A. Reynolds, and S. Marino, *Biochemistry* **13**, 2369 (1974).
[12] U. K. Laemmli, *Nature* (*London*) **227**, 680 (1970).
[13] J. Ruiz-Herrera and J. A. DeMoss, *J. Bacteriol.* **99**, 720 (1969).

Section V

Isomerases, Epimerases, and Mutases

[92] D-Ribose Isomerase[1]

By ALAN D. ELBEIN and KEN IZUMORI

$$\text{D-Ribose} \rightleftharpoons \text{D-ribulose}$$

Assay Method

Principle. The amount of ketose, D-ribulose, formed from D-ribose was determined by the cysteine-carbazole method.[2]

Reagents

Tris buffer, 50 mM, pH 7.5
$MnCl_2$, 50 mM
D-Ribose, 100 mM
H_2SO_4 solution: 19 ml of H_2O + 45 ml of concentrated H_2SO_4
0.12% Carbazole in 95% ethanol
1.5% Cysteine-HCl (prepared fresh)

Procedure. The assay mixture contained 100 μl of 50 mM Tris buffer, 50 μl of 50 mM $MnCl_2$, 50 μl of 100 mM D-ribose, and an appropriate amount of enzyme, all in a final volume of 1 ml. The reaction was started by the addition of the D-ribose. After an incubation of 10 min at 37°, the reaction was terminated by the addition of 0.1 ml of 10% trichloroacetic acid. The formation of ribulose was measured by the cysteine-carbazole method.

Definition of Unit and Specific Activity. One unit of enzyme is defined as the amount that will catalyze the conversion of 1 μmol of ribose to ribulose in 1 min. Specific activity is expressed as units per milligram of protein. Protein was measured by the method of Lowry *et al.*[3]

Purification Procedure

Step 1. Growth of the Bacteria. Since the D-ribose isomerase is an inducible enzyme, it is necessary to culture the bacteria in a medium containing D-ribose. Thus, the organism is routinely grown in nutrient broth containing 0.1% D-ribose. *Mycobacterium smegmatis* is maintained on slants of trypticase soy agar. A starter culture is prepared by inoculat-

[1] K. Izumori, A. W. Rees, and A. D. Elbein, *J. Biol. Chem.* **250**, 8085 (1975).
[2] Z. Dische and E. Borenfreund, *J. Biol. Chem.* **192**, 583 (1951).
[3] O. H. Lowry, N. J. Rosebough, A. L. Farr, and R. J. Randall, *J. Biol. Chem.* **193**, 265 (1951).

ing 125-ml Erlenmeyer flasks containing 25 ml of the nutrient broth-ribose medium with a loopful of bacteria from a slant. This starter culture is grown for 2 days at 37° on a rotary shaker and then used to inoculate 2-liter flasks containing 1 liter of the above medium. This 1-liter culture is used to inoculate 20 more flasks for the production of sufficient bacteria for the purification of the enzyme. The final cultures are grown for 15 hr at 37°, and the bacteria are harvested on a Büchner funnel and washed well with 50 mM Tris buffer, pH 7.5.

Step 2. Preparation of the Crude Extract. Cell-free extracts were prepared by suspending the *Mycobacterium smegmatis* in 50 mM Tris buffer containing 0.5 mM of MnCl$_2$ to the consistency of a thick paste. The cells were then ruptured by grinding them with alumina powder. The paste was then diluted with Tris buffer and centrifuged at 17,000 g for 20 min. The supernatant liquid was saved for the enzyme purification. All operations were performed at 0–5°.

Step 3. Mn^{2+} and Polyethylene Glycol Treatments. A 1.0 M solution of MnCl$_2$ is added slowly, with stirring, to the crude extract to a final concentration of 0.05 M. The solution is allowed to stand in ice for 15 min and then centrifuged at 17,000 g for 20 min. Most of the enzymic activity is recovered in the supernatant liquid. Since ammonium sulfate fractionation results in poor recovery of activity at this stage, polyethylene glycol is used to precipitate the enzyme. To the Mn^{2+} supernatant fluid, solid polyethylene glycol 6000 is added to a final concentration of 6%, and the mixture is gently stirred for 20 min in ice. The mixture is centrifuged at 17,000 g for 20 min; the precipitate is discarded. Solid polyethylene glycol 6000 is added to this supernatant fluid to a final concentration of 20%. The resulting precipitate is collected by centrifugation at 17,000 g for 20 min, dissolved in a minimum of Tris buffer (50 mM containing 0.5 mM MnCl$_2$). Any undissolved precipitate is removed by centrifugation, and the supernatant is further fractionated as indicated below.

Step 4. DEAE-Cellulose Chromatography. The 6–20% polyethylene glycol fraction is adsorbed on a 1.8 × 20 cm column of DEAE-cellulose (Cl$^-$) that has been equilibrated with the Tris-MnCl$_2$ buffer. The column is washed with 150 ml of the above buffer, and the enzyme is eluted with a linear gradient of 0 to 2 M KCl in the same buffer. A total of 400 ml of the KCl solution is used to elute the column, and most of the activity elutes between 1.1 and 1.5 M KCl. Active fractions are pooled and concentrated with an Amicon Diaflo apparatus using an XM-50 filter.

Step 5. Sephadex G-200 Chromatography. The concentrated enzyme from the above step is placed on a 2.5 × 63 cm column of Sephadex G-200 that has been equilibrated with the Tris-Mn^{2+} buffer. The enzyme is eluted with the same buffer. Fractions are collected and assayed for enzymic

PURIFICATION OF D-RIBOSE ISOMERASE

Step	Volume (ml)	Total protein (mg)	Total activity (units)	Specific activity (units/mg protein)	Recovery (%)
Preparation of crude extract	540	2260	50.4	0.022	100
Mn treatment and PEG[a] precipitation	50	550	38.3	0.070	76
DEAE-cellulose	2.7	14.3	18.7	1.31	37
Sephadex G-200	1.8	9.3	16.3	1.75	32
Crystallization	1.0	5.0	9.2	1.84	18

[a] PEG, polyethylene glycol 6000.

activity and for protein. Active fractions are pooled and concentrated with the Amicon filter.

Step 6. Crystallization of the Enzyme. The concentrated enzyme from step 5 (specific activity 1.75 units per milligram of protein) containing 5.1 mg of protein per milliliter is used for crystallization. Saturated ammonium sulfate, pH 7.5, is added slowly with stirring to the enzyme to a final concentration of 44%. A faint precipitate is removed by centrifugation. The resulting clear solution is allowed to stand at 5°. Hexagonal crystals begin to appear after about 16 hr, and crystallization is complete after about 2 days. Recrystallization is done by dissolving the crystals in buffer and subjecting them to another ammonium sulfate treatment.

A summary of a typical purification is given in the table. The enzyme was purified 85-fold with a recovery of 18%. The purified enzyme was homogeneous by disc gel electrophoresis and ultracentrifugal analysis.

Properties of the Enzyme

Purity. The crystalline enzyme was homogeneous when examined by disc gel electrophoresis and gave a single, symmetrical peak in the analytical ultracentrifuge.

Molecular Weight of the Enzyme and Its Subunits. In the analytical ultracentrifuge, the enzyme was calculated to have a molecular weight of between 145,000 and 174,000. Treatment of the isomerase with sodium dodecyl sulfate in the presence of 5 mM $MnCl_2$ resulted in dissociation to a single band which migrated with a molecular weight of 42,000 to 44,000. This indicates that the native enzyme contains four identical subunits.

Stability. The crystalline enzyme was stable for at least 1 month in 44% ammonium sulfate at 5°. The enzyme from the Sephadex G-200 column slowly lost activity when stored at 5°. However, both enzyme preparations appeared to be sensitive to freezing and thawing.

pH Optimum. The pH range for the enzyme was rather broad and showed an optimum between 7.5 and 8.5 in Tris or glycine buffers.

Substrate Specificity and Affinity. The isomerase was specific for the configuration at carbons 1–3 of the sugar. Thus, the enzyme isomerized L-lyxose, D-allose, and L-rhamnose as well as D-ribose. The K_m for D-ribose was estimated to be 4 mM, while that for L-lyxose was 5.3 mM. A number of other sugars including D- and L-arabinose, D- and L-xylose, D-lyxose, and the common hexoses were not isomerized by the enzyme.

Metal Ion Requirement. The enzyme required a divalent cation for activity, Mn^{2+} being the best activator. The K_m for Mn^{2+} was found to be $1 \times 10^{-7} M$, while that for Co^{2+} was $4 \times 10^{-7} M$ and for Mg^{2+} it was $1.8 \times 10^{-5} M$.

Effect of Polyols. Since polyols have been reported to inhibit some pentose isomerases, we tested a number of polyols on the ribose isomerase. At $0.05 M$ none of the following polyols showed any inhibitory effect: xylitol, D- and L-arabitol, ribitol, sorbitol, mannitol, dulcitol, glycerol, L-erythritol, and *i*-inositol.

Equilibrium. At equilibrium, at 37° in 0.025 M Tris buffer, pH 7.5 containing 0.0025 M MnCl$_2$, the ratio of D-ribose to D-ribulose, starting with either substrate, was 0.30.

[93] Glucosephosphate Isomerase[1] from Catfish Muscle and Liver and from Mammalian Tissues

By ROBERT W. GRACY

D-Glucose 6-Phosphate $\rightleftharpoons$ D-Fructose 6-Phosphate

Assay Method

The enzyme catalytic activity can easily be measured in either direction by coupling to the appropriate dehydrogenases and monitoring the oxidation or reduction of NADH or NADP spectrophotometrically. The isomerase activity is more conveniently assayed in the direction fructose 6-phosphate $\rightarrow$ glucose 6-phosphate by coupling with glucose-6-phosphate dehydrogenase and monitoring the reduction of NADP at 340 nm.

[1] Glucose-6-phosphate ketol-isomerase, EC 5.3.1.9.

The conditions of these and alternative assays have been detailed in this series.[1a]

Isozymes and Allozymes

Mammalian tissues contain a single form of glucosephosphate isomerase, thus making isolation of the enzyme relatively simple. It was initially observed by Tilley et al.[2] that the isomerase from human tissues could be isolated by the specific substrate elution of the enzyme from cellulose phosphate. Subsequently it was found that slight modifications of this procedure also could be used for isolating the enzyme from rabbit[3] or swine-muscle.[4] Gearhart and Oster-Granite[5] have also shown that this method can be used to isolate the enzyme from mouse tissues. Glucosephosphate isomerase exists as a dimer, which is the result of a single gene locus in vertebrates, with the notable exception of the fishes (see below). Payne et al.[6] and Gracy[7] showed that in mammals there are no tissue-specific isozymes of glucosephosphate isomerase, and the multiple electrophoretic forms occasionally observed are due to oxidation of sulfhydryl groups.[7,8] Genetic studies[9-11] have also shown that mammals possess a single glucosephosphate isomerase gene locus.

In the case of heterozygous individuals, tissues typically exhibit a three-banded electrophoretic pattern consisting of the normal dimer, the variant dimer, and a hybrid consisting of a normal and a variant subunit. We have shown that these allozymes can be resolved by a substrate gradient elution of the enzyme from carboxymethyl BioGel.[12]

A single gene locus exists for all vertebrates and invertebrates thus far examined, except for fishes. The lampreys and teleostean fish have two or

[1a] R. W. Gracy and B. E. Tilley, this series, Vol. 41, p. 392.

[2] B. E. Tilley, R. W. Gracy, and S. G. Welch, J. Biol. Chem. 249, 4571 (1974).

[3] T. L. Phillips, J. M. Talent, and R. W. Gracy, Biochim. Biophys. Acta 429, 624 (1976).

[4] H. S. Lu and R. W. Gracy, J. Biol. Chem. 256, 785 (1981).

[5] J. Gearhart and M. L. Oster-Granite, J. Histochem. Cytochem. 28, 245 (1980).

[6] D. M. Payne, D. W. Porter, and R. W. Gracy, Arch. Biochem. Biophys. 151, 122 (1972).

[7] R. W. Gracy, in "Isozymes" (C. L. Markert, ed.), Vol. 1, pp. 471–487. Academic Press, New York, 1975.

[8] M. N. Blackburn, J. N. Chirgwin, G. T. James, T. D. Kempe, T. F. Parsons, A. M. Register, K. D. Schnackerz, and E. A. Noltmann, J. Biol. Chem. 247, 1170 (1972).

[9] J. C. Detter, P. O. Ways, E. R. Giblett, M. A. Baughan, D. A. Hopkinson, S. Porey, and H. Harris, Ann. Hum. Genet. 31, 329 (1968).

[10] R. J. DeLorenzo and F. H. Ruddle, Biochem. Genet. 3, 151 (1969).

[11] F. A. McMorris, T. R. Chen, F. Ricciuti, J. Tischfield, R. Creagan, and F. Ruddle, Science 179, 1129 (1973).

[12] D. R. Gibson, J. M. Talent, R. W. Gracy, K. D. Schnackerz, and G. Ishimoto, Clin. Chim. Acta 89, 355 (1978).

more loci coding for the isomerase, and the isozymes exhibit a characteristic tissue-specific distribution.[13-15] Most tissues contain the more acidic isozyme referred to as "GPI-A" or "liver-type" isozyme, but skeletal muscle (and to a lesser extent the eye) exhibits the "GPI-B" or "muscle isozyme."[16]

Isolation of the Mammalian Isomerases by Substrate Elution

The overall procedure, regardless of source or tissue, consists of (a) extraction of the enzyme from cells followed by dialysis and adjustment of pH; (b) adsorption of the enzyme to cellulose phosphate; (c) removal of contaminating proteins from the ion exchanger; and (d) specific elution of the enzyme by its substrate.

The critical differences in isolating the enzyme from various sources are primarily in the choice of pH and buffer concentrations so that the maximum amount of the contaminating proteins are removed from the ion exchanger without "leaking" the enzyme from the column. The enzyme is adsorbed to the ion exchanger at a pH such that the initial binding is due to general ion exchange as well as by affinity of the enzyme for the phosphorylated glucose polymer. Subsequently the column is washed with a buffer of slightly higher pH to the point where the binding of the enzyme to the cellulose phosphate is primarily due to affinity binding. After the effluent is free of protein, and the pH has changed to the higher value, substrate (either glucose 6-phosphate or fructose 6-phosphate) is added to the eluent buffer, which causes the release of the enzyme in a small volume. The enzyme can be purified to homogeneity by this method with overall purifications of up to 30,000-fold[2] and recoveries of 65–75%.[1,2] The specific details of each of the procedures for isolating the enzyme from human placenta, muscle, and erythrocytes and from muscle from rabbit or swine are summarized in Table I.

Extraction of Tissues. For most mammalian tissues the enzyme can be extracted successfully by simply homogenizing finely chopped tissue in a blender. Table I lists the buffers and conditions that we have employed. It is important to adjust the pH immediately after homogenization. After extraction, the debris is sedimented by centrifugation. The supernatant solution is collected and passed through a glass-wool filter, and the pH again is measured and adjusted if necessary. At this stage the enzyme is dialyzed against the appropriate buffer (Table I) to assure complete bind-

[13] J. C. Avise and G. B. Kitto, *Biochem. Genet.* **8,** 113 (1973).
[14] P. R. Dando, *Comp. Biochem. Physiol. B* **47B,** 663 (1974).
[15] P. R. Dando, *Comp. Biochem. Physiol.* **66B,** 373 (1980).
[16] Y. Mo, C. D. Young, R. W. Gracy, N. D. Carter, and P. R. Dando, *J. Biol. Chem.* **250,** 6747 (1975).

CONDITIONS FOR PURIFYING MAMMALIAN GLUCOSEPHOSPHATE ISOMERASES BY SUBSTRATE ELUTION[a]

Source	Extraction conditions	Dialysis buffer	Binding buffer	Washing buffer	pH prior to adding substrate	Elution buffer	Column size	Specific activity (units/mg protein)	References[b]
Human muscle	370 g of tissue/750 ml of buffer 1; centrifuge; pH to 7.5	Buffer 2	Buffer 2	Buffer 3	8.2	Buffer 4	2 × 60 cm	850–1000	(1)
Human erythrocyte[c]	450 ml of blood; sediment, resuspend in 500 ml of 0.145 M NaCl, freeze/thaw; pH to 7.0	Buffer 2	Buffer 2	Buffer 3	8.2	Buffer 4	1.5 × 30 cm	850–1000	(2)
Human placenta	1200 g in 1600 ml of buffer 2; centrifuge; pH to 7.5	Buffer 2	Buffer 2	Buffer 3	8.2	Buffer 4	4.5 × 100 cm	850–1000	(18)
Rabbit muscle	200 g of tissue in 400 ml of buffer 2; centrifuge, pH to 6.6	Buffer 5	Buffer 5	Buffer 6	7.5	Buffer 7	8 × 19 cm	850–1000	(3)
Swine muscle[d]	600 g in 1500 ml of buffer 2; centrifuge; pH to 6.6	Buffer 5	Buffer 5	Buffer 3	8.0	Buffer 4	4.5 × 100 cm	800–900	(4)

[a] Buffers: 1. 10 mM triethanolamine (HCl), 1 mM EDTA, 10 mM KCl pH 11; 2. 10 mM triethanolamine (HCl), 1 mM EDTA, 0.1% v/v 2-mercaptoethanol, pH 7.5; 3. 25 mM triethanolamine (HCl), 0.1% 2-mercaptoethanol, pH 8.2; 4. 25 mM triethanolamine (HCl), 0.1% 2-mercaptoethanol containing 7 mM fructose 6-phosphate, pH 8.2; 5. 10 mM imidazole, 1 mM EDTA, 0.1% 2-mercaptoethanol, pH 6.6; 6. 10 mM imidazole, 1 mM EDTA, 0.1% 2-mercaptoethanol, pH 7.2; 7. 10 mM imidazole, 1 mM EDTA, 0.1% 2-mercaptoethanol containing 7 mM fructose 6-phosphate, pH 7.2.

[b] Numbers refer to text footnotes.

[c] Hemoglobin removal with $CHCl_3/CH_3OH$ is preferable.

[d] This procedure is used to isolate the basic isozyme. The other acidic isozymes can be isolated by substrate elution using buffer 4 at a lower pH.

ing to the cellulose phosphate. While this procedure is satisfactory with most tissues (e.g., liver, kidney, brain, skeletal or heart muscle, placenta), in the case of erythrocytes it is advantageous first to precipitate the hemoglobin. This is achieved by mixing 100-ml portions of hemolysate with 60-ml portions of 1 : 1 v/v chloroform : methanol and rapidly stirring at 0° for 120 sec followed by centrifugation.[2]

Adsorption of the Enzyme. Since various sources of cellulose phosphate differ in their ability to bind the enzyme, it is advisable first to establish the necessary amount of the ion exchanger by carrying out a pilot test. A small amount of weighed cellulose phosphate (e.g., 5–10 g), which has been washed by cycling through acid and base[2] and equilibrated at the proper pH, is added to a 15-ml sintered-glass funnel. Aliquots of the dialyzed enzyme are added to the cellulose, and the effluent is checked for isomerase activity. Additional aliquots are added until the column becomes saturated and the enzyme begins to "leak" through the ion exchanger. Knowing the amount of enzyme added and the amount of cellulose used, one thus determines the amount of cellulose required to bind all the enzyme.

The enzyme is then adsorbed to the ion exchanger in a batch operation. The proper amount of equilibrated cellulose is added to the enzyme solution, stirred for 15–45 min and poured into a 2-liter (e.g., 16 cm high × 14 cm in diameter) sintered-glass funnel. The effluent is assayed for isomerase activity to assure that binding of the enzyme has been 95–100% complete.

Washing. The cellulose pad is then washed with the same buffer (binding buffer) until the effluent is free of protein (i.e., absorbance at 280 nm < 0.01). The cellulose is then resuspended in a second washing buffer of higher pH, stirred and filtered as before. This washing is continued until the effluent is protein-free, and the pH of the effluent has reached the higher pH (see Table I). The entire procedure of binding and washing can be carried out in 1 day. At this point the enzyme is ready to be eluted with substrate.

Substrate Elution. Although the enzyme can be eluted from the cellulose in a batch operation, it is more desirable to elute the enzyme from a column. This simplifies collection of fractions and allows the enzyme to be obtained in a smaller volume. The enzyme-bound cellulose is thus packed into a column (e.g., 70 cm × 5.5 cm), and the elution buffer (Table I) containing substrate is pumped through the column. Fractions are collected, and the enzyme is eluted in a sharp peak.

The entire process described above can easily be completed in 2–4 days (allowing for overnight dialysis, preparation and equilibration of the ion exchangers and buffers, and assaying of fractions for enzyme activity).

Specific activities of 850–1000 units/mg are usually obtained, and the enzyme is homogeneous by the usual criteria. If desired, the enzyme can be crystallized,[2,3] but usually this is not necessary. In the event the enzyme contains contaminants, these can be removed by rechromatography with substrate elution from a small column of cellulose phosphate.

Resolution of Allozymes

When the enzyme is isolated from heterozygous individuals, it is necessary to separate the three allozymes. In some cases this can be achieved by preparative isoelectric focusing.[2,17,18] However, a more successful method is to utilize a substrate gradient elution from carboxymethyl BioGel.[12] For example, five different human genetic variant forms of glucosephosphate isomerase were resolved from heterozygous individuals by this method.

Isolation of Glucosephosphate Isomerase Isozymes from Catfish Muscle and Liver

As described above, most fish have tissue-specific isozymes of glucosephosphate isomerase. The isozymes are most conveniently isolated from muscle and liver tissues where GPI-B_2 and GPI-A_2 isozymes predominate, respectively. Unfortunately the substrate elution method that is so successful in isolating the enzyme from mammals cannot be used for isolating the enzyme from fish. This appears to be due, at least in part, to the lower isoelectric points of the fish enzymes, which interfere with their binding to the cellulose phosphate. Nevertheless, the two isozymes can be isolated to homogeneity by a series of two chromatographic steps on DEAE-cellulose followed by preparative isoelectric focusing.

Extraction of Tissues. Catfish, *Ictalarus punctatus* (Rafinesque), are obtained locally. Tissues are removed from freshly killed fish and extracted by blending for 30 sec with four volumes of 10 mM triethanolamine buffer, pH 9.0 at 4°. The extracts are centrifuged at 20,000 g for 1 hr, and the pellets are discarded. The supernatant solutions are filtered through glass-wool and dialyzed overnight against buffer A [10 mM triethanolamine, pH 9.0, containing 1 mM EDTA and 0.1% (v/v) 2-mercaptoethanol].

First DEAE-Sephadex Chromatography. The dialyzed extracts are first subjected to chromatography on DEAE-Sephadex columns (8 × 60 cm)

---

[17] P. M. Yuan, M. R. Zaun, M. V. Kester, C. E. Snider, M. Johnson, and R. W. Gracy, *Clin. Chim. Acta* **92**, 481 (1979).
[18] K. L. Purdy, H. H. Tai, and R. W. Gracy, *Arch. Biochem. Biophys.* **200**, 485 (1980).

which have been preequilibrated in buffer A. One liter of buffer A containing 0.01 M NaCl is then pumped through the column to elute a large amount of contaminating proteins.

For the isolation of the catfish muscle isozyme, a linear sodium chloride gradient of 2 liters with ionic strength ranging from 0.02 to 0.08 is applied, and the isozyme elutes in a sharp peak at an ionic strength of 0.04. When the liver extract is chromatographed, it is necessary to employ a linear gradient of higher ionic strength (0.01 to 0.15) since the isomerase from liver elutes only at a higher ionic strength (0.11). This higher affinity of the liver isozyme for the anion exchanger is consistent with its lower

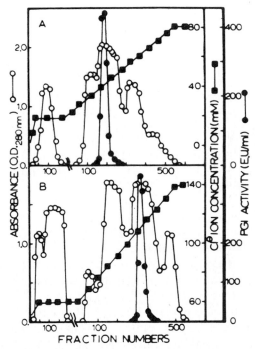

FIG. 1. Chromatography of liver and muscle isozymes on DEAE-Sephadex. (A) Partially purified catfish muscle glucosephosphate isomerase (2.8 g of protein in a volume of 50 ml; specific activity = 13 units/mg) was applied to a column, 3 × 100 cm, of DEAE-Sephadex as described in the text. The column was washed at a flow rate of 50 ml/hr with 20 mM NaCl, 10 mM triethanolamine, pH 9.2. Then, linear NaCl gradient (1 liter of 20 mM NaCl to 1 liter of 80 mM NaCl) was applied to elute the enzyme. (B) Partially purified catfish liver glucosephosphate isomerase (287 mg in a volume of 10 ml; specific activity = 43 units/mg) was applied to a column, 3 × 100 cm, of DEAE-Sephadex as described above. The column was washed with 60 mM salt, then the linear gradient (60 to 140 mM NaCl) was established. Glucosephosphate isomerase (PGI) activities (●), protein concentrations (○), and chloride concentrations (■) were determined.

TABLE II

Isolation of Catfish Muscle and Liver Glucosephosphate Isomerases

Fractions	Total activity (units)	Total protein (mg)	Specificity activity (units/mg protein)	Purification overall (fold)	Recovery overall (%)
fish muscle (400 g of tissue)					
I. Homogenate	41,000	39,300	1.03	(1.0)	(100)
I. 1st DEAE	36,700	2,810	13.1	12.7	89.5
I. 2nd DEAE	35,200	178	198	192	85.8
V. Isoelectric focusing	21,400	52	412	400	52.2
fish liver (100 g of tissue)					
I. Homogenate	13,400	11,700	1.15	(1.0)	(100)
I. 1st DEAE	12,300	287	42.7	37.1	91.8
I. 2nd DEAE	10,700	47	226	196	79.9
V. Isoelectric focusing	4,650	10	465	404	34.7

isoelectric point. In the case of the muscle isozyme, the first chromato-graphic step yields enzyme with a specific activity of 10–15 units/mg. This step in the purification from liver is more effective, yielding specific ac-tivities of 40–45 units/mg. In both cases the recoveries of enzyme activity are approximately 90%.

Second DEAE-Sephadex Chromatography. Fractions containing glucosephosphate isomerase activity are pooled, dialyzed, and re-chromatographed on DEAE-Sephadex columns, which are developed at pH 9.2 with linear salt gradients. For the muscle enzyme approximately 2.8 g of protein are available at this stage, whereas only about one-tenth that amount of protein is available for the partially purified liver enzyme. In either case the enzyme solutions are dialyzed against 10 mM triethanolamine containing 1 mM EDTA and 0.1% 2-mercaptoethanol at pH 9.2. After dialysis, the solutions are added to columns (3 × 100 cm) of DEAE-Sephadex equilibrated in the above buffer. The columns are ini-tially washed with the above buffer containing NaCl (20 mM in the case of the muscle isozyme and 60 mM for the liver isozyme). This washing procedure (approximately 1 liter) removes a substantial amount of con-taminating proteins. Thereafter linear NaCl gradients (1 liter of 20 mM NaCl to 1 liter of 80 mM NaCl for the muscle enzyme; and 1 liter of 60 mM NaCl to 1 liter of 140 mM NaCl for the liver enzyme) are utilized to elute the isomerases. Flow rates are maintained at 50 ml/hr, and fractions of 10 ml each are collected and assayed for glucosephosphate isomerase activ-ity. Figure 1 shows typical results of this chromatography. After the second DEAE chromatography, the enzymes exhibit specific activities of approximately 200 units/mg and are approximately 45% pure. Some fractions are approximately 90% pure. Overall recoveries at this stage are generally 80–85% pure.

Isoelectric Focusing. The glucosephosphate isomerase isozymes are purified further by isoelectric focusing. The isozymes from the previous chromatographic steps are electrofocused in 2% narrow range ampholines at 600 V and 4° for 72 hr. The pH range for the separation is 6–8 for the muscle isozyme. For the liver isozyme, pH 5–7 is used. Upon completion of electrofocusing, liquids in the columns are removed, as 1.0-ml fractions and assayed for isomerase activity. The catfish muscle isozyme exhibits a major component of glucosephosphate isomerase activity with an isoelectric pH of 7.0 and a minor component at pH 6.6. Isoelectric focusing of liver glucosephosphate isomerase shows a major component at pH 6.2 and a minor component at a pH of 5.6. These minor fractions seem to be due to sulfhydryl oxidation of the native isozymes, since addition of 2-mercaptoethanol or dithiothreitol greatly reduces the amounts of the minor components.

After isoelectric focusing, the isozymes from catfish muscle and liver are obtained with specific activities of approximately 400–450 units/mg and are judged to be homogeneous by a variety of criteria.[16] Rechromatography on DEAE-cellulose, Sephadex G-200, or additional isoelectric focusing do not increase the specific activity of the isozymes. Both the catfish liver and muscle isozymes yield single bands after polyacrylamide gel electrophoresis in the presence or the absence of sodium dodecyl sulfate. Table II summarizes the results of typical isolations of glucosephosphate isomerase from catfish muscle and liver.

Acknowledgment

This work was supported in part from grants from the National Institutes of Health (AM14638), (AG01274), and the R. A. Welch Foundation (B-502).

[94] Phosphoglucose Isomerase[1] from Mouse and *Drosophila melanogaster*

By CHI-YU LEE

Fructose 6-phosphate ⇌ glucose 6-phosphate

An 8-(6-Aminohexyl)amino-ATP-Sepharose column carries a multi-negatively charged ligand and acts like a cation exchange column under conditions of low pH and low ionic strength. Thus the positively charged phosphoglucose isomerase can be adsorbed on the ATP-Sepharose column mainly by electrostatic interactions. Elution with negatively charged substrate such as glucose 6-phosphate results in changes in protein conformation or surface charge of this enzyme and a subsequent desorption of the enzyme.[2–4]

Assay Method

Phosphoglucose isomerase from mouse and *Drosophila melanogaster* is routinely assayed spectrophotometrically at 340 nm at 25° by following the increase in absorbance due to the oxidation of glucose 6-phosphate by glucose-6-phosphate dehydrogenase. The assay mixture in a total volume of 1 ml contains $0.1\ M$ Tris-HCl, pH 8.0, 2 mM EDTA, 0.5 mM NADP$^+$, 1 mM fructose 6-phosphate, and 1 unit of glucose-6-phosphate dehydrogenase from yeast. A suitable amount of phosphoglucose isomerase is added to cause an increase in absorbance at 340 nm of 0.05 to 0.1 A per minute. One unit of enzyme activity is defined as the amount of enzyme that catalyzes the conversion of 1 μmol of fructose 6-phosphate to glucose 6-phosphate per minute under the above assay conditions.[5]

Affinity Column

An 8-(6-aminohexyl)amino-ATP-Sepharose column is used as the cation-exchange column for the purification of phosphoglucose isomerase. The preparation, maintenance, and commercial sources of this affinity column have been described elsewhere in this volume.[6,7]

[1] D-Glucose-6-phosphate ketol-isomerase, EC 5.3.1.9.
[2] C.-Y. Lee, D. Charles, and D. Bronson, *J. Biol. Chem.* **254,** 6375 (1979).
[3] D. Charles and C.-Y. Lee, *Biochem. Genet.* **18,** 153 (1980).
[4] D. Charles and C.-Y. Lee, *Mol. Cell. Biochem.* **28,** 11 (1980).
[5] R. W. Gracy and B. E. Tilley, this series, Vol. 41, p. 392.
[6] D. W. Niesel, G. Bewley, C.-Y. Lee, and F. B. Armstrong, this volume [51].
[7] C.-Y. Lee, J. H. Yuan, and E. Goldberg, this volume [61].

METHODS IN ENZYMOLOGY, VOL. 89

Drosophila Culture

Adult *Drosophila* (*Samarkand* and *cn bw; ri e*) are cultured and collected as described in this volume.[6]

Purification Procedure

An identical procedure is used for the purification of phosphoglucose isomerase from mouse and *Drosophila melanogaster* as well as their genetic variants.[2-4] The purification of this enzyme from the DBA/2J strain of mice and the *4/4* variant of *Drosophila* is presented here in detail. Unless otherwise indicated, all operations are performed at 4°, and phosphate buffers used are potassium salts.

Step 1. Preparation of Crude Homogenate. The muscle (~100 g) of whole mouse bodies is homogenized with a meat grinder in 500 ml of 10 mM phosphate, pH 6.5, containing 5 mM dithiothreitol and 1 mM EDTA (buffer A). After 1 hr, the homogenate is centrifuged at 27,000 g for 30 min, and the supernatant is filtered through glass wool. For purification of the *Drosophila* enzyme, 20 g of frozen flies are homogenized with a Polytron

TABLE I
PURIFICATION OF MOUSE AND *Drosophila* PHOSPHOGLUCOSE ISOMERASE

Step	Total protein[a] (mg)	Total activity (units)	Specific activity (units/mg protein)	Yield (%)
1. Preparation of crude extract				
Mouse	13,500	20,500	1.5	100
Drosophila	2,500	4,400	1.8	100
2. Ammonium sulfate fractionations				
Mouse	8,700	17,500	2	85
Drosophila	550	3,500	6.4	80
3. 8-(6-Aminohexyl)amino-ATP-Sepharose column				
Mouse	70	14,000	200	70
Drosophila	11	2,600	240	60
4. Isoelectric focusing				
Mouse	9.3	7,000	750	35
Drosophila	1.8	1,800	1,000	40

[a] Using bovine serum albumin as the protein standard, protein was determined by fluorescamine assays of P. Böhlen, S. Stein, W. Dairman, and S. Udenfriend, *Arch. Biochem. Biophys.* **155**, 213 (1973).

TABLE II
BIOCHEMICAL PROPERTIES OF MOUSE AND *Drosophila*
PHOSPHOGLUCOSE ISOMERASE

Property	Mouse DBA/2J	*Drosophila* (*cn bw;ri e*) (4/4)
Sedimentation coefficient (S)	7.0 ± 0.5	7.1 ± 0.5
Stokes' radius (nm)	4.46	4.54
Molecular weight		
Native	120,000 ± 7000	127,000 ± 10,000
Subunit	55,000 ± 2000	55,000 ± 2,000
Isoelectric point	8.4 ± 0.1	6.3 ± 0.2
Michaelis constant ($\times 10^5 M$)		
(fructose 6-phosphate)	4.0 ± 0.4	10.0 ± 0.8
Specific activity (units/mg)	750	1000

homogenizer (Brinkmann Instrument Co., Westbury, New York) in 150 ml of 10 mM phosphate, pH 6.0, containing 5 mM dithiothreitol and 1 mM EDTA (buffer B). After centrifugation, the supernatant is filtered through glass wool to remove suspended lipid.

Step 2. Ammonium Sulfate Fractionation. To the clear supernatant extract, ammonium sulfate is added to a final concentration of 24.3 g/100 ml. One hour later, the precipitate is removed by centrifugation (27,000 g for 20 min). More ammonium sulfate is added to the supernatant (28.5 g/100 ml of the supernatant). After 1 hr, the precipitate is recovered by centrifugation as described. The pellet is redissolved in 20 ml of buffer A for the mouse enzyme or buffer B for the fly enzyme and dialyzed against 4 liters each of the appropriate buffer.

Step 3. 8-(6-Aminohexyl)amino-ATP-Sepharose Chromatography. After dialysis, the denatured protein is removed by centrifugation, and the supernatant is loaded on the ATP-Sepharose column (2.5 × 20 cm) equilibrated with buffer A or buffer B. After washing with 1 liter of the appropriate buffer, the mouse enzyme is eluted with a 0 to 10 mM glucose 6-phosphate linear gradient (300 ml by 300 ml) in buffer A, while the fly enzyme is readily eluted with 300 ml of 5 mM glucose 6-phosphate in buffer B. Optimal results are obtained with a flow rate of 50–100 ml/hr.

Step 4. Isoelectric Focusing. Fractions containing phosphoglucose isomerase activity are pooled, concentrated to 5 ml, and dialyzed overnight with 1 liter of 1% glycine solution (pH 7.0). The dialyzed enzyme (in ~5 ml) is subjected to a preparative isoelectric focusing on a LKB 8101 column (110 ml) in the presence of 2% ampholyte generating a pH 3.5 to 10 gradient. After 18 hr at 1600 V, fractions of 1.2 ml are collected and assayed for enzyme activity. Peak fractions containing pure enzyme are pooled and stored at $-20°$. No attempt is made to remove ampholyte from the purified enzyme.

A summary of the purifications from mouse and *Drosophila melanogaster* is presented in Table I.

Properties

Some of the biochemical properties of mouse and *Drosophila* phosphoglucose isomerase are shown in Table II. The mouse and the *Drosophila* enzymes, in general, have quite similar molecular properties.[3,4] Antisera raised in rabbits against pure enzyme showed no mutual cross-reactivity between the mouse and the fly enzyme. Amino acid composition analysis revealed a certain degree of homology between the same enzyme from these two species.[4]

[95] D-Galactose-6-phosphate Isomerase

By RICHARD L. ANDERSON and DONALD L. BISSETT

D-Galactose 6-phosphate $\rightleftharpoons$ D-tagatose 6-phosphate

This inducible enzyme is instrumental in the catabolism of lactose and D-galactose in *Staphylococcus aureus*.[1-3]

Assay Method[3]

Principle. The continuous spectrophotometric assay is based on the following sequence of reactions:

$$\text{D-Galactose-6-P} \xrightarrow[\text{isomerase}]{\text{D-galactose-6-P}} \text{D-tagatose-6-P}$$

$$\text{D-Tagatose-6-P} + \text{MgATP} \xrightarrow[\text{kinase}]{\text{D-fructose-6-P}} \text{D-tagatose-1,6-P}_2 + \text{MgADP}$$

$$\text{MgADP} + \text{phosphoenolpyruvate} \xrightarrow[\text{kinase}]{\text{pyruvate}} \text{MgATP} + \text{pyruvate}$$

$$\text{Pyruvate} + \text{NADH} \xrightarrow[\text{dehydrogenase}]{\text{lactate}} \text{lactate} + \text{NAD}^+$$

[1] D. L. Bissett and R. L. Anderson, *Biochem. Biophys. Res. Commun.* **52**, 641 (1973).

[2] D. L. Bissett and R. L. Anderson, *J. Bacteriol.* **119**, 698 (1974).

[3] D. L. Bissett, W. C. Wenger, and R. L. Anderson, *J. Biol. Chem.* **255**, 8740 (1980).

METHODS IN ENZYMOLOGY, VOL. 89

With the three coupling enzymes in excess, the rate of D-galactose 6-phosphate isomerization is equivalent to the rate of NADH oxidation, which is measured by the absorbance decrease at 340 nm.

Reagents

Sodium Bicine[4] buffer, 0.25 M, pH 8.2
$MgCl_2$, 0.2 M
ATP, 0.1 M
Phosphoenolpyruvate, 0.1 M
NADH, 0.01 M
D-Galactose 6-phosphate, 0.2 M
Crystalline D-fructose-6-phosphate kinase (rabbit muscle)
Crystalline pyruvate kinase
Crystalline lactate dehydrogenase

Procedure. The following are added to a microcuvette with a 1.0-cm light path: 40 μl of Bicine buffer, 5 μl of $MgCl_2$, 5 μl of ATP, 5 μl of phosphoenolpyruvate, 5 μl of NADH, 40 μl of D-galactose 6-phosphate, non-rate-limiting amounts of D-fructose-6-phosphate kinase (500 mU), pyruvate kinase (500 mU), and lactate dehydrogenase (750 mU), a rate-limiting amount of D-galactose-6-phosphate isomerase, and water to a volume of 0.15 ml. The reaction is initiated by the addition of D-galactose-6-phosphate isomerase. A control cuvette minus D-galactose 6-phosphate measures ATPase and NADH oxidase, which must be subtracted from the total rate. The rates are conveniently measured with a Gilford multiple-sample absorbance-recording spectrophotometer. The cuvette compartment should be thermostatted at 30°. Care should be taken to confirm that the rates are constant with time and proportional to the D-galactose-6-phosphate isomerase concentration.

Definition of Unit and Specific Activity. One unit is defined as the amount of enzyme that catalyzes the isomerization of 1 μmol of D-galactose 6-phosphate to D-tagatose 6-phosphate per minute. Specific activity (units per milligram of protein) is based on protein determinations by the Lowry procedure.

Alternative Assay Procedure. The reaction may be measured in the reverse direction by the use of a colorimetric assay.[3]

Purification Procedure

Organism and Growth Conditions. Staphylococcus aureus NCTC 8511 is grown at 37° in Fernbach flasks (1500 ml per flask) on a rotary shaker. The medium is the induction medium of McClatchy and Rosenblum[5] supple-

---

[4] *N,N*-Bis(2-hydroxyethyl)glycine.
[5] J. K. McClatchy and E. D. Rosenblum, *J. Bacteriol.* **86,** 1211 (1963).

PURIFICATION OF D-GALACTOSE-6-PHOSPHATE ISOMERASE FROM *Staphylococcus aureus*

Fraction	Volume (ml)	Total protein (mg)	Total activity (units)	Specific activity (units/mg protein)	Recovery (%)
Cell extract	98	505	115	0.228	(100)
Bentonite	88	284	107	0.337	93
Hydroxyapatite	104	146	102	0.699	89
DEAE-cellulose I	36	17.2	57.6	3.35	50
DEAE-cellulose II	6.0	5.06	39.0	7.71	34
Sephadex G-100	8.0	1.25	18.9	15.1	16

mented with 1% D-galactose (autoclaved separately). The inoculum is 7 ml of an overnight culture in the same medium, except that no carbohydrate is added and the peptone concentration is increased to 2%. The cells are harvested 9 hr after inoculation, washed once by suspension in 0.85% (w/v) NaCl, and collected by centrifugation. The yield is about 4 g (wet weight) of cells per liter of medium.

Preparation of Cell Extracts. Cells (30 g) are suspended in 107 ml of buffer A [40 mM Tris-HCl (pH 8.5), 15% (v/v) glycerol, 0.2% (v/v) 2-thioethanol, and 0.2 mM EDTA]. The cells are broken by treating the suspension at 0 to 2° for 10 min (in 1-min bursts) with the 1.27-cm (diameter) horn of a Heat Systems-Ultrasonic W-185C sonifier at 100 W, in the presence of twice the packed-cell volume of glass beads (88–125 μm in diameter). The cell extract is the supernatant fluid resulting from a 10-min centrifugation at 40,000 g. The pH of the extract is 7.5.

General. The following procedures are performed at 0 to 4°. A summary of the purification is shown in the table.

Bentonite Treatment. To the cell extract 3.9 g of bentonite (Fisher Scientific Co.) are added with stirring. After 10 min, the suspension is clarified by centrifugation at 14,000 g for 10 min, and the sediment is discarded.

Negative Adsorption on Hydroxyapatite. A hydroxyapatite (Bio-Rad Laboratories, Richmond, California) column (4.7 × 2.9 cm) is equilibrated with buffer B [40 mM Tris-HCl (pH 7.5), 15% (v/v) glycerol, 0.2% (v/v) 2-thioethanol, and 0.2 mM EDTA]. The supernatant solution from the bentonite step is applied to the column, which is then washed (10 ml/hr) with the same buffer. Fractions (13 ml) are collected, and those that contain isomerase activity (fractions 3 through 10) are combined.

DEAE-Cellulose Chromatography I. A DEAE-cellulose column (2.6 × 6.6 cm) is equilibrated with buffer B. The combined fractions from the preceding step are applied to the column, which is then washed with 200 ml of buffer B. The protein is eluted with a linear gradient (350 ml; 60 ml/hr) of 0.1 to 0.8 M KCl in the same buffer. Forty-eight 7.3-ml fractions are collected, and those that contain most of the isomerase activity (fractions 19 through 23) are combined.

DEAE-Cellulose Chromatography II. A DEAE-cellulose column (1.2 × 4.5 cm) is equilibrated with buffer B. The combined fractions from the preceding step are diluted to 100 ml with the same buffer and applied to the column, which is then washed with 50 ml of the same buffer. The protein is eluted with a linear gradient (50 ml; 20 ml/hr) of 0.1 to 0.5 M KCl in the same buffer. Fifty 1.0-ml fractions are collected, and those that contain most of the isomerase activity (fractions 16 through 21) are combined.

Sephadex G-100 Chromatography. A Sephadex G-100 column (1.4 × 86 cm) is equilibrated with buffer B. The combined fractions from the preceding step are applied to the column and chromatographed with the same buffer. Two-milliliter fractions are collected, and those that contain the highest specific activity of isomerase (fractions 28 through 31) are combined. The enzyme is 67-fold purified with an overall recovery of 16% and is free from the constitutive D-glucose-6-phosphate isomerase.

Properties[3]

Stability. The enzyme in the Sephadex G-100 fractions was stable for several weeks during storage at $-20°$.

pH Optimum. The optimal pH in Bicine buffer was about 8.2.

Substrate Specificity. D-Galactose 6-phosphate (K_m = 9.6 mM) and D-tagatose 6-phosphate (K_m = 1.9 mM) are the only known substrates. D-Glucose 6-phosphate, D-mannose 6-phosphate, and D-fructose 6-phosphate at 33 mM were not isomerized (<1% of the rate with D-galactose 6-phosphate).

Effect of Divalent Metal Ions. No activation was observed with 1 mM $MgCl_2$, $MnCl_2$, $CoCl_2$, $FeSO_4$, or $CaCl_2$. Also, 10 mM EDTA did not affect the activity.

Molecular Weight. The molecular weight was estimated to be about 99,000–100,000, by both sucrose gradient centrifugation and Sephadex chromatography.

[96] 6-Phospho-3-Ketohexulose Isomerase from *Methylomonas* (*Methylococcus*) *capsulatus*

By J. Rodney Quayle

D-*arabino*-3-Ketohexulose 6-phosphate $\rightleftharpoons$ D-fructose 6-phosphate

This enzyme, in concert with hexulosephosphate synthase,[1] catalyzes a key step in the assimilation of carbon during growth of some bacteria on reduced C_1 compounds. The purification described below has been the subject of a previous report.[2] A similar enzyme has also been purified from *Methylomonas aminofaciens* 77*a* by Kato *et al.*[3]

Assay Methods

Principle. Two assay methods can be used, both of which depend on the prior preparation of substrate amounts of D-*arabino*-3-ketohexulose 6-phosphate (hexulose phosphate) by the use of hexulose phosphate synthase, purified as described elsewhere.[1] Methods 1 and 2 involve, respectively, the discontinuous or continuous measurement of the rate of production of fructose 6-phosphate from hexulose phosphate.

The preparation of hexulose phosphate described below has been the subject of a previous report.[4]

Preparation of Hexulose Phosphate. D-Ribose 5-phosphate (1 mmol, sodium salt) is incubated with 120 μg of phosphoriboisomerase (Calbiochem Ltd., London) in 98.5 ml of 50 mM imidazole–HCl buffer, pH 7.0, for 30 min at 37°. Hexulosephosphate synthase purified from *M. capsulatus* (60 units) is then added together with a further 120 μg of phosphoriboisomerase, 0.5 mmol of $MgCl_2$, and 1.5 mmol of formaldehyde to a total volume of 100 ml. After 10 and 20 min, further 0.35-mmol quantities of formaldehyde are added. The reaction is stopped after 30 min, when approximately 70% of the ribose phosphate has disappeared, by precipitation of the sugar phosphates as barium salts as described by Horecker.[5] The barium salts of the sugar phosphates are converted into the soluble sodium salts by dissolving in 2.5 ml of 0.2 M

---

[1] J. R. Quayle, see this series, Vol. 90 [51].
[2] T. Ferenci, T. Strøm, and J. R. Quayle, *Biochem. J.* **144**, 477 (1974).
[3] N. Kato, H. Ohashi, T. Hori, Y. Tani, and K. Ogata, *Agric. Biol. Chem.* **41**, 1133 (1977).
[4] T. Strøm, T. Ferenci, and J. R. Quayle, *Biochem. J.* **144**, 465 (1974).
[5] B. L. Horecker, see this series, Vol. 3, p. 190.

H_2SO_4, adding 0.5 mmol of Na_2SO_4 and water to a total volume of 20 ml. The resulting precipitate of $BaSO_4$ is collected by centrifugation at 38,000 g for 15 min and extracted twice with water; the precipitate is discarded. The combined supernatant and extracts are combined, and the volume is adjusted to 50 ml. A sample containing up to 500 μmol of sugar phosphates is taken and diluted to 50 ml; the pH is adjusted to 8.0 with aqueous NH_3. This solution is added, at a flow rate of 2 ml/min, to a column (1.5 cm × 5 cm) of Dowex 1 (X8; Cl^- form) at 3°, and the sugar phosphates are eluted by using the borate complexing method of Khym and Cohn.[6] A linear gradient of 20 to 40 mM NH_4Cl in 0.04 mM $Na_2B_4O_7$, pH 8.3, in a total volume of 1 liter, is used first, followed by 40 mM NH_4Cl in 0.01 mM $Na_2B_4O_7$, pH 8.3, at a flow rate of 2 ml/min. Fractions (10 ml) are collected, and D-ribulose 5-phosphate is first eluted in the effluent volume between 400 and 700 ml, followed by the hexulose phosphate in the effluent volume between 500 and 1200 ml. The total yield (based on starting ribose 5-phosphate) in all fractions containing hexulose phosphate should be 70–80%; the fractions from 710 to 1200 ml, free of ribulose 5-phosphate, are pooled. After precipitation with $Ba(OH)_2$ and conversion of the sugar phosphates into the acid form by using Amberlite CG-120 (H^+ form), borate is removed by distillation under reduced pressure in a rotary evaporator at 30° in the present of methanol, as described by Zill et al.[7] The evaporation is taken to near dryness, and the syrup is appropriately diluted with water. The pH of the solution is adjusted to 3.0 with NaOH and stored at 3°, under which conditions it is stable for at least 3 months. Hexulose phosphate is much less stable at pH values greater than 3. The final yield from pentose phosphate should be 25–30%.

Reagents for Method 1

Tris-HCl buffer, 0.2 M, pH 8.3
Hexulose phosphate, 4 mM
EDTA, 10 mM

Reagents for Method 2

Tris-HCl buffer, 0.2 M, pH 8.6
Hexulose phosphate 3.5 mM
EDTA, 10 mM
Glucose-6-phosphate dehydrogenase (grade II, from yeast, Boehringer Corporation, London Ltd.)
Phosphoglucoisomerase (from yeast, Boehringer, London)
NADP 2.5 mM

[6] J. X. Khym and W. E. Cohn, *J. Am. Chem. Soc.* **74**, 1153 (1953).
[7] L. P. Zill, J. X. Khym, and C. M. Chenise, *J. Am. Chem. Soc.* **75**, 1339 (1953).

Procedure for Method 1. The assay mixture contains in a final volume of 1 ml: Tris-HCl buffer, 0.5 ml; hexulose phosphate, 0.25 ml; EDTA 0.1 ml. Extract containing up to 0.2 unit of phosphohexuloisomerase is added, incubation is carried out at 30°, and the reaction is stopped at known times, up to 5 min, by the addition of 1 ml of 1 M perchloric acid. The perchlorate is removed from the assay mixture by discarding the precipitate formed after neutralizing with 0.5 ml of 2 M potassium carbonate at 0–4°. The supernatant is assayed for D-fructose 6-phosphate by the enzymic method of Hohorst.[8] Nonenzymic formation of fructose phosphate from hexulose phosphate is corrected for by running assays in the absence of phosphohexuloisomerase.

Procedure for Method 2. This assay has previously been used by van Dijken *et al.*[9] The assay mixture in a cuvette (10-mm light path) contains, in a final volume of 1 ml: Tris-HCl buffer, 0.25 ml; hexulose phosphate, 0.2 ml; EDTA, 0.1 ml; glucose phosphate dehydrogenase, 0.7 unit; phosphoglucoisomerase, 0.7 unit; NADP, 0.1 ml. After a suitable time interval at 30°, extract containing phosphohexuloisomerase is added, and the absorbance change at 340 nm is recorded. The resulting rate minus the phosphohexuloisomerase-independent rate is taken to be due to phosphohexuloisomerase activity.

Units. One unit of phosphohexuloisomerase is defined as the amount of enzyme that catalyzes the formation of 1 μmol of fructose 6-phosphate per minute at 30°. Specific activity is expressed as units per milligram of protein.

Purification Procedure

The organism used is *Methylomonas capsulatus* ATCC No. 19069 (*Methylococcus capsulatus*), and its growth is described elsewhere.[1] The first two purification steps described below are identical to those used for purification of hexulosephosphate synthase[1]; the particulate fraction from step 2 can then be used as a source of hexulosephosphate synthase, and the supernatant fraction from step 2 is used for phosphohexuloisomerase purification. If it is intended to purify phosphohexuloisomerase only, the 5 mM $MgCl_2$ used in step 1 should be replaced by 5 mM EDTA.

Step 1. Preparation of Cell-Free Extract. Cell paste (24 g wet weight) is suspended in 75 ml of 20 mM sodium potassium phosphate buffer, pH 7.0, containing 5 mM $MgCl_2$. The suspension is disrupted by ultrasonication

[8] H.-J. Hohorst, *in* "Methods of Enzymatic Analysis" (H. U. Bergmeyer, ed.), p. 134. Academic Press, New York, 1965.
[9] J. P. van Dijken, W. Harder, A. J. Beardsmore, and J. R. Quayle, *FEMS Microbiol. Lett.* **4**, 97 (1978).

for 6 min at 0–4° in an MSE ultrasonic disintegrator (150 W) and centrifuged at 6000 g for 10 min, yielding the cell-free extract as supernatant.

Step 2. Preparation of Membrane-Free Supernatant. The supernatant from step 1 is further centrifuged at 38,000 g for 60 min (the pellet can be used for purification of hexulosephosphate synthase)[1] and the resulting supernatant is retained. An apparent increase in the total activity of the enzyme is encountered in step 2. This is probably due to removal of inhibitor(s) present in the crude cell-free extract. Higher activities are observed in step 1 if the 5 mM MgCl$_2$ is replaced by 5 mM EDTA, however, the extract cannot then be used as a simultaneous source of hexulosephosphate synthase owing to instability of the latter in the absence of divalent metal ions.

Step 3. Protamine Sulfate Treatment. To the supernatant from step 2 is added EDTA to a final concentration of 1 mM, and to 40 ml of the resulting solution protamine sulfate (2%, w/v) is added slowly, to a final concentration of 1 mg of protamine sulfate to 10 mg of protein. After 20 min, the precipitate is removed by centrifugation at 25,000 g for 15 min and discarded. In all subsequent operations it is essential to include EDTA in the buffers and solutions used.

Step 4. Ammonium Sulfate Fractionation. To the supernatant from step 3, solid (NH$_4$)$_2$SO$_4$ is added with stirring to a concentration of 2.05 M. The resulting precipitate is removed by centrifugation and discarded. Further (NH$_4$)$_2$SO$_4$ is added to the supernatant to a final concentration of 3.9 M; the precipitate is collected by centrifugation and dissolved in 4 ml of 10 mM Tris-HCl buffer, pH 7.5, containing 1 mM EDTA.

Step 5. Gel Filtration. The protein solution from step 4 is applied to a column (70 cm × 2.5 cm) of Sephadex G-150 previously equilibrated with 10 mM Tris-HCl buffer, pH 7.5, containing 1 mM EDTA. Phosphohexuloisomerase activity is eluted from the column with the equilibration buffer between 220 ml and 280 ml (void volume of the column being 135 ml).

Step 6. DEAE-Cellulose Chromatography. Sodium chloride is added to the combined fractions from step 5 to a final concentration of 0.1 M, and the pH is adjusted to 7.5 with sodium hydroxide. The solution is applied to a column (20 cm × 1.5 cm) of DEAE-cellulose (Whatman DE-52) previously equilibrated with 10 mM Tris-HCl buffer–1 mM EDTA–0.1 M NaCl, pH 7.5.

The phosphohexuloisomerase is eluted from the column with 500 ml of a linear gradient of 0.1 M to 0.5 M NaCl in 10 mM Tris HCl buffer–1 mM EDTA, pH 7.5. The enzyme appears in the effluent volume between 310 and 410 ml.

Step 7. Second DEAE-Cellulose Chromatography. One-third of the

PURIFICATION OF PHOSPHOHEXULOISOMERASE FROM *Methylomonas capsulatus*

Step	Fraction	Total volume (ml)	Total protein (mg)	Total activity[a] (units)	Specific activity (units/mg protein)
1	Cell-free extract	47	1015	3150	3.12
2	Membrane-free supernatant	40	312	5520[b]	17.7
3	Protamine sulfate fraction	40	304	5480	18.0
4	(NH$_4$)$_2$SO$_4$ fraction (2.05–3.9 M)	5	135	5720	42.3
5	Pooled fractions after gel filtration	60	28.8	4650	162
6	Pooled fraction after first DEAE-cellulose chromatography	105	7.2	4450	1060
7	Pooled fractions after second DEAE-cellulose chromatography	10[c]	0.55	860[d]	1560

[a] Results obtained by Assay Method 1.
[b] See text for explanation of apparent increase in total activity.
[c] Volume after concentration of pooled column fractions.
[d] Step 7 was performed with one third of the enzyme solution from step 6.

pooled fractions from step 6 are diluted to 100 ml with 10 mM Tris-HCl buffer–1 mM EDTA, pH 7.5, to lower the NaCl concentration to below 0.15 M and applied to a second column (10 × 1.5 cm) of DEAE-cellulose (Whatman DE-52) previously equilibrated with 10 mM Tris-HCl buffer–1 mM EDTA–0.15 M NaCl, pH 7.5. The phosphohexuloisomerase is eluted from the column with 500 ml of a linear gradient of 0.15 M to 0.2 M in 10 mM Tris-HCl buffer–1 mM EDTA, pH 7.5. The enzyme appears in the effluent volume between 290 and 390 ml. The protein in the pooled fractions is concentrated by ultrafiltration through a Diaflo PM-30 membrane (Amicon Corp., Lexington, Massachusetts) to a volume of 1.0 ml and this is diluted to 10 ml with 10 mM-Tris-HCl buffer–1 mM EDTA, pH 7.5. This procedure of 10-fold concentration (from 10 ml to 1 ml) and dilution (1 ml to 10 ml) is repeated once more to lower further the concentration of NaCl. The final solution of phosphohexuloisomerase is stored at $-15°$.

A typical purification is summarized in the table.

Properties

Purity. As judged by polyacrylamide gel electrophoresis, the preparation of phosphohexuloisomerase is over 90% pure. The enzyme obtained from step 6 is contaminated with phosphoriboisomerase and phosphoglucoisomerase at 0.8% and 0.1%, respectively, of the activity of phosphohexuloisomerase. The second DEAE-cellulose chromatography in step 7 removes the phosphoriboisomerase and reduces the phosphoglucoisomerase activity to less than 0.01% that of the phosphohexuloisomerase. There is no detectable hexulosephosphate synthase activity.

Substrate Specificity. As far as has been tested, the enzyme is specific for hexulose phosphate, it is inactive with D-ribulose 5-phosphate, D-xylulose 5-phosphate, or D-allulose 6-phosphate.

Effect of Metal Ions. The enzyme is inhibited by divalent metal ions in the order $Cu^{2+} > Ni^{2+} > Co^{2+} > Zn^{2+} > Mg^{2+} > Ca^{2+}$. The inhibition by 1 m$M$ Cu^{2+} is 92% and that by 1 mM Mg^{2+} is 25%.

Stability. The purified enzyme can be stored in 10 mM Tris-HCl buffer–1 mM EDTA, pH 7.5, for 4 months at 0–4° and at −15° without detectable loss of activity. All activity is lost within 10 min at 60°.

pH Optimum. The enzyme has optimal activity at pH 8.3.

Michaelis Constants. In the forward reaction, the apparent K_m for hexulose phosphate at pH 8.3 is $1.0 \times 10^{-4} M$, and in the reverse direction for D-fructose 6-phosphate at pH 7.0 it is $1.1 \times 10^{-3} M$.

Molecular Weight. The molecular weight, as measured by gel permeation through Sephadex G-100,[10] is approximately 67,000.

Equilibrium Constant. The equilibrium constant of the reaction in the forward direction at pH 7.0 and 30° is 188.

[10] P. Andrews, *Biochem. J.* **96**, 595 (1965).

[97] Bacterial Ribosephosphate Isomerase

By R. D. MacElroy and C. R. Middaugh

D-Ribose 5-phosphate ⟷ D-ribulose 5-phosphate

Ribosephosphate isomerase (phosphoribose isomerase) [ribose-5-phosphate ketol-isomerase, EC 5.3.1.6] (RPI) catalyzes the interconversion of D-ribose 5-phosphate and ribulose 5-phosphate. The enzyme has been found in both heterotrophic and autotrophic bacteria, and its perceived role in metabolism differs significantly depending on whether it is considered to be an anabolic or a catabolic enzyme. In heterotrophic organisms it functions catabolically as a member of the pentose pathway. It is probable that the pentose shunt in heterotrophs is primarily involved in production of NADPH and for supplying ribose 5-phosphate for the formation of 5-phosphoribosyl pyrophosphate and the subsequent synthesis of nucleic acids. In autotrophic bacteria the enzyme occupies a position in the anabolic pentose (Calvin–Benson–Bassham) pathway, which serves as the major source of reduced carbon for these organisms.

Distribution. Among heterotrophic bacteria, ribosephosphate isomerase has been found in *Escherichia coli,*[1] *Bacillus caldolyticus,*[2] *Brevibacterium,*[3] and *Mycobacterium.*[3] Autotrophs containing the enzyme include *Pseudomonas* (*Hydrogenomonas*) *facilis,*[4] *Desulfovibrio vulgaris,*[5] *Rhodospirillum rubrum,*[6] *Chromatium* D,[7] *Nitrosomonas europaea,*[8] *Thiobacillus thioparus* and *T. neapolitanus.*[9]

Assay Methods

Principles. Several assay procedures are available for detecting ribosephosphate isomerase activity. Of these, the most direct and reliable for impure enzymes detects the appearance (or disappearance) of ribulose 5-phosphate, which produces a purple color after reaction with cysteine-carbazole.[10] The procedure is appropriate at all stages of enzyme purification and can be used for kinetic studies as well. Continuous assays have been described[11] that involve the use of coupling enzymes (phosphoribulokinase, phosphoenolpyruvate kinase, and lactate dehydrogenase); in these assays NADH use in the reduction of pyruvate to lactate is continuously monitored at 340 nm. Direct spectrophotometric assays have also been described in which the absorbance of ribulose is monitored at 280 nm[12] or at 290 nm.[13]

Reagents

Ribose 5-phosphate, sodium salt, 0.03 M. Frozen ribose 5-phosphate solutions are reasonably stable, although a slow isomerization to ribulose 5-phosphate occurs.

[1] J. David and H. Wiesmeyer, *Biochim. Biophys. Acta* **79**, 56 (1970).
[2] C. R. Middaugh and R. D. MacElroy, *J. Biochem.* (*Tokyo*) **79**, 1331 (1976).
[3] N. V. Loginova and Y. A. Trotsenko, *Mikrobiologiya* **47**, 939 (1978).
[4] R. D. MacElroy, E. J. Johnson, and M. K. Johnson, *Arch. Biochem. Biophys.* **131**, 272 (1969).
[5] M. Alvarez and L. L. Barton, *J. Bacteriol.* **131**, 133 (1977).
[6] L. Anderson, L. E. Worthen, and R. C. Fuller, *in* "Comparative Biochemistry and Biophysics of Photosynthesis" (K. Shibata, A. Takamiya, A. T. Jagendorf, and R. C. Fuller, eds.), p. 379. University of Tokyo Press, Tokyo, Japan, 1968.
[7] J. Gibson and B. Hart, *Biochemistry* **8**, 2737 (1969).
[8] P. S. Rao and D. J. D. Nicholas, *Biochim. Biophys. Acta* **124**, 221 (1966).
[9] E. J. Johnson and H. D. Peck, *J. Bacteriol.* **89**, 1041 (1965).
[10] G. F. Domagk and K. M. Doering, this series, Vol. 41 [90].
[11] R. C. Huffaker, R. L. Orendorf, C. J. Keller, and G. E. Kleinkopf, *Plant Physiol.* **41**, 913 (1966).
[12] F. C. Knowles, J. D. Chanley, and N. G. Pon, *Arch. Biochem. Biophys.* **202**, 106 (1980).
[13] T. Wood, *Anal. Biochem.* **33**, 297 (1970).

Tris-HCl buffer, 0.10 M, pH 7.4, containing 1.0 mM EDTA and 1.0 mM 2-mercaptoethanolamine. The pH of Tris buffer changes markedly with temperature, and the pH of the buffer should be adjusted at the temperature at which the assay is to be done.

Cysteine, 0.03 M. Dissolve 79 mg of cysteine hydrochloride in 15 ml of distilled water; prepare fresh on day of assay.

H_2SO_4, 75% solution. Slowly add 75 ml of concentrated H_2SO_4 to 25 ml of water and cool in ice bath

Carbazole, 0.1% solution. Carbazole should be recrystallized from xylene before use. The stock solution is made by dissolving 50 mg in 50 ml of absolute ethanol.

Procedure. Place 0.3 ml of Tris buffer, 0.1 ml of ribose 5-phosphate, and up to 0.1 ml of dilute enzyme in a large (e.g., 25 ml) test tube. The final volume of the assay mixture should be 0.5 ml. Incubate at assay temperature, usually 25–37°. If higher incubation temperatures are used, cover with a glass ball of appropriate size to prevent evaporation. With enzyme preparations of unknown activity, it is appropriate to stop the incubation at various times after 2–15 min, or to do parallel assays with various dilutions of the enzyme.

After incubation add 0.5 ml of cysteine, followed immediately by 5 ml of H_2SO_4. Mix vigorously by swirling, and develop color for 30 min at 37°. Use a substrate blank (substrate incubated without enzyme).

Cool tubes and measure color density within 5 min after color development. The density of color is determined with a spectrophotometer at 540 nm using a 1-cm cuvette, or with a colorimeter using a green filter. Reproducible readings are possible with absorptivities of up to 0.4. The absorption value of the blank is subtracted from the test assays; it should not exceed 0.05 when freshly prepared substrate is used.

Definition of Unit and Specific Activity. One unit of enzyme activity can be defined, for routine purposes, as the amount necessary to convert sufficient ribose 5-phosphate to ribulose 5-phosphate to produce an absorption of 1.0 at 540 nm in 1 min.

For conversion to standard units (micromoles per minute), the equilibrium constant for ribulose 5-phosphate formation (0.323 at 37°)[13a] is used. To determine a conversion factor, a known amount of ribose 5-phosphate (0.250 μmol) is incubated with an enzyme preparation containing no other enzymic activities for which ribose 5-phosphate or ribulose 5-phosphate are substrates. After 90 min at 37°, ribulose 5-phosphate concentration is determined with the cysteine–carbazole test, and the absorption value of the known (Aa), minus that of the blank (Ab), is used to determine the absorption per micromole.

[13a] T. Wood, *Arch. Biochem. Biophys.* **160**, 40 (1974).

$$[Aa - Ab]/0.323 \times 0.250 = \text{absorption per } \mu\text{mol}$$

Alternatively, a coupled spectrophotometric assay system[11] can be used to determine a standard curve for the assay based on a known stoichiometric relationship between ribose 5-phosphate isomerization and the extinction coefficient of NADPH or NADH.

For most purposes the extinction coefficient of Ashwell and Hickman[14] can be used: 0.1 μmol of ribulose 5-phosphate produces an absorption of 0.140 at 540 nm after color development at 37° for 30 min.

Purification Procedures

General Comments

The methods of preparing ribose-5-phosphate isomerase take advantage of general characteristics of the enzyme: its resistance to heat inactivation, its relatively small size, and precipitability by $(NH_4)_2SO_4$. The final stages of purification are most often accomplished by ion-exchange chromatography,[2] gel filtration,[2,7] or affinity chromatography.[15] References cited in footnotes 2, 6, 10, and 15 give several detailed preparative procedures.

Procedure for Bacillus caldolyticus

Growth of Bacillus caldolyticus (thermophilic heterotroph). Bacillus caldolyticus[16,17] is grown on a complex medium containing, per liter, 3 g of glucose, 3 g of tryptone, and 0.3 g of $MgSO_4 \cdot 7 H_2O$. After sterilization of the medium, 1 ml of a 0.5% solution of ferrous sulfate ($FeSO_4 \cdot 7 H_2O$; Mallinckrodt) is added; the medium is brought to 65°, with the pH adjusted to 7.5–7.6, and is inoculated with approximately 10^8 organisms grown overnight on petri plates. The characteristics of the organism, and the various strains of it, have been described by Johnson.[17] The organisms are harvested by centrifugation and washed twice with 0.1 M buffer (Tris-HCl, pH 7.5 containing 1 mM each of ethylenediaminetetracetic acid and 2-mercaptoethanol).

[14] G. Ashwell and J. Hickman, *J. Biol. Chem.* **226**, 65 (1957).
[15] H. Horitso, Y. Banno, S. Kimura, H. Shumizu, and M. Tomoyeda, *Agric. Biol. Chem.* **42**, 2195 (1978).
[16] U. J. Heinen and W. Heinen, *Arch. Mikrobiol.* **76**, 2 (1972).
[17] E. J. Johnson, *in* "Strategies of Microbial Life in Extreme Environments" (M. Shilo, ed.), p. 471. Verlag Chemie International, Weinheim, 1979.

Preparation of Cell-free Extracts. For preparation of a partially purified enzyme fraction, fresh or frozen cells are suspended to 50% (w/v) in buffer and passed through a French pressure cell at 1400 kg/cm². The homogenate is centrifuged at 27,000 g for 60 min; the supernatant is decanted and centrifuged again at 144,000 g for 120 min.

Ammonium Sulfate Fractionation. Solid (NH₄)₂SO₄ (22.6 g per 100 ml of solution) is added to the supernatant fluid and stirred for 30 min at 4°; the precipitate is removed by centrifugation at 20,000 g for 30 min. The supernatant fraction is decanted, and 12.0 g of solid (NH₄)₂SO₄ are added per 100 ml. The precipitate is collected, dissolved in a minimum volume of buffer, and desalted by passage over a column of Sephadex G-25 (coarse) (2.5 cm × 45 cm). The fractions containing protein are identified by their absorption at 280 nm in a spectrophotometer; ammonium sulfate can be detected by the formation of a white precipitate of barium sulfate when a drop of eluent is mixed with a small volume of 5% barium chloride in 0.1 N HCl. The salt-free eluent is concentrated by ultrafiltration to 10% of its original volume with an Amicon (UM-10) filter.

Protamine Sulfate Treatment. A 2% protamine sulfate solution is added (0.2 volume), and after 60 min the precipitate is removed by centrifugation (15,000 g, 30 min).

Heat Treatment. The supernatant fraction is then brought to 95° for 3 min; after cooling on ice, the coagulated protein is removed by centrifugation (15,000 g, 30 min).

Sephadex G-200 Chromatography. The supernatant fraction is passed over a G-200 Sephadex column (2.4 × 50 cm) with a flow rate of 50–75 ml/hr; 5-ml fractions are collected. The fractions containing activity are pooled and reduced by ultrafiltration to 20% of the volume originally applied to the column.

Second Ammonium Sulfate Fractionation. The (NH₄)₂SO₄ fractionation step is repeated. The precipitate is again desalted and concentrated.

DEAE-Sephadex (A-25) Chromatography. After equilibration overnight by dialysis against 0.01 M buffer (pH 8.0) the material is applied to a DEAE-Sephadex A-25 column (2.4 × 45 cm) and developed with a linear, 800-ml Tris-HCl gradient ranging from 0.01 to 0.5 M and containing 1 mM each of EDTA and 2-mercaptoethanol. The enzyme is eluted as a pair of overlapping peaks of activity at approximately 0.35 M and 0.38 M buffer. The flow rate of the column is approximately 50 ml/hr, and the activity emerges at between 500 ml and 700 ml. The resolution of the peaks is increased by chromatography at 25° rather than 4°. Rechromatography of either of the peaks of activity results in elution of two similar peaks, and no further purification.

Procedure for Thiobacillus thioparus

Preparation of Cell-free Extracts from Thiobacillus thioparus (mesophilic autotroph). *Thiobacillus thioparus* (ATCC 8158) is grown as described elsewhere,[18] and partially purified enzyme fractions are prepared as described elsewhere[2] and below. Fresh or frozen cells are suspended to 20% (w/v) in buffer and passed through a French pressure cell at 1400 kg/cm². The homogenate is centrifuged at 105,000 g, first for 30 min; the precipitate is discarded, and the homogenate is centrifuged again as before for 300 min.

Ammonium Sulfate Fractionation. The supernatant fluid is treated with 29.1 g of solid $(NH_4)_2SO_4$ per 100 ml, and the precipitate that forms in 30 min at 4° is discarded after collection by centrifugation at 15,000 g for 30 min. An additional 15.9 g of solid ammonium sulfate per 100 ml is added to the supernatant, and the precipitate is collected.

Streptomycin Sulfate Fractionation and Heat Treatment. The precipitate is dissolved in a minimum volume of buffer and desalted on a Sephadex G-25 column (2.5 × 45 cm). The active eluent fractions are concentrated by ultrafiltration, as described above, to 10% of the original volume; 0.1 volume of 3% streptomycin sulfate is added, and the temperature of the solution is brought to 60° for 5 min. After cooling on ice, the precipitate is removed by centrifugation (30 min at 15,000 g).

Sephadex G-75 Chromatography. The supernatant fraction collected after centrifugation is applied to a Sephadex G-75 column (2.5 × 50 cm). Fractions of 1 ml are collected; active fractions emerge after a volume of 115 ml and are pooled and concentrated to about 20% of the original volume.

Typical purifications of *B. caldolyticus* and *T. thioparus* are summarized in Table I.

Properties

Stability and Storage. Both the *B. caldolyticus* and the *T. thioparus* enzymes are freeze-labile and were stored at 4°. The *B. caldolyticus* enzyme retained appreciable activity for 2 weeks; the *T. thioparus* enzyme retained full activity for 6–8 weeks.

Physical Properties. The estimated size of ribosephosphate isomerase from bacteria ranges from 30,000 to 60,000 daltons. Table II compares molecular weights of ribosephosphate isomerases from several bacteria.

[18] R. D. MacElroy, E. J. Johnson, and M. K. Johnson, *Arch. Biochem. Biophys.* **127**, 310 (1968).

TABLE I

PURIFICATION OF RIBOSE-5-PHOSPHATE ISOMERASE FROM *Bacillus caldolyticus* AND *Thiobacillus thioparus*[a]

Step	Total units	Total protein (mg)	Specific activity (μmol/min mg^{-1})	Recovery (%)
B. caldolyticus				
Crude extract	780	4500	0.17	100
First $(NH_4)_2SO_4$	750	1950	0.39	96
Protamine sulfate	730	900	0.88	94
Heat treatment	700	170	4.1	90
G-200 Sephadex	610	60	10.2	78
Second $(NH_4)_2SO_4$	590	42	12.2	76
DEAE Sephadex A-25	580	11	52.7	74
T. thioparus				
Crude extract	47,000	9200	5.1	100
$(NH_4)_2SO_4$	8,000	470	17.0	17
Streptomycin sulfate	7,800	290	26.9	17
Heat treatment	5,200	42	124.0	11
Sephadex G-75	4,900	9	544.0	10

[a] From Middaugh and MacElroy.[2]

Regulation and Inhibition

The key position of ribose-5-phosphate isomerase in the pathway leading to the reductive assimilation of CO_2 has led to the suggestion that it may play a role in the regulation of metabolism, particularly in *Rhodospirillum rubrum*.[6] In cell-free extracts the enzyme from this organism exhibited significant inhibition of rate in the presence of citrate, NAD$^+$, and AMP when these were present at concentrations of 10 mM. Ribulose diphosphate inhibited strongly at concentrations of 1.0 mM. David and Wiesmeyer[1] have reported that 6-phosphogluconate, fructose 6-phosphate, and AMP all cause inhibition of the *E. coli* enzyme. *Bacillus caldolyticus* and *T. thioparus* were significantly inhibited by 6-phosphogluconate and AMP,[2] whereas AMP at 5.0 mM did not inhibit the enzyme from *Pseudomonas* (*Hydrogenomonas*) *facilis*.[4]

The response of phosphoribose isomerase to metal ions, metal ion complexing reagents, sulfhydryl-complexing reagents, and reducing agents varies according to the source of the enzyme. For example, the enzymes from *T. thioparus* and *B. caldolyticus*[2] were both stimulated by ethylenediaminetetracetic acid (EDTA) and by mercaptoethanol at concentrations between 0.5 and 1.0 mM. No stimulation of activity was ob-

TABLE II

MOLECULAR WEIGHTS OF RIBOSEPHOSPHATE ISOMERASES FROM
SEVERAL BACTERIA

Source	Estimated molecular weight	Reference[a]
Escherichia coli (constitutive)	45,000	(14)
E. coli (inducible)	32,000	(14)
Chromatium D	54,000	(7)
Rhodospirillum rubrum	57,000	(6)
Thiobacillus neapolitanus	40,000	(2)
Bacillus caldolyticus	40,000	(2)

[a] Numbers refer to text footnotes.

served upon addition of metal ions, and Mg^{2+}, Ca^{2+}, Co^{2+} and Ag^+ inhibited the activity of both enzymes. In contrast data on multiple enzyme forms from the yeast Candida utilis [19] indicate that each of the three forms behaves differently in the presence of ions, but that a divalent cation is necessary for enzymic activity.

David and Wiesmeyer[1] have published evidence for two distinct ribosephosphate isomerases in E. coli. The observation was confirmed by Skinner and Cooper[20] and extended by Essenberg and Cooper,[21] who characterized the two enzymes. These authors also identified one enzyme (A) as constitutive and the other (B) as inducible. Evidence was also presented to suggest that two distinct genes are responsible for the appearance of the enzymes. Middaugh and MacElroy[2] reported evidence that multiple forms of the enzyme are present in extracts of B. caldolyticus; however, the data suggest that these forms are the consequence of an equilibrium between different tertiary or quaternary forms of the same protein. Wood[13a] has reported multiple forms of mammalian and spinach enzymes, and Horitsu et al.[19] have reported three forms of the enzyme in extracts of Candida utilis.

Kinetics and Temperature Effects. The kinetics of all the ribosephosphate isomerases thus far reported follow standard Lineweaver–Burk kinetics. The K_m of the enzyme varies with the source, but is generally of the order 0.9 to 5×10^{-3}. The K_m varies with temperature, and in the case of B. caldolyticus and T. thioparus, plots of the pK_m versus reciprocal

[19] H. Horitsu, I. Sasaki, T. Kikuchi, H. Suzuki, M. Sumida, and M. Tomoyeda, Agric. Biol. Chem. 40, 257 (1976).
[20] A. J. Skinner and R. A. Cooper, J. Bacteriol. 118, 1183 (1973).
[21] M. K. Essenberg and R. A. Cooper, Eur. J. Biochem. 55, 323 (1975).

temperature (van 't Hoff plots) show biphasic linear slopes.[2] Similar discontinuities appear in Arrhenius plots (log of activity versus reciprocal temperature). These data have been interpreted to indicate that the isomerases from these two sources may undergo temperature-dependent conformational changes, and that more than one substrate binding site is found on the enzyme. Both salts and nonpolar solvents significantly affect the K_m and V_{max} of each enzyme and variously affect their thermal stability.[2]

[98] Triosephosphate Isomerase from Chicken and Rabbit Muscle

By M. P. ESNOUF, R. P. HARRIS, and J. D. McVITTIE

D-Glyceraldehyde 3-phosphate $\rightleftharpoons$ dihydroxyacetone phosphate

Assay Method

Principle. Triosephosphate isomerase (EC 5.3.1.1) activity may be measured by linking the isomerization reaction to either of the NAD^+-$NADH_2$ dependent enzymic reactions:

D-Glyceraldehyde 3-phosphate + NAD^+ $\xrightarrow{\text{G-3PDH}}$ 3-phosphoglyceric acid + NADH + H^+

or

Dihydroxyacetone phosphate + NADH + H^+ $\xrightarrow[\text{dehydrogenase}]{\alpha\text{-glycerophosphate}}$ α-glycerophosphate + NAD^+

Both of these approaches have been available for a number of years[1,2] and the D-glyceraldehyde 3-phosphate/α-glycerophosphate dehydrogenase procedure has been described in previous volumes of this series.[3,4] Both assays have been studied in detail by Knowles and co-workers,[5,6] but a number of problems remain.

[1] O. Warburg and W. Christian, *Biochem. Z.* **314,** 159 (1942).
[2] E. Meyer-Arendt, G. Beisenherz, and T. Bücher, *Naturwissenschaften* **2,** 59 (1953).
[3] G. Beisenherz, this series, Vol. 1, p. 387.
[4] F. C. Hartman, this series, Vol. 25, p. 661.
[5] S. J. Putman, A. F. W. Coulson, I. R. T. Farley, B. Riddleston, and J. R. Knowles, *Biochem. J.* **129,** 301 (1972).
[6] B. Plaut and J. R. Knowles, *Biochem. J.* **129,** 311 (1972).

The D-glyceraldehyde 3-phosphate/α-glycerophosphate dehydrogenase procedure exhibits substrate inhibition at high concentrations of D-glyceraldehyde 3-phosphate, and furthermore both triosephosphate isomerase[7] and α-glycerophosphate dehydrogenase are inhibited by phosphate and sulfate ions. In the dihydroxyacetone phosphate/glyceraldehyde-3-phosphate dehydrogenase procedure the triosephosphate isomerase is inhibited by arsenate ions, an essential component of the secondary enzyme reaction system. Thus, in neither instance can the isomerase activity be measured easily under optimum conditions. The assay conditions outlined here are those that have been used in our preparative work to date, but no advantages over alternative procedures are claimed.

Reagents

Triethanolamine-HCl buffer, 25 mM, pH 7.9
D-Glyceraldehyde 3-phosphate, 10 mM
NADH, disodium salt, 7.0 mM
α-Glycerophosphate dehydrogenase, 10 mg/ml
Triosephosphate isomerase, ca. 250 mg/ml

All reagents were purchased from Sigma Chemical Co.

Procedure. To a 3-ml cuvette with a 1-cm light path, add and mix 2.8 ml of triethanolamine, HCl buffer, 0.1 ml of glyceraldehyde phosphate, 0.05 ml of NADH, 0.01 ml of α-glycerophosphate dehydrogenase, and 0.02–0.04 ml of suitably diluted triosephosphate isomerase.

The reaction is started by addition of triosephosphate isomerase. The decrease in absorbance of the solution at 340 nm is monitored at 25° for 3–5 min, and the specific activity of the enzyme is calculated in the usual manner.

Specific activity (units/mg) = $\Delta A/6.3 \times V \times D/C$, where ΔA is the change of absorbance per minute, V is the total reaction volume in the cuvette (2.98–3.00 ml), D is the factor by which the stock enzyme has been diluted, and C is the protein concentration in milligrams per milliliter.

Purification Procedure for Chicken Muscle

Commercial frozen chickens for domestic consumption may be used as the source of the enzyme. The initial extraction and ammonium sulfate fractionation should be carried out at temperatures below 10°, but chromatographic procedures may be carried out at room temperatures (20°). Sephadex materials are obtainable from Pharmacia Fine Chemicals.

---

[7] P. M. Burton and S. G. Waley, *Biochim. Biophys. Acta* **151**, 714 (1968).

Preparation of Initial Extract. Deep-frozen chicken breast muscle (500 g) is thawed, minced in a top-drive macerator, and stirred for 30 min with 500 ml of 1.5 mM ethylenediaminetetracetic acid (pH 7.0) containing 0.2% (v/v) 2-mercaptoethanol. The homogenate is centrifuged (1200 g for 45 min), and the supernatant is filtered through muslin to remove any particles of fat. The sediment is reextracted by stirring for 30 min with 250 ml of the same solvent and then centrifuged as before. This supernatant is also filtered through muslin and combined with the supernatant obtained previously.

Ammonium Sulfate Fractionation. Ammonium sulfate (43 g) is added to each 100 ml of the combined supernatants. The resulting precipitate is centrifuged (5000 g for 60 min) and discarded. A further 16.8 g of ammonium sulfate are then added to each 100 ml of this supernatant; after standing for 2 hr, the suspension is centrifuged (5500 g for 75 min). The sediment at this stage is dissolved in the minimum volume of 0.02 M Tris-HCl (pH 7.2).

Column Chromatography. Ammonium sulfate is removed from the redissolved pellet by gel filtration on a Sephadex G-25 (coarse grade) column (100 × 3.8 cm) equilibrated with 0.02 M Tris-HCl (pH 7.2). This column can be run at a downward flow rate of 25 ml/cm² per hour, and 25-ml fractions should be collected. The column eluate is monitored by measuring the absorbance at 280 nm, and the high molecular weight fractions are combined. These are applied to a QAE-Sephadex A-50 column (55 × 5.1 cm) equilibrated with the same solvent. Protein can be eluted from this column by the application of a 2 × 1200 ml 0.02 to 0.10 M Tris-HCl (pH 7.2) linear gradient run at a downward flow rate of 3–5 ml/cm² per hour and detected as before by monitoring the absorbance of the eluate at 280 nm; 50-ml fractions should be collected. Triosephosphate isomerase elutes as a single peak at about 0.065 M Tris-HCl and can be precipitated from the fractions containing maximum activity by the addition of ammonium sulfate. The enzyme may be stored as an ammonium sulfate suspension at 0°.

Since this method was first developed, changes in the ion-exchange binding properties of QAE-Sephadex have been observed. The most successful preparations have been obtained using QAE-Sephadex of capacity 3.0 ± 0.4 meq/g. It has been noticed with some batches of QAE-Sephadex that the triosephosphate isomerase is eluted earlier (at about 0.04 M Tris-HCl), and there has been slight contamination with another protein. In such cases, increasing the pH of the eluting buffer to 7.8 has been found to give good separation. Alternatively, the contaminating protein can be removed by rechromatographing the enzymically active fractions on CM-Sephadex (25 × 2.0 cm) equilibrated with 0.02 M Tris-HCl (pH 7.0) and eluted by a 2 × 1500 ml 0 to 0.2 M KCl linear gradient. The column should be run at a flow rate of 3–5 ml/cm² per hour, and triosephosphate isomerase elutes at about 0.15 M KCl.

During the last 10 years over 50 g of chicken muscle triosephosphate isomerase of very reproducible quality have been produced by the authors using these procedures. Analysis of the enzyme by polyacrylamide gel or SDS-polyacrylamide gel electrophoresis shows only one component even at high (200 μg) protein loading.

Purification Procedure for Rabbit Muscle

The procedure outlined below is based on a method designed to recover triosephosphate isomerase from the preparation of myokinase, phosphorylase b, and aldolase from rabbit muscle. The initial stages of the extraction are identical to those described above for chicken muscle except that 2-mercaptoethanol has been found to be unnecessary in the extraction solvent.

Column Chromatography. After preparation of an ammonium sulfate suspension of crude enzyme, the ammonium sulfate is removed by gel filtration on Sephadex G-25 (coarse grade) as described previously. The high molecular weight fractions are then submitted to ion exchange chromatography on CM-Sephadex (55 × 5.1 cm) equilibrated with 0.02 M Tris-HCl (pH 7.0) and eluted by a 2 × 1500 ml linear gradient of 0 to 0.4 M KCl run at a downward flow rate of 3–5 ml/cm² per hour. Triosephosphate isomerase elutes at about 0.3 M KCl. 50 ml fractions should be collected. The active fractions are concentrated by precipitation with ammonium sulfate (600 mg/ml) and desalted by further gel filtration on Sephadex G-25 (the same column as in stage 1 may be used). The eluate from this stage is then applied to a column of QAE-Sephadex (55 × 5.1 cm) equilibrated with 0.02 M Tris-HCl (pH 7.5) and eluted by a 2 × 1200 ml linear gradient of 0.02 to 0.20 M Tris-HCl (pH 7.5) run at a downward flow rate of 3–5 ml/cm³ per hour. Triosephosphate isomerase activity is detected in fractions (50 ml) eluting in the range 0.12–0.15 M buffer. These active fractions are again concentrated by precipitation with ammonium sulfate and finally are submitted to gel filtration on Sephadex G-100 (120 × 4.2 cm) equilibrated with 0.02 M Tris-HCl (pH 7.5) and pumped at 15 ml/cm² per hour; 30-ml fractions are collected.

This procedure, summarized in the table, is admittedly more complex than that for the preparation of the chicken muscle enzyme but has proved to be quite reproducible over the years. In 5 years some 10 g of rabbit muscle triosephosphate isomerase have been prepared in this way. Electrophoresis of the product indicates one main component and a further one (or sometimes two) minor component(s). All components appear to exhibit triosephosphate isomerase activity.

PURIFICATION OF TRIOSEPHOSPHATE ISOMERASE

	Chicken muscle			Rabbit muscle		
		Activity			Activity	
Fraction	Protein (mg/500 g muscle)	Units/500 g of muscle ($\times 10^6$)	Units/mg protein	Protein (mg/500 g muscle)	Units/500 g of muscle ($\times 10^6$)	Units/mg protein
rude homogenate	20,000	4.0	200	25000	3.3	130
t $(NH_4)_2SO_4$ supernatant	7,900	3.1	390	8300	2.0	240
d $(NH_4)_2SO_4$ precipitate	5,300	2.6	490	3000	1.6	530
f G-25 column	5,100	2.45	480	—	—	—
f CM-Sephadex	NA[a]	NA	NA	1000	1.4	1400
f QAE-Sephadex	340	2.3	6700	200	1.1	5500
f G-100 column	NA	NA	NA	160	1.1	7000

[a] NA, not applicable.

Properties

In a survey of enzyme distribution in the muscle tissues of 13 widely different species, Scopes[8] showed that although most species exhibited multiple forms of triosephosphate isomerase, chicken muscle was almost unique in possessing only two isozymes. The second isozyme was present in very low amounts. Up to eight isozymes of rabbit muscle triosephosphate isomerase have been reported.[9]

The structure–function relationships of triosephosphate isomerase from both chicken and rabbit muscle have been widely investigated. The molecule is dimeric, each subunit having a molecular weight of approximately 25,000. Only the dimeric form is enzymically active. The amino acid sequence is known,[10,11] as is the crystallographic structure at 2.5 Å resolution.[12] The enzyme's extreme efficiency as a catalyst has been explained,[13] and there have been a number of studies of the structure of enzyme–substrate complexes.[14]

[8] R. K. Scopes, *Biochem. J.* **107**, 139 (1968).

[9] I. L. Norton, P. Pfuderer, C. D. Stringer, and F. C. Hartman, *Biochemistry* **9**, 4952 (1970).

[10] P. H. Corran and S. G. Waley, *FEBS Lett.* **30**, 97 (1973).

[11] A. J. Furth, J. D. Milman, J. D. Priddle, and R. E. Offord, *Biochem. J.* **139**, 11 (1974).

[12] D. W. Bonner, A. C. Bloomer, G. A. Petsko, D. C. Phillips, C. I. Pogson, I. A. Wilson, P. H. Corran, A. J. Furth, J. D. Milman, R. E. Offord, and S. G. Waley, *Nature (London)* **255**, 609 (1975).

[13] W. J. Albery and J. R. Knowles, *Biochemistry* **15**, 5631 (1976).

[14] I. D. Campbell, R. B. Jones, P. A. Kiener, and S. G. Waley, *Biochem. J.* **179**, 607 (1979).

[99] Uridine Diphosphate Glucose-4-epimerase and Galactose-1-phosphate Uridylyltransferase from *Saccharomyces cerevisiae*

By T. Fukasawa, T. Segawa, and Y. Nogi

$$\text{UDPG} + \text{Gal-1-P} \underset{}{\overset{\text{transferase}}{\rightleftharpoons}} \text{UDPGal} + \text{G-1-P}$$
$$\text{UDPGal} \underset{\text{4-epimerase}}{\rightleftharpoons} \text{UDPG}$$

$$\text{Gal-1-P} \rightleftharpoons \text{G-1-P}$$

Galactose-1-phosphate uridylyltransferase (EC 2.7.7.10) and UDPG-4-epimerase (EC 5.1.3.2) together with galactokinase and phosphoglucomutase comprise the group of enzymes that promote the catabolism of galactose by the common pathway for glucose.

Galactose-1-phosphate Uridylyltransferase

This enzyme has been purified from *Escherichia coli*[1] and from human tissue[2,3] as well as from *Saccharomyces cerevisiae*.[4]

Assay Method

Principle. The enzyme activity is measured with a spectrophotometer from NADPH formation in a coupled reaction of phosphoglucomutase and G-6-P dehydrogenase. The following is a slight modification of the method of Maxwell *et al.*,[5] which is applicable to crude extracts as well as to fractionated preparations.

Reagents and Enzymes

Solution A: 3.0 ml of 1 M glycine–NaOH buffer (pH 8.7), 1.5 ml of 0.1 M MgCl$_2$, 0.5 ml of 0.1 M dithiothreitol
Solution B: 0.5 ml of 10 mM UDPG, 0.5 ml of 0.2 mM G-1,6-P, 1.0 ml of 50 mM NADP

[1] S. Saito, M. Ozutsumi, and K. Kurahashi, *J. Biol. Chem.* **252**, 1162 (1967).
[2] T. A. Tedesco, *J. Biol. Chem.* **247**, 6631 (1972).
[3] G. L. Dale and G. Popják, *J. Biol. Chem.* **251**, 1057 (1976).
[4] T. Segawa and T. Fukasawa, *J. Biol. Chem.* **254**, 10707 (1979).
[5] E. S. Maxwell, K. Kurahashi, and H. M. Kalckar, this series, Vol. 5 [20].

Solution C: 0.5 ml of yeast G-6-P dehydrogenase (10 units/ml), 0.5 ml of rabbit phosphoglucomutase. Both enzymes as obtained from Boehringer Mannheim are diluted with dilution buffer (see below) to give the indicated concentrations. One unit of each coupling enzyme is defined as that amount of enzyme that catalyzes the reduction of 1 μmol of NADP per minute under the assay conditions for Gal-1-P uridylyltransferase except that G-6-P (for dehydrogenase) or G-1-P (for phosphoglucomutase) are used as substrates. For the calculation of the enzyme unit, see below.

Solution D: 1.0 ml of 20 mM Gal-1-P.[6]

Dilution buffer: 10 mM Tris-HCl (pH 7.5), 0.1% bovine serum albumin (fraction V of Sigma), 1 mM dithiothreitol, 1 mM EDTA-3 Na.

The amounts of solutions A–D given above suffice for 50 assays or one purification run. The solutions are stored at $-20°$.

Procedure. Into a quartz cuvette having a light path of 1 cm and a volume of 1 ml or less, are added 430 μl of water, 100 μl of solution A, 40 μl of solution B, 20 μl of solution C, and 10 μl of Gal-1-P uridylyltransferase to be assayed. The absorbance change is determined at 340 nm in a spectrophotometer at 1-min intervals after addition of 20 μl of solution D per cuvette. The total volume becomes 620 μl. At the beginning of the reaction, a short lag period is normally observed, and then the absorbance increases linearly with time. The linear part of the curve is taken for the calculation of the activity. In order to get maximum activity, the enzyme sample should be diluted with dilution buffer to give an absorbance change at 340 nm between 0.01 to 0.04 per minute. The spectrophotometer should preferably be equipped with a thermospacer to keep the cuvettes at a constant temperature, and an automatic recorder.

Definition and Calculation of the Enzyme Unit. One unit of Gal-1-P uridylyltransferase catalyzes the incorporation of 1 μmol of Gal-1-P into uridine nucleotide per minute and thus causes the formation of 1 μmol of NADPH per minute. Each nanomole of NADP converted to NADPH corresponds to an absorbance change of 0.010 at 340 nm.

Other Assay Methods. The two-step methods are sometimes suitable, especially for the determination of low uridylyltransferase activity in crude extracts or for the study of the catalytic properties of the enzyme. These include the enzymic method, described previously[4] and the radioactive method,[7] which is a modification from Buttin.[8]

---

[6] The final concentration of Gal-1-P thus becomes 0.32 mM, which is rather low considering the K_m value of the purified enzyme (see the text), and therefore should be increased to obtain a maximum activity of Gal-1-P uridylyltransferase in a given sample. However, for the sake of economy, we have used Gal-1-P at this concentration.

[7] Y. Nogi and T. Fukasawa, *Curr. Genet.* **2**, 115 (1980).

[8] G. Buttin, *J. Mol. Biol.* **7**, 164 (1963).

Purification Procedure

Growth of Yeast. A strain of *Saccharomyces cerevisiae,* 106-3D (ATCC44216), bears mutations of *gal 80, his 1,* and *ura 1.* The syntheses of Leloir pathway enzymes (galactokinase, Gal-1-P uridylyltransferase, UDPG-4-epimerase) are partially constitutive due to the mutation of *gal 80* and are fully induced during growth in the presence of galactose. One liter of growth medium contains 20 g of Polypeptone (Daigo Eiyo; this can be replaced with Bacto Peptone from Difco Co. Detroit, Michigan), 10 g of Bacto Yeast Extract (Difco), and 20 g of D-galactose (Sigma) (YPGal) or 3% v/v glycerol (YPGly). A fully grown preculture in YPGly (200 ml) is inoculated into 20 liters of YPGal. Each liter of culture is contained in a 3-liter Erlenmeyer flask. Cells are grown at 30° for 20 hr with vigorous shaking (logarithmic phase) and harvested by continuous-flow centrifugation at 17,000 g. Each liter of culture normally yields about 12 g of cells, wet weight. The cells are divided into 50-g batches and stored at $-80°$ until use. All operations described below, unless otherwise indicated, are carried out in the cold (0–5°). The following procedures up to the DEAE-cellulose column chromatography step are essentially the same as those described by Schell and Wilson for purification of galactokinase from *Saccharomyces cerevisiae.*[9]

Disruption of Cells. Fifty grams of frozen cells are thawed and resuspended in 35 ml of buffer K (20 mM triethanolamine acetate, pH 8.0, 1 mM EDTA, 1 mM dithiothreitol) to give a final volume of 80 ml, and disrupted in a Vibrogen Cell Mill (Edmond Bühler Tübingen, F.R.G.) by vibration for 3 min with 280 g of acid-washed glass beads (0.5 mm in diameter). The supernatant solution and a washing of the glass beads with about 150 ml of buffer K are combined and subjected to centrifugation at 18,000 g for 15 min. The supernatant fraction is decanted and saved. The pellet is resuspended in 10 ml of fresh buffer K and treated as above in the Vibrogen Cell Mill. The supernatant fluid combined with a washing of the glass beads is centrifuged as above. The supernatant fluids from two rounds of breakage are combined and diluted to 500 ml with buffer K (fraction 1).

Ammonium Sulfate Fractionation. To fraction 1 are added 115 g of ammonium sulfate with stirring over a period of 7 min; the stirring is continued for an additional 15 min, and the solution is allowed to sit for 20 min, without stirring; the solution is centrifuged at 10,000 g for 20 min. An additional 100 g of ammonium sulfate are added to the supernatant fraction as before, and then the pellet is dissolved in buffer K to give a final volume of 100 ml (fraction 2).

[9] M. A. Schell and D. B. Wilson, *J. Biol. Chem.* **252,** 1162 (1977).

Streptomycin Precipitation. Streptomycin sulfate (25 ml; 10% w/v) is added dropwise to fraction 2, and the solution is dialyzed against 2 liters of buffer A (10 mM Tris-HCl, pH 7.4, 1 mM EDTA, 1 mM dithiothreitol, 10% (v/v) glycerol) for 6 hr. The dialysis buffer is changed twice with an interval of 6 hr. The precipitated material is removed by centrifugation at 8000 g for 30 min, then the supernatant fraction is decanted and saved. Ammonium sulfate solution (1 M) is added to give a final concentration of 7 mM (fraction 3).

DEAE-Cellulose Column Chromatography. Fraction 3 is loaded onto a DEAE-cellulose (DE-52 Whatman Co., Clifton, New Jersey) column (5 × 25 cm) equilibrated with buffer A containing 7 mM ammonium sulfate at a flow rate of 120 ml/hr, and the column is washed with the same buffer at 88 ml/hr. Twenty-milliliter fractions are collected; fractions with activity greater than 3 units/ml are pooled (ca. 120 ml) and concentrated to 30 ml in an Amicon ultrafiltration apparatus with a PM-10 membrane. The concentrated fraction is dialyzed against 2 liters of buffer T (50 mM sodium phosphate, pH 6.9, 1 mM EDTA, 1 mM dithiothreitol, and 10% (v/v) glycerol) for at least 6 hr. The dialysis buffer is changed, and dialysis is continued for at least another 6 hr (fraction 4).

Hydroxyapatite Column Chromatography. Fraction 4 is loaded onto a column (2.6 × 10 cm) of hydroxyapatite (Hypatite C, Clarkson Chemical Co., Williamsport, Pennsylvania) equilibrated with buffer T at a flow rate of 45 ml/hr. The column is washed with about 500 ml of buffer T at a flow rate of 88 ml/hr until the $A_{280\,nm}$ of the effluent remains below 0.1. The bound transferase is eluted with 700 ml of a linear gradient of 50 to 200 mM sodium phosphate (pH 6.9) containing 1 mM EDTA, 1 mM dithiothreitol, and 10% (v/v) glycerol at a flow rate of 39 ml/hr. The 20-ml fractions of the highest specific activity eluting around 120 mM sodium phosphate are pooled (ca 300 ml) and concentrated to 30 ml in an Amicon ultrafiltration apparatus with PM-10 membrane. The concentrated sample is dialyzed once against 2 liters of buffer T for at least 6 hr and the insoluble material is removed by centrifugation at 8000 g for 15 min (fraction 5).

Phosphocellulose Column Chromatography. Fraction 5 is loaded onto a column (2.6 × 10 cm) of phosphocellulose (P-11, Whatman) equilibrated with buffer T at a flow rate of 38 ml/hr. The column is washed with 200 ml of buffer T at a flow rate of 43 ml/hr. The transferase adsorbed to the exchanges is eluted with 200 ml of a linear gradient of 50 to 200 mM sodium phosphate (pH 6.9) containing 1 mM EDTA, 1 mM dithiothreitol, and 10% (v/v) glycerol at a flow rate of 43 ml/hr and 4-ml fractions are collected. Transferase is eluted in the first $A_{280\,nm}$ peak, which emerges

TABLE I
PURIFICATION OF GALACTOSE-1-PHOSPHATE URIDYLYLTRANSFERASE

Fraction No.	Total volume (ml)	Total units[a]	Recovery of units (%)	Total protein (mg)	Specific activity (units/mg protein)
1	500	620	74.5	4900	0.1
2	100	800	96.2	2750	0.3
3	126	832	(100)	1714	0.5
4	29.5	694	83.4	340	2.0
5	28.5	459	55.2	94	4.9
6	6.2	192	23.0	5.5	34.9
7	9.8	157	18.9	2.9	54.1

[a] Micromoles of G-1-P formed per minute at 27°.

from the column at about 120 mM sodium phosphate. The pooled fraction is concentrated to 6 ml by PM-10 Amicon membrane as above (fraction 6).

BioGel A-0.5m Gel Filtration. Fraction 6 is applied to a column of BioGel A-0.5m (200–400 mesh, Bio-Rad, Richmond, California) (1.6 × 90 cm); the column is washed with buffer T at a pressure of about 90 cm$_{H_2O}$, and 2.5-ml fractions are collected. The fractions of the highest specific activity are pooled and stored at −20° (fraction 7).

A summary of the purification is presented in Table I.

Properties of the Purified Enzyme[4]

Molecular Weight. The enzyme in the native state has a molecular weight of 86,100 ± 900, as estimated by sedimentation equilibrium centrifugation. In SDS–polyacrylamide gel electrophoresis, the enzyme preparation exhibits a single band at a position of M_r 38,000. Threonine is the only terminal residue. Thus the native enzyme consists of two identical subunits.

Purity. The enzyme in fraction 7 of typical preparations should be more than 95% pure as judged by SDS–polyacrylamide (8%) gel electrophoresis and staining with Coomassie Brilliant Blue.

Kinetic Constants. The K_m values for UDPG and Gal-1-P are 0.26 mM and 4.0 mM, respectively. The V_{max} value is 0.688 μmol/min for 1 μg of the enzyme, and the calculated turnover number is 59,200 molecules of phosphate group transferred to glucose per minute per enzyme molecule.

Optimum pH. Maximum activity is observed between pH 8.4 and 8.8. At pH 7.1 or 9.4, the enzyme exhibits 56% and 83% of the maximum activity, respectively.

Activation Energy. Studies of the effect of temperature on the rate of Gal-1-P uridylyltransferase over the range of 18° to 39° indicate an activation energy to be 8.7 kcal/mol.

UDPG-4-Epimerase

This enzyme has been extensively purified from calf liver,[10] *Escherichia coli,*[11] and *Saccharomyces fragilis,*[12] in addition to *Saccharomyces cerevisiae.*[13] The *S. fragilis* is now classified in the genus *Kluyveromyces* and differs from a true *Saccharomyces* both physiologically and morphologically.

Assay Method

Principle. The enzyme activity is measured with a spectrophotometer from NADH formation in a coupled reaction with UDPG dehydrogenase. This method, essentially according to Maxwell *et al.,*[5] can be applied to crude extracts as well as to fractionated preparations.

Reagents and Enzymes

UDPGal, 4.5 mM
NAD, 25 mM
Glycine-NaOH, 1.0 M pH 8.7
UDPG dehydrogenase (10 units/ml). Enzyme preparations, either purified from calf liver up to fraction V according to Wilson[14] or purchased from Boehringer Mannheim, are equally usable. One unit of dehydrogenase activity catalyzes the oxidation of 1 μmol of UDPG per minute, and therefore the reduction of 2 μmol of NAD per minute under the assay conditions for UDPG-4-epimerase, except that UDPG is used in place of UDPGal. Each nanomole of UDPG oxidized corresponds to an absorbance change of 0.020 at 340 nm.
Dilution buffer: 0.01 M Tris-HCl (pH 7.4), 1% bovine serum albumin, 1 mM dithiothreitol, 1 mM EDTA
Procedure. Into a quartz cuvette of 1-cm light path and a volume of 1 ml or less, are added 480 μl of water, 60 μl of glycine–NaOH (pH 8.7), 20 μl of NAD, 40 μl of UDPGal, and 10 μl of the UDPG-4-epimerase sample

[10] R. Langer and L. Gealer, *J. Biol. Chem.* **249**, 1126 (1974).
[11] D. B. Wilson and D. S. Hogness, *J. Biol. Chem.* **239**, 2469 (1964).
[12] R. A. Darrow and R. Rodstrom, *Biochemistry* **7**, 1645 (1968).
[13] T. Fukasawa, K. Obonai, T. Segawa, and Y. Nogi, *J. Biol. Chem.* **255**, 2705 (1980).
[14] D. B. Wilson, *Anal. Biochem.* **10**, 472 (1965).

to be assayed. The reaction is started by the addition of 10 μl of UDPG dehydrogenase. The total volume of the reaction mixture becomes 620 μl. The absorbance change is determined at 340 nm at 1-min intervals. The sample of UDPG-4-epimerase should be diluted with dilution buffer to give an absorbance change between 0.010 and 0.03.

Definition and Calculation of the Enzyme Unit. One unit of UDPG-4-epimerase causes the conversion of 1 μmol of UDPGal to UDPG and thus the formation of 2 μmol of NDPH per minute. Each nanomole of UDPGal converted to UDPG corresponds to an absorbance change of 0.020 at 340 nm.

Two-Step Assay. The two-step assay[13] is highly recommended for determination of the low activity found in crude extracts. The incubation period or the amount of the enzyme to be added can be increased so that the sensitivity of the assay is increased.

Purification Procedure

The procedure for cell growth, preparation of crude extract (fraction 1), ammonium sulfate fractionation (fraction 2), and streptomycin treatment (fraction 3) are exactly like those described above for the purification of Gal-1-P uridylyltransferase. Thus starting from the same batch of cells (50 g wet weight), transferase and epimerase are purified through the common procedure for the first 3 steps without loss of either enzyme.

First DEAE-Cellulose Column Chromatography. A DEAE-cellulose (DE-52, Whatman) column (5 × 25 cm) was equilibrated with buffer A (10 mM Tris-HCl, pH 7.4, 1 mM EDTA, 1 mM DTT, 10% (v/v) glycerol) containing 7 mM ammonium sulfate and is loaded with fraction 3 at a flow rate of about 120 ml/hr. The column is then washed with the same buffer at a flow rate of 88 ml/hr until $A_{280\,nm}$ of the effluent becomes constant, which requires about 500 ml of the buffer. At this step, transferase and galactokinase are completely washed out from the exchanger. The column is then eluted by a linear gradient consisting of 1 liter of buffer A containing 7 mM ammonium sulfate and 1 liter of buffer A containing 0.4 M NaCl, collecting 20 ml fractions. The peak of epimerase activity appears at about 0.2 M NaCl. Fractions of the highest activity are pooled and concentrated to 50 ml in an Amicon ultrafiltration apparatus with a PM-10 membrane. The concentrated sample is dialyzed against 1 liter of buffer B (5 mM sodium phosphate (pH 6.9), 1 mM EDTA, 1 mM DTT, and 10% (v/v) glycerol). The dialysis buffer is changed, and dialysis is continued for an additional 6 hr (fraction 4).

Hydroxyapatite Chromatography. A column (2.6 × 15 cm) of hydroxyapatite (Hypatite C, Clarkson) is prepared and washed with 200 ml of buffer B. Fraction 4 is applied to the column at a flow rate of 120 ml/hr.

The column is washed with 200 ml of buffer B at a flow rate of 88 ml/hr, and the bound epimerase is eluted with a linear gradient consisting of 700 ml of buffer B and 700 ml of the same buffer with 250 mM sodium phosphate. Twenty-milliliter fractions are collected. Fractions with specific activities higher than 5.0 units per milligram of protein are pooled and concentrated to a final volume of 60 ml. (Further concentration sometimes give rise to heavy precipitation accompanied by a considerable loss of epimerase activity.) The sample is dialyzed against 2 liters of buffer A for more than 6 hr. The buffer is changed, and dialysis is continued for an additional 6 hr (fraction 5).

Second DEAE-Cellulose Column Chromatography. A column of DEAE-cellulose (2.6 × 15 cm) is prepared and equilibrated with buffer A. Fraction 5 is loaded onto the column at a flow rate of 60 ml/hr. The column is then washed successively with 200 ml of buffer A and 200 ml of buffer A containing 0.1 M NaCl at a flow rate of 88 ml/hr. The bound epimerase is eluted with a linear gradient consisting of 400 ml of buffer A with 0.1 M NaCl and 400 ml of that buffer containing 0.2 M NaCl. Fractions with specific activities of more than 15 units of protein per milligram are pooled and concentrated to 5 ml by an Amicon ultrafiltration apparatus with PM-10 membrane (fraction 6).

Gel Filtration. Fraction 6 is applied to a column (1.6 × 90 cm) of BioGel A-0.5 m (200–400 mesh, Bio-Rad, Richmond, California), and the column is eluted with buffer E (5 mM potassium phosphate buffer, pH 6.9, containing 1 mM EDTA, 1 mM DTT, and 10% (v/v) glycerol) at a pressure of about 90 cm$_{H_2O}$, collecting 2.5-ml fractions. The fractions of the highest specific activity are pooled (fraction 7) and stored at −80° without detectable loss of activity for at least 1 month.

A typical purification is summarized in Table II.

TABLE II

PURIFICATION OF URIDINE DIPHOSPHATE GLUCOSE (UDPG)-4-EPIMERASE

Fraction and step	Total volume (ml)	Total units[a]	Recovery of units (%)	Total protein (mg)	Specific activity (units/mg protein)
Crude extract	500	700	55.5	4800	0.2
Ammonium sulfate fractionation	100	1160	91.8	3320	0.3
Streptomycin supernatant	126	1140	89.7	1640	0.7
1st DEAE-cellulose chromatography	54	1260	(100)	260	4.9
Hydroxyapatite chromatography	65	660	52.2	80	8.3
2nd DEAE-cellulose chromatography	5	430	33.9	19.4	22.1
Gel filtration	12.5	420	33.0	13.2	31.8

[a] Micromoles of UDPG formed per minute at 25°.

Properties of the Purified Enzyme[13]

Purity. In typical preparations of fraction 7, the enzyme is more than 95% pure judged by SDS-gradient (6 to 16%) polyacrylamide gel electrophoresis and Coomassie Brilliant Blue staining.

Molecular Weight. The molecular weight of the purified native enzyme averages 183,000 or 187,000 as judged by sucrose gradient centrifugation or gel-filtration analyses, respectively. In the SDS–gradient polyacrylamide gel electrophoresis, the molecular weight of the enzyme under dissociating condition is estimated to be 78,000. Analyses of dansyl- or phenylthiohydantoin derivatives of NH_2-terminal residues indicate that threonine is the only NH_2-terminal residue. Thus the native UDPG-4-epimerase may consist of two identical subunits.

Kinetic Constant. The K_m value for UDPGal is 0.22 mM; V_{max} calculated from these data is 21 μmol/min per milligram of protein, which corresponds to a turnover number of 3890 molecules of UDPGal converted to UDPG per minute per enzyme molecule, assuming the molecular weight of epimerase to be 183,000. The equilibrium constant (K = UDPG/UDPGal) is estimated to be 3.5 ± 0.1 whether the starting substrate is UDPGal or UDPG.

pH Optimum. The effective pH range is rather broad; at final pH values of 6.65, 7.0, 7.3, and 8.05, the relative activities at 25° are 0.91, 0.97, 1.00, and 0.88, respectively.

Activators and Inhibitors. The purified epimerase from *S. cerevisiae* as well as other microbial epimerases requires neither NAD^+ nor metal ions like MnCl or $MgCl_2$. Unlike *S. fragilis* enzyme,[12] no stimulation of the enzyme activity is observed by addition of NaCl, $MgCl_2$, or spermidine-HCl to the reaction mixture. p-Chloromercuribenzoate (0.5 mM) completely inhibits *S. cerevisiae* enzyme, as it does the *S. fragilis* enzyme.

Thermostability. UDPG-4-epimerase from *S. cerevisiae* is highly thermolabile: The activity in fraction 7 decreases to 47% or 15% within 20 min of incubation at 30° or 37°, respectively. The inactivation of the enzyme is partially protected by addition of NAD^+ to a concentration of 7 mM.

Enzyme-Bound NAD. Each dimeric molecule of UDPG-4-epimerase from *S. cerevisiae* contains one molecule of NAD^+ tightly bound to the enzyme protein, as in case of the other microbial epimerases.[11,12]

In a remarkable contrast to *S. fragilis* epimerase, however, UDPG-4-epimerase from *S. cerevisiae* does not exhibit any detectable fluorescence at any step in the purification: The former enzyme has a unique fluorescence with an excitation maximum at 360 nm and emission at 433 nm, which closely resembles the fluorescence of free NADH.[15]

[15] E. S. Maxwell and H. de Robichon-Szulmajster, *J. Biol. Chem.* **235**, 308 (1960).

[100] Inositol Epimerase– Inosose Reductase from Bovine Brain[1]

By PAUL P. HIPPS, KAREN E. ACKERMANN, and WILLIAM R. SHERMAN

Inositol epimerase from bovine brain catalyzes the $NADP^+$-dependent epimerization of *myo*-inositol at the C-2 or C-5 position.[2] It is thought that this reaction occurs by oxidation of the inositol to an inosose followed by reduction to give an epimeric inositol. The inositol epimerase is accompanied by inosose reductase activity, and these activities copurify through ammonium sulfate fractionation, DEAE-cellulose chromatography and G-100 gel filtration.[3] It is, therefore, possible that a single enzyme is responsible for both processes. The reductase uses *myo*-inosose-2 as a substrate to produce *myo*- and *scyllo*-inositol with NADPH as cofactor;[4] i.e., it appears to be the half-reaction of the epimerase. A similar enzyme reaction has been observed, in cockroach fat-body extracts, that uses NAD^+ exclusively for the epimerization of inositols and uses either NADH or NADPH for the inosose reductase activity.[5] The cockroach activity has a lower substrate specificity than the bovine enzyme preparation.

[1] Supported by NIH Grants NS-05159, RR-00954, AM-20579.

[2] P. P. Hipps, W. H. Holland, and W. R. Sherman, *Biochem. Biophys. Res. Commun.* **77,** 340 (1977).

[3] W. R. Sherman, P. P. Hipps, L. A. Mauck, and A. Rasheed, *in* "Cyclitols and Phosphoinositides" (W. W. Wells and F. Eisenberg, Jr., eds.), p. 270. Academic Press, New York, 1978.

[4] P. P. Hipps, M. R. Eveland, M. H. Laird, and W. R. Sherman, *Biochem. Biophys. Res. Commun.* **68,** 1133 (1976).

[5] P. P. Hipps, R. K. Sehgal, W. H. Holland, and W. R. Sherman, *Biochemistry* **12,** 4705 (1973).

Assay Methods

Principle. In the bovine epimerase assay, an inositol substrate reacts with enzyme and NADP$^+$ to produce an epimeric inositol. In the bovine reductase assay, *myo*-inosose-2 reacts with the enzyme and NADPH to form *myo*- and *scyllo*-inositol. Analysis of the inositol products is performed by gas chromatography (GC) or by gas chromatography–mass spectrometry (GC-MS) using selected ion monitoring (SIM).

To facilitate analysis of the epimerase-reductase activity in column eluates, a fluorometric assay of the reductase is used. In this procedure *myo*-inosose-2, reduced cofactor, buffer, and enzyme are incubated together. After incubation, HCl is added to stop the reaction with destruction of the reduced cofactor. Alkaline hydrogen peroxide is then added to convert the oxidized cofactor into a fluorescent compound and to destroy unreacted inosose, which quenches the fluorescence.

Epimerase Activity

Reagents

myo-Inositol, 40 mM, in buffer; *scyllo*-inositol (Calbiochem, La Jolla, California) or *neo*-inositol[6] may also be used.

NADP$^+$, 1 mM in water (alkali labile)

Tris-glycine, 50 mM, pH 9.5 containing 1.0 mM dithiothreitol (DTT) and 3 mM (0.02%) sodium azide

Zinc sulfate, 30 mM with 3 mM (0.02%) sodium azide

A mixture of 2-[^{18}O]*myo*- and [^{18}O]*scyllo*-inositol, 0.57 mM (*myo* : *scyllo*-inositol ratio, 10 : 1)[7]

Procedure. Epimerase activity is measured by the incubation of 50 μl of the inositol substrate (10 mM final concentration), 50 μl of NADP$^+$ (0.25 mM final), and enzyme in pH 9.5 Tris-glycine buffer to give a total volume of 200 μl.

After incubation at 37° for 30 min, the reaction is stopped by heating for 10 min in a boiling water bath (or 100° dry bath). If GC–MS analysis is to be performed, 100 μl of [^{18}O]*myo*-/*scyllo*-inositol then is added as an internal standard. Protein is precipitated by adding 500 μl of 30 mM ZnSO$_4$. After centrifugation a 600-μl aliquot is lyophilized, and the trimethylsilyl (TMS) ethers of the products are prepared by reaction for

---

[6] S. J. Angyal and N. K. Matheson, *J. Am. Chem. Soc.* **77**, 4343 (1955).

[7] Prepared by the overnight room-temperature exchange of 30 mg of *myo*-inosose-2 in 1 ml of >96% enriched H$_2$^{18}O (requires warming to 80° to effect solution) followed by reduction with hydrogen gas (1 atmosphere) using 10 mg of 5% Ru/C catalyst (Aldrich, Milwaukee, Wisconsin). Filtration to remove the catalyst gives a solution of *myo*- and *scyllo*-inositols in about a 10 : 1 ratio. The H$_2$^{18}O may be recovered by lyophilization and trapping.

24 hr at room temperature with 100 μl of pyridine–N,O-bis(trimethyl-silyl)trifluoroacetamide–trimethylchlorosilane (1 : 1 : 0.2, v/v/v).

GC Analysis of Epimerase. A dilution curve of TMS *myo-, scyllo-,* and *neo*-inositols, as required, is prepared. A 1.8 meter × 0.6 cm o.d. silanized glass column packed with 3% SE-30 on 80/100 mesh Gas Chrom Q is used (Applied Science Laboratories, State College, Pennsylvania). The column is operated with He carrier, flow 30 ml/min, at a temperature (about 170°) such that TMS *myo*-inositol elutes at about 15 min. Retention times, relative to TMS *myo*-inositol, are TMS *neo-*, 0.44; TMS *chiro-*, 0.58; TMS *scyllo-*, 0.80. The GC analysis is limited in sensitivity when compared with the GC–MS assay, largely because of a large solvent front that makes *neo*-inositol analysis difficult. Nevertheless, the GC method is satisfactory for many experiments. For example, a 2-μl injection into the GC from an enzyme reaction using a 100-μl aliquot from a 10-ml G-100 column fraction gave peak heights for TMS *neo*-inositol of 4 cm, TMS *scyllo*-inositol of 2.2 cm, and TMS *myo*-inositol (substrate) of 300 cm, calculating the peaks at attenuation 4 range 10^{-11} A/mV, which is a 40-fold attenuation of the full electrometer gain on the gas chromatograph used.

GC–MS Analysis of Epimerase Products. This analysis is more sensitive and subject to less interference from contaminating substances than is GC alone. In particular, the solvent front is not a problem. The principle of the GC–MS analysis (SIM) is to assay the inositol by measuring the intensity of an ion from the spectrum of ions from its TMS derivative as it elutes from the GC column. This method uses m/z 432 and 507 for the inositols and m/z 434 and 509 for the [^{18}O]inositol internal standard. These ions are specific for inositols under the described conditions. The specificity can be tested in any assay by examining the ratios of m/z 432 : 507 in sample and standard: they should be the same, as should the retention times. The standard curve is constructed to contain the same amount of internal standard as the samples, and a plot of the ratio m/z 432 (507) to 434 (509) against the amount of the inositol is used for quantitation. Because the substrate, in each case, is present in much larger concentration than the product(s), the instrument must be protected against overload of the multiplier by either diverting the effluent or reducing the multiplier voltage until the substrate has eluted.

The mass spectrometer (Finnigan 3200) is operated at 70 eV ionization potential with the filament under emission control.

Reductase Activity

Reagents for Fluorometric Assay of Reductase

myo-Inosose-2 (Calbiochem, La Jolla, California) 4.2 mM in H_2O
NADPH, 50 mM in 50 mM sodium carbonate buffer, pH 10.5

Buffer: 100 mM Tris-HCl, pH 7.2, containing 3 mM (0.02%) sodium azide

Hydrogen peroxide, 30%

Sodium hydroxide, 8 M

Hydrochloric acid, 2 N

Fluorescence developing solution: Prepare fresh daily; dilute 30% hydrogen peroxide to 3% with water, then add to 8 M NaOH to give a final concentration of 0.06% H_2O_2 (adding 30% H_2O_2 directly to the sodium hydroxide will result in a precipitate).

Substrate solution: Heat NADPH at 60° for 10 min to destroy any NADP$^+$ present. Mix 0.5 ml of NADPH, 5 ml of inosose, and 5 ml of Tris buffer together immediately before assay.

Procedure. In the fluorometric procedure, 100 μl of substrate are incubated with 25 μl of enzyme preparation for 30 min at 37°. The reaction is stopped with the addition of 100 μl of HCl. After standing for 10 min the mixture is centrifuged to remove precipitated protein, and 50 μl of incubation mixture supernatant is added to 1 ml of fluorescence-developing solution with immediate and thorough mixing. Fluorescence is developed by heating at 60° for 10 min. Tubes are cooled to room temperature in the dark for about 1 hr. Fluorescence is measured on a filter fluorometer using an excitation wavelength of 360 nm and an emission wavelength of 460 nm.[8]

Reagents for GC and GC/MS Assay of Reductase

myo-Inosose-2, 6 mM in water (alkali labile)

NADPH, 8 mM in pH 7.2 Tris-glycine buffer (acid labile)

Buffer: 50 mM Tris-glycine containing 1.0 mM dithiothreitol (DTT) and 3 mM (0.02%) sodium azide, pH 7.2

Zinc sulfate, 0.03 M, with 3 mM (0.02%) sodium azide

[^{18}O]Inositol mixture, 0.14 mM (*myo*:*scyllo*-inositol ratio 10:1) in water with 3 mM (0.02%) sodium azide

Procedure. Reductase activity is measured as described for the epimerase except that 1.5 mM *myo*-inosose-2 is incubated with 2.0 mM NADPH in standard Tris-glycine buffer at pH 7.2.

GC and GC–MS Analysis of Reductase Products. The GC analysis of the reductase is similar to that of the epimerase: TMS *myo*-inosose-2 has a retention time on SE-30 of 0.44 relative to TMS *myo*-inositol and the product inositols are measured at their respective retention times as TMS derivatives. In the GC–MS assay a dilution of the same internal standard and the same mass values are used. Substrate is again diverted from the instrument or the multiplier voltage reduced.

[8] O. H. Lowry and J. V. Passonneau, "A Flexible System of Enzymatic Analysis," p. 8. Academic Press, New York, 1972.

TYPICAL PURIFICATION OF INOSITOL EPIMERASE–INOSOSE REDUCTASE

Step	Epimerase activity[a] (nmol/min mg^{-1} protein)		Reductase activity[b] (nmol/min mg^{-1} protein)	
	neo-Inositol	scyllo-Inositol	myo-Inositol	scyllo-Inositol
mmonium sulfate fraction	0.07	0.05	0.12	0.68
EAE-cellulose chromatography	0.59	0.39	0.33	2.06
ephadex G-100 gel filtration	0.62	0.39	0.69	3.30

[a] myo-Inositol (10 mM) as substrate.
[b] myo-Inosose-2 (1.5 mM) as substrate.

Definition of Unit. One unit is defined as the amount of enzyme that produces 1 nmol of inositol product per minute in this assay. Specific activity is defined as units per milligram of protein where protein (mg/ml) is determined[9] by $1.45\ A_{280} - 0.74\ A_{260}$.

Purification

The results of the purification procedure given below as summarized in the table. The isolation is carried out at 2–6° with centrifugation at 15,000 g for 1 hr. All buffers contained 1.0 mM dithiothreitol (DTT) and 0.02% NaN$_3$. Omission or decrease in the DTT results in partial or total inactivation of the epimerase activities, but has no effect on the reductase activity. The procedure has been performed many times with similar results. Either fresh or frozen bovine brain tissue is used for the preparation. After being weighed, the tissue is homogenized in a Waring blender using a volume of pH 7.2, 50 mM Tris-glycine buffer equal to the tissue weight. The homogenate is then centrifuged, and the supernatant is made 2.1 M with solid ammonium sulfate. After stirring for 2 hr, the preparation is centrifuged and the pellet is discarded. The supernatant is then made 3.2 M with solid ammonium sulfate, and after stirring for 2 hr it is allowed to stand without stirring overnight. After centrifugation, the supernatant is discarded and the pellet is dissolved in a minimum volume of standard buffer (about 100–150 ml for the ammonium sulfate pellet obtained from 6–7 liters of homogenate). After pressure dialysis using an Amicon ultrafiltration apparatus, the enzyme preparation (3000 mg) is pumped onto a DEAE-cellulose column (5 × 13 cm) and eluted with standard buffer. Neither epimerase nor reductase activities are retained by the column, both are recovered in the wash volume. The active fraction from the DEAE-cellulose column is combined and concentrated 5-fold in an Amicon ultrafiltration unit. The enzyme solution containing 100 mg of protein

[9] H. M. Kalckar, *J. Biol. Chem.* **167**, 461 (1947).

is then placed on a Sephadex G-100 column (2.5 × 80 cm) in the standard buffer (pH 7.2) containing 0.1 M HCl, eluted with this buffer, and then concentrated 10-fold by ultrafiltration.

Properties

Stability. The DEAE eluate can be stored frozen for months without loss in activity. The G-100 enzyme has been stored several months with no loss in activity; however, for unknown reasons, some preparations have been found to be very labile. DTT must always accompany the bovine epimerase preparations, but is not needed in the unpurified cockroach enzyme. Reductase activity is not labile to storage or freeze-thaw even in the absence of DTT.

Cofactor Specificity. The bovine epimerase is specific for $NADP^+$, and no activity is observed with NAD^+.[2] Inosose reductase activity has similar specificity for NADPH; no activity being observed with NADH. This is in contrast with the activities of the less pure cockroach epimerase, which has the same high degree of specificity, but for NAD^+, not $NADP^+$.[5] The cockroach reductase preparation lacks specificity and retains 73% of the NADPH activity using NADH.

Molecular Weight. The epimerase and reductase activities from bovine brain coelute from G-100 Sephadex columns at an elution volume that corresponds to a molecular weight of about 54,000.[5]

pH Optima. The bovine epimerase exhibits a sharp pH optimum at pH 9.0–9.5 with 40% of the activity remaining at pH 8.5 and 10.0.[2] In contrast, the bovine reductase activity has a broad maximum between 5.5 and 9.5, with 90% of the activity remaining at the extreme points.[2] This is similar to the crude cockroach preparation, which has a broad reductase pH optimum between pH 4 and 8 and a sharp epimerase maximum at pH 7.5–8.0.[5]

Substrate Specificity and Product Ratio. Bovine reductase effects the conversion of *myo*-inosose-2 to *scyllo*- and *myo*-inositol in a ratio of 3:1 using NADPH.[4] No activity is observed with NADH. This contrasts with the crude cockroach reductase, which reduces *myo*-inosose-2 to *scyllo*- and *myo*-inositol with a product ratio of 7:1 using NADPH, but using NADH it gives a *scyllo-myo* ratio of 1:2.[5]

Bovine epimerase converts *myo*- to *neo*- and *scyllo*-inositols (0.59 and 0.39 nmol/min per milligram, respectively), but does not convert *chiro*-inositol to *myo*-inositol.[10] *scyllo*- and *neo*-Inositol are epimerized to *myo*-inositol, but each is a less effective substrate than *myo*-inositol. This differs from the crude cockroach epimerase, which epimerized *scyllo*->D-*chiro*->epi->*myo*->L-*chiro*->*neo*-inositol.[5]

---

[10] P. P. Hipps, K. Ackermann, and W. R. Sherman, unpublished data.

[101] Phosphoglucomutase from Yeast[1]

By J. G. Joshi

$$EP + \text{glucose-1-P} \rightleftarrows \left\{ \begin{array}{c} EP \cdot \text{glucose-6-P} \\ \Updownarrow \\ EP \cdot \text{glucose-1-P} \\ \Updownarrow \\ E \cdot \text{glucose-1,6-P}_2 \end{array} \right\} \rightleftarrows EP + \text{glucose-6-P}$$

$$\Updownarrow$$

$$E \cdot \ + \ \text{glucose-1,6-P}_2$$

Phosphoglucomutase (E.C. 2.7.5.1) catalyzes an apparent intramolecular transfer of phosphate between C-1 and C-6 of glucose. This enzyme from diverse sources[1a-6] has a molecular weight between 60,000 and 65,000 and is optimally active at pH 7.5 in the presence of a chelating agent, magnesium, and glucose 1,6-biphosphate. Only the phosphoenzyme (EP) is catalytically active. The complex, $E \cdot \text{glucose 1,6-P}_2$, occasionally dissociates to yield inactive dephosphoenzyme and glucose-1,6-P$_2$. This is prevented by adding excess glucose-1,6-P$_2$ in the assay mixture.

Principle. The activity is measured by determining the rate of glucose-6-P formation using a coupled assay involving glucose-6-P dehydrogenase and NADP$^+$.

The coupled assay is convenient and also accurate provided that (*a*) the concentration of phosphoglucomutase is limiting; (*b*) only the initial rates are recorded; and (*c*) the change in the absorbance at 340 nm is less than 1.0 per minute. These conditions equate to the formation of less than 0.15 μmol of glucose-6-P per minute or less than 8% conversion of added glucose-1-P.

---

[1] Abbreviations used: glucose-1-P, glucose 1-phosphate; glucose-1,6-P$_2$, glucose 1,6-bisphosphate; glucose-6-P, glucose 6-phosphate; NADP$^+$, nicotinamide adenine dinucleotide phosphate; EDTA, ethylenediaminetetracetic acid; PGM, phosphoglucomutase.

[1a] J. G. Joshi, J. Hooper, T. Kuwaki, T. Sakurada, J. R. Swanson, and P. Handler, *Proc. Natl Acad. Sci. U.S.A.* **57**, 1482 (1967).

[2] J. G. Joshi and P. Handler, *J. Biol. Chem.* **239**, 2741 (1964).

[3] J. G. Joshi, T. Hashimoto, K. Hanabusa, H. W. Dougherty, and P. Handler, *in* "Evolving Genes and Proteins" (V. Bryson and H. J. Vogel, eds.), pp. 207–219. Academic Press, New York, 1965.

[4] K. Hanabusa, H. W. Dougherty, C. Del Rio, T. Hashimoto, and P. Handler, *J. Biol. Chem.* **241**, 3930 (1966).

[5] T. Hashimoto and P. Handler, *J. Biol. Chem.* **241**, 3940 (1966).

[6] J. G. Joshi and P. Handler, *J. Biol. Chem.* **244**, 3343 (1969).

Fleischmann's dry yeast, *Saccharomyces cerevisiae*, type 20-40 (lot No. 797), is purchased from Standard Brands Inc. (New York). All other reagents are purchased from Sigma Chemical Co. (St. Louis, Missouri). The standard reaction mixture contains 2.0 μmol of glucose-1-P, 7.9 nmol of glucose-1,6-P_2, 0.5 μmol of NADP$^+$, 5.0 μmol of MgSO$_4$, 40 μmol of imidazole-HCl (pH 7.4), 1 unit of glucose-6-P dehydrogenase (yeast), and phosphoglucomutase in a total volume of 1.0 ml. Activity at room temperature is determined by measuring the rate of increase in absorbance at 340 nm using a spectrophotometer, preferably equipped with an automatic recorder. One unit of enzyme is the amount that catalyzes the conversion of 1 μmol of glucose-1-P to glucose-6-P in 1 min under the assay conditions; specific activity is expressed as units per milligram of protein. Protein is routinely measured by its absorbance at 280 nm assuming $A_{1\,cm}^{1\%}$ to be 10.0.

The procedure described here[7] evolved from a combination of those described by McCoy and Najjar,[8] Tsoi and Douglas,[9] Sakurada *et al.,*[10] and Hirose *et al.*[11] The modifications separate isozymes[9,10] and improve the yield of the enzyme.

This procedure is very sensitive to variations in pH, and extra care is required during the ion-exchange chromatography. Thus, no appreciable enzyme is adsorbed by CM-cellulose at pH 6.1 or above, whereas at pH levels lower than 6.0 adsorption occurs and elution requires very high ionic strength with attendant elution of impurities.

Unless stated otherwise, all steps are performed at 4° and in the presence of 0.1 mM EDTA.

Step 1. Autolysis. In a typical run (see the table), 1000 g of Fleischmann's yeast (type 20-40), is added over a period of 20 min with continuous stirring to 2500 ml of 70.0 mM Na$_2$HPO$_4$, plus 1 ml of toluene. Autolysis is conducted for 4 hr at 38° with continuous gentle stirring. The autolyzate is centrifuged at 9000 g for 15 min, and the supernatant is set aside. The residue is extracted at room temperature with 1250 ml of 70.0 mM Na$_2$HPO$_4$ and centrifuged at 11,000 g for 20 min; the residue is discarded, and the supernatants are pooled. The second extraction yields about 50% additional units.

Step 2. Heat Treatment. The turbid, dark-brown supernatant has a pH between 6.30 and 6.35. The pH is adjusted at room temperature to 5.5 by

[7] J. P. Daugherty, W. F. Kraemer, and J. G. Joshi, *Eur. J. Biochem.* **57**, 115 (1975).

[8] E. E. McCoy and V. A. Najjar, *J. Biol. Chem.* **234**, 3017 (1959).

[9] A. Tsoi and H. C. Douglas, *Biochim. Biophys. Acta* **92**, 513 (1964).

[10] T. Sakurada, J. G. Joshi, and P. Handler, *Abstr. 155th Natl. Meet. Am. Chem. Soc.,* p. 6062 (1968).

[11] M. Hirose, E. Sugimoto, R. Sasaki, and H. Chiba, *J. Biochem.* (*Tokyo*) **68**, 440 (1970).

gradual addition of 0.15 M acetic acid. Solid ammonium sulfate (3.30 g/100 ml of acidified solution) is added. After 15 min the mixture is heated to 57° in a boiling-water bath, held for 5 min at 57°, transferred to an ice bath, and cooled to room temperature. After 30 min, the precipitate is removed by centrifugation at 11,000 g for 20 min and the supernatant is recovered.

Step 3. Ammonium Sulfate Fractionation. Solid ammonium sulfate (24.7 g/100 ml of original acidified solution) is added to the turbid, light-brown, heat-treated supernatant over a period of 15 min with continuous stirring. After 30 min the solution is centrifuged at 11,000 g for 20 min, and the precipitate is discarded. Insoluble material that floats on the top of the supernatant is separated by filtration through glass wool at room temperature. To the clear, dark-brown filtrate is added solid ammonium sulfate (17.5 g/100 ml of original solution) with continuous stirring over a period of 20 min. After 12 hr at 4°, the mixture is centrifuged at 11,000 g for 20 min, and the supernatant is discarded. The precipitate is dissolved in ice-cold water and dialyzed for 20 hr against 12 liters of 5.0 mM sodium acetate, pH 5.5, with one change after 8 hr. The dialyzate is centrifuged at 11,000 g for 30 min to remove any precipitate formed during dialysis.

Step 4. CM-Cellulose Column Chromatography. The supernatant is applied to a CM-cellulose column (4.5 × 44.0 cm) that has been previously equilibrated with 5.0 mM Na$_2$HPO$_4$-HCl, pH 6.0. The column is washed with 500 ml of equilibration buffer, and the activity is eluted with a linear sodium gradient generated with 1000 ml of 5.0 mM Na$_2$HPO$_4$-HCl, pH 6.0, in the mixing chamber, and 1000 ml of 250 mM Na$_2$HPO$_4$-HCl, pH 6.0, in the reservoir. Fractions collected are assayed for protein and activity. Phosphoglucomutase activity is eluted in two peaks. The first peak represents 15% of the total activity (Fig. 1A). Fractions from the major peak II, containing phosphoglucomutase of a specific activity of at least 18, are pooled; solid ammonium sulfate is added (70 g per 100 ml) over a period of 30 min with continuous stirring, and the mixture is stored at 4°.

For convenience, steps 1 through 4 are repeated on another 1 kg of yeast. The light-brown precipitate from both runs is dissolved in ice-cold water and dialyzed against 6 liters of 5.0 mM sodium acetate-HCl, pH 5.5, with one change after 8 hr. The dialyzate is centrifuged at 11,000 g for 30 min to remove protein, which precipitates during dialysis.

Step 5. DEAE-Cellulose Column Chromatography. The pH of the combined supernatants from two runs above is adjusted to 7.3 at room temperature with 5.0 mM Na$_2$HPO$_4$. The clear, light-yellow solution is applied to a DEAE-cellulose column (2.8 × 60.0 cm) previously equilibrated with 5.0 mM Na$_2$HPO$_4$-HCl, pH 7.3. The column is washed with 500 ml of the equilibration buffer, and the activity is eluted with a linear phosphate gradient generated with 500 ml of 5.0 mM Na$_2$HPO$_4$-HCl, pH

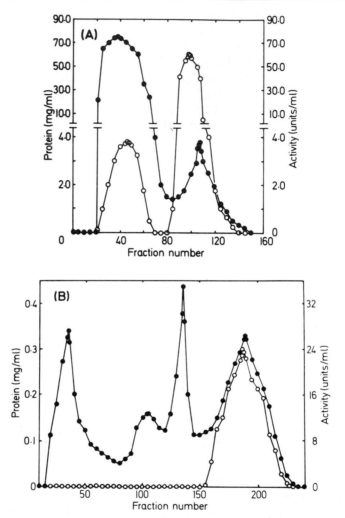

FIG. 1. (A) CM-cellulose column chromatography of yeast phosphoglucomutase (PGM). (B) DEAE-cellulose column chromatography and (C) Sephadex G-100 column chromatography of the major isozyme of yeast PGM. Fractions in (A) were 21.4 ml, (B) 5.3 ml, (C) 1.6 ml. ●——●, Protein; ○——○, enzyme activity.

7.3, in the mixing chamber and 500 ml of 50 mM Na$_2$HPO$_4$-HCl, pH 7.3, in the reservoir (Fig. 1B). The protein in fractions 152–220 is pooled, the enzyme is precipitated with ammonium sulfate, centrifuged, and dissolved in 2.5 ml of 0.1 M phosphate buffer, pH 7.0.

Step 6. Sephadex G-100 Column Chromatography. The clear supernatant (approximately 2.5 ml) is applied to a Sephadex G-100 column (2.8 ×

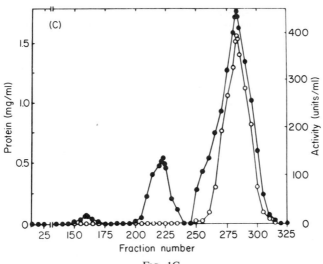

FIG. 1C.

100 cm) previously equilibrated with 0.10 M Na₂HPO₄-HCl, pH 7.0, containing 1.0 mM magnesium acetate and 0.10 mM EDTA. The protein is eluted with the equilibration buffer at a flow rate of 36 ml/hr. Fractions having a constant specific activity are pooled, solid ammonium sulfate is added (70 g/100 ml) with continuous stirring over a period of 60 min, and the enzyme is stored as such. Prior to use, an aliquot of suspension is removed and centrifuged at 20,000 g for 30 min; the supernatant is discarded. The precipitate is dissolved in the appropriate solution and desalted either by dialysis against an appropriate buffer or by passage through a Sephadex G-25 column equilibrated with an appropriate buffer. As described above, rechromatography of this preparation yields only one superimposable symmetrical peak of protein and activity.

The final preparation has a specific activity of 205 units per milligram of protein at room temperature or 400 units per milligram of protein at 30° and represents 2500-fold purification with 22% yield. It should be noted that these assays are conducted in the presence of 1 mM Mg²⁺. At 5 mM Mg²⁺, the specific activity is 490 units per milligram of protein. Only such preparations were used for further studies.

The procedure for purification is summarized in the table.

Properties

Purity and Molecular Weight. The protein as isolated is homogeneous as judged by ultracentrifugal analysis, sucrose density gradient centrifu-

PURIFICATION OF THE MAJOR ISOZYME OF PHOSPHOGLUCOMUTASE FROM FLEISCHMANN'S YEAST

Step	Volume (ml)	Total protein (mg)	Total activity (μmol/min)	Specific activity[b] (units/mg protein)	Recovery (%)
1. Autolysis	4076	436,200	35,240	0.081	100
2. Heat treatment	5376	321,000	35,200	0.11	99
3. Ammonium sulfate fractionation, precipitate dialyzed	880	53,280	31,780	0.60	90
4. CM-cellulose, peak II dialyzed	650	694	21,790	31.4	62
5. DEAE-cellulose fraction, precipitate dissolved	2.5	87	12,000	138	34
6. Sephadex G-100	36	38	7,785	205	22

[a] The values given are the averages of five separate purifications and are for 2.0 kg of starting material; however steps 1 through 4 were performed on 1.0-kg batches.
[b] These values are determined at room temperature and in the presence of 1 mM Mg^{2+}.

gation, and polyacrylamide gel electrophoresis. It has a molecular weight of about 65,600 and has 1 mol of phosphate bound per mole of enzyme.

Stability. Yeast phosphoglucomutase is quite stable at 4° for 3 months when stored as a precipitate in saturated ammonium sulfate in 0.1 M Na$_2$HPO$_4$-HCl, pH 7.0, containing 0.1 mM EDTA and 1 mM Mg^{2+}. It is also stable in solution for 5 days at 4° in 5 mM acetate or 10 mM citrate buffer, pH 5.5. In contrast, in 5 mM Tris buffer, pH 7.4 at 4°, more than 80% of the activity is lost overnight.

Effect of Temperature. Between 9.2° and 41° it has a Q$_{10}$ of 2.76 kcal (11.5 kJ) compared with about 3 kcal (12.5 kJ) reported for rabbit muscle phosphoglucomutase in the temperature range of 20° to 30°.[12]

Metals. Yeast PGM does not contain zinc. It requires a metal ion for activity, however. Mg^{2+}, 5.0 mM, is the best activator; the relative activities with other metals are Zn^{2+} (0.5 mM) 84%, Co^{2+} (2.0 mM) 95%, Mn^{2+} (0.5 mM) 62%. The specific activity in the presence of 5.0 mM Mg^{2+} and 0.5 mM Zn^{2+} is the same as that with 5.0 mM Mg alone. Beryllium is the only true inhibitor. Inactivation by Be^{2+} requires the presence of 0.1 mM EDTA or any other chelator. The inactivation is irreversible and time dependent. A completely inactivated enzyme has 1 g-atom of Be^{2+} bound per mole of enzyme.

[12] J. V. Passonneau, O. H. Lowry, D. W. Schulz, and J. G. Brown, *J. Biol. Chem.* **244**, 902 (1969).

Kinetics. The kinetic constants are determined by measuring the enzyme activity at varying concentrations of glucose-1,6-P_2 and glucose-1-P. The double-reciprocal plots of the data yields a family of parallel lines rather than converging lines conforming to the criterion of Ping-Pong kinetics rather than random sequential.[13] Secondary plots of $1/V$ against $1/[S]$ yield "true" K_m values of 2.15×0.0^{-5} M for glucose-1-P and of 2.37×10^{-6} M for glucose-1,6-P_2, respectively. These values are higher than those obtained for rabbit muscle enzyme.[14]

Chemical Probes. Yeast PGM has three free sulfhydryl groups, which can be titrated with p-chlormercuribenzoate or 5-5'-dithiobis(2-nitrobenzoic acid). More than 90% of the activity is abolished when an equivalent of one sulfhydryl has reacted.

Tetranitromethane at pH 8.0 or above reacts specifically with tyrosine to form nitrotyrosine. At pH 6.0 or below, it reacts specifically with cysteine to form nitroformate.[15] However, in the case of yeast PGM, cysteine, not tyrosine, is reactive at pH 8.0. This is established by amino acid analysis as well as spectrophometrically. Whether the tetranitromethane-reactive cysteine is identical to the one reacting with the sulfhydryl reagents above is not known.

Immunological Property. The amino acid compositions of PGM from yeast and rabbit muscle are significantly different. Antibodies made against rabbit muscle PGM in goats will not precipitate or inhibit yeast PGM.[6]

[13] W. W. Cleland, *Biochim. Biophys. Acta* **67**, 104 (1963).
[14] W. J. Ray, Jr. and G. A. Roscelli, *J. Biol. Chem.* **239**, 1228 (1964).
[15] J. F. Riordan, M. Sokolovsky, and B. L. Vallee, *Biochemistry* **7**, 3609 (1967).

Author Index

Numbers in parentheses are reference numbers and indicate that an author's work is referred to although the name is not cited in the text.

M

Subject Index

A

Liver
  bovine
    aldehyde dehydrogenase, 497–500
    diacetyl reductase, 517, 518, 521, 522
    D-erythrulose reductase, 232–237
    uridine diphosphate glucose-4-epimerase, 589
  catfish, glucosephosphate isomerase, 555–558
  chicken, D-erythrulose dehydrogenase, 237
  horse
    alcohol dehydrogenase
      affinity chromatography, 435–445
      NAD(H)-agarose complex, 457–468
      SS isozyme, 428–434, 435
    aldehyde dehydrogenase, isozymes, 474–479
  human
    aldehyde reductase, 508–510
    D-erythrulose dehydrogenase, 237
    glyceraldehyde-3-phosphate dehydrogenase, 301–305
  pig
    glyceraldehyde-3-phosphate dehydrogenase, 310–316
    2-keto--3-deoxy-L-fuconate dehydrogenase, 219–225
  pigeon, diacetyl reductase, 518, 519, 522
  rat
    2-ketoaldehyde dehydrogenase, 513–515
    sorbitol dehydrogenase, 135–140
  sheep, 2-ketoaldehyde dehydrogenase, 513, 515
Lobster, *see also Homarus americanus*
  glyceraldehyde-3-phosphate dehydrogenase, 334
Lung, human, glyceraldehyde-3-phosphate dehydrogenase, 303

M

Maltose, 26
  oxidation, by D-glucose dehydrogenase, 24
D-Mannitol, 26

D-Mannonate, 26
D-Mannonate dehydrogenase, *Escherichia coli*, 210–218
  assay, 210, 211
  colorimetry, 218
  coupling enzyme, in assay of uronate isomerase, 218
  effect of pH, 213
  enzyme unit, 211
  genetics, 218
  heat stability, 215
  inhibitors, 215, 216
  kinetic properties, 215
  properties, 213–218
  purification, 212–214
  specific activity, 211
  stability, 213
  substrate specificity, 215
Mannose, enantiomeric separation, as heptafluorobutyrate derivative, 7, 8
D-Mannose, 25, 26, 72
  conversion to D-glucose, 70, 72, 73
  oxidation, by D-glucose dehydrogenase, 24
D-[1-^{13}C]Mannose, conversion to D-[1-^{13}C]glucose, 90, 91
Methanol, metabolism, 424, 428
Methanol dehydrogenase, from *Hyphomicrobium* X, prosthetic group, 153, 457, 497
Methylgyoxal
  catabolism, D-lactate formation from, 35
  synthesis, 35
α-Methylglucoside, 24
Methylococcus capsulatus, see Methylomonas capsulatus
Methylomonas
  aminofaciens, 6-phospho-3-ketohexulose isomerase, 566
  capsulatus, 6-phospho-3-ketohexulose isomerase, 566–571
  culture, for glucose-6-phosphate dehydrogenase isolation, 272
  glucose-6-phosphate dehydrogenase, 271–275
  source, 272
Mevaldate reductase, 506
Micrococcus lactilyticus, see Veillonella alcalescens